Vice President, STEM: Ben Roberts
Publisher: Terri Ward
Senior Acquisitions Editor: Karen Carson
Marketing Manager: Tom DeMarco
Marketing Assistant: Cate McCaffery
Development Editor: Jorge Amaral
Senior Media Editor: Catriona Kaplan
Assistant Media Editor: Emily Tenenbaum
Director of Digital Production: Keri deManigold
Senior Media Producer: Alison Lorber
Associate Editor: Victoria Garvey
Editorial Assistant: Katharine Munz
Photo Editor: Cecilia Varas
Photo Researcher: Candice Cheesman
Director of Design, Content Management: Diana Blume
Text and Cover Designer: Blake Logan
Project Editor: Edward Dionne, MPS North America LLC
Illustrations: MPS North America LLC
Production Manager: Susan Wein
Composition: MPS North America LLC
Printing and Binding: LSC Communications
Cover Illustration: Drawing Water: Spring 2011 detail
(Midwest) by David Wicks
"Look Back" Arrow: NewCorner/Shutterstock

Library of Congress Control Number: 2016946039

Student Edition Hardcover:
ISBN-13: 978-1-319-01338-7
ISBN-10: 1-319-01338-4

Student Edition Loose-leaf:
ISBN-13: 978-1-319-01362-2
ISBN-10: 1-319-01362-7

Instructor Complimentary Copy:
ISBN-13: 978-1-319-01428-5
ISBN-10: 1-319-01428-3

W. H. Freeman and Company
One New York Plaza
Suite 4500
New York, NY 10004-1562
www.macmillanlearning.com

6.119, 6.124, 6.131, 6.135, 6.136, 6.137, 7.1, 7.2, 7.3, 7.6, 7.10, 7.11, 7.12, 7.13, 7.24, 7.25, 7.26, 7.27, 7.29, 7.67, 7.69, 7.76, 7.77, 7.81, 7.84, 7.102, 7.103, 7.115, 7.123, 7.124, 7.125, 7.128, 7.132, 7.135, 8.1, 8.3, 8.5, 8.6, 8.7, 8.8, 8.9, 8.16, 8.19, 8.32, 8.33, 8.34, 8.38, 8.40, 8.41, 8.44, 8.45, 8.50, 8.51, 8.52, 8.53, 8.54, 8.56, 8.57, 8.58, 8.59, 8.60, 8.61, 8.86, 8.87, 8.88, 9.26, 9.28, 10.10, 10.11, 10.12, 10.13, 10.14, 10.15, 10.16, 10.17, 10.38, 10.39, 10.40, 10.41, 10.45, 10.46, 11.11, 11.12, 11.18, 11.19, 11.20, 11.21, 11.23, 11.26, 11.27, 11.28, 11.29, 11.30, 11.31, 11.32, 11.57, 11.58, 11.59, 11.60, 11.61, 11.62, 11.63, 11.64, 11.65, 12.3, 12.4, 12.5, 12.6, 12.7, 12.8, 12.18, 12.20, 12.22, 12.26, 12.27, 12.28, 12.33, 12.34, 12.35, 12.36, 12.40, 12.41, 12.47, 12.61, 12.62, 12.67, 12.68, 12.72, 12.73, 12.74, 12.75, 12.79, 12.80, 12.83, 13.9, 13.11, 14.3, 14.4, 14.5, 14.6, 14.7, 14.8, 14.12, 14.30, 14.31, 14.32, 14.38, 14.39, 14.41, 14.44, 14.45, 14.46, 15.1, 15.2, 15.3, 15.4, 15.5, 15.6, 15.24, 15.25, 15.26, 15.27, 15.28, 15.29, 15.32, 15.33, 16.3, 16.4, 16.9, 16.14, 16.15, 16.22, 16.25, 16.34, 16.45, 16.47, 16.49, 16.55, 16.73, 16.75, 16.80, 16.81, 16.98, 16.99, 17.5, 17.6, 17.10

Demographics and characteristics of people:

1.75, 1.77, 1.78, 2.15, 2.34, 2.35, 2.36, 2.59, 2.75, 2.76, 2.77, 2.78, 2.79, 2.98, 2.99, 2.103, 2.111, 2.147, 2.148, 2.149, 4.52, 4.112, 4.113, 4.114, 5.1, 5.11, 5.17, 5.28, 5.29, 5.34, 5.43, 5.61, 5.63, 5.66, 5.75, 6.6, 6.7, 6.8, 6.11, 6.17, 6.18, 6.23, 6.24, 6.25, 6.27, 6.28, 6.57, 6.62, 6.66, 6.67, 6.74, 6.99, 6.121, 7.7, 7.8, 7.28, 7.30, 7.35, 7.93, 7.94, 7.95, 7.96, 7.104, 7.105, 7.110, 7.146, 7.148, 8.2, 8.4, 10.2, 10.5, 10.6, 10.10, 10.11, 10.12, 10.13, 10.14, 10.15, 10.16, 10.17, 10.29, 11.25, 12.4, 12.6, 12.7, 12.8, 12.35, 12.36, 12.40, 12.41, 12.68, 12.79, 12.80, 13.14, 13.15, 16.10, 16.24, 16.53, 16.57, 16.70, 16.75

Economics and Finance:
5.42, 6.132, 7.24, 10.27, 10.31, 10.32, 10.33, 10.42, 10.43, 10.44, 12.55, 12.56, 12.57

Education and child development:
1.5, 1.6, 1.7, 1.8, 1.11, 1.12, 1.13, 1.14, 1.16, 1.17, 1.20, 1.21, 1.22, 1.23, 1.24, 1.39, 1.40, 1.41, 1.44, 1.47, 1.48, 1.50, 1.51, 1.52, 1.53, 1.82, 1.93, 1.94, 1.95, 1.96, 1.97, 1.98, 1.99, 1.100, 1.104, 1.105, 1.106, 1.107, 1.108, 1.109, 1.123, 1.124, 1.125, 1.126, 1.127, 1.128, 1.129, 1.130, 1.131, 1.132, 1.133, 1.159, 1.164, 2.7, 2.15, 2.27, 2.49, 2.75, 2.76, 2.77, 2.78, 2.79, 2.86, 2.89, 2.98, 2.99, 2.111, 2.119, 2.123, 2.135, 2.140, 2.143,

21.53, 21.54, 2.155, 2.160, 2.161, 2.162, 2.167, 2.169, 3.1, 3.3, 3.7, 3.10, 3.20, 3.25, 3.44, 3.52, 3.53, 3.101, 4.29, 4.42, 4.45, 4.46, 4.47, 4.93, 4.94, 4.107, 4.108, 4.109, 4.110, 4.111, 4.112, 4.113, 4.114, 4.136, 5.7, 5.36, 5.74, 5.78, 5.94, 6.9, 6.10, 6.15, 6.21, 6.22, 6.55, 6.56, 6.61, 6.66, 6.67, 6.97, 6.125, 7.47, 7.48, 7.49, 7.50, 7.51, 7.64, 7.65, 7.66, 7.134, 7.143, 7.144, 7.147, 7.149, 8.31, 8.63, 8.80, 8.89, 8.90, 9.16, 9.18, 9.20, 9.22, 9.24, 9.25, 9.27, 9.39, 9.40, 9.41, 9.47, 9.48, 9.49, 9.50, 9.53, 9.56, 10.10, 10.11, 10.12, 10.13, 10.14, 10.15, 10.16, 10.17, 10.27, 10.30, 10.34, 10.36, 10.38, 10.39, 10.40, 10.41, 10.54, 10.58, 10.61, 10.62, 10.63, 11.1, 11.9, 11.13, 11.14, 11.15, 11.16, 11.18, 11.19, 11.20, 11.21, 11.22, 11.33, 11.34, 11.35, 11.36, 12.17, 12.19, 12.21, 12.31, 13.13, 13.36, 13.56, 13.57, 13.58, 13.59, 14.47, 14.48, 14.49, 14.50, 15.8, 15.9, 15.10, 15.11, 15.12, 15.13, 15.16, 15.17, 15.18, 15.19, 15.34, 15.46, 16.8, 16.18, 16.19, 16.26, 16.27, 16.40, 16.43, 16.48, 16.50, 16.51, 16.52, 16.53, 16.56, 16.57, 16.59, 16.65, 16.68, 16.70, 16.71, 16.72, 16.79, 16.83, 16.84, 16.85, 16.86

Environment:
1.32, 1.33, 1.36, 1.88, 1.142, 1.145, 1.146, 1.147, 2.21, 2.22, 2.23, 2.44, 2.45, 2.66, 2.67, 3.43, 3.45, 5.9, 5.37, 5.55, 5.56, 5.85, 5.91, 6.35, 6.70, 6.71, 6.116, 6.117, 6.140, 7.9, 7.33, 7.79, 7.80, 7.87, 7.89, 7.90, 7.112, 7.113, 7.133, 7.136, 9.55, 10.19, 10.20, 10.22, 10.48, 10.49, 10.50, 10.51, 10.52, 10.53, 10.64, 11.24, 11.46, 11.47, 11.48, 11.49, 11.50, 11.51, 11.52, 11.53, 11.54, 11.55, 11.56, 12.18, 12.20, 12.22, 13.16, 13.18, 13.46, 13.55, 15.24, 15.25, 15.26, 15.27, 15.28, 15.29, 15.45, 15.51, 16.28

Ethics:
3.71, 3.72, 3.73, 3.74, 3.75, 3.76, 3.77, 3.78, 3.79, 3.80, 3.81, 3.82, 3.83, 3.84, 3.85, 3.86, 3.87, 3.88, 3.89, 3.90, 3.91, 3.92, 3.104, 3.106, 3.107, 3.108, 3.109, 3.110, 5.4, 5.89, 6.44, 7.40, 7.115, 8.97, 16.98, 16.99

Health and nutrition:
1.15, 1.18, 1.19, 1.30, 1.31, 1.57, 1.58, 1.59, 1.60, 1.61, 1.62, 1.63, 1.64, 1.80, 1.89, 1.134, 1.135, 1.136, 1.141, 1.143, 1.144, 1.155, 1.156, 1.157, 1.158, 2.8, 2.18, 2.19, 2.20, 2.24, 2.25, 2.26, 2.37, 2.42, 2.43, 2.47, 2.48, 2.61, 2.62, 2.65, 2.68, 2.69, 2.70, 2.71, 2.72, 2.88, 2.92, 2.93, 2.94, 2.95, 2.106, 2.107, 2.113, 2.114, 2.115, 2.116, 2.117, 2.118, 2.121, 2.125, 2.126, 2.127, 2.128, 2.133, 2.134, 2.139, 2.141, 2.142, 2.166, 2.170, 2.174, 2.175, 2.176, 3.7, 3.11, 3.16, 3.17, 3.18, 3.19, 3.21, 3.22, 3.24, 3.26, 3.35, 3.39, 3.41, 3.42, 3.63, 3.71, 3.75, 3.76, 3.79, 3.81,

3.84, 3.86, 4.25, 4.26, 4.75, 4.78, 4.105, 4.106, 4.121, 4.122, 4.123, 5.9, 5.28, 5.29, 5.35, 5.40, 5.60, 5.66, 5.75, 5.85, 6.26, 6.29, 6.33, 6.34, 6.60, 6.62, 6.65, 6.69, 6.74, 6.94, 6.96, 6.99, 6.118, 6.125, 6.127, 6.133, 7.10, 7.11, 7.34, 7.36, 7.37, 7.38, 7.39, 7.45, 7.46, 7.53, 7.59, 7.60, 7.68, 7.71, 7.72, 7.73, 7.74, 7.83, 7.85, 7.86, 7.88, 7.91, 7.97, 7.98, 7.99, 7.109, 7.111, 7.116, 7.125, 7.131, 7.142, 7.148, 8.11, 8.15, 8.18, 8.20, 8.25, 8.26, 8.42, 8.43, 8.62, 8.63, 8.64, 8.65, 8.66, 8.76, 8.77, 8.78, 8.79, 8.91, 8.92, 8.96, 9.4, 9.5, 9.6, 9.7, 9.8, 9.14, 9.29, 9.30, 9.31, 9.42, 9.45, 9.46, 10.1, 10.2, 10.5, 10.6, 10.24, 10.28, 10.29, 10.37, 10.60, 10.66, 10.67, 11.25, 11.40, 11.41, 11.42, 11.43, 11.44, 11.45, 12.23, 12.25, 12.37, 12.46, 12.54, 12.55, 12.56, 12.57, 12.63, 12.65, 12.66, 12.69, 12.70, 12.71, 12.76, 12.77, 12.78, 12.82, 13.21, 13.24, 13.28, 13.29, 13.30, 13.31, 13.32, 13.37, 13.40, 13.41, 13.42, 13.45, 14.15, 14.16, 14.17, 14.18, 14.19, 14.20, 14.21, 14.23, 14.24, 14.25, 14.26, 14.27, 14.33, 14.40, 14.43, 14.45, 14.46, 15.21, 15.23, 15.35, 15.36, 15.41, 15.42, 15.43, 15.48, 15.49, 15.50, 15.52, 16.67, 16.76, 16.77, 16.82, 16.88, 16.94, 17.54

Humanities and social sciences:
1.10, 1.15, 1.27, 1.28, 1.29, 1.39, 1.40, 1.41, 1.110, 1.111, 1.150, 2.2, 2.4, 2.105, 2.122, 2.123, 2.136, 2.137, 2.172, 2.173, 3.6, 3.10, 3.50, 3.66, 3.70, 3.72, 3.73, 3.74, 3.83, 3.85, 3.87, 3.88, 3.92, 3.105, 3.109, 3.110, 4.28, 4.61, 4.107, 4.108, 4.138, 5.1, 5.43, 5.45, 5.61, 5.62, 5.63, 5.64, 5.65, 5.66, 5.69, 5.70, 5.71, 5.72, 5.77, 5.79, 5.88, 6.12, 6.13, 6.14, 6.17, 6.18, 6.29, 6.63, 6.64, 6.68, 6.72, 6.73, 6.97, 7.31, 7.36, 7.37, 7.61, 7.62, 7.63, 7.67, 7.84, 7.85, 7.129, 7.130, 7.132, 7.135, 7.137, 7.139, 7.140, 7.145, 8.27, 8.28, 8.29, 8.30, 8.35, 8.37, 8.75, 8.96, 8.97, 9.1, 9.2, 9.3, 9.10, 9.25, 9.27, 9.39, 9.40, 9.41, 9.48, 9.51, 9.56, 10.28, 10.65, 11.1, 11.9, 11.12, 11.17, 11.36, 11.37, 11.38, 11.39, 12.7, 12.8, 12.24, 12.27, 12.28, 12.35, 12.36, 12.40, 12.41, 12.46, 12.48, 12.49, 12.51, 12.52, 12.53, 12.58, 12.59, 12.60, 12.64, 12.68, 12.79, 12.81, 13.9, 13.13, 13.17, 13.19, 13.20, 13.22, 13.23, 13.25, 13.26, 13.27, 13.28, 13.29, 13.30, 13.31, 13.32, 13.35, 13.36, 13.39, 15.13, 15.14, 15.15, 15.31, 15.40, 15.44, 16.11, 16.23, 16.58, 16.66, 16.75, 16.80, 16.81, 16.89, 16.90, 16.91, 16.92

International:
1.111, 1.146, 1.147, 1.149, 1.152, 1.153, 2.21, 2.22, 2.23, 2.30, 2.31, 2.34, 2.35, 2.36, 2.44, 2.45, 2.53, 2.54, 2.55, 2.59, 2.66, 2.67, 2.83, 2.84, 2.103, 2.144, 2.145, 2.146, 2.147,

2.148, 2.149, 3.46, 3.47, 4.28, 5.1, 5.62, 5.64, 5.65, 5.69, 5.71, 5.81, 5.84, 6.64, 6.74, 6.99, 7.24, 7.34, 7.47, 7.58, 7.59, 7.60, 7.134, 8.25, 8.26, 8.64, 8.65, 8.66, 8.91, 8.92, 10.24, 10.31, 10.32, 10.33, 10.42, 10.55, 10.56, 10.57, 11.12, 11.37, 11.38, 11.39, 12.24, 12.47, 12.59, 12.60, 12.61, 12.67, 12.81, 13.28, 13.29, 13.30, 13.31, 13.32, 13.36, 14.26, 14.27, 14.39, 14.42, 15.7, 16.71, 16.72

Law and government data:
1.132, 1.133, 3.6, 3.83, 3.107, 4.52, 6.44, 7.24, 7.25, 7.26, 7.29, 7.119, 7.125, 8.96, 9.51, 10.19, 10.20, 10.22, 10.27, 11.24

Manufacturing, products, and processes:
2.164, 2.165, 3.5, 5.31, 5.33, 5.60, 5.83, 6.30, 6.31, 6.36, 6.39, 6.75, 6.140, 7.29, 7.32, 7.41, 7.42, 7.43, 7.45, 7.106, 7.107, 7.109, 7.116, 7.117, 7.118, 10.18, 13.9, 13.43, 13.44, 15.20, 15.21, 15.22, 15.23, 15.47, 16.31, 16.74, 16.82, 17.4, 17.5, 17.6, 17.7; 17.8, 17.9, 17.10, 17.11, 17.12, 17.13, 17.14, 17.15, 17.16, 17.17, 17.18, 17.19, 17.20, 17.21, 17.22, 17.23, 17.28, 17.29, 17.30, 17.31, 17.32, 17.33, 17.34, 17.35, 17.36, 17.37, 17.38, 17.39, 17.40, 17.41, 17.42, 17.43, 17.44, 17.45, 17.46, 17.47, 17.48, 17.49, 17.50, 17.53, 17.54, 17.55, 17.56, 17.57, 17.58, 17.59, 17.60, 17.61, 17.62, 17.63, 17.64, 17.65, 17.69, 17.70, 17.71, 17.72, 17.73, 17.74, 17.75, 17.76, 17.77, 17.78, 17.79, 17.80, 17.81, 17.82, 17.83, 17.84, 17.85, 17.86, 17.87, 17.88, 17.89, 17.90, 17.91

Marketing:
1.160, 3.23, 3.27, 3.28, 3.29, 3.30, 3.91, 3.95, 6.90, 6.91, 6.104, 7.78, 7.114, 7.137, 7.139, 7.140, 7.141, 8.2, 8.4, 8.40, 8.41, 12.45, 12.48, 12.49, 12.50, 13.1, 13.2, 13.19, 13.20, 13.25, 13.26, 13.27, 16.61, 16.62, 16.63, 16.64, 16.66

Science:
1.81, 1.90, 1.92, 2.32, 2.33, 2.50, 2.51, 2.73, 2.74, 2.96, 2.97, 5.46, 5.82, 5.90, 6.36, 6.39, 7.9, 7.10, 7.11, 7.100, 10.27, 10.65, 11.17, 11.40, 11.41, 11.42, 11.43, 11.44, 11.45, 11.57, 11.58, 11.59, 11.60, 11.61, 11.62, 11.63, 11.64, 11.65, 16.76, 16.77, 16.88

Sports and leisure:
1.25, 1.26, 1.42, 1.149, 1.150, 1.153, 1.154, 1.163, 2.1, 2.5, 2.24, 2.25, 2.26, 2.28, 2.29, 2.30, 2.31, 2.47, 2.48, 2.53, 2.54, 2.55, 2.68, 2.69, 2.70, 2.71, 2.72, 2.83, 2.84, 2.94, 2.95, 2.120, 2.125, 2.126, 2.137, 2.140, 2.142, 2.150, 3.27, 3.28, 3.29, 3.30, 3.32, 3.51, 3.99, 4.5, 4.7, 4.14, 4.16, 4.30, 4.32, 4.33, 4.54, 4.55, 4.56, 4.76, 4.77, 4.82, 4.89, 4.90, 4.91, 4.92, 4.128, 4.129, 4.131, 4.135, 4.137, 5.3, 5.27, 5.30, 5.32, 5.40, 5.48, 5.50, 5.73, 5.81, 5.87, 5.92, 5.93, 5.98, 6.16, 6.27, 6.28, 6.56, 6.126, 6.135, 7.7, 7.8, 7.13, 7.34, 7.58, 7.59, 7.60, 7.70, 7.75, 7.95, 7.96, 7.110, 7.138, 8.56, 8.57, 8.58, 8.59, 8.60, 8.61, 8.67, 8.68, 8.69, 8.70, 8.71, 9.1, 9.2, 9.3, 9.10, 9.15, 9.17, 9.19, 9.21, 9.23, 9.32, 9.33, 10.2, 10.5, 10.6, 10.43, 10.44, 10.47, 11.11, 11.25, 11.27, 11.28, 11.29, 11.30, 11.31, 11.32, 14.1, 14.2, 14.30, 14.31, 14.32, 14.38, 15.7, 15.30, 15.37, 15.38, 15.39, 16.9, 16.10, 16.12, 16.17, 16.22, 16.24, 16.46, 16.78, 16.87, 16.94, 16.95, 16.96, 16.97, 17.53

Students:
1.6, 1.7, 1.8, 1.11, 1.12, 1.14, 1.16, 1.17, 1.20, 1.21, 1.22, 1.23, 1.24, 1.44, 1.47, 1.48, 1.50, 1.51, 1.52, 1.53, 1.148, 1.154, 1.159, 1.164, 2.2, 2.4, 2.27, 2.49, 2.134, 3.1, 3.3, 3.15, 3.25, 3.57, 3.59, 3.63, 3.67, 4.29, 4.42, 4.45, 4.46, 4.47, 4.93, 4.94, 4.109, 4.110, 4.111, 4.136, 5.7, 5.9, 5.11, 5.28, 5.29, 5.40, 5.45, 5.59, 5.60, 5.95, 6.6, 6.7, 6.8, 6.9, 6.10, 6.11, 6.12, 6.13, 6.14, 6.21, 6.22, 6.24, 6.25, 6.27, 6.28, 6.38, 6.54, 6.55, 6.56, 6.57, 6.61, 6.63, 6.65, 6.73, 6.90, 6.119, 6.121, 6.123,

7.7, 7.8, 7.28, 7.30, 7.64, 7.65, 7.66, 7.93, 7.94, 7.95, 7.96, 7.104, 7.105, 7.110, 7.119, 7.123, 7.129, 7.130, 7.134, 8.10, 8.12, 8.13, 8.29, 8.30, 8.31, 8.32, 8.33, 8.34, 8.42, 8.43, 8.80, 8.86, 8.87, 8.88, 8.89, 8.90, 9.4, 9.5, 9.6, 9.7, 9.8, 10.2, 10.5, 10.6, 10.29, 10.30, 10.36, 10.37, 10.38, 10.39, 10.40, 10.41, 10.47, 10.60, 10.61, 10.62, 10.63, 11.1, 11.9, 11.13, 11.14, 11.15, 11.16, 11.18, 11.19, 11.20, 11.21, 11.22, 11.23, 11.25, 11.36, 12.7, 12.8, 12.17, 12.19, 12.21, 12.27, 12.28, 12.33, 12.34, 12.35, 12.36, 12.38, 12.39, 12.40, 12.41, 12.58, 12.68, 12.79, 13.13, 13.56, 13.57, 13.58, 13.59, 14.47, 14.48, 14.49, 14.50, 16.1, 16.10, 16.16, 16.19, 16.24, 16.26, 16.27, 16.30, 16.37, 16.43, 16.48, 16.50, 16.51, 16.52, 16.56, 16.75, 16.79, 16.83, 16.84, 16.85, 16.86, 17.4

Technology and the Internet:
1.25, 1.26, 1.150, 2.1, 2.5, 2.34, 2.35, 2.36, 2.59, 2.103, 2.135, 3.5, 3.10, 4.9, 4.23, 4.24, 4.34, 4.50, 4.51, 4.57, 5.6, 5.11, 5.17, 5.27, 5.30, 5.32, 5.34, 5.45, 5.62, 5.64, 5.65, 5.69, 5.71, 5.74, 5.84, 5.99, 5.102, 6.16, 6.54, 6.68, 6.123, 7.7, 7.8, 7.27, 7.28, 7.30, 7.35, 7.46, 7.64, 7.65, 7.66, 7.70, 7.75, 7.76, 7.77, 7.95, 7.96, 7.100, 8.1, 8.3, 8.14, 8.17, 8.21, 8.27, 8.28, 8.35, 8.55, 8.67, 8.68, 8.69, 8.70, 8.71, 8.74, 8.75, 8.81, 8.82, 8.83, 8.84, 8.85, 10.36, 12.7, 12.8, 12.18, 12.20, 12.22, 12.24, 12.27, 12.28, 12.35, 12.36, 12.38, 12.39, 12.40, 12.41, 12.45, 12.47, 12.68, 12.79, 13.11, 13.38, 14.9, 14.10, 14.11, 14.13, 14.14, 14.42, 14.44, 15.30, 15.37, 15.38, 15.39, 16.1, 16.10, 16.12, 16.16, 16.17, 16.24, 16.30, 16.37, 16.46, 16.75, 16.78, 16.87, 16.95, 16.96, 16.97

Introduction to the Practice of Statistics

NINTH EDITION

David S. Moore
George P. McCabe
Bruce A. Craig

Purdue University

w. h. freeman
Macmillan Learning
New York

Brief Contents

To Teachers: About This Book   xi

To Students: What Is Statistics?   xix

About the Authors   xxii

Data Table Index   xxiii

Beyond the Basics Index   xxiv

PART I   Looking at Data

CHAPTER 1	Looking at Data—Distributions	1
CHAPTER 2	Looking at Data—Relationships	79
CHAPTER 3	Producing Data	163

PART II   Probability and Inference

CHAPTER 4	Probability: The Study of Randomness	215
CHAPTER 5	Sampling Distributions	281
CHAPTER 6	Introduction to Inference	341
CHAPTER 7	Inference for Means	407
CHAPTER 8	Inference for Proportions	483

PART III   Topics in Inference

CHAPTER 9	Inference for Categorical Data	525
CHAPTER 10	Inference for Regression	555
CHAPTER 11	Multiple Regression	607

CHAPTER 12   One-Way Analysis of Variance   643

CHAPTER 13   Two-Way Analysis of Variance   697

Companion Chapters

(on the *IPS* website www.macmillanlearning.com/ips9e and in LaunchPad)

CHAPTER 14   Logistic Regression   14-1

CHAPTER 15   Nonparametric Tests   15-1

CHAPTER 16   Bootstrap Methods and Permutation Tests   16-1

CHAPTER 17   Statistics for Quality: Control and Capability   17-1

Tables   T-1
Answers to Odd-Numbered Exercises   A-1
Notes and Data Sources   N-1
Index   I-1

Contents

To Teachers: About This Book xi
To Students: What Is Statistics? xix
About the Authors xxii
Data Table Index xxiii
Beyond the Basics Index xxiv

PART I Looking at Data

CHAPTER 1 Looking at Data—Distributions 1

Introduction 1

1.1 Data 2
    Key characteristics of a data set 4
Section 1.1 Summary 7
Section 1.1 Exercises 7

1.2 Displaying Distributions with Graphs 8
    Categorical variables: Bar graphs
      and pie charts 9
    Quantitative variables: Stemplots
      and histograms 11
    Histograms 14
    Data analysis in action: Don't hang up
      on me 16
    Examining distributions 18
    Dealing with outliers 19
    Time plots 21
Section 1.2 Summary 23
Section 1.2 Exercises 23

1.3 Describing Distributions with Numbers 27
    Measuring center: The mean 28
    Measuring center: The median 30
    Mean versus median 31
    Measuring spread: The quartiles 32
    The five-number summary and
      boxplots 34
    The $1.5 \times IQR$ rule for suspected
      outliers 35
    Measuring spread: The standard
      deviation 38
    Properties of the standard deviation 40
    Choosing measures of center and spread 40
    Changing the unit of measurement 43
Section 1.3 Summary 46
Section 1.3 Exercises 47

1.4 Density Curves and Normal Distributions 51
    Density curves 53
    Measuring center and spread for
      density curves 54
    Normal distributions 56
    The 68–95–99.7 rule 57
    Standardizing observations 59
    Normal distribution calculations 61
    Using the standard Normal table 63
    Inverse Normal calculations 64
    Normal quantile plots 66
Beyond the Basics: Density estimation 68
Section 1.4 Summary 69
Section 1.4 Exercises 70
Chapter 1 Exercises 74

CHAPTER 2 Looking at Data—Relationships 79

Introduction 79

2.1 Relationships 79
    Examining relationships 81
Section 2.1 Summary 84
Section 2.1 Exercises 84

2.2 Scatterplots 85
    Interpreting scatterplots 88
    The log transformation 91
    Adding categorical variables to scatterplots 93
    Scatterplot smoothers 94
    Categorical explanatory variables 96
Section 2.2 Summary 96
Section 2.2 Exercises 96

2.3 Correlation 100
    The correlation r 101
    Properties of correlation 102
Section 2.3 Summary 104
Section 2.3 Exercises 105

2.4 Least-Squares Regression 107
    Fitting a line to data 108
    Prediction 110
    Least-squares regression 111
    Interpreting the regression line 113
    Facts about least-squares regression 114
    Correlation and regression 115
    Another view of r^2 117

v

Section 2.4 Summary 117
Section 2.4 Exercises 118

2.5 Cautions about Correlation and Regression 123
  Residuals 123
  Outliers and influential observations 127
  Beware of the lurking variable 129
  Beware of correlations based on averaged data 131
  Beware of restricted ranges 132
Beyond the Basics: Data mining 132
Section 2.5 Summary 133
Section 2.5 Exercises 133

2.6 Data Analysis for Two-Way Tables 136
  The two-way table 136
  Joint distribution 138
  Marginal distributions 139
  Describing relations in two-way tables 140
  Conditional distributions 140
  Simpson's paradox 143
Section 2.6 Summary 145
Section 2.6 Exercises 146

2.7 The Question of Causation 148
  Explaining association 149
  Establishing causation 150
Section 2.7 Summary 152
Section 2.7 Exercises 153
Chapter 2 Exercises 154

CHAPTER 3 Producing Data 163

Introduction 163

3.1 Sources of Data 164
  Anecdotal data 164
  Available data 165
  Sample surveys and experiments 167
Section 3.1 Summary 170
Section 3.1 Exercises 170

3.2 Design of Experiments 171
  Comparative experiments 173
  Randomization 175
  Randomized comparative experiments 177
  How to randomize 177
  Randomization using software 178
  Randomization using random digits 179
  Cautions about experimentation 181
  Matched pairs designs 182
  Block designs 183
Section 3.2 Summary 185
Section 3.2 Exercises 186

3.3 Sampling Design 188
  Simple random samples 191
  How to select a simple random sample 191
  Stratified random samples 194
  Multistage random samples 195
  Cautions about sample surveys 196
Beyond the Basics: Capture-recapture sampling 199
Section 3.3 Summary 200
Section 3.3 Exercises 200

3.4 Ethics 203
  Institutional review boards 204
  Informed consent 205
  Confidentiality 206
  Clinical trials 207
  Behavioral and social science experiments 209
Section 3.4 Summary 211
Section 3.4 Exercises 211
Chapter 3 Exercises 212

PART II Probability and Inference

CHAPTER 4 Probability: The Study of Randomness 215

Introduction 215

4.1 Randomness 215
  The language of probability 217
  Thinking about randomness 218
  The uses of probability 219
Section 4.1 Summary 219
Section 4.1 Exercises 220

4.2 Probability Models 220
  Sample spaces 221
  Probability rules 223
  Assigning probabilities: Finite number of outcomes 225
  Assigning probabilities: Equally likely outcomes 227
  Independence and the multiplication rule 228
  Applying the probability rules 231
Section 4.2 Summary 232
Section 4.2 Exercises 232

4.3 Random Variables 235
  Discrete random variables 236
  Continuous random variables 239
  Normal distributions as probability distributions 242
Section 4.3 Summary 243
Section 4.3 Exercises 244

4.4 Means and Variances of Random Variables 246
The mean of a random variable 246
Statistical estimation and the law of
large numbers 250
Thinking about the law of large numbers 251
Beyond the Basics: More laws of large numbers 253
Rules for means 253
The variance of a random variable 255
Rules for variances and standard deviations 257
Section 4.4 Summary **261**
Section 4.4 Exercises **262**

4.5 General Probability Rules 264
General addition rules 264
Conditional probability 267
General multiplication rules 270
Tree diagrams 271
Bayes's rule 273
Independence again 274
Section 4.5 Summary **274**
Section 4.5 Exercises **275**
Chapter 4 Exercises **278**

CHAPTER 5 Sampling Distributions 281
Introduction 281
5.1 Toward Statistical Inference 282
Sampling variability 283
Sampling distributions 284
Bias and variability 287
Sampling from large populations 289
Why randomize? 290
Section 5.1 Summary **290**
Section 5.1 Exercises **291**

**5.2 The Sampling Distribution of a Sample
Mean 293**
The mean and standard deviation of $\bar{x}$ 296
The central limit theorem 298
A few more facts 304
Beyond the Basics: Weibull distributions **305**
Section 5.2 Summary **307**
Section 5.2 Exercises **307**

**5.3 Sampling Distributions for Counts
and Proportions 310**
The binomial distributions for sample counts 312
Binomial distributions in statistical sampling 314
Finding binomial probabilities 315
Binomial mean and standard deviation 317
Sample proportions 319
Normal approximation for counts and
proportions 321
The continuity correction 325

Binomial formula 326
The Poisson distributions 328
Section 5.3 Summary **332**
Section 5.3 Exercises **333**
Chapter 5 Exercises **338**

CHAPTER 6 Introduction to Inference 341
Introduction 341
Overview of inference **342**
6.1 Estimating with Confidence 343
Statistical confidence 344
Confidence intervals 346
Confidence interval for a population mean 348
How confidence intervals behave 352
Choosing the sample size 353
Some cautions 355
Section 6.1 Summary **356**
Section 6.1 Exercises **357**

6.2 Tests of Significance 361
The reasoning of significance tests 361
Stating hypotheses 363
Test statistics 364
P-values 365
Statistical significance 367
Tests for a population mean 371
Two-sided significance tests and
confidence intervals 375
The *P*-value versus a statement of
significance 377
Section 6.2 Summary **379**
Section 6.2 Exercises **379**

6.3 Use and Abuse of Tests 384
Choosing a level of significance 384
What statistical significance does not mean 385
Don't ignore lack of significance 386
Statistical inference is not valid for all
sets of data 387
Beware of searching for significance 388
Section 6.3 Summary **389**
Section 6.3 Exercises **389**

6.4 Power and Inference as a Decision 391
Power 391
Increasing the power 395
Inference as decision 396
Two types of error 396
Error probabilities 397
The common practice of testing hypotheses 399
Section 6.4 Summary **400**
Section 6.4 Exercises **400**
Chapter 6 Exercises **402**

CHAPTER 7 Inference for Means 407

Introduction 407

7.1 Inference for the Mean of a Population 408
The t distributions 408
The one-sample t confidence interval 410
The one-sample t test 412
Matched pairs t procedures 419
Robustness of the t procedures 423
Beyond the Basics: The bootstrap **424**
Section 7.1 Summary **425**
Section 7.1 Exercises **426**

7.2 Comparing Two Means 432
The two-sample z statistic 434
The two-sample t procedures 436
The two-sample t confidence interval 436
The two-sample t significance test 439
Robustness of the two-sample procedures 442
Inference for small samples 444
Software approximation for the
degrees of freedom 447
The pooled two-sample t procedures 448
Section 7.2 Summary **453**
Section 7.2 Exercises **454**

7.3 Additional Topics on Inference 460
Choosing the sample size 461
Inference for non-Normal populations 470
Section 7.3 Summary **474**
Section 7.3 Exercises **474**
Chapter 7 Exercises **476**

CHAPTER 8 Inference for Proportions 483

Introduction 483

8.1 Inference for a Single Proportion 484
Large-sample confidence interval for a
single proportion 485
Beyond the Basics: The plus four confidence
interval for a single proportion **489**
Significance test for a single proportion 491
Choosing a sample size for a confidence
interval 494
Choosing a sample size for a significance
test 498
Section 8.1 Summary **500**
Section 8.1 Exercises **501**

8.2 Comparing Two Proportions 505
Large-sample confidence interval for a
difference in proportions 506
Beyond the Basics: The plus four confidence
interval for a difference in proportions **509**

Significance test for a difference in
proportions 511
Choosing a sample size for two
sample proportions 514
Beyond the Basics: Relative risk **518**
Section 8.2 Summary **519**
Section 8.2 Exercises **520**
Chapter 8 Exercises **522**

PART III Topics in Inference

CHAPTER 9 Inference for Categorical Data 525

Introduction 525

9.1 Inference for Two-Way Tables 526
The hypothesis: No association 532
Expected cell counts 533
The chi-square test 534
Computations 536
Computing conditional distributions 537
The chi-square test and the z test 540
Beyond the Basics: Meta-analysis **542**
Section 9.1 Summary **543**
Section 9.1 Exercises **544**

9.2 Goodness of Fit 545
Section 9.2 Summary **550**
Section 9.2 Exercises **550**
Chapter 9 Exercises **551**

CHAPTER 10 Inference for Regression 555

Introduction 555

10.1 Simple Linear Regression 556
Statistical model for linear regression 556
Preliminary data analysis and inference
considerations 558
Estimating the regression parameters 561
Checking model assumptions 565
Confidence intervals and significance tests 567
Confidence intervals for mean response 570
Prediction intervals 572
Transforming variables 574
Beyond the Basics: Nonlinear regression **576**
Section 10.1 Summary **577**
Section 10.1 Exercises **578**

**10.2 More Detail about Simple Linear
Regression 582**
Analysis of variance for regression 583
The ANOVA F test 585
Calculations for regression inference 588
Inference for correlation 593

Section 10.2 Summary **596**
Section 10.2 Exercises **597**
Chapter 10 Exercises **598**

CHAPTER 11 Multiple Regression **607**
Introduction 607

11.1 Inference for Multiple Regression 608
Population multiple regression equation 608
Data for multiple regression 609
Multiple linear regression model 610
Estimation of the multiple regression
   parameters 611
Confidence intervals and significance tests
   for regression coefficients 612
ANOVA table for multiple regression 613
Squared multiple correlation R^2 615
Section 11.1 Summary **615**
Section 11.1 Exercises **616**

11.2 A Case Study 618
Preliminary analysis 619
Relationships between pairs of variables 620
Regression on high school grades 622
Interpretation of results 624
Examining the residuals 624
Refining the model 625
Regression on SAT scores 626
Regression using all variables 627
Test for a collection of regression coefficients 631
Beyond the Basics: Multiple logistic regression **632**
Section 11.2 Summary **633**
Section 11.2 Exercises **634**
Chapter 11 Exercises **636**

CHAPTER 12 One-Way Analysis of Variance **643**
Introduction 643

**12.1 Inference for One-Way Analysis of
   Variance 644**
Data for one-way ANOVA 644
Comparing means 645
The two-sample t statistic 647
An overview of ANOVA 647
The ANOVA model 651
Estimates of population parameters 653
Testing hypotheses in one-way ANOVA 656
The ANOVA table 658
The F test 660
Software 663
Beyond the Basics: Testing the equality of spread **665**
Section 12.1 Summary **666**
Section 12.1 Exercises **667**

12.2 Comparing the Means 670
Contrasts 670
Multiple comparisons 677
Power 681
Section 12.2 Summary **685**
Section 12.2 Exercises **685**
Chapter 12 Exercises **687**

CHAPTER 13 Two-Way Analysis of Variance **697**
Introduction 697

13.1 The Two-Way ANOVA Model 698
Advantages of two-way ANOVA 698
The two-way ANOVA model 702
Main effects and interactions 703

13.2 Inference for Two-Way ANOVA 708
The ANOVA table for two-way ANOVA 708
Chapter 13 Summary **713**
Chapter 13 Exercises **714**

Companion Chapters

(on the *IPS* website www.macmillanlearning.com/
ips9e and in LaunchPad)

CHAPTER 14 Logistic Regression **14-1**
Introduction 14-1

14.1 The Logistic Regression Model 14-2
Binomial distributions and odds 14-2
Odds for two groups 14-3
Model for logistic regression 14-5
Fitting and interpreting the logistic
   regression model 14-6

14.2 Inference for Logistic Regression 14-9
Confidence intervals and significance tests 14-9
Multiple logistic regression 14-16
Chapter 14 Summary **14-18**
Chapter 14 Exercises **14-19**
Chapter 14 Notes and Data Sources **14-26**

CHAPTER 15 Nonparametric Tests **15-1**
Introduction 15-1

15.1 The Wilcoxon Rank Sum Test 15-3
The rank transformation 15-4
The Wilcoxon rank sum test 15-5
The Normal approximation 15-7
What hypotheses does Wilcoxon test? 15-9
Ties 15-10
Rank, t, and permutation tests 15-13

Section 15.1 Summary **15-14**
Section 15.1 Exercises **15-15**

15.2 The Wilcoxon Signed Rank Test 15-17
The Normal approximation 15-21
Ties 15-22
Testing a hypothesis about the median of
a distribution 15-23
Section 15.2 Summary **15-24**
Section 15.2 Exercises **15-24**

15.3 The Kruskal-Wallis Test 15-26
Hypotheses and assumptions 15-27
The Kruskal-Wallis test 15-28
Section 15.3 Summary **15-30**
Section 15.3 Exercises **15-31**
Chapter 15 Exercises **15-33**
Chapter 15 Notes and Data Sources **15-34**

CHAPTER 16 Bootstrap Methods and
Permutation Tests **16-1**
Introduction 16-1
Software 16-2

16.1 The Bootstrap Idea 16-3
The big idea: Resampling and the bootstrap
distribution 16-3
Thinking about the bootstrap idea 16-8
Using software 16-9
Section 16.1 Summary **16-10**
Section 16.1 Exercises **16-10**

16.2 First Steps in Using the Bootstrap 16-12
Bootstrap *t* confidence intervals 16-13
Bootstrapping to compare two groups 16-16
Beyond the Basics: The bootstrap for a scatterplot
smoother **16-19**
Section 16.2 Summary **16-21**
Section 16.2 Exercises **16-22**

16.3 How Accurate Is a Bootstrap Distribution? 16-24
Bootstrapping small samples 16-26
Bootstrapping a sample median 16-28
Section 16.3 Summary **16-30**
Section 16.3 Exercises **16-30**

16.4 Bootstrap Confidence Intervals 16-31
Bootstrap percentile confidence intervals 16-31
A more accurate bootstrap confidence
interval: BCa 16-32
Confidence intervals for the correlation 16-35
Section 16.4 Summary **16-37**
Section 16.4 Exercises **16-37**

**16.5 Significance Testing Using Permutation
Tests 16-41**
Using software 16-44
Permutation tests in practice 16-45
Permutation tests in other settings 16-47
Section 16.5 Summary **16-50**
Section 16.5 Exercises **16-51**
Chapter 16 Exercises **16-54**
Chapter 16 Notes and Data Sources **16-56**

CHAPTER 17 Statistics for Quality: Control and
Capability **17-1**
Introduction 17-1
Use of data to assess quality **17-2**

**17.1 Processes and Statistical Process
Control 17-3**
Describing processes 17-3
Statistical process control 17-6
$\bar{x}$ charts for process monitoring 17-8
s charts for process monitoring 17-12
Section 17.1 Summary **17-17**
Section 17.1 Exercises **17-18**

17.2 Using Control Charts 17-22
$\bar{x}$ and R charts 17-23
Additional out-of-control rules 17-24
Setting up control charts 17-26
Comments on statistical control 17-31
Don't confuse control with capability! 17-34
Section 17.2 Summary **17-35**
Section 17.2 Exercises **17-36**

17.3 Process Capability Indexes 17-40
The capability indexes C_p and C_{pk} 17-43
Cautions about capability indexes 17-46
Section 17.3 Summary **17-48**
Section 17.3 Exercises **17-48**

**17.4 Control Charts for Sample
Proportions 17-51**
Control limits for p charts 17-52
Section 17.4 Summary **17-56**
Section 17.4 Exercises **17-56**
Chapter 17 Exercises **17-57**
Chapter 17 Notes and Data Sources **17-59**

Tables T-1
Answers to Odd-Numbered Exercises A-1
Notes and Data Sources N-1
Index I-1

To Teachers: About This Book

Statistics is the science of data. *Introduction to the Practice of Statistics (IPS)* is an introductory text based on this principle. We present methods of basic statistics in a way that emphasizes working with data and mastering statistical reasoning. *IPS* is elementary in mathematical level but conceptually rich in statistical ideas. After completing a course based on our text, we would like students to be able to think objectively about conclusions drawn from data and use statistical methods in their own work.

In *IPS*, we combine attention to basic statistical concepts with a comprehensive presentation of the elementary statistical methods that students will find useful in their work. *IPS* has been successful for several reasons:

1. *IPS* examines the nature of modern statistical practice at a level suitable for beginners. We focus on the production and analysis of data as well as the traditional topics of probability and inference.

2. *IPS* has a logical overall progression, so data production and data analysis are a major focus, while inference is treated as a tool that helps us draw conclusions from data in an appropriate way.

3. *IPS* presents data analysis as more than a collection of techniques for exploring data. We emphasize systematic ways of thinking about data. Simple principles guide the analysis: always plot your data; look for overall patterns and deviations from them; when looking at the overall pattern of a distribution for one variable, consider shape, center, and spread; for relations between two variables, consider form, direction, and strength; always ask whether a relationship between variables is influenced by other variables lurking in the background. We warn students about pitfalls in clear cautionary discussions.

4. *IPS* uses real examples to drive the exposition. Students learn the technique of least-squares regression and how to interpret the regression slope. But they also learn the conceptual ties between regression and correlation and the importance of looking for influential observations.

5. *IPS* is aware of current developments both in statistical science and in teaching statistics. Brief, optional Beyond the Basics sections give quick overviews of topics such as density estimation, scatterplot smoothers, data mining, nonlinear regression, and meta-analysis. Chapter 16 gives an elementary introduction to the bootstrap and other computer-intensive statistical methods.

The title of the book expresses our intent to introduce readers to statistics as it is used in practice. Statistics in practice is concerned with drawing conclusions from data. We focus on problem solving rather than on methods that may be useful in specific settings.

GAISE The College Report of the Guidelines for Assessment and Instruction in Statistics Education (GAISE) Project (www.amstat.org/education/gaise/) was funded by the American Statistical Association to make recommendations for how introductory statistics courses should be taught. This report and its update contain many interesting teaching suggestions, and we strongly recommend that you read it. The philosophy and approach of *IPS* closely reflect the GAISE recommendations. Let's examine each of the latest recommendations in the context of *IPS*.

1. *Teach statistical thinking.* Through our experiences as applied statisticians, we are very familiar with the components that are needed for the appropriate use of statistical methods. We focus on formulating questions, collecting and finding data, evaluating the quality of data, exploring the relationships among variables, performing statistical analyses, and drawing conclusions. In examples and exercises throughout the text, we emphasize putting the analysis in the proper context and translating numerical and graphical summaries into conclusions.

2. *Focus on conceptual understanding.* With the software available today, it is very easy for almost anyone to apply a wide variety of statistical procedures, both simple and complex, to a set of data. Without a firm grasp of the concepts, such applications are frequently meaningless. By using the methods that we present on real sets of data, we believe that students will gain an excellent understanding of these concepts. Our emphasis is on the input (questions of interest, collecting or finding data, examining data) and the output (conclusions) for a statistical analysis. Formulas are given only where they will provide some insight into concepts.

3. *Integrate real data with a context and a purpose.* Many of the examples and exercises in *IPS* include data that we have obtained from collaborators or consulting clients. Other data sets have come from research related to these activities. We have also used the Internet as a data source, particularly for data related to social media and other topics of interest to undergraduates. Our emphasis on real data, rather than artificial data chosen to illustrate a calculation, serves to motivate students and help them see the usefulness of statistics in everyday life. We also frequently encounter interesting statistical issues that we explore. These include outliers and nonlinear relationships. All data sets are available from the text website.

4. *Foster active learning in the classroom.* As we mentioned earlier, we believe that statistics is exciting as something to do rather than something to talk about. Throughout the text, we provide exercises in Use Your Knowledge sections that ask the students to perform some relatively simple tasks that reinforce the material just presented. Other exercises are particularly suited to being worked on and discussed within a classroom setting.

5. *Use technology for developing concepts and analyzing data.* Technology has altered statistical practice in a fundamental way. In the past, some of the calculations that we performed were particularly difficult and tedious. In other words, they were not fun. Today, freed from the burden of computation by software, we can concentrate our efforts on the big picture: what questions are we trying to address with a study and what can we conclude from our analysis?

6. *Use assessments to improve and evaluate student learning.* Our goal for students who complete a course based on *IPS* is that they are able to design and carry out a statistical study for a project in their capstone course or other setting. Our exercises are oriented toward this goal. Many ask about the design of a statistical study and the collection of data. Others ask for a paragraph summarizing the results of an analysis. This recommendation includes the use of projects, oral presentations, article critiques, and written reports. We believe that students using this text will be well prepared to undertake these kinds of activities. Furthermore, we view these activities not only as assessments but also as valuable tools for learning statistics.

Teaching Recommendations We have used *IPS* in courses taught to a variety of student audiences. For general undergraduates from mixed disciplines, we recommend covering Chapters 1 through 8 and Chapters 9, 10, or 12. For a quantitatively strong audience—sophomores planning to major in actuarial science or statistics—we recommend moving more quickly. Add Chapters 10 and 11 to the core material in Chapters 1 through 8. In general, we recommend deemphasizing the material on probability because these students will take a probability course later in their program. For beginning graduate students in such fields as education, family studies, and retailing, we recommend that the students read the entire text (Chapters 11 and 13 lightly), again with reduced emphasis on Chapter 4 and some parts of Chapter 5. In all cases, beginning with data analysis and data production (Part I) helps students overcome their fear of statistics and builds a sound base for studying inference. We believe that *IPS* can easily be adapted to a wide variety of audiences.

The Ninth Edition: What's New?

- Chapter 1 now begins with a short section giving an overview of data.

- "Toward Statistical Inference" (previously Section 3.3), which introduces the concepts of statistical inference and sampling distributions, has been moved to Section 5.1 to better assist with the transition from a single data set to sampling distributions.

- Coverage of mosaic plots as a visual tool for relationships between two categorical variables has been added to Chapters 2 and 9.

- Chapter 3 now begins with a short section giving a basic overview of data sources.

- Coverage of equivalence testing has been added to Chapter 7.

- There is a greater emphasis on sample size determination using software in Chapters 7 and 8.

- Resampling and bootstrapping are now introduced in Chapter 7 rather than Chapter 6.

- "Inference for Categorical Data" is the new title for Chapter 9, which includes goodness of fit as well as inference for two-way tables.

- There are more JMP screenshots and updated screenshots of Minitab, Excel, and SPSS outputs.

- **Design** A new design incorporates colorful, revised figures throughout to aid the students' understanding of text material. Photographs related to chapter examples and exercises make connections to real-life applications and provide a visual context for topics. More figures with software output have been included.

- **Exercises and Examples** More than 30% of the exercises are new or revised, and there are more than 1700 exercises total. Exercise sets have been added at the end of sections in Chapters 9 through 12. To maintain the attractiveness of the examples to students, we have replaced or updated a large number of them. More than 30% of the 430 examples are new or revised. A list of exercises and examples categorized by application area is provided on the inside of the front cover.

In addition to the new ninth edition enhancements, *IPS* has retained the successful pedagogical features from previous editions:

- **Look Back** At key points in the text, Look Back margin notes direct the reader to the first explanation of a topic, providing page numbers for easy reference.

- **Caution** Warnings in the text, signaled by a caution icon, help students avoid common errors and misconceptions.

- **Challenge Exercises** More challenging exercises are signaled with an icon. Challenge exercises are varied: some are mathematical, some require open-ended investigation, and others require deeper thought about the basic concepts.

- **Applets** Applet icons are used throughout the text to signal where related interactive statistical applets can be found on the *IPS* website and in LaunchPad.

- **Use Your Knowledge Exercises** We have found these exercises to be a very useful learning tool. They appear throughout each section and are listed, with page numbers, before the section-ending exercises.

- **Technology output screenshots** Most statistical analyses rely heavily on statistical software. In this book, we discuss the use of Excel 2013, JMP 12, Minitab 17, SPSS 23, CrunchIt, R, and a TI-83/-84 calculator for conducting statistical analysis. As specialized statistical packages, JMP, Minitab, and SPSS are the most popular software choices both in industry and in colleges and schools of business. R is an extremely powerful statistical environment that is free to anyone; it relies heavily on members of the academic and general statistical communities for support. As an all-purpose spreadsheet program, Excel provides a limited set of statistical analysis options in comparison. However, given its pervasiveness and wide acceptance in industry and the computer world at large, we believe it is important to give Excel proper attention. It should be noted that for users who want more statistical capabilities but want to work in an Excel environment, there are a number of commercially available add-on packages (if you have JMP, for instance, it can be invoked from within Excel). Finally, instructions are provided for the TI-83/-84 calculators.

Even though basic guidance is provided in the book, it should be emphasized that *IPS* is not bound to any of these programs. Computer output from statistical packages is very similar, so you can feel quite comfortable using any one these packages.

Acknowledgments

We are pleased that the first eight editions of *Introduction to the Practice of Statistics* have helped to move the teaching of introductory statistics in a direction supported by most statisticians. We are grateful to the many colleagues and students who have provided helpful comments, and we hope that they will find this new edition another step forward. In particular, we would like to thank the following colleagues who offered specific comments on the new edition:

Ali Arab, *Georgetown University*

Tessema Astatkie, *Dalhousie University*

Fouzia Baki, *McMaster University*

Lynda Ballou, *New Mexico Institute of Mining and Technology*

Sanjib Basu, *Northern Illinois University*

David Bosworth, *Hutchinson Community College*

Max Buot, *Xavier University*

Nadjib Bouzar, *University of Indianapolis*

Matt Carlton, *California Polytechnic State University–San Luis Obispo*

Gustavo Cepparo, *Austin Community College*

Pinyuen Chen, *Syracuse University*

Dennis L. Clason, *University of Cincinnati–Blue Ash College*

Tadd Colver, *Purdue University*

Chris Edwards, *University of Wisconsin–Oshkosh*

Irina Gaynanova, *Texas A&M University*

Brian T. Gill, *Seattle Pacific University*

Mary Gray, *American University*

Gary E. Haefner, *University of Cincinnati*

Susan Herring, *Sonoma State University*

Lifang Hsu, *Le Moyne College*

Tiffany Kolba, *Valparaiso University*

Lia Liu , *University of Illinois at Chicago*

Xuewen Lu, *University of Calgary*

Antoinette Marquard, *Cleveland State University*

Frederick G. Schmitt, *College of Marin*

James D. Stamey, *Baylor University*

Engin Sungur, *University of Minnesota–Morris*

Anatoliy Swishchuk, *University of Calgary*

Richard Tardanico, *Florida International University*

Melanee Thomas, *University of Calgary*

Terri Torres, *Oregon Institute of Technology*

Mahbobeh Vezvaei, *Kent State University*

Yishi Wang, *University of North Carolina–Wilmington*

John Ward, *Jefferson Community and Technical College*

Debra Wiens, *Rocky Mountain College*

Victor Williams, *Paine College*

Christopher Wilson, *Butler University*

Anne Yust, *Birmingham-Southern College*

Biao Zhang, *The University of Toledo*

Michael L. Zwilling, *University of Mount Union*

The professionals at Macmillan, in particular, Terri Ward, Karen Carson, Jorge Amaral, Emily Tenenbaum, Ed Dionne, Blake Logan, and Susan Wein, have contributed greatly to the success of *IPS*. In addition, we would like to thank Tadd Colver at Purdue University for his valuable contributions to the ninth edition, including authoring the back-of-book answers, solutions, and Instructor's Guide. We'd also like to thank Monica Jackson at American University for accuracy reviewing the back-of-book answers and solutions and for authoring the test bank. Thanks also to Michael Zwilling at University of Mount Union for accuracy reviewing the test bank, Christopher Edwards at University of Wisconsin Oshkosh for authoring the lecture slides, and James Stamey at Baylor University for authoring the Clicker slides.

Most of all, we are grateful to the many friends and collaborators whose data and research questions have enabled us to gain a deeper understanding of the science of data. Finally, we would like to acknowledge the contributions of John W. Tukey, whose contributions to data analysis have had such a great influence on us as well as a whole generation of applied statisticians.

Media and Supplements

LaunchPad, our online course space, combines an interactive e-Book with high-quality multimedia content and ready-made assessment options, including LearningCurve adaptive quizzing. Content is easy to assign or adapt with your own material, such as readings, videos, quizzes, discussion groups, and more. LaunchPad also provides access to a Gradebook that offers a window into your students' performance—either individually or as a whole. Use LaunchPad on its own or integrate it with your school's learning management system so your class is always on the same page. To learn more about LaunchPad for *Introduction to the Practice of Statistics*, **Ninth Edition**, or to request access, go to launchpadworks.com.

Assets integrated into LaunchPad include:

Interactive e-Book. Every LaunchPad e-Book comes with powerful study tools for students, video and multimedia content, and easy customization for instructors. Students can search, highlight, and bookmark, making it easier to study and access key content. And teachers can ensure that their classes get just the book they want to deliver: customize and rearrange chapters; add and share notes and discussions; and link to quizzes, activities, and other resources.

LearningCurve provides students and instructors with powerful adaptive quizzing, a game-like format, direct links to the e-Book, and instant feedback. The quizzing system features questions tailored specifically to the text and adapts to students' responses, providing material at different difficulty levels and topics based on student performance.

JMP Student Edition (developed by SAS) is easy to learn and contains all the capabilities required for introductory statistics. JMP is the leading commercial data analysis software of choice for scientists, engineers, and analysts at companies throughout the world (for Windows and Mac). Register inside Launch-Pad at no additional cost.

CrunchIt!® is a Web-based statistical program that allows users to perform all the statistical operations and graphing needed for an introductory statistics course and more. It saves users time by automatically loading data from *IPS*, 9e, and it provides the flexibility to edit and import additional data.

StatBoards Videos are brief whiteboard videos that illustrate difficult topics through additional examples, written and explained by a select group of statistics educators.

Stepped Tutorials are centered on algorithmically generated quizzing with step-by-step feedback to help students work their way toward the correct solution. These exercise tutorials (two to three per chapter) are easily assignable and assessable.

Statistical Video Series consists of StatClips, StatClips Examples, and Statistically Speaking "Snapshots." View animated lecture videos, whiteboard lessons, and documentary-style footage that illustrate key statistical concepts and help students visualize statistics in real-world scenarios.

Video Technology Manuals, available for TI-83/84 calculators, Minitab, Excel, JMP, SPSS, R, Rcmdr, and CrunchIt!®, provide brief instructions for using specific statistical software.

StatTutor Tutorials offer multimedia tutorials that explore important concepts and procedures in a presentation that combines video, audio, and interactive features. The newly revised format includes built-in, assignable assessments and a bright new interface.

Statistical Applets give students hands-on opportunities to familiarize themselves with important statistical concepts and procedures in an interactive setting that allows them to manipulate variables and see the results graphically. Icons in the textbook indicate when an applet is available for the material being covered. Applets are assessable and assignable in LaunchPad.

Stats@Work Simulations put students in the role of the statistical consultant, helping them better understand statistics interactively within the context of real-life scenarios.

EESEE Case Studies (Electronic Encyclopedia of Statistical Examples and Exercises), developed by The Ohio State University Statistics Department, teach students to apply their statistical skills by exploring actual case studies using real data.

SolutionMaster offers an easy-to-use web-based version of the instructor's solutions, allowing instructors to generate a solution file for any set of homework exercises.

Data files are available in JMP, ASCII, Excel, TI, Minitab, SPSS (an IBM Company)*, R, and CSV formats.

Student Solutions Manual provides solutions to the odd-numbered exercises in the text and is available as a print supplement and electronically in LaunchPad.

Instructor's Guide with Full Solutions includes teaching suggestions, chapter comments, and detailed solutions to all exercises and is available electronically in LaunchPad.

Test Bank offers hundreds of multiple-choice questions and is available in LaunchPad.

Lecture Slides offer a customizable, detailed lecture presentation of statistical concepts covered in each chapter of *IPS*, 9e. **Image slides** contain all textbook figures and tables. Lecture slides and images slides are available in LaunchPad.

WebAssign offers algorithmic questions from *IPS*, 9e, in a powerful online instructional system. WebAssign lets you easily create assignments, grade homework,

---

*SPSS was acquired by IBM in October 2009

and give your students instant feedback. Along with flexible features, class and question-level analytics are available for instructors and students. WebAssign Premium also includes the following resources described above: e-Book, data files, LearningCurve, StatTutor Tutorials, Statistical Videos, Video Technology Manuals, solutions manuals, lecture and image slides, i-Clicker slides, test bank, and practice quizzes.

Additional Resources Available with *IPS*, 9e

Special Software Package A student version of JMP is available for packaging with the printed text. JMP is also available inside LaunchPad at no additional cost.

i-clicker. **i-Clicker** is a two-way radio-frequency classroom response solution developed by educators for educators. Each step of i-Clicker's development has been informed by teaching and learning.

To Students: What Is Statistics?

Statistics is the science of collecting, organizing, and interpreting numerical facts, which we call *data*. We are bombarded by data in our everyday lives. The news mentions movie box-office sales, the latest poll of the president's popularity, and the average high temperature for today's date. Advertisements claim that data show the superiority of the advertiser's product. All sides in public debates about economics, education, and social policy argue from data. A knowledge of statistics helps separate sense from nonsense in this flood of data.

The study and collection of data are also important in the work of many professions, so training in the science of statistics is valuable preparation for a variety of careers. Each month, for example, government statistical offices release the latest numerical information on unemployment and inflation. Economists and financial advisers, as well as policymakers in government and business, study these data in order to make informed decisions. Doctors must understand the origin and trustworthiness of the data that appear in medical journals. Politicians rely on data from polls of public opinion. Business decisions are based on market research data that reveal consumer tastes and preferences. Engineers gather data on the quality and reliability of manufactured products. Most areas of academic study make use of numbers and, therefore, also make use of the methods of statistics. This means it is extremely likely that your undergraduate research projects will involve, at some level, the use of statistics.

Learning from Data

The goal of statistics is to learn from data. To learn, we often perform calculations or make graphs based on a set of numbers. But to learn from data, we must do more than calculate and plot because data are not just numbers; they are numbers that have some context that helps us learn from them.

More than two-thirds of Americans are overweight or obese according to the Centers for Disease Control and Prevention (CDC) website (www.cdc.gov/nchs/nhanes.htm). What does it mean to be obese or to be overweight? To answer this question, we need to talk about body mass index (BMI). Your weight in kilograms divided by the square of your height in meters is your BMI. A man who is 6 feet tall (1.83 meters) and weighs 180 pounds (81.65 kilograms) will have a BMI of $81.65/(1.83)^2 = 24.4$ kg/m^2. How do we interpret this number? According to the CDC, a person is classified as overweight if his or her BMI is between 25 and 29.9 kg/m^2 and as obese if his or her BMI is 30 kg/m^2 or more. Therefore, more than two-thirds of Americans have a BMI of 25 kg/m^2 or more. The man who weighs 180 pounds and is 6 feet tall is not overweight or obese, but if he gains 5 pounds, his BMI would increase to 25.1, and he would be classified as overweight.

When you do statistical problems, even straightforward textbook problems, don't just graph or calculate. Think about the context and state your conclusions in the specific setting of the problem. As you are learning how to do statistical calculations and graphs, remember that the goal of statistics is not calculation for its own sake but gaining understanding from numbers. The calculations and graphs can be automated by a calculator or software, but you must supply the understanding. This book presents only the most common specific procedures for statistical analysis. A thorough grasp of the principles of statistics will enable you to quickly learn more advanced methods as needed. On the other hand, a fancy computer analysis carried out without attention to basic principles will often produce elaborate nonsense. As you read, seek to understand the principles as well as the necessary details of methods and recipes.

The Rise of Statistics

Historically, the ideas and methods of statistics developed gradually as society grew interested in collecting and using data for a variety of applications. The earliest origins of statistics lie in the desire of rulers to count the number of inhabitants or measure the value of taxable land in their domains. As the physical sciences developed in the seventeenth and eighteenth centuries, the importance of careful measurements of weights, distances, and other physical quantities grew. Astronomers and surveyors striving for exactness had to deal with variation in their measurements. Many measurements should be better than a single measurement, even though they vary among themselves. How can we best combine many varying observations? Statistical methods that are still important were invented in order to analyze scientific measurements.

By the nineteenth century, the agricultural, life, and behavioral sciences also began to rely on data to answer fundamental questions. How are the heights of parents and children related? Does a new variety of wheat produce higher yields than the old, and under what conditions of rainfall and fertilizer? Can a person's mental ability and behavior be measured just as we measure height and reaction time? Effective methods for dealing with such questions developed slowly and with much debate.

As methods for producing and understanding data grew in number and sophistication, the new discipline of statistics took shape in the twentieth century. Ideas and techniques that originated in the collection of government data, in the study of astronomical or biological measurements, and in the attempt to understand heredity or intelligence came together to form a unified "science of data." That science of data—statistics—is the topic of this text.

The Organization of This Book

Part I of this book, called simply "Looking at Data," concerns data analysis and data production. The first two chapters deal with statistical methods for organizing and describing data. These chapters progress from simpler to more complex data. Chapter 1 examines data on a single variable; Chapter 2 is devoted to relationships among two or more variables. You will learn both how to examine data produced by others and how to organize and summarize your own data. These summaries will first be graphical, then numerical, and then, when appropriate, in the form of a mathematical model that gives a compact description of the overall pattern of the data. Chapter 3 outlines arrangements (called designs) for producing data that answer specific questions. The principles presented in this chapter will help you to design proper samples and experiments for your research projects and to evaluate other such investigations in your field of study.

Part II, consisting of Chapters 4 through 8, introduces statistical inference—formal methods for drawing conclusions from properly produced data. Statistical inference uses the language of probability to describe how reliable its conclusions are, so some basic facts about probability are needed to understand inference. Probability is the subject of Chapters 4 and 5. Chapter 6, perhaps the most important chapter in the text, introduces the reasoning of statistical inference. Effective inference is based on good procedures for producing data (Chapter 3), careful examination of the data (Chapters 1 and 2), and an understanding of the nature of statistical inference as discussed in Chapter 6. Chapters 7 and 8 describe some of the most common specific methods of inference, for drawing conclusions about means and proportions from one and two samples.

The five shorter chapters in Part III introduce somewhat more advanced methods of inference, dealing with relations in categorical data, regression and correlation, and analysis of variance. Four supplementary chapters, available from the text website, present additional statistical topics.

What Lies Ahead

Introduction to the Practice of Statistics is full of data from many different areas of life and study. Many exercises ask you to express briefly some understanding gained from the data. In practice, you would know much more about the background of the data you work with and about the questions you hope the data will answer. No textbook can be fully realistic. But it is important to form the habit of asking, "What do the data tell me?" rather than just concentrating on making graphs and doing calculations.

You should have some help in automating many of the graphs and calculations. You should certainly have a calculator with basic statistical functions. Look for keywords such as "two-variable statistics" or "regression" when you shop for a calculator. More advanced (and more expensive) calculators will do much more, including some statistical graphs. You may be asked to use software as well. There are many kinds of statistical software, from spreadsheets to large programs for advanced users of statistics. The kind of computing available to learners varies a great deal from place to place—but the big ideas of statistics don't depend on any particular level of access to computing.

Because graphing and calculating are automated in statistical practice, the most important assets you can gain from the study of statistics are an understanding of the big ideas and the beginnings of good judgment in working with data. Ideas and judgment can't (at least yet) be automated. They guide you in telling the computer what to do and in interpreting its output. This book tries to explain the most important ideas of statistics, not just teach methods. Some examples of big ideas that you will meet are "always plot your data," "randomized comparative experiments," and "statistical significance."

You learn statistics by doing statistical problems. "Practice, practice, practice." Be prepared to work problems. The basic principle of learning is persistence. Being organized and persistent is more helpful in reading this book than knowing lots of math. The main ideas of statistics, like the main ideas of any important subject, took a long time to discover and take some time to master. The gain will be worth the pain.

About the Authors

David S. Moore is Shanti S. Gupta Distinguished Professor of Statistics, Emeritus, at Purdue University and was 1998 president of the American Statistical Association. He received his AB from Princeton and his PhD from Cornell, both in mathematics. He has written many research papers in statistical theory and served on the editorial boards of several major journals.

Professor Moore is an elected fellow of the American Statistical Association and of the Institute of Mathematical Statistics and is an elected member of the International Statistical Institute. He has served as program director for statistics and probability at the National Science Foundation.

In recent years, Professor Moore has devoted his attention to the teaching of statistics. He was the content developer for the Annenberg/Corporation for Public Broadcasting college-level telecourse, *Against All Odds: Inside Statistics*, and for the series of video modules, *Statistics: Decisions through Data,* intended to aid the teaching of statistics in schools. He is the author of influential articles on statistics education and of several leading texts. Professor Moore has served as president of the International Association for Statistical Education and has received the Mathematical Association of America's national award for distinguished college or university teaching of mathematics.

George P. McCabe is Associate Dean for Academic Affairs in the College of Science and Professor of Statistics at Purdue University. In 1966, he received a BS degree in mathematics from Providence College and in 1970 a PhD in mathematical statistics from Columbia University. His entire professional career has been spent at Purdue, with sabbaticals at Princeton University, the Commonwealth Scientific and Industrial Research Organization (CSIRO) in Melbourne (Australia), the University of Berne (Switzerland), the National Institute of Standards and Technology (NIST) in Boulder, Colorado, and the National University of Ireland in Galway. Professor McCabe is an elected fellow of the American Association for the Advancement of Science and of the American Statistical Association; he was 1998 chair of its section on Statistical Consulting. In 2008–2010, he served on the Institute of Medicine Committee on Nutrition Standards for the National School Lunch and Breakfast Programs. He has served on the editorial boards of several statistics journals. He has consulted with many major corporations and has testified as an expert witness on the use of statistics in several cases.

Professor McCabe's research interests have focused on applications of statistics. Much of his recent work has focused on problems in nutrition, including nutrient requirements, calcium metabolism, and bone health. He is the author or coauthor of more than 190 publications in many different journals.

Bruce A. Craig is Professor of Statistics and Director of the Statistical Consulting Service at Purdue University. He received his BS in mathematics and economics from Washington University in St. Louis and his PhD in statistics from the University of Wisconsin–Madison. He is an elected fellow of the American Association for the Advancement of Science and of the American Statistical Association and was chair of its section on Statistical Consulting in 2009. He has also been an active member of the Eastern North American Region of the International Biometrics Society and was elected by the voting membership to the Regional Committee between 2003 and 2006.

Professor Craig has served on the editorial board of several statistical journals and has been a member of several data and safety monitoring boards, including Purdue's institutional review board.

Professor Craig's research interests focus on the development of novel statistical methodology to address research questions in the life sciences. Areas of current interest are diagnostic testing, inter-rater agreement, and abundance estimation. He is an author or coauthor of more than 100 papers in more than 50 different journals. In 2005, he was named Purdue University Faculty Scholar.

Data Table Index

TABLE 1.1   IQ test scores for 60 randomly chosen fifth-grade students 14

TABLE 1.2   Service times (seconds) for calls to a customer service center 17

TABLE 1.3   Educational data for 78 seventh-grade students 26

TABLE 2.1   Four data sets for exploring correlation and regression 122

TABLE 2.2   Two measures of glucose level in diabetics 127

TABLE 2.3   Dwelling permits, sales, and production for 21 countries 154

TABLE 2.4   World record times for the 10,000-meter run 156

TABLE 5.1   Length (in minutes) of 60 visits to a statistics help room 295

TABLE 7.1   Monthly rates of return on a portfolio (%) 415

TABLE 7.2   Parts measurements using optical software 419

TABLE 7.3   DRP scores for third-graders 437

TABLE 7.4   Seated systolic blood pressure (mm Hg) 450

TABLE 7.5   Length (in seconds) of audio files sampled from an iPod 470

TABLE 10.1   Annual number of tornadoes in the United States between 1953 and 2014 581

TABLE 10.2   In-state tuition and fees (in dollars) for 33 public universities 600

TABLE 10.3   Sales price and assessed value (in thousands of $) of 35 homes in a midwestern city 602

TABLE 10.4   Watershed area (km^2), percent forest, and index of biotic integrity 603

TABLE 13.1   Iron content (mg/100 g) of food cooked in different pots 722

TABLE 13.2   Tool diameter data 723

TABLE 16.1   Degree of reading power scores for third-graders 16-42

TABLE 16.2   Aggressive behaviors of dementia patients 16-48

TABLE 16.3   Serum retinol levels (μmol/l in two groups of children 16-53

TABLE 17.1   Twenty control chart samples of water resistance (depth in mm) 17-10

TABLE 17.2   Control chart constants 17-14

TABLE 17.3   Twenty samples of size 3, with $\bar{x}$ and s 17-19

TABLE 17.4   Three sets of $\bar{x}$'s from 20 samples of size 4 17-20

TABLE 17.5   Twenty samples of size 4, with $\bar{x}$ and s 17-21

TABLE 17.6 $\bar{x}$ and s for 24 samples of elastomer viscosity (in Mooneys) 17-27

TABLE 17.7 $\bar{x}$ and s for 24 samples of label placement (in inches) 17-36
TABLE 17.8 $\bar{x}$ and s for 24 samples of label placement (in inches) 17-36
TABLE 17.9   Hospital losses for 15 samples of DRG 209 patients 17-37
TABLE 17.10   Daily calibration samples for a lunar bone densitometer 17-38
TABLE 17.11 $\bar{x}$ and s for samples of bore diameter 17-39
TABLE 17.12   Fifty control chart samples of call center response times 17-50
TABLE 17.13   Proportions of workers absent during four weeks 17-55
TABLE 17.14 $\bar{x}$ and s for samples of film thickness (mm $\times 10^{-4}$) 17-58

Beyond the Basics Index

Chapter 1   Density estimation 68
Chapter 2   Data mining 132
Chapter 3   Capture-recapture sampling 199
Chapter 4   More laws of large numbers 253
Chapter 5   Weibull distributions 305
Chapter 7   The bootstrap 424
Chapter 8   The plus four confidence interval for a single proportion 489
Chapter 8   The plus four confidence interval for a difference in
                 proportions 509
Chapter 8   Relative risk 518
Chapter 9   Meta-analysis 542
Chapter 10   Nonlinear regression 576
Chapter 11   Multiple logistic regression 632
Chapter 12   Testing the equality of spread 665
Chapter 16   The bootstrap for a scatterplot smoother 16-19

iStock/Creatista/Getty Images Plus

Looking at Data—Distributions

Introduction

Statistics is the science of learning from data. Data are numerical or qualitative descriptions of the objects that we want to study. In this chapter, we will master the art of examining data.

We begin in Section 1.1 with some basic ideas about data. We will learn about the different types of data that are collected and how data sets are organized.

Section 1.2 starts our process of learning from data by looking at graphs. These visual displays give us a picture of the overall patterns in a set of data. We have excellent software tools that help us make these graphs. However, it takes a little experience and a lot of judgment to study the graphs carefully and to explain what they tell us about our data.

Section 1.3 continues our process of learning from data by computing numerical summaries. These sets of numbers describe key characteristics of the patterns that we saw in our graphical summaries.

The final section in this chapter helps us make the transition from data summaries to statistical models that are used to draw conclusions and to make predictions. Specifically, we learn about using density curves to describe a set of data and are introduced to the Normal distributions. These distributions can be used to describe many sets of data that we will encounter. They also play a fundamental role in many of the methods of statistical analysis.

1.1 Data

1.2 Displaying Distributions with Graphs

1.3 Describing Distributions with Numbers

1.4 Density Curves and Normal Distributions

1.1 Data

When you complete this section, you will be able to:

- Give examples of cases in a data set.
- Identify the variables in a data set.
- Demonstrate how a label can be used as a variable in a data set.
- Identify the values of a variable.
- Classify variables as categorical or quantitative.
- Describe the key characteristics of a set of data.
- Explain how a rate is the result of adjusting one variable to create another.

A statistical analysis starts with a set of data. We construct a set of data by first deciding what *cases,* or units, we want to study. For each case, we record information about characteristics that we call *variables*.

CASES, LABELS, VARIABLES, AND VALUES

Cases are the objects described by a set of data. Cases may be customers, companies, subjects in a study, units in an experiment, or other objects.

A **label** is a special variable used in some data sets to distinguish the different cases.

A **variable** is a characteristic of a case.

Different cases can have different **values** of the variables.

EXAMPLE 1.1

COUPONS

Restaurant discount coupons. A website offers coupons that can be used to get discounts for various items at local restaurants. Coupons for food are very popular. Figure 1.1 gives information for seven restaurant coupons that were available for a recent weekend. These are the cases. Data for each coupon are listed on a different line, and the first column has the coupons numbered from 1 to 7. The remaining columns gives the type of restaurant, the name of the restaurant, the item being discounted, the regular price, and the discount price.

FIGURE 1.1 Spreadsheet of food discount coupons, Example 1.1.

	A	B	C	D	E	F
1	ID	Type	Name	Item	RegPrice	DiscPrice
2	1	Italian	Domo's	Pizza	20	10
3	2	Italian	Mama Rita's	Pizza	20	12
4	3	BBQ	Smokey McSween's	Barbecue	30	17
5	4	BBQ	Smokey Grill	Ribs	20	11
6	5	Mexican	Dos Amigos	Tacos	16	8
7	6	Mexican	Holy Guacamole	Steak fajitas	13	8
8	7	Seafood	Sea Grille	Shrimp platter	20	11

Some variables, like the type of restaurant, the name of the restaurant, and the item simply place coupons into categories. The regular price and discount price columns have numerical values for which we can do arithmetic. It makes sense to give an average of the regular prices, but it does not make sense to give an "average" type of restaurant. We can, however, do arithmetic to compare the regular prices classified by type of restaurant.

CATEGORICAL AND QUANTITATIVE VARIABLES

A **categorical variable** places a case into one of several groups or categories.

A **quantitative variable** takes numerical values for which arithmetic operations such as adding and averaging make sense.

EXAMPLE 1.2

COUPONS

Categorical and quantitative variables for coupons. The restaurant discount coupon file has six variables: coupon number, type of restaurant, name of restaurant, item, regular price, and discount price. The two price variables are quantitative variables. Coupon number, type of restaurant, name of restaurant, and item are categorical variables.

An appropriate label for your cases should be chosen carefully. In our food coupon example, a natural choice of a label would be the name of the restaurant. However, if there are two or more coupons available for a particular restaurant, or if a restaurant is a chain with different discounts offered at different locations, then the name of the restaurant would not uniquely label each of the coupons. In the restaurant discount coupon file, the first variable, ID, is a unique label for each coupon.

spreadsheet   The display in Figure 1.1 is from an Excel **spreadsheet**. Spreadsheets are very useful for doing the kind of simple computations that you will do in Exercise 1.2. You can type in a formula and have the same computation performed for each row.

Note that the names we have chosen for the variables in our spreadsheet do not have spaces. For example, instead of "Restaurant Name" for the name of the restaurant, we simply use Name. *In some statistical software packages, however, spaces are not allowed in variable names.* For this reason, when creating spreadsheets for eventual use with statistical software, it is best to avoid spaces in variable names. Another convention is to use an underscore (_) where you would normally use a space. For our data set, we could have used Regular_Price and Discount_Price for the two price variables.

USE YOUR KNOWLEDGE

1.1 Read the spreadsheet. Refer to Figure 1.1. Give the regular price and the discount price for the Smokey Grill ribs coupon.

1.2 How much is the discount worth? Refer to Example 1.1. Consider adding another column to the spreadsheet that gives the coupon savings. Explain how you would compute the entries in this column. Does the new column contain values for a categorical variable or for a quantitative variable? Explain your answer.

unit of measurement

Another important part of the description of any quantitative variable is its unit of measurement. For both RegPrice and DiscPrice, the **unit of measurement** is clearly dollars. In other settings, it may not be as obvious. For example, if we were measuring heights of children, we might choose to use either inches or centimeters. The units of measurement are an important part of the description of a quantitative variable.

Key characteristics of a data set

In practice, any set of data is accompanied by background information that helps us understand the data. When you plan a statistical study or explore data from someone else's work, ask yourself the following questions:

1. **Who?** What **cases** do the data describe? **How many** cases does the data set contain?

2. **What?** How many **variables** do the data contain? What are the **exact definitions** of these variables? What are the units of measurement for each quantitative variable?

3. **Why?** **What purpose** do the data have? Do we hope to answer some specific questions? Do we want to draw conclusions about cases other than the ones we actually have data for? Are the variables that are recorded suitable for the intended purpose?

EXAMPLE 1.3

Statistics class data. Suppose that you are a teaching assistant for a statistics class and one of your jobs is to keep track of the grades for students in two sections of the course. The cases are the students in the class. There are weekly homework assignments, two exams during the semester, and a final exam. Each of these components is given a numerical score, and the components are added to get a total score that can range from 0 to 1000. Cutoffs of 900, 800, 700, etc., are used to assign letter grades of A, B, C, etc.

The spreadsheet for this course will have seven variables:

- An identifier for each student.

- The number of points earned for homework.

- The number of points earned for the first exam.

- The number of points earned for the second exam.

- The number of points earned for the final exam.

- The total number of points earned.

- The letter grade earned.

The student identifier is a label and the letter grade earned is a categorical variable. All the other variables are measured in "points." Because we can do arithmetic with their values, these variables are quantitative variables.

In our example of statistics class data, the possible values for the grade variable are A, B, C, D, and F. When computing grade point averages, many colleges and universities translate these letter grades into numbers using A = 4, B = 3, C = 2, D = 1, and F = 0. The transformed variable with numeric values is considered to be quantitative because we can average the numerical values across different courses to obtain a grade point average.

Sometimes, experts argue about numerical scales such as this. They ask whether or not the difference between an A and a B is the same as the difference between a D and an F. Similarly, many questionnaires ask people to respond on a 1 to 5 scale, with 1 representing strongly agree, 2 representing agree, etc. Again we could ask whether or not the five possible values for this scale are equally spaced in some sense. From a practical point of view, the averages that can be computed when we convert categorical scales such as these to numerical values frequently provide a very useful way to summarize data.

EXAMPLE 1.4

Who, what, and why for the statistics class data. The data set in Example 1.3 was constructed to keep track of the grades for students in an introductory statistics course. The cases are the students in the class. There are seven variables in this data set. These include a label for each student and scores for the various course requirements. There are no units for the label and grade. The other variables all have "points" as the unit.

USE YOUR KNOWLEDGE

1.3 Who, what, and why? For the restaurant discount coupon data of Example 1.1 (page 2), what cases do the data describe? How many cases are there? How many variables are there? What are their definitions and units of measurement? What purpose do the data have?

EXAMPLE 1.5

Statistics class data for a different purpose. Suppose that the data for the students in the introductory statistics class were also to be used to study relationships between student characteristics and success in the course. Here, we have decided to focus on the TotalPoints and Grade as the outcomes of interest. Other variables of interest would have been included—for example, Sex, PrevStat (whether or not the student has taken a statistics course previously), and Year (student classification as first, second, third, or fourth year). ID is a categorical variable, TotalPoints is a quantitative variable, and the remaining variables are all categorical.

USE YOUR KNOWLEDGE

1.4 Apartment rentals. A data set lists apartments available for students to rent. Information provided includes the monthly rent, whether or not cable is included free of charge, whether or not pets are allowed, the number of bedrooms, and the distance to the campus. Describe the cases in the data set, give the number of variables, and specify whether each variable is categorical or quantitative.

Often, the variables in a statistical study are easy to understand: height in centimeters, study time in minutes, and so on. But each area of work also has its own special variables. A psychologist uses the Minnesota Multiphasic Personality Inventory (MMPI), and a physical fitness expert measures "VO2 max" (the volume of oxygen consumed per minute while exercising at your maximum capacity). Both of these variables are measured with special **instruments**. VO2 max is measured by exercising while breathing into a mouthpiece connected to an apparatus that measures oxygen consumed. Scores on the MMPI are based on a long questionnaire, which is also called an instrument.

instrument

Part of mastering your field of work is learning what variables are important and how they are best measured. Because details of particular measurements usually require knowledge of the particular field of study, we will say little about them.

Be sure that each variable really does measure what you want it to. A poor choice of variables can lead to misleading conclusions. Often, for example, the **rate** at which something occurs is a more meaningful measure than a simple count of occurrences.

rate

EXAMPLE 1.6

Comparing colleges based on graduates. Think about comparing colleges based on the numbers of graduates. This view tells you something about the relative sizes of different colleges. However, if you are interested in how well colleges succeed at graduating students they admit, it would be better to use a rate. For example, you can find data on the Internet on the six-year graduation rates of different colleges. These rates are computed by examining the progress of first-year students who enroll in a given year. Suppose that at College A there were 1000 first-year students in a particular year, and 800 graduated within six years. The graduation rate is

$$\frac{800}{1000} = 0.80$$

or 80%. College B has 2000 students who entered in the same year, and 1200 graduated within six years. The graduation rate is

$$\frac{1200}{2000} = 0.60$$

or 60%. How do we compare these two colleges? College B has more graduates but College A has a better graduation rate.

In Example 1.6, when we computed the graduation rate, we used the total number of students to adjust the number of graduates. We constructed a new variable by dividing the number of graduates by the total number of students. Computing a rate is just one of several ways of **adjusting one variable to create another**. We often divide one variable by another to compute a more meaningful variable to study. Example 1.20 (page 20) is another type of adjustment.

adjusting one variable to
create another

USE YOUR KNOWLEDGE

1.5 How should you express the change? Between the first exam and the second exam in your statistics course, you increased the amount of time that you spent working exercises. Which of the following three ways would you choose to express the results of your increased work: (a) give the grades on the two exams, (b) give the ratio of the grade on the second exam divided by the grade on the first exam, (c) take the difference between the grade on the second exam and the grade on the first exam, and express this as a percent of the grade on the first exam. Give reasons for your answer.

1.6 Which variable would you choose? Refer to Example 1.6 on colleges and their graduates.

(a) Give a setting in which you would prefer to evaluate the colleges based on the numbers of graduates. Give a reason for your choice.

(b) Give a setting in which you would prefer to evaluate the colleges based on the graduation rates. Give a reason for your choice.

Exercises 1.5 and 1.6 illustrate an important point about presenting the results of your statistical calculations. *Always consider how to best communicate your results to a general audience.* For example, the numbers produced by your calculator or by statistical software frequently contain more digits that are needed. Be sure that you do not include extra information generated by software that will distract from a clear explanation of what you have found.

SECTION 1.1 SUMMARY

• A data set contains information on a number of **cases**. Cases may be customers, companies, subjects in a study, units in an experiment, or other objects.

• For each case, the data give values for one or more **variables**. A variable describes some characteristic of a case, such as a person's height, gender, or salary. Variables can have different **values** for different cases.

• A **label** is a special variable used to identify cases in a data set.

• Some variables are **categorical** and others are **quantitative**. A categorical variable places each individual into a category, such as male or female. A quantitative variable has numerical values that measure some characteristic of each case, such as height in centimeters or annual salary in dollars.

• The **key characteristics** of a data set answer the questions Who?, What?, and Why?

SECTION 1.1 EXERCISES

For Exercises 1.1 and 1.2, see page 3; for Exercise 1.3, see page 5; for Exercise 1.4, see page 5; and for Exercises 1.5 and 1.6, see page 6.

1.7 How do you do online research? A study of 552 first-year college students asked about their favorite choice for doing online research. Possible choices were "Google or Google Scholar," "Library database or website," "Wikipedia or online encyclopedia," and "Other." Names of the students were not recorded, but the students were numbered from 1 to 552 in the data file. The researchers also recorded age, sex, and major area of study for each student.

(a) What are the cases?

(b) Identify the variables and their possible values.

(c) Classify each variable as categorical or quantitative. Be sure to include at least one of each.

(d) Was a label used? Explain your answer.

(e) Summarize the key characteristics of your data set.

1.8 Summer jobs. You are collecting information about summer jobs that are available for college students in your area. Describe a data set that you could use to organize the information that you collect.

(a) What are the cases?

(b) Identify the variables and their possible values.

(c) Classify each variable as categorical or quantitative. Be sure to include at least one of each.

(d) Use a label and explain how you chose it.

(e) Summarize the key characteristics of your data set.

1.9 Employee application data. The personnel department keeps records on all employees in a company. Here is the information that they keep in one of their data files: employee identification number, last name, first name, middle initial, department, number of years with the company, salary, education (coded as high school, some college, or college degree), and age.

(a) What are the cases for this data set?

(b) Describe each type of information as a label, a quantitative variable, or a categorical variable.

(c) Set up a spreadsheet that could be used to record the data. Give appropriate column headings and five sample cases.

1.10 How would you rank cities? Various organizations rank cities and produce lists of the 10 or the 100 best based on various measures. Create a list of criteria that you would use to rank cities. Include at least eight variables, and give reasons for your choices. Say whether each variable is quantitative or categorical.

1.11 Survey of students. A survey of students in an introductory statistics class asked the following questions: (1) age; (2) do you like to sing? (Yes, No); (3) can you play a musical instrument (not at all, a little, pretty well); (4) how much did you spend on food last week (in dollars); (5) height.

(a) Classify each of these variables as categorical or quantitative and give reasons for your answers.

(b) For each variable give the possible values.

1.12 What questions would you ask? Refer to the previous exercise. Make up your own survey with at least six questions. Include at least two categorical variables and at least two quantitative variables. Tell which variables are categorical and which are quantitative. Give reasons for your answers. For each variable, give the possible values.

1.13 How would you rate colleges? Popular magazines rank colleges and universities on their "academic quality" in serving undergraduate students. Describe five variables that you would like to see measured for each college if you were choosing where to study. Give reasons for each of your choices.

1.14 Attending college in your state or in another state. The U.S. Census Bureau collects a large amount of information concerning higher education.[1] For example, the bureau provides a table that includes the following variables: state, number of students from the state who attend college, number of students who attend college in their home state.

(a) What are the cases for this set of data?

(b) Is there a label variable? If yes, what is it?

(c) Identify each variable as categorical or quantitative.

(d) Explain how you might use each of the quantitative variables to explain something about the states.

(e) Consider a variable computed as the number of students in each state who attend college in the state divided by the total number of students from the state who attend college. Explain how you would use this variable to explain something about the states.

1.15 Alcohol-impaired driving fatalities. A report on drunk-driving fatalities in the United States gives the number of alcohol-impaired driving fatalities for each state.[2] Discuss at least three different ways that these numbers could be converted to rates. Give the advantages and disadvantages of each.

1.2 Displaying Distributions with Graphs

When you complete this section, you will be able to:

- Analyze the distribution of a categorical variable using a bar graph.
- Analyze the distribution of a categorical variable using a pie chart.
- Analyze the distribution of a quantitative variable using a stemplot.
- Analyze the distribution of a quantitative variable using a histogram.
- Examine the distribution of a quantitative variable with respect to the overall pattern of the data and deviations from that pattern.
- Identify the shape, center, and spread of the distribution of a quantitative variable.
- Identify and describe any outliers in the distribution of a quantitative variable.
- Use a time plot to describe the distribution of a quantitative variable that is measured over time.

exploratory data analysis

Statistical tools and ideas help us examine data to describe their main features. This examination is called **exploratory data analysis**. Like an explorer crossing unknown lands, we want first to simply describe what we see. Here are two basic strategies that help us organize our exploration of a set of data:

• Begin by examining each variable by itself. Then move on to study the relationships among the variables.

• Begin with a graph or graphs. Then add numerical summaries of specific aspects of the data.

We follow these principles in organizing our learning. This chapter presents methods for describing a single variable. We will study relationships among several variables in Chapter 2. Within each chapter, we will begin with graphical displays, then add numerical summaries for a more complete description.

Categorical variables: Bar graphs and pie charts

distribution of a
categorical variable

count
percent
proportion

The values of a categorical variable are labels for the categories, such as "yes" and "no." The **distribution of a categorical variable** lists the categories and gives either the **count** or the **percent** of cases that fall in each category. An alternative to the percent is the **proportion**, the count divided by the sum of the counts. Note that the percent is simply the proportion times 100.

EXAMPLE 1.7

ONLINE

© Carl Skepper/Alamy

How do you do online research? A study of 552 first-year college students asked about their preferences for online resources. One question asked them to pick their favorite.[3] Here are the results:

Resource	Count (n)
Google or Google Scholar	406
Library database or website	75
Wikipedia or online encyclopedia	52
Other	19
Total	552

Resource is the categorical variable in this example, and the values are the names of the online resources.

Note that the last value of the variable resource is "Other," which includes all other online resources that were given as selection options. For data sets that have a large number of values for a categorical variable, we often create a category such as this that includes categories that have relatively small counts or percents. *Careful judgment is needed when doing this.* You don't want to cover up some important piece of information contained in the data by combining data in this way.

EXAMPLE 1.8

ONLINE

Favorites as percents. When we look at the online resources data set, we see that Google is the clear winner. We see that 406 reported Google or Google Scholar as their favorite. To interpret this number, we need to know that the total number of students polled was 552. When we say that Google is the winner, we can describe this win by saying that 73.6% (406 divided by 552, expressed as a percent) of the students reported Google as their favorite. Here is a table of the preference percents:

Resource	Percent (%)
Google or Google Scholar	73.6
Library database or website	13.6
Wikipedia or online encyclopedia	9.4
Other	3.4
Total	100.0

The use of graphical methods allows us to see this information and other characteristics of the data easily. We now examine two types of graphs.

EXAMPLE 1.9

bar graph

ONLINE

Bar graph for the online resource preference data. Figure 1.2 displays the online resource preference data using a **bar graph**. The heights of the four bars show the percents of the students who reported each of the resources as their favorite.

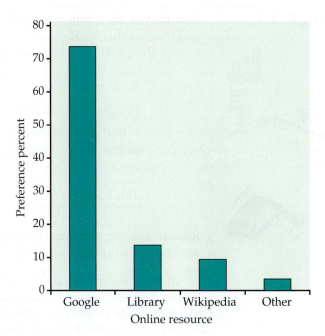

FIGURE 1.2 Bar graph for the online resource preference data, Example 1.9.

The categories in a bar graph can be put in any order. In Figure 1.2, we ordered the resources based on their preference percents. For other data sets, an alphabetical ordering or some other arrangement might produce a more useful graphical display.

You should always consider the best way to order the values of the categorical variable in a bar graph. Choose an ordering that will be useful to you. If you have difficulty, ask a friend if your choice communicates what you expect. Note that a bar graph using counts will look the same as a bar graph using percents. A pie chart naturally uses percents.

EXAMPLE 1.10

pie chart

ONLINE

Pie chart for the online resource preference data. The **pie chart** in Figure 1.3 helps us see what part of the whole each group forms. Here it is very easy to see that Google is the favorite for about three-quarters of the students.

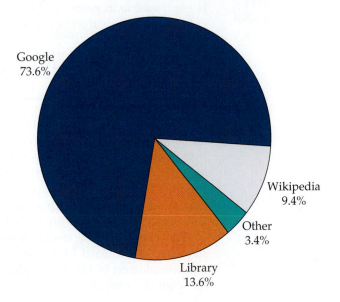

FIGURE 1.3 Pie chart for the online resource preference data, Example 1.10.

USE YOUR KNOWLEDGE

ONLINE

1.16 Compare the bar graph with the pie chart. Refer to the bar graph in Figure 1.2 and the pie chart in Figure 1.3 for the online resource preference data. Which graphical display does a better job of describing the data? Give reasons for your answer.

To make a pie chart, you must include all the categories that make up a whole. A category such as "Other" in this example can be used, but the sum of the percents for all the categories should be 100%. This constraint makes bar graphs more flexible.

Quantitative variables: Stemplots and histograms

A *stemplot* (also called a stem-and-leaf plot) gives a quick picture of the shape of a distribution while including the actual numerical values in the graph. Stemplots work best for small numbers of observations that are all greater than 0.

STEMPLOT

To make a **stemplot**,

1. Separate each observation into a **stem** consisting of all but the final (rightmost) digit and a **leaf**, the final digit. Stems may have as many digits as needed, but each leaf contains only a single digit.

2. Write the stems in a vertical column with the smallest at the top, and draw a vertical line at the right of this column.

3. Write each leaf in the row to the right of its stem, in increasing order out from the stem.

EXAMPLE 1.11

SCF

Soluble corn fiber and calcium. Soluble corn fiber (SCF) has been promoted for various health benefits. One study examined the effect of SCF on the absorption of calcium of adolescent boys and girls. Calcium absorption is expressed as a percent of calcium in the diet. Here are the data for the condition where subjects consumed 12 grams per day (g/d) of SCF.[4]

50	43	43	44	50	44	35	49	54	76	31	48
61	70	62	47	42	45	43	59	53	53	73	

To make a stemplot of these data, use the first digits as stems and the second digits as leaves. Figure 1.4 shows the steps in making the plot, We use the first digit of each value as the stem. Figure 1.4(a) shows the stems that have values 3, 4, 5, 6, and 7. The first entry in our data set is 50. This appears in Figure 1.4(b) on the 5 stem with a leaf of 0. Similarly, the second value, 43, appears in the 4 stem with a leaf of 3. The stemplot is completed in Figure 1.4(c), where the leaves are ordered from smallest to largest.

The center of the distribution is in the 40s, and the data are more stretched out toward high values than low values (the highest value is 76, while the lowest is 31). In the plot, we do not see any extreme values that lie far from the remaining data.

```
3               3 | 5 1            3 | 1 5
4               4 | 3 3 4 4 9 8 7 2 5 3   4 | 2 3 3 3 4 4 5 7 8 9
5               5 | 0 0 4 9 3 3    5 | 0 0 3 3 4 9
6               6 | 1 2            6 | 1 2
7               7 | 6 0 3          7 | 0 3 6

    (a)             (b)                    (c)
```

FIGURE 1.4 Making a stemplot of the data in Example 1.11. (a) Write the stems. (b) Go through the data and write each leaf on the proper stem. For example, the values on the 3-stem are 35 and 31 in the order given in the display for the example. (c) Arrange the leaves on each stem in order out from the stem. The 3-stem now has leaves 1 and 5.

USE YOUR KNOWLEDGE

STAT

1.17 Make a stemplot. Here are the scores on the first exam in an introductory statistics course for 30 students in one section of the course:

82	73	92	82	75	98	94	57	80	90	92	80	87	91	65
73	70	85	83	61	70	90	75	75	59	68	85	78	80	94

Use these data to make a stemplot. Then use the stemplot to describe the distribution of the first-exam scores for this course.

back-to-back stemplot

When you wish to compare two related distributions, a **back-to-back stemplot** with common stems is useful. The leaves on each side are ordered out from the common stem.

EXAMPLE 1.12

SCF

Soluble corn fiber and calcium. Refer to Example 1.11, which gives the data for subjects consuming 12 g/d of SCF. Here are the data for subjects under control conditions (0 g/d of SCF):

42	33	41	49	42	47	48	47	53	72	47	63
68	59	35	46	43	55	38	49	51	51	66	

Figure 1.5 gives the back-to-back stemplot for the SCF and control conditions. The values on the left give absorption for the control condition, while the values on the right give absorption when SCF was consumed. The values for SCF appear to be somewhat higher than the controls.

FIGURE 1.5 A back-to-back stemplot to compare the distributions of calcium absorption under control and SCF conditions, Example 1.12.

```
        8 5 3 | 3 | 1 5
9 9 8 7 7 7 6 3 2 2 1 | 4 | 2 3 3 3 4 4 5 7 8 9
            9 5 3 1 1 | 5 | 0 0 3 3 4 9
              8 6 3 | 6 | 1 2
                  2 | 7 | 0 3 6
```

splitting stems

trimming

There are two modifications of the basic stemplot that can be helpful in different situations. You can double the number of stems in a plot by **splitting each stem** into two: one with leaves 0 to 4 and the other with leaves 5 through 9. When the observed values have many digits, it is often best to **trim** the numbers by removing the last digit or digits before making a stemplot. If you are using software, you can round the data, which is what was done for the data given in Example 1.11.

You must use your judgment in deciding whether to split stems and whether to trim or round, though statistical software will often make these choices for you. Remember that the purpose of a stemplot is to display the shape of a distribution. If there are many stems with no leaves or only one leaf, trimming will reduce the number of stems. Let's take a look at the effect of splitting the stems for our SCF data.

EXAMPLE 1.13

SCF

Stemplot with split stems for SCF. Figure 1.6 presents the data from Example 1.12 in a stemplot with split stems.

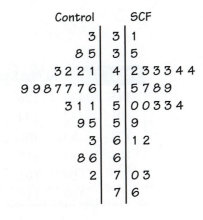

```
        Control        SCF
              3 | 3 | 1
            8 5 | 3 | 5
          3 2 2 1 | 4 | 2 3 3 3 4 4
      9 9 8 7 7 7 6 | 4 | 5 7 8 9
            3 1 1 | 5 | 0 0 3 3 4
              9 5 | 5 | 9
                3 | 6 | 1 2
              8 6 | 6 |
                2 | 7 | 0 3
                  | 7 | 6
```

FIGURE 1.6 A back-to-back stemplot with split stems to compare the distributions of calcium absorption under control and SCF conditions, Example 1.13.

USE YOUR KNOWLEDGE

1.18 Which stemplot do you prefer? Look carefully at the stemplots for the SCF data in Figures 1.5 and 1.6. Which do you prefer? Give reasons for your answer.

1.19 Why should you keep the space? Suppose that you had a data set similar to the one given in Example 1.12, but in which the control values of 66 and 68 were both changed to 64.

(a) Make a stemplot of these data using split stems.

(b) Should you use one stem or two stems for the 60s? Give a reason for your answer. (*Hint:* How would your choice reveal or conceal a potentially important characteristic of the data?)

Histograms

Stemplots display the actual values of the observations. This feature makes stemplots awkward for large data sets. Moreover, the picture presented by a stemplot divides the observations into groups (stems) determined by the number system rather than by judgment.

histogram   Histograms do not have these limitations. A **histogram** breaks the range of values of a variable into classes and displays only the count or percent of the observations that fall into each class. You can choose any convenient number of classes, but you should choose classes of equal width.

Making a histogram by hand requires more work than a stemplot. Histograms do not display the actual values observed. For these reasons, we prefer stemplots for small data sets.

The construction of a histogram is best shown by example. Most statistical software packages will make a histogram for you.

EXAMPLE 1.14

IQ

Distribution of IQ scores. You have probably heard that the distribution of scores on IQ tests is supposed to be roughly "bell-shaped." Let's look at some actual IQ scores. Table 1.1 displays the IQ scores of 60 fifth-grade students chosen at random from one school.

1. Divide the range of the data into classes of equal width. Let's use

$$75 \leq \text{IQ score} < 85$$
$$85 \leq \text{IQ score} < 95$$
$$\vdots$$
$$145 \leq \text{IQ score} < 155$$

TABLE 1.1		IQ Test Scores for 60 Randomly Chosen Fifth-Grade Students							
145	139	126	122	125	130	96	110	118	118
101	142	134	124	112	109	134	113	81	113
123	94	100	136	109	131	117	110	127	124
106	124	115	133	116	102	127	117	109	137
117	90	103	114	139	101	122	105	97	89
102	108	110	128	114	112	114	102	82	101

Be sure to specify the classes precisely so that each individual falls into exactly one class. A student with IQ 84 would fall into the first class, but IQ 85 falls into the second.

2. Count the number of individuals in each class. These counts are called **frequencies**, and a table of frequencies for all classes is a **frequency table**.

Class	Count	Class	Count
75 ≤ IQ score < 85	2	115 ≤ IQ score < 125	13
85 ≤ IQ score < 95	3	125 ≤ IQ score < 135	10
95 ≤ IQ score < 105	10	135 ≤ IQ score < 145	5
105 ≤ IQ score < 115	16	145 ≤ IQ score < 155	1

3. Draw the histogram. First, on the horizontal axis mark the scale for the variable whose distribution you are displaying. That's the IQ score. The scale runs from 75 to 155 because that is the span of the classes we chose. The vertical axis contains the scale of counts. Each bar represents a class. The base of the bar covers the class, and the bar height is the class count. There is no horizontal space between the bars unless a class is empty, so its bar has height zero. Figure 1.7 is our histogram. It does look roughly "bell-shaped."

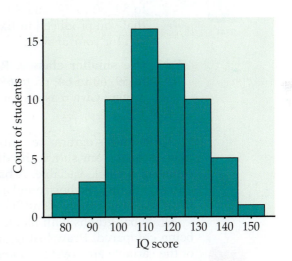

FIGURE 1.7 Histogram of the IQ scores of 60 fifth-grade students, Example 1.14.

Large sets of data are often reported in the form of frequency tables when it is not practical to publish the individual observations. In addition to the frequency (count) for each class, we may be interested in the fraction or percent of the observations that fall in each class. A histogram of percents looks just like a frequency histogram such as Figure 1.7. Simply relabel the vertical scale to read in percents. *Use histograms of percents for comparing several distributions that have different numbers of observations.*

USE YOUR KNOWLEDGE

STAT

1.20   Make a histogram. Refer to the first-exam scores from Exercise 1.17 (page 12). Use these data to make a histogram with classes 50 to 59, 60 to 69, etc. Compare the histogram with the stemplot as a way of describing this distribution. Which do you prefer for these data?

Our eyes respond to the *area* of the bars in a histogram. Because the classes are all the same width, area is determined by height and all classes are fairly represented. There is no one right choice of the classes in a histogram. Too few classes will give a "skyscraper" graph, with all values in a few classes with tall bars. Too many will produce a "pancake" graph, with most classes having one or no observations. Neither choice will give a good picture of the shape of the distribution. You must use your judgment in choosing classes to display the shape. Statistical software will choose the classes for you. The software's choice is often a good one, but you can change it if you want.

You should be aware that the appearance of a histogram can change when you change the classes. The histogram function in the *One-Variable Statistical Calculator* applet on the text website allows you to change the number of classes by dragging with the mouse, so that it is easy to see how the choice of classes affects the histogram.

USE YOUR KNOWLEDGE

STAT

1.21   Change the classes in the histogram. Refer to the first-exam scores from Exercise 1.17 (page 12) and the histogram that you produced in Exercise 1.20. Now make a histogram for these data using classes 40 to 59, 60 to 79, and 80 to 100. Compare this histogram with the one that you produced in Exercise 1.20. Which do you prefer? Give a reason for your answer.

1.22   Use smaller classes. Repeat the previous exercise using classes 55 to 59, 60 to 64, 65 to 69, etc. Of the three histograms, which do you prefer? Give reasons for your answer.

Although histograms resemble bar graphs, their details and uses are distinct. A histogram shows the distribution of counts or percents among the values of a single variable. A bar graph compares the counts or percents of different items. The horizontal axis of a bar graph need not have any measurement scale but simply identifies the items being compared.

Draw bar graphs with blank space between the bars to separate the items being compared. Draw histograms with no space, to indicate that all values of the variable are covered. *Some spreadsheet programs, which are not primarily intended for statistics, will draw histograms as if they were bar graphs, with space between the bars.* Often, you can tell the software to eliminate the space to produce a proper histogram.

Data analysis in action: Don't hang up on me

Many businesses operate call centers to serve customers who want to place an order or make an inquiry. Customers want their requests handled thoroughly. Businesses want to treat customers well, but they also want to avoid wasted time on the phone. They therefore monitor the length of calls and encourage their representatives to keep calls short.

TABLE 1.2	Service Times (Seconds) for Calls to a Customer Service Center						
77	289	128	59	19	148	157	203
126	118	104	141	290	48	3	2
372	140	438	56	44	274	479	211
179	1	68	386	2631	90	30	57
89	116	225	700	40	73	75	51
148	9	115	19	76	138	178	76
67	102	35	80	143	951	106	55
4	54	137	367	277	201	52	9
700	182	73	199	325	75	103	64
121	11	9	88	1148	2	465	25

EXAMPLE 1.15

CALLS80

How long are customer service center calls? We have data on the lengths of all 31,492 calls made to the customer service center of a small bank in a month. Table 1.2 displays the lengths of the first 80 calls.[5]

Take a look at the data in Table 1.2. In this data set, the *cases* are calls made to the bank's call center. The *variable* recorded is the length of each call. The *units* are seconds. We see that the call lengths vary a great deal. The longest call lasted 2631 seconds, almost 44 minutes. More striking is that 8 of these 80 calls lasted less than 10 seconds.

We started our study of the customer service center data by examining a few cases, the ones displayed in Table 1.2. It would be very difficult to examine all 31,492 cases in this way. How can we do this? Let's try a histogram.

EXAMPLE 1.16

CALLS

Histogram for customer service center call lengths. Figure 1.8 is a histogram of the lengths of all 31,492 calls. We did not plot the few lengths greater than 1200 seconds (20 minutes). As expected, the graph shows that most calls last between about 1 and 5 minutes, with some lasting much longer when customers have complicated problems. More striking is the fact that 7.6% of all calls are no more than 10 seconds long.

FIGURE 1.8 The distribution of call lengths for 31,492 calls to a bank's customer service center, Example 1.16. The data show a surprising number of very short calls. These are mostly due to representatives deliberately hanging up in order to bring down their average call length.

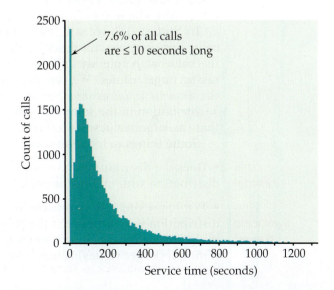

It turned out that the bank penalized representatives whose average call length was too long—so some representatives just hung up on customers to bring their average length down. Neither the customers nor the bank were happy about this. The bank changed its policy, and later data showed that calls under 10 seconds had almost disappeared.

tails

The extreme values of a distribution are in the **tails** of the distribution. The high values are in the upper, or right, tail and the low values are in the lower, or left, tail. The overall pattern in Figure 1.8 is made up of the many moderate call lengths and the long right tail of more lengthy calls. The striking deviation from the overall pattern is the surprising number of very short calls in the left tail.

Our examination of the call center data illustrates some important principles:

• After you understand the background of your data (cases, variables, units of measurement), the first thing to do is **plot** your data.

• When you look at a plot, look for an **overall pattern** and also for any **striking deviations** from the pattern.

Examining distributions

Making a statistical graph is not an end in itself. The purpose of the graph is to help us understand the data. After you make a graph, always ask, "What do I see?" Once you have displayed a distribution, you can see its important features as follows.

EXAMINING A DISTRIBUTION

In any graph of data, look for the **overall pattern** and for striking **deviations** from that pattern.

You can describe the overall pattern of a distribution by its **shape, center**, and **spread**.

An important kind of deviation is an **outlier**, an individual value that falls outside the overall pattern.

In Section 1.3, we will learn how to describe center and spread numerically. For now, we can describe the center of a distribution by its *midpoint*, the value with roughly half the observations taking smaller values and half taking larger values. We can describe the spread of a distribution by giving the *smallest and largest values*. Stemplots and histograms display the shape of a distribution in the same way. Just imagine a stemplot turned on its side so that the larger values lie to the right.

Some things to look for in describing shape are

modes
unimodal

• Does the distribution have one or several major peaks, called **modes**? A distribution with one major peak is called **unimodal**.

symmetric
skewed

• Is it approximately symmetric or is it skewed in one direction? A distribution is **symmetric** if the pattern of values smaller and larger than its midpoint are mirror images of each other. It is **skewed to the right** if the right tail (larger values) is much longer than the left tail (smaller values).

Some variables commonly have distributions with predictable shapes. Many biological measurements on specimens from the same species and sex—lengths of bird bills, heights of young women—have symmetric distributions. Money amounts, on the other hand, usually have right-skewed distributions. There are many moderately priced houses, for example, but the few very expensive mansions give the distribution of house prices a strong right-skew.

EXAMPLE 1.17

IQ

Examine the histogram of IQ scores. What does the histogram of IQ scores (Figure 1.7, page 15) tell us?

Shape: The distribution is *roughly symmetric* with a *single peak* in the center. We don't expect real data to be perfectly symmetric, so in judging symmetry, we are satisfied if the two sides of the histogram are roughly similar in shape and extent.

Center: You can see from the histogram that the midpoint is not far from 110. Looking at the actual data shows that the midpoint is 114.

Spread: The histogram has a spread from 75 to 155. Looking at the actual data shows that the spread is from 81 to 145. There are no outliers or other strong deviations from the symmetric, unimodal pattern.

EXAMPLE 1.18

Examine the histogram of call lengths. The distribution of call lengths in Figure 1.8 (page 17), on the other hand, is strongly *skewed to the right*. The midpoint, the length of a typical call, is about 115 seconds, or just under 2 minutes. The spread is very large, from 1 second to 28,739 seconds.

The longest few calls are outliers. They stand apart from the long right tail of the distribution, though we can't see this from Figure 1.8, which omits the largest observations. The longest call lasted almost 8 hours—that may well be due to equipment failure rather than an actual customer call.

USE YOUR KNOWLEDGE

STAT

1.23 Describe the first-exam scores. Refer to the first-exam scores from Exercise 1.17 (page 12). Use your favorite graphical display to describe the shape, the center, and the spread of these data. Are there any outliers?

Dealing with outliers

In data sets smaller than the service call data, you can spot outliers by looking for observations that stand apart (either high or low) from the overall pattern of a histogram or stemplot. *Identifying outliers is a matter for judgment. Look for points that are clearly apart from the body of the data, not just the most extreme observations in a distribution. You should search for an explanation for any outlier.* Sometimes outliers point to errors made in recording the data. In other cases, the outlying observation may be caused by equipment failure or other unusual circumstances.

EXAMPLE 1.19

COLLEGE

College students. How does the number of undergraduate college students vary by state? Figure 1.9 is a histogram of the numbers of undergraduate students in each of the states.[6] Notice that more than 50% of the states are included in the first bar of the histogram. These states have fewer than 300,000 undergraduates. The next bar includes another 30% of the states. These have between 300,000 and 600,000 students. The bar at the far right of the histogram corresponds to the state of California, which has 2,685,893 undergraduates. California certainly stands apart from the other states for this variable. It is an outlier.

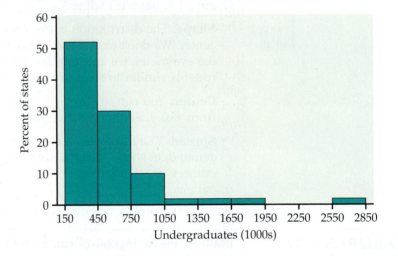

FIGURE 1.9 The distribution of the numbers of undergraduate college students for the 50 states, Example 1.19.

The state of California is an outlier in the previous example because it has a very large number of undergraduate students. California has the largest population of all the states, so we might expect it to have a large number of college students. Let's look at these data in a different way.

EXAMPLE 1.20

COLLEGE

College students per 1000. To account for the fact that there is large variation in the populations of the states, for each state we divide the number of undergraduate students by the population and then multiply by 1000. This gives the undergraduate college enrollment expressed as the number of students per 1000 people in each state. Figure 1.10 gives a stemplot of the distribution. California has 60 undergraduate students per 1000 people. This is one of the higher values in the distribution, but it is clearly not an outlier.

FIGURE 1.10 Stemplot of the numbers of undergraduate college students per 1000 people in each of the 50 states, Example 1.20.

```
3 | 8
4 | 1 1 1 1 1 3 3
4 | 5 5 5 5 6 6 6 7 7 7 7 7 8 8 8 8 9
5 | 0 0 1 1 1 2 2 2 4 4 4 4
5 | 5 6 6
6 | 0 0 0 0 1
6 | 7 9
7 | 1 2
7 | 7
```

USE YOUR KNOWLEDGE

COLLEGE

1.24 Four states with large populations. There are four states with populations greater than 15 million.

(a) Examine the data file and report the names of these four states.

(b) Find these states in the distribution of number of undergraduate students per 1000 people. To what extent do these four states influence the distribution of number of undergraduate students per 1000 people?

In Example 1.19, we looked at the distribution of the number of undergraduate students, while in Example 1.20, we adjusted these data by expressing the counts as number per 1000 people in each state. Which way is correct? The answer depends upon why you are examining the data.

If you are interested in marketing a product to undergraduate students, the unadjusted numbers would be of interest because you want to reach the most people. On the other hand, if you are interested in comparing states with respect to how well they provide opportunities for higher education to their residents, the population-adjusted values would be more suitable. *Always think about why you are doing a statistical analysis, and this will guide you in choosing an appropriate analytic strategy.*

Here is an example with a different kind of outlier.

EXAMPLE 1.21

PTH

Healthy bones and PTH. Bones are constantly being built up (bone formation) and torn down (bone resorption). Young people who are growing have more formation than resorption. When we age, resorption increases to the point where it exceeds formation. (The same phenomenon occurs when astronauts travel in space.) The result is osteoporosis, a disease associated with fragile bones that are more likely to break. The underlying mechanisms that control these processes are complex and involve a variety of substances. One of these is parathyroid hormone (PTH). Here are the values of PTH measured on a sample of 29 boys and girls aged 12 to 15 years:[7]

```
 1 | 9
 2 | 5 8 8 8 9
 3 | 0 1 1 1 3 5 8 9
 4 | 0 5 6 8 9 9
 5 | 0 0 9 9
 6 | 3 4
 7 | 1 1
 8 |
 9 |
10 |
11 |
12 | 7
```

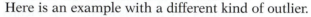

| 39 | 59 | 30 | 48 | 71 | 31 | 25 | 31 | 71 | 50 | 38 | 63 | 49 | 45 | 31 |
| 33 | 28 | 40 | 127 | 49 | 59 | 50 | 64 | 28 | 46 | 35 | 28 | 19 | 29 | |

The data are measured in picograms per milliliter (pg/ml) of blood. The original data were recorded with one digit after the decimal point. They have been rounded to simplify our presentation here. Figure 1.11 gives a stemplot of the data.

The observation 127 clearly stands out from the rest of the distribution. A PTH measurement on this individual taken on a different day was similar to the rest of the values in the data set. We conclude that this outlier was caused by a laboratory error or a recording error, and we are confident in discarding it for any additional analysis.

FIGURE 1.11 Stemplot of the values of PTH, Example 1.21.

Time plots

Whenever data are collected over time, it is a good idea to plot the observations in time order. *Displays of the distribution of a variable that ignore time order, such as stemplots and histograms, can be misleading when there is systematic change over time.*

TIME PLOT

A **time plot** of a variable plots each observation against the time at which it was measured. Always put time on the horizontal scale of your plot and the variable you are measuring on the vertical scale.

EXAMPLE 1.22

VITDS

Seasonal variation in vitamin D. Although we get some of our vitamin D from food, most of us get about 75% of what we need from the sun. Cells in the skin make vitamin D in response to sunlight. If people do not get enough exposure to the sun, they can become deficient in vitamin D, resulting in weakened bones and other health problems. The elderly, who need more vitamin D than younger people, and people who live in northern areas, where there is relatively little sunlight in the winter, are particularly vulnerable to these problems.

Figure 1.12 is a plot of the serum levels of vitamin D versus time of year for samples of subjects from Switzerland.[8] The units measuring Vitamin D are nanomoles per liter (nmol/l) of blood. The observations are grouped into periods of two months for the plot. Means are marked by filled-in circles and are connected by a line in the plot. The effect of the lack of sunlight in the winter months on vitamin D levels is clearly evident in the plot.

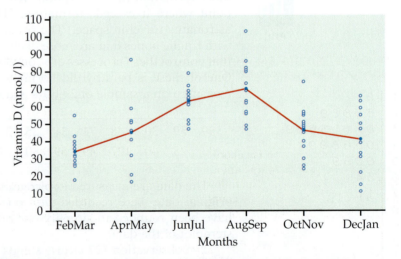

FIGURE 1.12 Plot of vitamin D versus months of the year, Example 1.22.

The data described in the preceding example are based on a subset of the subjects in a study of 248 subjects. The researchers were particularly concerned about subjects whose levels were deficient, defined as a serum vitamin D level of less than 50 nmol/l. They found that there was a 3.8-fold higher deficiency rate in February–March than in August–September: 91.2% versus 24.3%. To ensure that individuals from this population have adequate levels of vitamin D, some form of supplementation is needed, particularly during certain times of the year.

SECTION 1.2 SUMMARY

- **Exploratory data analysis** uses graphs and numerical summaries to describe the variables in a data set and the relations among them.

- The **distribution** of a variable tells us what values it takes and how often it takes these values.

- **Bar graphs** and **pie charts** display the distributions of categorical variables. These graphs use the counts or percents of the categories.

- **Stemplots** and **histograms** display the distributions of quantitative variables. Stemplots separate each observation into a **stem** and a one-digit **leaf**. Histograms plot the **frequencies** (counts) or the percents of equal-width classes of values.

- When examining a distribution, look for **shape, center**, and **spread** and for clear **deviations** from the overall shape.

- Some distributions have simple shapes, such as **symmetric** or **skewed**. The number of **modes** (major peaks) is another aspect of overall shape. Not all distributions have a simple overall shape, especially when there are few observations.

- **Outliers** are observations that lie outside the overall pattern of a distribution. Always look for outliers and try to explain them.

- When observations on a variable are taken over time, make a **time plot** that graphs time horizontally and the values of the variable vertically. A time plot can reveal changes over time.

SECTION 1.2 EXERCISES

For Exercise 1.16, see page 11; for Exercise 1.17, see page 12; for Exercises 1.18 and 1.19, see page 14; for Exercise 1.20, see page 16; for Exercises 1.21 and 1.22, see page 16; for Exercise 1.23, see page 19; and for Exercise 1.24, see page 21.

1.25 Your Facebook app can generate a million dollars a month. A report on Facebook suggests that Facebook apps can generate large amounts of money, as much as $1 million a month.[9] The following table gives the numbers of Facebook users by country for the top 10 countries based on the number of users:[10] 📊 FACEBK

Country	Facebook users (in millions)
Brazil	29.30
India	37.38
Mexico	29.80
Germany	21.46
France	23.19
Philippines	26.87
Indonesia	40.52
United Kingdom	30.39
United States	155.74
Turkey	30.63

(a) Use a bar graph to describe the numbers of users in these countries.

(b) Do you think that the United States is an outlier in this data set? Explain your answer.

(c) Describe the major features of your graph in a short paragraph.

1.26 Facebook use increases by country. Refer to the previous exercise. The report also gave the increases in the number of Facebook users for a one-month period for the same countries: 📊 FACEBK

Country	Increase in users (in millions)
Brazil	2.47
India	1.75
Mexico	0.84
Germany	0.51
France	0.38
Philippines	0.38
Indonesia	0.37
United Kingdom	0.22
United States	0.65
Turkey	0.09

(a) Use a bar graph to describe the increase in users in these countries.

(b) Describe the major features of your graph in a short paragraph.

(c) Do you think a stemplot would be a better graphical display for these data? Give reasons for your answer.

(d) Write a short paragraph about possible business opportunities suggested by the data you described in this exercise and the previous one.

1.27 The *Titanic* and class. On April 15, 1912, on her maiden voyage, the *Titanic* collided with an iceberg and sank. The ship was luxurious but did not have enough lifeboats for the 2224 passengers and crew. As a result of the collision, 1502 people died.[11] The ship had three classes of passengers. The level of luxury and the price of the ticket varied with the class, with first class being the most luxurious. There were 323 passengers in first class, 277 in second class, and 709 in third class.[12] TITANIC

(a) Make a bar graph of these data.

(b) Give a short summary of how the number of passengers varied with class.

(c) If you made a bar graph of the percent of passengers in each class, would the general features of the graph differ from the one you made in part (a)? Explain your answer.

1.28 Another look at the *Titanic* and class. Refer to the previous exercise. TITANIC

(a) Make a pie chart to display the data.

(b) Compare the pie chart with the bar graph. Which do you prefer? Give reasons for your answer.

1.29 Who survived? Refer to the two previous exercises. The number of first-class passengers who survived was 200. For second and third class, the numbers were 119 and 181, respectively. Create a graphical summary that shows how the survival of passengers depended on class. TITANIC

1.30 Potassium from potatoes. The 2015 Dietary Guidelines for Americans[13] notes that the average potassium (K) intake for U.S. adults is about half of the recommended amount. A major source of potassium is potatoes. Nutrients in the diet can have different absorption depending on the source. One study looked at absorption of potassium from different sources. Participants ate a controlled diet for five days, and the amount of potassium absorbed was measured. Data for a diet that included 40 milliequivalents (mEq) of potassium were collected from 27 adult subjects.[14] KPOT40

(a) Make a stemplot of the data.

(b) Describe the pattern of the distribution.

(c) Are there any outliers? If yes, describe them and explain why you have declared them to be outliers.

(d) Describe the shape, center, and spread of the distribution.

1.31 Potassium from a supplement. Refer to the previous exercise. Data were also recorded for 29 subjects who received a potassium salt supplement with 40 mEq of potassium. Answer the questions in the previous exercise for the supplemented subjects. KSUP40

1.32 Energy consumption. The U.S. Energy Information Administration reports data summaries of various energy statistics. Let's look at the total amount of energy consumed, in quadrillions of British thermal units (Btu), for each month in a recent year. Here are the data:[15] ENERGY

Month	Energy (quadrillion Btu)	Month	Energy (quadrillion Btu)
January	9.58	July	8.23
February	8.46	August	8.21
March	8.56	September	7.64
April	7.56	October	7.78
May	7.66	November	8.19
June	7.79	December	8.82

(a) Look at the table and describe how the energy consumption varies from month to month.

(b) Make a time plot of the data and describe the patterns.

(c) Suppose you wanted to communicate information about the month-to-month variation in energy consumption. Which would be more effective, the table of the data or the graph? Give reasons for your answer.

1.33 Energy consumption in a different year. Refer to the previous exercise. Here are the data for the previous year: ENERGY

Month	Energy (quadrillion Btu)	Month	Energy (quadrillion Btu)
January	8.99	July	8.27
February	8.02	August	8.17
March	8.38	September	7.64
April	7.52	October	7.72
May	7.62	November	8.14
June	7.72	December	9.08

(a) Analyze these data using the questions in the previous exercise as a guide.

(b) Compare the patterns across the two years. Describe any similarities and differences.

1.34 Favorite colors. What is your favorite color? One survey produced the following summary of responses to that question: blue, 42%; green, 14%; purple, 14%; red, 8%; black, 7%; orange, 5%; yellow, 3%; brown, 3%; gray, 2%; and white, 2%.[16] Make a bar graph of the percents and write a short summary of the major features of your graph. **FAVCOL**

1.35 Least-favorite colors. Refer to the previous exercise. The same study also asked people about their least-favorite color. Here are the results: orange, 30%; brown, 23%; purple, 13%; yellow, 13%; gray, 12%; green, 4%; white, 4%; red, 1%; black, 0%; and blue, 0%. Make a bar graph of these percents and write a summary of the results. **LFAVCOL**

1.36 Garbage. The formal name for garbage is "municipal solid waste." In the United States, approximately 250 million tons of garbage are generated in a year. Following is a breakdown of the materials that made up American municipal solid waste in 2012:[17] **GARBAGE**

Material	Weight (million tons)	Percent of total
Food scraps	36.4	14.5
Glass	11.6	4.6
Metals	22.4	8.9
Paper, paperboard	68.6	27.4
Plastics	31.7	12.7
Rubber, leather	7.5	3.0
Textiles	14.3	5.7
Wood	15.8	6.3
Yard trimmings	34.0	13.5
Other	8.5	3.4
Total	250.9	100.0

(a) Add the weights. The sum is not exactly equal to the value of 250.9 million tons given in the table. Why?

(b) Make a bar graph of the percents. The graph gives a clearer picture of the main contributors to garbage if you order the bars from tallest to shortest.

(c) Also make a pie chart of the percents. Comparing the two graphs, notice that it is easier to see the small differences among "Food scraps," "Plastics," and "Yard trimmings" in the bar graph.

1.37 Vehicle colors. Vehicle colors differ among regions of the world. Here are data on the most popular colors for vehicles in North America:[18] **VCOLOR**

Color	(percent)
White	24
Black	19
Silver	16
Gray	15
Red	10
Blue	7
Brown	5
Other	4

(a) Describe these data with a bar graph.

(b) Describe these data with a pie chart.

(c) Which graphical summary do you prefer. Give reasons for your answer.

1.38 Sketch a skewed distribution. Sketch a histogram for a distribution that is skewed to the left. Suppose that you and your friends emptied your pockets of coins and recorded the year marked on each coin. The distribution of dates would be skewed to the left. Explain why.

1.39 Grades and self-concept. Table 1.3 presents data on 78 seventh-grade students in a rural midwestern school.[19] The researcher was interested in the relationship between the students' "self-concept" and their academic performance. The data we give here include each student's grade point average (GPA), score on a standard IQ test, and gender, taken from school records. Gender is coded as F for female and M for male. The students are identified only by an observation number (OBS). The missing OBS numbers show that some students dropped out of the study. The final variable is each student's score on the Piers-Harris Children's Self-Concept Scale, a psychological test administered by the researcher. **SEVENGR**

(a) How many variables does this data set contain? Which are categorical variables and which are quantitative variables?

(b) Make a stemplot of the distribution of GPA, after rounding to the nearest tenth of a point.

(c) Describe the shape, center, and spread of the GPA distribution. Identify any suspected outliers from the overall pattern.

(d) Make a back-to-back stemplot of the rounded GPAs for female and male students. Write a brief comparison of the two distributions.

1.40 Describe the IQ scores. Make a graph of the distribution of IQ scores for the seventh-grade students in Table 1.3. Describe the shape, center, and spread of the

TABLE 1.3	Educational Data for 78 Seventh-Grade Students								
OBS	GPA	IQ	Gender	Self-concept	OBS	GPA	IQ	Gender	Self-concept
001	7.940	111	M	67	043	10.760	123	M	64
002	8.292	107	M	43	044	9.763	124	M	58
003	4.643	100	M	52	045	9.410	126	M	70
004	7.470	107	M	66	046	9.167	116	M	72
005	8.882	114	F	58	047	9.348	127	M	70
006	7.585	115	M	51	048	8.167	119	M	47
007	7.650	111	M	71	050	3.647	97	M	52
008	2.412	97	M	51	051	3.408	86	F	46
009	6.000	100	F	49	052	3.936	102	M	66
010	8.833	112	M	51	053	7.167	110	M	67
011	7.470	104	F	35	054	7.647	120	M	63
012	5.528	89	F	54	055	0.530	103	M	53
013	7.167	104	M	54	056	6.173	115	M	67
014	7.571	102	F	64	057	7.295	93	M	61
015	4.700	91	F	56	058	7.295	72	F	54
016	8.167	114	F	69	059	8.938	111	F	60
017	7.822	114	F	55	060	7.882	103	F	60
018	7.598	103	F	65	061	8.353	123	M	63
019	4.000	106	M	40	062	5.062	79	M	30
020	6.231	105	F	66	063	8.175	119	M	54
021	7.643	113	M	55	064	8.235	110	M	66
022	1.760	109	M	20	065	7.588	110	M	44
024	6.419	108	F	56	068	7.647	107	M	49
026	9.648	113	M	68	069	5.237	74	F	44
027	10.700	130	F	69	071	7.825	105	M	67
028	10.580	128	M	70	072	7.333	112	F	64
029	9.429	128	M	80	074	9.167	105	M	73
030	8.000	118	M	53	076	7.996	110	M	59
031	9.585	113	M	65	077	8.714	107	F	37
032	9.571	120	F	67	078	7.833	103	F	63
033	8.998	132	F	62	079	4.885	77	M	36
034	8.333	111	F	39	080	7.998	98	F	64
035	8.175	124	M	71	083	3.820	90	M	42
036	8.000	127	M	59	084	5.936	96	F	28
037	9.333	128	F	60	085	9.000	112	F	60
038	9.500	136	M	64	086	9.500	112	F	70
039	9.167	106	M	71	087	6.057	114	M	51
040	10.140	118	F	72	088	6.057	93	F	21
041	9.999	119	F	54	089	6.938	106	M	56

distribution, as well as any outliers. IQ scores are usually said to be centered at 100. Is the midpoint for these students close to 100, clearly above, or clearly below? SEVENGR

1.41 Describe the self-concept scores. Based on a suitable graph, briefly describe the distribution of self-concept scores for the students in Table 1.3. Be sure to identify any suspected outliers. SEVENGR

1.42 The Boston Marathon. Women were allowed to enter the Boston Marathon in 1972. Here are the times (in minutes, rounded to the nearest minute) for the winning women from 1972 to 2015.

Make a graph that shows change over time. What overall pattern do you see? Have times stopped improving in recent years? If so, when did improvement end?

📊 MARATH

Year	Time	Year	Time	Year	Time	Year	Time
1972	190	1983	143	1994	142	2005	145
1973	186	1984	149	1995	145	2006	143
1974	167	1985	154	1996	147	2007	149
1975	162	1986	145	1997	146	2008	145
1976	167	1987	146	1998	143	2009	152
1977	168	1988	145	1999	143	2010	146
1978	165	1989	144	2000	146	2011	142
1979	155	1990	145	2001	144	2012	151
1980	154	1991	144	2002	141	2013	146
1981	147	1992	144	2003	145	2014	139
1982	150	1993	145	2004	144	2015	145

1.3 Describing Distributions with Numbers

When you complete this section, you will be able to:

- Describe the center of a distribution by using the mean.
- Describe the center of a distribution by using the median.
- Compare the mean and the median as measures of center for a particular set of data.
- Describe the spread of a distribution by using quartiles.
- Describe a distribution by using the five-number summary.
- Describe a distribution by using a boxplot.
- Compare one or more sets of data measured on the same variable by using side-by-side boxplots.
- Identify outliers by using the $1.5 \times IQR$ rule.
- Describe the spread of a distribution by using the standard deviation.
- Choose measures of center and spread for a particular set of data.
- Compute the effects of a linear transformation on the mean, the median, the standard deviation, and the interquartile range.

We can begin our data exploration with graphs, but numerical summaries make our analysis more specific. For categorical variables, numerical summaries are the counts or percents that we use to construct pie charts or bar graphs. In this section, we focus on numerical summaries for quantitative variables. A brief description of the distribution of a quantitative variable should include its *shape* and numbers describing its *center* and *spread*. We describe the shape of a distribution based on inspection of a histogram or a stemplot. Now we will learn specific ways to use numbers to measure the center and spread of a distribution. We can calculate these numerical measures for any quantitative variable. But to interpret measures of center and spread, and to choose among the several measures we will learn, you must think about the shape of the distribution and the meaning of the data. The numbers, like graphs, are aids to understanding, not "the answer" in themselves.

EXAMPLE 1.23

TTS24

```
0 | 2455556678
1 | 01236799
2 | 45
3 | 28
4 | 9
5 | 3
```

FIGURE 1.13 Stemplot for the sample of 24 business start times, Example 1.23.

The distribution of business start times. An entrepreneur faces many bureaucratic and legal hurdles when starting a new business. The World Bank collects information about starting businesses throughout the world. They have determined the time, in days, to complete all the procedures required to start a business.[20] Data for 189 countries are included in the data set, TTS. For this section, we examine data, rounded to integers, for a sample of 24 of these countries. Here are the data:

16	4	5	6	5	7	12	19	10	2	25	19
38	5	24	8	6	5	53	32	13	49	11	17

The stemplot in Figure 1.13 shows us the *shape*, *center*, and *spread* of the business start times. The stems are tens of days and the leaves are days. The distribution is skewed to the right with a very long tail of high values. All but six of the times are less than 20 days. The center appears to be about 10 days, and the values range from 2 days to 53 days. There do not appear to be any outliers.

Measuring center: The mean

Numerical description of a distribution begins with a measure of its center or average. The two common measures of center are the *mean* and the *median*. The mean is the "average value" and the median is the "middle value." These are two different ideas for "center," and the two measures behave differently. We need precise recipes for the mean and the median.

THE MEAN $\bar{x}$

To find the **mean $\bar{x}$** of a set of observations, add their values and divide by the number of observations. If the n observations are $x_1, x_2, \ldots, x_n$, their mean is

$$\bar{x} = \frac{x_1 + x_2 + \cdots + x_n}{n}$$

or, in more compact notation,

$$\bar{x} = \frac{1}{n} \sum x_i$$

The Σ (capital Greek sigma) in the formula for the mean is short for "add them all up." The bar over the x indicates the mean of all the x-values. Pronounce the mean $\bar{x}$ as "x-bar." This notation is so common that writers who are discussing data use $\bar{x}$, $\bar{y}$, etc., without additional explanation. The subscripts on the observations x_i are a way of keeping the n observations separate.

EXAMPLE 1.24

TTS24

Mean time to start a business. The mean time to start a business is

$$\bar{x} = \frac{x_1 + x_2 + \cdots + x_n}{n}$$

$$= \frac{16 + 4 + \cdots + 17}{24}$$

$$= \frac{391}{24} = 16.292$$

The mean time to start a business for the 24 countries in our data set is 16.3 days. Note that we have rounded the answer. Our goal in using the mean to describe the center of a distribution is not to demonstrate that we can compute with great accuracy. The additional digits do not provide any additional useful information. In fact, they distract our attention from the important digits that are meaningful. Do you think it would be better to report the mean as 16 days?

The value of the mean will not necessarily be equal to the value of one of the observations in the data set. Our example of time to start a business illustrates this fact.

In practice, you can key the data into your calculator and hit the Mean key. You don't have to actually add and divide. But you should know that this is what the calculator is doing.

USE YOUR KNOWLEDGE

TTS25

1.43 Include the outlier. For Example 1.23, a random sample of 24 countries was selected from a data set that included 189 countries. The South American country of Suriname, where the start time is 208 days, was not included in the random sample. Consider the effect of adding Suriname to the original set. Show that the mean for the new sample of 25 countries has increased to 24 days. (This is a rounded number. You should report the mean with two digits after the decimal to show that you have performed this calculation.)

STAT

1.44 Find the mean. Here are the scores on the first exam in an introductory statistics course for 10 students:

83 74 93 85 75 97 93 55 92 81

Find the mean first-exam score for these students.

resistant measure

   Exercise 1.43 illustrates an important weakness of the mean as a measure of center: *the mean is sensitive to the influence of a few extreme observations.* These may be outliers, but a skewed distribution that has no outliers will also pull the mean toward its long tail. Because the mean cannot resist the influence of extreme observations, we say that it is not a **resistant measure** of center.

robust measure

   A measure that is resistant does more than limit the influence of outliers. Its value does not respond strongly to changes in a few observations, no matter how large those changes may be. The mean fails this requirement because we can make the mean as large as we wish by making a large enough increase in just one observation. A resistant measure is sometimes called a **robust measure**.

Measuring center: The median

We used the midpoint of a distribution as an informal measure of center in Section 1.2. The *median* is the formal version of the midpoint, with a specific rule for calculation.

THE MEDIAN *M*

The **median *M*** is the midpoint of a distribution. Half the observations are smaller than the median and the other half are larger than the median. Here is a rule for finding the median:

1. Arrange all observations in order of size, from smallest to largest.

2. If the number of observations n is odd, the median M is the center observation in the ordered list. Find the location of the median by counting $(n + 1)/2$ observations up from the bottom of the list.

3. If the number of observations n is even, the median M is the mean of the two center observations in the ordered list. The location of the median is again $(n + 1)/2$ from the bottom of the list.

   Note that the formula $(n + 1)/2$ does not give the median, just the location of the median in the ordered list. Medians require little arithmetic, so they are easy to find by hand for small sets of data. Arranging even a moderate number of observations in order is tedious, however, so that finding the median by hand for larger sets of data is unpleasant. Even simple calculators have an $\bar{x}$ button, but you will need computer software or a graphing calculator to automate finding the median.

EXAMPLE 1.25

TTS24

Median time to start a business. To find the median time to start a business for our 24 countries, we first arrange the data in order from smallest to largest:

2	4	5	5	5	5	6	6	7	8	10	11
12	13	16	17	19	19	24	25	32	38	49	53

The count of observations $n = 24$ is even. The median, then, is the average of the two center observations in the ordered list. To find the location of the center observations, we first compute

$$\text{location of } M = \frac{n+1}{2} = \frac{25}{2} = 12.5$$

Therefore, the center observations are the 12th and 13th observations in the ordered list. The median is

$$M = \frac{11+12}{2} = 11.5$$

Note that you can use the stemplot in Figure 1.13 (page 28) directly to compute the median. In the stemplot the cases are already ordered and you simply need to count from the top or the bottom to the desired location.

USE YOUR KNOWLEDGE

TTS25

CALLS80

STAT

1.45 **Include the outlier.** Include Suriname, where the start time is 208 days, in the data set, and show that the median is 12 days. Note that with this case included, the sample size is now 25 and the median is the 13th observation in the ordered list. Write out the ordered list and circle the outlier. Describe the effect of the outlier on the median for this set of data.

1.46 **Calls to a customer service center.** The service times for 80 calls to a customer service center are given in Table 1.2 (page 17). Use these data to compute the median service time.

1.47 **Find the median.** Here are the scores on the first exam in an introductory statistics course for 10 students:

    83   74   93   85   75   97   93   55   92   81

Find the median first-exam score for these students.

Mean versus median

Exercises 1.43 and 1.45 illustrate an important difference between the mean and the median. Suriname is an outlier. It pulls the mean time to start a business up from 16 days to 24 days. The median increased slightly, from 11.5 days to 12 days.

The median is more *resistant* than the mean. If the largest start time in the data set was 1200 days, the median for all 25 countries would still be 12 days. The largest observation just counts as one observation above the center, no matter how far above the center it lies. The mean uses the actual value of each observation and so will chase a single large observation upward.

The best way to compare the response of the mean and median to extreme observations is to use an interactive applet that allows you to place points on a line and then drag them with your computer's mouse. Exercises 1.83, 1.84, and 1.85 use the *Mean and Median* applet on the website for this text to compare the mean and the median.

The median and mean are the most common measures of the center of a distribution. The mean and median of a symmetric distribution are close together. If the distribution is exactly symmetric, the mean and median are exactly the same. In a skewed distribution, the mean is farther out in the long tail than is the median.

The endowment for a college or university is money set aside and invested. The income from the endowment is usually used to support various programs. The distribution of the sizes of the endowments of colleges and universities is strongly skewed to the right. Most institutions have modest endowments, but a few are very wealthy. The median endowment of colleges and universities in a recent year was $93 million—but the mean endowment was $498 million.[21] The few wealthy institutions pull the mean up but do not affect the median. *Don't confuse the "average" value of a variable (the mean) with its "typical" value, which we might describe by the median.*

We can now give a better answer to the question of how to deal with outliers in data. First, look at the data to identify outliers and investigate their causes. You can then correct outliers if they are wrongly recorded, delete them for good reason, or otherwise give them individual attention. The outlier in Example 1.21 (page 21) can be dropped from the data once we discover that it is an error. If you have no clear reason to drop outliers, you may want to use resistant measures in your analysis, so that outliers have little influence over your conclusions. The choice is often a matter for judgment.

Measuring spread: The quartiles

A measure of center alone can be misleading. Two countries with the same median family income are very different if one has extremes of wealth and poverty and the other has little variation among families. A drug manufactured with the correct mean concentration of active ingredient is dangerous if some batches are much too high and others much too low.

We are interested in the *spread* or *variability* of incomes and drug potencies as well as their centers. **The simplest useful numerical description of a distribution consists of both a measure of center and a measure of spread**.

We can describe the spread or variability of a distribution by giving several percentiles. The median divides the data in two; half of the observations are above the median and half are below the median. We could call the median **quartile**   the 50th percentile. The upper **quartile** is the median of the upper half of the data. Similarly, the lower quartile is the median of the lower half of the data. With the median, the quartiles divide the data into four equal parts; 25% of the data are in each part.

percentile   We can do a similar calculation for any percent. The ***p*th percentile** of a distribution is the value that has p percent of the observations fall at or below it. To calculate a percentile, arrange the observations in increasing order and count up the required percent from the bottom of the list.

Our definition of percentiles is a bit inexact because there is not always a value with exactly p percent of the data at or below it. We will be content to take the nearest observation for most percentiles, but the quartiles are important enough to require an exact rule.

THE QUARTILES Q_1 AND Q_3

To calculate the quartiles:

1. Arrange the observations in increasing order and locate the median M in the ordered list of observations.

2. The **first quartile Q_1** is the median of the observations whose positions in the ordered list are to the left of the location of the overall median.

3. The **third quartile Q_3** is the median of the observations whose positions in the ordered list are to the right of the location of the overall median.

Here is an example.

EXAMPLE 1.26

TTS24

Finding the quartiles. Here is the ordered list of the times to start a business in our sample of 24 countries:

2	4	5	5	5	5	6	6	7	8	10	11
12	13	16	17	19	19	24	25	32	38	49	53

The count of observations $n = 24$ is even, so the median is at position $(24 + 1)/2 = 12.5$, that is, between the 12th and the 13th observation in the ordered list. There are 12 cases above this position and 12 below it. The first quartile is the median of the first 12 observations, and the third quartile is the median of the last 12 observations. Check that $Q_1 = 5.5$ and $Q_3 = 21.5$.

Notice that the quartiles are resistant. For example, Q_3 would have the same value if the highest start time was 530 days rather than 53 days.

Be careful when several observations take the same numerical value. Write down all the observations and apply the rules just as if they all had distinct values.

USE YOUR KNOWLEDGE

STAT

1.48 Find the quartiles. Here are the scores on the first exam in an introductory statistics course for 10 students:

83   74   93   85   75   97   93   55   92   81

Find the quartiles for these first-exam scores.

There are several rules for calculating quartiles, which often give slightly different values. The differences are generally small. For describing data, just report the values that your software gives.

The five-number summary and boxplots

In Section 1.2, we used the smallest and largest observations to indicate the spread of a distribution. These single observations tell us little about the distribution as a whole, but they give information about the tails of the distribution that is missing if we know only Q_1, M, and Q_3. To get a quick summary of both center and spread, use all five numbers.

THE FIVE-NUMBER SUMMARY

The **five-number summary** of a set of observations consists of the smallest observation, the first quartile, the median, the third quartile, and the largest observation, written in order from smallest to largest. In symbols, the five-number summary is

Minimum Q_1 M Q_3 Maximum

EXAMPLE 1.27

CALLS80

Service center call lengths. Table 1.2 (page 17) gives the service center call lengths for the sample of 80 calls that we discussed in Example 1.15. The five-number summary for these data is 1.0, 54.5, 103.5, 200, and 2631. The distribution is highly skewed. The mean is 197 seconds, a value that is very close to the third quartile.

USE YOUR KNOWLEDGE

CALLS80

STAT

1.49 **Verify the calculations.** Refer to the five-number summary and the mean for service center call lengths given in Example 1.27. Verify these results. Do not use software for this exercise and be sure to show all your work.

1.50 **Find the five-number summary.** Here are the scores on the first exam in an introductory statistics course for 10 students:

83   74   93   85   75   97   93   55   92   81

Find the five-number summary for these first-exam scores.

The five-number summary leads to another visual representation of a distribution, the *boxplot*.

BOXPLOT

A **boxplot** is a graph of the five-number summary.

- A central box spans the quartiles Q_1 and Q_3.
- A line in the box marks the median M.
- Lines extend from the box out to the smallest and largest observations.

whiskers
box-and-whisker plots

The lines extending to the smallest and largest observations are sometimes called **whiskers**, and boxplots are sometimes called **box-and-whisker plots**. Software provides many varieties of boxplots, some of which use different choices for the placement of the whiskers.

When you look at a boxplot, first locate the median, which marks the center of the distribution. Then look at the spread. The quartiles show the spread of the middle half of the data, and the extremes (the smallest and largest observations) show the spread of the entire data set.

EXAMPLE 1.28

IQ

IQ scores. In Example 1.14 (page 14), we used a histogram to examine the distribution of a sample of 60 IQ scores. A boxplot for these data is given in Figure 1.14. Note that the mean is marked with a "+" and appears very close to the median. The two quartiles are each approximately the same distance from the median, and the two whiskers are approximately the same distance from the corresponding quartiles. All these characteristics are consistent with a symmetric distribution, as illustrated by the histogram in Figure 1.7.

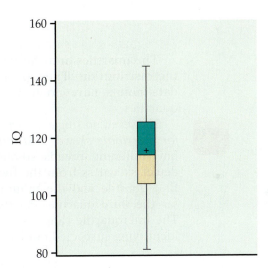

FIGURE 1.14 Boxplot for sample of 60 IQ scores, Example 1.28.

USE YOUR KNOWLEDGE

STAT

1.51 Make a boxplot. Here are the scores on the first exam in an introductory statistics course for 10 students:

83   74   93   85   75   97   93   55   92   81

Make a boxplot for these first-exam scores.

The 1.5 × *IQR* rule for suspected outliers

If we look at the data in Table 1.2 (page 17), we can spot a clear outlier, a call lasting 2631 seconds, more than twice the length of any other call. How can we describe the spread of this distribution? The smallest and largest observations are extremes that do not describe the spread of the majority of the data.

The distance between the quartiles (the range of the center half of the data) is a more resistant measure of spread than the range. This distance is called the *interquartile range*.

THE INTERQUARTILE RANGE *IQR*

The **interquartile range *IQR*** is the distance between the first and third quartiles,

$$IQR = Q_3 - Q_1$$

EXAMPLE 1.29

***IQR* for service center call length data.** In Exercise 1.49 (page 34) you verified that the five-number summary for our data on service center call lengths was 1.0, 54.5, 103.5, 200, and 2631. Therefore, we calculate

$$IQR = Q_3 - Q_1$$
$$IQR = 200 - 54.5$$
$$= 145.5$$

The quartiles and the *IQR* are not affected by changes in either tail of the distribution. They are resistant, therefore, because changes in a few data points have no further effect once these points move outside the quartiles.

However, *no single numerical measure of spread, such as IQR, is very useful for describing skewed distributions.* The two sides of a skewed distribution have different spreads, so one number can't summarize them. We can often detect skewness from the five-number summary by comparing how far the first quartile and the minimum are from the median (left tail) with how far the third quartile and the maximum are from the median (right tail). The interquartile range is mainly used as the basis for a rule of thumb for identifying suspected outliers.

THE 1.5 × *IQR* RULE FOR OUTLIERS

Call an observation a suspected outlier if it falls more than 1.5 × *IQR* above the third quartile or below the first quartile.

EXAMPLE 1.30

CALLS80

Suspected outliers for call length data. For the call length data in Table 1.2 (page 17),

$$1.5 \times IQR = 1.5 \times 145.5 = 218.25$$

Any values below $54.5 - 218.25 = -163.75$ or above $200 + 218.25 = 418.25$ are flagged as possible outliers. There are no low outliers, but the eight longest calls are flagged as possible high outliers. Their lengths are

438   465   479   700   700   951   1148   2631

It is difficult to imagine calls lasting this long.

1.52 Find the *IQR*. Here are the scores on the first exam in an introductory statistics course for 10 students:

$$83 \quad 74 \quad 93 \quad 85 \quad 75 \quad 97 \quad 93 \quad 55 \quad 92 \quad 81$$

Find the interquartile range and use the $1.5 \times IQR$ rule to check for outliers. How low would the lowest score need to be for it to be an outlier according to this rule?

modified boxplot

Two variations on the basic boxplot can be very useful. The first, called a **modified boxplot**, uses the $1.5 \times IQR$ rule. The lines that extend out from the quartiles are terminated in whiskers that are $1.5 \times IQR$ in length. Points beyond the whiskers are plotted individually and are classified as outliers according to the $1.5 \times IQR$ rule.

side-by-side boxplots

The other variation is to use two or more boxplots in the same graph to compare groups measured on the same variable. These are called **side-by-side boxplots**. The following example illustrates these two variations.

EXAMPLE 1.31

POETS

Do poets die young? According to William Butler Yeats, "She is the Gaelic muse, for she gives inspiration to those she persecutes. The Gaelic poets die young, for she is restless, and will not let them remain long on earth." One study designed to investigate this issue examined the age at death for writers from different cultures and genders.[22]

Three categories of writers examined were novelists, poets, and nonfiction writers. We examine the ages at death for female writers in these categories from North America. Figure 1.15 shows modified side-by-side boxplots for the three categories of writers.

Displaying the boxplots for the three categories of writers lets us compare the three distributions. We see that nonfiction writers tend to live the longest, followed by novelists. The poets do appear to die young! There is one outlier among the nonfiction writers, which is plotted individually along with the value of its label. This writer died at the age of 40, young for a nonfiction writer, but not for a novelist or a poet!

FIGURE 1.15 Modified side-by-side boxplots for the data on writers' age at death, for Example 1.31.

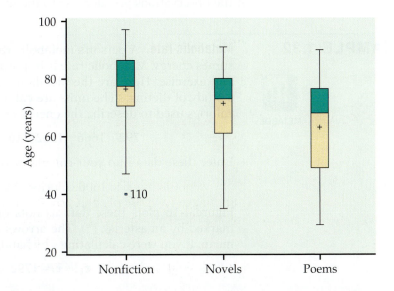

Measuring spread: The standard deviation

The five-number summary is not the most common numerical description of a distribution. That distinction belongs to the combination of the mean to measure center and the *standard deviation* to measure spread, or variability. The standard deviation measures spread by looking at how far the observations are from their mean.

THE STANDARD DEVIATION s

The **variance s^2** of a set of observations is the average of the squares of the deviations of the observations from their mean. In symbols, the variance of n observations $x_1, x_2, \ldots, x_n$ is

$$s^2 = \frac{(x_1 - \bar{x})^2 + (x_2 - \bar{x})^2 + \cdots + (x_n - \bar{x})^2}{n - 1}$$

or, in more compact notation,

$$s^2 = \frac{1}{n - 1} \sum (x_i - \bar{x})^2$$

The **standard deviation s** is the square root of the variance s^2:

$$s = \sqrt{\frac{1}{n - 1} \sum (x_i - \bar{x})^2}$$

The idea behind the variance and the standard deviation as measures of spread is as follows: The deviations $x_i - \bar{x}$ display the spread of the values x_i about their mean $\bar{x}$. Some of these deviations will be positive and some negative because some of the observations fall on each side of the mean. In fact, *the sum of the deviations of the observations from their mean will always be zero*. Squaring the deviations makes the negative deviations positive so that observations far from the mean in either direction have large positive squared deviations. The variance is the average squared deviation. Therefore, s^2 and s will be large if the observations are widely spread about their mean and small if the observations are all close to the mean.

EXAMPLE 1.32

METABOL

Metabolic rate. A person's metabolic rate is the rate at which the body consumes energy. Metabolic rate is important in studies of weight gain, dieting, and exercise. Here are the metabolic rates of seven men who took part in a study of dieting. (The units are calories per 24 hours. These are the same calories used to describe the energy content of foods.)

1792   1666   1362   1614   1460   1867   1439

Enter these data into your calculator or software and verify that

$$\bar{x} = 1600 \text{ calories} \qquad s = 189.24 \text{ calories}$$

Figure 1.16 plots these data as dots on the calorie scale, with their mean marked by an asterisk (*). The arrows mark two of the deviations from the mean. If you were calculating s by hand, you would find the first deviation as

$$x_1 - \bar{x} = 1792 - 1600 = 192$$

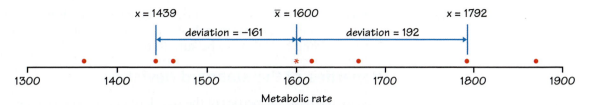

FIGURE 1.16 Metabolic rates for seven men, with the mean (*) and the deviations of two observations from the mean, Example 1.32.

Exercise 1.80 asks you to calculate the seven deviations from Example 1.32, square them, and find s^2 and s directly from the deviations. Working one or two short examples by hand helps you understand how the standard deviation is obtained. In practice, you will use either software or a calculator that will find s.

USE YOUR KNOWLEDGE

STAT

1.53 Find the variance and the standard deviation. Here are the scores on the first exam in an introductory statistics course for 10 students:

83   74   93   85   75   97   93   55   92   81

Find the variance and the standard deviation for these first-exam scores.

The idea of the variance is straightforward: it is the average of the squares of the deviations of the observations from their mean. The details we have just presented, however, raise some questions.

Why do we square the deviations?

- First, the sum of the squared deviations of any set of observations from their mean is the smallest that the sum of squared deviations from any number can possibly be. This is not true of the unsquared distances. So squared deviations point to the mean as center in a way that distances do not.

- Second, the standard deviation turns out to be the natural measure of spread for a particularly important class of symmetric unimodal distributions, the *Normal distributions*. We will meet the Normal distributions in the next section.

Why do we emphasize the standard deviation rather than the variance?

- One reason is that s, not s^2, is the natural measure of spread for Normal distributions, which are introduced in the next section.

- There is also a more general reason to prefer s to s^2. Because the variance involves squaring the deviations, it does not have the same unit of measurement as the original observations. The variance of the metabolic rates, for example, is measured in squared calories. Taking the square root gives us a description of the spread of the distribution in the original measurement units.

Why do we average by dividing by n − 1 rather than n in calculating the variance?

- Because the sum of the deviations is always zero, the last deviation can be found once we know the other $n - 1$. So we are not averaging n unrelated numbers. Only $n - 1$ of the squared deviations can vary freely, and we average by dividing the total by $n - 1$.

degrees of freedom

- The number $n - 1$ is called the **degrees of freedom** of the variance or standard deviation. Many calculators offer a choice between dividing by n and dividing by $n - 1$, so be sure to use $n - 1$.

Properties of the standard deviation

Here are the basic properties of the standard deviation s as a measure of spread.

PROPERTIES OF THE STANDARD DEVIATION

- s measures spread about the mean and should be used only when the mean is chosen as the measure of center.

- $s = 0$ only when there is *no spread*. This happens only when all observations have the same value. Otherwise, $s > 0$. As the observations become more spread out about their mean, s gets larger.

- s, like the mean $\bar{x}$, is not resistant. A few outliers can make s very large.

USE YOUR KNOWLEDGE

1.54 A standard deviation of zero. Construct a data set with 6 cases that has a variable with $s = 0$.

The use of squared deviations renders s even more sensitive than $\bar{x}$ to a few extreme observations. For example, when we add Suriname to our sample of 24 countries for the analysis of the time to start a business (Exercise 1.43, page 29, and Exercise 1.45, page 31), we increase the standard deviation from 14.2 to 40.8! Distributions with outliers and strongly skewed distributions have standard deviations that do not give much helpful information about such distributions.

USE YOUR KNOWLEDGE

TTS24, TTS25

1.55 Effect of an outlier on the *IQR*. Find the *IQR* for the time to start a business with and without Suriname. What do you conclude about the sensitivity of this measure of spread to the inclusion of an outlier?

Choosing measures of center and spread

How do we choose between the five-number summary and $\bar{x}$ and s to describe the center and spread of a distribution? Because the two sides of a strongly skewed distribution have different spreads, no single number such as s describes the spread well. The five-number summary, with its two quartiles and two extremes, does a better job.

CHOOSING A SUMMARY

The five-number summary is usually better than the mean and standard deviation for describing a skewed distribution or a distribution with strong outliers. Use $\bar{x}$ and s for reasonably symmetric distributions that are free of outliers.

Remember that a graph gives the best overall picture of a distribution. Numerical measures of center and spread report specific facts about a distribution, but they do not describe its shape. Numerical summaries do not disclose the presence of multiple modes or gaps, for example. **Always plot your data**.

EXAMPLE 1.33

TTS24

FIGURE 1.17 Graphical and numerical summaries from Minitab: (a) boxplot, (b) histogram, and (c) numerical summaries for the time to start a business, Example 1.33.

Results from software. We prefer to examine the numerical summaries and graphical summaries together. Figure 1.17 gives (a) a boxplot, (b) a histogram, and (c) numerical summaries for the time to start a business from Example 1.23 (page 28) using Minitab. Similar displays are given for SPSS in Figure 1.18 (a), (b), and (c) and for JMP

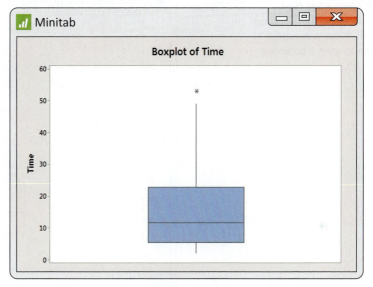

(a) Minitab boxplot

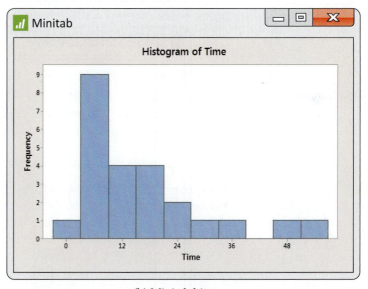

(b) Minitab histogram

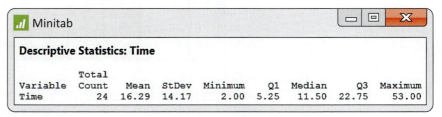

(c) Minitab numerical summaries

in Figure 1.19. Examine and compare the outputs carefully. Notice that they give different numbers of significant digits for some of these numerical summaries. There are also variations in how they make the boxplots and how they define classes for the histograms.

FIGURE 1.18 Graphical and numerical summaries from SPSS: (a) boxplot, (b) histogram, and (c) numerical summaries for the time to start a business, Example 1.33.

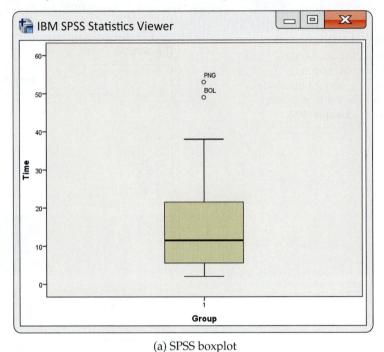

(a) SPSS boxplot

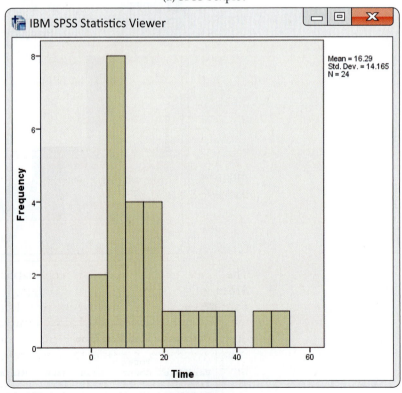

(b) SPSS histogram

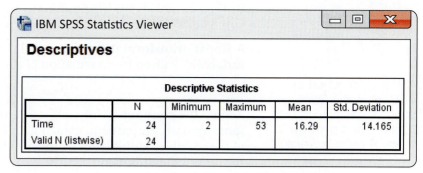

FIGURE 1.18 *Continued*   (c) SPSS numerical summaries

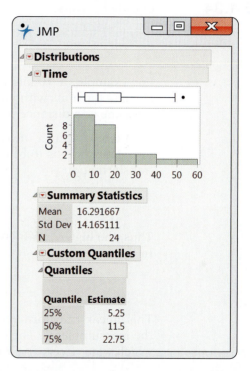

FIGURE 1.19 Graphical and numerical summaries from JMP for the time to start a business, Example 1.33.

Changing the unit of measurement

The same variable can be recorded in different units of measurement. Americans commonly record distances in miles and temperatures in degrees Fahrenheit, while the rest of the world measures distances in kilometers and temperatures in degrees Celsius. Fortunately, it is easy to convert numerical descriptions of a distribution from one unit of measurement to another. This is true because a change in the measurement unit is a *linear transformation* of the measurements.

LINEAR TRANSFORMATIONS

A **linear transformation** changes the original variable x into the new variable x_{new} given by an equation of the form

$$x_{new} = a + bx$$

Adding the constant a shifts all values of x upward or downward by the same amount. In particular, such a shift changes the origin (zero point) of the variable. Multiplying by the positive constant b changes the size of the unit of measurement.

EXAMPLE 1.34

Change the units.

(a) If a distance x is measured in kilometers, the same distance in miles is

$$x_{new} = 0.62x$$

For example, a 10-kilometer race covers 6.2 miles. This transformation changes the units without changing the origin—a distance of 0 kilometers is the same as a distance of 0 miles.

(b) A temperature x measured in degrees Fahrenheit must be reexpressed in degrees Celsius to be easily understood by the rest of the world. The transformation is

$$x_{new} = \frac{5}{9}(x - 32) = -\frac{160}{9} + \frac{5}{9}x$$

Thus, the high of 95°F on a hot American summer day translates into 35°C. In this case,

$$a = -\frac{160}{9} \quad \text{and} \quad b = \frac{5}{9}$$

This linear transformation changes both the unit size and the origin of the measurements. The origin in the Celsius scale (0°C, the temperature at which water freezes) is 32° in the Fahrenheit scale.

Linear transformations do not change the shape of a distribution. If measurements on a variable x have a right-skewed distribution, any new variable x_{new} obtained by a linear transformation $x_{new} = a + bx$ (for $b > 0$) will also have a right-skewed distribution. If the distribution of x is symmetric and unimodal, the distribution of x_{new} remains symmetric and unimodal.

Although a linear transformation preserves the basic shape of a distribution, the center and spread will change. Because linear changes of measurement scale are common, we must be aware of their effect on numerical descriptive measures of center and spread. Fortunately, the changes follow a simple pattern.

EXAMPLE 1.35

Use scores to find the points. In an introductory statistics course, homework counts for 300 points out of a total of 1000 possible points for all course requirements. During the semester, there were 12 homework assignments, and each was given a grade on a scale of 0 to 100. The maximum total score for the 12 homework assignments is therefore 1200. To convert the homework scores to final grade points, we need to convert the scale of 0 to 1200 to a scale of 0 to 300. We do this by multiplying the homework scores by 300/1200. In other words, we divide the homework scores by 4. Here are the homework scores and the corresponding final grade points for five students:

Student	1	2	3	4	5
Score	1056	1080	900	1164	1020
Points	264	270	225	291	255

These two sets of numbers measure the same performance on homework for the course. Because we obtained the points by dividing the scores by 4, the mean of the points will be the mean of the scores divided by 4. Similarly, the standard deviation of points will be the standard deviation of the scores divided by 4.

USE YOUR KNOWLEDGE

1.56 Calculate the points for a student. Use the setting of Example 1.35 to find the points for a student whose score is 950.

Here is a summary of the rules for linear transformations:

EFFECT OF A LINEAR TRANSFORMATION

To see the effect of a linear transformation on measures of center and spread, apply these rules:

• Multiplying each observation by a positive number b multiplies both measures of center (mean and median) and measures of spread (interquartile range and standard deviation) by b.

• Adding the same number a (either positive or negative) to each observation adds a to measures of center and to quartiles and other percentiles but does not change measures of spread.

In Example 1.35, when we converted from score to points, we described the transformation as dividing by 4. The multiplication part of the summary of the effect of a linear transformation applies to this case because division by 4 is the same as multiplication by 0.25. Similarly, the second part of the summary applies to subtraction as well as addition because subtraction is simply the addition of a negative number.

The measures of spread IQR and s do not change when we add the same number a to all the observations because adding a constant changes the location of the distribution but leaves the spread unaltered. You can find the effect of a linear transformation $x_{new} = a + bx$ by combining these rules. For example, if x has mean $\bar{x}$, the transformed variable x_{new} has mean $a + b\bar{x}$.

SECTION 1.3 SUMMARY

- A numerical summary of a distribution should report its **center** and its **spread** or **variability**.

- The **mean** $\bar{x}$ and the **median** M describe the center of a distribution in different ways. The mean is the arithmetic average of the observations, and the median is their midpoint.

- When you use the median to describe the center of a distribution, describe its spread by giving the **quartiles**. The **first quartile** Q_1 has one-fourth of the observations below it, and the **third quartile** Q_3 has three-fourths of the observations below it.

- The **interquartile range** is the difference between the quartiles. It is the spread of the center half of the data. The **1.5 × _IQR_ rule** flags observations more than $1.5 \times IQR$ beyond the quartiles as possible outliers.

- The **five-number summary** consisting of the median, the quartiles, and the smallest and largest individual observations provides a quick overall description of a distribution. The median describes the center, and the quartiles and extremes show the spread.

- **Boxplots** based on the five-number summary are useful for comparing several distributions. The box spans the quartiles and shows the spread of the central half of the distribution. The median is marked within the box. Lines extend from the box to the extremes and show the full spread of the data. In a **modified boxplot**, points identified by the $1.5 \times IQR$ rule are plotted individually. **Side-by-side boxplots** can be used to display boxplots for more than one group on the same graph.

- The **variance** s^2 and especially its square root, the **standard deviation** s, are common measures of spread about the mean as center. The standard deviation s is zero when there is no spread and gets larger as the spread increases.

- A **resistant measure** of any aspect of a distribution is relatively unaffected by changes in the numerical value of a small proportion of the total number of observations, no matter how large these changes are. The median and quartiles are resistant, but the mean and the standard deviation are not.

- The mean and standard deviation are good descriptions for symmetric distributions without outliers. They are most useful for the Normal distributions introduced in the next section. The five-number summary is a better exploratory description for skewed distributions.

- **Linear transformations** have the form $x_{new} = a + bx$. A linear transformation changes the origin if $a \neq 0$ and changes the size of the unit of measurement if $b > 0$. Linear transformations do not change the overall shape of a distribution. A linear transformation multiplies a measure of spread by b and changes a percentile or measure of center m into $a + bm$.

- Numerical measures of particular aspects of a distribution, such as center and spread, do not report the entire shape of most distributions. In some cases, particularly distributions with multiple peaks and gaps, these measures may not be very informative.

SECTION 1.3 EXERCISES

For Exercises 1.43 and 1.44, see page 29; for Exercises 1.45 to 1.47, see page 31; for Exercise 1.48, see page 33; for Exercises 1.49 and 1.50, see page 34; for Exercise 1.51, see page 35; for Exercise 1.52, see page 37; for Exercise 1.53, see page 39; for Exercise 1.54, see page 40; for Exercise 1.55, see page 40; and for Exercise 1.56, see page 45.

1.57 Potassium from potatoes. Refer to Exercise 1.30 (page 24) where you examined the potassium absorption of a group of 27 adults who ate a controlled diet that included 40 mEq of potassium from potatoes for five days. [III] KPOT40

(a) Compute the mean for these data.

(b) Compute the median for these data.

(c) Which measure do you prefer for describing the center of this distribution? Explain your answer. (You may include a graphical summary as part of your explanation.)

1.58 Potassium from a supplement. Refer to Exercise 1.31 (page 24) where you examined the potassium absorption of a group of 29 adults who ate a controlled diet that included 40 mEq of potassium from a supplement for five days. [III] KSUP40

(a) Compute the mean for these data.

(b) Compute the median for these data.

(c) Which measure do you prefer for describing the center of this distribution? Explain your answer. (You may include a graphical summary as part of your explanation.)

1.59 Potassium from potatoes. Refer to Exercise 1.30 (page 24) where you examined the potassium absorption of a group of 27 adults who ate a controlled diet that included 40 mEq of potassium from potatoes for five days. [III] KPOT40

(a) Compute the standard deviation for these data.

(b) Compute the quartiles for these data.

(c) Give the five-number summary and explain the meaning of each of the five numbers.

(d) Which numerical summaries do you prefer for describing the distribution, the mean, and the standard deviation of the five-number summary? Explain your answer. (You may include a graphical summary as part of your explanation.)

1.60 Potassium from a supplement. Refer to Exercise 1.31 (page 24) where you examined the potassium absorption of a group of 29 adults who ate a controlled diet that included 40 mEq of potassium from a supplement for five days. [III] KSUP40

(a) Compute the standard deviation for these data.

(b) Compute the quartiles for these data.

(c) Give the five-number summary and explain the meaning of each of the five numbers.

(d) Which numerical summaries do you prefer for describing the distribution, the mean, and the standard deviation of the five-number summary? Explain your answer. (You may include a graphical summary as part of your explanation.)

1.61 Potassium from potatoes. Refer to Exercise 1.30 (page 24) where you examined the potassium absorption of a group of 27 adults who ate a controlled diet that included 40 mEq of potassium from potatoes for five days. In Exercise 1.30, you used a stemplot to examine the distribution of the potassium absorption. [III] KPOT40

(a) Make a histogram and use it to describe the distribution of potassium absorption.

(b) Make a boxplot and use it to describe the distribution of potassium absorption.

(c) Compare the stemplot, the histogram, and the boxplot as graphical summaries of this distribution. Which do you prefer? Give reasons for your answer.

1.62 Potassium from a supplement. Refer to Exercise 1.31 (page 24) where you examined the potassium absorption of a group of 29 adults who ate a controlled diet that included 40 mEq of potassium from a supplement for five days. In Exercise 1.31, you used a stemplot to examine the distribution of the potassium absorption. [III] KSUP40

(a) Make a histogram and use it to describe the distribution of potassium absorption.

(b) Make a boxplot and use it to describe the distribution of potassium absorption.

(c) Compare the stemplot, the histogram, and the boxplot as graphical summaries of this distribution. Which do you prefer? Give reasons for your answer.

1.63 Compare the potatoes with the supplement. Refer to Exercises 1.30 and 1.31 (page 24). Use a back-to-back stemplot to display the data for the two sources of potassium. Use the stemplot to compare the two distributions and write a short summary of your findings. [III] KPS40

1.64 Potassium sources. Refer to Exercises 1.30 and 1.31 (page 24). Use side-by-side boxplots in to describe the distributions. [III] KPS40

(a) Summarize what you see in the boxplots and compare it with what you saw in the stemplots.

(b) For comparing these two distributions, do you prefer back-to-back stemplots or side-by-side boxplots? Give reasons for your answer.

1.65 Gosset's data on double stout sales. William Sealy Gosset worked at the Guinness Brewery in Dublin and made substantial contributions to the practice of statistics.[23] In his work at the brewery, he collected and analyzed a great deal of data. Archives with Gosset's handwritten tables, graphs, and notes have been preserved at the Guinness Storehouse in Dublin.[24] In one study, Gosset examined the change in the double stout market before and after World War I (1914–1918). For various regions in England and Scotland, he calculated the ratio of sales in 1925, after the war, as a percent of sales in 1913, before the war. Here are the data: STOUT

Bristol	94	Glasgow	66
Cardiff	112	Liverpool	140
English Agents	78	London	428
English O	68	Manchester	190
English P	46	Newcastle-on-Tyne	118
English R	111	Scottish	24

(a) Compute the mean for these data.

(b) Compute the median for these data.

(c) Which measure do you prefer for describing the center of this distribution? Explain your answer. (You may include a graphical summary as part of your explanation.)

1.66 Measures of spread for the double stout data. Refer to the previous exercise. STOUT

(a) Compute the standard deviation for these data.

(b) Compute the quartiles for these data.

(c) Which measure do you prefer for describing the spread of this distribution? Explain your answer. (You may include a graphical summary as part of your explanation.)

1.67 Are there outliers in the double stout data? Refer to the previous two exercises. STOUT

(a) Find the IQR for these data.

(b) Use the $1.5 \times IQR$ rule to identify and name any outliers.

(c) Make a boxplot for these data and describe the distribution using only the information in the boxplot.

(d) Make a modified boxplot for these data and describe the distribution using only the information in the boxplot.

(e) Make a stemplot for these data.

(f) Compare the boxplot, the modified boxplot, and the stemplot. Evaluate the advantages and disadvantages of each graphical summary for describing the distribution of the double stout data.

1.68 Smolts. Smolts are young salmon at a stage when their skin becomes covered with silvery scales and they start to migrate from freshwater to the sea. The reflectance of a light shined on a smolt's skin is a measure of the smolt's readiness for the migration. Here are the reflectances, in percents, for a sample of 50 smolts:[25] SMOLTS

57.6	54.8	63.4	57.0	54.7	42.3	63.6	55.5	33.5	63.3
58.3	42.1	56.1	47.8	56.1	55.9	38.8	49.7	42.3	45.6
69.0	50.4	53.0	38.3	60.4	49.3	42.8	44.5	46.4	44.3
58.9	42.1	47.6	47.9	69.2	46.6	68.1	42.8	45.6	47.3
59.6	37.8	53.9	43.2	51.4	64.5	43.8	42.7	50.9	43.8

(a) Find the mean reflectance for these smolts.

(b) Find the median reflectance for these smolts.

(c) Do you prefer the mean or the median as a measure of center for these data? Give reasons for your preference.

1.69 Measures of spread for smolts. Refer to the previous exercise. SMOLTS

(a) Find the standard deviation of the reflectance for these smolts.

(b) Find the quartiles of the reflectance for these smolts.

(c) Do you prefer the standard deviation or the quartiles as a measure of spread for these data? Give reasons for your preference.

1.70 Are there outliers in the smolt data? Refer to the previous two exercises. SMOLTS

(a) Find the IQR for the smolt data.

(b) Use the $1.5 \times IQR$ rule to identify any outliers.

(c) Make a boxplot for the smolt data and describe the distribution using only the information in the boxplot.

(d) Make a modified boxplot for these data and describe the distribution using only the information in the boxplot.

(e) Make a stemplot for these data.

(f) Compare the boxplot, the modified boxplot, and the stemplot. Evaluate the advantages and disadvantages of each graphical summary for describing the distribution of the smolt reflectance data.

1.71 Potatoes. A quality product is one that is consistent and has very little variability in its characteristics. Controlling variability can be more difficult with agricultural products than with those that are manufactured. The following table gives the weights, in ounces, of the 25 potatoes sold in a 10-pound bag:
▥ POTATO

7.6	7.9	8.0	6.9	6.7	7.9	7.9	7.9	7.6	7.8	7.0	4.7	7.6
6.3	4.7	4.7	4.7	6.3	6.0	5.3	4.3	7.9	5.2	6.0	3.7	

(a) Summarize the data graphically and numerically. Give reasons for the methods you chose to use in your summaries.

(b) Do you think that your numerical summaries do an effective job of describing these data? Why or why not?

(c) There appear to be two distinct clusters of weights for these potatoes. Divide the sample into two subsamples based on the clustering. Give the mean and standard deviation for each subsample. Do you think that this way of summarizing these data is better than a numerical summary that uses all the data as a single sample? Give a reason for your answer.

1.72 The alcohol content of beer. Brewing beer involves a variety of steps that can affect the alcohol content. A website gives the percent alcohol for 159 domestic brands of beer.[26] ▥ BEER

(a) Use graphical and numerical summaries of your choice to describe the data. Give reasons for your choice.

(b) The data set contains an outlier. Explain why this particular beer is unusual.

(c) For the outlier, give a short description of how you think this particular beer should be marketed.

1.73 Outlier for alcohol content of beer. Refer to the previous exercise. ▥ BEER

(a) Calculate the mean with and without the outlier. Do the same for the median. Explain how these values change when the outliers is excluded.

(b) Calculate the standard deviation with and without the outlier. Do the same for the quartiles. Explain how these values change when the outlier is excluded.

(c) Write a short paragraph summarizing what you have learned in this exercise.

1.74 Calories in beer. Refer to the previous two exercises. The data set also lists calories per 12 ounces of beverage. ▥ BEER

(a) Analyze the data and summarize the distribution of calories for these 159 brands of beer.

(b) In the previous exercise, you identified one brand of beer as an outlier. To what extent is this brand an outlier in the distribution of calories? Explain your answer.

(c) Does the distribution of calories suggest marketing strategies for this brand of beer? Describe some marketing strategies.

1.75 Median versus mean for net worth. A report on the assets of American households says that the median net worth of U.S. families is $81,200. The mean net worth of these families is $534,600.[27] What explains the difference between these two measures of center?

1.76 Create a data set. Create a data set with seven observations for which the median would change by a large amount if the smallest observation were deleted.

1.77 Mean versus median. A small accounting firm pays each of its seven clerks $55,000, three junior accountants $80,000 each, and the firm's owner $650,000. What is the mean salary paid at this firm? How many of the employees earn less than the mean? What is the median salary?

1.78 Be careful about how you treat the zeros. In computing the median income of any group, some federal agencies omit all members of the group who had no income. Give an example to show that the reported median income of a group can go down even though the group becomes economically better off. Is this also true of the mean income?

1.79 How does the median change? The firm in Exercise 1.77 gives no raises to the clerks and junior accountants, while the owner's take increases to $900,000. How does this change affect the mean? How does it affect the median?

1.80 Metabolic rates. Calculate the mean and standard deviation of the metabolic rates in Example 1.32 (page 38), showing each step in detail. First find the mean $\bar{x}$ by summing the seven observations and dividing by 7. Then find each of the deviations $x_i - \bar{x}$ and their squares. Check that the deviations have sum 0. Calculate the variance as an average of the squared deviations (remember to divide by $n - 1$). Finally, obtain s as the square root of the variance. ▥ METABOL

⚙ **1.81 Earthquakes.** Each year there are about 900,000 earthquakes of magnitude 2.5 or less that are usually not felt. In contrast, there are about 10 of magnitude 7.0 that cause serious damage.[28] Explain why the average magnitude of earthquakes is not a good measure of their impact.

1.82 IQ scores. Many standard statistical methods that you will study in Part II of this book are intended for use with distributions that are symmetric and have no outliers. These methods start with the mean and standard deviation, $\bar{x}$ and s. For example, standard methods would typically be used for the IQ and GPA data in Table 1.3 (page 26). **IQ**

(a) Find $\bar{x}$ and s for the IQ data. In large populations, IQ scores are standardized to have mean 100 and standard deviation 15. In what way does the distribution of IQ among these students differ from the overall population?

(b) Find the median IQ score. It is, as we expect, close to the mean.

(c) Find the mean and median for the GPA data. The two measures of center differ a bit. What feature of the data (see your stemplot in Exercise 1.39 or make a new stemplot) explains the difference?

1.83 Mean and median for two observations. The *Mean and Median* applet allows you to place observations on a line and see their mean and median visually. Place two observations on the line by clicking below it. Why does only one arrow appear?

1.84 Mean and median for four observations. In the *Mean and Median* applet, place four observations on the line by clicking below it, three close together near the center of the line and one somewhat to the right of these two.

(a) Pull the single rightmost observation out to the right. (Place the cursor on the point, hold down a mouse button, and drag the point.) How does the mean behave? How does the median behave? Explain briefly why each measure acts as it does.

(b) Now drag the rightmost point to the left as far as you can. What happens to the mean? What happens to the median as you drag this point past the other two (watch carefully)?

1.85 Mean and median for seven observations. Place seven observations on the line in the *Mean and Median* applet by clicking below it.

(a) Add one additional observation *without changing the median*. Where is your new point?

(b) Use the applet to convince yourself that when you add yet another observation (there are now nine in all), the median does not change no matter where you put the seventh point. Explain why this must be true.

1.86 Imputation. Various problems with data collection can cause some observations to be missing. Suppose a

data set has 20 cases. Here are the values of the variable x for 10 of these cases: **IMPUTE**

17 6 12 14 20 23 9 12 16 21

The values for the other 10 cases are missing. One way to deal with missing data is called **imputation**. The basic idea is that missing values are replaced, or imputed, with values that are based on an analysis of the data that are not missing. For a data set with a single variable, the usual choice of a value for imputation is the mean of the values that are not missing. The mean for this data set is 15.

(a) Verify that the mean is 15 and find the standard deviation for the 10 cases for which x is not missing.

(b) Create a new data set with 20 cases by setting the values for the 10 missing cases to 15. Compute the mean and standard deviation for this data set.

(c) Summarize what you have learned about the possible effects of this type of imputation on the mean and the standard deviation.

1.87 A standard deviation contest. This is a standard deviation contest. You must choose four numbers from the whole numbers 10 to 20, with repeats allowed.

(a) Choose four numbers that have the smallest possible standard deviation.

(b) Choose four numbers that have the largest possible standard deviation.

(c) Is more than one choice possible in either part (a) or part (b)? Explain.

1.88 Longleaf pine trees. The Wade Tract in Thomas County, Georgia, is an old-growth forest of longleaf pine trees (*Pinus palustris*) that has survived in a relatively undisturbed state since before the settlement of the area by Europeans. A study collected data on 584 of these trees.[29] One of the variables measured was the diameter at breast height (DBH). This is the diameter of the tree at 4.5 feet and the units are centimeters (cm). Only trees with DBH greater than 1.5 cm were sampled. Here are the diameters of a random sample of 40 of these trees: **PINES**

10.5	13.3	26.0	18.3	52.2	9.2	26.1	17.6	40.5	31.8
47.2	11.4	2.7	69.3	44.4	16.9	35.7	5.4	44.2	2.2
4.3	7.8	38.1	2.2	11.4	51.5	4.9	39.7	32.6	51.8
43.6	2.3	44.6	31.5	40.3	22.3	43.3	37.5	29.1	27.9

(a) Find the five-number summary for these data.

(b) Make a boxplot.

(c) Make a histogram.

(d) Write a short summary of the major features of this distribution. Do you prefer the boxplot or the histogram for these data?

1.89 Weight gain. A study of diet and weight gain deliberately overfed 15 volunteers for eight weeks. The mean increase in fat was $\bar{x} = 2.41$ kilograms, and the standard deviation was $s = 1.25$ kilograms. What are $\bar{x}$ and s in pounds? (A kilogram is 2.2 pounds.)

1.90 Changing units from inches to centimeters. Changing the unit of length from inches to centimeters multiplies each length by 2.54 because there are 2.54 centimeters in an inch. This change of units multiplies our usual measures of spread by 2.54. This is true of *IQR* and the standard deviation. What happens to the variance when we change units in this way?

1.91 A different type of mean. The **trimmed mean** is a measure of center that is more resistant than the mean but uses more of the available information than the median. To compute the 10% trimmed mean, discard the highest 10% and the lowest 10% of the observations and compute the mean of the remaining 80%. Trimming eliminates the effect of a small number of outliers. Compute the 10% trimmed mean of the service time data in Table 1.2 (page 17). Then compute the 20% trimmed mean. Compare the values of these measures with the median and the ordinary untrimmed mean.

1.92 Changing units from centimeters to inches. Refer to Exercise 1.88 (page 50). Change the measurements from centimeters to inches by multiplying each value by 0.39. Answer the questions from that exercise and explain the effect of the transformation on these data.

1.4 Density Curves and Normal Distributions

When you complete this section, you will be able to:

- Compare the mean and the median for symmetric and skewed distributions.
- Sketch a Normal distribution for any given mean and standard deviation.
- Apply the 68–95–99.7 rule to find proportions of observations within one, two, and three standard deviations of the mean for any Normal distribution.
- Transform values of a variable from a general Normal distribution to the standard Normal distribution.
- Compute areas under a Normal curve using software or Table A.
- Perform inverse Normal calculations to find values of a Normal variable corresponding to various areas.
- Assess the extent to which the distribution of a set of data can be approximated by a Normal distribution.

We now have a kit of graphical and numerical tools for describing distributions. What is more, we have a clear strategy for exploring data on a single quantitative variable:

1. Always plot your data: make a graph, usually a stemplot or a histogram.

2. Look for the overall pattern and for striking deviations such as outliers.

3. Calculate an appropriate numerical summary to briefly describe center and spread.

Technology has expanded the set of graphs that we can choose for Step 1. It is possible, though painful, to make histograms by hand. Using software, clever algorithms can describe a distribution in a way that is not feasible by hand, by fitting a smooth curve to the data in addition to or instead of a histogram. The curves used are called **density curves**. Before we examine density curves in detail, here is an example of what software can do.

density curves

EXAMPLE 1.36

TTS

TITANIC

Density curves for times to start a business and *Titanic* passenger ages.
Figure 1.20 illustrates the use of density curves along with histograms to describe distributions. Figure 1.20(a) shows the distribution of the times to start a business for 189 countries (see Example 1.23. page 28). The outlier, Suriname, described in Exercise 1.43 (page 29) has been deleted from the data set. The distribution is highly skewed to the right. Most of the data are in the first two classes, with 40 or fewer days to start a business.

Exercise 1.27 (page 24) describes data on the class of the ticket of the *Titanic* passengers, and Figure 1.20(b) shows the distribution of the ages of these passengers. It has a single mode, a long right tail, and a relatively short left tail.

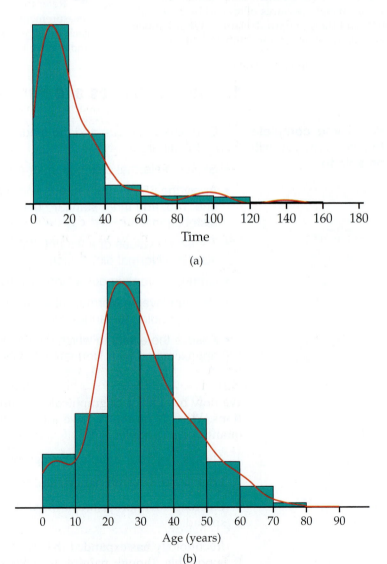

(a)

(b)

FIGURE 1.20 (a) The distribution of the time to start a business, Example 1.36. The distribution is pictured with both a histogram and a density curve. (b) The distribution of the ages of the *Titanic* passengers, Example 1.36. These distributions have a single mode with tails of two different lengths.

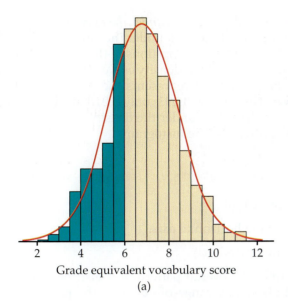
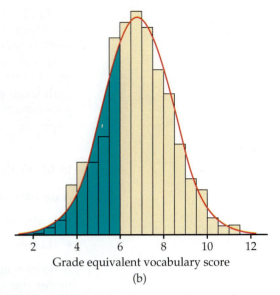

Grade equivalent vocabulary score
(a)

Grade equivalent vocabulary score
(b)

FIGURE 1.21 (a) The distribution of Iowa Test vocabulary scores for Gary, Indiana, seventh-graders, Example 1.37. The shaded bars in the histogram represent scores less than or equal to 6.0. (b) The shaded area under the Normal density curve also represents scores less than or equal to 6.0. This area is 0.293, close to the true 0.303 for the actual data.

A smooth density curve is an idealization that gives the overall pattern of the data but ignores minor irregularities. We first discuss density curves in general and then focus on a special class of density curves, the bell-shaped Normal curves.

Density curves

One way to think of a density curve is as a smooth approximation to the irregular bars of a histogram. Figure 1.21 shows a histogram of the scores of all 947 seventh-grade students in Gary, Indiana, on the vocabulary part of the Iowa Test of Basic Skills. Scores of many students on this national test have a very regular distribution. The histogram is symmetric, and both tails fall off quite smoothly from a single center peak. There are no large gaps or obvious outliers. The curve drawn through the tops of the histogram bars in Figure 1.21 is a good description of the overall pattern of the data.

EXAMPLE 1.37

Vocabulary scores. In a histogram, the *areas* of the bars represent either counts or proportions of the observations. In Figure 1.21(a), we shaded the bars that represent students with vocabulary scores 6.0 or lower. There are 287 such students, who make up the proportion 287/947 = 0.303 of all Gary seventh-graders. The shaded bars in Figure 1.21(a) make up *proportion* 0.303 of the total area under all the bars. If we adjust the scale so that the total area of the bars is 1, the *area of the shaded bars* will also be 0.303.

In Figure 1.21(b), we shaded the *area under the curve* to the left of 6.0. If we adjust the scale so that the total area under the curve is exactly 1, areas under the curve will then represent proportions of the observations. That is, *area = proportion*. The curve is then a density curve. The shaded area under the density curve in Figure 1.21(b) represents the proportion of students with score 6.0 or lower. This area is 0.293, only 0.010 away from the histogram result. You can see that areas under the density curve give quite good approximations of areas given by the histogram.

DENSITY CURVE

A **density curve** is a curve that

- Is always on or above the horizontal axis.

- Has area exactly 1 underneath it.

A density curve describes the overall pattern of a distribution. The area under the curve and above any range of values is the proportion of all observations that fall in that range.

The density curve in Figure 1.21 is a *Normal curve*. Density curves, like distributions, come in many shapes. Figure 1.22 shows two density curves, a symmetric Normal density curve and a right-skewed curve.

We will discuss Normal density curves in detail in this section because of the important role that they play in statistics. There are, however, many applications where the use of other families of density curves are essential.

A density curve of an appropriate shape is often an adequate description of the overall pattern of a distribution. Outliers, which are deviations from the overall pattern, are not described by the curve.

Measuring center and spread for density curves

Our measures of center and spread apply to density curves as well as to actual sets of observations, but only some of these measures are easily seen from the curve. A *mode* of a distribution described by a density curve is a peak point of the curve, the location where the curve is highest. Because

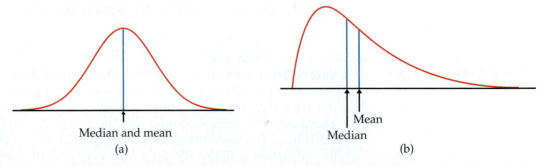

FIGURE 1.22 (a) A symmetric Normal density curve with its mean and median marked. (b) A right-skewed density curve with its mean and median marked.

FIGURE 1.23 The mean of a density curve is the point at which it would balance.

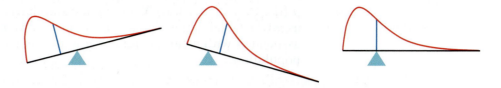

areas under a density curve represent proportions of the observations, the *median* is the point with half the total area on each side. You can roughly locate the *quartiles* by dividing the area under the curve into quarters as accurately as possible by eye. The *IQR* is the distance between the first and third quartiles. There are mathematical ways of calculating areas under curves. These allow us to locate the median and quartiles exactly on any density curve.

What about the mean and standard deviation? The mean of a set of observations is their arithmetic average. If we think of the observations as weights strung out along a thin rod, the mean is the point at which the rod would balance. This fact is also true of density curves. The mean is the point at which the curve would balance if it were made out of solid material. Figure 1.23 illustrates this interpretation of the mean.

A symmetric curve, such as the Normal curve in Figure 1.22(a), balances at its center of symmetry. Half the area under a symmetric curve lies on either side of its center, so this is also the median.

For a right-skewed curve, such as those shown in Figures 1.22(b) and 1.23, the small area in the long right tail tips the curve more than the same area near the center. The mean (the balance point), therefore, lies to the right of the median. It is hard to locate the balance point by eye on a skewed curve. There are mathematical ways of calculating the mean for any density curve, so we are able to mark the mean as well as the median in Figure 1.22(b). The standard deviation can also be calculated mathematically, but it can't be located by eye on most density curves.

MEDIAN AND MEAN OF A DENSITY CURVE

The **median** of a density curve is the equal-areas point, the point that divides the area under the curve in half.

The **mean** of a density curve is the balance point, at which the curve would balance if made of solid material.

The median and mean are the same for a symmetric density curve. They both lie at the center of the curve. The mean of a skewed curve is pulled away from the median in the direction of the long tail.

A density curve is an idealized description of a distribution of data. For example, the density curve in Figure 1.21 (page 53) is exactly symmetric, but the histogram of vocabulary scores is only approximately symmetric. We therefore need to distinguish between the mean and standard deviation of the density curve and the numbers $\bar{x}$ and s computed from the actual observations. The usual notation for the mean of an idealized distribution is

mean μ
standard deviation σ

μ (the Greek letter mu). We write the standard deviation of a density curve as σ (the Greek letter sigma). In Chapter 5, we refer to $\bar{x}$ and s as statistics associated with a sample and to μ and σ as parameters associated with a population.

Normal distributions

One particularly important class of density curves has already appeared in Figures 1.21 and 1.22(a). These density curves are symmetric, unimodal,

Normal curves
Normal distributions

and bell-shaped. They are called **Normal curves**, and they describe **Normal distributions**. All Normal distributions have the same overall shape.

The exact density curve for a particular Normal distribution is specified by giving the distribution's mean μ and its standard deviation σ. The mean is located at the center of the symmetric curve and is the same as the median. Changing μ without changing σ moves the Normal curve along the horizontal axis without changing its spread.

The standard deviation σ controls the spread of a Normal curve. Figure 1.24 shows two Normal curves with different values of σ. The curve with the larger standard deviation is more spread out.

The standard deviation σ is the natural measure of spread for Normal distributions. Not only do μ and σ completely determine the shape of a Normal curve, but we can locate σ by eye on the curve. Here's how. As we move out in either direction from the center μ, the curve changes from falling ever more steeply

to falling ever less steeply

The points at which this change of curvature takes place are located at distance σ on either side of the mean μ. You can feel the change as you run your finger along a Normal curve, and so find the standard deviation. Remember that μ and σ alone do not specify the shape of most distributions, and that the shape

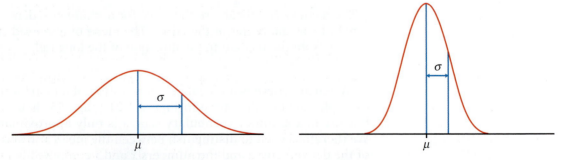

FIGURE 1.24 Two Normal curves, showing the mean μ and the standard deviation σ.

of density curves in general does not reveal σ. These are special properties of Normal distributions.

There are other symmetric bell-shaped density curves that are not Normal. The Normal density curves are specified by a particular equation. The height of the density curve at any point x is given by

$$\frac{1}{\sigma\sqrt{2\pi}}e^{-\frac{1}{2}\left(\frac{x-\mu}{\sigma}\right)^2}$$

We will not make direct use of this fact, although it is the basis of mathematical work with Normal distributions. Notice that the equation of the curve is completely determined by the mean μ and the standard deviation σ.

Why are the Normal distributions important in statistics? Here are three reasons.

1. Normal distributions are good descriptions for some distributions of *real data*. Distributions that are often close to Normal include scores on tests taken by many people (such as the Iowa Test of Figure 1.21, page 53), repeated careful measurements of the same quantity, and characteristics of biological populations (such as lengths of baby pythons and yields of corn).

2. Normal distributions are good approximations to the results of many kinds of *chance outcomes*, such as tossing a coin many times.

3. Many *statistical inference* procedures based on Normal distributions work well for other roughly symmetric distributions.

 However, *even though many sets of data follow a Normal distribution, many do not*. Most income distributions, for example, are skewed to the right and so are not Normal. Non-Normal data, like nonnormal people, not only are common but are also sometimes more interesting than their Normal counterparts.

The 68–95–99.7 rule

Although there are many Normal curves, they all have common properties. Here is one of the most important.

THE 68–95–99.7 RULE

In the Normal distribution with mean μ and standard deviation σ:

- Approximately **68%** of the observations fall within σ of the mean μ.
- Approximately **95%** of the observations fall within 2σ of μ.
- Approximately **99.7%** of the observations fall within 3σ of μ.

Figure 1.25 illustrates the 68–95–99.7 rule. By remembering these three numbers, you can think about Normal distributions without constantly making detailed calculations.

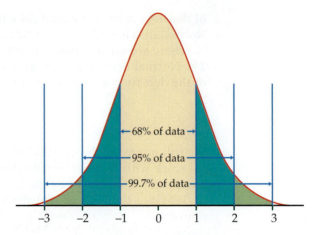

FIGURE 1.25 The 68–95–99.7 rule for Normal distributions.

EXAMPLE 1.38

Heights of young women. The distribution of heights of young women aged 18 to 24 is approximately Normal with mean $\mu = 64.5$ inches and standard deviation $\sigma = 2.5$ inches. Figure 1.26 shows what the 68–95–99.7 rule says about this distribution.

Two standard deviations equals five inches for this distribution. The 95 part of the 68–95–99.7 rule says that the middle 95% of young women are between $64.5 - 5$ and $64.5 + 5$ inches tall, that is, between 59.5 and 69.5 inches. This fact is exactly true for an exactly Normal distribution. It is approximately true for the heights of young women because the distribution of heights is approximately Normal.

The other 5% of young women have heights outside the range from 59.5 to 69.5 inches. Because the Normal distributions are symmetric, half of these women are on the tall side. So the tallest 2.5% of young women are taller than 69.5 inches.

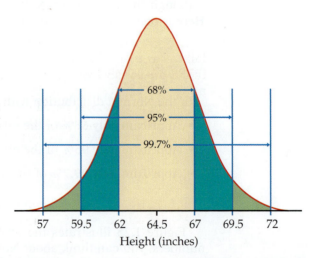

FIGURE 1.26 The 68–95–99.7 rule applied to the heights of young women, Example 1.38.

Because we will mention Normal distributions often, a short notation is helpful. We abbreviate the Normal distribution with mean μ and standard deviation σ as $N(\mu, \sigma)$. For example, the distribution of young women's heights is $N(64.5, 2.5)$.

$N(\mu, \sigma)$

USE YOUR KNOWLEDGE

1.93 Test scores. Many states assess the skills of their students in various grades. One program that is available for this purpose is the National Assessment of Educational Progress (NAEP).[30] One of the tests provided by the NAEP assesses the reading skills of 12th-grade students. In a recent year, the national mean score was 288 and the standard deviation was 38. Assuming that these scores are approximately Normally distributed, $N(288, 38)$, use the 68–95–99.7 rule to give a range of scores that includes 95% of these students.

1.94 Use the 68–95–99.7 rule. Refer to the previous exercise. Use the 68–95–99.7 rule to give a range of scores that includes 99.7% of these students.

Standardizing observations

As the 68–95–99.7 rule suggests, all Normal distributions share many properties. In fact, all Normal distributions are the same if we measure in units of size σ about the mean μ as center. Changing to these units is called *standardizing*. To standardize a value, subtract the mean of the distribution and then divide by the standard deviation.

STANDARDIZING AND z-SCORES

If x is an observation from a distribution that has mean μ and standard deviation σ, the **standardized value** of x is

$$z = \frac{x - \mu}{\sigma}$$

A standardized value is often called a **z-score**.

A z-score tells us how many standard deviations the original observation falls away from the mean, and in which direction. Observations larger than the mean are positive when standardized, and observations smaller than the mean are negative.

To compare scores based on different measures, z-scores can be very useful. For example, see Exercise 1.124 (page 73), where you are asked to compare an SAT score with an ACT score.

EXAMPLE 1.39

Find some z-scores. The heights of young women are approximately Normal with $\mu = 64.5$ inches and $\sigma = 2.5$ inches. The z-score for height is

$$z = \frac{\text{height} - 64.5}{2.5}$$

A woman's standardized height is the number of standard deviations by which her height differs from the mean height of all young women. A woman 68 inches tall, for example, has z-score

$$z = \frac{68 - 64.5}{2.5} = 1.4$$

or 1.4 standard deviations above the mean. Similarly, a woman 5 feet (60 inches) tall has z-score

$$z = \frac{60 - 64.5}{2.5} = -1.8$$

or 1.8 standard deviations less than the mean height.

USE YOUR KNOWLEDGE

1.95   Find the z-score. Consider the NAEP scores (see Exercise 1.93, page 59), which we assume are approximately Normal, $N(288, 38)$. Give the z-score for a student who received a score of 350.

1.96   Find another z-score. Consider the NAEP scores, which we assume are approximately Normal, $N(288, 38)$. Give the z-score for a student who received a score of 240. Explain why your answer is negative even though all the test scores are positive.

We need a way to write variables, such as "height" in Example 1.38, that follow a theoretical distribution such as a Normal distribution. We use capital letters near the end of the alphabet for such variables. If X is the height of a young woman, we can then shorten "the height of a young woman is less than 68 inches" to "$X < 68$." We will use lowercase x to stand for any specific value of the variable X.

We often standardize observations from symmetric distributions to express them in a common scale. We might, for example, compare the heights of two children of different ages by calculating their z-scores. The standardized heights tell us where each child stands in the distribution for his or her age group.

Standardizing is a linear transformation that transforms the data into the standard scale of z-scores. We know that a linear transformation does not change the shape of a distribution, and that the mean and standard deviation change in a simple manner. In particular, *the standardized values for any distribution always have mean 0 and standard deviation 1.*

If the variable we standardize has a Normal distribution, standardizing does more than give a common scale. It makes all Normal distributions into a single distribution, and this distribution is still Normal. Standardizing a variable that has any Normal distribution produces a new variable that has the *standard Normal distribution.*

THE STANDARD NORMAL DISTRIBUTION

The **standard Normal distribution** is the Normal distribution $N(0, 1)$ with mean 0 and standard deviation 1.

If a variable X has any Normal distribution $N(\mu, \sigma)$ with mean μ and standard deviation σ, then the standardized variable

$$Z = \frac{X - \mu}{\sigma}$$

has the standard Normal distribution.

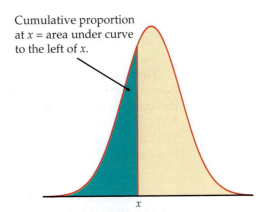

Cumulative proportion at x = area under curve to the left of x.

FIGURE 1.27 The cumulative proportion for a value x is the proportion of all observations from the distribution that are less than or equal to x. This is the area to the left of x under the Normal curve.

Normal distribution calculations

Areas under a Normal curve represent proportions of observations from that Normal distribution. There is no formula for areas under a Normal curve. Calculations use either software that calculates areas or a table of areas. The table and most software calculate one kind of area: **cumulative proportions**. A cumulative proportion is the proportion of observations in a distribution that lie at or below a given value. When the distribution is given by a density curve, the cumulative proportion is the area under the curve to the left of a given value. Figure 1.27 shows the idea more clearly than words do.

cumulative proportion

The key to calculating Normal proportions is to match the area you want with areas that represent cumulative proportions. Then get areas for cumulative proportions either from software or (with an extra step) from a table. The following examples show the method in pictures.

EXAMPLE 1.40

Mitchell Layton/Getty Images

NCAA eligibility for competition. To be eligible to compete in their first year of college, the National Collegiate Athletic Association (NCAA) requires Division I athletes to meet certain academic standards. These are based on their grade point average (GPA) in certain courses and combined scores on the SAT Critical Reading and Mathematics sections or the ACT composite score.[31]

For a student with a 3.0 GPA, the combined SAT score must be 800 or higher. Based on the distribution of SAT scores for college-bound students, we assume that the distribution of the combined Critical Reading and Mathematics scores is approximately Normal with mean 1010 and standard deviation 225.[32] What proportion of college-bound students have SAT scores of 800 or more?

Here is the calculation in pictures: the proportion of scores above 800 is the area under the curve to the right of 800. That's the total area under the curve (which is always 1) minus the cumulative proportion up to 800.

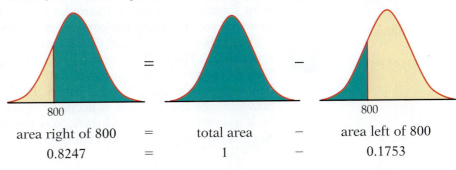

area right of 800	=	total area	−	area left of 800
0.8247	=	1	−	0.1753

That is, the proportion of college-bound SAT takers with a 3.0 GPA who are eligible to compete is 0.8247, or about 82%.

There is *no* area under a smooth curve that is exactly over the point 800. Consequently, the area to the right of 800 (the proportion of scores > 800) is the same as the area at or to the right of this point (the proportion of scores ≥ 800). The actual data may contain a student who scored exactly 800 on the SAT. That the proportion of scores exactly equal to 800 is 0 for a Normal distribution is a consequence of the idealized smoothing of Normal distributions for data.

EXAMPLE 1.41

NCAA eligibility for aid and practice. The NCAA has a category of eligibility in which a first-year student may not compete but is still eligible to receive an athletic scholarship and to practice with the team. The requirements for this category are a 3.0 GPA and combined SAT Critical Reading and Mathematics scores of at least 620.

What proportion of college-bound students who take the SAT would be eligible to receive an athletic scholarship and to practice with the team but would not be eligible to compete? That is, what proportion have scores between 620 and 800? Here are the pictures:

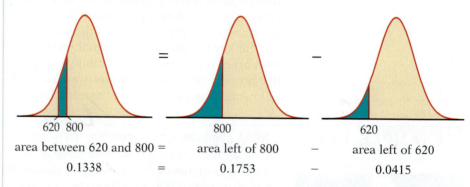

| area between 620 and 800 = | area left of 800 | − | area left of 620 |
| 0.1338 | = | 0.1753 | − | 0.0415 |

About 13% of college-bound students with a 3.0 GPA have SAT scores between 620 and 800.

How do we find the numerical values of the areas in Examples 1.40 and 1.41? If you use software, just plug in mean 1010 and standard deviation 225. Then ask for the cumulative proportions for 800 and for 620. (Your software will probably refer to these as "cumulative probabilities." We will learn in Chapter 4 why the language of probability fits.) Sketches of the areas that you want similar to the ones in Examples 1.40 and 1.41 are very helpful in making sure that you are doing the correct calculations.

You can use the *Normal Curve* applet on the text website to find Normal proportions. The applet is more flexible than most software—it will find any Normal proportion, not just cumulative proportions. The applet is an excellent way to understand Normal curves. But, because of the limitations of web browsers, the applet is not as accurate as statistical software.

If you are not using software, you can find cumulative proportions for Normal curves from a table. That requires an extra step, as we now explain.

Using the standard Normal table

The extra step in finding cumulative proportions from a table is that we must first standardize to express the problem in the standard scale of z-scores. This allows us to get by with just one table, a table of *standard Normal cumulative proportions*. Table A in the back of the book gives standard Normal probabilities. The picture at the top of the table reminds us that the entries are cumulative proportions, areas under the curve to the left of a value z.

EXAMPLE 1.42

Find the proportion from z. What proportion of observations on a standard Normal variable Z take values less than 1.47? We need to find the area to the left of 1.47; locate 1.4 in the left-hand column of Table A and then locate the remaining digit 7 as .07 in the top row. The entry opposite 1.4 and under .07 is 0.9292. This is the cumulative proportion we seek. Figure 1.28 illustrates this area.

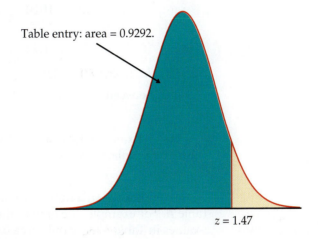

Table entry: area = 0.9292.

$z = 1.47$

FIGURE 1.28 The area under a standard Normal curve to the left of the point z = 1.47 is 0.9292, Example 1.42.

Now that you see how Table A works, let's redo the NCAA Examples 1.40 and 1.41 using the table.

EXAMPLE 1.43

Find the proportion from x. What proportion of college-bound students who take the SAT have scores of at least 800? The picture that leads to the answer is exactly the same as in Example 1.40. The extra step is that we first standardize to read cumulative proportions from Table A. If X is SAT score, we want the proportion of students for which $X \geq x$, where $x = 800$.

1. *Standardize.* Subtract the mean, then divide by the standard deviation, to transform the problem about X into a problem about a standard Normal Z:

$$X \geq 800$$

$$\frac{X - 1010}{225} \geq \frac{800 - 1010}{225}$$

$$Z \geq -0.93$$

2. *Use the table.* Look at the pictures in Example 1.40. From Table A, we see that the proportion of observations less than -0.93 is 0.1762. The area to the right of -0.93 is therefore $1 - 0.1762 = 0.8238$. This is about 82%.

The area from the table in Example 1.43 (0.8238) is slightly less accurate than the area from software in Example 1.40 (0.8247) because we must round z to two places when we use Table A. The difference is rarely important in practice.

EXAMPLE 1.44

Eligibility for aid and practice. What proportion of all students who take the SAT would be eligible to receive athletic scholarships and to practice with the team but would not be eligible to compete in the eyes of the NCAA? That is, what proportion of students have SAT scores between 620 and 800? First, sketch the areas, exactly as in Example 1.41. We again use X as shorthand for an SAT score.

1. *Standardize.*

$$620 \leq X < 800$$
$$\frac{620 - 1010}{225} \leq \frac{X - 1010}{225} < \frac{800 - 1010}{225}$$
$$-1.73 \leq Z < -0.93$$

2. *Use the table.*

area between -1.73 and -0.93 = (area left of -0.93) − (area left of -1.73)
$$= 0.1762 - 0.0418 = 0.1344$$

As in Example 1.41, about 13% of students would be eligible to receive athletic scholarships and to practice with the team.

Sometimes we encounter a value of z more extreme than those appearing in Table A. For example, the area to the left of $z = -4$ is not given in the table. The z-values in Table A leave only area 0.0002 in each tail unaccounted for. For practical purposes, we can act as if there is zero area outside the range of Table A.

USE YOUR KNOWLEDGE

1.97 **Find the proportion.** Consider the NAEP scores, which are approximately Normal, $N(288, 38)$. Find the proportion of students who have scores less than 350. Find the proportion of students who have scores greater than or equal to 350. Sketch the relationship between these two calculations using pictures of Normal curves similar to the ones given in Example 1.40 (page 61).

1.98 **Find another proportion.** Consider the NAEP scores, which are approximately Normal, $N(288, 38)$. Find the proportion of students who have scores between 300 and 350. Use pictures of Normal curves similar to the ones given in Example 1.41 (page 62) to illustrate your calculations.

Inverse Normal calculations

Examples 1.40 to 1.44 illustrate the use of Normal distributions to find the proportion of observations in a given event, such as "SAT score between 620 and 800." We may instead want to find the observed value corresponding to a given proportion.

Statistical software will do this directly. Without software, use Table A backward, finding the desired proportion in the body of the table and then reading the corresponding z from the left column and top row.

EXAMPLE 1.45

How high for the top 10%? Scores for college-bound students on the SAT Critical Reading test in recent years follow approximately the $N(500, 120)$ distribution.[33] How high must a student score to place in the top 10% of all students taking the SAT?

Again, the key to the problem is to draw a picture. Figure 1.29 shows that we want the score x with an area of 0.10 above it. That's the same as area below x equal to 0.90.

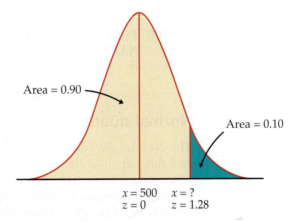

Area = 0.90

Area = 0.10

$x = 500$ $x = ?$
$z = 0$ $z = 1.28$

FIGURE 1.29 Locating the point on a Normal curve with area 0.10 to its right, Example 1.45.

Statistical software has a function that will give you the x for any cumulative proportion you specify. The function often has a name such as "inverse cumulative probability." Plug in mean 500, standard deviation 120, and cumulative proportion 0.9. The software tells you that $x = 653.786$. We see that a student must score at least 654 to place in the highest 10%.

Without software, first find the standard score z with cumulative proportion 0.9, then "unstandardize" to find x. Here is the two-step process:

1. *Use the table.* Look in the body of Table A for the entry closest to 0.9. It is 0.8997. This is the entry corresponding to $z = 1.28$. So $z = 1.28$ is the standardized value with area 0.9 to its left.

2. *Unstandardize* to transform the solution from z back to the original x scale. We know that the standardized value of the unknown x is $z = 1.28$. So x itself satisfies

$$\frac{x - 500}{120} = 1.28$$

Solving this equation for x gives

$$x = 500 + (1.28)(120) = 653.6$$

This equation should make sense: it finds the x that lies 1.28 standard deviations above the mean on this particular Normal curve. That is the "unstandardized" meaning of $z = 1.28$. The general rule for unstandardizing a z-score is

$$x = \mu + z\sigma$$

USE YOUR KNOWLEDGE

1.99 **What score is needed to be in the top 20%?** Consider the NAEP scores, which are approximately Normal, $N(288, 38)$. How high a score is needed to be in the top 20% of students who take this exam?

1.100 **Find the score that 75% of students will exceed.** Consider the NAEP scores, which are approximately Normal, $N(288, 38)$. Seventy-five percent of the students will score above x on this exam. Find x.

Normal quantile plots

The Normal distributions provide good descriptions of some distributions of real data, such as the Iowa Test vocabulary scores. The distributions of some other common variables are usually skewed and therefore distinctly non-Normal. Examples include economic variables such as personal income and gross sales of business firms, the survival times of cancer patients after treatment, and the service lifetime of mechanical or electronic components. While experience can suggest whether or not a Normal distribution is plausible in a particular case, it is risky to assume that a distribution is Normal without actually inspecting the data.

A histogram or stemplot can reveal distinctly non-Normal features of a distribution, such as outliers, pronounced skewness, or gaps and clusters. If the stemplot or histogram appears roughly symmetric and unimodal, however, we need a more sensitive way to judge the adequacy of a Normal model.

Normal quantile plot   The most useful tool for assessing Normality is another graph, the **Normal quantile plot**.

Here is the basic idea of a Normal quantile plot. The graphs produced by software use more sophisticated versions of this idea. It is not practical to make Normal quantile plots by hand.

1. Arrange the observed data values from smallest to largest. Record what percentile of the data each value occupies. For example, the smallest observation in a set of 20 is at the 5% point, the second smallest is at the 10% point, and so on.

2. Do Normal distribution calculations to find the values of z corresponding to these same percentiles. For example, $z = -1.645$ is the 5% point of the standard Normal distribution, and $z = -1.282$ is the 10% point. We call

Normal scores   these values of Z **Normal scores**.

3. Plot each data point x against the corresponding Normal score. If the data distribution is close to any Normal distribution, the plotted points will lie close to a straight line.

   Any Normal distribution produces a straight line on the plot because standardizing turns any Normal distribution into a standard Normal distribution. Standardizing is a linear transformation that can change the slope and intercept of the line in our plot but cannot turn a line into a curved pattern.

USE OF NORMAL QUANTILE PLOTS

If the points on a **Normal quantile plot** lie close to a straight line, the plot indicates that the data are Normal. Systematic deviations from a straight line indicate a non-Normal distribution. Outliers appear as points that are far away from the overall pattern of the plot. An optional line can be drawn on the plot that corresponds to the Normal distribution with mean equal to the mean of the data and standard deviation equal to the standard deviation of the data.

Figures 1.30 and 1.31 are Normal quantile plots for data we have met earlier. The data x are plotted vertically against the corresponding standard Normal z-score plotted horizontally. The z-score scale generally extends from -3 to 3 because almost all of a standard Normal curve lies between these values. These figures show how Normal quantile plots behave.

EXAMPLE 1.46

IQ

IQ scores are approximately Normal. Figure 1.30 is a Normal quantile plot of the 60 fifth-grade IQ scores from Table 1.1 (page 14). The points lie very close to the straight line drawn on the plot. We conclude that the distribution of IQ data is approximately Normal.

FIGURE 1.30 Normal quantile plot of IQ scores, Example 1.46. This distribution is approximately Normal.

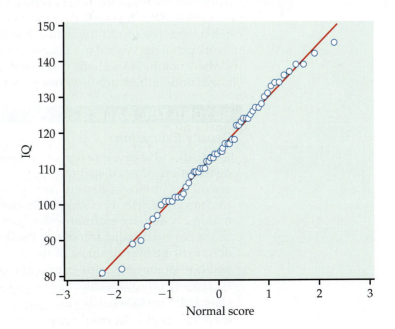

EXAMPLE 1.47

TTS

Times to start a business are skewed. Figure 1.31 is a Normal quantile plot of the data on times to start a business from Example 1.23. We have excluded Suriname, the outlier that you examined in Exercise 1.43 (page 29). The line drawn on the plot shows clearly that the plot of the data is curved. We conclude that these data are not Normally distributed. The shape of the curve is what we typically see with a distribution that is strongly skewed to the right.

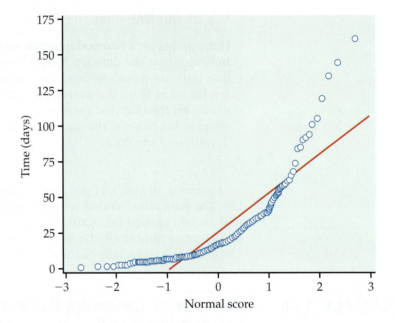

FIGURE 1.31 Normal quantile plot of 188 times to start a business, with the outlier, Suriname, excluded, Example 1.47. This distribution is highly skewed.

Real data often show some departure from the theoretical Normal model. *When you examine a Normal quantile plot, look for shapes that show clear departures from Normality. Don't overreact to minor wiggles in the plot.* When we discuss statistical methods that are based on the Normal model, we are interested in whether or not the data are sufficiently Normal for these procedures to work properly. We are not concerned about minor deviations from Normality. Many common methods work well as long as the data are approximately Normal and outliers are not present.

BEYOND THE BASICS

Density Estimation

A density curve gives a compact summary of the overall shape of a distribution. Many distributions do not have the Normal shape. There are other families of density curves that are used as mathematical models for various distribution shapes. Modern software offers more flexible options. A **density estimator** does not start with any specific shape, such as the Normal shape. It looks at the data and draws a density curve that describes the overall shape of the data. Density estimators join stemplots and histograms as useful graphical tools for exploratory data analysis.

Density estimates can capture other unusual features of a distribution. Here is an example.

density estimator

EXAMPLE 1.48

STUBHUB

StubHub! StubHub! is a website where fans can buy and sell tickets to sporting events. Ticket holders wanting to sell their tickets provide the location of their seats and the selling price. People wanting to buy tickets can choose from among the tickets offered for a given event.[34]

Tickets for the 2015 NCAA women's basketball tournament were available from StubHub! in a package deal that included the semifinal games and the championship game. On June 28, 2014, StubHub! listed 518 tickets for sale. A histogram of the distribution of ticket prices with a density estimate is given in Figure 1.32. The distribution has three peaks: one around $700, another around $2800, and the third around $4650. This is the identifying characteristic of a trimodal distribution. There appears to be three types of tickets. How would you name the three types?

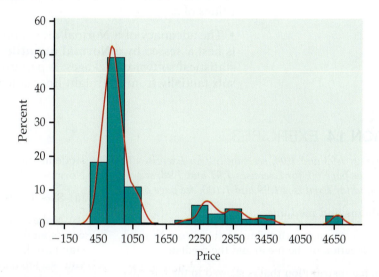

FIGURE 1.32 Histogram of StubHub! price per seat for tickets to the 2015 NCAA Women's Semifinal and Championship games, with a density estimate, Example 1.48.

Many distributions that we have met have a single peak, or mode. The distribution described in Example 1.48 has three modes and is called a **trimodal distribution**. A distribution that has two modes is called a **bimodal distribution**.

trimodal distribution
bimodal distribution

The previous example reminds of a continuing theme for data analysis. We looked at a histogram and a density estimate and saw something interesting. This led us to speculation. Additional data on the type and location of the seats may explain more about the prices than we see in Figure 1.32.

SECTION 1.4 SUMMARY

• The overall pattern of a distribution can often be described compactly by a **density curve**. A density curve has total area 1 underneath it. Areas under a density curve give proportions of observations for the distribution.

• The **mean** μ (balance point), the **median** (equal-areas point), and the **quartiles** can be approximately located by eye on a density curve. The **standard deviation** σ cannot be located by eye on most density curves. The mean and median are equal for symmetric density curves, but the mean of a skewed curve is located farther toward the long tail than is the median.

• The **Normal distributions** are described by bell-shaped, symmetric, unimodal density curves. The mean μ and standard deviation σ completely specify the Normal distribution $N(\mu, \sigma)$. The mean is the center of symmetry, and σ is the distance from μ to the change-of-curvature points on either side. All Normal distributions satisfy the **68–95–99.7 rule**.

- To **standardize** any observation x, subtract the mean of the distribution and then divide by the standard deviation. The resulting **z-score** $z = (x - \mu)/\sigma$ says how many standard deviations x lies from the distribution mean. All Normal distributions are the same when measurements are transformed to the standardized scale.

- If X has the $N(\mu, \sigma)$ distribution, then the standardized variable $Z = (X - \mu)/\sigma$ has the **standard Normal distribution** $N(0, 1)$. Proportions for any Normal distribution can be calculated by software or from the **standard Normal table** (Table A), which gives the **cumulative proportions** of $Z < z$ for many values of z.

- The adequacy of a Normal model for describing a distribution of data is best assessed by a **Normal quantile plot**, which is available in most statistical software packages. A pattern on such a plot that deviates substantially from a straight line indicates that the data are not Normal.

SECTION 1.4 EXERCISES

For Exercises 1.93 and 1.94, see page 59; for Exercises 1.95 and 1.96, see page 60; for Exercises 1.97 and 1.98, see page 64; and for Exercises 1.99 and 1.100, see page 66.

1.101 Means and medians.

(a) Sketch a symmetric distribution that is *not* Normal. Mark the location of the mean and the median.

(b) Sketch a distribution that is skewed to the left. Mark the location of the mean and the median.

1.102 The effect of changing the standard deviation.

(a) Sketch a Normal curve that has mean 30 and standard deviation 8.

(b) On the same x axis, sketch a Normal curve that has mean 30 and standard deviation 12.

(c) How does the Normal curve change when the standard deviation is varied but the mean stays the same?

1.103 The effect of changing the mean.

(a) Sketch a Normal curve that has mean 30 and standard deviation 8.

(b) On the same x axis, sketch a Normal curve that has mean 40 and standard deviation 8.

(c) How does the Normal curve change when the mean is varied but the standard deviation stays the same?

1.104 NAEP music scores. In Exercise 1.93 (page 59) we examined the distribution of NAEP scores for the 12th-grade reading skills assessment. For eighth-grade students, the average music score is approximately Normal with mean 150 and standard deviation 35.

(a) Sketch this Normal distribution.

(b) Make a table that includes values of the scores corresponding to plus or minus one, two, and three standard deviations from the mean. Mark these points on your sketch along with the mean.

(c) Apply the 68–95–99.7 rule to this distribution. Give the ranges of reading score values that are within one, two, and three standard deviations of the mean.

1.105 NAEP U.S. history scores. Refer to the previous exercise. The scores for 12th-grade students on the U.S. history assessment are approximately $N(288,32)$ Answer the questions in the previous exercise for this assessment.

1.106 Standardize some NAEP music scores. The NAEP music assessment scores for eighth-grade students are approximately $N(150,35)$. Find z-scores by standardizing the following scores: 150, 140, 100, 180, 230.

1.107 Compute the percentile scores. Refer to the previous exercise. When scores such as the NAEP assessment scores are reported for individual students, the actual values of the scores are not particularly meaningful. Usually, they are transformed into percentile scores. The percentile score is the proportion of students who would score less than or equal to the score for the individual student. Compute the percentile scores for the five scores in the previous exercise. State whether you used software or Table A for these computations.

1.108 Are the NAEP U.S. history scores approximately Normal? In Exercise 1.105, we assumed that the NAEP U.S history scores for 12th-grade students are approximately Normal with the reported mean and standard deviation, $N(288,32)$. Let's check that assumption. In addition to means and standard deviations, you can find selected percentiles for the NAEP assessments (see previous exercise). For the 12th-grade U.S. history scores, the following percentiles are reported:

Percentile	Score
10%	246
25%	276
50%	290
75%	311
90%	328

Use these percentiles to assess whether or not the NAEP U.S History scores for 12th-grade students are approximately Normal. Write a short report describing your methods and conclusions.

1.109 Are the NAEP mathematics scores approximately Normal? Refer to the previous exercise. For the NAEP mathematics scores for 12th-graders, the mean is 153 and the standard deviation is 34. Here are the reported percentiles:

Percentile	Score
10%	110
25%	130
50%	154
75%	177
90%	197

Is the $N(153,34)$ distribution a good approximation for the NAEP mathematics scores? Write a short report describing your methods and conclusions.

1.110 Do women talk more? Conventional wisdom suggests that women are more talkative than men. One study designed to examine this stereotype collected data on the speech of 42 women and 37 men in the United States.[35] TALK

(a) The mean number of words spoken per day by the women was 14,297 with a standard deviation of 6441. Use the 68–95–99.7 rule to describe this distribution.

(b) Do you think that applying the rule in this situation is reasonable? Explain your answer.

(c) The men averaged 14,060 words per day with a standard deviation of 9056. Answer the questions in parts (a) and (b) for the men.

(d) Do you think that the data support the conventional wisdom? Explain your answer. Note that in Section 7.2 we will learn formal statistical methods to answer this type of question.

1.111 Data from Mexico. Refer to the previous exercise. A similar study in Mexico was conducted with 31 women and 20 men. The women averaged 14,704 words per day with a standard deviation of 6215. For men the mean was 15,022 and the standard deviation was 7864. TALKM

(a) Answer the questions from the previous exercise for the Mexican study.

(b) The means for both men and women are higher for the Mexican study than for the U.S. study. What conclusions can you draw from this observation?

1.112 A uniform distribution. If you ask a computer to generate "random numbers" between 0 and 1, you will get observations from a **uniform distribution**. Figure 1.33 graphs the density curve for a uniform distribution. Use areas under this density curve to answer the following questions.

(a) Why is the total area under this curve equal to 1?

(b) What proportion of the observations lie above 0.44?

(c) What proportion of the observations lie between 0.44 and 0.70?

1.113 Use a different range for the uniform distribution. Many random number generators allow users to specify the range of the random numbers to be produced. Suppose that you specify that the outcomes are to be distributed uniformly between 0 and 4. Then the density curve of the outcomes has constant height between 0 and 4, and height 0 elsewhere.

(a) What is the height of the density curve between 0 and 4? Draw a graph of the density curve.

(b) Use your graph from part (a) and the fact that areas under the curve are proportions of outcomes to find the proportion of outcomes that are more than 1.

(c) Find the proportion of outcomes that lie between 1.5 and 2.5.

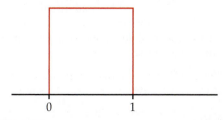

FIGURE 1.33 The density curve of a uniform distribution, Exercise 1.122.

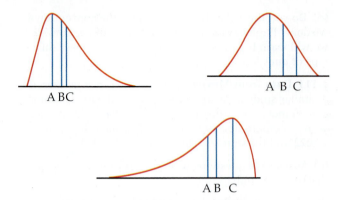

FIGURE 1.34 Three density curves, Exercise 1.115.

1.114 Find the mean, the median, and the quartiles. What are the mean and the median of the uniform distribution in Figure 1.33? What are the quartiles?

1.115 Three density curves. Figure 1.34 displays three density curves, each with three points marked on it. At which of these points on each curve do the mean and the median fall?

1.116 Use the *Normal Curve* applet. Use the *Normal Curve* applet for the standard Normal distribution to say how many standard deviations above and below the mean the quartiles of any Normal distribution lie.

1.117 Use the *Normal Curve* applet. The 68–95–99.7 rule for Normal distributions is a useful approximation. You can use the *Normal Curve* applet on the text website to see how accurate the rule is. Drag one flag across the other so that the applet shows the area under the curve between the two flags.

(a) Place the flags one standard deviation on either side of the mean. What is the area between these two values? What does the 68–95–99.7 rule say this area is?

(b) Repeat for locations two and three standard deviations on either side of the mean. Again compare the 68–95–99.7 rule with the area given by the applet.

1.118 Find some proportions. Using either Table A or your calculator or software, find the proportion of observations from a standard Normal distribution that satisfies each of the following statements. In each case, sketch a standard Normal curve and shade the area under the curve that is the answer to the question.

(a) $Z > 1.75$

(b) $Z < 1.75$

(c) $Z > -0.80$

(d) $-0.80 < Z < 1.75$

1.119 Find more proportions. Using either Table A or your calculator or software, find the proportion of observations from a standard Normal distribution for each of the following events. In each case, sketch a standard Normal curve and shade the area representing the proportion.

(a) $Z \leq -1.4$

(b) $Z \geq -1.4$

(c) $Z > 2.0$

(d) $-1.4 < Z < 2.0$

1.120 Find some values of z. Find the value z of a standard Normal variable Z that satisfies each of the following conditions. (If you use Table A, report the value of z that comes closest to satisfying the condition.) In each case, sketch a standard Normal curve with your value of z marked on the axis.

(a) 38% of the observations fall below z

(b) 70% of the observations fall above z

1.121 Find more values of z. The variable Z has a standard Normal distribution.

(a) Find the number z that has cumulative proportion 0.88.

(b) Find the number z such that the event $Z > z$ has proportion 0.12.

1.122 Find some values of z. The Wechsler Adult Intelligence Scale (WAIS) is the most common IQ test. The scale of scores is set separately for each age group, and the scores are approximately Normal with mean 100 and standard deviation 15. People with WAIS scores below 70 are considered developmentally disabled when, for example, applying for Social Security disability benefits. What percent of adults are developmentally disabled by this criterion?

1.123 High IQ scores. The Wechsler Adult Intelligence Scale (WAIS) is the most common IQ test. The scale of scores is set separately for each age group, and the scores are approximately Normal with mean 100 and standard deviation 15. The organization MENSA, which calls itself "the high-IQ society," requires a WAIS score of 130 or higher for membership. What percent of adults would qualify for membership?

There are two major tests of readiness for college, the ACT and the SAT. ACT scores are reported on a scale from 1 to 36. The distribution of ACT scores is approximately Normal with mean $\mu = 21.5$ and standard deviation $\sigma = 5.4$. SAT scores are reported on a scale from 600 to 2400. The distribution of SAT scores is approximately Normal with mean $\mu = 1498$ and standard deviation $\sigma = 316$. Exercises 1.124 through 1.133 are based on this information.

1.124 Compare an SAT score with an ACT score. Jessica scores 1830 on the SAT. Ashley scores 27 on the ACT. Assuming that both tests measure the same thing, who has the higher score? Report the z-scores for both students.

1.125 Make another comparison. Joshua scores 16 on the ACT. Anthony scores 1050 on the SAT. Assuming that both tests measure the same thing, who has the higher score? Report the z-scores for both students.

1.126 Find the ACT equivalent. Jorge scores 2090 on the SAT. Assuming that both tests measure the same thing, what score on the ACT is equivalent to Jorge's SAT score?

1.127 Find the SAT equivalent. Alyssa scores 30 on the ACT. Assuming that both tests measure the same thing, what score on the SAT is equivalent to Alyssa's ACT score?

1.128 Find an SAT percentile. Reports on a student's ACT or SAT results usually give the percentile as well as the actual score. The percentile is just the cumulative proportion stated as a percent: the percent of all scores that were lower than or equal to this one. Renee scores 2050 on the SAT. What is her percentile?

1.129 Find an ACT percentile. Reports on a student's ACT or SAT results usually give the percentile as well as the actual score. The percentile is just the cumulative proportion stated as a percent: the percent of all scores that were lower than or equal to this one. Joshua scores 19 on the ACT. What is his percentile?

1.130 How high is the top 12%? What SAT scores make up the top 12% of all scores?

1.131 How low is the bottom 12%? What SAT scores make up the bottom 12% of all scores?

1.132 Find the ACT quintiles. The quintiles of any distribution are the values with cumulative proportions 0.20, 0.40, 0.60, and 0.80. What are the quintiles of the distribution of ACT scores?

1.133 Find the SAT quartiles. The quartiles of any distribution are the values with cumulative proportions 0.25 and 0.75. What are the quartiles of the distribution of SAT scores?

1.134 Do you have enough "good cholesterol?" High-density lipoprotein (HDL) is sometimes called the "good cholesterol" because low values are associated with a higher risk of heart disease. According to the American Heart Association, people over the age of 20 years should have at least 40 milligrams per deciliter (mg/dl) of HDL cholesterol.[36] U.S. women aged 20 and over have a mean HDL of 55 mg/dl with a standard deviation of 15.5 mg/dl. Assume that the distribution is Normal.

(a) What percent of women have low values of HDL (40 mg/dl or less)?

(b) HDL levels of 60 mg/dl and higher are believed to protect people from heart disease. What percent of women have protective levels of HDL?

(c) Women with more than 40 mg/dl but less than 60 mg/dl of HDL are in the intermediate range, neither very good or very bad. What proportion are in this category?

1.135 Men and HDL cholesterol. HDL cholesterol levels for men have a mean of 46 mg/dl with a standard deviation of 13.6 mg/dl. Answer the questions given in the previous exercise for the population of men.

1.136 Diagnosing osteoporosis. Osteoporosis is a condition in which the bones become brittle due to loss of minerals. To diagnose osteoporosis, an elaborate apparatus measures bone mineral density (BMD). BMD is usually reported in standardized form. The standardization is based on a population of healthy young adults. The World Health Organization (WHO) criterion for osteoporosis is a BMD 2.5 standard deviations below the mean for young adults. BMD measurements in a population of people similar in age and sex roughly follow a Normal distribution.

(a) What percent of healthy young adults have osteoporosis by the WHO criterion?

(b) Women aged 70 to 79 are of course not young adults. The mean BMD in this age is about -2 on the standard scale for young adults. Suppose that the standard deviation is the same as for young adults. What percent of this older population has osteoporosis?

1.137 Deciles of Normal distributions. The **deciles** of any distribution are the 10th, 20th, ..., 90th percentiles. The first and last deciles are the 10th and 90th percentiles, respectively.

(a) What are the first and last deciles of the standard Normal distribution?

(b) The weights of 9-ounce potato chip bags are approximately Normal with mean 9.12 ounces and standard deviation 0.15 ounce. What are the first and last deciles of this distribution?

1.138 Quartiles for Normal distributions. The quartiles of any distribution are the values with cumulative proportions 0.25 and 0.75.

(a) What are the quartiles of the standard Normal distribution?

(b) Using your numerical values from part (a), write an equation that gives the quartiles of the $N(\mu, \sigma)$ distribution in terms of μ and σ.

1.139 IQR for Normal distributions. Continue your work from the previous exercise. The interquartile range *IQR* is the distance between the first and third quartiles of a distribution.

(a) What is the value of the *IQR* for the standard Normal distribution?

(b) There is a constant *c* such that $IQR - c\sigma$ for any Normal distribution $N(\mu, \sigma)$. What is the value of *c*?

1.140 Outliers for Normal distributions. Continue your work from the previous two exercises. The percent of the observations that are suspected outliers according to the $1.5 \times IQR$ rule is the same for any Normal distribution. What is this percent?

1.141 Deciles of HDL cholesterol. The **deciles** of any distribution are the 10th, 20th,..., 90th percentiles. Refer to Exercise 1.134 where we assumed that the distribution of HDL cholesterol in U.S. women aged 20 and over is Normal with mean 55 mg/dl and standard deviation 15.5 mg/dl. Find the deciles for this distribution.

1.142 Longleaf pine trees. Exercise 1.88 (page 50) gives the diameter at breast height (DBH) for 40 longleaf pine trees from the Wade Tract in Thomas County, Georgia.

Make a Normal quantile plot for these data and write a short paragraph interpreting what it describes. ▥ PINES

1.143 Potassium from potatoes. Refer to Exercise 1.30 (page 24) where you used a stemplot to examine the potassium absorption of a group of 27 adults who ate a controlled diet that included 40 mEq of potassium from potatoes for five days. In Exercise 1.61 (page 47), you compared the stemplot, the histogram, and the boxplot as graphical summaries of this distribution. ▥ KPOT40

(a) Generate these three graphical summaries.

(b) Make a Normal quantile plot and interpret it.

1.144 Potassium from a supplement. Refer to Exercise 1.31 (page 24) where you used a stemplot to examine where you examined the potassium absorption of a group of 29 adults who ate a controlled diet that included 40 mEq of potassium from a supplement for five days. In Exercise 1.62 (page 47), you compared the stemplot, the histogram, and the boxplot as graphical summaries of this distribution. ▥ KSUP40

(a) Generate these three graphical summaries.

(b) Make a Normal quantile plot and interpret it.

CHAPTER 1 EXERCISES

1.145 Sources of energy consumed. Energy consumed in the United States can be classified as coming from one of three sources: fossil fuels, nuclear and electric power, and renewable energy. In 2014, the energy from these three sources was 80.3, 8.3, and 9.6 quadrillion Btu, respectively. In 2004, the corresponding amounts were 85.8, 8.2, and 6.1.[37] Write a description of the changes from 2004 to 2014 expressed in these data. Illustrate your summary with appropriate graphical summaries. Be sure to discuss both the amounts of energy from each source as well as the percents.

1.146 CO₂ emissions in vehicles. Natural Resources Canada tests new vehicles each year and reports several variables related to fuel consumption for vehicles in different classes.[38] For 2015, it provides data for 526 vehicles that use regular fuel. Two variables reported are carbon dioxide (CO_2) emissions and highway fuel consumption. CO_2 is measured in grams per kilometer (g/km), and highway fuel consumption measured in liters per 100 kilometers (L/km). Use graphical and numerical summaries to describe the distribution of CO_2 emissions for these vehicles. Be sure to justify your choice of summaries. ▥ CANFREG

1.147 Highway fuel consumption. Refer to the previous exercise. Use graphical and numerical summaries to describe the distribution of highway fuel consumption

for these vehicles. Be sure to justify your choice of summaries. ▥ CANFREG

1.148 Jobs for business majors. What types of jobs are available for students who graduate with a business degree? The website **careerbuilder.com** lists job opportunities classified in a variety of ways. A recent posting had 25,120 jobs. The following table gives types of jobs and the numbers of postings listed under the classification "business administration" on a recent day:[39] ▥ BUSJOBS

Type	Number
Management	10916
Sales	5981
Information technology	4605
Customer service	4116
Marketing	3821
Finance	2339
Health care	2231
Accounting	2175
Human resources	1685

Describe these data using the methods you learned in this chapter, and write a short summary about jobs that are available for those who have a business degree. Include comments on the limitations that should be kept in mind when interpreting this particular set of data.

1.149 Flopping in the 2014 World Cup. Soccer players are often accused of spending an excessive amount of time dramatically falling to the ground followed by other activities, suggesting that a possible injury is very serious. It has been suggested that these tactics are often designed to influence the call of a referee or to take extra time off the clock. Recordings of the first 32 games of the 2014 World Cup were analyzed, and there were 302 times when the referee interrupted the match because of a possible injury. The number of injuries and the total time, in minutes, spent flopping for each of the 32 teams who participated in these matches was recorded.[40] Here are the data: **Ⅲ FLOPS**

Country	Injuries	Time
Brazil	17	3.30
Chile	16	6.97
Honduras	15	7.67
Nigeria	15	6.42
Mexico	15	3.97
Costa Rica	13	3.80
USA	12	6.40
Ecuador	12	4.55
France	10	7.32
South Korea	10	4.52
Algeria	10	4.05
Iran	9	5.43
Russia	9	5.27
Ivory Coast	9	4.63
Croatia	9	4.32
Colombia	9	4.32
Uruguay	9	4.12
Greece	9	2.65
Cameroon	8	3.15
Germany	8	1.97
Spain	8	1.82
Belgium	7	3.38
Japan	7	2.08
Italy	7	1.60
Switzerland	7	1.35
England	7	3.13
Argentina	6	2.80
Ghana	6	1.85
Australia	6	1.83
Portugal	4	1.82
Netherlands	4	1.65
Bosnia and Herzegovina	2	0.40

Describe these data using the methods you learned in this chapter, and write a short summary about flopping in the 2014 World Cup based on your analysis.

1.150 Twitter accounts. Twitter has more than 52,900,000 million users in the United States. A study of Twitter accounts classified users by age. Here are the numbers of users (in millions) for six age groups:[41] **Ⅲ TWIT**

Age	Number	Age	Number
18–24	11.7	45–54	6.7
25–34	13.3	55–64	4.1
35–44	8.7	65 and over	2.7

Describe these data using the methods you learned in this chapter, and write a short summary about the age distribution of Twitter users based on your analysis.

1.151 What graph would you use? What type of graph or graphs would you plan to make in a study of each of the following issues?

(a) What makes of cars do students drive? How old are their cars?

(b) How many hours per week do students study? How does the number of study hours change during a semester?

(c) Which radio stations are most popular with students?

(d) When many students measure the concentration of the same solution for a chemistry course laboratory assignment, do their measurements follow a Normal distribution?

1.152 Canadian international trade. The government organization Statistics Canada provides data on many topics related to Canada's population, resources, economy, society, and culture. Go to the web page **statcan.gc.ca/start-debut-eng.html**. Under the "Subject" tab, choose "International trade." Pick some data from the resources listed and use the methods that you learned in this chapter to create graphical and numerical summaries. Write a report summarizing your findings that includes supporting evidence from your analyses.

1.153 Travel and tourism in Canada. Refer to the previous exercise. Under the "Subject" tab, choose "Travel and tourism." Pick some data from the resources listed and use the methods that you learned in this chapter to create graphical and numerical summaries. Write a report summarizing your findings that includes supporting evidence from your analyses.

1.154 Leisure time for college students. You want to measure the amount of "leisure time" that college students enjoy. Write a brief discussion of two issues:

(a) How will you define "leisure time"?

(b) Once you have defined leisure time, how will you measure it?

1.155 How much vitamin C do you need?
The U.S. Food and Nutrition Board of the Institute of Medicine, working in cooperation with scientists from Canada, have used scientific data to answer this question for a variety of vitamins and minerals.[42] Their methodology assumes that needs, or requirements, follow a distribution. They have produced guidelines called dietary reference intakes for different gender-by-age combinations. For vitamin C, there are three dietary reference intakes: the estimated average requirement (EAR), which is the mean of the requirement distribution; the recommended dietary allowance (RDA), which is the intake that would be sufficient for 97% to 98% of the population; and the tolerable upper level (UL), the intake that is unlikely to pose health risks. For women aged 19 to 30 years, the EAR is 60 milligrams per day (mg/d), the RDA is 75 mg/d, and the UL is 2000 mg/d.[43]

(a) The researchers assumed that the distribution of requirements for vitamin C is Normal. The EAR gives the mean. From the definition of the RDA, let's assume that its value is the 97.72 percentile. Use this information to determine the standard deviation of the requirement distribution.

(b) Sketch the distribution of vitamin C requirements for 19- to 30-year-old women. Mark the EAR, the RDA, and the UL on your plot.

1.156 How much vitamin C do men need?
Refer to the previous exercise. For men aged 19 to 30 years, the EAR is 75 milligrams per day (mg/d), the RDA is 90 mg/d, and the UL is 2000 mg/d. Answer the questions in the previous exercise for this population.

1.157 How much vitamin C do women consume?
To evaluate whether or not the intake of a vitamin or mineral is adequate, comparisons are made between the intake distribution and the requirement distribution. Here is some information about the distribution of vitamin C intake, in milligrams per day, for women aged 19 to 30 years:[44]

					Percentile (mg/d)				
Mean	1st	5th	19th	25th	50th	75th	90th	95th	99th
84.1	31	42	48	61	79	102	126	142	179

(a) Use the 5th, the 50th, and the 95th percentiles of this distribution to estimate the mean and standard deviation of this distribution assuming that the distribution is Normal. Explain your method for doing this.

(b) Sketch your Normal intake distribution on the same graph with a sketch of the requirement distribution that you produced in part (b) of Exercise 1.155.

(c) Do you think that many women aged 19 to 30 years are getting the amount of vitamin C that they need? Explain your answer.

1.158 How much vitamin C do men consume?
To evaluate whether or not the intake of a vitamin or mineral is adequate, comparisons are made between the intake distribution and the requirement distribution. Here is some information about the distribution of vitamin C intake, in milligrams per day, for men aged 19 to 30 years:

					Percentile (mg/d)				
Mean	1st	5th	19th	25th	50th	75th	90th	95th	99th
122.2	39	55	65	85	114	150	190	217	278

(a) Use the 5th, the 50th, and the 95th percentiles of this distribution to estimate the mean and standard deviation of this distribution assuming that the distribution is Normal. Explain your method for doing this.

(b) Sketch your Normal intake distribution on the same graph with a sketch of the requirement distribution that you produced in Exercise 1.156.

(c) Do you think that many men aged 19 to 30 years in the United States are getting the amount of vitamin C that they need? Explain your answer.

1.159 Time spent studying. Do women study more than men? We asked the students in a large first-year college class how many minutes they studied on a typical weeknight. Here are the responses of random samples of 30 women and 30 men from the class: **STUDY**

Women					Men				
170	120	180	360	240	80	120	30	90	200
120	180	120	240	170	90	45	30	120	75
150	120	180	180	150	150	120	60	240	300
200	150	180	150	180	240	60	120	60	30
120	60	120	180	180	30	230	120	95	150
90	240	180	115	120	0	200	120	120	180

(a) Examine the data. Why are you not surprised that most responses are multiples of 10 minutes? We eliminated one student who claimed to study 30,000 minutes per night. Are there any other responses that you consider suspicious?

(b) Make a back-to-back stemplot of these data. Report the approximate midpoints of both groups. Does it appear that women study more than men (or at least claim that they do)?

(c) Make side-by-side boxplots of these data. Compare the boxplots with the stemplot you made in part (b). Which to you prefer? Give reasons for your answer.

1.160 Product preference. Product preference depends in part on the age, income, and gender of the consumer. A market researcher selects a large sample of potential car buyers. For each consumer, she records gender, age, household income, and automobile preference. Which of these variables are categorical and which are quantitative?

1.161 Two distributions. If two distributions have exactly the same mean and standard deviation, must their histograms have the same shape? If they have the same five-number summary, must their histograms have the same shape? Explain.

1.162 Spam filters. A university department installed a spam filter on its computer system. During a 21-day period, 6693 messages were tagged as spam. How much spam you get depends on what your online habits are. Here are the counts for some students and faculty in this department (with log-in IDs changed, of course):

ID	Count	ID	Count	ID	Count	ID	Count
AA	1818	BB	1358	CC	442	DD	416
EE	399	FF	389	GG	304	HH	251
II	251	JJ	178	KK	158	LL	103

All other department members received fewer than 100 spam messages. How many did the others receive in total? Make a graph and comment on what you learn from these data. SPAM

1.163 Phish. One of the most favored songs of the band Phish is "Divided Sky." The band plays this song at many of their concerts. Frequently, after the main theme, Trey, the guitarist, pauses before playing the resolving note.[45] The data file PHISH gives the date of each concert where "Divided Sky" was played, the venue, and the length of the pause for 366 concerts. Analyze the data and write a report summarizing what you have found. Be sure to include graphical and numerical summaries. Include the rationale for decisions that you made in performing your analysis. For example, did you give any consideration to the relatively large number of zeros? PHISH

1.164 Visits to a help room for statistics. A help room staffed by graduate students provides assistance to students taking statistics courses. To justify the cost of providing this service, extensive records are kept. Each time a student visits the help room, the student signs a sheet with several variables. These include the date of the visit, the course number that they are taking, the time they arrived at the room, and the time that they left the room. The length of time that the each student spent in the help room is computed from the two time variables. Data for 1268 visits are given in the file HELP.[46] Analyze the data and write a report summarizing what you have found. Be sure to include graphical and numerical summaries. Include the rationale for the choices of methods that you chose for your analysis. There are some missing course numbers. How did you handle these? HELP

1.165 Blueberries and anthocyanins. Anthocyanins are compounds that have been associated with health benefits associated with the heart, bones, and the brain. Blueberries are a good source of many different anthocyanins. Researchers at the Piedmont Research Station of North Carolina State University have assembled a database giving the concentrations of 18 different anthocyanins for 267 varieties of blueberries.[47] Four of the anthocyanins measured are delphinidin-3-arabinoside, malvidin-3-arabinoside, cyanidin-3-galactoside, and delphinidin-3-glucoside, all measured in units of mg/100g of berries. In the data file, we have simplified the names of these anthocyanins to Antho1, Antho2, Antho3, and Antho4. Figure 1.35 gives graphical and numeric summaries from JMP for Antho1. Use this output to write a summary of the distribution of Antho1 using the methods and ideas that you learned in this chapter. BERRIES

1.166 Blueberries and anthocyanins, Antho2. Refer to the previous exercise. Generate your own output for the analysis of Antho2 and use your output to write a summary of the distribution of Antho2 using the methods and ideas that you learned in this chapter. BERRIES

1.167 Blueberries and anthocyanins, Antho3. Refer to Exercise 1.165. Figure 1.36 gives the JMP output for Antho3. Use this output to write a summary of the distribution of Antho3 using the methods and ideas that you learned in this chapter. BERRIES

1.168 Blueberries and anthocyanins, Antho4. Refer to Exercise 1.165. Generate your own output for the analysis of Antho4 and use your output to write a summary of the distribution of Antho4 using the methods and ideas that you learned in this chapter. BERRIES

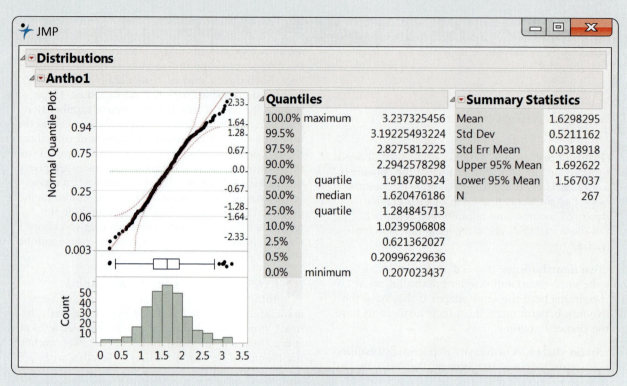

FIGURE 1.35 JMP descriptive statistics for Antho1, Exercise 1.165.

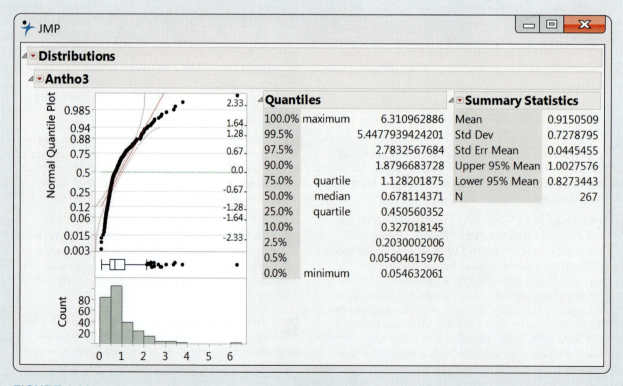

FIGURE 1.36 JMP descriptive statistics for Antho3, Exercise 1.167.

Yellow Dog Productions/Getty Images

Looking at Data—Relationships

Introduction

In Chapter 1, we learned to use graphical and numerical methods to describe the distribution of a single variable. Many of the interesting examples of the use of statistics involve relationships between pairs of variables. Learning ways to describe relationships with graphical and numerical methods is the focus of this chapter.

In Section 2.2, we focus on graphical descriptions. The scatterplot is our fundamental graphical tool for displaying the relationship between two quantitative variables. Sections 2.3 and 2.4 move on to numerical summaries for these relationships. Cautions about the use of these methods are discussed in Section 2.5. Graphical and numerical methods for describing the relationship between two categorical variables are presented in Section 2.6. We conclude with Section 2.7, a brief overview of issues related to the distinction between associations and causation.

2.1 Relationships

2.2 Scatterplots

2.3 Correlation

2.4 Least-Squares Regression

2.5 Cautions about Correlation and Regression

2.6 Data Analysis for Two-Way Tables

2.7 The Question of Causation

2.1 Relationships

When you complete this section, you will be able to:	• Identify the key characteristics of a data set to be used to explore a relationship between two variables.
	• Categorize variables as response variables or explanatory variables.

In Chapter 1 (page 2), we discussed the key characteristics of a data set. Cases are the objects described by a set of data, and a variable is a characteristic of a case. We also learned to categorize variables as categorical or quantitative. For Chapter 2, we focus on data sets that have pairs of variables that we want to study together. Here is an example.

EXAMPLE 2.1

College students cope with stress. Stress is a common problem for college students. Exploring factors that are associated with stress may lead to strategies that will help students to relieve some of the stress that they experience. A recent study found that students who experienced greater stress had less access to resources that would help them to cope with their stress.[1] The two variables involved in the relationship here are perceived stress and resources to cope. The cases are the 97 students who are the subjects for a particular study.

When we study relationships between two variables, it is not sufficient to collect data on the two variables. *A key idea for this chapter is that both variables must be measured on the same cases.*

USE YOUR KNOWLEDGE

2.1 Facebook friends. Do people who have more Facebook friends spend more time on Facebook? In an introductory statistics class of 38 students, there were 32 users of Facebook. Each of these students was asked to report how many Facebook friends they had and the average amount of time that they spent on Facebook per week.

(a) Who are the cases for this study?

(b) What are the variables?

(c) Are the variables quantitative or categorical? Explain your answer.

We use the term *associated* to describe the relationship between two variables, such as stress and access to resources to cope in Example 2.1. Here is another example where two variables are associated.

EXAMPLE 2.2

Anthony Behar/Sipa USA/Sipa via AP Images

Size and price of a coffee beverage. You visit a local Starbucks to buy a Mocha Frappuccino®. The barista explains that this blended coffee beverage comes in three sizes and asks if you want a Tall, a Grande, or a Venti. The prices are $3.95, $4.45, and $4.95, respectively. There is a clear association between the size of the Mocha Frappuccino® and its price.

ASSOCIATION BETWEEN VARIABLES

Two variables measured on the same cases are **associated** if knowing the values of one of the variables tells you something about the values of the other variable.

In the Mocha Frappuccino® example, knowing the size tells you the exact price, so the association here is very strong. Many statistical associations, however, are simply overall tendencies that allow exceptions. For example, it's likely that some students in Example 2.1 are highly stressed and have a high level of resources to cope. Others experience little stress and have a low level of resources to cope. The association in that example is much weaker than the one in the Mocha Frappuccino® example.

Examining relationships

To examine the relationship between two or more variables, we first need to know some basic characteristics of the data.

EXAMPLE 2.3

Stress and resources to cope. Refer to Example 2.1. The study asked 97 first-year college students about their stress (perceived stress) and the availability of resources to deal with stress (resources to cope).[2] Perceived stress is based on responses to 10 questions that are summarized in a single variable. Therefore, we will treat the perceived stress as a quantitative variable. Resources to cope is constructed in a similar way summarizing the responses to 20 questions. We treat resources to cope as a quantitative variable also.

In many situations, we measure a collection of categorical variables and then combine them in a scale that can be viewed as a quantitative variable. The perceived stress and resources to cope are examples. We can also turn the tables in the other direction. Here is an example.

EXAMPLE 2.4

Hemoglobin and anemia. Hemoglobin is a measure of iron in the blood. The units are grams of hemoglobin per deciliter of blood (g/dl). Typical values depend on age and sex. Adult women typically have values between 12 and 16 g/dl.

Anemia is a major problem in developing countries, and many studies have been designed to address the problem. In these studies, computing the mean hemoglobin is not particularly useful. For studies like these, it is more appropriate to use a definition of severe anemia (a hemoglobin of less than 8 g/dl). Thus, for example, researchers can compare the proportions of subjects who are severely anemic for two treatments rather than the difference in the mean hemoglobin levels. In this situation, the categorical variable, severely anemic or not, is much more useful than the quantitative variable, hemoglobin.

When analyzing data to draw conclusions, it is important to carefully consider the best way to summarize the data. *Just because a variable is measured as a quantitative variable, it does not necessarily follow that the best summary is based on the mean (or the median).* As the previous example illustrates, converting a quantitative variable to a categorical variable is a very useful option to keep in mind.

2.2 Create a categorical variable from a quantitative variable. Consider the study described in Example 2.3. Suppose that we order the students based on the values of resources to cope from smallest to largest. Then, we define three resource groups: low resources, the first 32 students; medium resources, the next 33 students; and high resources, the remaining 32 students. If we compare the perceived stress of the three resource groups, are we using resource group as a quantitative variable or as a categorical variable? Explain your answer and describe some advantages to using the groups versus the original variable in explaining the results of a study such as this.

2.3 Replace names by ounces. In the Mocha Frappuccino® example, the variable size is categorical, with Tall, Grande, and Venti as the possible values. Suppose that you converted these values to the number of ounces: Tall is 12 ounces, Grande is 16 ounces, and Venti is 24 ounces. For studying the relationship between ounces and price, describe the cases and the variables and state whether each is quantitative or categorical.

When you examine the relationship between two variables, a new question becomes important:

Is your purpose simply to explore the nature of the relationship, or do you hope to show that one of the variables can explain variation in the other? In other words, is one of the variables a *response variable* and the other an *explanatory variable*?

RESPONSE VARIABLE, EXPLANATORY VARIABLE

A **response variable** measures an outcome of a study. An **explanatory variable** explains or causes changes in the response variable.

EXAMPLE 2.5

Stress and resources to cope. Refer to the study of stress and resources to cope in Example 2.3. Here, the explanatory variable is resources to cope and the response variable is perceived stress.

2.4 Stress and resources or resources and stress? Consider the scenario described in the previous example. Note that the variable, resources to cope, is constructed by summarizing the responses to 20 questions that include items measuring the skills that the student has developed to reduce stress. Make an argument for treating stress as the explanatory variable and resources to cope as the response variable.

In some studies, it is easy to identify explanatory and response variables. The following example illustrates one situation where this is true: when we actually set values of one variable to see how it affects another variable.

EXAMPLE 2.6

How much calcium do you need? Adolescence is a time when bones are growing very actively. If young people do not have enough calcium, their bones will not grow properly. How much calcium is enough? Research designed to answer this question has been performed for many years at events called "Camp Calcium."[3] At these camps, subjects eat controlled diets that are identical except for the amount of calcium. The amount of calcium retained by the body is the major response variable of interest. Because the amount of calcium consumed is controlled by the researchers, this variable is the explanatory variable.

When you don't set the values of either variable but just observe both variables, there may or may not be explanatory and response variables. Whether there are depends on how you plan to use the data.

EXAMPLE 2.7

Student loans. A college student aid officer looks at the findings of the National Student Loan Survey. She notes data on the amount of debt of recent graduates, their current income, and how stressful they feel about college debt. She isn't interested in predictions but is simply trying to understand the situation of recent college graduates.

A sociologist looks at the same data with an eye to using amount of debt and income, along with other variables, to explain the stress caused by college debt. Now, amount of debt and income are explanatory variables, and stress level is the response variable.

In many studies, the goal is to show that changes in one or more explanatory variables actually *cause* changes in a response variable. But many explanatory-response relationships do not involve direct causation. The SAT scores of high school students help predict the students' future college grades, but high SAT scores certainly don't cause high college grades.

KEY CHARACTERISTICS OF DATA FOR RELATIONSHIPS

A description of the key characteristics of a data set that will be used to explore a relationship between two variables should include

- **Cases.** Identify the cases and how many there are in the data set.

- **Categorical or quantitative.** Classify each variable as categorical or quantitative.

- **Values.** Identify the possible values for each variable.

- **Explanatory or response.** If appropriate, classify each variable as explanatory or response.

- **Label.** Identify what is used as a label variable if one is present.

Some of the statistical techniques in this chapter require us to distinguish explanatory from response variables; others make no use of this distinction. independent variable / dependent variable / You will often see explanatory variables called **independent variables** variable and response variables called **dependent variables**. These terms express mathematical ideas; they are not statistical terms. The concept that underlies this language is that the response *depends* on explanatory variables. Because the words "independent" and "dependent" have other meanings in statistics that are unrelated to the explanatory-response distinction, we prefer to avoid those words.

Most statistical studies examine data on more than one variable. Fortunately, statistical analysis of several-variable data builds on the tools used for examining individual variables. The principles that guide our work also remain the same:

- Start with a graphical display of the data.

- Look for overall patterns and deviations from those patterns.

- Based on what you see, use numerical summaries to describe specific aspects of the data.

SECTION 2.1 SUMMARY

- To study relationships between variables, we must measure the variables on the same cases. It is also important to determine the number of cases and to classify each variable as categorical or quantitative.

- Two variables measured on the same cases are **associated** if knowing the values of one variable tells you something about the values of the other variable. If we think that a variable x may explain or even cause changes in another variable y, we call x an **explanatory variable** and y a **response variable**.

SECTION 2.1 EXERCISES

For Exercise 2.1, see page 80; for Exercises 2.2 and 2.3, see page 82; and for Exercise 2.4, see page 82.

2.5 High click counts on Twitter. A study was done to identify variables that might produce high click counts on Twitter. You and nine of your friends collect data on all of your tweets for a week. You record the number of click counts, the time of day, the day of the week, the sex of the person posting the tweet, and the length of the tweet.

(a) What are the cases for this study?

(b) Classify each of the variables as categorical or quantitative.

(c) Classify each of the variables as explanatory, response, or neither. Explain your answers.

2.6 Explanatory or response? For each of the following scenarios, classify each of the pair of variables as explanatory or response or neither. Give reasons for your answers.

(a) Whether or not a person likes to sing and whether or not a person likes to dance.

(b) The number of pages in a textbook and the cost of a new copy of the textbook.

(c) The number of alcoholic drinks consumed and the blood alcohol content.

(d) In a study of adolescents, the dose of vitamin D given each day for a year (50, 100, or 200 international units) and the change in total bone mineral content from the beginning of the study to the end of the study.

2.7 Buy and sell prices of used textbooks. Think about a study designed to compare the prices of textbooks for third- and fourth-year college courses in five different majors. For the five majors, you want to examine the relationship between the difference in the price that you pay for a used textbook and the price that the seller gives back to you when you return the textbook. Describe a data set that could be used for this study, and give the key characteristics of the data.

2.8 Protein and fat. Think about a study designed to examine the relationship between protein intake and fat intake in the diets of first-year college students. Describe a data set that could be used for this study, and give the key characteristics of the data.

2.9 Soccer tickets and performance. For the teams in the Big Ten Conference last year, plan a study of the relationship between the average number of tickets sold for home soccer games and the percentage of games won. Give the key characteristics of the data that could be used for your study.

2.2 Scatterplots

When you complete this section, you will be able to:

- Make a scatterplot to examine a relationship between two variables.
- Describe the overall pattern in a scatterplot and any striking deviations from that pattern.
- Use a scatterplot to describe the form, direction, and strength of a relationship.
- Use a scatterplot to identify outliers.
- Identify a linear pattern in a scatterplot.
- Explain the effect of a change of units on a scatterplot.
- Use a log transformation to change a curved relationship into a linear relationship.
- Use different plotting symbols to include information about a categorical variable in a scatterplot.

EXAMPLE 2.8

LAUNDRY

Laundry detergents. Consumers Union provides ratings on a large variety of consumer products. They use sophisticated testing methods as well as surveys of their members to create these ratings. The ratings are published in their magazine, *Consumer Reports*.[4]

One recent study rated 53 laundry detergents on a scale from 1 to 100. The scale summarizes washing performance under a variety of conditions. Price per load is given in cents.[5] We will examine the relationship between rating and price per load for these laundry detergents. We expect that the higher-priced detergents will tend to have higher ratings.

USE YOUR KNOWLEDGE

LAUNDRY

2.10 **Examine the spreadsheet.** Examine the spreadsheet of the laundry detergent data.

(a) How many cases are in the data set?

(b) Describe the labels, variables, and values.

(c) Which columns represent quantitative variables? Which columns give categorical variables.

(d) Is there an explanatory variable? A response variable? Explain your answer.

LAUNDRY

2.11 **Use the data set.** Using the data set from the previous exercise, create graphical and numerical summaries for the rating and for the price per load.

The most common way to display the relationship between two quantitative variables is a *scatterplot*.

SCATTERPLOT

A **scatterplot** shows the relationship between two quantitative variables measured on the same cases. The values of one variable appear on the horizontal axis, and the values of the other variable appear on the vertical axis. Each case in the data appears as the point in the plot determined by the values of both variables for that case.

EXAMPLE 2.9

LAUNDRY

Laundry detergents. A higher price for a product should be associated with a better product. Therefore, let's treat price per load as the explanatory variable and rating as the response variable in our examination of the relationship between these two variables. We begin with a graphical display.

Figure 2.1 gives a scatterplot that displays the relationship between the response variable, rating, and the explanatory variable, price per load. The most striking feature that we see in the plot is a case that appears to be very different from the others. One of the laundry detergents has a rating that is about average (51), but the price per load (56 cents) is almost double that of the other products.

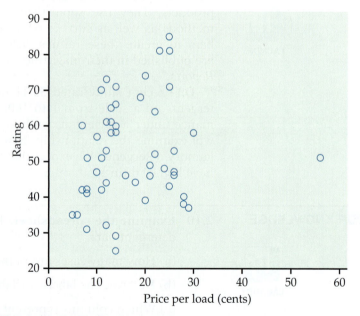

FIGURE 2.1 Scatterplot of price per load (in cents) versus rating for 53 laundry detergents, Example 2.9.

Cases that fall well outside the general pattern of the relationship are called outliers. We provide a more detailed description of these in Section 2.5. For now, we remove this case and focus on the relationship of the remaining data.

Figure 2.2 gives the scatterplot with the outlier removed. The relationship is weak. Paying a high price for your laundry detergent will not guarantee that you have selected a highly rated product.

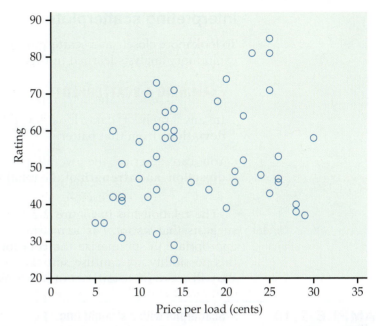

FIGURE 2.2 Scatterplot of price per load (in cents) versus rating for 52 laundry detergents (with the outlier removed), Example 2.9.

Always plot the explanatory variable, if there is one, on the horizontal axis (the *x* axis) of a scatterplot. We usually call the explanatory variable *x* and the response variable *y*. If there is no explanatory-response distinction, either variable can go on the horizontal axis. Time plots such as the one in Figure 1.12 (page 22) are special scatterplots where the explanatory variable *x* is a measure of time.

USE YOUR KNOWLEDGE

LAUND

jitter

LAUNDRY

2.12 Make a scatterplot. Let's consider the laundry data with the outlier removed.

(a) Make a scatterplot similar to Figure 2.2.

(b) Two of the laundry detergents cost 14 cents per load with a rating of 60. Mark the location of these items on your plot.

(c) Cases with identical values for both variables are generally indistinguishable in a scatterplot. To what extent do you think that this could give a distorted picture of the relationship between two variables for a data set that has a large number of duplicate values? Explain your answer.

(d) An option called **jitter** is available with some statistical software that will add a little noise to each point so that points with identical values will appear to be different. If you have software that includes this option, apply it to your plot and summarize the effect of the jittering.

2.13 Change the units. Refer to the laundry data with the outlier.

(a) Create a spreadsheet with the price per load expressed in dollars.

(b) Make a scatterplot for the data in your spreadsheet.

(c) Describe how this scatterplot differs from Figure 2.2.

Interpreting scatterplots

To look more closely at a scatterplot such as Figure 2.2, apply the strategies of exploratory analysis learned in Chapter 1.

> **EXAMINING A SCATTERPLOT**
>
> In any graph of data, look for the **overall pattern** and for striking **deviations** from that pattern.
>
> You can describe the overall pattern of a scatterplot by the **form**, **direction**, and **strength** of the relationship.

linear relationship   The relationship in Figure 2.2 is difficult to see. Looking at it carefully suggests that its *form* is approximately **linear**. In other words, it may be appropriate to summarize the **relationship** with a straight line. To explore this possibility, we can use software to put a straight line through the data. We will see more details about how this is done in Section 2.4.

EXAMPLE 2.10

LAUND

Scatterplot with a straight line.   Figure 2.3 plots the laundry detergent data with a straight line. The line helps us to see and to evaluate the linear form of the relationship. In Section 2.4 (page 107), we will learn how to determine this line.

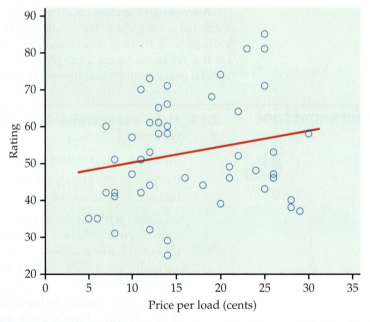

FIGURE 2.3 Scatterplot of rating versus price per load (in cents), with a fitted straight line, Example 2.10.

There is a large amount of scatter about the line. We see that there are eight laundry detergents with a price of 14 cents per load. For these products, the variation in ratings is substantial, from 25 to 71. We do not see any additional outliers in this plot.

Although it is very weak, the relationship in Figure 2.3 has a *direction*, laundry detergents that cost more have somewhat higher ratings. This is a *positive association* between the two variables.

POSITIVE ASSOCIATION, NEGATIVE ASSOCIATION

Two variables are **positively associated** when above-average values of one tend to accompany above-average values of the other and below-average values also tend to occur together.

Two variables are **negatively associated** when above-average values of one tend to accompany below-average values of the other, and vice versa.

The *strength* of a relationship in a scatterplot is determined by how closely the points follow a clear form. The overall relationship in Figure 2.3 is weak. Here is an example of a stronger linear relationship.

EXAMPLE 2.11

EDSPEND

benchmarking

Education spending and population: Benchmarking. We expect that states with larger populations would spend more on education than states with smaller populations.[6] What is the nature of this relationship? Can we use this relationship to evaluate whether some states are spending more than we expect or less than we expect? This type of exercise is called **benchmarking**. The basic idea is to compare processes or procedures of an organization with those of similar organizations.

Figure 2.4 is a spreadsheet giving the education spending and the populations of the 50 U.S. states for 2015. Figure 2.5 is a scatterplot of the education spending versus the population with a straight line. The scatterplot shows a strong positive relationship between these two variables.

FIGURE 2.4 State spending (in billions of dollars) and population (in millions) for the 50 U.S. states, Example 2.11.

	A	B	C
1	State	Spending	Population
2	Alabama	14.9	4.9
3	Alaska	3.8	0.7
4	Arizona	14.8	6.8
5	Arkansas	8.5	3.0
6	California	110.7	39.2
7	Colorado	14.3	5.4
8	Connecticut	13.1	3.6
9	Delaware	4.0	0.9
10	Florida	40.1	20.2
11	Georgia	26.4	10.2
12	Hawaii	3.3	1.4
13	Idaho	3.4	1.7
14	Illinois	38.0	12.9
15	Indiana	17.5	6.6
16	Iowa	10.8	3.1
17	Kansas	8.6	2.9
18	Kentucky	12.8	4.4
19	Louisiana	13.1	4.7
20	Maine	4.2	1.3
21	Maryland	18.6	6.0
22	Massachusetts	21.2	6.8
23	Michigan	24.0	9.9
24	Minnesota	17.7	5.5
25	Mississippi	8.1	3.0
26	Missouri	16.3	6.1

	A	B	C
27	Montana	2.7	1.0
28	Nebraska	6.5	1.9
29	Nevada	5.7	2.9
30	New Hampshire	4.2	1.3
31	New Jersey	36.6	9.0
32	New Mexico	6.6	2.1
33	New York	74.3	19.8
34	North Carolina	30.2	10.0
35	North Dakota	3.0	0.8
36	Ohio	32.1	11.6
37	Oklahoma	10.8	3.9
38	Oregon	11.3	4.0
39	Pennsylvania	37.5	12.8
40	Rhode Island	3.5	1.1
41	South Carolina	10.8	4.9
42	South Dakota	2.1	0.9
43	Tennessee	15.9	6.6
44	Texas	89.4	27.4
45	Utah	10.0	3.0
46	Vermont	2.5	0.6
47	Virginia	24.9	8.4
48	Washington	21.2	7.2
49	West Virginia	6.0	1.8
50	Wisconsin	18.9	5.8
51	Wyoming	2.7	0.6
52			

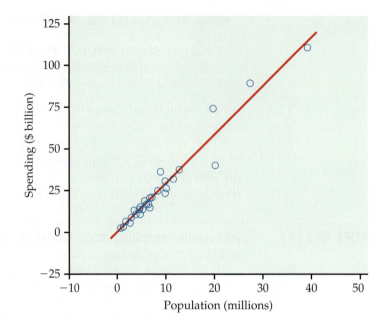

FIGURE 2.5 Scatterplot of state spending (in billions of dollars) versus population for the 50 U.S. states, with a fitted straight line, Example 2.11.

USE YOUR KNOWLEDGE

2.14 **Make a scatterplot.** In our Mocha Frappuccino® example, the 12-ounce drink costs $3.95, the 16-ounce drink costs $4.45, and the 24-ounce drink costs $4.95. Explain which variable should be used as the explanatory variable, and make a scatterplot and include the fitted straight line if your software includes this option. Describe the scatterplot and the association between these two variables.

Of course, not all relationships are linear. Here is an example where a relationship is described by a curve.

EXAMPLE 2.12

CALCIUM

Calcium retention. Our bodies need calcium to build strong bones. How much calcium do we need? Does the amount that we need depend on our age? Questions like these are studied by nutrition researchers. One series of studies used the amount of calcium retained by the body as a response variable and the amount of calcium consumed as an explanatory variable.[7]

Figure 2.6 is a scatterplot of calcium retention in milligrams per day (mg/d) versus calcium intake (mg/d) for 56 children aged 11 to 15 years.

FIGURE 2.6 Scatterplot of calcium retention (mg/d) versus calcium intake (mg/d) for 56 children, with a fitted curve, Example 2.12. There is a positive relationship between these two variables, but it is not linear.

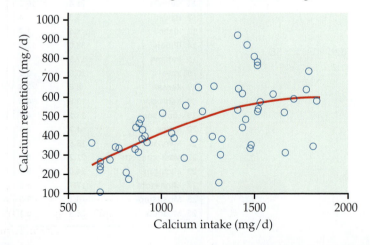

A smooth curve generated by software helps us see the relationship between the two variables.

There is clearly a relationship here. As calcium intake increases, the body retains more calcium. However, the relationship is not linear. The curve is approximately linear for low values of intake, but then the line curves more and becomes almost level.

transformation

There are many kinds of curved relationships like that in Figure 2.6. For some of these, we can apply a **transformation** to the data that will make the relationship approximately linear. To do this, we replace the original values with the transformed values and then use the transformed values for our analysis.

Transforming data is common in statistical practice. There are systematic principles that describe how transformations behave and guide the search for transformations that will, for example, make a distribution more Normal or a curved relationship more linear.

The log transformation

log transformation

The most important transformation that we will use is the **log transformation**. This transformation can be used for variables that have positive values only. Occasionally, we use it when there are zeros, but in this case we first replace the zero values by some small value, often one-half of the smallest positive value in the data set.

You have probably encountered logarithms in one of your high school mathematics courses as a way to do certain kinds of arithmetic. Logarithms are a powerful tool when used in statistical analyses. We will use natural logarithms. Statistical software and statistical calculators generally provide easy ways to perform this transformation.

Let's try a log transformation on our calcium retention data. Here are the details.

EXAMPLE 2.13

CALCIUM

Calcium retention with logarithms. Figure 2.7 is a scatterplot of the log of calcium retention versus calcium intake. The plot includes a fitted straight line to help us see the relationship. We see that the transformation has worked. Our relationship is now approximately linear.

FIGURE 2.7 Scatterplot of log calcium retention versus calcium intake, with a fitted line, for 56 children, Example 2.13. The relationship is approximately linear.

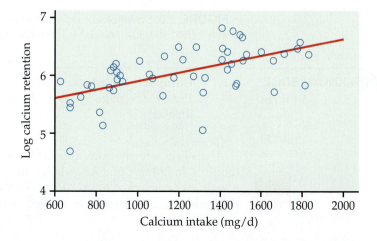

Our analysis of the calcium retention data in Examples 2.12 and 2.13 reminds us of an important issue when describing relationships. In Example 2.12, we noted that the relationship appeared to become approximately flat. Biological processes are consistent with this observation. There is probably a point where additional intake does not result in any additional retention. With our transformed relationship in Figure 2.7, however, there is no leveling off as we saw in Figure 2.6, even though we appear to have a good fit to the data. The relationship and fit apply to the range of data that are analyzed. *We cannot assume that the relationship extends beyond the range of the data.*

For the calcium data, we used a log transformation to describe the curved relationship in Figure 2.6 as the linear relationship in Figure 2.7. Here is another application of a log transformation.

EXAMPLE 2.14

EDSPEND

Education spending and population with logarithms. Let's examine the relationship between spending and population using logs for both variables. Figure 2.8 gives the plot with the fitted line.

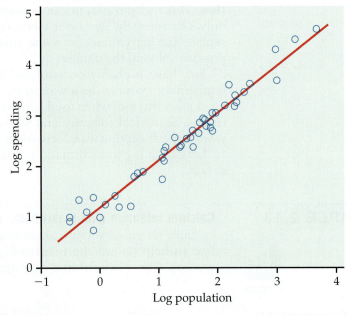

FIGURE 2.8 Scatterplot of log spending versus log population for the 50 U.S. states, with a fitted line, Example 2.14. The relationship is approximately linear.

USE YOUR KNOWLEDGE

EDSPEND

2.15 Compare the plots. Compare the plot in Figure 2.8 with the one in Figure 2.5 (page 90). Which one do you prefer? Give reasons for your answer.

Use of transformations and the interpretation of scatterplots are an art that requires judgment and knowledge about the variables that we are studying. *Always ask yourself if the relationship that you see makes sense.* If it does not, then additional analyses are needed to understand the data.

Adding categorical variables to scatterplots

In Figure 2.3 (page 88), we looked at the relationship between the rating and the price per load for 52 laundry detergents. A more detailed look at the data shows that there are two different types of laundry detergent included in this data set, liquid and powder. Let's examine where these two types of laundry detergents are in our plot.

CATEGORICAL VARIABLES IN SCATTERPLOTS

To add a categorical variable to a scatterplot, use a different plot color or symbol for each category.

EXAMPLE 2.15

LAUND

Rating versus price and type of laundry detergent. In our scatterplot, we use the color blue for liquids and the color red for powders. The scatterplot is given in Figure 2.9. Separate lines are given for each type of laundry detergent. Most of the laundry detergents are liquids. There are three powders with somewhat low prices and four powders with relatively high prices. The prices of the powders are similar to the prices of the liquids.

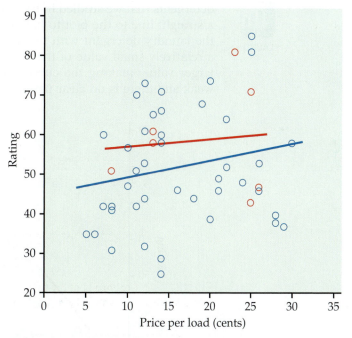

FIGURE 2.9 Scatterplot of rating versus price per load (in cents), with fitted straight lines, for 52 laundry detergents, Example 2.15. The type of detergent is indicated by the color: blue for liquid and red for powder.

In this example, we used a categorical variable, type, to distinguish the two types of laundry detergents in our plot. Suppose that the additional variable that we want to investigate is quantitative. In this situation, we sometimes can combine the values into ranges of the quantitative variable—such as high, medium, and low—to create a categorical variable.

Careful judgment is needed in using this graphical method. Don't be discouraged if your first attempt is not very successful. In performing a good data analysis, you will often produce several plots before you find the one that you believe to be the most effective in describing the data.[8]

Scatterplot smoothers

In Figure 2.6 (page 90), we added a curve to our scatterplot to better understand the relationship between calcium retention and calcium intake. This curve helped us to see that the amount of calcium retained tends to level off as the intake increases. The method that we used to construct the curve is called **smoothing**.

smoothing

Today, most statistical software includes options to perform the calculations needed for smoothing. The technical details vary, but the basic idea is that there is a smoothing parameter that controls the degree to which the relationship is smoothed. Here is another example.

EXAMPLE 2.16

LAUND

Laundry rating versus price with a smooth fit. Figure 2.2 (page 87) gives the scatterplot for rating versus price for the remaining 52 laundry detergents that we studied in Example 2.9. In Figure 2.3 (page 88), we added a straight line to the plot to help us see the relationship. Figure 2.10 shows the laundry detergent with two different smooth curves. The first (a) used a relatively small value of the smoothing parameter. The second (b) used a larger value, making the curve smoother. Overall, the relationship is very weak and there is no clear pattern in the plot.

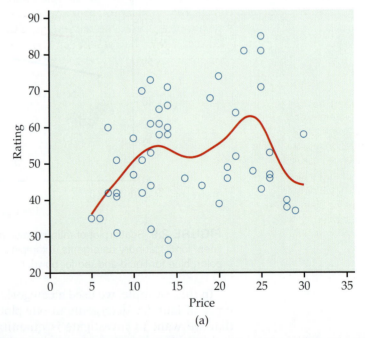

(a)

FIGURE 2.10 Scatterplot of rating versus price per load (in cents), with smooth curves, Example 2.16: (a) with a small value of the smoothing parameter; (b) with a higher value of the smoothing parameter.

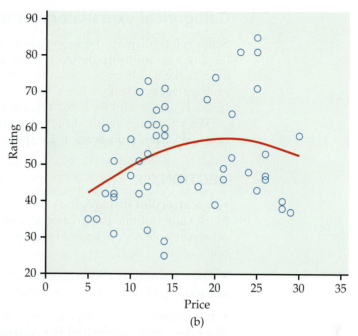

(b)

FIGURE 2.10 *Continued*

Scatterplot smoothers can help you to learn about relationships between two quantitative variables. They can confirm that there is a linear relationship, or they can suggest other features that are not evident in a casual look at the scatterplot. Here is an example of the latter scenario.

EXAMPLE 2.17

EDSPEND

A smooth fit for education spending and population with logs. Figure 2.11 gives the scatterplot of log education spending versus log population with a smooth curve. The curve suggests that the relationship is approximately linear except for states with relatively small populations. For these, the spending is relatively flat.

FIGURE 2.11 Scatterplot of log spending versus log population, with a smooth curve fitted to the data, for 50 U.S. states, Example 2.17. This smooth curve fits the data very well and suggests that the relationship is generally linear except for states with small populations.

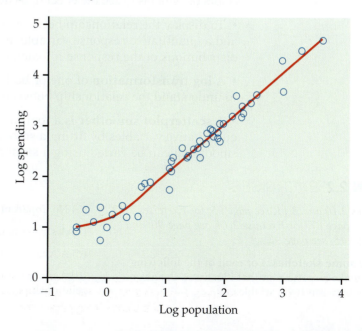

Categorical explanatory variables

Scatterplots display the association between two quantitative variables. To display a relationship between a categorical variable and a quantitative variable, make a side-by-side comparison of the distributions of the response for each category. Back-to-back stemplots (page 12) and side-by-side boxplots (page 37) are useful tools for this purpose.

We will study methods for describing the association between two categorical variables in Section 2.6 (page 136).

SECTION 2.2 SUMMARY

• A **scatterplot** displays the relationship between two quantitative variables. Mark values of one variable on the horizontal axis (x axis) and values of the other variable on the vertical axis (y axis). Plot each individual's data as a point on the graph.

• Always plot the explanatory variable, if there is one, on the x axis of a scatterplot. Plot the response variable on the y axis.

• In examining a scatterplot, look for an overall pattern showing the **form**, **direction**, and **strength** of the relationship, and then for **outliers** or other deviations from this pattern.

• **Form: Linear relationships**, where the points show a straight-line pattern, are an important form of relationship between two variables. Curved relationships are other forms to watch for.

• **Direction:** If the relationship has a clear direction, we speak of either **positive association** (high values of the two variables tend to occur together) or **negative association** (high values of one variable tend to occur with low values of the other variable).

• **Strength:** The **strength** of a relationship is determined by how close the points in the scatterplot lie to a simple form such as a line. Plot points with different colors or symbols to see the effect of a categorical variable in a scatterplot.

• To display the relationship between a categorical explanatory variable and a quantitative response variable, make a graph that compares the distributions of the response for each category of the explanatory variable.

• A **log transformation** of one or both variables in a scatterplot can help us to understand the relationship between two quantitative variables.

• A **scatterplot smoother** is a tool to examine the relationship between two quantitative variables by fitting a smooth curve to the data. The amount of smoothing can be varied using a **smoothing parameter**.

SECTION 2.2 EXERCISES

For Exercises 2.10 and 2.11, see page 85; for Exercises 2.12 and 2.13, see page 87; for Exercise 2.14, see page 90; and for Exercise 2.15, see page 92.

2.16 Make some sketches. For each of the following situations, make a scatterplot that illustrates the given relationship between two variables.

(a) No apparent relationship.

(b) A strong negative linear relationship.

(c) A weak positive relationship that is not linear.

(d) A more complicated relationship. Explain the relationship.

2.17 What's wrong? Explain what is wrong with each of the following:

(a) If two variables are negatively associated, then low values of one variable are associated with low values of the other variable.

(b) A stemplot can be used to examine the relationship between two variables.

(c) In a scatterplot, we put the response variable on the x axis and the explanatory variable on the y axis.

2.18 Blueberries and anthocyanins. Anthocyanins are compounds that have been associated with health benefits associated with the heart, bones, and the brain. Blueberries are a good source of many different anthocyanins. Researchers at the Piedmont Research Station of North Carolina State University have assembled a database giving the concentrations of 18 different anthocyanins for 267 varieties of blueberries.[9] Four of the anthocyanins measured are delphinidin-3-arabinoside, malvidin-3-arabinoside, cyanidin-3-galactoside, and delphinidin-3-glucoside, all measured in units of mg per 100g of berries (dry weight). In the data file, we have simplified the names of these anthocyanins to Antho1, Antho2, Antho3, and Antho4. In Exercises 1.167 and 1.168 (page 77), you examined the distributions of each Antho3 and Antho4. **BERRIES**

(a) Make a scatterplot of the data with Antho3 on the x axis and Antho4 on the y axis.

(b) Describe the form, direction, and strength of the relationship.

(c) Are there any outliers or unusual observations?

(d) Is it useful to add a straight line to your scatterplot? Explain your answer.

(e) If you have access to the appropriate software, explore the use of a scatterplot smoother to understand this relationship. Summarize what you have found using this method.

2.19 Blueberries and anthocyanins with logs. Refer to the previous exercise. Transform each of the variables with a log, make a scatterplot and answer the questions in the previous exercise for the transformed data. **BERRIES**

2.20 Blueberries and anthocyanins: Raw data or logs. Refer to Exercises 2.18 and 2.19.

(a) Compare your results from the two exercises.

(b) For exploring and explaining the relationship between Antho4 and Antho3, do you prefer the analysis you performed in Exercise 2.18 or the one you performed in Exercise 2.19? Give reasons for your answer. **BERRIES**

2.21 Fuel consumption. Natural Resources Canada tests new vehicles each year and reports several variables related to fuel consumption for vehicles in different classes.[10] For 2015 they provide data for 527 vehicles that use regular fuel. Two variables reported are carbon dioxide (CO_2) emissions and highway fuel consumption. CO_2 is measured in grams per kilometer (g/km) and highway fuel consumption measured in liters per 100 kilometers (L/100km). **CANFREG**

(a) Make a scatterplot of the data with highway fuel consumption on the x axis and CO_2 emissions on the y axis.

(b) Describe the form, direction, and strength of the relationship.

(c) Are there any outliers or unusual observations?

(d) Is it useful to add a straight line to your scatterplot? Explain your answer.

(e) If you have access to the appropriate software, explore the use of a scatterplot smoother to understand this relationship. Summarize what you find using this method.

2.22 Fuel consumption with a line. Refer to the previous exercise. **CANFREG**

(a) Add a line to the plot. To what extent to you think that the line does a good job of summarizing the relationship?

(b) If your have the appropriate software, use smooth curves to examine the relationship. Does your analysis support the idea of using a straight line to summarize the relationship? Explain your answer.

2.23 Fuel consumption for different types of vehicles. Refer to the previous two exercises. Those exercises examined data for vehicles that used regular fuel. Data are also available for vehicles that use several other types of fuel. There are 1067 vehicles in total. The variable Fuel has four different possible values: X, for regular fuel; Z, for premium fuel; D, for diesel; and E, for ethanol. **CANFUEL**

(a) Make a scatterplot of all of the data using different symbols or colors for the different fuel types.

(b) Does the relationship between CO_2 and highway fuel consumption depend upon the type of fuel that the vehicle uses? Explain your answer.

2.24 Bone strength. Osteoporosis is a condition where bones become weak. It affects more than 200 million people worldwide. Exercise is one way to produce strong bones and to prevent osteoporosis. Because we use our dominant arm (the right arm for most people) more than our nondominant arm, we expect the bone in our dominant arm to be stronger than the bone in our nondominant arm. By comparing the strengths, we can get an idea of the effect that exercise can have on bone strength. Here are some data on the strength of bones, measured in Newton meters divided by 1000 (Nm/1000), for the arms of 15 young men:[11] 📊 ARMSTR

ID	Nondominant	Dominant	ID	Nondominant	Dominant
1	15.7	16.3	9	15.9	20.1
2	25.2	26.9	10	13.7	18.7
3	17.9	18.7	11	17.7	18.7
4	19.1	22.0	12	15.5	15.2
5	12.0	14.8	13	14.4	16.2
6	20.0	19.8	14	14.1	15.0
7	12.3	13.1	15	12.3	12.9
8	14.4	17.5			

Before attempting to compare the arm strengths of the nondominant and dominant arms, let's take a careful look at the data for these two variables.

(a) Make a scatterplot of the data with the nondominant arm strength on the *x* axis and the dominant arm strength on the *y* axis.

(b) Describe the overall pattern in the scatterplot and any striking deviations from the pattern.

(c) Describe the form, direction, and strength of the relationship.

(d) Identify any outliers.

(e) Is the relationship approximately linear?

2.25 Bone strength for baseball players. Refer to the previous exercise. The study collected arm bone strength information for two groups of young men. The data in the previous exercise were for a control group. The second group in the study comprised men who played baseball. We know that these baseball players use their dominant arm in throwing (those who throw with their nondominant arm were excluded), so they get more arm exercise than the controls. Here are the data for the baseball players:

ID	Nondominant	Dominant	ID	Nondominant	Dominant
16	17.0	19.3	24	15.1	19.4
17	16.9	19.0	25	13.5	20.4
18	17.7	25.2	26	13.6	17.1
19	21.2	37.7	27	20.3	26.5
20	21.0	40.3	28	17.3	30.3
21	14.6	20.8	29	14.6	17.4
22	31.5	36.9	30	22.6	35.0
23	14.9	21.2			

Answer the questions in the previous exercise for the baseball players. 📊 ARMSTR

2.26 Compare the baseball players with the controls. Refer to the previous two exercises. 📊 ARMSTR

(a) Plot the data for the two groups on the same graph using different symbols for the baseball players and the controls.

(b) Use your plot to describe and compare the relationships for the two variables. Write a short paragraph summarizing what you have found.

2.27 Parents' income and student loans. How well does the income of a college student's parents predict how much the student will borrow to pay for college? We have data on parents' income and college debt for a sample of 1200 recent college graduates. What are the explanatory and response variables? Are these variables categorical or quantitative? Do you expect a positive or negative association between these variables? Why?

2.28 What's in the beer? The website **beer100.com** advertises itself as "Your Place for All Things Beer." One of their "things" is a list of 159 domestic beer brands with the percent alcohol, calories per 12 ounces, and carbohydrates per 12 ounces (in grams).[12] 📊 BEERD

(a) Figure 2.12 gives a scatterplot of calories versus percent alcohol. Give a short summary of what can be learned from the plot.

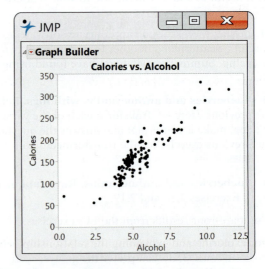

FIGURE 2.12 Scatterplot of calories versus percent alcohol for 159 domestic brands of beer, Exercise 2.28.

(b) One of the points is an outlier. Find the brand of the outlier. How is this brand of beer different from the other brands?

(c) Remove the outlier from the data set and generate a scatterplot of the remaining data.

(d) Describe the relationship between calories and percent alcohol based on what you see in your scatterplot.

2.29 More beer. Refer to the previous exercise.
BEERD

(a) Make a scatterplot of calories versus percent alcohol using the data set without the outlier.

(b) Describe the relationship between these two variables. If your software is capable, use a line and smoothers to explore the relationship.

2.30 Imported beer. The beer100 website also gives data for imported beers. Describe the relationship between calories and percent alcohol for these imported beers.
BEERI

2.31 Compare domestic with imported. Plot calories versus percent alcohol for domestic and imported beers on the same scatterplot. Use different colors or symbols for the two types of beers. Summarize what this plot tells you about the relationship and the difference between the two types of beer. In particular, note any characteristics that are better shown in this plot relative to what was learned in Exercises 2.28, 2.29, and 2.30. **BEER**

2.32 Decay of a radioactive element. Barium-137m is a radioactive form of the element barium that decays very rapidly. It is easy and safe to use for lab experiments in schools and colleges.[13] In a typical experiment, the radioactivity of a sample of barium-137m is measured for one minute. It is then measured for three additional one-minute periods, separated by two minutes. So data are recorded at one, three, five, and seven minutes after the start of the first counting period. The measurement units are counts. Here are the data for one of these experiments:[14] **DECAY**

Time	1	3	5	7
Count	578	317	203	118

(a) Make a scatterplot of the data. Give reasons for the choice of which variables to use on the x and y axes.

(b) Describe the overall pattern in the scatterplot and any striking deviations from the pattern.

(c) Describe the form, direction, and strength of the relationship.

(d) Identify any outliers.

(e) Is the relationship approximately linear?

2.33 Use a log for the radioactive decay. Refer to the previous exercise. Transform the counts using a log transformation. Then repeat parts (a) through (e) for the transformed data and compare your results with those from the previous exercise. **DECAY**

2.34 Internet use and babies. The World Bank collects data on many variables related to world development for countries throughout the world. Two of these are Internet use, in number of users per 100 people, and birthrate, in births per 1000 people.[15] Figure 2.13 is a scatterplot of birthrate versus Internet use for the 106 countries that have data available for both variables. **INBIRTH**

(a) Describe the relationship between these two variables.

(b) A friend looks at this plot and concludes that using the Internet will decrease the number of babies born. Write a short paragraph explaining why the association seen in the scatterplot does not provide a reason to draw this conclusion.

2.35 Try a log. Refer to the previous exercise. **INBIRTH**

(a) Make a scatterplot of the log of births per 1000 people versus Internet users per 100 people.

(b) Describe the relationship that you see in this plot and compare it with Figure 2.13.

(c) Which plot do you prefer? Give reasons for your answer.

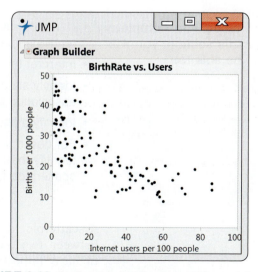

FIGURE 2.13 Scatterplot of births (per 1000 people) versus Internet users (per 100 people) for 106 countries, Exercise 2.34.

2.36 Make another plot. Refer to Exercise 2.34.

In INBIRTH

(a) Make a new data set that has Internet users expressed as users per 10,000 people and births as births per 10,000 people.

(b) Explain why these transformations to give new variables are linear transformations. (*Hint:* See linear transformations on page 44.)

(c) Make a scatterplot using the transformed variables.

(d) Compare your new plot with the one in Figure 2.13.

(e) Why do you think that the analysts at the World Bank chose to express births as births per 1000 people and Internet users as users per 100 people?

2.37 Body mass and metabolic rate. Metabolic rate, the rate at which the body consumes energy, is important in studies of weight gain, dieting, and exercise. The following table gives data on the lean body mass and resting metabolic rate for 12 women and 7 men who are subjects in a study of dieting. Lean body mass, given in kilograms, is a person's weight leaving out all fat. Metabolic rate is measured in calories burned per 24 hours, the same calories used to describe the energy content of foods. The researchers believe that lean body mass is an important influence on metabolic rate. **In** BMASS

Subject	Sex	Mass	Rate	Subject	Sex	Mass	Rate
1	M	62.0	1792	11	F	40.3	1189
2	M	62.9	1666	12	F	33.1	913
3	F	36.1	995	13	M	51.9	1460
4	F	54.6	1425	14	F	42.4	1124
5	F	48.5	1396	15	F	34.5	1052
6	F	42.0	1418	16	F	51.1	1347
7	M	47.4	1362	17	F	41.2	1204
8	F	50.6	1502	18	M	51.9	1867
9	F	42.0	1256	19	M	46.9	1439
10	M	48.7	1614				

(a) Make a scatterplot of the data, using different symbols or colors for men and women.

(b) Is the association between these variables positive or negative? What is the form of the relationship? How strong is the relationship?

(c) Does the pattern of the relationship differ for women and men? How do the male subjects as a group differ from the female subjects as a group?

2.3 Correlation

When you complete this section, you will be able to:	• Use a correlation to describe the direction and strength of a linear relationship between two quantitative variables.
	• Interpret the sign of a correlation.
	• Identify situations in which the correlation is not a good measure of association between two quantitative variables.
	• Identify a linear pattern in a scatterplot.
	• For describing the relationship between two quantitative variables, identify the roles of the correlation, a numerical summary, and the scatterplot, a graphical summary.

A scatterplot displays the form, direction, and strength of the relationship between two quantitative variables. Linear (straight-line) relations are particularly important because a straight line is a simple pattern that is quite common. We say a linear relationship is strong if the points lie close to a straight line and weak if they are widely scattered about a line. Our eyes are not good judges of how strong a relationship is. The two scatterplots in Figure 2.14 depict exactly the same data, but the plot on the right is drawn smaller in a large field. The plot on the right seems to show a stronger relationship.

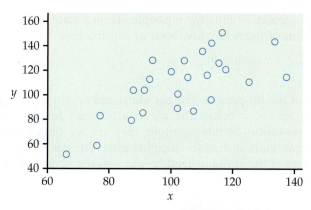 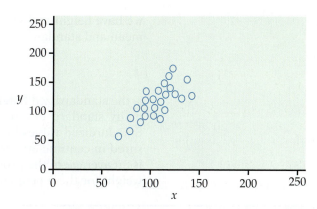

FIGURE 2.14 Two scatterplots of the same data. The linear pattern in the plot on the right appears stronger because of the surrounding space.

Our eyes can be fooled by changing the plotting scales or the amount of white space around the cloud of points in a scatterplot.[16] We need to follow our strategy for data analysis by using a numerical measure to supplement the graph. *Correlation* is the measure we use.

The correlation *r*

We have data on variables x and y for n individuals. Think, for example, of measuring height and weight for n people. Then x_1 and y_1 are your height and your weight, x_2 and y_2 are my height and my weight, and so on. For the ith individual, height x_i goes with weight y_i. Here is the definition of correlation.

CORRELATION

The **correlation** measures the direction and strength of the linear relationship between two quantitative variables. Correlation is usually written as r.

Suppose that we have data on variables x and y for n individuals. The means and standard deviations of the two variables are $\bar{x}$ and s_x for the x-values, and $\bar{y}$ and s_y for the y-values. The correlation r between x and y is

$$r = \frac{1}{n-1} \sum \left(\frac{x_i - \bar{x}}{s_x} \right) \left(\frac{y_i - \bar{y}}{s_y} \right)$$

As always, the summation sign Σ means "add these terms for all the individuals." The formula for the correlation r is a bit complex. It helps us see what correlation is but is not convenient for actually calculating r. In practice, you should use software or a calculator that computes r from the values of x and y pairs.

LOOK BACK

standardize, p. 59

The formula for r begins by standardizing the observations. Suppose, for example, that x is height in centimeters and y is weight in kilograms and that we have height and weight measurements for n people. Then $\bar{x}$ and s_x are the mean and standard deviation of the n heights, both in centimeters. The value

$$\frac{x_i - \bar{x}}{s_x}$$

is the standardized height of the ith person. The standardized height says how many standard deviations above or below the mean a person's height lies. Standardized values have no units—in this example, they are no longer measured in centimeters. You can standardize the weights also. The correlation r is an average of the products of the standardized height and the standardized weight for the n people.

USE YOUR KNOWLEDGE

LAUNDRY

LAUNDRY

2.38 Laundry detergents. Example 2.8 (page 85) describes data on the rating and price per load for 53 laundry detergents. Use these data to compute the correlation between rating and the price per load.

2.39 Change the units. Refer to the previous exercise. Express the price per load in dollars.

(a) Is the transformation from cents to dollars a linear transformation? Explain your answer.

(b) Compute the correlation between rating and price per load expressed in dollars.

(c) How does the correlation that you computed in part (b) compare with the one you computed in the previous exercise?

(d) What can you say in general about the effect of changing units using linear transformations on the size of the correlation?

Properties of correlation

The formula for correlation helps us see that r is positive when there is a positive association between the variables. Height and weight, for example, have a positive association. People who are above average in height tend to also be above average in weight. Both the standardized height and the standardized weight for such a person are positive. People who are below average in height tend also to have below-average weight. Then both standardized height and standardized weight are negative. In both cases, the products in the formula for r are mostly positive, so r is positive. In the same way, we can see that r is negative when the association between x and y is negative. More detailed study of the formula gives more detailed properties of r.

Here is what you need to know to interpret correlation:

• Correlation makes no use of the distinction between explanatory and response variables. It makes no difference which variable you call x and which you call y in calculating the correlation.

• *Correlation requires that both variables be quantitative.* For example, we cannot calculate a correlation between the incomes of a group of people and what city they live in because city is a categorical variable.

• Because r uses the standardized values of the observations, r does not change when we change the units of measurement (a linear transformation) of x, y, or both. Measuring height in inches rather than centimeters and weight in pounds rather than kilograms does not change the correlation between height and weight. The correlation r itself has no unit of measurement; it is just a number.

- Positive r indicates positive association between the variables, and negative r indicates negative association.

- The correlation r is always a number between -1 and 1. Values of r near 0 indicate a very weak linear relationship. The strength of the relationship increases as r moves away from 0 toward either -1 or 1. Values of r close to -1 or 1 indicate that the points lie close to a straight line. The extreme values $r = -1$ and $r = 1$ occur only when the points in a scatterplot lie exactly along a straight line.

- Correlation measures the strength of only the linear relationship between two variables. *Correlation does not describe curved relationships between variables, no matter how strong they are.*

- *Like the mean and standard deviation, the correlation is not resistant: r is strongly affected by a few outlying observations.* Use r with caution when outliers appear in the scatterplot.

LOOK BACK

resistant,
p. 30

The scatterplots in Figure 2.15 illustrate how values of r closer to 1 or -1 correspond to stronger linear relationships. To make the essential meaning of

FIGURE 2.15 How the correlation r measures the direction and strength of a linear association.

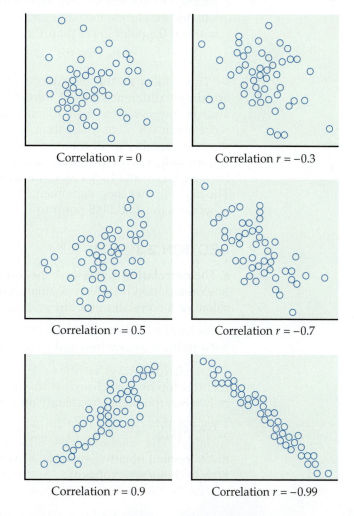

Correlation $r = 0$

Correlation $r = -0.3$

Correlation $r = 0.5$

Correlation $r = -0.7$

Correlation $r = 0.9$

Correlation $r = -0.99$

r clear, the standard deviations of both variables in these plots are equal, and the horizontal and vertical scales are the same. In general, it is not so easy to guess the value of *r* from the appearance of a scatterplot. Remember that changing the plotting scales in a scatterplot may mislead our eyes, but it does not change the standardized values of the variables and, therefore, cannot change the correlation. To explore how extreme observations can influence *r*, use the *Correlation and Regression* applet available on the text website. Also, see Exercises 2.56 and 2.57 (page 106).

Finally, remember that **correlation is not a complete description of two-variable data,** even when the relationship between the variables is linear. You should give the means and standard deviations of both *x* and *y* along with the correlation. (Because the formula for correlation uses the means and standard deviations, these measures are the proper choices to accompany a correlation.) Conclusions based on correlations alone may require rethinking in the light of a more complete description of the data.

EXAMPLE 2.18

Scoring of figure skating in the Olympics. Until a scandal at the 2002 Olympics brought change, figure skating was scored by judges on a scale from 0.0 to 6.0. The scores were often controversial. We have the scores awarded by two judges, Pierre and Elena, to many skaters. How well do they agree? We calculate that the correlation between their scores is *r* = 0.9. But the mean of Pierre's scores is 0.8 point lower than Elena's mean.

These facts in the example above do not contradict each other. They are simply different kinds of information. The mean scores show that Pierre awards lower scores than Elena. But because Pierre gives *every* skater a score about 0.8 point lower than Elena, the correlation remains high. Adding the same number to all values of either *x* or *y* does not change the correlation. If both judges score the same skaters, the competition is scored consistently because Pierre and Elena agree on which performances are better than others. The high *r* shows their agreement. But if Pierre scores some skaters and Elena others, we must add 0.8 point to Pierre's scores to arrive at a fair comparison.

SECTION 2.3 SUMMARY

- The **correlation *r*** measures the direction and strength of the linear (straight line) association between two quantitative variables *x* and *y*. Although you can calculate a correlation for any scatterplot, *r* measures only linear relationships.

- Correlation indicates the direction of a linear relationship by its sign: *r* > 0 for a positive association and *r* < 0 for a negative association.

- Correlation always satisfies −1 ≤ *r* ≤ 1 and indicates the strength of a relationship by how close it is to −1 or 1. Perfect correlation, *r* = ±1, occurs only when the points lie exactly on a straight line.

- Correlation ignores the distinction between explanatory and response variables. The value of *r* is not affected by changes in the unit of measurement of either variable. Correlation is not resistant, so outliers can greatly change the value of *r*.

SECTION 2.3 EXERCISES

For Exercises 2.38 and 2.39, see page 102.

2.40 Correlations and scatterplots. Explain why you should always look at a scatterplot when you want to use a correlation to describe the relationship between two quantitative variables.

2.41 Interpret some correlations. For each of the following correlations, describe the relationship between the two quantitative variables in terms of the direction and the strength of the linear relationship.

(a) $r = 0.9$.

(b) $r = -0.9$.

(c) $r = -0.3$.

(d) $r = 0.0$.

2.42 Blueberries and anthocyanins. In Exercise 2.18 (page 97), you examined the relationship between Antho4 and Antho3, two anthocyanins found in blueberries. 📊 BERRIES

(a) Find the correlation between these two anthocyanins.

(b) Look at the scatterplot for these data that you made in part (a) of Exercise 2.18 (or make one if you did not do that exercise). Is the correlation a good numerical summary of the graphical display in the scatterplot? Explain your answer.

(c) Does the size of the correlation suggest that the amounts of these two anthocyanins is approximately equal in these blueberries? Explain why or why not.

2.43 Blueberries and anthocyanins with logs. In Exercise 2.19 (page 97), you examined the relationship between Antho4 and Antho3, two anthocyanins found in blueberries, using logs for both variables. Answer the questions in the previous exercise for the variables transformed in this way. 📊 BERRIES

2.44 Fuel consumption. In Exercise 2.21 (page 97), you examined the relationship between CO_2 emissions and highway fuel consumption for 527 vehicles that use regular fuel. Find the correlation between these two variables. Write a short paragraph describing the relationship using the scatterplot and the correlation. 📊 CANFREG

2.45 Fuel consumption for different types of vehicles. In Exercise 2.23 (page 97), you examined the relationship between CO_2 emissions and highway fuel consumption for 1067 vehicles that use four different types of fuel. Find the correlations between CO_2 and highway fuel consumption for each of these four categories of vehicle. Summarize your results explaining similarities and differences in the relationships among the four types of fuel. 📊 CANFUEL

2.46 Strong association but no correlation. Here is a data set that illustrates an important point about correlation: 📊 CORR

X	25	35	45	55	65
Y	10	30	50	30	10

(a) Make a scatterplot of Y versus X.

(b) Describe the relationship between Y and X. Is it weak or strong? Is it linear?

(c) Find the correlation between Y and X.

(d) What important point about correlation does this exercise illustrate?

2.47 Bone strength. Exercise 2.24 (page 97) gives the bone strengths of the dominant and the nondominant arms for 15 men who were controls in a study. 📊 ARMSTR

(a) Find the correlation between the bone strength of the dominant arm and the bone strength of the nondominant arm.

(b) Look at the scatterplot for these data that you made in part (a) of Exercise 2.24 (or make one if you did not do that exercise). Is the correlation a good numerical summary of the graphical display in the scatterplot? Explain your answer.

2.48 Bone strength for baseball players. Refer to the previous exercise. Similar data for baseball players are given in Exercise 2.25 (page 98). Answer parts (a) and (b) of the previous exercise for these data. 📊 ARMSTR

2.49 Student ratings of teachers. A college newspaper interviews a psychologist about student ratings of the teaching of faculty members. The psychologist says, "The evidence indicates that the correlation between the research productivity and teaching rating of faculty members is close to zero." The paper reports this as "Professor McDaniel said that good researchers tend to be poor teachers, and vice versa." Explain why the paper's report is wrong. Write a statement in plain language (don't use the word "correlation") to explain the psychologist's meaning.

2.50 Decay of a radioactive element. Data for an experiment on the decay of barium-137m is given in Exercise 2.32 (page 99). 📊 DECAY

(a) Find the correlation between the radioactive counts and the time after the start of the first counting period.

(b) Does the correlation give a good numerical summary of the relationship between these two variables? Explain your answer.

2.51 Decay in the log scale. Refer to the previous exercise and to Exercise 2.33 (page 99), where the counts were transformed by a log. ▥ DECAY

(a) Find the correlation between the log counts and the time after the start of the first counting period.

(b) Does the correlation give a good numerical summary of the relationship between these two variables? Explain your answer.

(c) Compare your results for this exercise with those from the previous exercise.

2.52 Brand names and generic products.

(a) If a store always prices its generic "store brand" products at 80% of the brand name products' prices, what would be the correlation between the prices of the brand name products and the store brand products? (*Hint:* Draw a scatterplot for several prices.)

(b) If the store always prices its generic products $2 less than the corresponding brand name products, then what would be the correlation between the prices of the brand name products and the store brand products?

2.53 Alcohol and calories in beer. Figure 2.12 (page 98) gives a scatterplot of the calories versus percent alcohol for 159 brands of domestic beer. ▥ BEERD

(a) Compute the correlation for these data.

(b) Does the correlation do a good job of describing the direction and strength of this relationship? Explain your answer.

2.54 Alcohol and calories in beer revisited. Refer to the previous exercise. The data that you used to compute the correlation includes an outlier. ▥ BEERD

(a) Remove the outlier and recompute the correlation.

(b) Write a short paragraph about the possible effects of outliers on a correlation using this example to illustrate your ideas.

2.55 Compare domestic with imported. In Exercise 2.31 (page 99), you compared domestic beers with imported beers with respect to the relationship between calories and percent alcohol. In that exercise, you used scatterplots to make the comparison. Compute the correlations for these two categories of beer and write a new summary of the comparison using correlations in addition to the scatterplots. ▥ BEERD, BEERI

2.56 Use the applet. Go to the *Correlation and Regression* applet. Click on the scatterplot to create a group of 12 points in the lower-right corner of the scatterplot with a strong straight-line negative pattern (correlation about −0.9).

(a) Add one point at the upper left that is in line with the first 12. How does the correlation change?

(b) Drag this last point down until it is opposite the group of 12 points. How small can you make the correlation? Can you make the correlation positive? *A single outlier can greatly strengthen or weaken a correlation. Always plot your data to check for outlying points.*

2.57 Use the applet. You are going to use the *Correlation and Regression* applet to make different scatterplots with 12 points that have correlation close to 0.8. *Many patterns can have the same correlation. Always plot your data before you trust a correlation.*

(a) Stop after adding the first two points. What is the value of the correlation? Why does it have this value no matter where the two points are located?

(b) Make a lower-left to upper-right pattern of 12 points with correlation about $r = 0.8$. (You can drag points up or down to adjust r after you have 12 points.) Make a rough sketch of your scatterplot.

(c) Make another scatterplot, this time with 11 points in a vertical stack at the left of the plot. Add one point far to the right and move it until the correlation is close to 0.8. Make a rough sketch of your scatterplot.

(d) Make yet another scatterplot, this time with 12 points in a curved pattern that starts at the lower left, rises to the right, then falls again at the far right. Adjust the points up or down until you have a quite smooth curve with correlation close to 0.8. Make a rough sketch of this scatterplot also.

2.58 An interesting set of data. Make a scatterplot of the following data: ▥ INTER

X	1	2	3	4	10	10
Y	1	3	3	5	1	10

Verify that the correlation is about 0.5. What feature of the data is responsible for reducing the correlation to this value despite a strong straight-line association between x and y in most of the observations?

2.59 Internet use and babies. Figure 2.13 (page 99) is a scatterplot of the number of births per 1000 people rate versus Internet users per 100 people for 106 countries. In Exercise 2.34 (page 99), you described this relationship. ▥ INBIRTH

(a) Make a plot of the data similar to Figure 2.13 and report the correlation.

(b) Is the correlation a good numerical summary for this relationship? Explain your answer.

2.60 What's wrong? Each of the following statements contains a blunder. Explain in each case what is wrong.

(a) There is a high correlation between the age of American workers and their occupation.

(b) We found a high correlation ($r = 1.19$) between students' ratings of faculty teaching and ratings made by other faculty members.

(c) The correlation between the sex of a group of students and the color of their cell phone was $r = 0.23$.

2.4 Least-Squares Regression

When you complete this section, you will be able to:

- Draw a straight line on a scatterplot of a set of data, given the equation of the line.
- Predict a value of the response variable y for a given value of the explanatory variable x using a regression equation.
- Explain the meaning of the term *least squares*.
- Calculate the equation of a least-squares regression line from the means and standard deviations of the explanatory and response variables and their correlation.
- Read the output of statistical software to find the equation of the least-squares regression line and the value of r^2.
- Explain the meaning of r^2 in the regression setting.

Correlation measures the direction and strength of the linear (straight-line) relationship between two quantitative variables. If a scatterplot shows a linear relationship, we would like to summarize this overall pattern by drawing a line on the scatterplot. A *regression line* summarizes the relationship between two variables, but only in a specific setting: when one of the variables helps explain or predict the other. That is, regression describes a relationship between an explanatory variable and a response variable.

REGRESSION LINE

A **regression line** is a straight line that describes how a response variable y changes as an explanatory variable x changes. We often use a regression line to **predict** the value of y for a given value of x. Regression, unlike correlation, requires that we have an explanatory variable and a response variable.

EXAMPLE 2.19

FIDGET

Fidgeting and fat gain. Does fidgeting keep you slim? Some people don't gain weight even when they overeat. Perhaps fidgeting and other "nonexercise activity" (NEA) explains why—the body might spontaneously increase nonexercise activity when fed more. Researchers deliberately overfed 16 healthy young adults for eight weeks. They measured fat gain (in kilograms) and, as an explanatory variable, increase in energy use (in calories) from activity other than deliberate exercise—fidgeting, daily living, and the like. Here are the data:[17]

NEA increase (cal)	−94	−57	−29	135	143	151	245	355
Fat gain (kg)	4.2	3.0	3.7	2.7	3.2	3.6	2.4	1.3
NEA increase (cal)	392	473	486	535	571	580	620	690
Fat gain (kg)	3.8	1.7	1.6	2.2	1.0	0.4	2.3	1.1

Figure 2.16 is a scatterplot of these data. The plot shows a moderately strong negative linear association with no outliers. The correlation is $r = -0.7786$. People with larger increases in nonexercise activity do indeed gain less fat. A line drawn through the points will describe the overall pattern well.

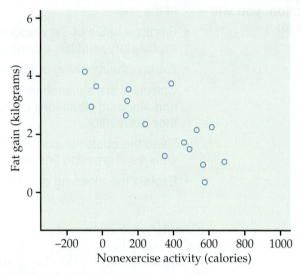

FIGURE 2.16 Fat gain after eight weeks of overeating plotted against the increase in nonexercise activity over the same period, Example 2.19.

Fitting a line to data

fitting a line

When a scatterplot displays a linear pattern, we can describe the overall pattern by drawing a straight line through the points. Of course, no straight line passes exactly through all the points. **Fitting a line** to data means drawing a line that comes as close as possible to the points. The equation of a line fitted to the data gives a concise description of the relationship between the response variable y and the explanatory variable x. It is the numerical summary that supports the scatterplot, our graphical summary.

STRAIGHT LINES

Suppose that y is a response variable (plotted on the vertical axis) and x is an explanatory variable (plotted on the horizontal axis). A straight line relating y to x has an equation of the form

$$y = b_0 + b_1 x$$

In this equation, b_1 is the **slope**, the amount by which y changes when x increases by one unit. The number b_0 is the **intercept**, the value of y when $x = 0$.

In practice, we will use software to obtain values of b_0 and b_1 for a given set of data.

EXAMPLE 2.20

FIDGET

Regression line for fat gain. Any straight line describing the nonexercise activity data has the form

$$\text{fat gain} = b_0 + (b_1 \times \text{NEA increase})$$

In Figure 2.17, we have drawn the regression line with the equation

$$\text{fat gain} = 3.505 - (0.00344 \times \text{NEA increase})$$

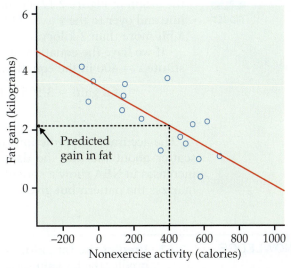

FIGURE 2.17 A regression line fitted to the nonexercise activity data and used to predict fat gain for an NEA increase of 400 calories, Examples 2.20 and 2.21.

The figure shows that this line fits the data well. The slope $b_1 = -0.00344$ tells us that fat gained goes down by 0.00344 kilogram for each added calorie of NEA increase.

The slope b_1 of a line $y = b_0 + b_1 x$ is the *average change* in the response y as the explanatory variable x changes. The slope of a regression line is an important numerical description of the relationship between the two variables. For Example 2.20, the intercept, $b_0 = 3.505$ kilograms. This value is the estimated fat gain if NEA does not change. When we substitute the value zero for the NEA increase, the regression equation gives 3.505 (the intercept) as the predicted value of the fat gain.

USE YOUR KNOWLEDGE

FIDGET

2.61 Plot the line. Make a plot of the data in Example 2.19 and plot the line

$$\text{fat gain} = 4.505 - (0.00344 \times \text{NEA increase})$$

on your sketch. Explain why this line does not give a good fit to the data.

Prediction

prediction   We can use a regression line to **predict** the response y for a specific value of the explanatory variable x. We can interpret the prediction as the *average* value of y corresponding to a collection of cases at the particular value of x or as our best guess at the value of y for an *individual* with the particular value of x.

EXAMPLE 2.21

FIDGET

Prediction for fat gain. Based on the linear pattern, we want to predict the fat gain for an individual whose NEA increases by 400 calories when she overeats. To use the fitted line to predict fat gain, go "up and over" on the graph in Figure 2.17. From 400 calories on the x axis, go up to the fitted line and over to the y axis. The graph shows that the predicted gain in fat is a bit more than 2 kilograms.

If we have the equation of the line, it is faster and more accurate to substitute $x = 400$ in the equation. The predicted fat gain is

$$\text{fat gain} = 3.505 - (0.00344 \times 400) = 2.13 \text{ kilograms}$$

The accuracy of predictions from a regression line depends on how much scatter about the line the data show. In Figure 2.17, fat gains for similar increases in NEA show a spread of 1 or 2 kilograms. The regression line summarizes the pattern but gives only roughly accurate predictions.

USE YOUR KNOWLEDGE

2.62 Predict the fat gain. Use the regression equation in Example 2.20 to predict the fat gain for a person whose NEA increases by 250 calories.

EXAMPLE 2.22

Is this prediction reasonable? Can we predict the fat gain for someone whose nonexercise activity increases by 1500 calories when she overeats? We can certainly substitute 1500 calories into the equation of the line. The prediction is

$$\text{fat gain} = 3.505 - (0.00344 \times 1500) = -1.66 \text{ kilograms}$$

That is, we predict that this individual loses fat when she overeats. This prediction is not trustworthy. Look again at Figure 2.17. An NEA increase of 1500 calories is far outside the range of our data. We can't say whether increases this large ever occur, or whether the relationship remains linear at such extreme values. Predicting fat gain when NEA increases by 1500 calories *extrapolates* the relationship beyond what the data show.

EXTRAPOLATION

Extrapolation is the use of a regression line for prediction far outside the range of values of the explanatory variable x used to obtain the line. Such predictions are often not accurate and should be avoided.

USE YOUR KNOWLEDGE

2.63 Would you use the regression equation to predict? Consider the following values for NEA increase: −300, 300, 600, 800. For each, decide whether you would use the regression equation in Example 2.20 to predict fat gain or whether you would be concerned that the prediction would not be trustworthy because of extrapolation. Give reasons for your answers.

Least-squares regression

Different people might draw different lines by eye on a scatterplot. This is especially true when the points are widely scattered. We need a way to draw a regression line that doesn't depend on our guess as to where the line should go. No line will pass exactly through all the points, but we want one that is as close as possible. We will use the line to predict y from x, so we want a line that is as close as possible to the points in the *vertical* direction. That's because the prediction errors we make are errors in y, which is the vertical direction in the scatterplot.

The line in Figure 2.17 predicts 2.13 kilograms of fat gain for an increase in nonexercise activity of 400 calories. If the actual fat gain turns out to be 2.3 kilograms, the error is

$$\text{error} = \text{observed gain} - \text{predicted gain}$$
$$= 2.3 - 2.13 = 0.17 \text{ kilogram}$$

Errors are positive if the observed response lies above the line and negative if the response lies below the line. We want a regression line that makes these prediction errors as small as possible. Figure 2.18 illustrates the idea. For clarity, the plot shows only three of the points from Figure 2.17, along with the line, on an expanded scale. The line passes below two of the points and above one of them. The vertical distances of the data points from the line appear as vertical line segments. A "good" regression line makes these distances as small

FIGURE 2.18 The least-squares idea: make the errors in predicting y as small as possible by minimizing the sum of their squares.

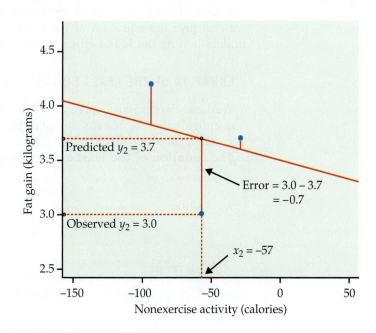

as possible. There are many ways to make "as small as possible" precise. The most common is the *least-squares* idea. The line in Figures 2.17 and 2.18 is, in fact, the least-squares regression line.

LEAST-SQUARES REGRESSION LINE

The **least-squares regression line of y on x** is the line that makes the sum of the squares of the vertical distances of the data points from the line as small as possible.

Here is the least-squares idea expressed as a mathematical problem. We represent n observations on two variables x and y as

$$(x_1, y_1), (x_2, y_2), \ldots, (x_n, y_n)$$

If we draw a line $y = b_0 + b_1 x$ through the scatterplot of these observations, the line predicts the value of y corresponding to x_i as $\hat{y}_i = b_0 + b_1 x_i$. We write $\hat{y}$ (read "y-hat") in the equation of a regression line to emphasize that the line gives a *predicted* response $\hat{y}$ for any x. The predicted response will usually not be exactly the same as the actually *observed* response y. The method of least squares chooses the line that makes the sum of the squares of these errors as small as possible. To find this line, we must find the values of the intercept b_0 and the slope b_1 that minimize

$$\sum (\text{error})^2 = \sum (y_i - b_0 - b_1 x_i)^2$$

for the given observations x_1 and y_1. For the NEA data, for example, we must find the b_0 and b_1 that minimize

$$(4.2 - b_0 + 94 b_1)^2 + (3.0 - b_0 + 57 b_1)^2 + \cdots + (1.1 - b_0 - 690 b_1)^2$$

These values are the intercept and slope of the least-squares line.

You will use software or a calculator with a regression function to find the equation of the least-squares regression line from data on x and y. Therefore, we will give the equation of the least-squares line in a form that helps our understanding but is not efficient for calculation.

EQUATION OF THE LEAST-SQUARES REGRESSION LINE

We have data on an explanatory variable x and a response variable y for n individuals. The means and standard deviations of the sample data are $\bar{x}$ and s_x for x and $\bar{y}$ and s_y for y, and the correlation between x and y is r. The **equation of the least-squares regression line** of y on x is

$$\hat{y} = b_0 + b_1 x$$

with **slope**

$$b_1 = r \frac{s_y}{s_x}$$

and **intercept**

$$b_0 = \bar{y} - b_1 \bar{x}$$

EXAMPLE 2.23

Check the calculations. Verify from the data in Example 2.19 that the mean and standard deviation of the 16 increases in NEA are

$$\bar{x} = 324.8 \text{ calories} \quad \text{and} \quad s_x = 257.66 \text{ calories}$$

The mean and standard deviation of the 16 fat gains are

$$\bar{y} = 2.388 \text{ kg} \quad \text{and} \quad s_y = 1.1389 \text{ kg}$$

The correlation between fat gain and NEA increase is $r = -0.7786$. Therefore, the least-squares regression line of fat gain y on NEA increase x has slope

$$b_1 = r\frac{s_y}{s_x} = -0.7786\frac{1.1389}{257.66}$$
$$= -0.00344 \text{ kg per calorie}$$

and intercept

$$b_0 = \bar{y} - b_1\bar{x}$$
$$= 2.388 - (-0.00344)(324.8) = 3.505 \text{ kg}$$

The equation of the least-squares line is

$$\hat{y} = 3.505 - 0.00344x$$

When doing calculations like this by hand, you may need to carry extra decimal places in the preliminary calculations to get accurate values of the slope and intercept. Using software or a calculator with a regression function eliminates this worry.

Interpreting the regression line

The slope $b_1 = -0.00344$ kilograms per calorie in Example 2.23 is the change in fat gain as NEA increases. The units "kilograms of fat gained per calorie of NEA" come from the units of y (kilograms) and x (calories). Although the correlation does not change when we change the units of measurement, the equation of the least-squares line does change. The slope in grams per calorie would be 1000 times as large as the slope in kilograms per calorie because there are 1000 grams in a kilogram. The small value of the slope, $b_1 = -0.00344$, does not mean that the effect of increased NEA on fat gain is small—it just reflects the choice of kilograms as the unit for fat gain. *The slope and intercept of the least-squares line depend on the units of measurement—you can't conclude anything from their size.*

EXAMPLE 2.24

FIDGET

Regression using software. Figure 2.19 displays the basic regression output for the nonexercise activity data from three statistical software packages. Other software produces very similar output. You can find the slope and intercept of the least-squares line, calculated to more decimal places than we need, in each output. The software also provides information that we do not yet need, including some that we trimmed from Figure 2.19.

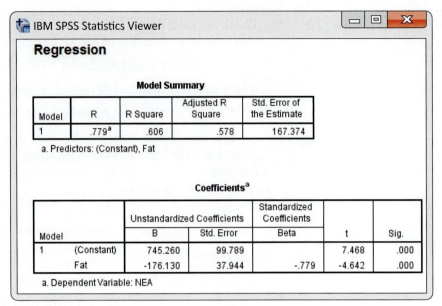

(a) SPSS

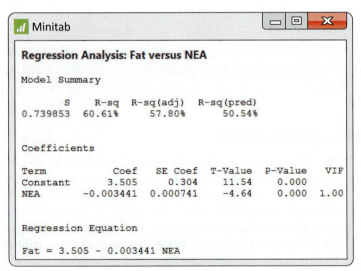

(b) Minitab

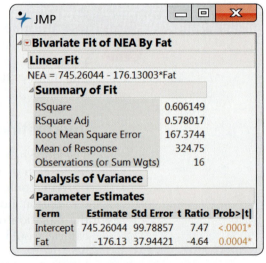

(c) JMP

FIGURE 2.19 Regression results for the nonexercise activity data from three statistical software packages: (a) SPSS; (b) Minitab; (c) JMP. Other software produces similar output.

Part of the art of using software is to ignore the extra information that is almost always present. Look for the results that you need. Once you understand a statistical method, you can read output from almost any software.

Facts about least-squares regression

Regression is one of the most common statistical settings, and least squares is the most common method for fitting a regression line to data. Here are some facts about least-squares regression lines.

Fact 1. There is a close connection between correlation and the slope of the least-squares line. The slope is

$$b_1 = r\frac{s_y}{s_x}$$

This equation says that along the regression line, **a change of one standard deviation in x corresponds to a change of r standard deviations in y.** When the variables are perfectly correlated ($r = 1$ or $r = -1$), the change in the predicted response $\hat{y}$ is the same (in standard deviation units) as the change in x. Otherwise, because $-1 \leq r \leq 1$, the change in $\hat{y}$ is less than the change in x. As the correlation grows less strong, the prediction y moves less in response to changes in x. Note that if the correlation is zero, then the slope of the least-squares regression line will be zero.

Fact 2. The least-squares regression line always passes through the point $(\overline{x}, \overline{y})$ on the graph of y against x. So, the least-squares regression line of y on x is the line with slope rs_y/s_x that passes through the point $(\overline{x}, \overline{y})$. We can describe regression entirely in terms of the basic descriptive measures $\overline{x}$, s_x, $\overline{y}$, s_y, and r.

Fact 3. The distinction between explanatory and response variables is essential in regression. Least-squares regression looks at the distances of the data points from the line only in the y direction. If we reverse the roles of the two variables, we get a different least-squares regression line.

Correlation and regression

Least-squares regression looks at the distances of the data points from the line only in the y direction. So the two variables x and y play different roles in regression.

EXAMPLE 2.25

FIDGET

Fidgeting and fat gain. Figure 2.20 is a scatterplot of the fidgeting and fat gain data described in Example 2.19 (page 107). There is a negative linear relationship. The two lines on the plot are the two least-squares regression

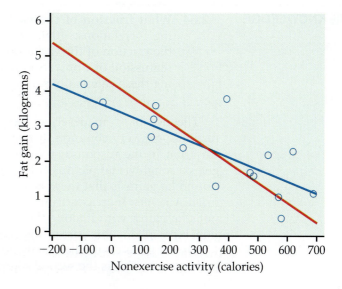

FIGURE 2.20 Scatterplot of fat gain versus nonexercise activity for 16 subjects from Example 2.19. The two lines are the two least-squares regression lines, using nonexercise activity to predict fat gain (blue) and using fat gain to predict nonexercise activity (red), Example 2.25.

lines. The regression line using nonexercise activity to predict fat gain is blue. The regression line using fat gain to predict nonexercise activity is red. *Regression of fat gain on nonexercise activity and regression of nonexercise activity on fat gain give different lines*. In the regression setting, you must decide which variable is explanatory.

Even though the correlation r ignores the distinction between explanatory and response variables, there is a close connection between correlation and regression. We saw that the slope of the least-squares line involves r. Another connection between correlation and regression is even more important. In fact, the numerical value of r as a measure of the strength of a linear relationship is best interpreted by thinking about regression. Here is the fact we need.

r^2 IN REGRESSION

The **square of the correlation, r^2**, is the fraction of the variation in the values of y that is explained by the least-squares regression of y on x.

The correlation between NEA increase and fat gain for the 16 subjects in Example 2.19 (page 107) is $r = -0.7786$. Because $r^2 = 0.6062$, the straight-line relationship between NEA and fat gain explains about 61% of the vertical scatter in fat gains in Figure 2.17 (page 109).

When you report a regression, give r^2 as a measure of how successfully the regression explains the response. All three software outputs in Figure 2.19 include r^2, either in decimal form or as a percent.

When you see a correlation, square it to get a better feel for the strength of the association. Perfect correlation ($r = -1$ or $r = 1$) means that the points lie exactly on a line. Then $r^2 = 1$ and all the variation in one variable is accounted for by the linear relationship with the other variable. If $r = -0.7$ or $r = 0.7$, $r^2 = 0.49$ and about half the variation is accounted for by the linear relationship. In the r^2 scale, correlation ±0.7 is about halfway between 0 and ±1.

USE YOUR KNOWLEDGE

2.64 What fraction of the variation is explained? Consider the following correlations: -0.9, -0.5, -0.2, 0, 0.2, 0.5, and 0.9. For each, give the fraction of the variation in y that is explained by the least-squares regression of y on x. Summarize what you have found from performing these calculations.

The use of r^2 to describe the success of regression in explaining the response y is very common. It rests on the fact that there are two sources of variation in the responses y in a regression setting. Figure 2.17 (page 109) gives a rough visual picture of the two sources. The first reason for the variation in fat gains is that there is a relationship between fat gain y and increase in NEA x. As x increases from -94 to 690 calories among the 16 subjects, it pulls fat gain y with it along the regression line in the figure. The linear relationship explains this part of the variation in fat gains.

The fat gains do not lie exactly on the line, however, but are scattered above and below it. This is the second source of variation in y, and the regression line tells us nothing about how large it is. The dashed lines in Figure 2.17 show a rough average for y when we fix a value of x. We use r^2 to measure variation along the line as a fraction of the total variation in the fat gains. In Figure 2.17, about 61% of the variation in fat gains among the 16 subjects is due to the straight-line relationship between y and x. The remaining 39% is vertical scatter in the observed responses remaining after the line has fixed the predicted responses.

Another view of r^2

Here is a more specific interpretation of r^2. The fat gains y in Figure 2.17 range from 0.4 to 4.2 kilograms. The variance of these responses, a measure of how variable they are, is

$$\text{variance of observed values } y = 1.297$$

Much of this variability is due to the fact that as x increases from -94 to 690 calories, it pulls y along with it. If the only variability in the observed responses were due to the straight-line dependence of fat gain on NEA, the observed gains would lie exactly on the regression line. That is, they would be the same as the predicted gains $\hat{y}$. We can compute the predicted gains by substituting the NEA values for each subject into the equation of the least-squares line. Their variance describes the variability in the predicted responses. The result is

$$\text{variance of predicted values } \hat{y} = 0.786$$

This is what the variance would be if the responses fell exactly on the line; that is, if the linear relationship explained 100% of the observed variation in y. Because the responses don't fall exactly on the line, the variance of the predicted values is smaller than the variance of the observed values. Here is the fact we need:

$$r^2 = \frac{\text{variance of predicted values } \hat{y}}{\text{variance of observed values } y}$$

$$= \frac{0.786}{1.297} = 0.606$$

This fact is always true. The squared correlation gives the variance that the responses would have if there were no scatter about the least-squares line as a fraction of the variance of the actual responses. This is the exact meaning of "fraction of variation explained" as an interpretation of r^2.

These connections with correlation are special properties of least-squares regression. They are not true for other methods of fitting a line to data. One reason that least squares is the most common method for fitting a regression line to data is that it has many convenient special properties.

SECTION 2.4 SUMMARY

• A **regression line** is a straight line that describes how a response variable y changes as an explanatory variable x changes.

• The most common method of fitting a line to a scatterplot is least squares. The **least-squares regression line** is the straight line $\hat{y} = b_0 + b_1 x$ that minimizes the sum of the squares of the vertical distances of the observed y-values from the line.

• You can use a regression line to **predict** the value of y for any value of x by substituting this x into the equation of the line. **Extrapolation** beyond the range of x-values spanned by the data is risky.

• The **slope** b_1 of a regression line $\hat{y} = b_0 + b_1 x$ is the rate at which the predicted response $\hat{y}$ changes along the line as the explanatory variable x changes. Specifically, b_1 is the change in $\hat{y}$ when x increases by 1. The numerical value of the slope depends on the units used to measure x and y.

• The **intercept** b_0 of a regression line $\hat{y} = b_0 + b_1 x$ is the predicted response $\hat{y}$ when the explanatory variable $x = 0$. This prediction is not particularly useful unless x can actually take values near 0.

• The least-squares regression line of y on x is the line with slope $b_1 = r s_y / s_x$ and intercept $b_0 = \bar{y} - b_1 \bar{x}$. This line always passes through the point $(\bar{x}, \bar{y})$.

• **Correlation and regression** are closely connected. The correlation r is the slope of the least-squares regression line when we measure both x and y in standardized units. The square of the correlation r^2 is the fraction of the variance of one variable that is explained by least-squares regression on the other variable.

SECTION 2.4 EXERCISES

For Exercise 2.61, see page 109; for Exercise 2.62, see page 110; for Exercise 2.63, see page 111; and for Exercise 2.64, see page 116.

2.65 Blueberries and anthocyanins. In Exercise 2.18 (page 97), you examined the relationship between Antho4 and Antho3, two anthocyanins found in blueberries. In Exercise 2.42 (page 105), you found the correlation between these two variables. [📊] BERRIES

(a) Find the equation of the least-squares regression line for predicting Antho4 from Antho3.

(b) Make a scatterplot of the data with the fitted line.

(c) How well does the line fit the data? Explain your answer.

(d) Use the line to predict the value of Antho4 when Antho3 is equal to 1.5.

2.66 Fuel consumption. In Exercise 2.21 (page 97), you examined the relationship between CO_2 emissions and highway fuel consumption for 527 vehicles that use regular fuel. In Exercise 2.44 (page 105), you found the correlation between these two variables. [📊] CANFREG

(a) Find the equation of the least-squares regression line for predicting CO_2 emissions from highway fuel consumption.

(b) Make a scatterplot of the data with the fitted line.

(c) How well does the line fit the data? Explain your answer.

(d) Use the line to predict the value of CO_2 for vehicles that consume 8.0 liters per kilometer (L/km).

2.67 Fuel consumption for different types of vehicles. In Exercise 2.23 (page 97), you examined the relationship between CO_2 emissions and highway fuel consumption for 1067 vehicles. You used different plotting symbols for the four different types of fuel used by these vehicles: regular, premium, diesel, and ethanol. [📊] CANFUEL

(a) Find the least-squares equation for predicting CO_2 emissions from highway fuel consumption for all 1067 vehicles.

(b) Make a scatterplot of the data with the fitted line.

(c) Based on what you learned from Example 2.23, do you think that a single least-squares regression line provides a good fit for all four types of vehicles? Explain your answer.

2.68 Bone strength. Exercise 2.24 (page 97), gives the bone strengths of the dominant and the nondominant arms for 15 men who were controls in a study. [📊] ARMSTR

(a) Plot the data. Use the bone strength in the nondominant arm as the explanatory variable and bone strength in the dominant arm as the response variable.

(b) The least-squares regression line for these data is

dominant = 2.74 + (0.936 × nondominant)

Add this line to your plot.

(c) Use the scatterplot (a graphical summary), with the least-squares line (a graphical display of a numerical summary) to write a short paragraph describing this relationship.

2.69 Bone strength for baseball players. Refer to the previous exercise. Similar data for baseball players are given in Exercise 2.25 (page 98). Here is the equation of the least-squares line for the baseball players:

dominant = 0.886 + (1.373 × nondominant)

Answer parts (a) and (c) of the previous exercise for these data. [Im] ARMSTR

2.70 Predict the bone strength. Refer to Exercise 2.68. A young male who is not a baseball player has a bone strength of 16.0 $cm^4/1000$ in his nondominant arm. Predict the bone strength in the dominant arm for this person. [Im] ARMSTR

2.71 Predict the bone strength for a baseball player. Refer to Exercise 2.69. A young male who is a baseball player has a bone strength of 16.0 $cm^4/1000$ in his nondominant arm. Predict the bone strength in the dominant arm for this person. [Im] ARMSTR

2.72 Compare the predictions. Refer to the two previous exercises. You have predicted two dominant-arm bone strengths, one for a baseball player and one for a person who is not a baseball player. The nondominant bone strengths are both 16.0 $cm^4/1000$. [Im] ARMSTR

(a) Compare the two predictions by computing the difference in means, baseball player minus control.

(b) Explain how the difference in the two predictions is an estimate of the effect of baseball throwing exercise on the strength of arm bones.

(c) For nondominant arm strengths of 12 $cm^4/1000$ and 20 $cm^4/1000$, repeat your predictions and take the differences. Make a table of the results of all three calculations (for 12, 16, and 20 $cm^4/1000$).

(d) Write a short summary of the results of your calculations for the three different nondominant-arm strengths. Be sure to include an explanation of why the differences are not the same for the three nondominant-arm strengths.

2.73 Least-squares regression for radioactive decay. Refer to Exercise 2.32 (page 99) for the data on radioactive decay of barium-137m. Here are the data: [Im] DECAY

Time	1	3	5	7
Count	578	317	203	118

(a) Using the least-squares regression equation

count = 602.8 − (74.7 × time)

find the predicted values for the counts.

(b) Compute the differences, observed count minus predicted count. How many of these are positive; how many are negative?

(c) Square and sum the differences that you found in part (b).

(d) Repeat the calculations that you performed in parts (a), (b), and (c) using the equation

count = 500 − (100 × time)

(e) In a short paragraph, explain the least-squares idea using the calculations that you performed in this exercise.

2.74 Least-squares regression for the log counts. Refer to Exercise 2.33 (page 99), where you analyzed the radioactive decay of barium-137m data using log counts. Here are the data: [Im] DECAY

Time	1	3	5	7
Log count	6.35957	5.75890	5.31321	4.77068

(a) Using the least-squares regression equation

log count = 6.593 − (0.2606 × time)

find the predicted values for the log counts.

(b) Compute the differences, observed count minus predicted count. How many of these are positive; how many are negative?

(c) Square and sum the differences that you found in part (b).

(d) Repeat the calculations that you performed in parts (a), (b), and (c) using the equation

log count = 7 − (0.2 × time)

(e) In a short paragraph, explain the least-squares idea using the calculations that you performed in this exercise.

2.75 College students by state. How well does the population of a state predict the number of undergraduates? The National Center for Education

Statistics collects data for each of the 50 U.S. states that we can use to address this question.[18] 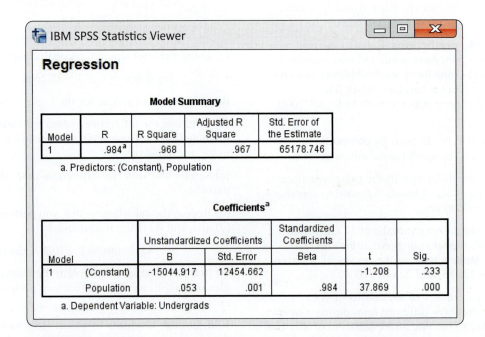 COLLEGE

(a) Make a scatterplot with population on the x axis and number of undergraduates on the y axis.

(b) Describe the form, direction, and strength of the relationship. Are there any outliers?

(c) For the number of undergraduates, the mean is 302,136 and the standard deviation is 358,460, and for population, the mean is 5,955,551 and the standard deviation is 6,620,733. The correlation between the number of undergraduates and the population is 0.98367. Use this information to find the least-squares regression line. Show your work.

(d) Add the least-squares line to your scatterplot.

2.76 College students by state without the four largest states. Refer to the previous exercise. Let's eliminate the four largest states, which have populations greater than 15 million. Here are the numerical summaries: for number of undergraduate college students, the mean is 220,134 and the standard deviation is 165,270; for population, the mean is 4,367,448 and the standard deviation is 3,310,957. The correlation between the number of undergraduate college students and the population is 0.97081. Use this information to find the least-squares regression line. Show your work. COL46

2.77 Make predictions and compare. Refer to the two previous exercises. Consider a state with a population of 4 million (this value is approximately the median population for the 50 states). COLLEGE

(a) Using the least-squares regression equation for all 50 states, find the predicted number of undergraduate college students.

(b) Do the same using the least-squares regression equation for the 46 states with populations less than 15 million.

(c) Compare the predictions that you made in parts (a) and (b). Write a short summary of your results and conclusions. Pay particular attention to the effect of including the four states with the largest populations in the prediction equation for a median-sized state.

2.78 College students by state. Refer to Exercise 2.75, where you examined the relationship between the number of undergraduate college students and the populations for the 50 states. Figure 2.21 gives the output from a software package for the regression. Use this output to answer the following questions: COLLEGE

(a) What is the equation of the least-squares regression line?

(b) What is the value of r^2?

(c) Interpret the value of r^2.

(d) Does the software output tell you that the relationship is linear and not, for example, curved? Explain your answer.

IBM SPSS Statistics Viewer

Regression

Model Summary

Model	R	R Square	Adjusted R Square	Std. Error of the Estimate
1	.984[a]	.968	.967	65178.746

a. Predictors: (Constant), Population

Coefficients[a]

Model		Unstandardized Coefficients B	Std. Error	Standardized Coefficients Beta	t	Sig.
1	(Constant)	-15044.917	12454.662		-1.208	.233
	Population	.053	.001	.984	37.869	.000

a. Dependent Variable: Undergrads

FIGURE 2.21 SPSS output for predicting number of undergraduate college students using population for the 50 states, Exercise 2.78.

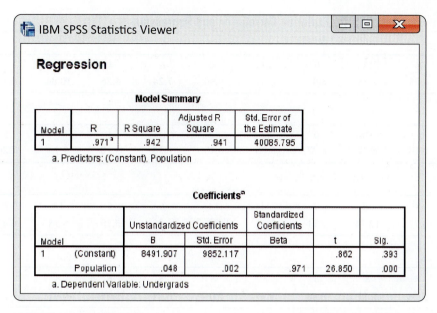

FIGURE 2.22 SPSS output for predicting number of undergraduate college students using population, with the four largest states deleted, Exercise 2.79.

2.79 College students by state without the four largest states. Refer to Exercise 2.76, where you eliminated the four largest states that have populations greater than 15 million. Figure 2.22 gives software output for these data. Answer the questions in the previous exercise for the data set with the 46 states. COL46

2.80 Data generated by software. The following 20 observations on Y and X were generated by a computer program. GENDATA

X	Y	X	Y
22.06	34.38	17.75	27.07
19.88	30.38	19.96	31.17
18.83	26.13	17.87	27.74
22.09	31.85	20.20	30.01
17.19	26.77	20.65	29.61
20.72	29.00	20.32	31.78
18.10	28.92	21.37	32.93
18.01	26.30	17.31	30.29
18.69	29.49	23.50	28.57
18.05	31.36	22.02	29.80

(a) Make a scatterplot and describe the relationship between Y and X.

(b) Find the equation of the least-squares regression line and add the line to your plot.

(c) What percent of the variability in Y is explained by X?

(d) Summarize your analysis of these data in a short paragraph.

2.81 Add an outlier. Refer to Exercise 2.80. Add an additional observation with $y = 25$ and $x = 35$ to the data set. Repeat the analysis that you performed in Exercise 2.80 and summarize your results, paying particular attention to the effect of this outlier. GEN21A

2.82 Add a different outlier. Refer to Exercise 2.80 and the previous exercise. Add an additional observation with $y = 36$ and $x = 30$ to the original data set. GEN21B

(a) Repeat the analysis that you performed in Exercise 2.80 and summarize your results, paying particular attention to the effect of this outlier.

(b) In this exercise and in the previous one, you added an outlier to the original data set and reanalyzed the data. Write a short summary of the changes in correlations that can result from different kinds of outliers.

2.83 Alcohol and calories in beer. Figure 2.12 (page 98) gives a scatterplot of calories versus percent alcohol in 159 brands of domestic beer. BEERD

(a) Find the equation of the least-squares regression line for these data.

(b) Find the value of r^2 and interpret it in the regression context.

TABLE 2.1	Four Data Sets for Exploring Correlation and Regression										
Data Set A											
x	10	8	13	9	11	14	6	4	12	7	5
y	8.04	6.95	7.58	8.81	8.33	9.96	7.24	4.26	10.84	4.82	5.68
Data Set B											
x	10	8	13	9	11	14	6	4	12	7	5
y	9.14	8.14	8.74	8.77	9.26	8.10	6.13	3.10	9.13	7.26	4.74
Data Set C											
x	10	8	13	9	11	14	6	4	12	7	5
y	7.46	6.77	12.74	7.11	7.81	8.84	6.08	5.39	8.15	6.42	5.73
Data Set D											
x	8	8	8	8	8	8	8	8	8	8	19
y	6.58	5.76	7.71	8.84	8.47	7.04	5.25	5.56	7.91	6.89	12.50

(c) Write a short report on the relationship between calories and percent alcohol in beer. Include graphical and numerical summaries for each variable separately as well as graphical and numerical summaries for the relationship in your report.

2.84 Alcohol and calories in beer revisited. Refer to the previous exercise. The data that you used includes an outlier. [IIi] BEERD

(a) Remove the outlier and answer parts (a), (b), and (c) for the new set of data.

(d) Write a short paragraph about the possible effects of outliers on a least-squares regression line and the value of r^2, using this example to illustrate your ideas.

2.85 Always plot your data! Table 2.1 presents four sets of data prepared by the statistician Frank Anscombe to illustrate the dangers of calculating without first plotting the data.[19] [IIi] ANSC

(a) Without making scatterplots, find the correlation and the least-squares regression line for all four data sets. What do you notice? Use the regression line to predict y for $x = 10$.

(b) Make a scatterplot for each of the data sets and add the regression line to each plot.

(c) In which of the four cases would you be willing to use the regression line to describe the dependence of y on x? Explain your answer in each case.

2.86 Progress in math scores. Every few years, the National Assessment of Educational Progress asks a

national sample of eighth-graders to perform the same math tasks. The goal is to get an honest picture of progress in math. Here are the last few national mean scores, on a scale of 0 to 500:[20] [IIi] NAEP

Year	1990	1992	1996	2000	2003	2005	2008	2011	2013
Score	263	268	272	273	278	279	281	283	285

(a) Make a time plot of the mean scores, by hand. This is just a scatterplot of score against year. There is a slow linear increasing trend.

(b) Find the regression line of mean score on time step-by-step. First calculate the mean and standard deviation of each variable and their correlation (use a calculator with these functions). Then find the equation of the least-squares line from these. Draw the line on your scatterplot. What percent of the year-to-year variation in scores is explained by the linear trend?

(c) Now use software or the regression function on your calculator to verify your regression line.

2.87 The regression equation. The equation of a least-squares regression line is $y = 15 - 2x$.

(a) What is the value of y for $x = 4$?

(b) If x increases by one unit, what is the corresponding change in y?

(c) What is the intercept for this equation?

2.88 Metabolic rate and lean body mass. Compute the mean and the standard deviation of the metabolic rates and lean body masses in Exercise 2.37 (page 100) and the correlation between these two variables. Use these values to find the slope of the regression line of metabolic rate on lean body mass. Also find the slope of the regression line of lean body mass on metabolic rate. What are the units for each of the two slopes? BMASS

2.89 Use an applet for progress in math scores.
Go to the *Two-Variable Statistical Calculator*. Enter the data for the progress in math scores from Exercise 2.86 using the "User-entered data" option in the "Data" tab. Explore the data by clicking the other tabs in the applet. Using only the only the results provided by the applet, write a short report summarizing the analysis of these data.

2.90 A property of the least-squares regression line. Use the equation for the least-squares regression line to show that this line always passes through the point $(\bar{x}, \bar{y})$.

2.91 Class attendance and grades. A study of class attendance and grades among first-year students at a state university showed that, in general, students who missed a higher percent of their classes earned lower grades. Class attendance explained 16% of the variation in grade index among the students. What is the numerical value of the correlation between percent of classes attended and grade index?

2.5 Cautions about Correlation and Regression

When you complete this section, you will be able to:	• Calculate the residuals for a set of data using the equation of the least-squares regression line and the observed values of the explanatory variable.
	• Use a plot of the residuals versus the explanatory variable to assess the fit of a regression line.
	• Identify outliers and influential observations by examining scatterplots and residual plots.
	• Identify lurking variables that can influence the interpretation of relationships between two variables.
	• Explain the difference between association and causality when interpreting the relationship between two variables.

Correlation and regression are among the most common statistical tools. They are used in more elaborate form to study relationships among many variables, a situation in which we cannot see the essentials by studying a single scatterplot. We need a firm grasp of the use and limitations of these tools, both now and as a foundation for more advanced statistics.

Residuals

A regression line describes the overall pattern of a linear relationship between an explanatory variable and a response variable. Deviations from the overall pattern are also important. In the regression setting, we see deviations by looking at the scatter of the data points about the regression line. The vertical distances from the points to the least-squares regression line are as small as possible in the sense that they have the smallest possible sum of squares. Because they represent "leftover" variation in the response after fitting the regression line, these distances are called *residuals*.

> **RESIDUALS**
>
> A **residual** is the difference between an observed value of the response variable and the value predicted by the regression line. That is,
>
> $$\text{residual} = \text{observed } y - \text{predicted } y$$
> $$= y - \hat{y}$$

EXAMPLE 2.26

FIDGET

Residuals for fat gain. Example 2.19 (page 107) describes measurements on 16 young people who volunteered to overeat for eight weeks. Those whose nonexercise activity (NEA) spontaneously rose substantially gained less fat than others. Figure 2.23(a) is a scatterplot of these data. The pattern is linear. The least-squares line is

$$\text{fat gain} = 3.505 - (0.00344 \times \text{NEA increase})$$

One subject's NEA rose by 135 calories. That subject gained 2.7 kilograms of fat. The predicted gain for 135 calories is

$$\hat{y} = 3.505 - (0.00344 \times 135) = 3.04 \text{ kg}$$

The residual for this subject is, therefore,

$$\text{residual} = \text{observed } y - \text{predicted } y$$
$$= y - \hat{y}$$
$$= 2.7 - 3.04 = -0.34 \text{ kg}$$

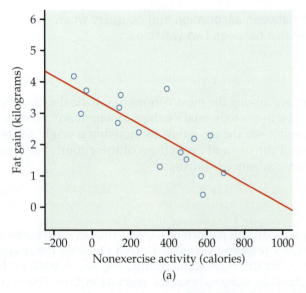

(a)

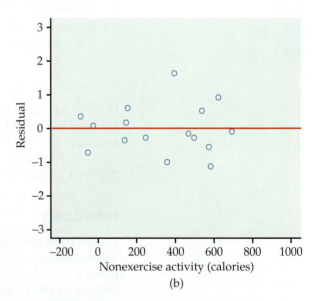

(b)

FIGURE 2.23 (a) Scatterplot of fat gain versus increase in nonexercise activity, with the least-squares regression line, Example 2.26. (b) Residual plot for the regression displayed in panel (a); the line at $y = 0$ marks the mean of the residuals.

Most regression software will calculate and store residuals for you.

USE YOUR KNOWLEDGE

2.92 Find the predicted value and the residual. Let's say that we have an individual in the NEA data set who has NEA increase equal to 250 calories and fat gain equal to 2.4 kg. Find the predicted value of fat gain for this individual and then calculate the residual. Explain why this residual is negative.

Because the residuals show how far the data fall from our regression line, examining the residuals helps us assess how well the line describes the data. Although residuals can be calculated from any model fit to the data, the residuals from the least-squares line have a special property: **the mean of the least-squares residuals is always zero**.

USE YOUR KNOWLEDGE

FIDGET

2.93 Find the sum of the residuals. Here are the 16 residuals for the NEA data rounded to two decimal places:

0.37	−0.70	0.10	−0.34	0.19	0.61	−0.26	−0.98
1.64	−0.18	−0.23	0.54	−0.54	−1.11	0.93	−0.03

Find the sum of these residuals. Note that the sum is not exactly zero because of roundoff error.

You can see the residuals in the scatterplot of Figure 2.23(a) by looking at the vertical deviations of the points from the line. The *residual plot* in Figure 2.23(b) makes it easier to study the residuals by plotting them against the explanatory variable, increase in NEA.

RESIDUAL PLOTS

A **residual plot** is a scatterplot of the regression residuals against the explanatory variable. Residual plots help us assess the fit of a regression line.

Because the mean of the residuals is always zero, the horizontal line at zero in Figure 2.23(b) helps orient us. This line (residual = 0) corresponds to the fitted line in Figure 2.23(a). The residual plot magnifies the deviations from the line to make patterns easier to see. If the regression line catches the overall pattern of the data, there should be *no pattern* in the residuals. That is, the residual plot should show an unstructured horizontal band centered at zero. The residuals in Figure 2.23(b) do have this irregular scatter.

You can see the same thing in the scatterplot of Figure 2.23(a) and the residual plot of Figure 2.23(b). It's just a bit easier in the residual plot. Deviations from an irregular horizontal pattern point out ways in which the regression line fails to catch the overall pattern. Here is an example.

EXAMPLE 2.27

INBIRTH

Patterns in birthrate and Internet user residuals. In Exercise 2.34 (page 99) we used a scatterplot to study the relationship between birthrate and Internet users for 106 countries. In this scatterplot, Figure 2.13, we see that there are many countries with low numbers of Internet users. In addition, the relationship between births and Internet users appears to be curved. For low values of Internet users, there is a clear relationship, while for higher values, the curve becomes relatively flat.

Figure 2.24(a) gives the data with the least-squares regression line, and Figure 2.24(b) plots the residuals. Look at the right part of Figure 2.24(b), where the values of Internet users are high. Here we see that the residuals tend to be positive.

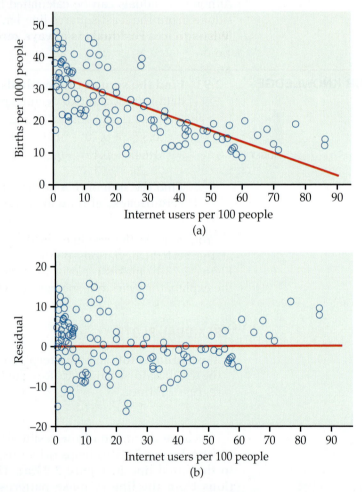

FIGURE 2.24 (a) Scatterplot of birthrate versus Internet users, with the least-squares regression line, Example 2.27. (b) Residual plot for the regression displayed in panel (a); the line at $y = 0$ marks the mean of the residuals.

The residual pattern in Figure 2.24(b) is characteristic of a simple curved relationship. *There are many ways in which a relationship can deviate from a linear pattern.* We now have an important tool for examining these deviations. Use it frequently and carefully when you study relationships.

TABLE 2.2		Two Measures of Glucose Level in Diabetics						
Subject	HbA1c (%)	FPG (mg/ml)	Subject	HbA1c (%)	FPG (mg/ml)	Subject	HbA1c (%)	FPG (mg/ml)
1	6.1	141	7	7.5	96	13	10.6	103
2	6.3	158	8	7.7	78	14	10.7	172
3	6.4	112	9	7.9	148	15	10.7	359
4	6.8	153	10	8.7	172	16	11.2	145
5	7.0	134	11	9.4	200	17	13.7	147
6	7.1	95	12	10.4	271	18	19.3	255

Outliers and influential observations

When you look at scatterplots and residual plots, look for striking individual points as well as for an overall pattern. Here is an example of data that contain some unusual cases.

EXAMPLE 2.28

HBA1C

Diabetes and blood sugar. People with diabetes must manage their blood sugar levels carefully. They measure their fasting plasma glucose (FPG) several times a day with a glucose meter. Another measurement, made at regular medical checkups, is called HbA1c. This is roughly the percent of red blood cells that have a glucose molecule attached. It measures average exposure to glucose over a period of several months.

This diagnostic test is becoming widely used and is sometimes called A1c by health care professionals. Table 2.2 gives data on both HbA1c and FPG for 18 diabetics five months after they completed a diabetes education class.[21]

Because both FPG and HbA1c measure blood glucose, we expect a positive association. The scatterplot in Figure 2.25(a) shows a

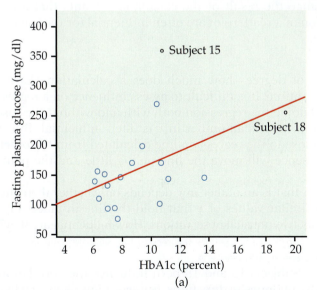

(a)

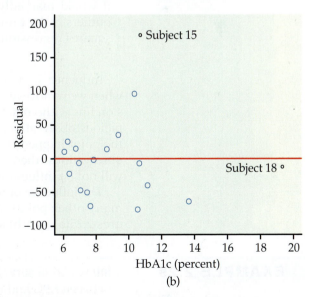

(b)

FIGURE 2.25 (a) Scatterplot of fasting plasma glucose against HbA1c (which measures long-term blood glucose), with the least-squares regression line, Example 2.28. (b) Residual plot for the regression of fasting plasma glucose on HbA1c. Subject 15 is an outlier in fasting plasma glucose. Subject 18 is an outlier in HbA1c that may be influential but does not have a large residual.

surprisingly weak relationship, with correlation $r = 0.4819$. The line on the plot is the least-squares regression line for predicting FPG from HbA1c. Its equation is

$$\hat{y} = 66.4 + 10.41x$$

It appears that one-time measurements of FPG can vary quite a bit among people with similar long-term levels, as measured by HbA1c. This is why A1c is an important diagnostic test.

Two unusual cases are marked in Figure 2.25(a). Subjects 15 and 18 are unusual in different ways. Subject 15 has dangerously high FPG and lies far from the regression line in the y direction. Subject 18 is close to the line but far out in the x direction. The residual plot in Figure 2.25(b) confirms that Subject 15 has a large residual and that Subject 18 does not.

Points that are outliers in the x direction, like Subject 18, can have a strong influence on the position of the regression line. Least-squares lines make the sum of squares of the vertical distances to the points as small as possible. A point that is extreme in the x direction with no other points near it pulls the line toward itself.

OUTLIERS AND INFLUENTIAL OBSERVATIONS IN REGRESSION

An **outlier** is an observation that lies outside the overall pattern of the other observations. Points that are outliers in the y direction of a scatterplot have large regression residuals, but other outliers need not have large residuals.

An observation is **influential** for a statistical calculation if removing it would markedly change the result of the calculation. Points that are outliers in the x direction of a scatterplot are often influential for the least-squares regression line.

Influence is a matter of degree—how much does a calculation change when we remove an observation? It is difficult to assess influence on a regression line without actually doing the regression both with and without the suspicious observation. A point that is an outlier in x is often influential. But if the point happens to lie close to the regression line calculated from the other observations, then its presence will move the line only a little and the point will not be influential.

The influence of a point that is an outlier in y depends on whether there are many other points with similar values of x that hold the line in place. Figures 2.25(a) and (b) identify two unusual observations. How influential are they?

EXAMPLE 2.29

HBA1C

Influential observations. Subjects 15 and 18 both influence the correlation between FPG and HbA1c, in opposite directions. Subject 15 weakens the linear pattern; if we drop this point, the correlation increases from $r = 0.4819$ to $r = 0.5684$. Subject 18 extends the linear pattern; if we omit this subject, the correlation drops from $r = 0.4819$ to $r = 0.3837$.

To assess influence on the least-squares line, we recalculate the line leaving out a suspicious point. Figure 2.26 shows three least-squares lines. The solid line is the regression line of FPG on HbA1c based on all 18 subjects. This is the same line that appears in Figure 2.25(a). The dotted line is calculated from all subjects except Subject 18. You see that point 18 does pull the line down toward itself. But the influence of Subject 18 is not very large—the dotted and solid lines are close together for HbA1c values between 6 and 14, the range of all except Subject 18.

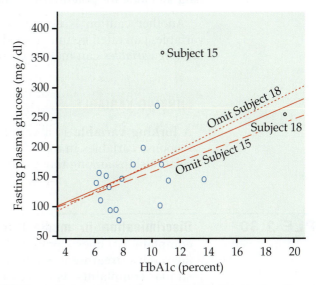

FIGURE 2.26 Three regression lines for predicting fasting plasma glucose from HbA1c, Example 2.29. The solid line uses all 18 subjects. The dotted line leaves out Subject 18. The dashed line leaves out Subject 15. "Leaving one out" calculations are the surest way to assess influence.

The dashed line omits Subject 15, the outlier in y. Comparing the solid and dashed lines, we see that Subject 15 pulls the regression line up. The influence is again not large, but it exceeds the influence of Subject 18.

We did not need the distinction between outliers and influential observations in Chapter 1. A single large salary that pulls up the mean salary $\bar{x}$ for a group of workers is an outlier because it lies far above the other salaries. It is also influential because the mean changes when it is removed. In the regression setting, however, not all outliers are influential. Because influential observations draw the regression line toward themselves, we may not be able to spot them by looking for large residuals.

Beware of the lurking variable

Correlation and regression are powerful tools for measuring the association between two variables and for expressing the dependence of one variable on the other. These tools must be used with an awareness of their limitations. We have seen that

• Correlation measures *only linear association*, and fitting a straight line makes sense only when the overall pattern of the relationship is linear. Always plot your data before calculating.

• *Extrapolation* (using a fitted model far outside the range of the data that we used to fit it) often produces unreliable predictions.

• Correlation and least-squares regression are *not resistant*. Always plot your data and look for potentially influential points.

Another caution is even more important: the relationship between two variables can often be understood only by taking other variables into account. *Lurking variables* can make a correlation or regression misleading.

LURKING VARIABLE

A **lurking variable** is a variable that is not among the explanatory or response variables in a study and yet may influence the interpretation of relationships among those variables.

EXAMPLE 2.30

Discrimination in medical treatment? Studies show that men who complain of chest pain are more likely to get detailed tests and aggressive treatment such as bypass surgery than are women with similar complaints. Is this association between sex and treatment due to discrimination?

Perhaps not. Men and women develop heart problems at different ages—women are, on the average, between 10 and 15 years older than men. Aggressive treatments are more risky for older patients, so doctors may hesitate to recommend them. Lurking variables—the patient's age and condition—may explain the relationship between sex and doctors' decisions.

Here is an example of a different type of lurking variable.

EXAMPLE 2.31

Gas and electricity bills. A single-family household receives bills for gas and electricity each month. The 12 observations for a recent year are plotted with the least-squares regression line in Figure 2.27. We have arbitrarily chosen to put the electricity bill on the *x* axis and the gas bill on the *y* axis. There is a clear negative association. Does this mean that a high electricity bill causes the gas bill to be low and vice versa?

To understand the association in this example, we need to know a little more about the two variables. In this household, heating is done by gas and cooling is done by electricity. Therefore, in the winter months the gas bill will be relatively high and the electricity bill will be relatively low. The pattern is reversed in the summer months. The association that we see in this example is due to a lurking variable: time of year.

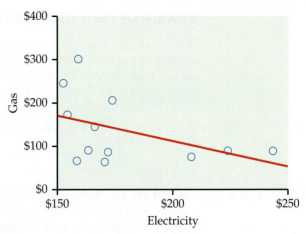

FIGURE 2.27 Scatterplot with least-squares regression line for predicting a household's monthly charges for gas using its monthly charges for electricity, Example 2.31.

Correlations that are due to lurking variables are sometimes called "nonsense correlations." The correlation is real. What is nonsense is the suggestion that the variables are directly related so that changing one of the variables *causes* changes in the other. The question of causation is important enough to merit separate treatment in Section 2.7. For now, just *remember that an association between two variables x and y can reflect many types of relationships among x, y, and one or more lurking variables*.

ASSOCIATION DOES NOT IMPLY CAUSATION

An association between an explanatory variable x and a response variable y, even if it is very strong, is not by itself good evidence that changes in x actually cause changes in y.

Lurking variables sometimes create a correlation between x and y, as in Examples 2.30 and 2.31. *When you observe an association between two variables, always ask yourself if the relationship that you see might be due to a lurking variable*. As in Example 2.31, time is often a likely candidate.

Beware of correlations based on averaged data

Regression or correlation studies sometimes work with averages or other measures that combine information from many individuals. For example, if we plot the average height of young children against their age in months, we will see a very strong positive association with correlation near 1. But individual children of the same age vary a great deal in height. A plot of height against age for individual children will show much more scatter and lower correlation than the plot of average height against age.

A correlation based on averages over many individuals is usually higher than the correlation between the same variables based on data for individuals. This fact reminds us again of the importance of noting exactly what variables a statistical study involves.

Beware of restricted ranges

The range of values for the explanatory variable in a regression can have a large impact on the strength of the relationship. For example, if we use age as a predictor of reading ability for a sample of students in the third grade, we will probably see little or no relationship. However, if our sample includes students from grades 1 through 8, we would expect to see a relatively strong relationship. We call this phenomenon the **restricted-range problem**.

restricted-range problem

EXAMPLE 2.32

A test for job applicants. Your company gives a test of cognitive ability to job applicants before deciding whom to hire. Your boss has asked you to use company records to see if this test really helps predict the performance ratings of employees. The restricted-range problem may make it difficult to see a strong relationship between test scores and performance ratings. The current employees were selected by a mechanism that is likely to result in scores that tend to be higher than those of the entire pool of applicants.

BEYOND THE BASICS

Data Mining

Chapters 1 and 2 of this text are devoted to the important aspect of statistics called *exploratory data analysis* (EDA). We use graphs and numerical summaries to examine data, searching for patterns and paying attention to striking deviations from the patterns we find. In discussing regression, we advanced to using the pattern we find (in this case, a linear pattern) for prediction.

Suppose now that we have a truly enormous database, such as all purchases recorded by the cash register scanners of a national retail chain during the past week. Surely this treasure chest of data contains patterns that might guide business decisions. If we could see clearly the types of activewear preferred in large California cities and compare the preferences of small Midwest cities—right now, not at the end of the season—we might improve profits in both parts of the country by matching stock with demand. This sounds much like EDA, and indeed it is. Exploring really large databases in the hope of finding useful patterns is called **data mining**. Here are some distinctive features of data mining:

data mining

- When you have terrabytes of data, even straightforward calculations and graphics become very time-consuming. So efficient algorithms are very important.

- The structure of the database and the process of storing the data (the fashionable term is *data warehousing*), perhaps by unifying data scattered across many departments of a large corporation, require careful consideration.

- Data mining requires automated tools that work based on only vague queries by the user. The process is too complex to do step-by-step as we have done in EDA.

All these features point to the need for sophisticated computer science as a basis for data mining. Indeed, data mining is often viewed as a part of computer science. Yet many statistical ideas and tools—mostly tools for dealing with multidimensional data, not the sort of thing that appears in a first statistics course—are very helpful. Like many other modern developments, data mining crosses the boundaries of traditional fields of study.

Do remember that the perils we associate with blind use of correlation and regression are yet more perilous in data mining, where the fog of an immense database can prevent clear vision. Extrapolation, ignoring lurking variables, and confusing association with causation are traps for the unwary data miner.

SECTION 2.5 SUMMARY

- You can examine the fit of a regression line by plotting the **residuals**, which are the differences between the observed and predicted values of y. Be on the lookout for points with unusually large residuals and also for nonlinear patterns and uneven variation about the line.

- Also look for **influential observations**, individual points that substantially change the regression line. Influential observations are often outliers in the x direction, but they need not have large residuals.

- Correlation and regression must be **interpreted with caution**. Plot the data to be sure that the relationship is roughly linear and to detect outliers and influential observations.

- **Lurking variables** may explain the relationship between the explanatory and response variables. Correlation and regression can be misleading if you ignore important lurking variables.

- We cannot conclude that there is a cause-and-effect relationship between two variables just because they are strongly associated. **High correlation does not imply causation**.

- **A correlation based on averages** is usually higher than if we used data for individuals.

SECTION 2.5 EXERCISES

For Exercise 2.92, see page 125; and for Exercise 2.93 see page 125.

2.94 Bone strength. Exercise 2.24 (page 97) gives the bone strengths of the dominant and the nondominant arms for 15 men who were controls in a study. The least-squares regression line for these data is

dominant = 2.74 + (0.936 × nondominant)

Here are the data for four cases:

ID	Nondominant	Dominant	ID	Nondominant	Dominant
5	12.0	14.8	7	12.3	13.1
6	20.0	19.8	8	14.4	17.5

Calculate the residuals for these four cases. ARMSTR

2.95 Bone strength for baseball players. Refer to the previous exercise. Similar data for baseball players is given in Exercise 2.25 (page 98). The equation of the least-squares line for the baseball players is

dominant = 0.886 + (1.373 × nondominant)

Here are the data for the first four cases:

ID	Nondominant	Dominant	ID	Nondominant	Dominant
20	21.0	40.3	22	31.5	36.9
21	14.6	20.8	23	14.9	21.2

Calculate the residuals for these four cases. ARMSTR

2.96 Least-squares regression for radioactive decay. Refer to Exercise 2.32 (page 99) for the data on radioactive decay of barium-137m. Here are the data: [Iltl] DECAY

Time	1	3	5	7
Count	578	317	203	118

(a) Using the least-squares regression equation

$$\text{count} = 602.8 - (74.7 \times \text{time})$$

and the observed data, find the residuals for the counts.

(b) Plot the residuals versus time.

(c) Write a short paragraph assessing the fit of the least-squares regression line to these data based on your interpretation of the residual plot.

2.97 Least-squares regression for the log counts. Refer to Exercise 2.33 (page 99), where you analyzed the radioactive decay of barium-137m data using log counts. Here are the data: [Iltl] DECAY

Time	1	3	5	7
Log count	6.35957	5.75890	5.31321	4.77068

(a) Using the least-squares regression equation

$$\log \text{count} = 6.593 - (0.2606 \times \text{time})$$

and the observed data, find the residuals for the counts.

(b) Plot the residuals versus time.

(c) Write a short paragraph assessing the fit of the least-squares regression line to these data based on your interpretation of the residual plot.

2.98 College students by state. Refer to Exercise 2.75 (page 119), where you examined the relationship between the number of undergraduate college students and the populations for the 50 states. [Iltl] COLLEGE

(a) Make a scatterplot of the data with the least-squares regression line.

(b) Plot the residuals versus population.

(c) Focus on California, the state with the largest population. Is this state an outlier when you consider only the distribution of population? Explain your answer and describe what graphical and numerical summaries you used as the basis for your conclusion.

(d) Is California an outlier in the distribution of undergraduate college students? Explain your answer and describe what graphical and numerical summaries you used as the basis for your conclusion.

(e) Is California an outlier when viewed in terms of the relationship between number of undergraduate college students and population? Explain your answer and describe what graphical and numerical summaries you used as the basis for your conclusion.

(f) Is California influential in terms of the relationship between number of undergraduate college students and population? Explain your answer and describe what graphical and numerical summaries you used as the basis for your conclusion.

2.99 College students by state using logs. Refer to the previous exercise. Answer parts (a) through (f) for that exercise using the logs of both variables. Write a short paragraph summarizing your findings and comparing them with those from the previous exercise. [Iltl] COLLEGE

2.100 Make some scatterplots. For each of the following scenarios, make a scatterplot with 10 observations that show a moderate positive association, plus one that illustrates the unusual case. Explain each of your answers.

(a) An outlier in x that is not influential for the regression.

(b) An outlier in x that is influential for the regression.

(c) An influential observation that is not an outlier in x.

(d) An observation that is influential for the intercept but not for the slope.

2.101 What's wrong? Each of the following statements contains an error. Describe each error and explain why the statement is wrong.

(a) An influential observation will always have a large residual.

(b) High correlation is never present when there is causation.

(c) If we have data at values of x equal to 1, 2, 3, 4, and 5, and we try to predict the value of y for $x = 2.5$ using a least-squares regression equation, we are doing an extrapolation.

2.102 What's wrong? Each of the following statements contains an error. Describe each error and explain why the statement is wrong.

(a) If the residuals are all negative, this implies that there is a negative relationship between the response variable and the explanatory variable.

(b) A strong negative relationship does not imply that there is an association between the explanatory variable and the response variable.

(c) A lurking variable is always something that can be measured.

2.103 Internet use and babies. Exercise 2.34 (page 99) explores the relationship between Internet use and birthrate for 106 countries. Figure 2.13 (page 99) is a scatterplot of the data. It shows a negative association between these two variables. Do you think that this plot indicates that Internet use causes people to have fewer babies? Give another possible explanation for why these two variables are negatively associated. INBIRTH

2.104 A lurking variable. The effect of a lurking variable can be surprising when individuals are divided into groups. In recent years, the mean SAT score of all high school seniors has increased. But the mean SAT score has decreased for students at each level of high school grades (A, B, C, and so on). Explain how grade inflation in high school (the lurking variable) can account for this pattern. *A relationship that holds for each group within a population need not hold for the population as a whole. In fact, the relationship can even change direction.*

2.105 How's your self-esteem? People who do well tend to feel good about themselves. Perhaps helping people feel good about themselves will help them do better in their jobs and in life. For a time, raising self-esteem became a goal in many schools and companies. Can you think of explanations for the association between high self-esteem and good performance other than "Self-esteem causes better work"?

2.106 Are big hospitals bad for you? A study shows that there is a positive correlation between the size of a hospital (measured by its number of beds x) and the median number of days y that patients remain in the hospital. Does this mean that you can shorten a hospital stay by choosing a small hospital? Why?

2.107 Does herbal tea help nursing-home residents? A group of college students believes that herbal tea has remarkable powers. To test this belief, they make weekly visits to a local nursing home, where they visit with the residents and serve them herbal tea. The nursing-home staff reports that after several months many of the residents are healthier and more cheerful. We should commend the students for their good deeds but doubt that herbal tea helped the residents. Identify the explanatory and response variables in this informal study. Then explain what lurking variables account for the observed association.

2.108 Price and ounces. In Example 2.2 (page 80) and Exercise 2.3 (page 82), we examined the relationship between the price and the size of a Mocha Frappuccino®. The 12-ounce Tall drink costs $3.95, the 16-ounce Grande is $4.45, and the 24-ounce Venti is $4.95.

(a) Plot the data and describe the relationship. (Explain why you should plot size in ounces on the x axis.)

(b) Find the least-squares regression line for predicting the price using size. Add the line to your plot.

(c) Draw a vertical line from the least-squares line to each data point. This gives a graphical picture of the residuals.

(d) Find the residuals and verify that they sum to zero.

(e) Plot the residuals versus size. Interpret this plot.

2.109 Use the applet. It isn't easy to guess the position of the least-squares line by eye. Use the *Correlation and Regression* applet to compare a line you draw with the least-squares line. Click on the scatterplot to create a group of 15 points from lower left to upper right with a clear, positive straight-line pattern (correlation around 0.6). Click the "Draw line" button and use the mouse to draw a line through the middle of the cloud of points from lower left to upper right. Note the "thermometer" that appears above the plot. The red portion is the sum of the squared vertical distances from the points in the plot to the least-squares line. The green portion is the "extra" sum of squares for your line—it shows by how much your line misses the smallest possible sum of squares.

(a) You drew a line by eye through the middle of the pattern. Yet the right-hand part of the bar is probably almost entirely green. What does that tell you?

(b) Now click the "Show least-squares line" box. Is the slope of the least-squares line smaller (the new line is less steep) or larger (line is steeper) than that of your line? If you repeat this exercise several times, you will consistently get the same result. *The least-squares line minimizes the vertical distances of the points from the line. It is not the line through the "middle" of the cloud of points.* This is one reason it is hard to draw a good regression line by eye.

2.110 Use the applet. Go to the *Correlation and Regression* applet. Click on the scatterplot to create a group of 12 points in the lower-right corner of the scatterplot with a strong straight-line pattern (correlation about −0.8). Now click the "Show least-squares line" box to display the regression line.

(a) Add one point at the upper left that is far from the other 12 points but exactly on the regression line. Why does this outlier have no effect on the line even though it changes the correlation?

(b) Now drag this last point down until it is opposite the group of 12 points. You see that one end of the least-squares line chases this single point, while the other end remains near the middle of the original group of 12. What makes the last point so influential?

2.111 Education and income. There is a strong positive correlation between years of education and income for economists employed by business firms. (In particular, economists with doctorates earn more than economists with only a bachelor's degree.) There is also a strong positive correlation between years of education and income for economists employed by colleges and universities. But when all economists are considered, there is a *negative* correlation between education and income. The explanation for this is that business pays high salaries and employs mostly economists with bachelor's degrees, while colleges pay lower salaries and employ mostly economists with doctorates. Sketch a scatterplot with two groups of cases (business and academic) that illustrates how a strong positive correlation within each group and a negative overall correlation can occur together.

2.112 Dangers of not looking at a plot. Table 2.1 (page 122) presents four sets of data prepared by the statistician Frank Anscombe to illustrate the dangers of calculating without first plotting the data.[22] ANSC

(a) Use x to predict y for each of the four data sets. Find the predicted values and residuals for each of the four regression equations.

(b) Plot the residuals versus x for each of the four data sets.

(c) Write a summary of what the residuals tell you for each data set, and explain how the residuals help you to understand these data.

2.6 Data Analysis for Two-Way Tables

When you complete this section, you will be able to:
- Identify the row variable, the column variable, and the cells in a two-way table.
- Find and interpret the joint distribution in a two-way table.
- Find and interpret the marginal distributions in a two-way table.
- Use the conditional distributions to describe the relationship displayed in a two-way table.
- Determine the joint distribution, the marginal distributions, and the conditional distributions in a two-way table from software output.
- Interpret examples of Simpson's paradox.

LOOK BACK

quantitative and categorical variables, p. 3

When we study relationships between two variables, one of the first questions we ask is whether each variable is quantitative or categorical. For two quantitative variables, we use a scatterplot to examine the relationship, and we fit a line to the data if the relationship is approximately linear. If one of the variables is quantitative and the other is categorical, we can use the methods in Chapter 1 to describe the distribution of the quantitative variable for each value of the categorical variable. This leaves us with the situation where both variables are categorical. In this section, we discuss methods for studying these relationships.

Some variables—such as sex, race, and occupation—are inherently categorical. Other categorical variables are created by grouping values of a quantitative variable into classes. Published data are often reported in grouped form to save space. To describe categorical data, we use the *counts* (frequencies) or *percents* (relative frequencies) of individuals that fall into various categories.

The two-way table

two-way table

A key idea in studying relationships between two variables is that both variables must be measured on the same individuals or cases. When both variables are categorical, the raw data are summarized in a **two-way table** that gives counts of observations for each combination of values of the two categorical variables. Here is an example.

EXAMPLE 2.33

IOM

Is the calcium intake adequate? Young children need calcium in their diet to support the growth of their bones. The Institute of Medicine provides guidelines for how much calcium should be consumed for people of different ages.[23] One study examined whether or not a sample of children consumed an adequate amount of calcium, based on these guidelines. Because there are different requirements for children aged 5 to 10 years and for children aged 11 to 13 years of age, the children were classified into these two age groups. For each student, his or her calcium intake was classified as meeting or not meeting the requirement. There were 2029 children in the study. Here are the data:[24]

Two-way table for "met requirement" and age

	Age (years)	
Met requirement	5 to 10	11 to 13
No	194	557
Yes	861	417

We see that 194 children aged 5 to 10 did not meet the calcium requirement, and 861 children aged 5 to 10 years met the calcium requirement.

USE YOUR KNOWLEDGE

2.113 Read the table. Refer to the table in Example 2.33. How many children aged 11 to 13 met the requirement? How many did not?

For the calcium requirement example, we could view age as an explanatory variable and "met requirement" as a response variable. This is why we put age in the columns (like the *x* axis in a scatterplot) and "met requirement" in the rows (like the *y* axis in a scatterplot). We call "met requirement" the **row variable** because each horizontal row in the table describes whether or not the requirement was met. Age is the **column variable** because each vertical column describes one age group. Each combination of values for these two variables is called a **cell**. For example, the cell corresponding to children who are 5 to 10 years old and who have not met the requirement contains the number 194. This table is called a 2 × 2 table because there are two rows and two columns.

row variable

column variable

cell

To describe relationships between two categorical variables, we compute different types of percents. Our job is easier if we expand the basic two-way table by adding various totals. We illustrate the idea with our calcium requirement example.

EXAMPLE 2.34

IOM

Add the margins to the table. We expand the table in Example 2.33 by adding the totals for each row, for each column, and the total number of all the observations. Here is the result:

Two-way table for "met requirement" and age

	Age (years)		
Met requirement	5 to 10	11 to 13	Total
No	194	557	751
Yes	861	417	1278
Total	1055	974	2029

In this study there were 1055 children aged 5 to 10. The total number of children who did not meet the calcium requirement is 751, and the total number of children in the study is 2029.

USE YOUR KNOWLEDGE

2.114 Read the margins of the table. How many children aged 11 to 13 were subjects in the calcium requirement study? What is the total number of children who met the calcium requirement?

In this example, be sure that you understand how the table is obtained from the raw data. Think about a data file with one line per subject. There would be 2029 lines or records in this data set. In the two-way table, each individual is counted once and only once. As a result, the sum of the counts in the table is the total number of individuals in the data set. *Most errors in the use of categorical-data methods come from a misunderstanding of how these tables are constructed.*

Joint distribution

We are now ready to compute some proportions that help us understand the data in a two-way table. Suppose that we are interested in the children aged 5 to 10 years who do not meet the calcium requirement. The proportion of these is simply 194 divided by 2029, or 0.0956. We would estimate that 9.56% of children in the population from which this sample was drawn are 5- to 10-year-olds who do not meet the calcium requirement. For each cell, we can compute a proportion by dividing the cell entry by the total sample size. The collection

joint distribution

of these proportions is the **joint distribution** of the two categorical variables.

EXAMPLE 2.35

IOM

The joint distribution. For the calcium requirement example, the joint distribution of age "met requirement" and age is

Joint distribution of "met requirement" and age		
	Age (years)	
Met requirement	5 to 10	11 to 13
No	0.0956	0.2745
Yes	0.4243	0.2055

Because this is a distribution, the sum of the proportions should be 1. For this example the sum is 0.9999. The difference is due to roundoff error.

USE YOUR KNOWLEDGE

2.115 Explain the computation. Explain how the entry for the children aged 5 to 10 who met the calcium requirement in Example 2.35 is computed from the table in Example 2.34.

How might we use the information in the joint distribution for this example? Suppose that we were to develop an outreach unit to increase the consumption of calcium. The distribution suggests that the older students should be targeted if we have to make a choice because of limited funds. Children who are 11 to 13 years old and do not meet the calcium requirement are 27.45% of the total;

however, children who are 5 to 10 years old and do not meet the requirement are only 9.56% of the total. For other uses of these data, we may need to calculate different numerical summaries. Let's look at the distribution of age.

Marginal distributions

marginal distribution

When we examine the distribution of a single variable in a two-way table, we are looking at a **marginal distribution**. There are two marginal distributions, one for each categorical variable in the two-way table. They are very easy to compute.

EXAMPLE 2.36

IOM

The marginal distribution of age. Look at the table in Example 2.34. The total numbers of children aged 5 to 10 and children aged 11 to 13 are given in the bottom row, labeled "Total." Our sample has 1055 children aged 5 to 10 and 974 children aged 11 to 13. To find the marginal distribution of age, we simply divide these numbers by the total sample size, 2029. The marginal distribution of age is

Marginal distribution of age		
	5 to 10	11 to 13
Proportion	0.52	0.48

Note that the proportions sum to 1; there is no roundoff error.

Often, we prefer to use percents rather than proportions. Here is the marginal distribution of age described with percents:

Marginal distribution of age		
	5 to 10	11 to 13
Percent	52%	48%

Which form do you prefer?

The percent of children in each age group is approximately the same. This is interesting because the first category includes six ages (5, 6, 7, 8, 9, and 10); whereas the second includes only three ages (11, 12, and 13). Recall that the age categories were chosen in this way because the Institute of Medicine defined the calcium requirement differently for these age groups. In this study, the children were selected from grades 4, 5, and 6. The distribution of ages within these grades explains the marginal distribution of age for our sample.

The other marginal distribution for this example is the distribution of "met requirement."

EXAMPLE 2.37

IOM

The marginal distribution of "met requirement." Here is the marginal distribution of "met requirement," in percents:

Marginal distribution of "met requirement"		
	No	Yes
Percent	37.01%	62.99%

USE YOUR KNOWLEDGE

2.116 Explain the marginal distribution. Explain how the marginal distribution of "met requirement" given in Example 2.37 is computed from the entries in the table given in Example 2.34.

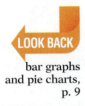

LOOK BACK

bar graphs and pie charts, p. 9

Each marginal distribution from a two-way table is a distribution for a single categorical variable. We can use a bar graph or a pie chart to display such a distribution. For our two-way table, we will be content with numerical summaries: for example, 52% of the children are aged 5 to 10, and 37% of the children are not meeting their calcium requirement. When we have more rows or columns, the graphical displays are particularly useful.

Describing relations in two-way tables

The table in Example 2.34 contains much more information than the two marginal distributions of age alone and "met requirement" alone. We need to do a little more work to examine the relationship. *Relationships among categorical variables are described by calculating appropriate percents from the counts given.* What percents do you think we should use to describe the relationship between age and meeting the calcium requirement?

EXAMPLE 2.38

IOM

Meeting the calcium requirement for children aged 5 to 10. What percent of the children aged 5 to 10 in our sample met the calcium requirement? This is the count of the children who are 5 to 10 years old and who met the calcium requirement as a percent of the number of children who are 5 to 10 years old:

$$\frac{861}{1055} = 0.8161 = 82\%$$

USE YOUR KNOWLEDGE

2.117 Find the percent. Refer to the table in Example 2.34 (page 137). Show that the percent of children 11 to 13 years old who met the calcium requirement is 43%.

Conditional distributions

In Example 2.38, we looked at the children aged 5 to 10 alone and examined the distribution of the other categorical variable, "met requirement." Another way to say this is that we *conditioned* on the value of age, 5 to 10 years old. Similarly, we can condition on the value of age being 11 to 13 years old. When we condition on the value of one variable and calculate the distribution of the other variable, we obtain a **conditional distribution**. Note that in Example 2.38, we calculated only the percent for children aged 5 to 10 years. The complete conditional distribution gives the proportions or percents for all possible values of the conditioning variable.

conditional distribution

EXAMPLE 2.39

Conditional distribution of "met requirement" for children aged 5 to 10. For children aged 5 to 10 years, the conditional distribution of the "met requirement" variable in terms of percents is

Conditional distribution of "met requirement" for children aged 5 to 10		
	No	**Yes**
Percent	18.39%	81.61%

Note that we have included the percents for both of the possible values, Yes and No, of the "met requirement" variable. These percents sum to 100%.

USE YOUR KNOWLEDGE

2.118 A conditional distribution. Perform the calculations to show that the conditional distribution of "met requirement" for children aged 11 to 13 years is

Conditional distribution of "met requirement" for children aged 11 to 13		
	No	**Yes**
Percent	57.19%	42.81%

Comparing the conditional distributions (Example 2.39 and Exercise 2.118) reveals the nature of the association between age and meeting the calcium requirement. In this set of data, the older children are more likely to fail to meet the calcium requirement.

Bar graphs can help us to see relationships between two categorical variables. No single graph (such as a scatterplot) portrays the form of the relationship between categorical variables, and no single numerical measure (such as the correlation) summarizes the strength of an association. Bar graphs are flexible enough to be helpful, but you must think about what comparisons you want to display. For numerical measures, we must rely on well-chosen percents or on more advanced statistical methods.[25]

A two-way table contains a great deal of information in compact form. Making that information clear almost always requires finding percents. You must decide which percents you need. Of course, we prefer to use software to compute the joint, marginal, and conditional distributions.

EXAMPLE 2.40

Software output. Figure 2.28 gives computer output for the data in Example 2.33 using Minitab, SPSS, and JMP. There are minor variations among software packages, but these outputs are typical of what is usually produced. Each cell in the 2 × 2 table has four entries. These are the count (the number of observations in the cell), the conditional distributions for rows and columns, and the joint distribution. Note that all of these are expressed as percents rather than proportions. Marginal totals and distributions are given in the rightmost column and the bottom row.

Most software packages order the row and column labels numerically or alphabetically. In general, it is better to use words rather than numbers for the column labels. This sometimes involves some additional work, but it avoids the kind of confusion that can result when you forget the real values associated with each numerical value. You should verify that the entries in Figure 2.28 correspond to the calculations that we performed in Examples 2.34 through 2.39. In addition, verify the calculations for the conditional distributions of age for each value of "met requirement."

FIGURE 2.28 Computer output for the calcium requirement study, Example 2.40: (a) Minitab; (b) SPSS; (c) JMP.

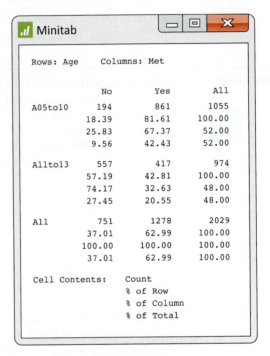

(a) Minitab

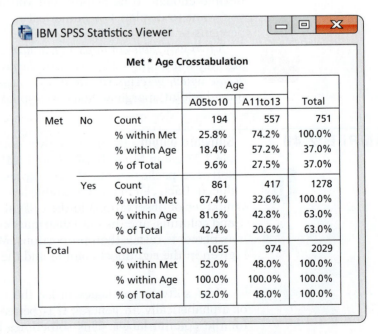

(b) SPSS

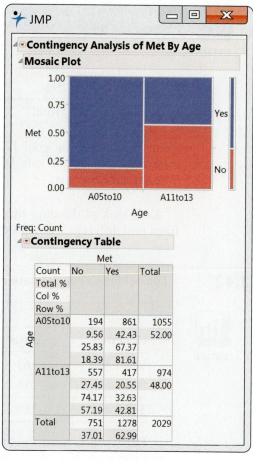

FIGURE 2.28 *Continued*   (c) JMP

The JMP output in Figure 2.28 includes a graphical display of the data called
mosaic plot   a **mosaic plot**. The sizes of the four boxes display the joint distribution. The
narrow bar to the right shows the marginal distribution of Met and the widths
of the vertical bars show the marginal distribution of Age. The conditional distribution of Met for each Age is represented in each of these vertical bars by the
heights of the blue and red sections. Notice that they always add to one.

Simpson's paradox

As is the case with quantitative variables, the effects of lurking variables can
strongly influence relationships between two categorical variables. Here is
an example that demonstrates the surprises that can await the unsuspecting
consumer of data.

EXAMPLE 2.41   **Which customer service representative is better?** A customer service center
has a goal of resolving customer questions in 10 minutes or less. Here are
the records for two representatives:

CUSTSER

Goal met	Representative	
	Alexis	Peyton
Yes	172	118
No	28	82
Total	200	200

Alexis has met the goal 172 times out of 200, a success rate of 86%. For Peyton, the success rate is 118 out of 200, or 59%. Alexis clearly has the better success rate.

Let's look at the data in a little more detail. The data summarized come from two different weeks in the year.

EXAMPLE 2.42

CUSTSER

Look at the data more carefully. Here are the counts broken down by week:

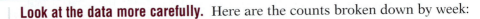

Goal met	Week 1		Week 2	
	Alexis	Peyton	Alexis	Peyton
Yes	162	19	10	99
No	18	1	10	81
Total	180	20	20	180

For Week 1, Alexis met the goal 90% of the time (162/180), while Peyton met the goal 95% of the time (19/20). Peyton had the better performance in Week 1. What about Week 2? Here, Alexis met the goal 50% of the time (10/20), while the success rate for Peyton was 55% (99/180). Peyton again had the better performance. How does this analysis compare with the analysis that combined the counts for the two weeks? That analysis clearly showed that Alexis had the better performance, 86% versus 59%.

These results can be explained by a lurking variable, Week. The first week was during a period when the product had been in use for several months. Most of the calls to the customer service center concerned problems that had been encountered before. The representatives were trained to answer these questions and usually had no trouble in meeting the goal of resolving the problems quickly. On the other hand, the second week occurred shortly after the release of a new version of the product. Most of the calls during this week concerned new problems that the representatives had not yet encountered. Many more of these questions took longer than the 10-minute goal to resolve.

Look at the totals in the bottom row of the detailed table. During the first week, when calls were easy to resolve, Alexis handled 180 calls and Peyton handled 20. The situation was exactly the opposite during the second week, when the calls were difficult to resolve. There were 20 calls for Alexis and 180 for Peyton.

The original two-way table, which did not take account of week, was misleading. This example illustrates *Simpson's paradox*.

SIMPSON'S PARADOX

An association or comparison that holds for all of several groups can reverse direction when the data are combined to form a single group. This reversal is called **Simpson's paradox**.

three-way table

aggregation

The lurking variables in our Simpson's paradox example, Week and problem difficulty, are categorical. That is, they break the observations into groups by workweek. *Simpson's paradox is an extreme form of the fact that observed associations can be misleading when there are lurking variables.*

The data in Example 2.42 are given in a **three-way table** that reports counts for each combination of three categorical variables: week, representative, and whether or not the goal was met. In our example, we constructed the three-way table by constructing two two-way tables for representative by goal, one for each week. The original table in Example 2.41 can be obtained by adding the corresponding counts for these two tables. This process is called **aggregating** the data. When we aggregated data in Example 2.41, we ignored the variable week, which then became a lurking variable. *Conclusions that seem obvious when we look only at aggregated data can become quite different when the data are examined in more detail.*

SECTION 2.6 SUMMARY

• A **two-way table** of counts organizes data about two categorical variables. Values of the **row variable** label the rows that run across the table, and values of the **column variable** label the columns that run down the table. Two-way tables are often used to summarize large amounts of data by grouping outcomes into categories.

• The **joint distribution** of the row and column variables is found by dividing the count in each cell by the total number of observations.

• The **row totals** and **column totals** in a two-way table give the **marginal distributions** of the two variables separately. It is clearer to present these distributions as percents of the table total. Marginal distributions do not give any information about the relationship between the variables.

• To find the **conditional distribution** of the row variable for one specific value of the column variable, look only at that one column in the table. Find each entry in the column as a percent of the column total.

• There is a conditional distribution of the row variable for each column in the table. Comparing these conditional distributions is one way to describe the association between the row and the column variables. It is particularly useful when the column variable is the explanatory variable. When the row variable is explanatory, find the conditional distribution of the column variable for each row and compare these distributions.

• **Bar graphs** are a flexible means of presenting categorical data. There is no single best way to describe an association between two categorical variables.

• We present data on three categorical variables in a **three-way table**, printed as separate two-way tables for each level of the third variable. A comparison between two variables that holds for each level of a third variable can be changed or even reversed when the data are **aggregated** by summing over all levels of the third variable. **Simpson's paradox** refers to the reversal of a comparison by aggregation. It is an example of the potential effect of lurking variables on an observed association.

SECTION 2.6 EXERCISES

For Exercise 2.113, see page 137; for 2.114, see page 138; for 2.115, see page 138; for 2.116, see page 140; for 2.117, see page 140; and for 2.118, see page 141.

2.119 Does driver's ed help? A study is planned to look at the effect of driver education programs on accidents. The driving records of all drivers under 18 in a given year will classify each driver as having taken a driver's education course or not. The drivers will also classified with respect to the number of accidents that they had in the year after they received their license. The categories are zero, one, and two or more accidents.

(a) There are two variables in this study. Do you think one is an explanatory variable and the other is a response variable? Explain your answer.

(b) Sketch a two-way table that could be used to organize the data. Which variable is the row variable? Which variable is the column variable?

(c) How many cells are in the table? Describe in words what each of the cells will contain when the data are collected.

2.120 Music and video games. You are planning a study of undergraduates in which you will examine the relationship between listening to music and playing video games. The study subjects will be asked how much time they spend in each of these activities during a typical day. The choices for both activities will be a half hour or less, more than a half hour but less than an hour, and more than an hour.

(a) There are two variables in this study. Do you think one is an explanatory variable and the other is a response variable? Explain your answer.

(b) Sketch a two-way table that could be used to organize the data. Which variable is the row variable? Which variable is the column variable?

(c) How many cells are in the table? Describe in words what each of the cells will contain when the data are collected.

2.121 Eight is enough. A healthy body needs good food, and healthy teeth are needed to chew our food so that it can nourish our bodies. The U.S. Army has recognized this fact and requires recruits to pass a dental examination. If you wanted to be a soldier in the Spanish American War, which took place in 1898, you needed to have at least eight teeth. Here is the statement of the requirement: [Ⅲ] TEETH

Unless an applicant has at least four sound double teeth, one above and one below on each side of the mouth, and so opposed as to serve the purpose of mastication, he should be rejected.

A study reported the rejection data for enlistment candidates classified by age. Here are the data:[26]

Rejected	Under 20	20 to 25	25 to 30	30 to 35	35 to 40	Over 40
Yes	68	647	1114	1783	2887	3801
No	58,884	77,992	55,597	43,994	47,569	39,985

Column group header: Age

(a) Which variable is the explanatory variable? Which variable is the response variable? Give reasons for your answer.

(b) Find the joint distribution. Write a brief summary explaining the major features of this distribution.

(c) Find the two marginal distributions. Write a brief summary explaining the major features of these distributions.

(d) Which conditional distribution would you choose to explain the relationship between these two variables? Explain your answer.

(e) Find the conditional distribution that you chose in part (d), and write a summary that includes your interpretation of the relationship based on this conditional distribution.

2.122 Survival and class on the *Titanic*. In Exercise 1.27 (page 24), you created a graphical summary of the number of passengers who survived classified by the accommodations that they had on the ship: first, second, or third class. Let's look at these data with a two-way table. [Ⅲ] TITANIC

(a) Create a two-way table that you could use to explore the relationship between survival and class.

(b) Which variable is the explanatory variable and which is the response variable? Give reasons for your answers.

(c) Find the two marginal distributions. Write a brief summary explaining the major features of these distributions.

(d) Which conditional distribution would you choose to explain the relationship between these two variables? Explain your answer.

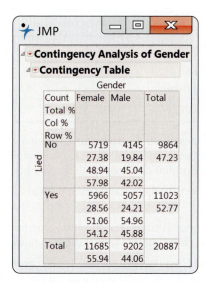

FIGURE 2.29 Computer output for the lying to a teacher data, Exercise 2.123.

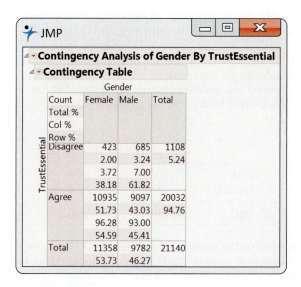

FIGURE 2.30 Computer output for the trust and honesty in the workplace data, Exercise 2.124.

(e) Find the conditional distribution that you chose in part (d) and, write a summary that includes your interpretation of the relationship based on this conditional distribution.

2.123 Lying to a teacher. One of the questions in a survey of high school students asked about lying to teachers.[27] The data set LYING gives the numbers of students who said that they lied to a teacher about something significant at least once during the past year, classified by sex. Figure 2.29 gives software output for these data. Use this output to analyze these data and write a report summarizing your work. Be sure to include a discussion of whether or not you consider this relationship to involve an explanatory variable and a response variable. LYING

2.124 Trust and honesty in the workplace. The students surveyed in the study described in the previous exercise were also asked whether they thought trust and honesty were essential in business and the workplace. Figure 2.30 gives software output for these data. Use this output to analyze these data and write a report summarizing your work. Be sure to include a discussion of whether or not you consider this relationship to involve an explanatory variable and a response variable. TRUST

2.125 Exercise and adequate sleep. A survey of 656 boys and girls, who were 13 to 18 years old, asked about adequate sleep and other health-related behaviors. The recommended amount of sleep is six to eight hours per night.[28] In the survey, 59% of the respondents reported that they got less than this amount of sleep on school nights. An exercise scale was developed and used to classify the students as above or below the median in this

domain. Here is the 2 × 2 table of counts with students classified as getting or not getting adequate sleep and by the exercise variable: SLEEP

Enough sleep	Exercise	
	High	Low
Yes	151	115
No	148	242

(a) Find the distribution of adequate sleep for the high exercisers.

(b) Do the same for the low exercisers.

(c) Summarize the relationship between adequate sleep and exercise using the results of parts (a) and (b).

2.126 Adequate sleep and exercise. Refer to the previous exercise. SLEEP

(a) Find the distribution of exercise for those who get adequate sleep.

(b) Do the same for those who do not get adequate sleep.

(c) Write a short summary of the relationship between adequate sleep and exercise using the results of parts (a) and (b).

(d) Compare this summary with your summary from part (c) of the previous exercise. Which do you prefer? Give a reason for your answer.

2.127 Which hospital is safer? Insurance companies and consumers are interested in the performance of hospitals. The government releases data about patient outcomes in hospitals that can be useful in making informed health care decisions. Here is a two-way table of data on the survival of patients after surgery in two hospitals. All patients undergoing surgery in a recent time period are included. "Survived" means that the patient lived at least six weeks following surgery.

HOSP

	Hospital A	Hospital B
Died	63	16
Survived	2037	784
Total	2100	800

What percent of Hospital A patients died? What percent of Hospital B patients died? These are the numbers one might see reported in the media.

2.128 Patients in "poor" or "good" condition. Refer to the previous exercise. Not all surgery cases are equally serious, however. Patients are classified as being in either "poor" or "good" condition before surgery. Here are the data broken down by patient condition. The entries in the original two-way table are just the sums of the "poor" and "good" entries in this pair of tables. **HOSP**

Good condition		
	Hospital A	Hospital B
Died	6	8
Survived	594	592
Total	600	600

Poor condition		
	Hospital A	Hospital B
Died	57	8
Survived	1443	192
Total	1500	200

(a) Find the death rate for Hospital A patients who were classified as "poor" before surgery. Do the same for Hospital B. In which hospital do "poor" patients fare better?

(b) Repeat part (a) for patients classified as "good" before surgery.

(c) What is your recommendation to someone facing surgery and choosing between these two hospitals?

(d) How can Hospital A do better in both groups, yet do worse overall? Look at the data and carefully explain how this can happen.

2.129 Complete the table. Here are the row and column totals for a two-way table with two rows and two columns:

$$\begin{array}{cc|c} a & b & 300 \\ c & d & 200 \\ \hline 300 & 200 & 500 \end{array}$$

Find *two different* sets of counts a, b, c, and d for the body of the table that give these same totals. This shows that the relationship between two variables cannot be obtained from the two individual distributions of the variables.

2.130 Construct a table with no association. Construct a 3×4 table of counts where there is no apparent association between the row and column variables.

2.7 The Question of Causation

When you complete this section, you will be able to:	• Identify the differences among causation, common response, and confounding in explaining an association.
	• Apply the five criteria for establishing causation.

In many studies of the relationship between two variables, the goal is to establish that changes in the explanatory variable *cause* changes in the response variable. Even when a strong association is present, however, the conclusion that this association is due to a causal link between the variables is often hard to justify. What ties between two variables (and others lurking in the background) can explain an observed association? What constitutes good evidence for causation? We begin our consideration of these questions with a set of observed associations. In each case, there is a clear association between variable x and variable y. Moreover, the association is positive whenever the direction makes sense.

Explaining association

EXAMPLE 2.43

Observed associations. Here are some examples of observed association between x and y:

1. $x =$ mother's body mass index

   $y =$ daughter's body mass index

2. $x =$ amount of the artificial sweetener saccharin in a rat's diet

   $y =$ count of tumors in the rat's bladder

3. $x =$ a student's SAT score as a high school senior

   $y =$ a student's first-year college grade point average

4. $x =$ monthly flow of money into stock mutual funds

   $y =$ monthly rate of return for the stock market

5. $x =$ whether a person regularly attends religious services

   $y =$ how long the person lives

6. $x =$ the number of years of education a worker has

   $y =$ the worker's income

Explaining association: Causation Figure 2.31 shows in outline form how a variety of underlying links between variables can explain association. The dashed double-arrow line represents an observed association between the variables x and y. Some associations are explained by a direct cause-and-effect link between these variables. The first diagram in Figure 2.31 shows "x causes y" by a solid arrow running from x to y.

 Items 1 and 2 in Example 2.43 are examples of direct causation. *Even when direct causation is present, very often it is not a complete explanation of an association between two variables.* The best evidence for causation comes from experiments that actually change x while holding all other factors fixed. If y changes, we have good reason to think that x caused the change in y.

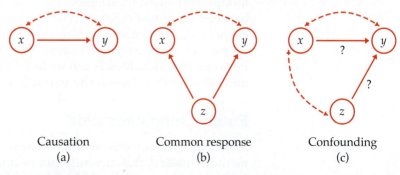

Causation	Common response	Confounding
(a)	(b)	(c)

FIGURE 2.31 Possible explanations for an observed association. The dashed double-arrow lines show an association. The solid arrows show a cause-and-effect link. The variable x is explanatory, y is a response variable, and z is a lurking variable.

common response

Explaining association: Common response "Beware of the lurking variable" is good advice when thinking about an association between two variables. The second diagram in Figure 2.31 illustrates **common response**. The observed association between the variables x and y is explained by a lurking variable z. Both x and y change in response to changes in z. This common response creates an association even though there may be no direct causal link between x and y.

The third and fourth items in Example 2.43 illustrate how common response can create an association. What would be a good candidate for the variable z in these two examples?

Explaining association: Confounding For the first item in Example 2.43, we expect that inheritance explains part of the association between the body mass indexes (BMIs) of daughters and their mothers. Can we use r or r^2 to say how much inheritance contributes to the daughters' BMIs? No. It may well be that mothers who are overweight also set an example of little exercise, poor eating habits, and lots of television. Their daughters pick up these habits to some extent, so the influence of heredity is mixed up with influences from the girls' environment. We call this mixing of influences *confounding*.

CONFOUNDING

Two variables are **confounded** when their effects on a response variable cannot be distinguished from each other. The confounded variables may be either explanatory variables or lurking variables or both.

When many uncontrolled variables are related to a response variable, you should always ask whether or not confounding of several variables prevents you from drawing conclusions about causation. The third diagram in Figure 2.31 illustrates confounding. Both the explanatory variable x and the lurking variable z may influence the response variable y. Because x is confounded with z, we cannot distinguish the influence of x from the influence of z. We cannot say how strong the direct effect of x on y is. In fact, it can be hard to say if x influences y at all.

The last two associations in Example 2.43 (Items 5 and 6) are explained in part by confounding. What would be a good candidate for the confounding variable z in these two examples?

Many observed associations are at least partly explained by lurking variables. Both common response and confounding involve the influence of a lurking variable (or variables) z on the response variable y. The distinction between these two types of relationship is less important than the common element, the influence of lurking variables. The most important lesson of these examples is one we have already emphasized: **even a very strong association between two variables is not by itself good evidence that there is a cause-and-effect link between the variables**.

Establishing causation

How can a direct causal link between x and y be established? The best method—indeed, the only fully compelling method—of establishing causation is to conduct a carefully designed experiment in which the effects of possible lurking variables are controlled. Chapter 3 explains how to design convincing experiments.

Many of the sharpest disputes in which statistics plays a role involve questions of causation that cannot be settled by experiment. Does gun control reduce violent crime? Does living near power lines cause cancer? Has "outsourcing" work to overseas locations reduced overall employment in the United States? All these questions have become public issues. All concern associations among variables. And all have this in common: they try to pinpoint cause and effect in a setting involving complex relations among many interacting variables. Common response and confounding, along with the number of potential lurking variables, make observed associations misleading. Experiments are not possible for ethical or practical reasons. We can't assign some people to live near power lines or compare the same nation with and without strong gun controls.

EXAMPLE 2.44

Power lines and leukemia. Electric currents generate magnetic fields. So living with electricity exposes people to magnetic fields. Living near power lines increases exposure to these fields. Really strong fields can disturb living cells in laboratory studies. Some people claim that the weaker fields we experience if we live near power lines cause leukemia in children.

It isn't ethical to do experiments that expose children to magnetic fields. It's hard to compare cancer rates among children who happen to live in more and less exposed locations because leukemia is rare and locations vary in many ways other than magnetic fields. We must rely on studies that compare children who have leukemia with children who don't.

A careful study of the effect of magnetic fields on children took five years and cost $5 million. The researchers compared 638 children who had leukemia and 620 who did not. They went into the homes and actually measured the magnetic fields in the children's bedrooms, in other rooms, and at the front door. They recorded facts about nearby power lines for the family home and also for the mother's residence when she was pregnant. Result: no evidence of more than a chance connection between magnetic fields and childhood leukemia.[29]

"No evidence" that magnetic fields are connected with childhood leukemia doesn't prove that there is no risk. It says only that a careful study could not find any risk that stands out from the play of chance that distributes leukemia cases across the landscape. Critics continue to argue that the study failed to measure some lurking variables or that the children studied don't fairly represent all children. Nonetheless, a carefully designed study comparing children with and without leukemia is a great advance over haphazard and sometimes emotional counting of cancer cases.

EXAMPLE 2.45

Smoking and lung cancer. Despite the difficulties, it is sometimes possible to build a strong case for causation in the absence of experiments. The evidence that smoking causes lung cancer is about as strong as nonexperimental evidence can be.

Doctors had long observed that most lung cancer patients were smokers. Comparison of smokers and similar nonsmokers showed a very strong association between smoking and death from lung cancer. Could the association be due to common response? Might there be, for example, a genetic factor that predisposes people both to nicotine addiction and to lung cancer? Smoking and lung cancer would then be positively associated even if smoking had no direct effect on the lungs. Or perhaps confounding is to blame. It might be that smokers live unhealthy lives in other ways (diet, alcohol, lack of exercise) and that some other habit confounded with smoking is a cause of lung cancer. How were these objections overcome?

Let's answer this question in general terms: what are the criteria for establishing causation when we cannot do an experiment?

- *The association is strong.* The association between smoking and lung cancer is very strong.

- *The association is consistent.* Many studies of different kinds of people in many countries link smoking to lung cancer. That reduces the chance that a lurking variable specific to one group or one study explains the association.

- *Higher doses are associated with stronger responses.* People who smoke more cigarettes per day or who smoke over a longer period get lung cancer more often. People who stop smoking reduce their risk.

- *The alleged cause precedes the effect in time.* Lung cancer develops after years of smoking.

- *The alleged cause is plausible.* Experiments show that tars from cigarette smoke cause cancer when applied to the backs of mice.

Medical authorities do not hesitate to say that smoking causes lung cancer. The U.S. Surgeon General states that cigarette smoking is "the largest avoidable cause of death and disability in the United States."[30] The evidence for causation is strong—but it is not as strong as the evidence provided by well-designed experiments.

SECTION 2.7 SUMMARY

- Some observed associations between two variables are due to a **cause-and-effect** relationship between these variables, but others are explained by **lurking variables**.

- The effect of lurking variables can operate through **common response** if changes in both the explanatory and the response variables are caused by changes in lurking variables. **Confounding** of two variables (either explanatory or lurking variables or both) means that we cannot distinguish their effects on the response variable.

- Establishing that an association is due to causation is best accomplished by conducting an **experiment** that changes the explanatory variable while controlling other influences on the response.

- In the absence of experimental evidence, be cautious in accepting claims of causation. Good evidence of causation requires (1) a strong association, (2) that appears consistently in many studies, (3) that has higher doses associated with stronger responses, (4) with the alleged cause preceding the effect in time, and (5) that is plausible.

SECTION 2.7 EXERCISES

2.131 Examples of association. Give three examples of association: one due to causation, one due to common response, and one due to confounding. Use your examples to write a short paragraph explaining the differences among these three explanations for an observed association.

2.132 The five criteria for establishing causation. Consider the five criteria for establishing causation. Explain how each of these, if not established seriously, weakens the case that an association is due to causation.

2.133 Iron and anemia. A lack of adequate iron in the diet is associated with anemia, a condition in which the body does not have enough red blood cells. However, anemia is also associated with malaria and infections with worms called helminths. Discuss these observed associations using the framework of Figure 2.31 (page 149).

2.134 Stress and lack of sleep in college students. Studies of college students have shown that stress and lack of sleep are associated. Do you think that lack of sleep causes stress or that stress causes lack of sleep? Write a short paragraph summarizing your opinions.

2.135 Online courses. Many colleges offer online versions of some courses that are also taught in the classroom. It often happens that the students who enroll in the online version do better than the classroom students on the course exams. This does not show that online instruction is more effective than classroom teaching because the people who sign up for online courses are often quite different from the classroom students. Suggest some student characteristics that you think could be confounded with online versus classroom. Use a diagram like Figure 2.31(c) to illustrate your ideas.

2.136 Marriage and income. Data show that men who are married, and also divorced or widowed men, earn quite a bit more than men who have never been married. This does not mean that a man can raise his income by getting married. Suggest several lurking variables that you think are confounded with marital status and that help explain the association between marital status and income. Use a diagram like Figure 2.31(c) to illustrate your ideas.

2.137 Exercise and self-confidence. A college fitness center offers an exercise program for staff members who choose to participate. The program assesses each participant's fitness, using a treadmill test, and also administers a personality questionnaire. There is a moderately strong positive correlation between fitness score and score for self-confidence. Is this good evidence that improving fitness increases self-confidence? Explain why or why not.

2.138 Computer chip manufacturing and miscarriages. A study showed that women who work in the production of computer chips have abnormally high numbers of miscarriages. The union claimed that exposure to chemicals used in production caused the miscarriages. Another possible explanation is that these workers spend most of their work time standing up. Illustrate these relationships in a diagram like one of those in Figure 2.31.

2.139 Hospital size and length of stay. A study shows that there is a positive correlation between the size of a hospital (measured by its number of beds x) and the median number of days y that patients remain in the hospital. Does this mean that you can shorten a hospital stay by choosing a small hospital? Use a diagram like one of those in Figure 2.31 to explain the association.

2.140 Watching TV and low grades. Children who watch many hours of television get lower grades in school, on average, than those who watch less TV. Explain clearly why this fact does not show that watching TV *causes* poor grades. In particular, suggest some other variables that may be confounded with heavy TV viewing and may contribute to poor grades.

2.141 Artificial sweeteners. People who use artificial sweeteners in place of sugar tend to be heavier than people who use sugar. Does this mean that artificial sweeteners cause weight gain? Give a more plausible explanation for this association.

2.142 Exercise and mortality. A sign in a fitness center says, "Mortality is halved for men over 65 who walk at least 2 miles a day."

(a) Mortality is eventually 100% for everyone. What do you think "mortality is halved" means?

(b) Assuming that the claim is true, explain why this fact does not show that exercise *causes* lower mortality.

2.143 Effect of a math skills refresher initiative. Students enrolling in an elementary statistics course take a pretest that assesses their math skills. Those who receive low scores are given the opportunity to take three one-hour refresher sessions designed to review the basic math skills needed for the statistics course. Those who took the refresher sessions performed worse than those who did not on the final exam in the statistics course. Can you conclude that the refresher course has a negative impact on performance in the statistics course? Explain your answer.

CHAPTER 2 EXERCISES

2.144 Dwelling permits and sales for 23 countries.
The Organisation for Economic Co-operation and
Development collects data on main economic indicators
(MEIs) for many countries. Each variable is recorded as
an index with the year 2000 serving as a base year. This
means that the variable for each year is reported as a
ratio of the value for the year divided by the value for
2000. Use of indices in this way makes it easier to
compare values for different countries. Table 2.3 gives the
values of three MEIs for 23 countries.[31] MEIS

(a) Make a scatterplot with sales as the response variable
and permits issued for new dwellings as the explanatory
variable. Describe the relationship. Are there any outliers
or influential observations?

(b) Find the least-squares regression line and add it to
your plot.

(c) Interpret the slope of the line in the context of this
exercise.

(d) Interpret the intercept of the line in the context of this
exercise. Explain whether or not this interpretation is useful
in explaining the relationship between these two variables.

(e) What is the predicted value of sales for a country that
has an index of 224 for dwelling permits?

(f) Canada has an index of 224 for dwelling permits. Find
the residual for this country.

(g) What percent of the variation in sales is explained by
dwelling permits?

2.145 Dwelling permits and production. Refer to the
previous exercise. MEIS

(a) Make a scatterplot with production as the response
variable and permits issued for new dwellings as the
explanatory variable. Describe the relationship. Are there
any outliers or influential observations?

(b) Find the least-squares regression line and add it to
your plot.

(c) Interpret the slope of the line in the context of this
exercise.

(d) Interpret the intercept of the line in the context of this
exercise. Explain whether or not this interpretation is useful
in explaining the relationship between these two variables.

(e) What is the predicted value of production for a
country that has an index of 224 for dwelling permits?

(f) Canada has an index of 224 for dwelling permits. Find
the residual for this country.

(g) What percent of the variation in production is
explained by dwelling permits? How does this value
compare with the value that you found in the previous
exercise for the percent of variation in sales that is
explained by building permits?

2.146 Sales and production. Refer to the previous two
exercises. MEIS

(a) Make a scatterplot with sales as the response variable
and production as the explanatory variable. Describe
the relationship. Are there any outliers or influential
observations?

(b) Find the least-squares regression line and add it to
your plot.

(c) Interpret the slope of the line in the context of this
exercise.

(d) Interpret the intercept of the line in the context of
this exercise. Explain whether or not this interpretation
is useful in explaining the relationship between these two
variables.

(e) What is the predicted value of sales for a country that
has an index of 109 for production?

(f) The Netherlands has an index of 109 for production.
Find the residual for this country.

(g) What percent of the variation in sales is explained
by production? How does this value compare with the
percents of variation that you calculated in the two
previous exercises?

TABLE 2.3	Dwelling Permits, Sales, and Production for 21 Countries		
Country	Dwelling permits	Sales	Production
Australia	116	137	109
Belgium	125	105	112
Canada	224	122	101
Czech Republic	178	134	162
Denmark	121	126	109
Finland	105	136	125
France	145	121	104
Germany	54	100	119
Greece	117	136	102
Hungary	109	140	155
Ireland	92	123	144
Japan	86	99	109
Korea	158	110	156
Luxembourg	145	161	118
Netherlands	160	107	109
New Zealand	127	139	112
Norway	125	136	94
Poland	163	139	159
Portugal	53	112	105
Spain	122	123	108
Sweden	180	142	116

	A	B	C	D
1	ProvinceOrTerritory	Population	Pct15&Under	Pct65&Over
2	Alberta	4121.7	18.3	11.4
3	British Columbia	4631.3	14.6	17.0
4	Manitoba	1282.0	18.7	14.6
5	New Brunswick	753.9	14.6	18.3
6	Newfoundland & Labrador	527.0	14.4	17.7
7	Northwest Territories	43.6	21.4	6.6
8	Nova Scotia	942.7	14.1	18.3
9	Nunavut	36.6	31.1	3.7
10	Ontario	13678.7	16.0	15.6
11	Prince Edward Island	146.3	15.9	17.9
12	Quebec	8214.7	15.4	17.1
13	Saskatchewan	1125.4	18.9	14.5
14	Yukon	36.5	16.6	10.5

FIGURE 2.32 Percent of the population over 65 years and percent of the population under 15 years in the 13 Canadian provinces and territories, Exercise 2.147.

2.147 Population in Canadian provinces and territories. Statistics Canada provides a great deal of demographic data organized in different ways.[32] Figure 2.32 gives the percent of the population aged 65 years and older and the percent aged 15 years and younger for each of the 13 Canadian provinces and territories. Figure 2.33 is a scatterplot of the percent of the population over 65 versus the percent under 15. CANADAP

(a) Write a short paragraph explaining what the plot tells you about these two demographic groups in the 13 Canadian provinces and territories.

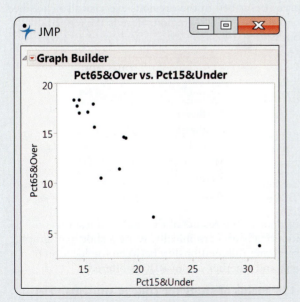

FIGURE 2.33 Scatterplot of percent of the population over 65 years versus percent of the population under 15 years for the 13 Canadian provinces and territories, Exercise 2.147.

(b) Find the correlation between the percent of the population over 65 and the percent under 15. Does the correlation give a good numerical summary of the strength of this relationship? Explain your answer.

2.148 Nunavut. Refer to the previous exercise and Figures 2.32 and 2.33. CANADAP

(a) Do you think that Nunavut is an outlier?

(b) Make a residual plot for these data. Comment on the size of the residual for Nunavut. Use this information to expand on your answer to part (a).

(c) Find the value of the correlation without Nunavut. How does this compare with the value you computed in part (b) of the previous exercise?

(d) Write a short paragraph about Nunavut based on what you have found in this exercise and the previous one.

2.149 Compare the provinces with the territories. Refer to the previous exercise. The three Canadian territories are the Northwest Territories, Nunavut, and the Yukon Territories. All the other entries in Figure 2.32 are provinces. CANADAP

(a) Generate a scatterplot of the Canadian demographic data similar to Figure 2.33 but with the points labeled "P" for provinces and "T" for territories.

(b) Use your new scatterplot to write a new summary of the demographics for the 13 Canadian provinces and territories.

2.150 Records for men and women in the 10K. Table 2.4 shows the progress of world record times (in seconds) for the 10,000-meter run for both men and women.[33] TENK

TABLE 2.4   World Record Times for the 10,000-Meter Run

Men				Women	
Record year	Time (seconds)	Record year	Time (seconds)	Record year	Time (seconds)
1912	1880.8	1963	1695.6	1967	2286.4
1921	1840.2	1965	1659.3	1970	2130.5
1924	1835.4	1972	1658.4	1975	2100.4
1924	1823.2	1973	1650.8	1975	2041.4
1924	1806.2	1977	1650.5	1977	1995.1
1937	1805.6	1978	1642.4	1979	1972.5
1938	1802.0	1984	1633.8	1981	1950.8
1939	1792.6	1989	1628.2	1981	1937.2
1944	1775.4	1993	1627.9	1982	1895.3
1949	1768.2	1993	1618.4	1983	1895.0
1949	1767.2	1994	1612.2	1983	1887.6
1949	1761.2	1995	1603.5	1984	1873.8
1950	1742.6	1996	1598.1	1985	1859.4
1953	1741.6	1997	1591.3	1986	1813.7
1954	1734.2	1997	1587.8	1993	1771.8
1956	1722.8	1998	1582.7		
1956	1710.4	2004	1580.3		
1960	1698.8	2005	1577.5		
1962	1698.2				

(a) Make a scatterplot of world record time against year, using separate symbols for men and women. Describe the pattern for each sex. Then compare the progress of men and women.

(b) Women began running this long distance later than men, so we might expect their improvement to be more rapid. Moreover, it is often said that men have little advantage over women in distance running as opposed to sprints, where muscular strength plays a greater role. Do the data appear to support these claims?

2.151 Remote deposit capture. The Federal Reserve has called remote deposit capture (RDC) "the most important development the [U.S.] banking industry has seen in years." This service allows users to scan checks and to transmit the scanned images to a bank for posting.[34] In its annual survey of community banks, the American Bankers Association asked banks whether or not they offered this service.[35] Here are the results classified by the asset size (in millions of dollars) of the bank: RDCSIZE

	Offer RDC	
Asset size	Yes	No
Under $100	63	309
$101 to $200	59	132
$201 or more	112	85

Summarize the results of this survey question numerically and graphically. Write a short paragraph explaining the relationship between the size of a bank, measured by assets, and whether or not RDC is offered.

2.152 How does RDC vary across the country? The survey described in the previous exercise also classified community banks by region. Here is the 6 × 2 table of counts:[36] RDCREG

	Offer RDC	
Region	Yes	No
Northeast	28	38
Southeast	57	61
Central	53	84
Midwest	63	181
Southwest	27	51
West	61	76

Summarize the results of this survey question numerically and graphically. Write a short paragraph explaining the relationship between the location of a bank and whether or not RDC is offered.

2.153 Fields of study for college students. The following table gives the number of students (in thousands) graduating from college with degrees in several fields of study for seven countries:[37] FOS

Field of study	Canada	France	Germany	Italy	Japan	U.K.	U.S.
Social sciences, business, law	64	153	66	125	250	152	878
Science, mathematics, engineering	35	111	66	80	136	128	355
Arts and humanities	27	74	33	42	123	105	397
Education	20	45	18	16	39	14	167
Other	30	289	35	58	97	76	272

(a) Calculate the marginal totals and add them to the table.

(b) Find the marginal distribution of country and give a graphical display of the distribution.

(c) Do the same for the marginal distribution of field of study.

2.154 Fields of study by country for college students. In the previous exercise you examined data on fields of study for graduating college students from seven countries. **FOS**

(a) Find the seven conditional distributions giving the distribution of graduates in the different fields of study for each country.

(b) Display the conditional distributions graphically.

(c) Write a paragraph summarizing the relationship between field of study and country.

2.155 Graduation rates. One of the factors used to evaluate undergraduate programs is the proportion of incoming students who graduate. This quantity, called the graduation rate, can be predicted by other variables such as the SAT or ACT scores and the high school records of the incoming students. One of the components that *U.S. News & World Report* uses when evaluating colleges is the difference between the actual graduation rate and the rate predicted by a regression equation.[38] In this chapter, we call this quantity the residual. Explain why the residual is a better measure to evaluate college graduation rates than the raw graduation rate.

2.156 Salaries and raises. For this exercise, we consider a hypothetical employee who starts working in Year 1 with a salary of $50,000. Each year her salary increases by approximately 5%. By Year 20, she is earning $126,000. The following table gives her salary for each year (in thousands of dollars): **RAISES**

Year	Salary	Year	Salary	Year	Salary	Year	Salary
1	50	6	63	11	81	16	104
2	53	7	67	12	85	17	109
3	56	8	70	13	90	18	114
4	58	9	74	14	93	19	120
5	61	10	78	15	99	20	126

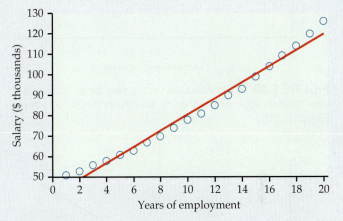

FIGURE 2.34 Plot of salary versus year for an individual who receives approximately a 5% raise each year for 20 years, with the least-squares regression line, Exercise 2.156.

(a) Figure 2.34 is a scatterplot of salary versus year, with the least-squares regression line. Describe the relationship between salary and year for this person.

(b) The value of r^2 for these data is 0.9832. What percent of the variation in salary is explained by year? Would you say that this is an indication of a strong linear relationship? Explain your answer.

2.157 Look at the residuals. Refer to the previous exercise. Figure 2.35 is a plot of the residuals versus year. **RAISES**

(a) Interpret the residual plot.

(b) Explain how this plot highlights the deviations from the least-squares regression line that you can see in Figure 2.34.

2.158 Try logs. Refer to the previous two exercises. Figure 2.36 is a scatterplot with the least-squares regression line for log salary versus year. For this model, $r^2 = 0.9995$. **RAISES**

(a) Compare this plot with Figure 2.34. Write a short summary of the similarities and the differences.

(b) Figure 2.37 is a plot of the residuals for the model using year to predict log salary. Compare this plot with Figure 2.35 and summarize your findings.

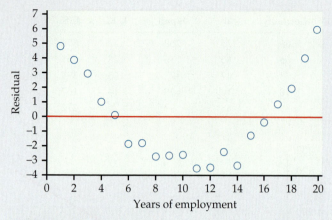

FIGURE 2.35 Plot of residuals versus year for an individual who receives approximately a 5% raise each year for 20 years, Exercise 2.157.

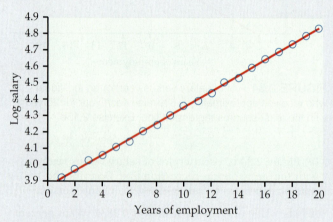

FIGURE 2.36 Plot of log salary versus year for an individual who receives approximately a 5% raise each year for 20 years, with the least-squares regression line, Exercise 2.158.

2.159 Make some predictions. The individual whose salary we have been studying wants to do some financial planning. Specifically, she would like to predict her salary five years into the future, that is, for Year 25. She is willing to assume that her employment situation will be stable for the next five years and that it will be similar to the last 20 years. **RAISES**

(a) Predict her salary for Year 25 using the least-squares regression equation constructed to predict salary from year.

(b) Predict her salary for Year 25 using the least-squares regression equation constructed to predict log salary from year. Note that you will need to take the predicted log salary and convert this value back to the predicted salary. Many calculators have a function that will perform this operation.

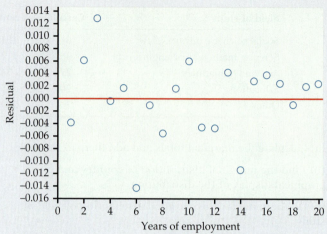

FIGURE 2.37 Plot of residuals, based on log salary, versus year for an individual who receives approximately a 5% raise each year for 20 years, Exercise 2.158.

(c) Which prediction do you prefer? Explain your answer.

(d) Someone looking at the numerical summaries and not the plots for these analyses says that because both models have very high values of r^2, they should perform equally well in doing this prediction. Write a response to this comment.

(e) Discuss the value of graphical summaries and the problems of extrapolation using what you have learned in studying these salary data.

2.160 Faculty salaries. Here are the salaries for a sample of professors in a mathematics department at a large midwestern university for the academic years 2014–2015 and 2015–2016. **FACULTY**

2014–2015 salary ($)	2015–2016 salary ($)	2014–2015 salary ($)	2015–2016 salary ($)
145,700	147,700	136,650	138,650
112,700	114,660	132,160	134,150
109,200	111,400	74,290	76,590
98,800	101,900	74,500	77,000
112,000	113,000	83,000	85,400
111,790	113,800	141,850	143,830
103,500	105,700	122,500	124,510
149,000	150,900	115,100	117,100

(a) Construct a scatterplot with the 2015–2016 salaries on the vertical axis and the 2014–2015 salaries on the horizontal axis.

(b) Comment on the form, direction, and strength of the relationship in your scatterplot.

(c) What proportion of the variation in 2015–2016 salaries is explained by 2014–2015 salaries?

2.161 Find the line and examine the residuals. Refer to the previous exercise. [≡] FACULTY

(a) Find the least-squares regression line for predicting 2015–2016 salaries from 2014–2015 salaries.

(b) Analyze the residuals, paying attention to any outliers or influential observations. Write a summary of your findings.

2.162 Bigger raises for those earning less. Refer to the previous two exercises. The 2014–2015 salaries do an excellent job of predicting the 2015–2016 salaries. Is there anything more that we can learn from these data? In this department, there is a tradition of giving higher-than-average percent raises to those whose salaries are lower. Let's see if we can find evidence to support this idea in the data. [≡] FACULTY

(a) Compute the percent raise for each faculty member. Take the difference between the 2015–2016 salary and the 2014–2015 salary, divide by the 2014–2015 salary, and then multiply by 100. Make a scatterplot with raise as the response variable and the 2014–2015 salary as the explanatory variable. Describe the relationship that you see in your plot.

(b) Find the least-squares regression line and add it to your plot.

(c) Analyze the residuals. Are there any outliers or influential cases? Make a graphical display and include this in a short summary of your conclusions.

(d) Is there evidence in the data to support the idea that greater percent raises are given to those with lower salaries? Include numerical and graphical summaries to support your conclusion.

2.163 Firefighters and fire damage. Someone says, "There is a strong positive correlation between the number of firefighters at a fire and the amount of damage the fire does. So sending lots of firefighters just causes more damage." Explain why this reasoning is wrong.

2.164 Predicting text pages. The editor of a statistics text would like to plan for the next edition. A key variable is the number of pages that will be in the final version. Text files are prepared by the authors using a word processor called LaTeX, and separate files contain figures and tables. For the previous edition of the text, the number of pages in the LaTeX files can easily be determined, as well as the number of pages in the final version of the text. Here are the data: [≡] TEXTP

Chapter	1	2	3	4	5	6	7	8	9	10	11	12	13
LaTeX pages	77	73	59	80	45	66	81	45	47	43	31	46	26
Text pages	99	89	61	82	47	68	87	45	53	50	36	52	19

(a) Plot the data and describe the overall pattern.

(b) Find the equation of the least-squares regression line and add the line to your plot.

(c) Find the predicted number of pages for the next edition if the number of LaTeX pages is 62.

(d) Write a short report for the editor explaining to her how you constructed the regression equation and how she could use it to estimate the number of pages in the next edition of the text.

2.165 Plywood strength. How strong is a building material such as plywood? To be specific, support a 24-inch by 2-inch strip of plywood at both ends and apply force in the middle until the strip breaks. The modulus of rupture (MOR) is the force needed to break the strip. We would like to be able to predict MOR without actually breaking the wood. The modulus of elasticity (MOE) is found by bending the wood without breaking it. Both MOE and MOR are measured in pounds per square inch. Here are data for 32 specimens of the same type of plywood:[39] [≡] MOEMOR

MOE	MOR	MOE	MOR	MOE	MOR	MOE	MOR
2,005,400	11,591	2,181,910	12,702	1,774,850	10,541	1,747,010	11,794
1,166,360	8,542	1,559,700	11,209	1,457,020	10,314	1,791,150	11,413
1,842,180	12,750	2,372,660	12,799	1,959,590	11,983	2,535,170	13,920
2,088,370	14,512	1,580,930	12,062	1,720,930	10,232	1,355,720	9,286
1,615,070	9,244	1,879,900	11,357	1,355,960	8,395	1,646,010	8,814
1,938,440	11,904	1,594,750	8,889	1,411,210	10,654	1,472,310	6,326
2,047,700	11,208	1,558,770	11,565	1,842,630	10,223	1,488,440	9,214
2,037,520	12,004	2,212,310	15,317	1,984,690	13,499	2,349,090	13,645

Can we use MOE to predict MOR accurately? Use the data to write a discussion of this question.

2.166 Distribution of the residuals. Some statistical methods require that the residuals from a regression line have a Normal distribution. The residuals for the nonexercise activity example are given in Exercise 2.93 (page 125). Is their distribution close to Normal? Make a Normal quantile plot to find out. [≡] FIDGET

2.167 An example of Simpson's paradox. Mountain View University has professional schools in business and law. Here is a three-way table of applicants to these professional schools, categorized by sex, school, and admission decision:[40] **ADMITS**

	Business			Law		
		Admit			Admit	
Sex	Yes	No	Sex	Yes	No
Male	400	200	Male	90	110
Female	200	100	Female	200	200

(a) Make a two-way table of sex by admission decision for the combined professional schools by summing entries in the three-way table.

(b) From your two-way table, compute separately the percents of male and female applicants admitted. Male applicants are admitted to Mountain View's professional schools at a higher rate than female applicants.

(c) Now compute separately the percents of male and female applicants admitted by the business school and by the law school.

(d) Explain carefully, as if speaking to a skeptical reporter, how it can happen that Mountain View appears to favor males when this is not true within each of the professional schools.

2.168 Simpson's paradox and regression. Simpson's paradox occurs when a relationship between variables within groups of observations reverses when all of the data are combined. The phenomenon is usually discussed in terms of categorical variables, but it also occurs in other settings. Here is an example: **SIMREG**

y	x	Group	y	x	Group
10.1	1	1	18.3	6	2
8.9	2	1	17.1	7	2
8.0	3	1	16.2	8	2
6.9	4	1	15.1	9	2
6.1	5	1	14.3	10	2

(a) Make a scatterplot of the data for Group 1. Find the least-squares regression line and add it to your plot. Describe the relationship between y and x for Group 1.

(b) Do the same for Group 2.

(c) Make a scatterplot using all 10 observations. Find the least-squares line and add it to your plot.

(d) Make a plot with all of the data using different symbols for the two groups. Include the three regression lines on the plot. Write a paragraph about Simpson's paradox for regression using this graphical display to illustrate your description.

2.169 Class size and class level. A university classifies its classes as either "small" (fewer than 40 students) or "large." A dean sees that 62% of Department A's classes are small, while Department B has only 40% small classes. She wonders if she should cut Department A's budget and insist on larger classes. Department A responds to the dean by pointing out that classes for third- and fourth-year students tend to be smaller than classes for first- and second-year students. The following three-way table gives the counts of classes by department, size, and student audience. Write a short report for the dean that summarizes these data. Start by computing the percents of small classes in the two departments and include other numerical and graphical comparisons as needed. Here are the numbers of classes to be analyzed: **CSIZE**

Year	Department A			Department B		
	Large	Small	Total	Large	Small	Total
First	2	0	2	18	2	20
Second	9	1	10	40	10	50
Third	5	15	20	4	16	20
Fourth	4	16	20	2	14	16

2.170 More smokers live at least 20 more years! You can see the headlines: "More smokers than nonsmokers live at least 20 more years after being contacted for study!" A medical study contacted randomly chosen people in a district in England. Here are data on the 1314 women contacted who were either current smokers or who had never smoked. The tables classify these women by their smoking status and age at the time of the survey and whether they were still alive 20 years later.[41] **SMOKERS**

	Age 18 to 44		Age 45 to 64		Age 65+	
	Smoker	Not	Smoker	Not	Smoker	Not
Dead	19	13	78	52	42	165
Alive	269	327	167	147	7	28

(a) From these data, make a two-way table of smoking (yes or no) by dead or alive. What percent of the smokers stayed alive for 20 years? What percent of the nonsmokers survived? It seems surprising that a higher percent of smokers stayed alive.

(b) The age of the women at the time of the study is a lurking variable. Show that within each of the three age groups in the data, a higher percent of nonsmokers remained alive 20 years later. This is another example of Simpson's paradox.

(c) The study authors give this explanation: "Few of the older women (over 65 at the original survey) were smokers, but many of them had died by the time of follow-up." Compare the percent of smokers in the three age groups to verify the explanation.

2.171 Recycled product quality. Recycling is supposed to save resources. Some people think recycled products are lower in quality than other products, a fact that makes recycling less practical. People who actually use a recycled product may have different opinions from those who don't use it. Here are data on attitudes toward coffee filters made of recycled paper among people who do and don't buy these filters:[42] **RECYCLE**

	Think the quality of the recycled product is:		
	Higher	The same	Lower
Buyers	20	7	9
Nonbuyers	29	25	43

(a) Find the marginal distribution of opinion about quality. Assuming that these people represent all users of coffee filters, what does this distribution tell us?

(b) How do the opinions of buyers and nonbuyers differ? Use conditional distributions as a basis for your answer. Include a mosaic plot if you have access to the needed software. Can you conclude that using recycled filters causes more favorable opinions? If so, giving away samples might increase sales.

2.172 Survival and sex on the *Titanic*. In Exercise 2.122, you examined the relationship between survival and class on the *Titanic*. The data file TITANIC contains data on the sex of the *Titanic* passengers. Examine the relationship between survival and sex and write a short summary of your findings. **TITANIC**

2.173 Survival, class, and sex on the *Titanic*. Refer to the previous exercise and Exercise 2.122 (page 146). When we looked at survival and class, we ignored sex. When we looked at survival and sex, we ignored class. Are we missing something interesting about these data when we choose this approach to the analysis? Here is one way to answer this question. **TITANIC**

(a) Create two separate two-way tables. One for survival and class for the women and another for survival and class for the men.

(b) Perform an analysis of the relationship between survival and class for the women. Summarize your findings.

(c) Perform an analysis of the relationship between survival and class for the men. Summarize your findings.

(d) Compare the analyses that you performed in parts (b) and (c). Write a short report on the relationship between survival and the two explanatory variables, class and sex.

2.174 Blueberries and anthocyanins. Refer to Exercises 1.165 and 1.166 (page 77). Figure 2.38 gives JMP output for examining the relationship between Antho2 and Antho1. Use this output to write a summary of this relationship using the methods and ideas that you learned in this chapter. **BERRIES**

2.175 Averaged date for blueberries and anthocyanins. Refer to the previous exercise where you examined the relationship between Antho2 and Antho1. The variables Antho2M and Antho1M were computed by averaging Antho2 and Antho1 for values of Antho1 in the intervals (0, 0.5), [0.5, 1.0], [1.0, 1.5], [1.5, 2.0], [2.0, 2.5), [2.5, 3.0), and [3.0, 3.5). Analyze the relationship between Antho2M and Antho1M, and compare your results that you found in the previous exercise using Antho2 and Antho1. Summarize what the comparison tells you about relationships with averaged data. **BERRIEM**

2.176 Restricting the range for blueberries and anthocyanins. Refer to Exercise 2.174 where you examined the relationship between Antho2 and Antho1. The data file BERRIER was created from the data file BERRIES by excluding cases with values of Antho1 that are less than 1.5 and cases with values of Antho1 that are greater than 3. Analyze the relationship between Antho2 and Antho1 for this restricted range data set, and compare your results that you found in Exercise 2.174 for the complete data set. Summarize what the comparison tells you about relationships with a restricted range. **BERRIER**

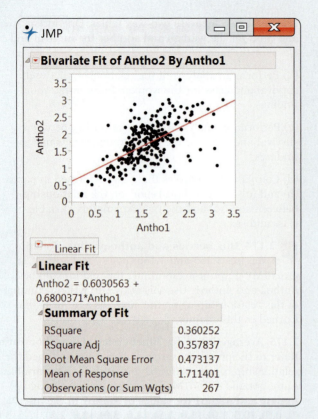

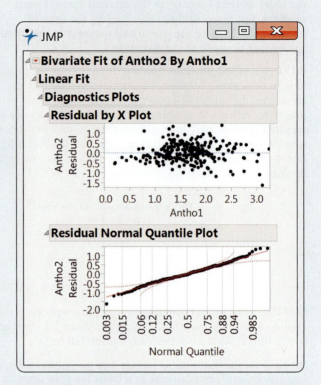

FIGURE 2.38 Selected JMP output for examining the relationship between Antho2 and Antho1, Exercise 2.174.

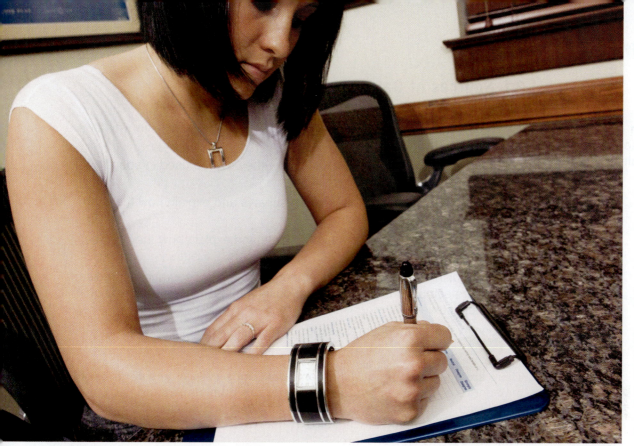
Thinkstock

Producing Data

3.1 Sources of Data

3.2 Design of Experiments

3.3 Sampling Design

3.4 Ethics

Introduction

In Chapters 1 and 2, we learned some basic tools of *data analysis*. We used graphs and numbers to describe data. When we do exploratory data analysis, we rely heavily on plotting the data. We look for patterns that suggest interesting conclusions or questions for further study. However, *exploratory analysis alone can rarely provide convincing evidence for its conclusions because striking patterns that we find in data can arise from many sources.*

The validity of the conclusions that we draw from an analysis of data depends not only on the use of the best methods to perform the analysis, but also on the quality of the data. Therefore, Section 3.1 begins this chapter with a short overview on sources of data.

The two main sources for quality data are designed experiments and sample surveys. We study these two sources in Sections 3.2 and 3.3, respectively.

Should an experiment or sample survey that could possibly provide interesting and important information always be performed? How can we safeguard the privacy of subjects in a sample survey? What constitutes the mistreatment of people or animals who are studied in an experiment? These are questions of **ethics**. In Section 3.4, we address ethical issues related to the design of studies and the analysis of data.

LOOK BACK

exploratory data analysis, p. 8

ethics

3.1 Sources of Data

When you complete this section, you will be able to:	• Identify anecdotal data and, using specific examples, explain why they have limited value.
	• Identify available data and explain how they can be used in specific examples.
	• Identify data collected from sample surveys and explain how they can be used in specific examples.
	• Identify data collected from experiments and explain how they can be used in specific examples.
	• Distinguish data that are from experiments, from observational studies that are sample surveys, and from observational studies that are not sample surveys.
	• Identify the treatment in an experiment.

There are many sources of data. Some data are very easy to collect, but they may not be very useful. Other data require careful planning and need professional staff to gather. These can be much more useful. Whatever the source, a good statistical analysis will start with a careful study of the source of the data. Here is one type of source.

Anecdotal data

It is tempting to simply draw conclusions from our own experience, making no use of more broadly representative data. A magazine article about Pilates says that men need this form of exercise even more than women do. The article describes the benefits that two men received from taking Pilates classes. A newspaper ad states that a particular brand of window is "considered to be the best" and says that "now is the best time to replace your windows and doors." These types of stories, or *anecdotes*, sometimes provide data. However, this type of data does not give us a sound basis for drawing conclusions.

ANECDOTAL DATA

Anecdotal data represent individual cases, which often come to our attention because they are striking in some way. These cases are not necessarily representative of any larger group of cases.

USE YOUR KNOWLEDGE

3.1 **The best instructor?** A friend tells you that the instructor in her statistics class is the best teacher in the college. Can you conclude that this teacher is better than all of the other instructors in the college? Explain your answer.

3.2 **Describe an anecdote.** Find an example from some recent experience where anecdotal evidence was used to draw a conclusion that is not justified. Describe the example and explain why the anecdote should not be used in this way.

3.3 **Opposition to a new requirement.** Your student newspaper ran a story describing interviews with three students who were strongly opposed to a proposed new requirement that all students take a course on ethics. Can you conclude that most students are opposed to this requirement? Explain your answer.

3.4 **Are all vehicles this good?** A friend has driven a Toyota Camry for more than 200,000 miles and with only the usual service maintenance expenses. Explain why not all Camry owners can expect this kind of performance.

Not all anecdotal data are bad. The experiences of an individual or a small group of individuals might suggest an interesting study that could be performed using more carefully collected data.

Available data

Occasionally, data are collected for a particular purpose but can also serve as the basis for drawing sound conclusions about other research questions. We use the term *available data* for this type of data.

AVAILABLE DATA

Available data are data that were produced for some other purpose but that may help answer a question of interest.

The library and the Internet can be good sources of available data. Because producing new data is expensive, we all use available data whenever possible. Here are two examples.

EXAMPLE 3.1

How Americans use their time. If you visit the U.S. Bureau of Labor Statistics website, **bls.gov**, you will find many interesting sets of data and statistical summaries. The American Time Use Survey[1] recently reported that men spend an average of 5.71 hours per day on leisure and sports activities, while women spend an average of 4.93 hours on these activities.

EXAMPLE 3.2

Math skills. At the website of the National Center for Education Statistics, **nces.ed.gov**, you will find full details about the math skills of schoolchildren as determined by the latest National Assessment of Educational Progress (Figure 3.1). Mathematics scores have slowly but steadily increased since 1990. Across all racial/ethnic groups, both boys and girls in most states are getting better in math.

FIGURE 3.1 Websites of government statistical offices are prime sources of data. Here is a page from the website of the National Center for Education Statistics. (Source: U.S. Department of Education Institute of Education Sciences National Center for Education Statistics)

Many nations have a single national statistical office, such as Statistics Canada (**statcan.gc.ca**) and Mexico's INEGI (**www.inegi.org.mx**). More than 70 different U.S. agencies collect data. You can reach most of them through the U.S. government's FedStats site (**fedstats.sites.usa.gov**).

USE YOUR KNOWLEDGE

3.5 What more do you need? A website claims that Millennial generation consumers are very loyal to the brands that they prefer. What additional information do you need to evaluate this claim?

A survey of college athletes is designed to estimate the percent who gamble. Do restaurant patrons give higher tips when their server repeats their order carefully? The validity of our conclusions from the analysis of data collected to address these issues rests on a foundation of carefully collected data.

In this chapter, we will develop the skills needed to produce trustworthy data and to judge the quality of data produced by others. The techniques for producing data that we will study require no formulas, but they are among the most important ideas in statistics. Statistical designs for producing data rely on either *sampling* or *experiments.*

Sample surveys and experiments

sample
surveys

How have the attitudes of Americans, on issues ranging from abortion to work, changed over time? **Sample surveys** are the usual tool for answering questions like these.

EXAMPLE 3.3

The General Social Survey. One of the most important sample surveys is the General Social Survey (GSS) conducted by the National Opinion Research Center (NORC), an organization affiliated with the University of Chicago.[2] The GSS interviews about 3000 adult residents of the United States every other year.

sample
population

The GSS selects a **sample** of adults to represent the larger **population** of all English-speaking adults living in the United States. The idea of *sampling* is to study a part in order to gain information about the whole. Data are often produced by sampling a population of people or things. Opinion polls, for example, may report the views of the entire country based on interviews with a sample of about 1000 people. Government reports on employment and unemployment are produced from a monthly sample of about 60,000 households. The quality of manufactured items is monitored by inspecting small samples each hour or each shift.

USE YOUR KNOWLEDGE

3.6 **Check out the General Social Survey.** Visit the General Social Survey website at **gss.norc.org**. Write a short summary of one of their reports, paying particular attention to the methods used to collect the data.

---

census

In all our examples, the expense of examining every item in the population makes sampling a practical necessity. Timeliness is another reason for preferring a sample to a **census**, which is an attempt to contact every individual in the population. We want information on current unemployment and public opinion next week, not next year. Moreover, a carefully conducted sample is often more accurate than a census. Accountants, for example, sample a firm's inventory to verify the accuracy of the records. Attempting to count every last item in the warehouse would be not only expensive, but also inaccurate. Bored people do not count carefully.

If conclusions based on a sample are to be valid for the population, a sound design for selecting the sample is required. Sampling designs are the topic of Section 3.3.

A sample survey collects information about a population by selecting and measuring a sample from the population. The goal is a picture of the population, disturbed as little as possible by the act of gathering information. Sample surveys are one kind of *observational study.*

OBSERVATION VERSUS EXPERIMENT

In an **observational study**, we observe individuals and measure variables of interest but do not attempt to influence the responses.

In an **experiment**, we deliberately impose some condition on individuals and we observe their responses.

EXAMPLE 3.4

Baseball players have strong bones in their throwing arms. A study of young baseball players measured the strength of the bones in their throwing arms. A control group of subjects who were matched with the baseball players based on age were also measured. This is an example of an observational study that is not a sample survey. The study reported that bone strength was 30% higher in the baseball players.[3]

What can we conclude from this study? If you start to play baseball, will you have stronger bones in your throwing arm?

EXAMPLE 3.5

Is there a cause-and-effect relationship? Example 3.4 describes an observational study. People choose to participate in baseball or not. Is it possible that those who choose to play baseball have stronger arms than those who do not? The study does not address this question.

We can imagine an experiment that would remove these difficulties. From a large group of subjects, require some to play baseball and forbid the rest from playing. This is an experiment because the condition (playing baseball or not) is imposed on the subjects. Of course, this particular experiment is neither practical nor ethical.

EXAMPLE 3.6

Baseball and bones. Example 3.4 compared the arm bone strengths of baseball players with those of age-matched controls. Although the study tells us something about baseball players, the results are particularly interesting because they suggest that certain kinds of exercise can help us to build strong bones.

USE YOUR KNOWLEDGE

3.7 **Available data.** Can available data be from an observational study? Can available data be from an experiment? Explain your answers.

3.8 **Picky eaters.** A study of 2049 children in grades 4 to 6 in 33 schools recorded their behaviors in the lunchroom. One of the conclusions of the study was that girls discarded more food than boys.[4] Is this an observational study or an experiment? Is it a sample survey? If it is an experiment, what is the condition? Explain your answers.

3.9 Automatic soap dispensers. A study compared several brands of automatic soap dispensers. For one test, the dispensers were run until their AA batteries failed. The times to failure were compared for the different brands.[5] Is this an observational study or an experiment? Is it a sample survey? If it is an experiment, what is the condition? Explain your answers.

An observational study, even one based on a carefully chosen sample, is a poor way to determine what will happen if we change something. The best way to see the effects of a change is to do an **intervention**—where we actually impose the change. When our goal is to understand cause and effect, experiments are the only source of fully convincing data.

Confounding occurs when an explanatory variable is related to one or more other variables that have an influence on the response variable. When this happens, we sometimes attribute a relationship to an explanatory when the effect is fully or partly due to the confounding variables.

In Example 3.4, the effect of baseball playing on arm bone strength is confounded with (mixed up with) other characteristics of the subjects in the study. Observational studies that examine the effect of a single variable on an outcome can be misleading when the effects of the explanatory variable are confounded with those of other variables.

Because experiments allow us to isolate the effects of specific variables, we generally prefer them. Here is an example.

intervention

LOOK BACK

confounding,
p. 150

explanatory
variable,
p. 82

EXAMPLE 3.7

Which web page design sells more? A company that sells products on the Internet wants to decide which of two possible web page designs to use. During a two-week period, they will use both designs and collect data on sales. They randomly select one of the designs to be used on the first day and then alternate the two designs on each of the following days. At the end of this period, they compare the sales for the two designs.

Experiments usually require some sort of randomization, as in this example. We begin the discussion of statistical designs for data collection in Section 3.2 with the principles underlying the design of experiments.

USE YOUR KNOWLEDGE

3.10 Software for teaching creative writing. An educational software company wants to compare the effectiveness of its computer animation for teaching creative writing with that of a textbook presentation. The company tests the creative-writing skills of a number of second-year college students and then randomly divides them into two groups. One group uses the animation, and the other studies the text. The company retests all the students and compares the increase in creative-writing skills in the two groups. Is this an experiment? Why or why not? What are the explanatory and response variables?

LOOK BACK

response
variable,
p. 82

3.11 Apples or apple juice? Food rheologists study different forms of foods and how the form of a food affects how full we feel when we eat it. One study prepared samples of apple juice and samples of apples with the same number of calories. Half of the subjects were fed apples on one day followed by apple juice on a later day; the other half received the apple juice followed by the apples. After eating, the subjects were asked about how full they felt. Is this an experiment? Why or why not? What are the explanatory and response variables?

SECTION 3.1 SUMMARY

- **Anecdotal data** come from stories or reports about cases that do not necessarily represent a larger group of cases.

- **Available data** are data that were produced for some other purpose but that may help answer a question of interest.

- A **sample survey** collects data from a **sample** of cases that represent some larger **population** of cases.

- A **census** collects data from all cases in the population of interest.

- In an **experiment**, a condition or intervention is imposed and the responses are recorded.

- **Confounding** occurs when the effects of two or more variables are related in such a way that we need to take care in assigning the effect on the response variable to one or to the other.

SECTION 3.1 EXERCISES

For Exercises 3.1 to 3.4, see pages 164–165; for Exercise 3.5, see page 166; for Exercise 3.6, see page 167; for Exercises 3.7 to 3.9, see pages 168–169; and for Exercises 3.10 and 3.11, see pages 169–170.

In several of the following exercises, you are asked to identify the type of data that is described. Possible answers include anecdotal data, available data, observational data that are from sample surveys, observational data that are not from sample surveys, and experimental data. It is possible for some data to be classified in more than one category.

3.12 Not enough tuna. You like to eat tuna fish sandwiches. Recently you have noticed that there does not seem to be as much tuna as you expect when you open the can. Identify the type of data that this represents, and describe how it can or cannot be used to reach a conclusion about the amount of tuna in cans of tuna fish. Is this an experiment? If yes, what is the treatment?

3.13 More about tuna. According to a story in *Consumer Reports*, three major producers of canned tuna agreed to pay $3,300,000 to settle claims in California that the amount of tuna in their cans was less than the amount printed on the label of the cans.[6] What kind of data do you think was used in this situation to convince the producers to pay this amount of money to settle the claims? Explain your answer fully.

3.14 What's wrong? Explain what is wrong in each of the following statements.

(a) Available data is always anecdotal.

(b) A census collects information on a subset of the population of interest.

(c) A sample survey usually involves a treatment.

3.15 Satisfaction with allocation of concert tickets. Your college sponsored a concert that sold out.

(a) After the concert, an article in the student newspaper reported interviews with three students who were unable to get tickets and were very upset. What kind of data does this represent? Explain your answer.

(b) A week later the student organization that sponsored the concert set up a website where students could rank their satisfaction with the way that the tickets were allocated using a 5-point scale with values "very satisfied," "satisfied," "neither satisfied nor unsatisfied," "dissatisfied," and "very dissatisfied." The website was open to any students who chose to provide their opinion. How would you classify these data? Give reasons for your answer.

(c) Suppose that the website in part (b) was changed so that only a sample of students from the college were invited by a text message to respond, and those who did not respond within three days were sent an additional text message reminding them to respond. How would your answer to part (b) change, if at all?

(d) Is the description in part (c) an experiment? If yes, what is the treatment?

(e) Write a short summary contrasting different types of data using your answers to parts (a), (b), (c), and (d) of this exercise.

3.16 Does echinacea reduce the severity of the common cold? In a study designed to evaluate the benefits of taking echinacea when you have a cold,

719 patients were randomly divided into four groups. The groups were (1) no pills, (2) pills that had no echinacea, (3) pills that had echinacea but the subjects did not know whether or not the pills contained echinacea, and (4) pills that had echinacea and the bottle containing the pills stated that the contents included echinacea. The outcome was a measure of the severity of the cold.[7]

(a) Identify the type of data collected in this study. Give reasons for your answer.

(b) Is this an experiment? If yes, what is the treatment?

3.2 Design of Experiments

When you complete this section, you will be able to:

- Identify experimental units, subjects, treatments, and outcomes for an experiment.
- Identify a comparative experiment.
- Describe a placebo effect in an experiment.
- Identify bias in an experiment.
- Explain the need for a control group in an experiment.
- Explain the need for randomization in an experiment.
- When evaluating an experiment, apply the basic principles of experimental design: compare, randomize, and repeat.
- Use a table of random digits to randomly assign experimental units to treatments in an experiment.
- Use software to randomly assign experimental units to treatments in an experiment.
- Identify a matched pairs design.
- Identify a block design.

An experiment is a study in which we actually do something to people, animals, or objects in order to observe the response. Here is the basic vocabulary of experiments.

EXPERIMENTAL UNITS, SUBJECTS, TREATMENTS, AND OUTCOMES

The individuals on which the experiment is done are the **experimental units**. When the units are human beings, they are called **subjects**. Experimental conditions applied to the units are called **treatments**. The **outcomes** are the measured variables that are used to compare the treatments.

Because the purpose of an experiment is to reveal the response of one variable to changes in one or more other variables, the distinction between explanatory and response variables is important. The explanatory variables

factors

level of a factor

in an experiment are often called **factors**. Many experiments study the joint effects of several factors. In such an experiment, each treatment is formed by combining a specific value (often called a **level**) of each of the factors.

EXAMPLE 3.8

Are smaller class sizes better? Do smaller classes in elementary school really benefit students in areas such as scores on standard tests, staying in school, and going on to college? We might do an observational study that compares students who happened to be in smaller classes with those who happened to be in larger classes in their early school years. Small classes are expensive, so they are more common in schools that serve richer communities. Students in small classes tend to also have other advantages: their schools have more resources, their parents are better educated, and so on. Confounding makes it impossible to isolate the effects of small classes.

The Tennessee STAR program was an experiment on the effects of class size. It has been called "one of the most important educational investigations ever carried out." The *subjects* were 6385 students who were beginning kindergarten. Each student was assigned to one of three *treatments*: regular class (22 to 25 students) with one teacher, regular class (22 to 25 students) with a teacher and a full-time teacher's aide, and small class (13 to 17 students). These treatments are levels of a single *factor*, the type of class. The students stayed in the same type of class for four years, then all returned to regular classes. In later years, students from the small classes had higher scores on the *outcomes*, standard tests. The benefits of small classes were greatest for minority students.[8]

LOOK BACK

lurking variables, p. 130

Example 3.8 illustrates the big advantage of experiments over observational studies. **In principle, experiments can give good evidence for causation.** In an experiment, we study the specific factors we are interested in while controlling the effects of lurking variables. All the students in the Tennessee STAR program followed the usual curriculum at their schools. Because students were assigned to different class types within their schools, school resources and family backgrounds were not confounded with class type. The only systematic difference was the type of class. When students from the small classes did better than those in the other two types, we can be confident that class size made the difference.

EXAMPLE 3.9

Repeated exposure to advertising. What are the effects of repeated exposure to an advertising message? The answer may depend both on the length of the ad and on how often it is repeated. An experiment investigated this question using undergraduate students as *subjects*. All subjects viewed a 40-minute television program that included ads for a digital camera. Some subjects saw a 30-second commercial; others, a 90-second version. The same commercial was shown either one, three, or five times during the program.

This experiment has two *factors*: length of the commercial, with two levels, and repetitions, with three levels. The six combinations of one level of each factor form six *treatments*. Figure 3.2 shows the layout of the treatments. After viewing the TV program, all the subjects answered questions about their recall of the ad, their attitude toward the camera, and their intention to purchase it. These are the *outcomes*.

		Factor B Repetitions		
		1 time	3 times	5 times
Factor A Length	30 seconds	1	2	3
	90 seconds	4	5	6

FIGURE 3.2 The treatments in the study of advertising, Example 3.9. Combining the levels of the two factors forms six treatments.

Example 3.9 shows how experiments allow us to study the combined effects of more than one factor. The interaction of several factors can produce effects that cannot be predicted from looking at the effects of each factor alone. Perhaps longer commercials increase interest in a product, and more commercials also increase interest, but if we both make a commercial longer and show it more often, viewers get annoyed and their interest in the product drops. The two-factor experiment in Example 3.9 will help us find out.

USE YOUR KNOWLEDGE

3.17 Calcium and bones. Calcium is important for the growth of bone for children. In a study designed to understand how calcium is processed by the body, 40 young girls attended a summer camp where they were fed a controlled diet. The camp ran for two 3-week periods. For one period, the diet included a low amount of calcium. For the other period, there was a high amount of calcium in the diet. The researchers recorded the amount of calcium retained in the body for each girl. Explain why this study is an experiment and identify the experimental units, the treatments, and the response variable. Describe the factor and its levels.

3.18 Does echinacea reduce the severity of the common cold? In a study designed to evaluate the benefits of taking echinacea when you have a cold, 719 patients were randomly divided into four groups. The groups were (1) no pills, (2) pills that had no echinacea, (3) pills that had echinacea but the subjects did not know whether or not the pills contained echinacea, and (4) pills that had echinacea and the bottle containing the pills stated that the contents included echinacea. The outcome was a measure of the severity of the cold.[9] Identify the experimental units, the treatments, and the outcome. Describe the factor and its levels. The study subjects were aged 12 to 80 years. To what extent do you think the results of this experiment can be generalized to young children?

Comparative experiments

Laboratory experiments in science and engineering often have a simple design with only a single treatment, which is applied to all experimental units. The design of such an experiment can be outlined as

Treatment ⟶ Observe response

For example, we may subject a beam to a load (treatment) and measure its deflection (observation). We rely on the controlled environment of the laboratory to protect us from lurking variables. When experiments are conducted outside the laboratory or with living subjects, such simple designs often yield invalid data. That is, we cannot tell whether the response was due to the treatment or to lurking variables.

EXAMPLE 3.10

Will writing about it reduce test anxiety? A study designed to reduce test anxiety had students write an essay about their feelings concerning an upcoming exam.[10] The scores on this exam, the second of the semester, were compared with those on the first exam in the course. The mean scores on the second exam were higher than the mean scores on the first exam.

$$\text{Write about feelings} \longrightarrow \text{Observe exam scores}$$

The test anxiety experiment of Example 3.10 was poorly designed to evaluate the effect of the writing exercise. Perhaps exam scores would have increased on the second exam because the students became more familiar with the exam style of this particular instructor even without the writing exercise. Another possible explanation is that people typically respond to the personal attention that the students received by the person who explained how to write about their feelings regarding the exam.

placebo effect   In medical settings, this phenomenon is called the **placebo effect**. In medicine, a placebo is a dummy treatment, such as a sugar pill. People respond favorably to personal attention or to any treatment that they hope will help them. On the other hand, the writing exercise may have been very effective in improving exam scores. For this experiment, we don't know whether the change was due to writing the essay, to the personal contacts with the study personnel, or to greater familiarity with the way the instructor designed exams.

The test anxiety experiment gave inconclusive results because the effect of writing the essay was confounded with other factors that could have had an effect on exam scores. The best way to avoid confounding is to do a *comparative experiment*   **comparative experiment**. Think about a study in which some students performed the writing exercise and others did not. A comparison of the exam scores of these two groups of students would provide an evaluation of the effect of the writing exercise.

control group   In medical settings, it is standard practice to randomly assign patients *treatment group*   either to a **control group** or a **treatment group**. All patients are treated the same in every way except that the treatment group receives the product that is being evaluated.

Uncontrolled experiments (that is, experiments that don't include a control group) in medicine and the behavioral sciences can be dominated by such influences as the details of the experimental arrangement, the selection of subjects, and the placebo effect. The result is often bias.

BIAS

The design of a study is **biased** if it systematically favors certain outcomes.

An uncontrolled study of a new medical therapy, for example, is biased in favor of finding the treatment effective because of the placebo effect. Uncontrolled studies in medicine give new therapies a much higher success rate than proper comparative experiments do. Well-designed experiments usually compare several treatments.

USE YOUR KNOWLEDGE

3.19 Does aspirin cure headaches? A study enrolled 100 college students who had frequent headaches to participate in a study to examine the effects of aspirin on their headaches. The students were instructed to take aspirin when they had a headache and to report whether there was a substantial relief from the headache pain within an hour.

(a) Explain why this study is biased.

(b) How would you change the study to remove the bias? Explain your answer.

3.20 Are the teacher evaluations biased? The evaluations of two instructors by their students are compared when it is time to decide raises for the coming year. One teacher always hands out the evaluation forms in class when the grades on the first exam are given to the students. The other instructor always hands out the evaluation forms at the end of a class in which a very interesting film clip is shown. Discuss the possibility of bias in this context.

Randomization

experimental design

The **design of an experiment** first describes the response variable or variables, the factors (explanatory variables), and the treatments, with comparison as the leading principle. Figure 3.2 (page 173) illustrates this aspect of the design of a study of response to advertising. The second aspect of experimental design is how the experimental units are assigned to the treatments. Comparison of the effects of several treatments is valid only when all treatments are applied to similar groups of experimental units. If one corn variety is planted on more fertile ground or if one cancer drug is given to more seriously ill patients, comparisons among treatments are meaningless. If groups assigned to treatments are quite different in a comparative experiment, we should be concerned that our experiment will be biased. How can we assign experimental units to treatments in a way that is fair to all treatments?

Experimenters often attempt to match groups by elaborate balancing acts. Medical researchers, for example, try to match the patients in a "new drug" experimental group and a "standard drug" control group by age, sex, physical matching   condition, smoker or not, and so on. **Matching** is helpful but not adequate—there are too many lurking variables that might affect the outcome. The experimenter is unable to measure some of these variables and will not think of others until after the experiment.

Some important variables, such as how advanced a cancer patient's disease is, are so subjective that they can't be measured. In other cases, an experimenter might unconsciously bias a study by assigning those patients who seemed the sickest to a promising new treatment in the (unconscious) hope that it would help them.

The statistician's remedy is to rely on chance to make an assignment that does not depend on any characteristic of the experimental units and that does not rely on the judgment of the experimenter in any way. The use of chance can be combined with matching, but the simplest experimental design creates groups by chance alone. Here is an example.

EXAMPLE 3.11

Which smartphone should be marketed? Two teams have each prepared a prototype for a new smartphone. Before deciding which one will be marketed, the smartphones will be evaluated by college students. Forty students will receive a new phone. They will use it for two weeks and then answer some questions about how well they like the phone. The 40 students will be randomized, with 20 receiving each phone.

This experiment has a single factor (prototype) with two levels. The researchers must divide the 40 student subjects into two groups of 20. To do this in a completely unbiased fashion, put the names of the 40 students in a hat, mix them up, and draw 20. These students will receive Phone 1, and the remaining 20 will receive Phone 2. Figure 3.3 outlines the design of this experiment.

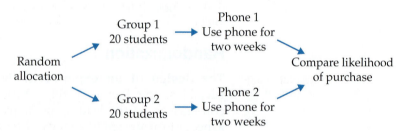

FIGURE 3.3 Outline of a randomized comparative experiment, Example 3.11.

The use of chance to divide experimental units into groups is called **randomization**. The design in Figure 3.3 combines comparison and randomization to arrive at the simplest randomized comparative design. This "flowchart" outline presents all the essentials: randomization, the sizes of the groups and which treatment they receive, and the response variable. There are, as we will see later, statistical reasons for using treatment groups that are about equal in size.

randomization

USE YOUR KNOWLEDGE

3.21 Diagram the echinacea experiment. Refer to Exercise 3.16 (page 171). Draw a diagram similar to Figure 3.3 that describes the experiment.

3.22 Diagram the aspirin experiment. Draw a diagram similar to Figure 3.3 that describes the experiment you suggested in part (b) of Exercise 3.19 (page 175).

Randomized comparative experiments

The logic behind the randomized comparative design in Figure 3.3 is as follows:

• Randomization produces two groups of subjects that we expect to be similar in all respects before the treatments are applied.

• Comparative design helps ensure that influences other than the characteristics of the smartphone operate equally on both groups.

• Therefore, differences in the satisfaction with the smartphone must be due either to the characteristics of the phone or to the chance assignment of subjects to the two groups.

That "either-or" deserves more comment. We cannot say that *all* the difference in the satisfaction with the two smartphones is caused by the characteristics of the phones. There would be some difference even if both groups used the same phone. Some students would be more likely to be highly favorable of any new phone. Chance can assign more of these students to one of the phones so that there is a chance difference between the groups. We would not trust an experiment with just one subject in each group, for example. The results would depend too much on which phone got lucky and received the subject who was more likely to be highly satisfied. If we assign many students to each group, however, the effects of chance will average out. There will be little difference in the satisfaction between the two groups unless the phone characteristics causes a difference. "Use enough subjects to reduce chance variation" is the third big idea of statistical design of experiments.

PRINCIPLES OF EXPERIMENTAL DESIGN

The basic principles of statistical design of experiments are

1. **Compare** two or more treatments. This will control the effects of lurking variables on the response.

2. **Randomize**—use chance to assign experimental units to treatments.

3. **Repeat** each treatment on many units to reduce chance variation in the results.

How to randomize

The idea of randomization is to assign subjects to treatments by drawing names from a hat. In practice, experimenters use software to carry out randomization. For example, most statistical software can choose five out of a list of 10 at random. The list might contain the names of 10 human subjects to be randomly assigned to two groups. The five chosen form one group, and the five that remain form the second group. The *Simple Random Sample* applet on the text website makes it particularly easy to choose treatment groups at random.

label

When we randomize, we first give a **label** to each in the collection of items to be randomized. The label could be the name of a subject in a clinical study or simply a numerical identification number. We then perform the randomization using software or a table of random numbers. To illustrate these methods, let's randomize 10 subjects for a study that will compare a treatment with a placebo control. We will randomly select the five subjects for the treatment group, and the remaining subjects will receive the placebo. We start by labeling the subjects with the numbers 1 through 10.

Randomization using software

LOOK BACK

uniform distribution, p. 71

Here is an example of one way to do the randomization using Excel. We start with a spreadsheet that has 10 rows corresponding to 10 subjects to be randomized to treatment or placebo.

The basic idea is that we generate a uniform random variable for each subject. In Excel, we use the RAND() function for this step. Then we sort the spreadsheet by the column with the uniform numbers and select the first five labels to be the treatment group and the remaining labels to be the placebo controls.

This process is essentially the same as writing the labels on a deck of 10 cards. We then shuffle the cards and deal five cards to form the treatment group.

EXAMPLE 3.12

Using software for the randomization. First create a data set with the numbers 1 to 10 in the first column. See Figure 3.4(a). Then we use RAND() to generate 10 random numbers in the second column. See Figure 3.4(b). Finally, we sort the data set based on the numbers in the second column. See Figure 3.4(c). The first five labels (7, 3, 4, 2, and 9) will receive the treatment. The remaining five labels (10, 6, 8, 1, and 5) will receive the placebo control.

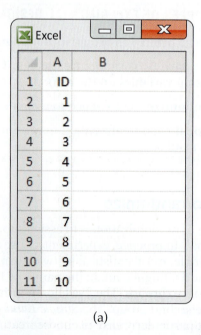

(a) (b)

FIGURE 3.4 Randomization of 10 experimental units using an Excel spreadsheet, Example 3.12: (a) labels; (b) random numbers; (c) sorted list of labels; (d) labels with group assignments.

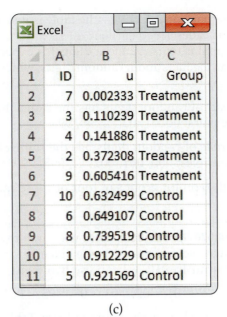

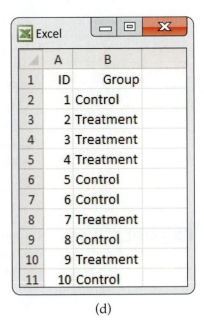

(c) (d)

FIGURE 3.4 *Continued*

If you want to save the uniform numbers that you generated in your file, you should copy them to another column using the "paste values" option before you perform the sort. Note that we have added a column called Group to the spreadsheet, which gives the group to which each subject is assigned. With this variable included, we can now sort the file on ID and delete the column with the random numbers. The result is shown in Figure 3.4(d). The spreadsheet in this form can now be used as a template for entering data.

Randomization using random digits

You can randomize without software by using a *table of random digits*. Thinking about random digits helps you to understand randomization even if you will use software in practice. Table B at the back of the book is a table of random digits.

RANDOM DIGITS

A **table of random digits** is a list of the digits 0, 1, 2, 3, 4, 5, 6, 7, 8, 9 that has the following properties:

1. The digit in any position in the list has the same chance of being any one of 0, 1, 2, 3, 4, 5, 6, 7, 8, 9.

2. The digits in different positions are independent in the sense that the value of one has no influence on the value of any other.

You can think of Table B as the result of asking an assistant (or a computer) to mix the digits 0 to 9 in a hat, draw one, then replace the digit drawn, mix again, draw a second digit, and so on. The assistant's mixing and drawing save us the work of mixing and drawing when we need to randomize. Table B begins with the digits 19223950340575628713. To make the table easier to read, the digits appear in groups of five and in numbered rows. The groups and rows have no meaning—the table is just a long list of digits having Properties 1 and 2 described earlier.

Our goal is to use random digits for experimental randomization. We need the following facts about random digits, which are consequences of Properties 1 and 2:

• Any *pair* of random digits has the same chance of being any of the 100 possible pairs: 00, 01, 02,…, 98, 99.

• Any *triple* of random digits has the same chance of being any of the 1000 possible triples: 000, 001, 002,…, 998, 999.

• …and so on for groups of four or more random digits.

EXAMPLE 3.13

Randomize the subjects. Let's use random digits to perform the randomization that we performed using Excel in Example 3.12. Because the labels range from 1 to 10, we can use two digits for our labels

01, 02, 03, 04, 05, 06, 07, 08, 09, 10

when we select random digits from Table B. We could also have changed our labels to 0 through 9 and then we would only need to use single digits from Table B.

Start anywhere in Table B and read two-digit groups. Suppose we begin at line 175, which is

80011   09937   57195   33906   94831   10056   42211   65491

The first 10 two-digit groups in this line are

80   01   10   99   37   57   19   53   39   06

Each of these two-digit groups is a label. The labels 00 and 11 to 99 are not used in this example, so we ignore them. The first 10 labels between 01 and 10 that we encounter in the table choose subjects who will receive the treatment. Of the first 10 labels in line 175, we ignore seven because they are too high (over 10). The others are 01, 10, and 06. Continue across line 175 and 176 and verify that the next two subjects selected correspond to labels 03 and 04. Our randomization has selected subjects 1, 3, 4, 6, and 10 to receive the treatment. The remaining subjects, 2, 5, 7, 8, and 9 will receive the placebo control.

---

completely randomized design

When all experimental units are allocated at random among all treatments, as in Examples 3.12 and 3.13, the experimental design is **completely randomized**. Completely randomized designs can compare any number of treatments. The treatments can be formed by levels of a single factor or by more than one factor.

EXAMPLE 3.14

Randomization for the TV commercial experiment. Figure 3.2 (page 173) displays six treatments formed by the two factors in an experiment on response to a TV commercial. Suppose that we have 150 students who are willing to serve as subjects. We must assign 25 students at random to each group. Figure 3.5 outlines the completely randomized design.

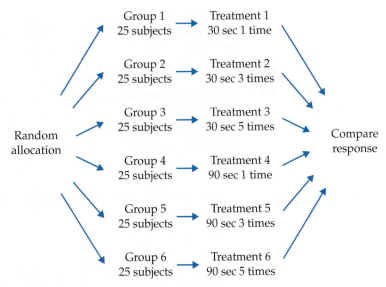

FIGURE 3.5 Outline of a completely randomized design comparing six treatments, Example 3.14.

To carry out the random assignment, label the 150 students 001 to 150. (Three digits are needed to label 150 subjects.) Using Excel, we would generate a uniform random variable for each label and sort the file as we did in Example 3.12. The first 25 students in this sorted file will receive Treatment 1, the next 25 will receive Treatment 2, etc.

Using random digits, we could enter Table B and read three-digit groups until you have selected 25 students to receive Treatment 1 (a 30-second ad shown once). If you start at line 140, the first few labels for Treatment 1 subjects are 129, 048, and 003.

Continue in Table B to select 25 more students to receive Treatment 2 (a 30-second ad shown three times). Then select another 25 for Treatment 3 and so on until you have assigned 125 of the 150 students to Treatments 1 through 5. The 25 students who remain get Treatment 6.

The randomization is straightforward but very tedious to do by using random digits. We strongly recommend that you use software, such as Excel or the *Simple Random Sample* applet. Exercise 3.37 (page 187) shows how to use the applet to do the randomization for this example.

USE YOUR KNOWLEDGE

3.23 Do the randomization. Use computer software to carry out the randomization in Example 3.14. Show your work by including the random uniform numbers in your final spreadsheet.

Cautions about experimentation

The logic of a randomized comparative experiment depends on our ability to treat all the experimental units identically in every way except for the actual treatments being compared. Good experiments, therefore, require careful attention

double-blind   to details. The ideal situation is where a study is **double-blind**—neither the subjects themselves nor the experimenters know which treatment any subject has received. The double-blind method avoids unconscious bias by, for example, a doctor who doesn't think that "just a placebo" can benefit a patient.

Many—perhaps most—experiments have some weaknesses in detail. The environment of an experiment can influence the outcomes in unexpected ways. Although experiments are the gold standard for evidence of cause and effect, really convincing evidence usually requires that a number of studies in different places with different details produce similar results. Here are some brief examples of what can go wrong.

EXAMPLE 3.15

Placebo for a marijuana experiment. A study of the effects of marijuana recruited young men who used marijuana. Some were randomly assigned to smoke marijuana cigarettes, while others were given placebo cigarettes. This failed: the control group recognized that their cigarettes were phony and complained loudly. It may be quite common for blindness to fail because the subjects can tell which treatment they are receiving.[11]

lack of realism

The most serious potential weakness of experiments is **lack of realism**. The subjects or treatments or setting of an experiment may not realistically duplicate the conditions we really want to study. Here is an example.

EXAMPLE 3.16

Layoffs and feeling bad. How do layoffs at a workplace affect the workers who remain on the job? To try to answer this question, psychologists asked student subjects to proofread text for extra course credit, then "let go" some of the workers (who were actually accomplices of the experimenters). Some subjects were told that those let go had performed poorly (Treatment 1). Others were told that not all could be kept and that it was just luck that they were kept and others let go (Treatment 2). We can't be sure that the reactions of the students are the same as those of workers who survive a layoff in which other workers lose their jobs. Many behavioral science experiments use student subjects in a campus setting. Do the conclusions apply to the real world?

Lack of realism can limit our ability to apply the conclusions of an experiment to the settings of greatest interest. Most experimenters want to generalize their conclusions to some setting wider than that of the actual experiment. *Statistical analysis of an experiment cannot tell us how far the results will generalize to other settings.* Nonetheless, the randomized comparative experiment, because of its ability to give convincing evidence for causation, is one of the most important ideas in statistics.

Matched pairs designs

Completely randomized designs are the simplest statistical designs for experiments. They illustrate clearly the principles of control, randomization, and repetition. However, completely randomized designs are often inferior to more elaborate statistical designs. In particular, matching the subjects in various ways can produce more precise results than simple randomization.

matched pairs design

The simplest use of matching is a **matched pairs design**, which compares just two treatments. The subjects are matched in pairs. For example, an experiment to compare two advertisements for the same product might use pairs of subjects with the same age, sex, and income. The idea is that matched subjects are more similar than unmatched subjects so that comparing responses within a number of pairs is more efficient than comparing the responses of groups of randomly assigned subjects. Randomization remains important: which one of a matched pair sees the first ad is decided at random. One common variation of the matched pairs design imposes both treatments on the same subjects so that each subject serves as his or her own control. Here is an example.

EXAMPLE 3.17

Matched pairs for the smartphone prototype experiment. Example 3.11 describes an experiment to compare two prototypes of a new smartphone. The experiment compared two treatments: Phone 1 and Phone 2. The response variable is the satisfaction of the college student participant with the new smartphone. In Example 3.11, 40 student subjects were assigned at random, 20 students to each phone. This is a completely randomized design, outlined in Figure 3.3. Subjects differ in how satisfied they are with smartphones in general. The completely randomized design relies on chance to create two similar groups of subjects.

If we wanted to do a matched pairs version of this experiment, we would have each college student use each phone for two weeks. An effective design would randomize the *order* in which the phones are evaluated by each student. This will eliminate bias due to the possibility that the first phone evaluated will be systematically evaluated higher or lower than the second phone evaluated.

cross-over

The completely randomized design uses chance to decide which subjects will evaluate each smartphone prototype. The matched pairs design uses chance to decide which 20 subjects will evaluate Phone 1 first. The other 20 will evaluate Phone 2 first. This experiment is called a **cross-over** experiment. Situations where there are more than two treatments and all subjects receive all treatments can also be performed in this way.

Block designs

The matched pairs design of Example 3.17 uses the principles of comparison of treatments, randomization, and repetition on several experimental units. However, the randomization is not complete (all subjects randomly assigned to treatment groups) but is restricted to assigning the order of the treatments for each subject. *Block designs* extend the use of "similar subjects" from pairs to larger groups.

BLOCK DESIGN

A **block** is a group of experimental units or subjects that are known before the experiment to be similar in some way that is expected to affect the response to the treatments. In a **block design**, the random assignment of units to treatments is carried out separately within each block.

Block designs can have blocks of any size. A block design combines the idea of creating equivalent treatment groups by matching with the principle of forming treatment groups at random. Blocks are another form of *control*. They control the effects of some outside variables by bringing those variables into the experiment to form the blocks. Here are some typical examples of block designs.

EXAMPLE 3.18

Blocking in a cancer experiment. The progress of a type of cancer differs in women and men. A clinical experiment to compare three therapies for this cancer then treats sex as a blocking variable. Two separate randomizations are done, one assigning the female subjects to the treatments and the other assigning the male subjects. Figure 3.6 outlines the design of this experiment. Note that there is no randomization involved in making up the blocks. They are groups of subjects who differ in some way (sex in this case) that is apparent before the experiment begins.

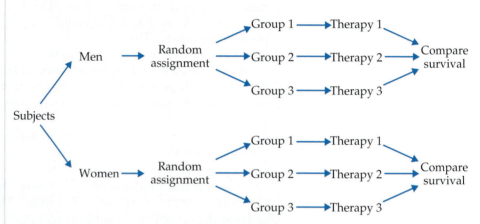

FIGURE 3.6 Outline of a block design, Example 3.18. The blocks consist of male and female subjects. The treatments are the three therapies for cancer.

EXAMPLE 3.19

Blocking in an agriculture experiment. The soil type and fertility of farmland differ by location. Because of this, a test of the effect of tillage type (two types) and pesticide application (three application schedules) on soybean yields uses small fields as blocks. Each block is divided into six plots, and the six treatments are randomly assigned to plots separately within each block.

EXAMPLE 3.20

Blocking in an education experiment. The Tennessee STAR class size experiment (Example 3.8, page 172) used a block design. It was important to compare different class types in the same school because the children in a school come from the same neighborhood, follow the same curriculum, and have the same school environment outside class. In all, 79 schools across Tennessee participated in the program. That is, there were 79 blocks. New kindergarten students were randomly placed in the three types of class separately within each school.

Blocks allow us to draw separate conclusions about each block, for example, about men and women in the cancer study in Example 3.18. Blocking also allows more precise overall conclusions because the systematic differences between men and women can be removed when we study the overall effects of the three therapies. The idea of blocking is an important additional principle of statistical design of experiments. A wise experimenter will form blocks based on the most important unavoidable sources of variability among the experimental units. Randomization will then average out the effects of the remaining variation and allow an unbiased comparison of the treatments.

SECTION 3.2 SUMMARY

• In an experiment, one or more **treatments** are imposed on the **experimental units** or **subjects**. Each treatment is a combination of **levels** of the explanatory variables, which we call **factors**. **Outcomes** are the measured variables that are used to compare the treatments.

• The **design** of an experiment refers to the choice of treatments and the manner in which the experimental units or subjects are assigned to the treatments.

• The basic principles of statistical design of experiments are **compare**, **randomization**, and **repetition**.

• The simplest form of control is **comparison**. Experiments should compare two or more treatments in order to prevent **confounding** the effect of a treatment with other influences, such as lurking variables.

• **Randomization** uses chance to assign subjects to the treatments. Randomization creates treatment groups that are similar (except for chance variation) before the treatments are applied. Randomization and comparison together prevent **bias**, or systematic favoritism, in experiments.

• You can carry out randomization by giving numerical labels to the experimental units and using a **table of random digits** to choose treatment groups.

• **Repetition** of the treatments on many units reduces the role of chance variation and makes the experiment more sensitive to differences among the treatments.

• Good experiments require attention to detail as well as good statistical design. Many behavioral and medical experiments are **double-blind**. **Lack of realism** in an experiment can prevent us from generalizing its results.

• In addition to comparison, a second form of control is to restrict randomization by forming **blocks** of experimental units that are similar in some way that is important to the response. Randomization is then carried out separately within each block.

• **Matched pairs** are a common form of blocking for comparing just two treatments. In some matched pairs designs, each subject receives both treatments in a random order. In others, the subjects are matched in pairs as closely as possible, and one subject in each pair receives each treatment.

SECTION 3.2 EXERCISES

For Exercises 3.17 and 3.18, see page 173; for Exercises 3.19 and 3.20, see page 175; for Exercises 3.21 and 3.22, see page 176; and for Exercise 3.23, see page 181.

3.24 Blueberries and bones. A study of the effects of blueberries on the bones of mice compared diets containing no blueberries, blueberries as 5% of the diet, and blueberries as 10% of the diet. Ten mice were randomly assigned to each diet. The mice were fed the diets for 30 days, and the total body bone mineral density (TBBMD) was measured at the end of the feeding period. What are the experimental units, the treatments, and the outcomes for this experiment? Would you use the term *subjects* for the experimental units? Explain your answers.

3.25 Online homework. Thirty students participated in a study designed to evaluate a new online homework system. None of the students had used an online homework system in the past. After using the system for a month, they were asked to rate their satisfaction with the system using a five-point scale.

(a) What are the experimental units, the treatment, and the outcome for this experiment? Can we use the term *subjects* for the experimental units? Explain your answers.

(b) Is this a comparative experiment? If your answer is Yes, explain why. If your answer is No, describe how you would change the design so that it would be a comparative experiment.

(c) Suggest some different outcomes that you think would be appropriate for this experiment.

3.26 Do magnets reduce pain? Some claim that magnets can be used to reduce pain. Design a double-blind experiment to test this claim. Write a proposal requesting funding for your study giving all the important details, including the number of subjects, issues concerning randomization, and how you will make the study double-blind.

3.27 Online sales of running shoes. A company that sells running shoes online wants to compare two new marketing strategies. They will test the strategies on 10 weekdays. In the morning of each day, a web page describing the comfort of the running shoes will be displayed. In the afternoon of each day, a web page describing the discounted price for the shoes will be displayed. Sales of the featured running shoes in the morning will be compared with sales in the afternoon at the end of the experiment.

(a) What are the experimental units, the treatments, and the outcomes for this experiment? Explain your answers.

(b) Is this a comparative experiment? Why or why not?

(c) Could the experiment be improved by using randomization? Explain your answer.

(d) Could the experiment be improved by using a placebo treatment? Explain your answer.

3.28 Online sales of running shoes. Refer to the previous exercise. Suppose that for each day, you randomized the web pages, showing one in the morning and the other in the afternoon.

(a) Can you view this experiment as a block design? Explain your answer.

(b) Do you prefer this experiment or the one in the previous exercise? Give reasons for your answer.

3.29 Online sales of running shoes. Refer to Exercise 3.27. Here is another way in which the experiment could be designed. Suppose that you alternate the display each time a customer visits the website. Can you view this experiment as a matched pairs design? Explain your answer.

3.30 Randomize the web pages for the running shoes. Refer to Exercise 3.28. Use software or Table B to randomize the treatments. Give a step-by-step detailed description of how you performed the randomization.

3.31 What is needed? Explain what is deficient in each of the following proposed experiments and explain how you would improve the experiment.

(a) Two product promotion offers are to be compared. The first, which offers two items for $2, will be used in a store on Friday. The second, which offers three items for $3, will be used in the same store on Saturday.

(b) A study compares two marketing campaigns to encourage individuals to eat more fruits and vegetables. The first campaign is launched in Florida at the same time that the second campaign is launched in Minnesota.

(c) You want to evaluate the effectiveness of a new investment strategy. You try the strategy for one year and evaluate the performance of the strategy.

3.32 The *Madden* curse. Some people believe that individuals who appear on the cover of the football game *Madden NFL* will soon have a serious injury. Can you evaluate this belief with an experiment? Explain your answer.

3.33 Evaluate a new orientation program. Your company runs a two-day orientation program Monday and Tuesday each week for new employees. A new program is to be compared with the current one. Set up an experiment to compare the new program with the old. Be sure to provide details regarding randomization and what outcome variables you will measure.

3.34 What is wrong? Explain what is wrong with each of the following randomization procedures, and describe how you would do the randomization correctly.

(a) Twenty students are to be used to evaluate a new treatment. Ten men are assigned to receive the treatment, and 10 women are assigned to be the controls.

(b) Ten subjects are to be assigned to two treatments, five to each. For each subject, a coin is tossed. If the coin comes up heads, the subject is assigned to the first treatment; if the coin comes up tails, the subject is assigned to the second treatment.

(c) An experiment will assign 40 rats to four different treatment conditions. The rats arrive from the supplier in batches of 10, and the treatment lasts two weeks. The first batch of 10 rats is randomly assigned to one of the four treatments, and data for these rats are collected. After a one-week break, another batch of 10 rats arrives and is assigned to one of the three remaining treatments. The process continues until the last batch of rats is given the treatment that has not been assigned to the three previous batches.

3.35 Calcium and vitamin D. Vitamin D is needed for the body to use calcium. An experiment is designed to study the effects of calcium and vitamin D supplements on the bones of first-year college students. The outcome measure is the total body bone mineral content (TBBMC), a measure of bone health. Three doses of calcium will be used: 0, 250, and 500 milligrams per day (mg/day). The doses of vitamin D will be 0, 75, and 150 international units (IU) per day. The calcium and vitamin D will be given in a single tablet. All tablets, including those with no calcium and no vitamin D, will look identical. Subjects for the study will be 45 men and 45 women.

(a) What are the factors and the treatments for this experiment?

(b) Draw a picture explaining how you would randomize the 90 college students to the treatments.

(c) Use a spreadsheet to carry out the randomization.

(d) Is there a placebo in this experiment? Explain your answer.

3.36 Use the *Simple Random Sample* applet. You can use the *Simple Random Sample* applet to choose a group at random once you have labeled the subjects. Example 3.12 (page 178) uses Excel to choose five students from a group of 10 to receive a treatment in an experiment. The remaining five students will receive a placebo control.

(a) Use the applet to choose five students. Which students were selected?

(b) Compare using Excel, as we did in Example 3.12, with the applet that you used for this exercise. Which do you prefer? Give reasons for your answer.

3.37 Use the *Simple Random Sample* applet. The *Simple Random Sample* applet allows you to randomly assign experimental units to more than two groups without difficulty. Consider a randomized comparative experiment in which 100 students are randomly assigned to four groups of 25.

(a) Use the applet to randomly choose 25 out of 100 students to form the first group. Which students are in this group?

(b) The "population hopper" now contains the 75 students who were not chosen, in scrambled order. Click "Sample" again to choose 25 of these remaining students to make up the second group. Which students were chosen?

(c) Click "Sample" one more time to choose the third group. Don't take the time to write this down. Check that there are only 25 students remaining in the "population hopper." These subjects get Treatment 4. Which students are they?

3.38 Use the *Simple Random Sample* applet. The *Simple Random Sample* applet can demonstrate how randomization works to create similar groups for comparative experiments. Suppose that (unknown to the experimenters) the 20 even-numbered students among the 40 subjects for the smartphone study in Example 3.11 (page 176) tend to send more text messages than the odd-numbered students. We would like the two groups to be similar with respect to text messaging. Use the applet to choose 10 samples of size 20 from the 40 students. (Be sure to click "Reset" after each sample.) Record the counts of even-numbered students in each of your 10 samples. You see that there is considerable chance variation but no systematic bias in favor of one or the other group in assigning the fast-reacting students. Larger samples from larger populations will, on the average, do a better job of making the two groups equivalent.

3.39 Health benefits of bee pollen. "Bee pollen is effective for combating fatigue, depression, cancer, and colon disorders." So says a website that offers the pollen for sale. We wonder if bee pollen really does prevent colon disorders. Here are two ways to study this question. Explain why the first design will produce more trustworthy data.

(a) Find 400 women who do not have colon disorders. Randomly assign 200 to take bee pollen capsules and the other 200 to take placebo capsules that are identical in appearance. Follow both groups for five years.

(b) Find 200 women who take bee pollen regularly. Match each with a woman of the same age, race, and occupation who does not take bee pollen. Follow both groups for five years.

3.40 Random digits. Table B is a table of random digits. Which of the following statements are true of a table of random digits, and which are false? Explain your answers.

(a) There are exactly four 0s in each row of 40 digits.

(b) Each pair of digits has chance 1/100 of being 00.

(c) The digits 0000 can never appear as a group because this pattern is not random.

3.41 Calcium and the bones of young girls. Calcium is important to the bone development of young girls. To study how the bodies of young girls process calcium, investigators used the setting of a summer camp. Calcium was given in punch at either a high or a low level. The camp diet was otherwise the same for all girls. Suppose that there are 30 campers.

(a) Outline a completely randomized design for this experiment.

(b) Use software or Table B to do the randomization. Explain in step-by-step detail how you carried out how you performed the randomization.

(c) Make a table giving the treatment that each camper will receive.

3.42 Calcium and the bones of young girls. Refer to the previous exercise.

(a) Outline a matched pairs design in which each girl receives both levels of calcium (with a "washout period" in which no calcium supplementation was given between the two treatment periods).

(b) What is the advantage of the matched pairs design over the completely randomized design?

(c) The same randomization can be used in different ways for both designs. Explain why this is true.

(d) Use software or Table B to do the randomization. Explain what each subject will do for the matched pairs design.

3.43 Measuring water quality in streams and lakes. Water quality of streams and lakes is an issue of concern to the public. Although trained professionals typically are used to take reliable measurements, many volunteer groups are gathering and distributing information based on data that they collect.[12] You are part of a team to train volunteers to collect accurate water quality data. Design an experiment to evaluate the effectiveness of the training. Write a summary of your proposed design to present to your team. Be sure to include all the details that they will need to evaluate your proposal.

3.3 Sampling Design

When you complete this section, you will be able to:

- Distinguish between a population and a sample.
- Use the response rate to evaluate a survey.
- Use Table B to generate a simple random sample (SRS).
- Use software to generate a simple random sample.
- Construct a stratified random sample using Table B or software to select the samples from the strata.
- Identify voluntary response samples, simple random samples, stratified random samples, and multistage random samples.
- Identify characteristics of samples that limit their usefulness, including undercoverage, nonresponse, response bias, and the wording of questions.

A political scientist wants to know what percent of college-age adults consider themselves conservatives. An automaker hires a market research firm to learn what percent of adults aged 18 to 35 recall seeing television advertisements for a new sports utility vehicle. Government economists inquire about average household income.

In all these cases, we want to gather information about a large group of individuals. We will not, as in an experiment, impose a treatment in order to observe the response. Also, time, cost, and inconvenience forbid contacting every individual. In such cases, we gather information about only part of the group—a *sample*—in order to draw conclusions about the whole. **Sample surveys** are an important kind of observational study.

sample survey

POPULATION AND SAMPLE

The entire group of individuals that we want information about is called the **population**.

A **sample** is a part of the population that we actually examine in order to gather information.

Notice that "population" is defined in terms of our desire for knowledge. If we wish to draw conclusions about all U.S. college students, that group is our population even if only local students are available for questioning. The sample is the part from which we draw conclusions about the whole. The *sample design* **design of a sample survey** refers to the method used to choose the sample from the population.

EXAMPLE 3.21

hartcreations/iStockphoto

The Reading Recovery program. The Reading Recovery (RR) program has specially trained teachers work one-on-one with at-risk first-grade students to help them learn to read. A study was designed to examine the relationship between the RR teachers' beliefs about their ability to motivate students and the progress of the students whom they teach.[13] The Reading Recovery International Data Evaluation Center website (**www.idecweb.us**) says that there are 13,823 RR teachers. The researchers send a questionnaire to a random sample of 200 of these. The population consists of all 13,823 RR teachers, and the sample is the 200 that were randomly selected.

Unfortunately, our idealized framework of population and sample does not exactly correspond to the situations that we face in many cases. In Example 3.21, the list of teachers was prepared at a particular time in the past. It is very likely that some of the teachers on the list are no longer working as RR teachers today. New teachers have been trained in RR methods and are not on the list. Despite these difficulties, we still view the list as the population. Also, we may have out-of-date addresses for some who are still working as RR teachers, and some teachers may choose not to respond to the survey questions.

In reporting the results of a sample survey, it is important to include all details regarding the procedures used. Follow-up mailings or phone calls to those who do not initially respond can help increase the response rate. The proportion of the original sample who actually provide usable data is called *response rate* the **response rate** and should be reported for all surveys. If only 150 of the teachers who were sent questionnaires provided usable data, the response rate would be 150/200, or 75%.

USE YOUR KNOWLEDGE

3.44 Are they satisfied? An educational research team wanted to examine the relationship between faculty participation in decision making and job satisfaction in Mongolian public universities. They are planning to randomly select 250 faculty members from a list of 2000 faculty members in these universities. The Job Descriptive Index will be used to measure job satisfaction, and the Conway Adaptation of the

Alutto-Belasco Decisional Participation Scale will be used to measure decision participation.

(a) Describe the population for this study.

(b) Describe the sample for this study.

(c) How would you determine the response rate for this study? Can you calculate it from the information given? If your answer is yes, calculate it.

3.45 What is the impact of the taxes? A study was designed to assess the impact of taxes on forest land usage in part of the Upper Wabash River Watershed in Indiana.[14] A survey was sent to 772 forest owners from this region, and 348 were returned.

(a) What is the sample for this study?

(b) What is the population for this study?

(c) How would you determine the response rate for this study? Can you calculate it from the information given? If your answer is yes, calculate it.

Poor sample designs can produce misleading conclusions. Here is an example.

EXAMPLE 3.22

Sampling pieces of steel. A mill produces large coils of thin steel for use in manufacturing home appliances. The quality engineer wants to submit a sample of 5-centimeter squares to detailed laboratory examination. She asks a technician to cut a sample of 10 such squares. Wanting to provide "good" pieces of steel, the technician carefully avoids the visible defects in the coil material when cutting the sample. The laboratory results are wonderful, but the customers complain about the material they are receiving.

LOOK BACK

bias,
p. 174

In Example 3.22, the sample was selected in a manner that guaranteed that it would not be representative of the entire population. This sampling scheme displays *bias*, or systematic error, in favoring some parts of the population over others.

Online polls use *voluntary response samples*, a particularly common form of biased sample. The sample who respond are not representative of the population at large. People who take the trouble to respond to an open invitation are not representative of the entire population.

VOLUNTARY RESPONSE SAMPLE

A **voluntary response sample** consists of people who choose themselves by responding to a general appeal. Voluntary response samples are biased because people with strong opinions, especially negative opinions, are most likely to respond.

The remedy for bias in choosing a sample is to allow chance to do the choosing so that there is neither favoritism by the sampler (Example 3.22) nor voluntary response (online opinion polls). Random selection of a sample eliminates bias by giving all individuals an equal chance to be chosen, just as randomization eliminates bias in assigning experimental units.

Simple random samples

The simplest sampling design amounts to placing names in a hat (the population) and drawing out a handful (the sample). This is *simple random sampling*.

> ### SIMPLE RANDOM SAMPLE
>
> A **simple random sample (SRS)** of size n consists of n individuals from the population chosen in such a way that every set of n individuals has an equal chance to be the sample actually selected.

Each treatment group in a completely randomized experimental design is an SRS drawn from the available experimental units. We select an SRS by labeling all the individuals in the population and using software or a table of random digits to select a sample of the desired size, just as in experimental randomization. Notice that an SRS not only gives every possible sample an equal chance to be chosen, but also gives each individual an equal chance to be chosen. There are other random sampling designs that give each individual, but not each sample, an equal chance. One such design, systematic random sampling, is described in Exercise 3.64 (page 202).

How to select a simple random sample

The basic ideas needed to select a simple random sample are very similar to those that we discussed when we randomized subjects to treatments (page 177). We first assign a label to each case in our population. Then we perform the randomization using software or random digits from Table B.

Selection of a simple random sample using software The World Bank collects information about starting businesses throughout the world. In Example 1.23 (page 26) and several other examples in Chapter 1, we examined the time to start a business in a subset of these countries. For those exercises, we used a subset of the data because it was easier to show some details about our calculations with a smaller amount of data.

Now, suppose we want to collect additional information about countries that would help us to understand the processes of starting a business. The complete data set contains entries for 189 countries, and the time required to collect the additional information on all these would be too much. Let's use Excel to select a sample of 25 countries for a more detailed examination of these countries.

EXAMPLE 3.23

TTS

Select an SRS of countries using Excel. The data file TTS includes columns for the country name and a three-letter country code for each of the 189 countries. We could use either of these for our label. We will use the three letter codes in our screen shots to save space.

Figure 3.7(a) shows the codes for the first ten countries. In Figure 3.7(b), we show the uniform numbers generated with the RAND() function (and

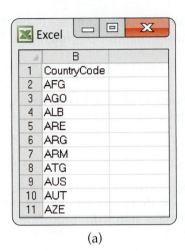
(a)

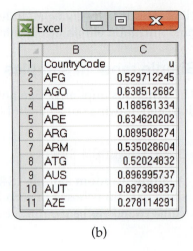
(b)

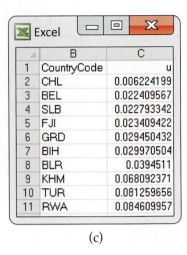
(c)

FIGURE 3.7 Selection of a simple random sample of countries from the population of 189 countries, Example 3.23.

then pasted into column C). The first three countries are Afghanistan (AFG), Angola (AGO), and Albania (ALB).

Figure 3.7(c) shows the file after we sort on the uniform random numbers in column C. Our sample is the first 25 countries in the sorted file. The first three selected are Chile (CHL), Belgium (BEL), and the Solomon Islands (SLB). Note that Excel does not display the last two digits for Belarus (BLR) because they are zero.

The *Simple Random Sample* applet on the text website is another convenient way to select an SRS.

USE YOUR KNOWLEDGE

TTS

3.46 Select an SRS. Use the *Simple Random Sample* applet or Excel to select an SRS of five countries from the TTS data file. Include a step-by-step detailed description of how you selected the countries.

Selection of a simple random sample using random digits We illustrate the procedure by selecting an SRS of countries from the population of 189 countries in the data file TTS. Recall that we used Excel to select such a sample in Example 3.23.

EXAMPLE 3.24

Select an SRS of countries using random digits. To use Table B, we need a numeric label. We could create such a label by adding a column to the data file TTS containing the numbers 1 to 189. An alternative requiring less work would be to use the numbers in the leftmost part of the spreadsheet. Notice in Figure 3.7(a), for example, that there is a 1 in the first row of the spreadsheet where we have entered the names of the variables in the columns. Therefore, the numbers corresponding to countries run from 2 through 190. We will use these numbers as our label.

We will examine the entries in Table B in sets of three. Three digit numbers between 2 and 190 will correspond to selected countries. We will ignore three digit numbers equal to 000, 001, or greater than 190. Let's start our selection at line 106 in Table B. The entries on this line are

68417   35013   15529   72765   85089   57067   50211   47487

If we arrange these into sets of three, we have

684   173   501   315   529   727   658   508   057   067   502   114   748   7

The selected labels from this set of random digits are 173, 057, and 067. Checking the spreadsheet, we see that these numbers correspond to Turkey, France, and Greece.

Note that we do not use the last digit on line 106 to select the country with the label 7. We should combine this single digit with the first two digits from line 107 of Table B. This gives us the three-digit number 782, which is a number that we ignore. We complete our selection of the additional 22 countries that we need in our SRS using additional lines from Table B as needed.

USE YOUR KNOWLEDGE

3.47 Find the next three countries to be selected. Continue the process described in Example 3.24 to select the next three countries for the SRS. Show your work.

3.48 Listen to three rock songs. The walk to your statistics class takes about 10 minutes, about the amount of time needed to listen to three songs on your iPod or smartphone. You decide to take a simple random sample of songs from the top 10 listed on a Billboard Hot Rock Songs.[15] Here is the list:

Shut Up and Dance	Uma Thurman	Renegades	Ex's & Oh's
Centuries	Cecilia and the Satellite	Tear in My Heart	Brother
Stressed Out	Shots		

Select your three hot rock songs using a simple random sample. Show your work.

3.49 Listen to three songs. Refer to the previous exercise. Suppose that you like to include more variety in your music, so you look at the Billboard Top 100 songs.[16] Here are the top 10 on this list:

Cheerleader	Can't Feel My Face	Watch Me	Lean On
The Hills	Good for You	Fight Song	679
Trap Queen	Shut Up and Dance		

Select the three songs for your iPod or smartphone using a simple random sample. Show your work.

EXAMPLE 3.25

Select an SRS of countries using JMP. Refer to Example 3.23, where we selected an SRS of countries from a population of 189 countries. Also see Example 3.24, where we used the random digits in Table B to select the SRS. We can also use JMP to select an SRS. The output for the first 17 countries selected is displayed in Figure 3.8. JMP provides a file with the selected countries with all columns in the original data file. Note that the selected files are listed in the order that they appear in the original file, which is alphabetically in this case.

	CountryName	CountryCode
1	Bulgaria	BGR
2	Bahrain	BHR
3	Bahamas, The	BHS
4	Bolivia	BOL
5	Barbados	BRB
6	Bhutan	BTN
7	Costa Rica	CRI
8	Dominican Republic	DOM
9	Micronesia, Fed. Sts.	FSM
10	Equatorial Guinea	GNQ
11	Jordan	JOR
12	Japan	JPN
13	Kiribati	KIR
14	Korea, Rep.	KOR
15	St. Lucia	LCA
16	Marshall Islands	MHL
17	Malaysia	MYS

FIGURE 3.8 JMP output for selecting an SRS of 25 countries from a population of 189 countries, Example 3.25. Only the first 17 are displayed.

Stratified random samples

The general framework for designs that use chance to choose a sample is a *probability sample*.

PROBABILITY SAMPLE

A **probability sample** is a sample chosen by chance. We must know what samples are possible and what chance, or probability, each possible sample has.

Some probability sampling designs (such as an SRS) give each member of the population an *equal* chance to be selected. This may not be true in more elaborate sampling designs. In every case, however, the use of chance to select the sample is the essential principle of statistical sampling.

Designs for sampling from large populations spread out over a wide area are usually more complex than an SRS. For example, it is common to sample important groups within the population separately, then combine these samples. This is the idea of a *stratified sample*.

STRATIFIED RANDOM SAMPLE

To select a **stratified random sample**, first divide the population into groups of similar individuals, called **strata**. Then choose a separate SRS in each stratum and combine these SRSs to form the full sample.

Choose the strata based on facts known before the sample is taken. For example, a population of election districts might be divided into urban, suburban, and rural strata.

A stratified design can produce more exact information than an SRS of the same size by taking advantage of the fact that individuals in the same stratum are similar to one another. Think of the extreme case in which all individuals in each stratum are identical: just one individual from each stratum is then enough to completely describe the population.

Strata for sampling are similar to blocks in experiments. We have two names because the idea of grouping similar units before randomizing arose separately in sampling and in experiments.

EXAMPLE 3.26

A stratified sample of countries. In Examples 3.23 and 3.24, we selected SRSs of size 25 from the population of 189 countries in the World Bank file with data on starting businesses. Let's think about using a stratified sample. You still want to select 25 companies to examine in detail.

Let's classify each of the countries as located in Asia, Africa, Europe, North or South America, and Other. We have five strata, and we want a total of 25 countries to examine in detail. Therefore, we need to sample five countries from each stratum. We take an SRS of size 5 from each of these strata.

Multistage random samples

multistage
random sample

Another common means of restricting random selection is to choose the sample in stages. These designs are called **multistage designs**. They are widely used in national samples of households or people. For example, data on employment and unemployment are gathered by the government's Current Population Survey, which conducts interviews in about 60,000 households each month. The cost of sending interviewers to the widely scattered households in an SRS would be too high. Moreover, the government wants data broken down by states and large cities.

Thus, the Current Population Survey uses a multistage random sampling design. The final sample consists of groups of nearby households, called clusters

clusters, that an interviewer can easily visit. Most opinion polls and other national samples are also multistage, though interviewing in most national samples today is done by telephone rather than in person, eliminating the economic need for clustering. The Current Population Survey sampling design is roughly as follows:[17]

Stage 1. Divide the United States into 2007 geographical areas called Primary Sampling Units, or PSUs. PSUs do not cross state lines. Select a sample of 754 PSUs. This sample includes the 428 PSUs with the largest population and a stratified sample of 326 of the others.

Stage 2. Divide each PSU selected into smaller areas called "blocks." Stratify the blocks using ethnic and other information, and take a stratified sample of the blocks in each PSU.

Stage 3. Sort the housing units in each block into clusters of four nearby units. Interview the households in a probability sample of these clusters.

Analysis of data from sampling designs more complex than an SRS takes us beyond basic statistics. But the SRS is the building block of more elaborate designs, and analysis of other designs differs more in complexity of detail than in fundamental concepts.

Cautions about sample surveys

Random selection eliminates bias in the choice of a sample from a list of the population. Sample surveys of large human populations, however, require much more than a good sampling design.[18] To begin, we need an accurate and complete list of the population. Because such a list is rarely available, most samples suffer from some degree of *undercoverage*. A sample survey of households, for example, will miss not only homeless people, but also prison inmates and students in dormitories. An opinion poll conducted by telephone will miss the large number of American households without residential phones. The results of national sample surveys, therefore, have some bias if the people not covered—who most often are poor people—differ from the rest of the population.

A more serious source of bias in most sample surveys is *nonresponse*, which occurs when a selected individual cannot be contacted or refuses to cooperate. Nonresponse to sample surveys often reaches 50% or more, even with careful planning and several callbacks. Because nonresponse is higher in urban areas, most sample surveys substitute other people in the same area to avoid favoring rural areas in the final sample. If the people contacted differ from those who are rarely at home or who refuse to answer questions, some bias remains.

UNDERCOVERAGE AND NONRESPONSE

Undercoverage occurs when some groups in the population are left out of the process of choosing the sample.

Nonresponse occurs when an individual chosen for the sample can't be contacted or does not cooperate.

EXAMPLE 3.27

Nonresponse in the Current Population Survey. How bad is nonresponse? The Current Population Survey (CPS) has the lowest nonresponse rate of any poll we know: only about 5% of the households in the CPS sample refuse to take part, and another 2% or 3% can't be contacted.[19] People are more likely to respond to a government survey such as the CPS, and the CPS contacts its sample in person before doing later interviews by phone.

The General Social Survey (Figure 3.9) is the nation's most important social science research survey. The GSS also contacts its sample in person, and it is run by a university. Despite these advantages, its most recent survey had a 30% rate of nonresponse.[20]

FIGURE 3.9 Part of the home page for the General Social Survey (GSS). The GSS has assessed attitudes on a wide variety of topics since 1972. Its continuity over time makes the GSS a valuable source for studies of changing attitudes. (Source: GSS)

What about polls done by the media and by market research and opinion-polling firms? Often, we don't know their rates of nonresponse because they won't say. That itself is a bad sign.

EXAMPLE 3.28

Change in nonresponse in Pew surveys. The Pew Research Center conducts research using surveys on a variety of issues, attitudes, and trends.[21] A study by the center examined the decline in the response rates to their surveys over time. The changes are dramatic, and there is a consistent pattern over time. Here are some data from the report:[22]

Year	1997	2000	2003	2006	2009	2012
Nonresponse rate	64%	72%	75%	79%	85%	91%

The center is devising alternative methods that show some promise of improving the response rates of their surveys.

Most sample surveys, and almost all opinion polls, are now carried out by telephone. This and other details of the interview method can affect the results. When presented with several options for a reply, such as "completely agree," "mostly agree," "mostly disagree," and "completely disagree," people tend to be a little more likely to respond to the first one or two options presented.

response bias   The behavior of the respondent or of the interviewer can cause **response bias** in sample results. Respondents may lie, especially if asked about illegal or unpopular behavior. The race or sex of the interviewer can influence responses to questions about race relations or attitudes toward feminism. Answers to questions that ask respondents to recall past events are often inaccurate because of faulty memory. For example, many people "telescope" events in the past, bringing them forward in memory to more recent time periods. "Have you visited a dentist in the last six months?" will often elicit a Yes from someone who last visited a dentist eight months ago.

wording of questions   The **wording of questions** is the most important influence on the answers given to a sample survey. Confusing or leading questions can introduce strong bias, and even minor changes in wording can change a survey's outcome. Here are some examples.

EXAMPLE 3.29

The form of the question is important. In response to the question "Are you heterosexual, homosexual, or bisexual?" in a social science research survey, one woman answered, "It's just me and my husband, so bisexual." The issue is serious, even if the example seems silly: reporting about sexual behavior is difficult because people understand and misunderstand sexual terms in many ways.

How do Americans feel about government help for the poor? Only 13% think we are spending too much on "assistance to the poor," but 44% think we are spending too much on "welfare." How do the Scots feel about the movement to become independent from England? Well, 51% would vote for "independence for Scotland," but only 34% support "an independent Scotland separate from the United Kingdom." It seems that "assistance to the poor" and "independence" are nice, hopeful words. "Welfare" and "separate" are negative words.[23]

The statistical design of sample surveys is a science, but this science is only part of the art of sampling. Because of nonresponse, response bias, and the difficulty of posing clear and neutral questions, you should hesitate to fully trust reports about complicated issues based on surveys of large human populations. *Insist on knowing the exact questions asked, the rate of nonresponse, and the date and method of the survey before you trust a poll result.*

BEYOND THE BASICS

Capture-Recapture Sampling

Sockeye salmon return to reproduce in the river where they were hatched four years earlier. How many salmon survived natural perils and heavy fishing to make it back this year? How many mountain sheep are there in Colorado? Are migratory songbird populations in North America decreasing or holding their own? These questions concern the size of animal populations. Biologists address them with a special kind of repeated sampling, called *capture-recapture sampling*.

EXAMPLE 3.30

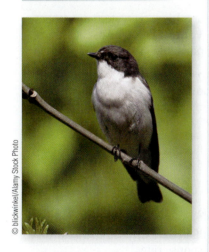

© blickwinkel/Alamy Stock Photo

Estimate the number of least flycatchers. You are interested in the number of least flycatchers migrating along a major route in the north-central United States. You set up "mist nets" that capture the birds but do not harm them. The birds caught in the net are fitted with a small aluminum leg band and released. Last year, you banded and released 200 least flycatchers. This year, you repeat the process. Your net catches 120 least flycatchers, 12 of which have tags from last year's catch.

The proportion of your second sample that have bands should estimate the proportion in the entire population that are banded. So if N is the unknown number of least flycatchers, we should have approximately

$$\text{proportion banded in sample} = \text{proportion banded in population}$$

$$\frac{12}{120} = \frac{200}{N}$$

Solve for N to estimate that the total number of flycatchers migrating while your net was up this year is approximately

$$N = 200 \times \frac{120}{12} = 2000$$

The capture-recapture idea extends the use of a sample proportion to estimate a population proportion. The idea works well if both samples are SRSs from the population and the population remains unchanged between samples. In practice, complications arise because, for example, some of the birds tagged last year died before this year's migration.

Variations on capture-recapture samples are widely used in wildlife studies and are now finding other applications. One way to estimate the census undercount in a district is to consider the census as "capturing and marking" the households that respond. Census workers then visit the district, take an SRS of households, and see how many of those counted by the census show up in the sample. Capture-recapture estimates the total count of households in the district. As with estimating wildlife populations, there are many practical pitfalls. Our final word is as before: the real world is less orderly than statistics textbooks imply.

SECTION 3.3 SUMMARY

- A sample survey selects a **sample** from the **population** of all individuals about which we desire information. We base conclusions about the population on data about the sample.

- The **design** of a sample refers to the method used to select the sample from the population. **Probability sampling designs** use impersonal chance to select a sample.

- The basic probability sample is a **simple random sample (SRS)**. An SRS gives every possible sample of a given size the same chance to be chosen.

- Choose an SRS using software. This can also be done using a **table of random digits** to select the sample.

- To choose a **stratified random sample**, divide the population into **strata**, groups of individuals that are similar in some way that is important to the response. Then choose a separate SRS from each stratum, and combine them to form the full sample.

- **Multistage random samples** select successively smaller groups within the population in stages, resulting in a sample consisting of clusters of individuals. Each stage may employ an SRS, a stratified sample, or another type of sample.

- Failure to use probability sampling often results in **bias**, or systematic errors in the way the sample represents the population. **Voluntary response** samples, in which the respondents choose themselves, are particularly prone to large bias.

- In human populations, even probability samples can suffer from bias due to **undercoverage** or **nonresponse**, from **response bias** due to the behavior of the interviewer or the respondent, or from misleading results due to **poorly worded questions**.

SECTION 3.3 EXERCISES

For Exercises 3.44 and 3.45, see pages 189–190; for Exercise 3.46, see page 192; and for Exercises 3.47, 3.48, and 3.49, see page 193.

3.50 How many text messages? You would like to know something about how many text messages you will receive in the next 100 days. Counting the number for each of the 100 days would take more time than you would like to spend on this project, so you randomly select 10 days from the hundred to count.

(a) Describe the population for this setting.

(b) What is the sample?

3.51 Response rate? A survey designed to assess satisfaction with food items sold at a college's football games was sent to 150 fans who had season tickets. The total number of fans who have season tickets is 5674. Responses to the survey were received from 98 fans.

(a) Describe the population for this survey.

(b) What is the sample?

(c) What is the response rate?

(d) What is the nonresponse rate?

(e) Suggest some ways that could be used in a future survey to increase the response rate.

3.52 Interview some students. You are a teaching assistant for an introductory statistics class. The instructor would like you to interview some of the students in the class to find out their opinion regarding some new interactive activities that she has introduced to the course. There are 123 students in the class, so you cannot interview all of them. You decide to select eight students to interview.

(a) What is the population for this setting?

(b) What is the sample?

(c) Make a spreadsheet with the numeric labels for the 123 students in the class.

(d) Use Excel to select the labels of the eight students to be interviewed from the spreadsheet.

(e) Explain the steps that you used in sufficient detail so that another person could repeat your work.

3.53 Interview some students. Refer to the previous exercise.

(a) Use Table B to select the students. Give details.

(b) Compare the use of Table B with software for selecting the students. Which do you prefer? Give reasons for your answer.

3.54 What kind of sample? In each of the following situations, identify the sample as an SRS, a stratified random sample, a multistage random sample, or a voluntary response sample. Explain your answers.

(a) There are seven sections of an introductory statistics course. A random sample of three sections is chosen, and then random samples of eight students from each of these sections are chosen.

(b) A student organization has 55 members. A table of random numbers is used to select a sample of five.

(c) An online poll asks people who visit this site to choose their favorite television show.

(d) Separate random samples of male and female first-year college students in an introductory psychology course are selected to receive a one-week alternative instructional method.

3.55 What's wrong? Explain what is wrong in each of the following scenarios.

(a) The population consists of all individuals selected in a simple random sample.

(b) In a poll of an SRS of residents in a local community, respondents are asked to indicate the level of their concern about the dangers of dihydrogen monoxide, a substance that is a major component of acid rain and that, in its gaseous state, can cause severe burns. (*Hint:* Ask a friend who is majoring in chemistry about this substance or search the Internet for information about it.)

(c) Students in a class are asked to raise their hands if they have cheated on an exam one or more times within the past year.

3.56 What's wrong? Explain what is wrong with each of the following random selection procedures, and explain how you would do the randomization correctly.

(a) To determine the reading level of an introductory statistics text, you evaluate all the written material in the third chapter.

(b) You want to sample student opinions about a proposed change in procedures for changing majors. You hand out questionnaires to 100 students as they arrive for class at 7:30 A.M.

(c) A population of subjects is put in alphabetical order, and a simple random sample of size 10 is taken by selecting the first 10 subjects in the list.

3.57 Importance of students as customers. A committee on community relations in a college town plans to survey local businesses about the importance of students as customers. From telephone book listings, the committee chooses 80 businesses at random. Of these, 46 return the questionnaire mailed by the committee.

(a) What is the population for this sample survey?

(b) What is the sample?

(c) What is the rate (percent) of nonresponse?

3.58 Identify the populations. For each of the following sampling situations, identify the population as exactly as possible. That is, say what kind of individuals the population consists of and say exactly which individuals fall in the population. If the information given is not complete, complete the description of the population in a reasonable way.

(a) A college has changed its core curriculum and wants to obtain detailed feedback information from the students during each of the first 12 weeks of the coming semester. Each week, a random sample of five students will be selected to be interviewed.

(b) The American Community Survey (ACS) replaced the census "long form" starting with the 2010 census. The ACS contacts 250,000 addresses by mail each month, with follow-up by phone and in person if there is no response. Each household answers questions about their housing, economic, and social status.

(c) An opinion poll contacts 1161 adults and asks them, "Which political party do you think has better ideas for leading the country in the twenty-first century?"

3.59 Interview residents of apartment complexes. You are planning a report on apartment living in a college town. You decide to select eight apartment complexes at random for in-depth interviews with residents. Select a simple random sample of eight of the following apartment complexes. If you use Table B, start at line 136. **RESIDEN**

Ashley Oaks	Country View	Mayfair Village
Bay Pointe	Country Villa	Nobb Hill
Beau Jardin	Crestview	Pemberly Courts
Bluffs	Del-Lynn	Peppermill
Brandon Place	Fairington	Pheasant Run
Briarwood	Fairway Knolls	Richfield
Brownstone	Fowler	Sagamore Ridge
Burberry	Franklin Park	Salem Courthouse
Cambridge	Georgetown	Village Manor
Chauncey Village	Greenacres	Waterford Court
Country Squire	Lahr House	Williamsburg

3.60 Using GIS to identify mint field conditions. A Geographic Information System (GIS) is to be used to distinguish different conditions in mint fields. Ground observations will be used to classify regions of each field as either healthy mint, diseased mint, or weed-infested mint. The GIS divides mint-growing areas into regions called pixels. An experimental area contains 100 pixels. For a random sample of 15 pixels, ground measurements will be made to determine the status of the mint, and these observations will be compared with information obtained by the GIS. Select the random sample. If you use Table B, start at line 132 and choose only the first 15 pixels in the sample.

3.61 Use the _Simple Random Sample_ applet. After you have labeled the individuals in a population, the _Simple Random Sample_ applet automates the task of choosing an SRS. Use the applet to choose the sample in the previous exercise.

3.62 Select a simple random sample. There are 38 active telephone area codes in California. You want to choose an SRS of 10 of these area codes for a study of available telephone numbers. Label the codes 01 to 38 and use the _Simple Random Sample_ applet, Table B, or software to choose your sample. (If you use Table B, start at line 131.)

3.63 Stratified samples for attitudes about alcohol. At a party, there are 25 students over age 21 and 15 students under age 21. You choose at random five of those over 21 and separately choose at random three of those under 21 to interview about attitudes toward alcohol. You have given every student at the party the same chance to be interviewed: what is that chance? Why is your sample not an SRS?

3.64 Systematic random samples. Systematic random samples are often used to choose a sample of apartments in a large building or dwelling units in a block at the last stage of a multistage sample. An example will illustrate the idea of a systematic sample. Suppose that we must choose five addresses out of 125. Because 125/5 = 25, we can think of the list as five lists of 25 addresses. Choose one of the first 25 at random, using software or Table B. The sample contains this address and the addresses 25, 50, 75, and 100 places down the list from it. If 13 is chosen, for example, then the systematic random sample consists of the addresses numbered 13, 38, 63, 88, and 113.

(a) A study of dating among college students wanted a sample of 200 of the 8000 single male students on campus. The sample consisted of every 40th name from a list of the 8000 students. Explain why the survey chooses every 40th name.

(b) Use software or Table B at line 112 to choose the starting point for this systematic sample.

3.65 Systematic random samples versus simple random samples. The previous exercise introduces systematic random samples. Explain carefully why a systematic random sample _does_ give every individual the same chance to be chosen but is _not_ a simple random sample.

3.66 Random digit telephone dialing. An opinion poll in California uses random digit dialing to choose telephone numbers at random. Numbers are selected separately within each California area code. The size of the sample in each area code is proportional to the population living there. 📊 AREACOD

(a) What is the name for this kind of sampling design?

(b) California area codes, in rough order from north to south, are

209	213	310	323	341	369	408	415	424	442
510	530	559	562	619	626	627	628	650	657
661	669	707	714	747	752	760	764	805	818
831	858	909	916	925	935	949	951		

Another California survey does not call numbers in all area codes but starts with an SRS of eight area codes. Choose such an SRS. If you use Table B, start at line 132.

3.67 Select club members to go to a convention. A club has 30 student members and 10 faculty members. The students are

Abel	Fisher	Huber	Moran	Reinmann
Carson	Golomb	Jimenez	Moskowitz	Santos
Chen	Griswold	Jones	Neyman	Shaw
David	Hein	Kiefer	O'Brien	Thompson
Deming	Hernandez	Klotz	Pearl	Utts
Elashoff	Holland	Liu	Potter	Vlasic

and the faculty members are

Andrews	Fernandez	Kim	Moore	Rabinowitz
Besicovitch	Gupta	Lightman	Phillips	Yang

The club can send seven students and three faculty members to a convention and decides to choose those who will go by random selection. Select a stratified random sample of seven students and three faculty members.

3.68 Stratified samples for accounting audits. Accountants use stratified samples during audits to verify a company's records of such things as accounts receivable. The stratification is based on the dollar amount of the item and often includes 100% sampling of the largest items. One company reports 5000 accounts receivable. Of these, 100 are in amounts over $50,000; 500 are in amounts between $1000 and $50,000; and the remaining 4400 are in amounts under $1000. Using these groups as strata, you decide to verify all of the largest accounts and to sample 5% of the midsize accounts and 1% of the small accounts. How would you label the two strata from which you will sample? Use software or Table B, starting at line 125, to select the first six accounts from each of these strata.

3.69 The sampling frame. The list of individuals from which a sample is actually selected is called the **sampling frame**. Ideally, the frame should list every individual in the population, but in practice this is often difficult. A frame that leaves out part of the population is a common source of undercoverage.

(a) Suppose that a sample of households in a community is selected at random from the telephone directory. What households are omitted from this frame? What types of people do you think are likely to live in these households? These people will probably be underrepresented in the sample.

(b) It is usual in telephone surveys to use random digit dialing equipment that selects the last four digits of a telephone number at random after being given the area code and the exchange. The exchange is the first three digits of the telephone number. Which of the households that you mentioned in your answer to part (a) will be included in the sampling frame by random digit dialing?

3.70 Survey questions. Comment on each of the following as a potential sample survey question. Is the question clear? Is it slanted toward a desired response?

(a) "Some cell phone users have developed brain cancer. Should all cell phones come with a warning label explaining the danger of using cell phones?"

(b) "Do you agree that a national system of health insurance should be favored because it would provide health insurance for everyone and would reduce administrative costs?"

(c) "In view of escalating environmental degradation and incipient resource depletion, would you favor economic incentives for recycling of resource-intensive consumer goods?"

3.4 Ethics

When you complete this section, you will be able to:	• Describe the purpose of an institutional review board and describe what kinds of expertise its members require.
	• Describe informed consent and evaluate whether or not it has been given in specific examples.
	• Determine when data have been kept confidential in a study.
	• Evaluate a clinical trial from the viewpoint of ethics.

The production and use of data, like all human endeavors, raise ethical questions. We won't discuss the telemarketer who begins a telephone sales pitch with "I'm conducting a survey." Such deception is clearly unethical. It enrages legitimate survey organizations, which find the public less willing to talk with them. Neither will we discuss those few researchers who, in the pursuit of professional advancement, publish fake data. There is no ethical question here—faking data to advance your career is just wrong. It will end your career when uncovered.

But just how honest must researchers be about real, unfaked data? Here is an example that suggests the answer is "More honest than they often are."

EXAMPLE 3.31

Provide all the critical information. Papers reporting scientific research are supposed to be short, with no extra baggage. But brevity can allow the researchers to avoid complete honesty about their data. Did they choose their subjects in a biased way? Did they report data on only some of their subjects? Did they try several statistical analyses and report only the ones that looked best?

The statistician John Bailar screened more than 4000 medical papers in more than a decade as consultant to the *New England Journal of Medicine.* He says, "When it came to the statistical review, it was often clear that critical information was lacking, and the gaps nearly always had the practical effect of making the authors' conclusions look stronger than they should have."[24] The situation is no doubt worse in fields that screen published work less carefully.

The most complex issues of data ethics arise when we collect data from people. The ethical difficulties are more severe for experiments that impose some treatment on people than for sample surveys that simply gather information. Trials of new medical treatments, for example, can do harm as well as good to their subjects. Here are some basic standards of data ethics that must be obeyed by any study that gathers data from human subjects, whether sample survey or experiment.

BASIC DATA ETHICS

The organization that carries out the study must have an **institutional review board** that reviews all planned studies in advance in order to protect the subjects from possible harm.

All individuals who are subjects in a study must give their **informed consent** before data are collected.

All individual data must be kept **confidential**. Only statistical summaries for groups of subjects may be made public.

The law requires that studies funded by the federal government obey these principles. But neither the law nor the consensus of experts is completely clear about the details of their application.

Institutional review boards

The purpose of an institutional review board is not to decide whether a proposed study will produce valuable information or whether it is statistically sound. The board's purpose is, in the words of one university's board, "to protect the rights and welfare of human subjects (including patients) recruited to participate in research activities." When protocols are greater than minimal risk, a statistician is often included on the board to help determine benefits.

The board reviews the plan of the study and can require changes. It reviews the consent form to be sure that subjects are informed about the nature of the study and about any potential risks. Once research begins, the board monitors its progress at least once a year.

The most pressing issue concerning institutional review boards is whether their workload has become so large that their effectiveness in protecting subjects drops. There are shorter review procedures for projects that involve only minimal risks to subjects, such as most sample surveys. When a board is overloaded, there is a temptation to put more proposals in the minimal-risk category to speed the work.

USE YOUR KNOWLEDGE

The exercises in this section on ethics are designed to help you think about the issues that we are discussing and to formulate some opinions. In general, there are no wrong or right answers, but you need to give reasons for your answers.

3.71 Do these proposals involve minimal risk? You are a member of your college's institutional review board. You must decide whether several research proposals qualify for lighter review because they involve only minimal risk to subjects. Federal regulations say that "minimal risk" means that the risks are no greater than "those ordinarily encountered in daily life or during the performance of routine physical or psychological examinations or tests." That's vague. Which of these do you think qualifies as "minimal risk"? Explain your choices.

(a) Draw a drop of blood by pricking a finger in order to measure blood sugar.

(b) Draw blood from the arm for a full set of blood tests.

(c) Insert a tube that remains in the arm so that blood can be drawn regularly.

3.72 Who should be on an institutional review board? Government regulations require that institutional review boards consist of at least five people, including at least one scientist, one nonscientist, and one person from outside the institution. Most boards are larger, but many contain just one outsider.

(a) Why should review boards contain people who are not scientists?

(b) Do you think that one outside member is enough? How would you choose that member? (For example, would you prefer a medical doctor? A member of the clergy? An activist for patients' rights?)

Informed consent

Both words in the phrase "informed consent" are important, and both can be controversial. Subjects must be *informed* in advance about the nature of a study and any risk of harm it may bring. In the case of a sample survey, physical harm is not possible. The subjects should be told what kinds of questions the survey will ask and about how much of their time it will take. Experimenters must tell subjects the nature and purpose of the study and outline possible risks. Subjects must then *consent* in writing.

EXAMPLE 3.32

Who can give informed consent? Are there some subjects who can't give informed consent? It was once common, for example, to test new vaccines on prison inmates who gave their consent in return for good-behavior credit. Now we worry that prisoners are not really free to refuse, and the law forbids medical experiments in prisons.

Young children can't give fully informed consent, so the usual procedure is to ask their parents. A study of new ways to teach reading is about to start at a local elementary school, so the study team sends consent forms home to parents. Many parents don't return the forms. Can their children take part in the study because the parents did not say No, or should we allow only children whose parents returned the form and said Yes?

What about research into new medical treatments for people with mental disorders? What about studies of new ways to help emergency room patients who may be unconscious or have suffered a stroke? In most cases, there is not time even to get the consent of the family. Does the principle of informed consent bar realistic trials of new treatments for unconscious patients?

These are questions without clear answers. Reasonable people differ strongly on all of them. There is nothing simple about informed consent.[25]

The difficulties of informed consent do not vanish even for capable subjects. Some researchers, especially in medical trials, regard consent as a barrier to getting patients to participate in research. They may not explain all possible risks; they may not point out that there are other therapies that might be better than those being studied; they may be too optimistic in talking with patients even when the consent form has all the right details.

On the other hand, mentioning every possible risk leads to very long consent forms that really are barriers. "They are like rental car contracts," one lawyer said. Some subjects don't read forms that run five or six printed pages. Others are frightened by the large number of possible (but unlikely) disasters that might happen and so refuse to participate. Of course, unlikely disasters sometimes happen. When they do, lawsuits follow and the consent forms become yet longer and more detailed.

Confidentiality

Ethical problems do not disappear once a study has been cleared by the review board, has obtained consent from its subjects, and has actually collected data about the subjects. **Confidentiality** means that only the researchers can identify responses of individual subjects. The report of an opinion poll may say what percent of the 1500 respondents felt that legal immigration should be reduced. It may not report what *you* said about this or any other issue.

confidentiality

anonymity

Confidentiality is not the same as **anonymity**. Anonymity means that subjects are anonymous—their names are not known even to the director of the study. Anonymity is rare in statistical studies. Even where anonymity is possible (mainly in surveys conducted by mail), it prevents any follow-up to improve nonresponse or inform subjects of results.

Any breach of confidentiality is a serious violation of data ethics. The best practice is to separate the identity of the subjects from the rest of the data at once. Sample surveys, for example, use the identification only to check on who did or did not respond. In an era of advanced technology, however, it is no longer enough to be sure that each individual set of data protects people's privacy.

The government, for example, maintains a vast amount of information about citizens in many separate databases—census responses, tax returns, Social Security information, data from surveys such as the Current Population Survey, and so on. Many of these databases can be searched by computers for statistical studies.

A clever computer search of several databases might be able, by combining information, to identify you and learn a great deal about you even if your name and other identification have been removed from the data available for search. A colleague from Germany once remarked that "female full professor of statistics with a PhD from the United States" was enough to identify her among all the citizens of Germany. Privacy and confidentiality of data are hot issues among statisticians in the computer age.

EXAMPLE 3.33

Data collected by the government. Citizens are required to give information to the government. Think of tax returns and Social Security contributions. The government needs these data for administrative purposes—to see if we paid the right amount of tax and how large a Social Security benefit we are owed when we retire. Some people feel that individuals should be able to forbid any other use of their data, even with all identification removed. This would prevent using government records to study, say, the ages, incomes, and household sizes of Social Security recipients. Such a study could well be vital to debates on reforming Social Security.

USE YOUR KNOWLEDGE

3.73 How can we obtain informed consent? A researcher suspects that traditional religious beliefs tend to be associated with an authoritarian personality. She prepares a questionnaire that measures authoritarian tendencies and also asks many religious questions. Write a description of the purpose of this research to be read by subjects in order to obtain their informed consent. You must balance the conflicting goals of not deceiving the subjects as to what the questionnaire will tell about them and of not biasing the sample by scaring off religious people.

3.74 Should we allow this personal information to be collected? In which of the following circumstances would you allow collecting personal information without the subjects' consent?

(a) A government agency takes a random sample of income tax returns to obtain information on the average income of people in different occupations. Only the incomes and occupations are recorded from the returns, not the names.

(b) A social psychologist attends public meetings of a religious group to study the behavior patterns of members.

(c) A social psychologist pretends to be converted to membership in a religious group and attends private meetings to study the behavior patterns of members.

Clinical trials

Clinical trials are experiments that study the effectiveness of medical treatments on actual patients. Medical treatments can harm as well as heal, so clinical trials spotlight the ethical problems of experiments with human subjects. Here are the starting points for a discussion:

• Randomized comparative experiments are the only way to see the true effects of new treatments. Without them, risky treatments that are no better than placebos will become common.

• Clinical trials produce great benefits, but most of these benefits go to future patients. The trials also pose risks, and these risks are borne by the subjects of the trial. So we must balance future benefits against present risks.

• Both medical ethics and international human rights standards say that "the interests of the subject must always prevail over the interests of science and society."

The quoted words are from the 1964 Helsinki Declaration of the World Medical Association, the most respected international standard. The most outrageous examples of unethical experiments are those that ignore the interests of the subjects.

National Archives and Records Administration NARA

EXAMPLE 3.34

The Tuskegee study. In the 1930s, syphilis was common among black men in the rural South, a group that had almost no access to medical care. The Public Health Service Tuskegee study recruited 399 poor black sharecroppers with syphilis and 201 others without the disease in order to observe how syphilis progressed when no treatment was given. Beginning in 1943, penicillin became available to treat syphilis. The study subjects were not treated. In fact, the Public Health Service prevented any treatment until word leaked out and forced an end to the study in the 1970s.

The Tuskegee study is an extreme example of investigators following their own interests and ignoring the well-being of their subjects. A 1996 review said, "It has come to symbolize racism in medicine, ethical misconduct in human research, paternalism by physicians, and government abuse of vulnerable people." In 1997, President Clinton formally apologized to the surviving participants in a White House ceremony.[26]

Because "the interests of the subject must always prevail," medical treatments can be tested in clinical trials only when there is reason to hope that they will help the patients who are subjects in the trials. Future benefits aren't enough to justify experiments with human subjects. Of course, if there is already strong evidence that a treatment works and is safe, it is unethical *not* to give it.

Here is the view of Dr. Charles Hennekens of the Harvard Medical School, who directed the large clinical trial that showed that aspirin reduces the risk of heart attacks:[27] A clinical trial is justified if there is some evidence that the treatment will be effective. This evidence, however, is not sufficiently strong, to conclude that we would be harming the subjects who would receive the placebo.

Why is it ethical to give a control group of patients a placebo? Well, we know that placebos often work. What is more, placebos have no harmful side effects. In fact, the placebo group may be getting a better treatment than the drug group. If we *knew* which treatment was better, we would give it to everyone. When we don't know, it is ethical to try both and compare them.

The idea of using a control or placebo is a fundamental principle to be considered in designing experiments. In many situations, deciding what to use as an appropriate control requires some careful thought.

The choice of the control can have a substantial impact on how the results of an experiment are interpreted. Here is an example.

EXAMPLE 3.35

Attentiveness improves by nearly 20%. The manufacturer of a breakfast cereal designed for children claims that eating this cereal has been clinically shown to improve attentiveness by nearly 20%. The study used two groups of children who were tested before and after breakfast. One group received the cereal for breakfast, while breakfast for the control group was water. The results of the tests taken three hours after breakfast were used in the claim.

The Federal Trade Commission investigated the marketing of this product. It charged that the claim was false and violated federal law. The charges were settled and the company agreed to not use misleading claims in its advertising.[28]

It is not sufficient to obtain appropriate controls. The data must be collected from all groups in the same way. Here is an example of this type of flawed design.

EXAMPLE 3.36

Accurate identification of ovarian cancer. Two scientists published a paper claiming to have developed an exciting new method to detect ovarian cancer using blood samples. When other scientists were unable to reproduce the results in different labs, the original work was examined more carefully. In the original study, there were samples for women with ovarian cancer and for healthy controls. The blood samples were all analyzed using a mass spectrometer. The control samples were analyzed on one day, and the cancer samples were analyzed on the next day. This design was flawed in that it could not control for changes over time in the measuring instrument.[29]

USE YOUR KNOWLEDGE

3.75 **Is this study ethical?** Researchers on aging proposed to investigate the effect of supplemental health services on the quality of life of older people. Eligible patients on the rolls of a large medical clinic were to be randomly assigned to treatment and control groups. The treatment group would be offered hearing aids, dentures, transportation, and other services not available without charge to the control group. The review board felt that providing these services to some but not other persons in the same institution raised ethical questions. Do you agree?

3.76 **Should the treatments be given to everyone?** Effective drugs for treating AIDS are very expensive, so most African nations cannot afford to give them to large numbers of people. Yet AIDS is more common in parts of Africa than anywhere else. Several clinical trials are looking at ways to prevent pregnant mothers infected with HIV from passing the infection to their unborn children, a major source of HIV infections in Africa. Some people say these trials are unethical because they do not give effective AIDS drugs to their subjects, as would be required in rich nations. Others reply that the trials are looking for treatments that can work in the real world in Africa and that they promise benefits at least to the children of their subjects. What do you think?

Behavioral and social science experiments

When we move from medicine to the behavioral and social sciences, the direct risks to experimental subjects are less acute, but so are the possible benefits to the subjects. Consider, for example, the experiments conducted by psychologists in their study of human behavior.

EXAMPLE 3.37

Personal space. Psychologists observe that people have a "personal space" and get annoyed if others come too close to them. We don't like strangers to sit at our table in a coffee shop if other tables are available, and we see people move apart in elevators if there is room to do so. Americans tend to require more personal space than people in most other cultures. Can violations of personal space have physical, as well as emotional, effects?

Investigators set up shop in a men's public restroom. They blocked off urinals to force men walking in to use either a urinal next to an experimenter (treatment group) or a urinal separated from the experimenter (control group). Another experimenter, using a periscope from a toilet stall, measured how long the subject took to start urinating and how long he kept at it.[30]

This personal space experiment illustrates the difficulties facing those who plan and review behavioral studies:

• There is no risk of harm to the subjects, although they would certainly object to being watched through a periscope. What should we protect subjects from when physical harm is unlikely? Possible emotional harm? Undignified situations? Invasion of privacy?

• What about informed consent? The subjects in Example 3.37 did not even know they were participating in an experiment. Many behavioral experiments rely on hiding the true purpose of the study. The subjects would change their behavior if told in advance what the investigators were looking for. Subjects are asked to consent on the basis of vague information. They receive full information only after the experiment.

The "Ethical Principles" of the American Psychological Association require consent unless a study merely observes behavior in a public place. They allow deception only when it is necessary to the study, does not hide information that might influence a subject's willingness to participate, and is explained to subjects as soon as possible. The personal space study (from the 1970s) does not meet current ethical standards.

We see that the basic requirement for informed consent is understood differently in medicine and psychology. Here is an example of another setting with yet another interpretation of what is ethical. The subjects get no information and give no consent. They don't even know that an experiment may be sending them to jail for the night.

EXAMPLE 3.38

Domestic violence. How should police respond to domestic violence calls? In the past, the usual practice was to remove the offender and order him to stay out of the household overnight. Police were reluctant to make arrests because the victims rarely pressed charges. Women's groups argued that arresting offenders would help prevent future violence even if no charges were filed. Is there evidence that arrest will reduce future offenses? That's a question that experiments have tried to answer.

A typical domestic violence experiment compares two treatments: arrest the suspect and hold him overnight, or warn the suspect and release him. When police officers reach the scene of a domestic violence call, they calm the participants and investigate. Weapons or death threats require an arrest. If the facts permit an arrest but do not require it, an officer radios headquarters for instructions. The person on duty opens the next envelope in a file prepared in advance by a statistician. The envelopes contain the treatments in random order. The police either arrest the suspect or warn and release him, depending on the contents of the envelope. The researchers then watch police records and visit the victim to see if the domestic violence reoccurs.

The first such experiment appeared to show that arresting domestic violence suspects does reduce their future violent behavior. As a result of this evidence, arrest has become the common police response to domestic violence.

The domestic violence experiments shed light on an important issue of public policy. Because there is no informed consent, the ethical rules that govern clinical trials and most social science studies would forbid these experiments.

They were cleared by review boards because, in the words of one domestic violence researcher, "These people became subjects by committing acts that allow the police to arrest them. You don't need consent to arrest someone."

SECTION 3.4 SUMMARY

- Approval of an **institutional review board** is required for studies that involve humans or animals as subjects.

- Human subjects must give **informed consent** if they are to participate in experiments.

- Data on human subjects must be kept **confidential**.

SECTION 3.4 EXERCISES

For Exercises 3.71 and 3.72, see pages 204–205; for Exercises 3.73 and 3.74, see page 207; and for Exercises 3.75 and 3.76, see page 209.

3.77 Did you give informed consent? You were asked to participate in a study by a friend who is recruiting subjects. You trust your friend and you tell her that you are willing to do whatever is needed for the study. Have you given informed consent? Explain your answer.

3.78 Are the data confidential? You have participated in a study, and the results were published in an article in a very prestigious journal. Only summary information was published. The policy of the journal requires that all data used in the articles it publishes be available to the public, and it archives the data on a website. When you examine the data, you realize that you have a unique set of characteristics that would allow someone who knows you very well to identify which data are from you. Someone who does not know you would not be able to do this. Are the data confidential? Explain your answer.

3.79 Is the IRB responsible? An institutional review board (IRB) approved an experimental cancer vaccine for use in a clinical trial. The subjects were patients who had advanced disease and had received standard treatments with no success. Of the 94 subjects who received the vaccine, 26 died during the study. Their deaths were not due to the vaccine. Some family members of the subjects sued the hospital, the study director, the company that made the vaccine, a university official, individual members of the IRB, and the university bioethicist who consulted with the IRB.[31] Discuss this case from the point of view of ethics. Include any additional information that you would need to form your opinion.

3.80 What is wrong? Explain what is wrong in each of the following scenarios.

(a) Clinical trials are always ethical as long as they randomly assign patients to the treatments.

(b) The job of an institutional review board is complete when it decides to allow a study to be conducted.

(c) A treatment that has no risk of physical harm to subjects is always ethical.

3.81 How should the samples have been analyzed? Refer to the ovarian cancer diagnostic test study in Example 3.36 (page 209). Describe how you would process the samples through the mass spectrometer.

3.82 The Vytorin controversy. Vytorin is a combination pill designed to lower cholesterol. It consists of a relatively inexpensive and widely used drug, Zocor, and a newer drug called Zetia. Early study results suggested that Vytorin was no more effective than Zetia. Critics claimed that the makers of the drugs tried to change the response variable for the study, and two congressional panels investigated why there was a two-year delay in the release of the results. Use the Internet to search for more information about this controversy and write a report about what you find. Include an evaluation in the framework of ethical use of experiments and data. A good place to start your search would be to look for the phrase "Vytorin's shortcomings."

3.83 The General Social Survey. One of the most important nongovernment surveys in the United States is the National Opinion Research Center's General Social Survey. The GSS regularly monitors public opinion on a wide variety of political and social issues. Interviews are conducted in person in the subject's home. Are a subject's responses to GSS questions anonymous, confidential, or both? Explain your answer.

3.84 Anonymity and confidentiality in health screening. Texas A&M, like many universities, offers free screening for HIV, the virus that causes AIDS. The announcement says, "Persons who sign up for the HIV Screening will be assigned a number so that they do not have to give their name." They can learn the results of the test by telephone, still without giving their name. Does this practice offer *anonymity* or just *confidentiality*?

3.85 Anonymity and confidentiality in mail surveys. Some common practices may appear to offer anonymity while actually delivering only confidentiality. Market researchers often use mail surveys that do not ask the respondent's identity but contain hidden codes on the questionnaire that identify the respondent. A false claim of anonymity is clearly unethical. If only confidentiality is promised, is it also unethical to say nothing about the identifying code, perhaps causing respondents to believe their replies are anonymous?

3.86 Use of stored blood. Long ago, doctors drew a blood specimen from you as part of treating minor anemia. Unknown to you, the sample was stored. Now researchers plan to use stored samples from you and many other people to look for genetic factors that may influence anemia. It is no longer possible to ask your consent. Modern technology can read your entire genetic makeup from the blood sample.

(a) Do you think it violates the principle of informed consent to use your blood sample if your name is on it but you were not told that it might be saved and studied later?

(b) Suppose that your identity is not attached. The blood sample is known only to come from (say) "a 20-year-old white female being treated for anemia." Is it now OK to use the sample for research?

(c) Perhaps we should use biological materials such as blood samples only from patients who have agreed to allow the material to be stored for later use in research. It isn't possible to say in advance what kind of research, so this falls short of the usual standard for informed consent. Is it nonetheless acceptable, given complete confidentiality and the fact that using the sample can't physically harm the patient?

3.87 Political polls. The presidential election campaign is in full swing, and the candidates have hired polling organizations to take regular polls to find out what the voters think about the issues.

(a) What information should the pollsters be required to give out?

(b) What does the standard of informed consent require the pollsters to tell potential respondents?

(c) The standards accepted by polling organizations also require giving respondents the name and address of the organization that carries out the poll. Why do you think this is required?

(d) The polling organization usually has a professional name such as "Samples Incorporated," so respondents don't know that the poll is being paid for by a political party or candidate. Would revealing the sponsor to respondents bias the poll? Should the sponsor always be announced whenever poll results are made public?

3.88 Should poll results be made public? Some people think that the law should require that all political results be made public. Otherwise, the possessors of poll results can use the information to their own advantage. They can act on the information, release only selected parts of it, or time the release for best effect. A candidate's organization replies that it is paying for the poll to gain information for its own use, not to amuse the public. Do you favor requiring complete disclosure of political poll results? What about other private surveys, such as market research surveys of consumer tastes?

3.89 Informed consent to take blood samples. Researchers from Yale, working with medical teams in Tanzania, wanted to know how common infection with the AIDS virus is among pregnant women in that country. To do this, they planned to test blood samples drawn from pregnant women.

Yale's institutional review board insisted that the researchers get the informed consent of each woman and tell her the results of the test. This is the usual procedure in developed nations. The Tanzanian government did not want to tell the women why blood was drawn or tell them the test results. The government feared panic if many people turned out to have an incurable disease for which the country's medical system could not provide care. The study was canceled. Do you think that Yale was right to apply its usual standards for protecting subjects?

CHAPTER 3 EXERCISES

3.90 Visit statistics and the news. STATS is an organization concerned about the appropriate reporting of statistical knowledge in the news media. Visit the website **stats.org/blog.** Some recent postings include discussions of deflategate, health claims for coffee, and the graph that launched a thousand news stories. Read one of the articles posted on this site and then write a short report summarizing the major ideas in the article.

3.91 Online behavioral advertising. The Federal Trade Commission Staff Report, "Self-Regulatory Principles for Online Behavioral Advertising," defines behavioral advertising as "the tracking of a consumer's online activities over time—including the searches the consumer has conducted, the Web pages visited and the content viewed—in order to deliver advertising targeted to the individual consumer's interests." The report suggests four governing concepts for their proposals. These are (1) transparency and control: when companies collect information from consumers for advertising, they should tell consumers how the data will be collected, and consumers should be given a choice about whether to allow the data to be collected; (2) security and data retention: data should be kept secure and should be retained only as long as they are needed; (3) privacy: before data are used in a way that differs from promises made when they were collected, consent should be obtained from the consumer; and (4) sensitive data: affirmative express consent should be obtained before using any sensitive data.[32] Write a report discussing your opinions concerning online behavioral advertising and the four governing concepts. Pay particular attention to issues related to the ethical collection and use of statistical data.

3.92 Confidentiality at NORC. The National Opinion Research Center conducts a large number of surveys and has established procedures for protecting the confidentiality of its survey participants. For its Survey of Consumer Finances, NORC provides a pledge to participants regarding confidentiality. This pledge is available at **scf.norc.org/confidentiality.html**. Review the pledge and summarize its key parts. Do you think that the pledge adequately addresses issues related to the ethical collection and use of data? Explain your answer.

3.93 Make it an experiment! In the following observational studies, describe changes that could be made to the data collection process that would result in an experiment rather than an observational study. Also, offer suggestions about unseen biases or lurking variables that may be present in the studies as they are described here.

(a) A friend of yours likes to play Texas hold 'em. Every time that he tells you about his playing, he says that he won.

(b) In an introductory statistics class, you notice that the students who sit in the first two rows of seats had higher scores on the first exam than the other students in the class.

3.94 Name the designs. What is the name for each of these study designs?

(a) A study to compare two methods of preserving wood started with boards of southern white pine. Each board was ripped from end to end to form two edge-matched specimens. One was assigned to Method A; the other, to Method B.

(b) A survey on youth and smoking contacted by telephone 300 smokers and 300 nonsmokers, all 14 to 22 years of age.

(c) Does air pollution induce DNA mutations in mice? Starting with 40 male and 40 female mice, 20 of each sex were housed in a polluted industrial area downwind from a steel mill. The other 20 of each sex were housed at an unpolluted rural location 30 kilometers away.

3.95 Price promotions and consumer expectations. A researcher studying the effect of price promotions on consumer expectations makes up two different histories of the store price of a hypothetical brand of laundry detergent for the past year. Students in a marketing course view one or the other price history on a computer. Some students see a steady price, while others see regular promotions that temporarily cut the price. The students are then asked what price they would expect to pay for the detergent. Is this study an experiment? Why? What are the explanatory and response variables?

3.96 Calcium and healthy bones. Adults need to eat foods or supplements that contain enough calcium to maintain healthy bones. Calcium intake is generally measured in milligrams per day (mg/d), and one measure of healthy bones is total body bone mineral density measured in grams per centimeter squared (TBBMD, g/cm^2). Suppose that you want to study the relationship between calcium intake and TBBMD.

(a) Design an observational study to study the relationship.

(b) Design an experiment to study the relationship.

(c) Compare the relative merits of your two designs. Which do you prefer? Give reasons for your answer.

3.97 Choose the type of study. Give an example of a question about pets and their owners, their behavior, or their opinions that would best be answered by

(a) a sample survey.

(b) an observational study that is not a sample survey.

(c) an experiment.

3.98 Compare the fries. Do consumers prefer the fries from Burger King or from McDonald's? Design a blind test in which the source of the fries is not identified. Describe briefly the design of a matched pairs experiment to investigate this question. How will you use randomization?

3.99 Bicycle gears. How does the time it takes a bicycle rider to travel 100 meters depend on which gear is used and how steep the course is? It may be, for example, that higher gears are faster on level ground, but lower gears are faster on steep inclines. Discuss the design of a two-factor experiment to investigate this issue, using one bicycle with three gears and one rider. How will you use randomization?

3.100 Design an experiment. The previous two exercises illustrate the use of statistically designed experiments to answer questions that arise in everyday life. Select a question of interest to you that an experiment might answer, and carefully discuss the design of an appropriate experiment.

3.101 Design a survey. You want to investigate the attitudes of students at your school about the faculty's commitment to teaching. The student government will pay the costs of contacting about 500 students.

(a) Specify the exact population for your study; for example, will you include part-time students?

(b) Describe your sample design. Will you use a stratified sample?

(c) Briefly discuss the practical difficulties that you anticipate; for example, how will you contact the students in your sample?

3.102 Compare two doses of a drug. A drug manufacturer is studying how a new drug behaves in patients. Investigators compare two doses: 5 milligrams (mg) and 10 mg. The drug can be administered by injection, by a skin patch, or by intravenous drip. Concentration in the blood after 30 minutes (the response variable) may depend both on the dose and on the method of administration.

(a) Make a sketch that describes the treatments formed by combining dose and method. Then use a diagram to outline a completely randomized design for this two-factor experiment.

(b) "How many subjects?" is a tough issue. We will explain the basic ideas in Chapter 6. What can you say now about the advantage of using larger groups of subjects?

3.103 Would the results be different for men and women? The drug that is the subject of the experiment in Exercise 3.102 may behave differently in men and women. How would you modify your experimental design to take this into account?

3.104 Informed consent. The requirement that human subjects give their informed consent to participate in an experiment can greatly reduce the number of available subjects. For example, a study of new teaching methods asks the consent of parents for their children to be randomly assigned to be taught by either a new method or the standard method. Many parents do not return the forms, so their children must continue to be taught by the standard method. Why is it not correct to consider these children as part of the control group along with children who are randomly assigned to the standard method?

3.105 Two ways to ask sensitive questions. Sample survey questions are usually read from a computer screen. In a Computer-aided personal interview (CAPI), the interviewer reads the questions and enters the responses. In a Computer-aided self interview (CASI), the interviewer stands aside and the respondent reads the questions and enters responses. One method almost always shows a higher percent of subjects admitting use of illegal drugs. Which method? Explain why.

3.106 Your institutional review board. Your college or university has an institutional review board that screens all studies that use human subjects. Get a copy of the document that describes this board (you can probably find it online).

(a) According to this document, what are the duties of the board?

(b) How are members of the board chosen? How many members are not scientists? How many members are not employees of the college? Do these members have some special expertise, or are they simply members of the "general public"?

3.107 Use of data produced by the government. Data produced by the government are often available free or at low cost to private users. For example, satellite weather data produced by the U.S. National Weather Service are available free to TV stations for their weather reports and to anyone on the Internet. *Opinion 1:* Government data should be available to everyone at minimal cost. *Opinion 2:* The satellites are expensive, and the TV stations are making a profit from their weather services, so they should share the cost. European governments, for example, charge TV stations for weather data. Which opinion do you support, and why?

3.108 Should we ask for the consent of the parents? The Centers for Disease Control and Prevention, in a survey of teenagers, asked the subjects if they were sexually active. Those who said Yes were then asked, "How old were you when you had sexual intercourse for the first time?" Should consent of parents be required to ask minors about sex, drugs, and other such issues, or is consent of the minors themselves enough? Give reasons for your opinion.

3.109 A theft experiment. Students sign up to be subjects in a psychology experiment. When they arrive, they are told that interviews are running late and are taken to a waiting room. The experimenters then stage a theft of a valuable object left in the waiting room. Some subjects are alone with the thief, and others are in pairs—these are the treatments being compared. Will the subject report the theft? The students had agreed to take part in an unspecified study, and the true nature of the experiment is explained to them afterward. Do you think this study is ethically OK?

3.110 A cheating experiment. A psychologist conducts the following experiment. She measures the attitude of subjects toward cheating and then has them play a game rigged so that winning without cheating is impossible. The computer that organizes the game also records—unknown to the subjects—whether or not they cheat. Then attitude toward cheating is retested. Subjects who cheat tend to change their attitudes to find cheating more acceptable. Those who resist the temptation to cheat tend to condemn cheating more strongly on the second test of attitude. These results confirm the psychologist's theory. This experiment tempts subjects to cheat. The subjects are led to believe that they can cheat secretly when, in fact, they are observed. Is this experiment ethically objectionable? Explain your position.

Probability: The Study of Randomness

Jgroup/Dreamstime.com

Introduction

The reasoning of statistical inference rests on asking, "How often would this method give a correct answer if I used it very many times?" When we produce data by random sampling or randomized comparative experiments, the laws of probability answer the question, "What would happen if we did this many times?" Games of chance like Texas hold 'em are exciting because the outcomes are determined by the rules of probability.

4.1 Randomness

4.2 Probability Models

4.3 Random Variables

4.4 Means and Variances of Random Variables

4.5 General Probability Rules

4.1 Randomness

When you complete this section, you will be able to:

- Identify random phenomena.
- Interpret the term *probability* for particular examples.
- Identify trials as independent or not.

Toss a coin, or choose a simple random sample (SRS). The result can't be predicted in advance because the result will vary when you toss the coin or choose the sample repeatedly. But there is, nonetheless, a regular pattern in the results, a pattern that emerges clearly only after many repetitions. This remarkable fact is the basis for the idea of probability.

EXAMPLE 4.1

Toss a coin 5000 times. When you toss a coin, there are only two possible outcomes, heads or tails. Figure 4.1 shows the results of tossing a coin 5000 times twice (Trial A and Trial B). For each number of tosses from 1 to 5000, we have plotted the proportion of those tosses that gave a head.

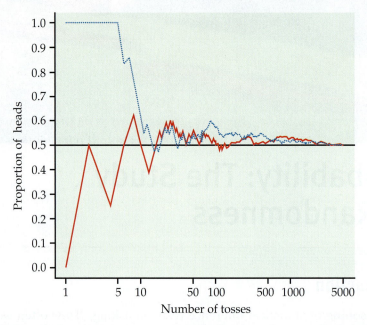

FIGURE 4.1 The proportion of tosses of a coin that give a head varies as we make more tosses. Eventually, however, the proportion approaches 0.5, the probability of a head. This figure shows the results of two trials of 5000 tosses each, Example 4.1.

Trial A (red line) begins tail, head, tail, tail. You can see that the proportion of heads for Trial A starts at 0 on the first toss, rises to 0.5 when the second toss gives a head, then falls to 0.33 and 0.25 as we get two more tails. Trial B (blue dotted line), on the other hand, starts with five straight heads, so the proportion of heads is 1 until the sixth toss.

The proportion of tosses that produce heads is quite variable at first. Trial A starts low and Trial B starts high. As we make more and more tosses, however, the proportion of heads for each trial gets close to 0.5 and stays there.

If we made yet a third trial at tossing the coin a great many times, the proportion of heads would again settle down to 0.5 in the long run. We say that 0.5 is the **probability** of a head. The probability 0.5 appears as a horizontal line on the graph.

probability

The *Probability* applet on the text website animates Figure 4.1. It allows you to choose the probability of a head and simulate any number of tosses of a coin with that probability. Try it. You will see that the proportion of heads gradually settles down close to the chosen probability. Equally important, you will also see that the proportion in a small or moderate number of tosses can be far from this probability. *Probability describes only what happens in the long run. Most people expect chance outcomes to show more short-term regularity than is actually true.*

EXAMPLE 4.2

Significance testing and Type I errors. In Chapter 6, we will learn about significance testing and Type I errors. When we perform a significance test, we have the possibility of making a Type I error under certain circumstances. The significance-testing procedure is set up so that the probability of making this kind of error is small, usually 5%. If we perform a large number of significance tests under this set of circumstances, the proportion of times that we will make a Type I error is 0.05.

fair coin

In the coin toss setting, the probability of a head is a characteristic of the coin being tossed. A coin is called **fair** if the probability of a head is 0.5; that is, it is equally likely to come up heads or tails. If we toss a coin five times and it comes up heads for all five tosses, we might suspect that the coin is not fair. Is this outcome likely if, in fact, the coin is fair? This is what happened in Trials A and B of Example 4.1. We will learn a lot more about significance testing in later chapters. For now, we are content with some very general ideas.

When the Type I error of a statistical significance procedure is set at 0.05, this probability is a characteristic of the procedure. If we roll a pair of dice once, we do not know whether the sum of the faces will be seven or not. Similarly, if we perform a significance test once, we do not know if we will make a Type I error or not. However, if the procedure is designed to have a Type I error probability of 0.05, then we are much less likely than not to make a Type I error.

The language of probability

"Random" in statistics is not a synonym for "unpredictable" but a description of a kind of order that emerges in the long run. We often encounter the unpredictable side of randomness in our everyday experience, but we rarely see enough repetitions of the same random phenomenon to observe the long-term regularity that probability describes. You can see that regularity emerging in Figure 4.1. In the very long run, the proportion of tosses that give a head is 0.5. This is the intuitive idea of probability. Probability 0.5 means "occurs half the time in a very large number of trials."

RANDOMNESS AND PROBABILITY

We call a phenomenon **random** if individual outcomes are uncertain but there is, nonetheless, a regular distribution of outcomes in a large number of repetitions.

The **probability** of any outcome of a random phenomenon is the proportion of times the outcome would occur in a very long series of repetitions.

Not all coins are fair. In fact, most real coins have bumps and imperfections that make the probability of heads a little different from 0.5. The probability might be 0.499999 or 0.500002. For our study of probability in this chapter, we will assume that we know the actual values of probabilities. Thus, we assume things like fair coins, even though we know that real coins are not exactly fair. We do this to learn what kinds of outcomes we are likely to see when we make such assumptions. When we study statistical inference in later chapters, we look at the situation from the opposite point of view: given that we have observed certain outcomes, what can we say about the probabilities that generated these outcomes?

USE YOUR KNOWLEDGE

4.1 Use Table B. We can use the random digits in Table B in the back of the book to simulate tossing a fair coin. Start at line 131 and read the numbers from left to right. If the number is 0, 2, 4, 6, or 8, you will say that the coin toss resulted in a head; if the number is a 1, 3, 5, 7, or 9, the outcome is tails. Use the first 10 random digits on line 131 to simulate 10 tosses of a fair coin. What is the actual proportion of heads in your simulated sample? Explain why you did not get exactly five heads.

Probability describes what happens in very many trials, and we must actually observe many trials to pin down a probability. In the case of tossing a coin, some diligent people have in fact made thousands of tosses.

EXAMPLE 4.3

Many tosses of a coin. The French naturalist Count Buffon (1707–1788) tossed a coin 4040 times. Result: 2048 heads, or proportion 2048/4040 = 0.5069 for heads.

Around 1900, the English statistician Karl Pearson heroically tossed a coin 24,000 times. Result: 12,012 heads, a proportion of 0.5005.

While imprisoned by the Germans during World War II, the South African statistician John Kerrich tossed a coin 10,000 times. Result: 5067 heads, proportion of heads 0.5067.

Thinking about randomness

That some things are random is an observed fact about the world. The outcome of a coin toss, the time between emissions of particles by a radioactive source, and the sexes of the next litter of lab rats are all random. So is the outcome of a random sample or a randomized experiment. Probability theory is the branch of mathematics that describes random behavior. Of course, we can never observe a probability exactly. We could always continue tossing the coin, for example. Mathematical probability is an idealization based on imagining what would happen in an indefinitely long series of trials.

The best way to understand randomness is to observe random behavior—not only the long-run regularity but the unpredictable results of short runs. You can do this with physical devices such as coins and dice, but software simulations of random behavior allow faster exploration. As you explore randomness, remember:

independence

- You must have a long series of **independent** trials. That is, the outcome of one trial must not influence the outcome of any other. Imagine a crooked gambling house where the operator of a roulette wheel can stop it where she chooses—she can prevent the proportion of "red" from settling down to a fixed number. These trials are not independent.

• The idea of probability is empirical. Simulations start with given probabilities and imitate random behavior, but we can estimate a real-world probability only by actually observing many trials.

• Nonetheless, simulations are very useful because we need long runs of trials. In situations such as coin tossing, the proportion of an outcome often requires several hundred trials to settle down to the probability of that outcome. The kinds of physical random devices suggested in the exercises are too slow to make performing so many trials practical. Short runs give only rough estimates of a probability.

The uses of probability

Probability theory originated in the study of games of chance. Tossing dice, dealing shuffled cards, and spinning a roulette wheel are examples of deliberate randomization. In that respect, they are similar to random sampling. Although games of chance are ancient, they were not studied by mathematicians until the sixteenth and seventeenth centuries.

It is only a mild simplification to say that probability as a branch of mathematics arose when seventeenth-century French gamblers asked the mathematicians Blaise Pascal and Pierre de Fermat for help. Gambling is still with us, in casinos and state lotteries. We will make use of games of chance as simple examples that illustrate the principles of probability.

Careful measurements in astronomy and surveying led to further advances in probability in the eighteenth and nineteenth centuries because the results of repeated measurements are random and can be described by distributions much like those arising from random sampling. Similar distributions appear in data on human life span (mortality tables) and in data on lengths or weights in a population of skulls, leaves, or cockroaches.[1]

Now, we employ the mathematics of probability to describe the flow of traffic through a highway system, the Internet, or a computer processor; the genetic makeup of individuals or populations; the energy states of subatomic particles; the spread of epidemics or tweets; and the rate of return on risky investments. Although we are interested in probability because of its usefulness in statistics, the mathematics of chance is important in many fields of study.

SECTION 4.1 SUMMARY

• A **random phenomenon** has outcomes that we cannot predict but that nonetheless have a regular distribution in very many repetitions.

• The **probability** of an event is the proportion of times the event occurs in many repeated trials of a random phenomenon.

• Trials are **independent** if the outcome of one trial does not influence the outcome of any other trial.

SECTION 4.1 EXERCISES

For Exercise 4.1, see page 218.

4.2 Are these phenomena random? Identify each of the following phenomena as random or not. Give reasons for your answers.

(a) The outside temperature in your town at noon on Groundhog Day, February 2.

(b) The first digit in your student identification number.

(c) You draw an ace from a well-shuffled deck of 52 cards.

4.3 Interpret the probabilities. Refer to the previous exercise. In each case, interpret the term *probability* for the phenomena that are random. For those that are not random, explain why the term *probability* does not apply.

4.4 Are the trials independent? For each of the following situations, identify the trials as independent or not. Explain your answers.

(a) You record the outside temperature in your town at noon on Groundhog Day, February 2, each year for the next five years.

(b) The number of tweets that you receive on the next 10 Mondays.

(c) Your grades in the five courses that you are taking this semester.

4.5 Winning at craps. The game of craps starts with a "come-out" roll, in which the shooter rolls a pair of dice. If the total of the "spots" on the up-faces is 7 or 11, the shooter wins immediately (there are ways that the shooter can win on later rolls if other numbers are rolled on the come-out roll). Roll a pair of dice 25 times and estimate the probability that the shooter wins immediately on the come-out roll. For a pair of perfectly made dice, the probability is 0.2222.

4.6 Use the *Probability* applet. The idea of probability is that the *proportion* of heads in many tosses of a balanced coin eventually gets close to 0.5. But does the actual *count* of heads get close to one-half the number of tosses? Let's find out. Set the "Probability of Heads" in the *Probability* applet to 0.5 and the number of tosses to 50. You can extend the number of tosses by clicking "Toss" again to get 50 more. Don't click "Reset" during this exercise.

(a) After 50 tosses, what is the proportion of heads? What is the count of heads? What is the difference between the count of heads and 25 (one-half the number of tosses)?

(b) Keep going to 200 tosses. Again record the proportion and count of heads and the difference between the count and 100 (half the number of tosses).

(c) Keep going. Stop at 300 tosses and again at 400 tosses to record the same facts. Although it may take a long time, the laws of probability say that the proportion of heads will always get close to 0.5 and also that the difference between the count of heads and half the number of tosses will always grow without limit.

4.7 A question about dice. Here is a question that a French gambler asked the mathematicians Fermat and Pascal at the very beginning of probability theory: what is the probability of getting at least one 6 in rolling four dice? The *Law of Large Numbers* applet allows you to roll several dice and watch the outcomes. (Ignore the title of the applet for now.) Because simulation—just like real random phenomena—often takes very many trials to estimate a probability accurately, let's simplify the question: is this probability clearly greater than 0.5, clearly less than 0.5, or quite close to 0.5? Use the applet to roll four dice until you can confidently answer this question. You will have to set "Rolls" to 1 so that you have time to look at the four up-faces. Keep clicking "Roll dice" to roll again and again. How many times did you roll four dice? What percent of your rolls produced at least one 6?

4.2 Probability Models

When you complete this section, you will be able to:	• Describe a sample space from a description of a random phenomenon.
	• Apply the four probability rules.
	• Identify random phenomena that have equally likely outcomes and distinguish them from those that do not.

The idea of probability as a proportion of outcomes in very many repeated trials guides our intuition but is hard to express in mathematical form. A

probability model

description of a random phenomenon in the language of mathematics is called a **probability model**. To see how to proceed, think first about a very simple random phenomenon, tossing a coin once. When we toss a coin, we cannot know the outcome in advance. What do we know? We are willing to say that the outcome will be either heads or tails. Because the coin appears to be balanced, we believe that each of these outcomes has probability 1/2. This description of coin tossing has two parts:

- A list of possible outcomes.

- A probability for each outcome.

This two-part description is the starting point for a probability model. We will begin by describing the outcomes of a random phenomenon and then learn how to assign probabilities to the outcomes.

Sample spaces

A probability model first tells us what outcomes are possible.

> ### SAMPLE SPACE
>
> The **sample space S** of a random phenomenon is the set of all possible outcomes.

The name "sample space" is natural in random sampling, where each possible outcome is a sample and the sample space contains all possible samples. To specify S, we must state what constitutes an individual outcome and then state which outcomes can occur. We often have some freedom in defining the sample space, so the choice of S is a matter of convenience as well as correctness. The idea of a sample space, and the freedom we may have in specifying it, are best illustrated by examples.

EXAMPLE 4.4

Sample space for tossing a coin. Toss a coin. There are only two possible outcomes, and the sample space is

$$S = \{\text{heads, tails}\}$$

or, more briefly, $S = \{\text{H, T}\}$.

EXAMPLE 4.5

Sample space for random digits. Let your pencil point fall blindly into Table B of random digits. Record the value of the digit it lands on. The possible outcomes are

$$S = \{0, 1, 2, 3, 4, 5, 6, 7, 8, 9\}$$

EXAMPLE 4.6

Sample space for tossing a coin four times. Toss a coin four times and record the results. That's a bit vague. To be exact, record the results of each of the four tosses in order. A typical outcome is then HTTH. Counting shows that there are 16 possible outcomes. The sample space S is the set of all 16 strings of four H's and T's.

Suppose that our only interest is the number of heads in four tosses. Now we can be exact in a simpler fashion. The random phenomenon is to toss a coin four times and count the number of heads. The sample space contains only five outcomes:

$$S = \{0, 1, 2, 3, 4\}$$

This example illustrates the importance of carefully specifying what constitutes an individual outcome.

Although these examples seem remote from the practice of statistics, the connection is surprisingly close. Suppose that in conducting an opinion poll you select four people at random from a large population and ask each if he or she favors reducing federal spending on low-interest student loans. The answers are Yes or No. The possible outcomes—the sample space—are exactly as in Example 4.6 if we replace heads by Yes and tails by No. Similarly, the possible outcomes of an SRS of 1500 people are the same, in principle, as the possible outcomes of tossing a coin 1500 times. One of the great advantages of mathematics is that the essential features of quite different phenomena can be described by the same probability model.

USE YOUR KNOWLEDGE

4.8 When were you born? A student is asked "In what month were you born? Set up an appropriate sample space for this setting.

The sample spaces described in Examples 4.4, 4.5, and 4.6 correspond to categorical variables where we can list all the possible values. Other sample spaces correspond to quantitative variables. Here is an example.

EXAMPLE 4.7

Using software. Most statistical software has a function that will generate a random number between 0 and 1. The sample space is

$$S = \{\text{all numbers between 0 and 1}\}$$

This S is a mathematical idealization. Any specific random number generator produces numbers with some limited number of decimal places so that, strictly speaking, not all numbers between 0 and 1 are possible outcomes. For example, Minitab generates random numbers like 0.736891, with six decimal places. The entire interval from 0 to 1 is easier to think about. It also has the advantage of being a suitable sample space for different software systems that produce random numbers with different numbers of digits.

USE YOUR KNOWLEDGE

4.9 How many hours do you text? You record the number of hours per week that a randomly selected student spends texting. What is the sample space?

A sample space S lists the possible outcomes of a random phenomenon. To complete a mathematical description of the random phenomenon, we must also give the probabilities with which these outcomes occur.

The true long-term proportion of any outcome—say, "exactly two heads in four tosses of a coin"—can be found only empirically, and then only approximately. How then can we describe probability mathematically? Rather than immediately attempting to give "correct" probabilities, let's confront the easier task of laying down rules that any assignment of probabilities must satisfy. We need to assign probabilities not only to single outcomes but also to sets of outcomes.

EVENT

An **event** is an outcome or a set of outcomes of a random phenomenon. That is, an event is a subset of the sample space.

EXAMPLE 4.8

Exactly one head in four tosses. Take the sample space S for four tosses of a coin to be the 16 possible outcomes in the form HTHH. Then "exactly one head" is an event. Call this event A. The event A expressed as a set of outcomes is

$$A = \{\text{HTTT, THTT, TTHT, TTTH}\}$$

In a probability model, events have probabilities. What properties must any assignment of probabilities to events have? Here are some basic facts about any probability model. These facts follow from the idea of probability as "the long-run proportion of repetitions on which an event occurs."

1. **Any probability is a number between 0 and 1.** Any proportion is a number between 0 and 1, so any probability is also a number between 0 and 1. An event with probability 0 never occurs, and an event with probability 1 occurs on every trial. An event with probability 0.5 occurs in half the trials in the long run.

2. **All possible outcomes together must have probability 1.** Because every trial will produce an outcome, the sum of the probabilities for all possible outcomes must be exactly 1.

3. **If two events have no outcomes in common, the probability that one or the other occurs is the sum of their individual probabilities.** If one event occurs in 40% of all trials, a different event occurs in 25% of all trials, and the two can never occur together, then one or the other occurs in 65% of all trials because 40% + 25% = 65%.

4. **The probability that an event does not occur is 1 minus the probability that the event does occur.** If an event occurs in (say) 70% of all trials, it fails to occur in the other 30%. The probability that an event occurs and the probability that it does not occur always add to 100%, or 1.

Probability rules

Formal probability uses mathematical notation to state Facts 1 through 4 more concisely. We use capital letters near the beginning of the alphabet to denote events. If A is any event, we write its probability as $P(A)$. Here are our probability facts in formal language. As you apply these rules, remember that they are just another form of intuitively true facts about long-run proportions.

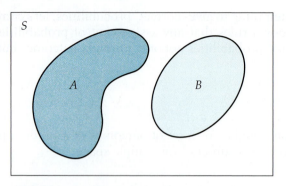

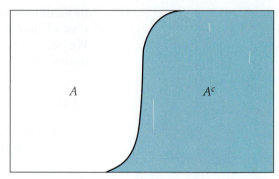

FIGURE 4.2 Venn diagram showing disjoint events A and B. Disjoint events have no common outcomes.

FIGURE 4.3 Venn diagram showing the complement A^c of an event A. The complement consists of all outcomes that are not in A.

PROBABILITY RULES

Rule 1. The probability $P(A)$ of any event A satisfies $0 \leq P(A) \leq 1$.

Rule 2. If S is the sample space in a probability model, then $P(S) = 1$.

Rule 3. Two events A and B are **disjoint** if they have no outcomes in common and so can never occur together. If A and B are disjoint,

$$P(A \text{ or } B) = P(A) + P(B)$$

This is the **addition rule for disjoint events**.

Rule 4. The **complement** of any event A is the event that A does not occur, written as A^c. The **complement rule** states that

$$P(A^c) = 1 - P(A)$$

You may find it helpful to draw a picture to remind yourself of the meaning of complements and disjoint events. A picture like Figure 4.2 that shows the sample space S as a rectangular area and events as areas within S is called a **Venn diagram**. The events A and B in Figure 4.2 are disjoint because they do not overlap. As Figure 4.3 shows, the complement A^c contains exactly the outcomes that are not in A.

Venn diagram

EXAMPLE 4.9

Favorite vehicle colors. What is your favorite color for a vehicle? Our preferences can be related to our personality, our moods, or particular objects. Here is a probability model for color preferences.[2]

Color	White	Black	Silver	Gray
Probability	0.24	0.19	0.16	0.15

Color	Red	Blue	Brown	Other
Probability	0.10	0.07	0.05	0.04

Each probability is between 0 and 1. The probabilities add to 1 because these outcomes together make up the sample space S. Our probability model corresponds to selecting a person at random and asking what is their favorite color.

Let's use the probability Rules 3 and 4 to find some probabilities for favorite vehicle colors.

EXAMPLE 4.10

Black or silver? What is the probability that a person's favorite vehicle color is black or silver? If the favorite is black, it cannot be silver, so these two events are disjoint. Using Rule 3, we find

$$P(\text{black or silver}) = P(\text{black}) + P(\text{silver})$$
$$= 0.19 + 0.16 = 0.35$$

There is a 35% chance that a randomly selected person will choose black or silver as their favorite color. Suppose that we want to find the probability that the favorite color is not blue.

EXAMPLE 4.11

Use the complement rule. To solve this problem, we could use Rule 3 and add the probabilities for white, black, silver, gray, red, brown and other. However, it is easier to use the probability that we have for blue and Rule 4. The event that the favorite is not blue is the complement of the event that the favorite is blue. Using our notation for events, we have

$$P(\text{not blue}) = 1 - P(\text{blue})$$
$$= 1 - 0.07 = 0.93$$

We see that 93% of people have a favorite vehicle color that is not blue.

USE YOUR KNOWLEDGE

4.10 Red or brown. Find the probability that the favorite color is red or brown.

4.11 White, black, silver, gray, or red. Find the probability that the favorite color is white, black, silver, gray, or red using Rule 4. Explain why this calculation is easier than finding the answer using Rule 3.

Assigning probabilities: Finite number of outcomes

The individual outcomes of a random phenomenon are always disjoint. So the addition rule provides a way to assign probabilities to events with more than one outcome: start with probabilities for individual outcomes and add to get probabilities for events. This idea works well when there are only a finite (fixed and limited) number of outcomes.

PROBABILITIES IN A FINITE SAMPLE SPACE

Assign a probability to each individual outcome. These probabilities must be numbers between 0 and 1 and must have sum 1.

The probability of any event is the sum of the probabilities of the outcomes making up the event.

EXAMPLE 4.12

Benford's law

Benford's law. Faked numbers in tax returns, payment records, invoices, expense account claims, and many other settings often display patterns that aren't present in legitimate records. Some patterns, such as too many round numbers, are obvious and easily avoided by a clever crook. Others are more subtle. It is a striking fact that the first digits of numbers in legitimate records often follow a distribution known as **Benford's law**. Here it is (note that a first digit can't be 0):[3]

First digit	1	2	3	4	5	6	7	8	9
Probability	0.301	0.176	0.125	0.097	0.079	0.067	0.058	0.051	0.046

Benford's law usually applies to the first digits of the sizes of similar quantities, such as invoices, expense account claims, and county populations. Investigators can detect fraud by comparing the first digits in records such as invoices paid by a business with these probabilities.

EXAMPLE 4.13

Find some probabilities for Benford's law. Consider the events

$$A = \{\text{first digit is 4}\}$$
$$B = \{\text{first digit is 7 or more}\}$$

From the table of probabilities in Example 4.12,

$$P(A) = P(4) = 0.097$$
$$P(B) = P(7) + P(8) + P(9)$$
$$= 0.058 + 0.051 + 0.046 = 0.155$$

Note that $P(B)$ is not the same as the probability that a first digit is strictly more than 7. The probability $P(7)$ that a first digit is 7 is included in "7 or more" but not in "more than 7."

USE YOUR KNOWLEDGE

4.12 Benford's law. Using the probabilities for Benford's law, find the probability that a first digit is anything other than 1.

4.13 Use the addition rule. Use the addition rule with the probabilities for the events A and B from Example 4.13 to find the probability that a first digit is either 4 or 7 or more.

Be careful to apply the addition rule only to disjoint events.

EXAMPLE 4.14

Find more probabilities for Benford's law. Check that the probability of the event C that a first digit is odd is

$$P(C) = P(1) + P(3) + P(5) + P(7) + P(9) = 0.609$$

The probability

$$P(B \text{ or } C) = P(1) + P(3) + P(5) + P(7) + P(8) + P(9) = 0.660$$

is *not* the sum of $P(B)$ and $P(C)$ because events B and C are not disjoint. Outcomes 7 and 9 are common to both events.

Assigning probabilities: Equally likely outcomes

Assigning correct probabilities to individual outcomes often requires long observation of the random phenomenon. In some circumstances, however, we are willing to assume that individual outcomes are equally likely because of some balance in the phenomenon. Ordinary coins have a physical balance that should make heads and tails equally likely, for example, and the table of random digits comes from a deliberate randomization.

EXAMPLE 4.15

First digits that are equally likely. You might think that first digits are distributed "at random" among the digits 1 to 9 in business records. The nine possible outcomes would then be equally likely. The sample space for a single digit is

$$S = \{1, 2, 3, 4, 5, 6, 7, 8, 9\}$$

Because the total probability must be 1, the probability of each of the nine outcomes must be 1/9. That is, the assignment of probabilities to outcomes is

First digit	1	2	3	4	5	6	7	8	9
Probability	1/9	1/9	1/9	1/9	1/9	1/9	1/9	1/9	1/9

The probability of the event B that a randomly chosen first digit is 7 or more is

$$P(B) = P(7) + P(8) + P(9)$$

$$= \frac{1}{9} + \frac{1}{9} + \frac{1}{9} = \frac{3}{9} = 0.333$$

Compare this with the Benford's law probability in Example 4.13. A person who fakes data by using "random" digits will end up with too many first digits that are 7 or more.

In Example 4.15, all outcomes have the same probability. Because there are nine equally likely outcomes, each must have probability 1/9. Because exactly three of the nine equally likely outcomes are 7 or more, the probability of this event is 3/9. In the special situation where all outcomes are equally likely, we have a simple rule for assigning probabilities to events.

EQUALLY LIKELY OUTCOMES

If a random phenomenon has k possible outcomes, all equally likely, then each individual outcome has probability $1/k$. The probability of any event A is

$$P(A) = \frac{\text{count of outcomes in } A}{\text{count of outcomes in } S}$$

$$= \frac{\text{count of outcomes in } A}{k}$$

Most random phenomena do not have equally likely outcomes, so the general rule for finite sample spaces (page 225) is more important than the special rule for equally likely outcomes.

USE YOUR KNOWLEDGE

4.14   Possible outcomes for rolling a die. A die has six sides with one to six spots on the sides. Give the probability distribution for the six possible outcomes that can result when a perfect die is rolled.

Independence and the multiplication rule

Rule 3, the addition rule for disjoint events, describes the probability that *one or the other* of two events *A* and *B* will occur in the special situation when *A* and *B* cannot occur together because they are disjoint. Our final rule describes the probability that *both* events *A* and *B* occur, again only in a special situation. More general rules appear in Section 4.5, but in our study of statistics, we will need only the rules that apply to special situations.

Suppose that you toss a fair coin twice. You are counting heads, so two events of interest are

$$A = \{\text{first toss is a head}\}$$
$$B = \{\text{second toss is a head}\}$$

The events *A* and *B* are not disjoint. They occur together whenever both tosses give heads. We want to compute the probability of the event {*A* and *B*} that *both* tosses are heads. The Venn diagram in Figure 4.4 illustrates the event {*A* and *B*} as the overlapping area that is common to both *A* and *B*.

The coin tossing of Buffon, Pearson, and Kerrich described in Example 4.3 makes us willing to assign probability 1/2 to a head when we toss a coin. So

$$P(A) = 0.5$$
$$P(B) = 0.5$$

What is *P*(*A* and *B*)? Our common sense says that it is 1/4. The first toss will give a head half the time and the second toss will give a head half the time, so both tosses will give heads on 1/2 × 1/2 = 1/4 of all trials in the long run. This reasoning assumes that the second toss still has probability 1/2 of a head after the first has given a head. This is true—we can verify it by tossing a coin twice many times and observing the proportion of heads on the second toss after the first toss has produced a head. We say that the events "head on the first toss" and "head on the second toss" are *independent*. Here is our final probability rule.

FIGURE 4.4 Venn diagram showing the event {*A* and *B*}. This event consists of outcomes common to *A* and *B*.

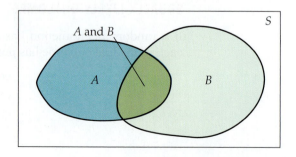

MULTIPLICATION RULE FOR INDEPENDENT EVENTS

Rule 5. Two events A and B are **independent** if knowing that one occurs does not change the probability that the other occurs. If A and B are independent,

$$P(A \text{ and } B) = P(A) \, P(B)$$

This is the **multiplication rule for independent events**.

Our definition of independence is rather informal. We will make this informal idea precise in Section 4.5. In practice, though, we rarely need a precise definition of independence because independence is usually *assumed* as part of a probability model when we want to describe random phenomena that seem to be physically unrelated to each other. Here is an example of independence.

EXAMPLE 4.16

Coins do not have memory. Because a coin has no memory, we assume that coin tosses are independent. For a fair coin, this means that the outcome of the first toss does not influence the outcome of any other toss.

USE YOUR KNOWLEDGE

4.15 A head and then a tail in two tosses. What is the probability of obtaining a head and then a tail on two tosses of a fair coin?

Here is an example of a situation where there are dependent events.

EXAMPLE 4.17

Dependent events in cards. The colors of successive cards dealt from the same deck are not independent. A standard 52-card deck contains 26 red and 26 black cards. For the first card dealt from a shuffled deck, the probability of a red card is $26/52 = 0.50$ because the 52 possible cards are equally likely. Once we see that the first card is red, we know that there are only 25 reds among the remaining 51 cards. The probability that the second card is red is therefore only $25/51 = 0.49$. Knowing the outcome of the first deal changes the probabilities for the second.

USE YOUR KNOWLEDGE

4.16 The probability of a second ace. A deck of 52 cards contains four aces, so the probability that a card drawn from this deck is an ace is $4/52$. If we know that the first card drawn is an ace, what is the probability that the second card drawn is also an ace? Using the idea of independence, explain why this probability is not $4/52$.

Here is another example of a situation where events are dependent.

EXAMPLE 4.18

Taking a test twice. If you take an IQ test or other mental test twice in succession, the two test scores are not independent. The learning that occurs on the first attempt influences your second attempt. If you learn a lot, then your second test score might be a lot higher than your first test score.

When independence is part of a probability model, the multiplication rule applies. Here is an example.

Profimedia.CZ a.s./Alamy

EXAMPLE 4.19

Mendel's peas. Gregor Mendel used garden peas in some of the experiments that revealed that inheritance operates randomly. The seed color of Mendel's peas can be either green or yellow. Two parent plants are "crossed" (one pollinates the other) to produce seeds.

Each parent plant carries two genes for seed color, and each of these genes has probability 0.5 of being passed to a seed. The two genes that the seed receives, one from each parent, determine its color. The parents contribute their genes independently of each other.

Suppose that both parents carry the G and the Y genes. The seed will be green if both parents contribute a G gene; otherwise, it will be yellow. If M is the event that the male contributes a G gene and F is the event that the female contributes a G gene, then the probability of a green seed is

$$P(M \text{ and } F) = P(M)\,P(F)$$
$$= (0.5)(0.5) = 0.25$$

In the long run, 1/4 of all seeds produced by crossing these plants will be green.

The multiplication rule applies only to independent events; you cannot use it if events are not independent. Here is a distressing example of misuse of the multiplication rule.

EXAMPLE 4.20

Sudden infant death syndrome. Sudden infant death syndrome (SIDS) causes babies to die suddenly (often in their cribs) with no explanation. Deaths from SIDS have been greatly reduced by placing babies on their backs, but as yet no cause is known.

When more than one SIDS death occurs in a family, the parents are sometimes accused. One "expert witness" popular with prosecutors in England told juries that there is only a 1 in 73 million chance that two children in the same family could have died from SIDS. Here's his calculation: the rate of SIDS in a nonsmoking middle-class family is 1 in 8500. So the probability of two deaths is

$$\frac{1}{8500} \times \frac{1}{8500} = \frac{1}{72{,}250{,}000}$$

Several women were convicted of murder on this basis, without any direct evidence that they harmed their children.

As the Royal Statistical Society said, this reasoning is nonsense. It assumes that SIDS deaths in the same family are independent events. The cause of SIDS is unknown: "There may well be unknown genetic or environmental factors that predispose families to SIDS, so that a second case within the family becomes much more likely."[4] The British government decided to review the cases of 258 parents convicted of murdering their babies.

The multiplication rule $P(A \text{ and } B) = P(A)P(B)$ holds if A and B are independent but not otherwise. The addition rule $P(A \text{ or } B) = P(A) + P(B)$ holds if

mosaic plot,
p. 143

A and *B* are disjoint but not otherwise. Resist the temptation to use these simple formulas when the circumstances that justify them are not present. *You must also be certain not to confuse disjointness and independence. Disjoint events cannot be independent.* If *A* and *B* are disjoint, then the fact that *A* occurs tells us that *B* cannot occur—look again at Figure 4.2 (page 224). Unlike disjointness or complements, independence cannot be pictured by a Venn diagram because it involves the probabilities of the events rather than just the outcomes that make up the events. However, it could be displayed in a mosaic plot.

Applying the probability rules

If two events *A* and *B* are independent, then their complements A^c and B^c are also independent and A^c is independent of B^c. Suppose, for example, that 75% of all registered voters in a suburban district are Republicans. If an opinion poll interviews two voters chosen independently, the probability that the first is a Republican and the second is not a Republican is $(0.75)(0.25) = 0.1875$.

The multiplication rule also extends to collections of more than two events, provided that all are independent. Independence of events *A*, *B*, and *C* means that no information about any one or any two can change the probability of the remaining events. The formal definition is a bit messy. Fortunately, independence is usually assumed in setting up a probability model. We can then use the multiplication rule freely.

By combining the rules we have learned, we can compute probabilities for rather complex events. Here is an example.

EXAMPLE 4.21

HIV testing. Many people who come to clinics to be tested for HIV, the virus that causes AIDS, don't come back to learn the test results. Clinics now use "rapid HIV tests" that give a result in a few minutes. The false-positive rate for a diagnostic test is the probability that a person with no disease will have a positive test result. For the rapid HIV tests, the Food and Drug Administration (FDA) has established 2% as the maximum false-positive rate allowed.[5] If a clinic uses a test that matches the FDA standard and tests 50 people who are free of HIV antibodies, what is the probability that at least one false-positive will occur?

It is reasonable to assume as part of the probability model that the test results for different individuals are independent. The probability that the test is positive for a single person is 0.02, so the probability of a negative result is $1 - 0.02 = 0.98$ by the complement rule. The probability of at least one false-positive among the 50 people tested is, therefore,

$$P(\text{at least 1 positive}) = 1 - P(\text{no positives})$$
$$= 1 - P(50 \text{ negatives})$$
$$= 1 - 0.98^{50}$$
$$= 1 - 0.3642 = 0.6358$$

There is approximately a 64% chance that at least 1 of the 50 people will test positive for HIV even though none of them has the virus.

Concern about excessive numbers of false-positives led the New York City Department of Health and Mental Hygiene to suspend the use of one particular rapid HIV test.[6]

SECTION 4.2 SUMMARY

• A **probability model** for a random phenomenon consists of a sample space *S* and an assignment of probabilities *P*.

• The **sample space *S*** is the set of all possible outcomes of the random phenomenon. Sets of outcomes are called **events**. *P* assigns a number *P(A)* to an event *A* as its probability.

• The **complement A^c** of an event *A* consists of exactly the outcomes that are not in *A*. Events *A* and *B* are **disjoint** if they have no outcomes in common. Events *A* and *B* are **independent** if knowing that one event occurs does not change the probability we would assign to the other event.

• Any assignment of probability must obey the rules that state the basic properties of probability:

Rule 1. $0 \le P(A) \le 1$ for any event *A*.

Rule 2. $P(S) = 1$.

Rule 3. Addition rule: If events *A* and *B* are **disjoint**, then $P(A \text{ or } B) = P(A) + P(B)$.

Rule 4. Complement rule: For any event *A*, $P(A^c) = 1 - P(A)$.

Rule 5. Multiplication rule: If events *A* and *B* are **independent**, then $P(A \text{ and } B) = P(A)P(B)$.

SECTION 4.2 EXERCISES

For Exercise 4.8, see page 222; for Exercise 4.9, see page 222; for Exercises 4.10 and 4.11, see page 225; for Exercises 4.12 and 4.13, see page 226; for Exercise 4.14, see page 228; for Exercise 4.15, see page 229; and for Exercise 4.16, see page 229.

4.17 What is the sample space? For each of the following questions, define a sample space for the associated random phenomenon. Explain your answers. Be sure to specify units if that is appropriate.

(a) Will it rain tomorrow?

(b) How many times do you tweet in a typical day?

(c) What is the average age of your Facebook friends?

(d) What are the majors for students at your college?

4.18 Probability rules. For each of the following situations, state the probability rule or rules that you would use and apply it or them. Write a sentence explaining how the situation illustrates the use of the probability rules.

(a) The probability of event *A* is 0.417. What is the probability that event *A* does not occur?

(b) A coin is tossed four times. The probability of zero heads is 1/16 and the probability of zero tails is 1/16.

What is the probability that all four tosses result in the same outcome?

(c) Refer to part (b). What is the probability that there is at least one head and at least one tail?

(d) The probability of event *A* is 0.4 and the probability of event *B* is 0.8. Events *A* and *B* are disjoint. Can this happen?

(e) Event *A* is very rare. Its probability is −0.04. Can this happen?

4.19 Equally likely events. For each of the following situations, explain why you think that the events are equally likely or not. Explain your answers.

(a) The outcome of the next tennis match for Sloane Stevens is either a win or a loss. (You might want to check the Internet for information about this tennis player.)

(b) You roll a fair die and get a 3 or a 4.

(c) You are observing turns at an intersection. You classify each turn as a right turn or a left turn.

(d) For college basketball games, you record the times that the home team wins and the number of times that the home team loses.

4.20 The multiplication rule for independent events. The probability that a randomly selected person prefers the vehicle color white is 0.24. Can you apply the multiplication rule for independent events in the situations described in parts (a) and (b)? If your answer is Yes, apply the rule.

(a) Two people are chosen at random from the population. What is the probability that both prefer white?

(b) Two people who are sisters are chosen. What is the probability that both prefer white?

(c) Write a short summary about the multiplication rule for independent events using your answers to parts (a) and (b) to illustrate the basic idea.

4.21 What's wrong? In each of the following scenarios, there is something wrong. Describe what is wrong and give a reason for your answer.

(a) If two events are disjoint, we can multiply their probabilities to determine the probability that they will both occur.

(b) If the probability of A is 0.7 and the probability of B is 0.5, the probability of both A and B happening is 1.2.

(c) If the probability of A is 0.45, then the probability of the complement of A is −0.45.

4.22 What's wrong? In each of the following scenarios, there is something wrong. Describe what is wrong and give a reason for your answer.

(a) If the sample space consists of two outcomes, then each outcome has probability 0.5.

(b) If we select a digit at random, then the probability of selecting a 3 is 0.3.

(c) If the probability of A is 0.3, the probability of B is 0.4, and the probability of A and B is 0.5, then A and B are independent.

4.23 Evaluating web page designs. You are a web page designer and you set up a page with four different links. A user of the page can click on one of the links or he or she can leave that page. Describe the sample space for the outcome of someone visiting your web page.

4.24 Record the length of time spent on the page. Refer to the previous exercise. You also decide to measure the length of time a visitor spends on your page. Give the sample space for this measure.

4.25 Distribution of blood types. All human blood can be "ABO-typed" as one of O, A, B, or AB, but the distribution of the types varies a bit among groups of people. Here is the distribution of blood types for a randomly chosen person in the United States:[7]

Blood type	A	B	AB	O
U.S. probability	0.42	0.11	?	0.44

(a) What is the probability of type AB blood in the United States?

(b) Maria has type B blood. She can safely receive blood transfusions from people with blood types O and B. What is the probability that a randomly chosen person from the United States can donate blood to Maria?

4.26 Blood types in Ireland. The distribution of blood types in Ireland differs from the U.S. distribution given in the previous exercise:

Blood type	A	B	AB	O
Ireland probability	0.35	0.10	0.03	0.52

Choose a person from the United States and a person from Ireland at random, independently of each other. What is the probability that both have type O blood? What is the probability that both have the same blood type?

4.27 Are the probabilities legitimate? In each of the following situations, state whether or not the given assignment of probabilities to individual outcomes is legitimate—that is, it satisfies the rules of probability. If not, give specific reasons for your answer.

(a) Choose a college student at random and record gender and enrollment status: P(female full-time) = 0.44, P(female part-time) = 0.56, P(male full-time) = 0.46, P(male part-time) = 0.54.

(b) Deal a card from a shuffled deck: P(clubs) = 16/52, P(diamonds) = 12/52, P(hearts) = 12/52, P(spades) = 12/52.

(c) Roll a die and record the count of spots on the up-face: $P(1) = 1/3$, $P(2) = 0$, $P(3) = 1/6$, $P(4) = 1/3$, $P(5) = 1/6$, $P(6) = 0$.

4.28 French and English in Canada. Canada has two official languages, English and French. Choose a Canadian at random and ask, "What is your mother tongue?" Here is the distribution of responses, combining many separate languages from the broad Asian/Pacific region:[8]

Language	English	French	Asian/Pacific	Other
Probability	0.59	?	0.07	0.11

(a) What probability should replace "?" in the distribution?

(b) What is the probability that a Canadian's mother tongue is not English? Explain how you computed your answer.

4.29 Education levels of young adults. Choose a young adult (age 25 to 34 years) at random. The probability is 0.12 that the person chosen did not complete high school, 0.31 that the person has a high school diploma but no further education, and 0.29 that the person has at least a bachelor's degree.

(a) What must be the probability that a randomly chosen young adult has some education beyond high school but does not have a bachelor's degree?

(b) What is the probability that a randomly chosen young adult has at least a high school education?

4.30 Loaded dice. There are many ways to produce crooked dice. To *load* a die so that 6 comes up too often and 1 (which is opposite 6) comes up too seldom, add a bit of lead to the filling of the spot on the 1 face. Because the spot is solid plastic, this works even with transparent dice. If a die is loaded so that 6 comes up with probability 0.24 and the probabilities of the 2, 3, 4, and 5 faces are not affected, what is the assignment of probabilities to the six faces?

4.31 Rh blood types. Human blood is typed as O, A, B, or AB and also as Rh-positive or Rh-negative. ABO type and Rh-factor type are independent because they are governed by different genes. In the American population, 84% of people are Rh-positive. Use the information about ABO type in Exercise 4.25 to give the probability distribution of blood type (ABO and Rh) for a randomly chosen American.

4.32 Roulette. A roulette wheel has 38 slots, numbered 0, 00, and 1 to 36. The slots 0 and 00 are colored green, 18 of the others are red, and 18 are black. The dealer spins the wheel and, at the same time, rolls a small ball along the wheel in the opposite direction. The wheel is carefully balanced so that the ball is equally likely to land in any slot when the wheel slows. Gamblers can bet on various combinations of numbers and colors.

(a) What is the probability that the ball will land in any one slot?

(b) If you bet on "red," you win if the ball lands in a red slot. What is the probability of winning?

(c) The slot numbers are laid out on a board on which gamblers place their bets. One column of numbers on the board contains all multiples of 3, that is, 3, 6, 9, . . . , 36. You place a "column bet" that wins if any of these numbers comes up. What is your probability of winning?

4.33 Winning the lottery. A state lottery's Pick 3 game asks players to choose a three-digit number, 000 to 999. The state chooses the winning three-digit number at random so that each number has probability 1/1000. You

win if the winning number contains the digits in your number, in any order.

(a) Your number is 059. What is your probability of winning?

(b) Your number is 223. What is your probability of winning?

4.34 PINs. The personal identification numbers (PINs) for automatic teller machines usually consist of four digits. You notice that most of your PINs have at least one 0, and you wonder if the issuers use lots of 0s to make the numbers easy to remember. Suppose that PINs are assigned at random, so that all four-digit numbers are equally likely.

(a) How many possible PINs are there?

(b) What is the probability that a PIN assigned at random has at least one 0?

4.35 Universal blood donors. People with type O-negative blood are universal donors. That is, any patient can receive a transfusion of O-negative blood. Only 7% of the American population have O-negative blood. If eight people appear at random to give blood, what is the probability that at least one of them is a universal donor?

4.36 Axioms of probability. Show that any assignment of probabilities to events that obeys Rules 2 and 3 on page 224 automatically obeys the complement rule (Rule 4). This implies that a mathematical treatment of probability can start from just Rules 1, 2, and 3. These rules are sometimes called *axioms* of probability.

4.37 Independence of complements. Show that if events A and B obey the multiplication rule, $P(A \text{ and } B) = P(A)P(B)$, then A and the complement B^c of B also obey the multiplication rule, $P(A \text{ and } B^c) = P(A)P(B^c)$. That is, if events A and B are independent, then A and B^c are also independent. (*Hint:* Start by drawing a Venn diagram and noticing that the events "A and B" and "A and B^c" are disjoint.)

Mendelian inheritance. Some traits of plants and animals depend on inheritance of a single gene. This is called Mendelian inheritance, after Gregor Mendel (1822–1884). Exercises 4.38 through 4.41 are based on the following information about Mendelian inheritance of blood type.

Each of us has an ABO blood type, which describes whether two characteristics, called A and B, are present. Every one of us has two blood type alleles (gene forms), one inherited from our mother and one from our father. Each of these alleles can be A, B, or O. Which two we inherit determines our blood type. Here is a table that shows what our blood type is for each combination of two alleles:

Alleles inherited	Blood type
A and A	A
A and B	AB
A and O	A
B and B	B
B and O	B
O and O	O

We inherit each of a parent's two alleles with probability 0.5. We inherit independently from our mother and father.

4.38 Blood types of children. Emily and Michael both have alleles O and O.

(a) What blood types can their children have?

(b) What is the probability that their next child has each of these blood types?

4.39 Parents with alleles B and O. Andreona and Caleb both have alleles B and O.

(a) What blood types can their children have?

(b) What is the probability that their next child has each of these blood types?

4.40 Two children. Samantha has alleles B and O. Dylan has alleles A and B. They have two children. What is the probability that both children have blood type A? What is the probability that both children have the same blood type?

4.41 Three children. Anna has alleles B and O. Nathan has alleles A and O.

(a) What is the probability that a child of these parents has blood type O?

(b) If Anna and Nathan have three children, what is the probability that all three have blood type O? What is the probability that the first child has blood type O and the next two do not?

4.3 Random Variables

When you complete this section, you will be able to:

- Describe the probability distribution of a discrete random variable.
- Use a probability histogram to provide a graphical description of the probability distribution of a discrete random variable.
- Use the distribution of a discrete random variable to calculate probabilities of events.
- Find probabilities of events for the uniform and normal distributions.

Sample spaces need not consist of numbers. When we toss a coin four times, we can record the outcome as a string of heads and tails, such as HTTH. In statistics, however, we are most often interested in numerical outcomes such as the count of heads in the four tosses. It is convenient to use a shorthand notation: Let X be the number of heads. If our outcome is HTTH, then $X = 2$. If the next outcome is TTTH, the value of X changes to $X = 1$. The possible values of X are 0, 1, 2, 3, and 4. Tossing a coin four times will give X one of these possible values. Tossing four more times will give X another and probably different value. We call X a *random variable* because its values vary when the coin tossing is repeated.

RANDOM VARIABLE

A **random variable** is a variable whose value is a numerical outcome of a random process.

In our earlier coin-tossing example, the process is the tossing of a coin four times. The random variable is the number of heads in the four tosses.

We usually denote random variables by capital letters near the end of the alphabet, such as X or Y. Of course, the random variables of greatest interest to us are outcomes such as the mean $\bar{x}$ of a random sample, for which we will keep the familiar notation.[9] As we progress from general rules of probability toward statistical inference, we will concentrate on random variables.

When a random variable X describes a random process, the sample space S just lists the possible values of the random variable. We usually do not mention S separately. There remains the second part of any probability model, the assignment of probabilities to events. There are two main ways of assigning probabilities to the values of a random variable. The two types of probability models that result will dominate our application of probability to statistical inference.

Discrete random variables

We have learned several rules of probability, but only one method of assigning probabilities: state the probabilities of the individual outcomes and assign probabilities to events by summing over the outcomes. The outcome probabilities must be between 0 and 1 and have sum 1. When the outcomes are numerical, they are values of a random variable. We will now attach a name to random variables having probability assigned in this way.[10]

DISCRETE RANDOM VARIABLE

A **discrete random variable** X has possible values that can be given in an ordered list. The **probability distribution** of X lists the values and their probabilities:

Value of X	x_1	x_2	x_3	$\cdots$
Probability	p_1	p_2	p_3	$\cdots$

The probabilities p_i must satisfy two requirements:

1. Every probability p_i is a number between 0 and 1.

2. $p_1 + p_2 + \cdots = 1$.

Find the probability of any event by adding the probabilities p_i of the particular values x_i that make up the event.

In most discrete random variable situations that we will study, the number of possible values is a finite number, k. For example, in our example on the number of heads in four tosses of a coin, there are $k = 5$ possible values: 0, 1, 2, 3, and 4.

There are, however, settings in which the number of possible values can be infinite. Think about tossing a fair coin until you get a head. The number of possible tosses is any positive integer.

EXAMPLE 4.22

Grade distributions. A liberal arts college posts the grade distributions for its courses. In a recent semester, students in one section of English 130 received 32% A's, 42% B's, 19% C's, 3% D's, and 4% F's. Choose an English 130 student at random. To "choose at random" means to give every student the same chance to be chosen. The student's grade on a five-point scale (with A = 4) is a random variable X.

The value of X changes when we repeatedly choose students at random, but it is always one of 0, 1, 2, 3, or 4. Here is the distribution of X:

Value of X	0	1	2	3	4
Probability	0.04	0.03	0.19	0.42	0.32

The probability that the student got a B or better is the sum of the probabilities of an A and a B. In the language of random variables,

$$P(X \geq 3) = P(X = 3) + P(X = 4)$$
$$= 0.42 + 0.32 = 0.74$$

USE YOUR KNOWLEDGE

4.42 Will the course satisfy the requirement? Refer to Example 4.22. Suppose that a grade of D or F in English 130 does not satisfy a requirement for a major in linguistics. What is the probability that a randomly selected student will not satisfy this requirement?

probability histogram

We can use histograms to show probability distributions as well as distributions of data. Figure 4.5 displays **probability histograms** that compare the probability model for equally likely random digits (Example 4.15, page 227) with the model given by Benford's law (Example 4.12, page 226). The height of each bar shows the probability of the outcome at its base. Because the heights

FIGURE 4.5 Probability histograms for (a) equally likely random digits 1 to 9 and (b) Benford's law. The height of each bar shows the probability assigned to a single outcome.

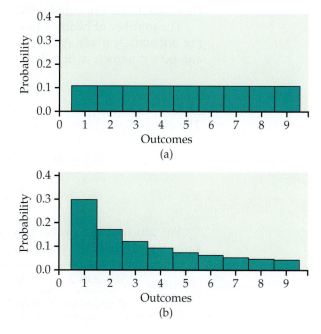

are probabilities, they add to 1. As usual, all the bars in a histogram have the same width. So the areas also display the assignment of probability to outcomes. Think of these histograms as idealized pictures of the results of very many trials. The histograms make it easy to quickly compare the two distributions.

EXAMPLE 4.23

Number of heads in four tosses of a coin. What is the probability distribution of the discrete random variable X that counts the number of heads in four tosses of a coin? We can derive this distribution if we make two reasonable assumptions:

- The coin is balanced, so it is fair and each toss is equally likely to give H or T.

- The coin has no memory, so tosses are independent.

The outcome of four tosses is a sequence of heads and tails such as HTTH. There are 16 possible outcomes in all. Figure 4.6 lists these outcomes along with the value of X for each outcome. The multiplication rule for independent events tells us that, for example,

$$P(\text{HTTH}) = \frac{1}{2} \times \frac{1}{2} \times \frac{1}{2} \times \frac{1}{2} = \frac{1}{16}$$

		HTTH		
		HTHT		
	HTTT	THTH	HHHT	
	THTT	HHTT	HHTH	
	TTHT	THHT	HTHH	
TTTT	TTTH	TTHH	THHH	HHHH
$X = 0$	$X = 1$	$X = 2$	$X = 3$	$X = 4$

FIGURE 4.6 Possible outcomes in four tosses of a coin, Example 4.23. The outcomes are arranged by the values of the random variable X, the number of heads.

Each of the 16 possible outcomes similarly has probability 1/16. That is, these outcomes are equally likely.

The number of heads X has possible values 0, 1, 2, 3, and 4. These values are *not* equally likely. As Figure 4.6 shows, there is only one way that $X = 0$ can occur: namely, when the outcome is TTTT. So

$$P(X = 0) = \frac{1}{16} = 0.0625$$

The event $\{X = 2\}$ can occur in six different ways, so that

$$P(X = 2) = \frac{\text{count of ways } X = 2 \text{ can occur}}{16}$$

$$= \frac{6}{16} = 0.375$$

We can find the probability of each value of X from Figure 4.6 in the same way. Here is the result:

Value of X	0	1	2	3	4
Probability	0.0625	0.25	0.375	0.25	0.0625

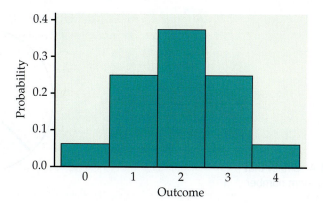

FIGURE 4.7 Probability histogram for the number of heads in four tosses of a coin (Example 4.23).

Figure 4.7 is a probability histogram for the distribution in Example 4.23. The probability distribution is exactly symmetric. The probabilities (bar heights) are idealizations of the proportions after very many tosses of four coins. The actual distribution of proportions observed would be nearly symmetric but is unlikely to be exactly symmetric.

EXAMPLE 4.24

Probability of at least three heads. Any event involving the number of heads observed can be expressed in terms of X, and its probability can be found from the distribution of X. For example, the probability of tossing at least three heads is

$$P(X \geq 3) = 0.25 + 0.0625 = 0.3125$$

The probability of at least one head is most simply found by use of the complement rule:

$$P(X \geq 1) = 1 - P(X = 0)$$
$$= 1 - 0.0625 = 0.9375$$

Recall that tossing a coin n times is similar to choosing an SRS of size n from a large population and asking a Yes or No question. We will extend the results of Example 4.23 when we return to sampling distributions in the next chapter.

USE YOUR KNOWLEDGE

4.43 Two tosses of a fair coin. Find the probability distribution for the number of heads that appear in two tosses of a fair coin.

Continuous random variables

When we use the table of random digits to select a digit between 0 and 9, the result is a discrete random variable. The probability model assigns probability 1/10 to each of the 10 possible outcomes. Suppose that we want to choose a number at random between 0 and 1, allowing *any* number between 0 and 1 as the outcome. Software random number generators will do this.

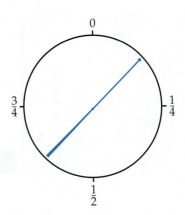

FIGURE 4.8 A spinner that generates a random number between 0 and 1.

You can visualize such a random number by thinking of a spinner (Figure 4.8) that turns freely on its axis and slowly comes to a stop. The pointer can come to rest anywhere on a circle that is marked from 0 to 1. The sample space is now an entire interval of numbers:

$$S = \{\text{all numbers } x \text{ such that } 0 \leq x \leq 1\}$$

LOOK BACK

density curve, p. 51

How can we assign probabilities to events such as $\{0.3 \leq x \leq 0.7\}$? As in the case of selecting a random digit, we would like all possible outcomes to be equally likely. But we cannot assign probabilities to each individual value of x and then sum because there are too many possible values. Instead, we use a new way of assigning probabilities directly to events—as *areas under a density curve*. Any density curve has area exactly 1 underneath it, corresponding to total probability 1.

EXAMPLE 4.25

uniform distribution

Uniform random numbers. The random number generator will spread its output uniformly across the entire interval from 0 to 1 as we allow it to generate a long sequence of numbers. The results of many trials are represented by the density curve of a **uniform distribution**.

This density curve appears in red in Figure 4.9. It has height 1 over the interval from 0 to 1, and height 0 everywhere else. The area under the density curve is 1: the area of a square with base 1 and height 1. The probability of any event is the area under the density curve and above the event in question.

FIGURE 4.9 Assigning probabilities for generating a random number between 0 and 1, Example 4.25. The probability of any interval of numbers is the area above the interval and under the density curve.

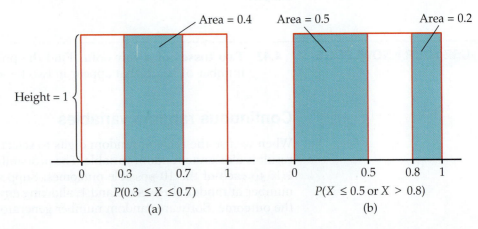

As Figure 4.9(a) illustrates, the probability that the random number generator produces a number X between 0.3 and 0.7 is

$$P(0.3 \leq X \leq 0.7) = 0.4$$

because the area under the density curve and above the interval from 0.3 to 0.7 is 0.4. The height of the density curve is 1, and the area of a rectangle is the product of height and length, so the probability of any interval of outcomes is just the length of the interval.

Similarly,

$$P(X \leq 0.5) = 0.5$$
$$P(X > 0.8) = 0.2$$
$$P(X \leq 0.5 \text{ or } X > 0.8) = 0.7$$

Notice that the last event consists of two nonoverlapping intervals, so the total area above the event is found by adding two areas, as illustrated by Figure 4.9(b). This assignment of probabilities obeys all of our rules for probability.

USE YOUR KNOWLEDGE

4.44 Find the probability. For the uniform distribution described in Example 4.25, find the probability that X is between 0.3 and 0.9.

Probability as area under a density curve is a second important way of assigning probabilities to events. Figure 4.10 illustrates this idea in general form. We call X in Example 4.25 a *continuous random variable* because its values are not isolated numbers but an entire interval of numbers.

CONTINUOUS RANDOM VARIABLE

A **continuous random variable** X takes all values in an interval of numbers. The **probability distribution** of X is described by a density curve. The probability of any event is the area under the density curve and above the values of X that make up the event.

The probability model for a continuous random variable assigns probabilities to intervals of outcomes rather than to individual outcomes. In fact,

FIGURE 4.10 The probability distribution of a continuous random variable assigns probabilities as areas under a density curve. The total area under any density curve is 1.

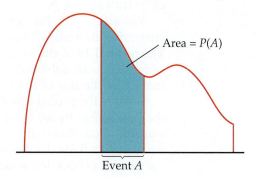

Area = $P(A)$

Event A

all continuous probability distributions assign probability 0 to every individual outcome. Only intervals of values have positive probability. To see that this is true, consider a specific outcome such as $P(X = 0.8)$ in the context of Example 4.25. The probability of any interval is the same as its length. The point 0.8 has no length, so its probability is 0.

Although this fact may seem odd, it makes intuitive, as well as mathematical, sense. The random number generator produces a number between 0.79 and 0.81 with probability 0.02. An outcome between 0.799 and 0.801 has probability 0.002. A result between 0.799999 and 0.800001 has probability 0.000002. You see that as we approach 0.8, the probability gets closer to 0.

To be consistent, the probability of an outcome *exactly* equal to 0.8 must be 0. Because there is no probability exactly at $X = 0.8$, the two events $\{X > 0.8\}$ and $\{X \geq 0.8\}$ have the same probability. *We can ignore the distinction between $>$ and $\geq$ when finding probabilities for continuous (but not discrete) random variables.*

Normal distributions as probability distributions

The density curves that are most familiar to us are the Normal curves. Because any density curve describes an assignment of probabilities, *Normal distributions are probability distributions*. Recall that $N(\mu, \sigma)$ is our shorthand for the Normal distribution having mean μ and standard deviation σ. In the language of random variables, if X has the $N(\mu, \sigma)$ distribution, then the standardized variable

$$Z = \frac{X - \mu}{\sigma}$$

is a standard Normal random variable having the distribution $N(0,1)$.

LOOK BACK

Normal distributions, p. 56

EXAMPLE 4.26

Texting while driving. Texting while driving can be dangerous, but young people want to remain connected. Suppose that 26% of teen drivers text while driving. If we take a sample of 500 teen drivers, what percent would we expect to say that they text while driving?[11]

The proportion $p = 0.26$ is a number that describes the population of teen drivers. The proportion $\hat{p}$ of the sample who say that they text while driving is used to estimate p. The proportion $\hat{p}$ is a random variable because repeating the SRS would give a different sample of 500 teen drivers and a different value of $\hat{p}$.

We will see in the next chapter that in this setting, with teen drivers answering honestly, $\hat{p}$ has approximately the $N(0.26, 0.0196)$ distribution. The mean 0.26 of this distribution is the same as the population proportion because $\hat{p}$ is an unbiased estimate of p. The standard deviation is controlled mainly by the size of the sample.

What is the probability that the survey result differs from the truth about the population by no more than 3 percentage points? We can use what we learned about Normal distribution calculations to answer this question. Because $p = 0.26$, the survey misses by no more than 3 percentage points if the sample proportion is between 0.23 and 0.29.

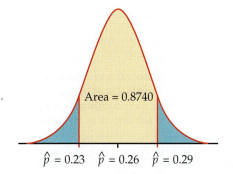

FIGURE 4.11 Probability as area under a Normal density curve, Example 4.26.

Figure 4.11 shows this probability as an area under a Normal density curve. You can find it by software or by standardizing and using Table A. From Table A,

$$P(0.23 \le \hat{p} \le 0.29) = P\left(\frac{0.23 - 0.26}{0.0196} \le \frac{\hat{p} - 0.26}{0.0196} \le \frac{0.29 - 0.26}{0.0196}\right)$$
$$= P(-1.53 \le Z \le 1.53)$$
$$= 0.9370 - 0.0630 = 0.8740$$

About 87% of the time, the sample $\hat{p}$ will be within 3 percentage points of the proportion p.

We began this chapter with a general discussion of the idea of probability and the properties of probability models. Two very useful specific types of probability models are distributions of discrete and continuous random variables. In our study of statistics, we will employ only these two types of probability models.

SECTION 4.3 SUMMARY

• A **random variable** is a variable taking numerical values determined by the outcome of a random phenomenon. The **probability distribution** of a random variable X tells us what the possible values of X are and how probabilities are assigned to those values.

• A random variable X and its distribution can be **discrete** or **continuous**.

• A **discrete random variable** has possible values that can be given in an ordered list. The probability distribution assigns each of these values a probability between 0 and 1 such that the sum of all the probabilities is exactly 1. The probability of any event is the sum of the probabilities of all the values that make up the event.

• A **continuous random variable** takes all values in some interval of numbers. A **density curve** describes the probability distribution of a continuous random variable. The probability of any event is the area under the curve and above the values that make up the event.

• **Uniform distributions** are continuous probability distributions that are very similar to equally likely discrete distributions.

• **Normal distributions** are one type of continuous probability distribution.

• You can picture a probability distribution by drawing a **probability histogram** in the discrete case or by graphing the density curve in the continuous case.

SECTION 4.3 EXERCISES

For Exercise 4.42, see page 237; for Exercise 4.43, see page 239; and for Exercise 4.44, see page 241.

4.45 How many courses? At a small liberal arts college, students can register for one to six courses. Let X be the number of courses taken in the fall by a randomly selected student from this college. In a typical fall semester, 5% take one course, 5% take two courses, 13% take three courses, 26% take four courses, 36% take five courses, and 15% take six courses. Let X be the number of courses taken in the fall by a randomly selected student from this college. Describe the probability distribution of this random variable.

4.46 Make a graphical display. Refer to the previous exercise. Use a probability histogram to provide a graphical description of the distribution of X.

4.47 Find some probabilities. Refer to Exercise 4.45.

(a) Find the probability that a randomly selected student takes three or fewer courses.

(b) Find the probability that a randomly selected student takes four or five courses.

(c) Find the probability that a randomly selected student takes eight courses.

4.48 Use the uniform distribution. Suppose that a random variable X follows the uniform distribution described in Example 4.25 (page 240). For each of the following events, find the probability and illustrate your calculations with a sketch of the density curve similar to the ones in Figure 4.9 (page 240).

(a) The probability that X is less than 0.2.

(b) The probability that X is greater than or equal to 0.7.

(c) The probability that X is less than 0.8 and greater than 0.4.

(d) The probability that X is 0.7.

4.49 What's wrong? In each of the following scenarios, there is something wrong. Describe what is wrong and give a reason for your answer.

(a) The possible values for a discrete random variable can't be negative.

(b) A continuous random variable can take any value between 0 and 1.

(c) Normal distributions are discrete random variables.

4.50 Use of Twitter. Suppose that the population proportion of Internet users who say that they use Twitter or another service to post updates about themselves or to see updates about others is 19%.[12] Think about selecting random samples from a population in which 19% are Twitter users.

(a) Describe the sample space for selecting a single person.

(b) If you select three people, describe the sample space.

(c) Using the results of part (b), define the sample space for the random variable that expresses the number of Twitter users in the sample of size 3.

(d) What information is contained in the sample space for part (b) that is not contained in the sample space for part (c)? Do you think this information is important? Explain your answer.

4.51 Use of Twitter. Refer to the previous exercise. Find the probabilities for the number of Twitter users in a sample of size 2.

4.52 Households and families in government data. In government data, a household consists of all occupants of a dwelling unit, while a family consists of two or more persons who live together and are related by blood or marriage. So all families form households, but some households are not families. Here are the distributions of household size and of family size in the United States:

Number of persons	1	2	3	4	5	6	7
Household probability	0.27	0.33	0.16	0.14	0.06	0.03	0.01
Family probability	0	0.44	0.22	0.20	0.09	0.03	0.02

Make probability histograms for these two discrete distributions, using the same scales. What are the most important differences between the sizes of households and families?

4.53 Discrete or continuous. In each of the following situations, decide whether the random variable is discrete or continuous and give a reason for your answer.

(a) Your web page has five different links, and a user can click on one of the links or can leave the page. You record the length of time that a user spends on the web page before clicking one of the links or leaving the page.

(b) You record the number of hits per day on your web page.

(c) You record the yearly income of a visitor to your web page.

4.54 Texas hold 'em. The game of Texas hold 'em starts with each player receiving two cards. Here is the probability distribution for the number of aces in two-card hands:

Number of aces	0	1	2
Probability	0.8507	0.1448	0.0045

(a) Verify that this assignment of probabilities satisfies the requirement that the sum of the probabilities for a discrete distribution must be 1.

(b) Make a probability histogram for this distribution.

(c) What is the probability that a hand contains at least one ace? Show two different ways to calculate this probability.

4.55 Tossing two dice. Some games of chance rely on tossing two dice. Each die has six faces, marked with one, two, . . . , six spots called pips. The dice used in casinos are carefully balanced so that each face is equally likely to come up. When two dice are tossed, each of the 36 possible pairs of faces is equally likely to come up. The outcome of interest to a gambler is the sum of the pips on the two up-faces. Call this random variable X.

(a) Write down all 36 possible pairs of up-faces.

(b) If all pairs have the same probability, what must be the probability of each pair?

(c) Write the value of X next to each pair of up-faces and use this information with the result of part (b) to give the probability distribution of X. Draw a probability histogram to display the distribution.

(d) One bet available in the game called craps wins if a 7 or an 11 comes up on the next roll of two dice. What is the probability of rolling a 7 or an 11 on the next roll?

(e) Several bets in craps lose if a 7 is rolled. If any outcome other than 7 occurs, these bets either win or continue to the next roll. What is the probability that anything other than a 7 is rolled?

4.56 Nonstandard dice. Nonstandard dice can produce interesting distributions of outcomes. You have two balanced, six-sided dice. One is a standard die, with faces having one, two, three, four, five, and six spots. The other die has three faces with one spot and three faces with six spots. Find the probability distribution for the total number of spots Y on the up-faces when you roll these two dice.

4.57 Spell-checking software. Spell-checking software catches "nonword errors," which are strings of letters that are not words, as when "the" is typed as "eth." When undergraduates are asked to write a 250-word essay (without spell-checking), the number X of nonword errors has the following distribution:

Value of X	0	1	2	3	4
Probability	0.1	0.3	0.3	0.2	0.1

(a) Sketch the probability distribution for this random variable.

(b) Write the event "at least one nonword error" in terms of X. What is the probability of this event?

(c) Describe the event $X \leq 3$ in words. What is its probability? What is the probability that $X < 3$?

4.58 Find the probabilities. Let the random variable X be a random number with the uniform density curve in Figure 4.9 (page 240). Find the following probabilities:

(a) $P(X \geq 0.35)$.

(b) $P(X = 0.35)$.

(c) $P(0.35 < X < 1.35)$.

(d) $P(0.18 \leq X \leq 0.25 \text{ or } 0.4 \leq X \leq 0.5)$.

(e) X is not in the interval 0.4 to 0.8.

4.59 Uniform numbers between 0 and 2. Many random number generators allow users to specify the range of the random numbers to be produced. Suppose that you specify that the range is to be all numbers between 0 and 2. Call the random number generated Y. Then the density curve of the random variable Y has constant height between 0 and 2, and height 0 elsewhere.

(a) What is the height of the density curve between 0 and 2? Draw a graph of the density curve.

(b) Use your graph from part (a) and the fact that probability is area under the curve to find $P(Y \leq 1.6)$.

(c) Find $P(0.5 < Y < 1.7)$.

(d) Find $P(Y \geq 0.95)$.

4.60 The sum of two uniform random numbers. Generate *two* random numbers between 0 and 1 and take Y to be their sum. Then Y is a continuous random variable that can take any value between 0 and 2. The density curve of Y is the triangle shown in Figure 4.12.

(a) Verify by geometry that the area under this curve is 1.

(b) What is the probability that Y is less than 1? [Sketch the density curve, shade the area that represents the probability, then find that area. Do this for part (c) also.]

(c) What is the probability that Y is greater than 1.5?

(d) What is the probability that Y is greater than 0.5?

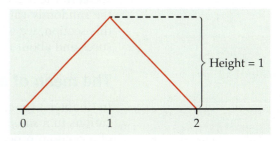

FIGURE 4.12 The density curve for the sum Y of two random numbers, Exercise 4.60.

4.61 How many close friends? How many close friends do you have? Suppose that the number of close friends adults claim to have varies from person to person with mean $\mu = 9$ and standard deviation $\sigma = 2.4$. An opinion poll asks this question of an SRS of 1100 adults. We will see in the next chapter that, in this situation, the sample mean response $\bar{x}$ has approximately the Normal distribution with mean 9 and standard deviation 0.0724. What is $P(8 \leq \bar{x} \leq 10)$, the probability that $\bar{x}$ estimates μ to within ± 1?

4.62 Normal approximation for a sample proportion. A sample survey contacted an SRS of 700 registered voters in Oregon shortly after an election and asked respondents whether they had voted. Voter records show that 56% of registered voters had actually voted. We will see in the next chapter that, in this situation, the proportion $\hat{p}$ of the sample who voted has approximately the Normal distribution with mean $\mu = 0.56$ and standard deviation $\sigma = 0.019$.

(a) If the respondents answer truthfully, what is $P(0.52 \leq \hat{p} \leq 0.60)$? This is the probability that $\hat{p}$ estimates 0.56 within plus or minus 0.04.

(b) In fact, 72% of the respondents said they had voted ($\hat{p} = 0.72$). If respondents answer truthfully, what is $P(\hat{p} \geq 0.72)$? This probability is so small that it is good evidence that some people who did not vote claimed that they did vote.

4.4 Means and Variances of Random Variables

When you complete this section, you will be able to:

- Use a probability distribution to find the mean of a discrete random variable.
- Apply the law of large numbers to describe the behavior of the sample mean as the sample size increases.
- Find means using the rules for means of linear transformations, sums, and differences.
- Use a probability distribution to find the variance and the standard deviation of a discrete random variable.
- Find variances and standard deviations using the rules for variances and standard deviations for linear transformations.
- Find variances and standard deviations using the rules for variances and standard deviations for sums of and differences between two random variables and for uncorrelated and for correlated random variables.

The probability histograms and density curves that picture the probability distributions of random variables resemble our earlier pictures of distributions of data. In describing data, we moved from graphs to numerical measures such as means and standard deviations. Now we will make the same move to expand our descriptions of the distributions of random variables. We can speak of the mean winnings in a game of chance or the standard deviation of the randomly varying number of calls a travel agency receives in an hour. In this section, we will learn more about how to compute these descriptive measures and about the laws they obey.

The mean of a random variable

In Chapter 1 (page 28), we learned that the mean $\bar{x}$ is the average of the observations in a *sample*. Recall that a random variable X is a numerical outcome of a random process. Think about repeating the random process many times and recording the resulting values of the random variable. You can think of the value of a random variable as the average of a very large sample where the relative frequencies of the values are the same as their probabilities.

If we think of the random process as corresponding to the population, then the mean of the random variable is a characteristic of this population. Here is an example.

EXAMPLE 4.27

The Tri-State Pick 3 lottery. Most states and Canadian provinces have government-sponsored lotteries. Here is a simple lottery wager from the Tri-State Pick 3 game that New Hampshire shares with Maine and Vermont. You choose a three-digit number, 000 to 999. The state chooses a three-digit winning number at random and pays you $500 if your number is chosen.

Because there are 1000 three-digit numbers, you have probability 1/1000 of winning. Taking X to be the amount your ticket pays you, the probability distribution of X is

Payoff X	$0	$500
Probability	0.999	0.001

The random process consists of drawing a three-digit number. The population consists of the numbers 000 to 999. Each of these possible outcomes is equally likely in this example. In the setting of sampling in Chapter 3 (page 191), we can view the random process as selecting an SRS of size 1 from the population. The random variable X is 1 if the selected number is equal to the one that you chose and 0 if it is not.

What is your average payoff from many tickets? The ordinary average of the two possible outcomes $0 and $500 is $250, but that makes no sense as the average because $500 is much less likely than $0. In the long run, you receive $500 once in every 1000 tickets and $0 on the remaining 999 of 1000 tickets. The long-run average payoff is

$$\$500 \, \frac{1}{1000} \; + \; \$0 \, \frac{999}{1000} = \$0.50$$

or 50 cents. That number is the mean of the random variable X. (Tickets cost $1, so in the long run, the state keeps half the money you wager.)

If you play Tri-State Pick 3 several times, we would—as usual—call the mean of the actual amounts you win $\bar{x}$. The mean in Example 4.27 is a different quantity—it is the long-run average winnings you expect if you play a very large number of times.

USE YOUR KNOWLEDGE

4.63 Find the mean of the probability distribution. You toss a fair coin. If the outcome is heads, you win $10.00; if the outcome is tails, you win nothing. Let X be the amount that you win in a single toss of a coin. Find the probability distribution of this random variable and its mean.

Just as probabilities are an idealized description of long-run proportions, the mean of a probability distribution describes the long-run average outcome. We can't call this mean $\bar{x}$, so we need a different symbol. The common mean μ symbol for the **mean of a probability distribution** is μ, the Greek letter mu. We used μ in Chapter 1 for the mean of a Normal distribution, so this is not a new notation. We will often be interested in several random variables, each having a different probability distribution with a different mean.

To remind ourselves that we are talking about the mean of X, we often write μ_X rather than simply μ. In Example 4.27, $\mu_X = \$0.50$. Notice that, as often happens, the mean is not a possible value of X. You will often find the

expected value   mean of a random variable X called the **expected value** of X. This term can be misleading because we don't necessarily expect one observation on X to be close to its expected value.

The mean of any discrete random variable is found just as in Example 4.27. It is an average of the possible outcomes, but a weighted average in which each outcome is weighted by its probability. Because the probabilities add to 1, we have total weight 1 to distribute among the outcomes. An outcome that occurs half the time has probability one-half and gets one-half the weight in calculating the mean. Here is the general definition.

MEAN OF A DISCRETE RANDOM VARIABLE

Suppose that X is a **discrete random variable** whose distribution is

Value of X	x_1	x_2	x_3	$\cdots$
Probability	p_1	p_2	p_3	$\cdots$

To find the **mean** of X, multiply each possible value by its probability, then add all the products:

$$\mu_X = x_1 p_1 + x_2 p_2 + \cdots$$
$$= \sum x_i p_i$$

EXAMPLE 4.28

The mean of equally likely first digits. If first digits in a set of data all have the same probability, the probability distribution of the first digit X is then

First digit X	1	2	3	4	5	6	7	8	9
Probability	1/9	1/9	1/9	1/9	1/9	1/9	1/9	1/9	1/9

The mean of this distribution is

$$\mu_X = 1 \times \frac{1}{9} + 2 \times \frac{1}{9} + 3 \times \frac{1}{9} + 4 \times \frac{1}{9} + 5 \times \frac{1}{9}$$
$$+ 6 \times \frac{1}{9} + 7 \times \frac{1}{9} + 8 \times \frac{1}{9} + 9 \times \frac{1}{9}$$
$$= 45 \times \frac{1}{9} = 5$$

Suppose that the random digits in Example 4.28 had a different probability distribution. In Example 4.12 (page 226), we described Benford's law as a probability distribution that describes first digits of numbers in many real situations. Let's calculate the mean for Benford's law.

EXAMPLE 4.29

The mean of first digits that follow Benford's law. Here is the distribution of the first digit for data that follow Benford's law. We use the letter V for this random variable to distinguish it from the one that we studied in Example 4.28. The distribution of V is

First digit V	1	2	3	4	5	6	7	8	9
Probability	0.301	0.176	0.125	0.097	0.079	0.067	0.058	0.051	0.046

The mean of V is

$$\mu_V = (1)(0.301) + (2)(0.176) + (3)(0.125) + (4)(0.097) + (5)(0.079)$$
$$+ (6)(0.067) + (7)(0.058) + (8)(0.051) + (9)(0.046)$$
$$= 3.441$$

The mean reflects the greater probability of smaller first digits under Benford's law than when first digits 1 to 9 are equally likely.

Figure 4.13 locates the means of X and V on the two probability histograms. Because the discrete uniform distribution of Figure 4.13(a) is symmetric, the mean lies at the center of symmetry. We can't locate the mean of the right-skewed distribution of Figure 4.13(b) by eye—calculation is needed.

What about continuous random variables? The probability distribution of a continuous random variable X is described by a density curve. Chapter 1 (page 54) showed how to find the mean of the distribution: it is the point at

FIGURE 4.13 Locating the mean of a discrete random variable on the probability histogram for (a) digits between 1 and 9 chosen at random; (b) digits between 1 and 9 chosen from records that obey Benford's law.

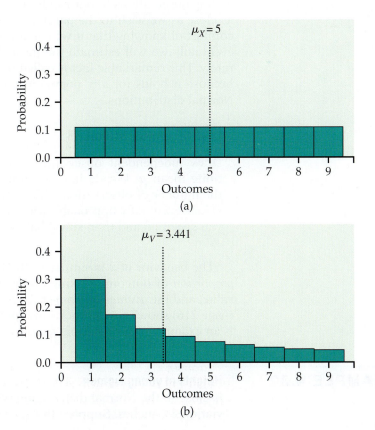

which the area under the density curve would balance if it were made out of solid material. The mean lies at the center of symmetric density curves such as the Normal curves. Exact calculation of the mean of a distribution with a skewed density curve requires advanced mathematics.[13] The idea that the mean is the balance point of the distribution applies to discrete random variables as well, but in the discrete case, we have a formula that gives us this point.

Statistical estimation and the law of large numbers

We would like to estimate the mean height μ of the population of all American women between the ages of 18 and 24 years. This μ is the mean μ_X of the random variable X obtained by choosing a young woman at random and measuring her height. To estimate μ, we choose an SRS of young women and use the sample mean $\bar{x}$ to estimate the unknown population mean μ. In the language of Section 5.1 (page 282), μ is a *parameter* and $\bar{x}$ is a *statistic*.

Statistics obtained from probability samples are random variables because their values vary in repeated sampling. The sampling distributions of statistics are just the probability distributions of these random variables.

It seems reasonable to use $\bar{x}$ to estimate μ. An SRS should fairly represent the population, so the mean $\bar{x}$ of the sample should be somewhere near the mean μ of the population. Of course, we don't expect $\bar{x}$ to be exactly equal to μ, and we realize that if we choose another SRS, the luck of the draw will probably produce a different $\bar{x}$.

If $\bar{x}$ is rarely exactly right and varies from sample to sample, why is it nonetheless a reasonable estimate of the population mean μ? If we keep on adding observations to our random sample, the statistic $\bar{x}$ is *guaranteed* to get as close as we wish to the parameter μ and then stay that close. We have the comfort of knowing that if we can afford to keep on measuring more women, eventually we will estimate the mean height of all young women very accurately. This remarkable fact is called the *law of large numbers*. It is remarkable because it holds for *any* population, not just for some special class such as Normal distributions.

LAW OF LARGE NUMBERS

Draw independent observations at random from any population with finite mean μ. Decide how accurately you would like to estimate μ. As the number of observations drawn increases, the mean $\bar{x}$ of the observed values eventually approaches the mean μ of the population as closely as you specified and then stays that close.

The behavior of $\bar{x}$ is similar to the idea of probability. In the long run, the *proportion* of outcomes taking any value gets close to the *probability* of that value, and the *average outcome* gets close to the distribution *mean*. Figure 4.1 (page 216) shows how proportions approach probability in one example. Here is an example of how sample means approach the distribution mean.

EXAMPLE 4.30

Heights of young women. The distribution of the heights of all young women is close to the Normal distribution with mean 64.5 inches and standard deviation 2.5 inches. Suppose that $\mu = 64.5$ were exactly true.

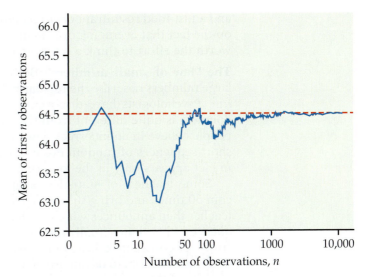

FIGURE 4.14 The law of large numbers in action, Example 4.30. As we take more observations, the sample mean always approaches the mean of the population.

Figure 4.14 shows the behavior of the mean height $\bar{x}$ of n women chosen at random from a population whose heights follow the $N(64.5, 2.5)$ distribution. The graph plots the values of $\bar{x}$ as we add women to our sample. The first woman drawn had height 64.21 inches, so the line starts there. The second had height 64.35 inches, so for $n = 2$ the mean is

$$\bar{x} = \frac{64.21 + 64.35}{2} = 64.28$$

This is the second point on the line in the graph.

At first, the graph shows that the mean of the sample changes as we take more observations. Eventually, however, the mean of the observations gets close to the population mean $\mu = 64.5$ and settles down at that value. The law of large numbers says that this *always* happens.

USE YOUR KNOWLEDGE

4.64 Use the *Law of Large Numbers* applet. The *Law of Large Numbers* applet animates a graph like Figure 4.14 for rolling dice. Use it to better understand the law of large numbers by making a similar graph.

The mean μ of a random variable is the average value of the variable in two senses. By its definition, μ is the average of the possible values, weighted by their probability of occurring. The law of large numbers says that μ is also the long-run average of many independent observations on the variable. The law of large numbers can be proved mathematically starting from the basic laws of probability.

Thinking about the law of large numbers

The law of large numbers says broadly that the average results of many independent observations are stable and predictable. The gamblers in a casino may win or lose, but the casino will win in the long run because the law of large numbers says what the average outcome of many thousands of bets will be. An insurance company deciding how much to charge for life insurance and a fast-food restaurant deciding how many beef patties to prepare also rely on the fact that averaging over many individuals produces a stable result. It is worth the effort to think a bit more closely about so important a fact.

The "law of small numbers" Both the rules of probability and the law of large numbers describe the regular behavior of chance phenomena *in the long run*. Psychologists have discovered that our intuitive understanding of randomness is quite different from the true laws of chance.[14] For example, most people believe in an incorrect "law of small numbers." That is, we expect even short sequences of random events to show the kind of average behavior that in fact appears only in the long run.

Some teachers of statistics begin a course by asking students to toss a coin 50 times and bring the sequence of heads and tails to the next class. The teacher then announces which students just wrote down a random-looking sequence rather than actually tossing a coin. The faked tosses don't have enough "runs" of consecutive heads or consecutive tails. Runs of the same outcome don't look random to us but are, in fact, common. For example, the probability of a run of three or more consecutive heads or tails in just 10 tosses is greater than 0.8.[15] The runs of consecutive heads or consecutive tails that appear in real coin tossing (and that are predicted by the mathematics of probability) seem surprising to us. Because we don't expect to see long runs, we may conclude that the coin tosses are not independent or that some influence is disturbing the random behavior of the coin.

EXAMPLE 4.31

The "hot hand" in basketball. Belief in the law of small numbers influences behavior. If a basketball player makes several consecutive shots, both the fans and her teammates believe that she has a "hot hand" and is more likely to make the next shot. This is doubtful.

Careful study suggests that runs of baskets made or missed are no more frequent in basketball than would be expected if each shot were independent of the player's previous shots. Baskets made or missed are just like heads and tails in tossing a coin. (Of course, some players make 30% of their shots in the long run and others make 50%, so a coin-toss model for basketball must allow coins with different probabilities of a head.) Our perception of hot or cold streaks simply shows that we don't perceive random behavior very well.[16]

Our intuition doesn't do a good job of distinguishing random behavior from systematic influences. This is also true when we look at data. We need statistical inference to supplement exploratory analysis of data because probability calculations can help verify that what we see in the data is more than a random pattern.

How large is a large number? The law of large numbers says that the actual mean outcome of many trials gets close to the distribution mean μ as more trials are made. It doesn't say how many trials are needed to guarantee a mean outcome close to μ. That depends on the *variability* of the random outcomes. The more variable the outcomes, the more trials are needed to ensure that the mean outcome $\bar{x}$ is close to the distribution mean μ. Casinos understand this: the outcomes of games of chance are variable enough to hold the interest of gamblers. Only the casino plays often enough to rely on the law of large numbers. Gamblers get entertainment; the casino has a business.

BEYOND THE BASICS

More Laws of Large Numbers

The law of large numbers is one of the central facts about probability. It helps us understand the mean μ of a random variable. It explains why gambling casinos and insurance companies make money. It assures us that statistical estimation will be accurate if we can afford enough observations. The basic law of large numbers applies to independent observations that all have the same distribution. Mathematicians have extended the law to many more general settings. Here are two of these.

Is there a winning system for gambling? Serious gamblers often follow a system of betting in which the amount bet on each play depends on the outcome of previous plays. You might, for example, double your bet on each spin of the roulette wheel until you win—or, of course, until your fortune is exhausted. Such a system tries to take advantage of the fact that you have a memory even though the roulette wheel does not. Can you beat the odds with a system based on the outcomes of past plays? No. Mathematicians have established a stronger version of the law of large numbers that says that, if you do not have an infinite fortune to gamble with, your long-run average winnings μ remain the same as long as successive trials of the game (such as spins of the roulette wheel) are independent.

What if observations are not independent? You are in charge of a process that manufactures video screens for computer monitors. Your equipment measures the tension on the metal mesh that lies behind each screen and is critical to its image quality. You want to estimate the mean tension μ for the process by the average $\bar{x}$ of the measurements. Alas, the tension measurements are not independent. If the tension on one screen is a bit too high, the tension on the next is more likely to also be high. Many real-world processes are like this—the process stays stable in the long run, but two observations made close together are likely to both be above or both be below the long-run mean. Again the mathematicians come to the rescue: as long as the dependence dies out fast enough as we take measurements farther and farther apart in time, the law of large numbers still holds.

Rules for means

You are studying flaws in the painted finish of refrigerators made by your firm. Dimples and paint sags are two kinds of surface flaw. Not all refrigerators have the same number of dimples: many have none, some have one, some two, and so on. You ask for the average number of imperfections on a refrigerator. The inspectors report finding an average of 0.7 dimple and 1.4 sags per refrigerator. How many total imperfections of both kinds (on the average) are there on a refrigerator? That's easy: if the average number of dimples is 0.7 and the average number of sags is 1.4, then counting both gives an average of $0.7 + 1.4 = 2.1$ flaws.

In more formal language, the number of dimples on a refrigerator is a random variable X that varies as we inspect one refrigerator after another. We know only that the mean number of dimples is $\mu_X = 0.7$. The number of paint sags is a second random variable Y having mean $\mu_X = 1.4$. (As usual, the subscripts keep straight which variable we are talking about.) The total number of both dimples and sags is another random variable, the sum $X + Y$. Its mean μ_{X+Y} is the average number of dimples and sags together. It is just the sum of the individual means μ_X and μ_Y. That's an important rule for how means of random variables behave.

Here's another rule. The crickets living in a field have mean length of 1.2 inches. What is the mean in centimeters? There are 2.54 centimeters in an inch, so the length of a cricket in centimeters is 2.54 times its length in inches. If we multiply every observation by 2.54, we also multiply their average by 2.54. The mean in centimeters must be 2.54 × 1.2, or about 3.05 centimeters. More formally, the length in inches of a cricket chosen at random from the field is a random variable X with mean μ_X. The length in centimeters is 2.54X, and this new random variable has mean 2.54μ_X.

The point of these examples is that means behave like averages. Here are the rules we need.

> **RULES FOR MEANS OF LINEAR TRANSFORMATIONS, SUMS, AND DIFFERENCES**
>
> **Rule 1.** If X is a random variable and a and b are fixed numbers, then
>
> $$\mu_{a+bX} = a + b\mu_X$$
>
> **Rule 2.** If X and Y are random variables, then
>
> $$\mu_{X+Y} = \mu_X + \mu_Y$$
>
> **Rule 3.** If X and Y are random variables, then
>
> $$\mu_{X-Y} = \mu_X - \mu_Y$$

LOOK BACK

linear transformation, p. 44

Note that $a + bX$ is a linear transformation of the random variable X.

skynesher/iStockphoto

EXAMPLE 4.32

How many courses? In Exercise 4.45 (page 244) you described the probability distribution of the number of courses taken in the fall by students at a small liberal arts college. Here is the distribution:

Courses in the fall	1	2	3	4	5	6
Probability	0.05	0.05	0.13	0.26	0.36	0.15

For the spring semester, the distribution is a little different.

Courses in the spring	1	2	3	4	5	6
Probability	0.06	0.08	0.15	0.25	0.34	0.12

For a randomly selected student, let X be the number of courses taken in the fall semester, and let Y be the number of courses taken in the spring semester. The means of these random variables are

$$\mu_X = (1)(0.05) + (2)(0.05) + (3)(0.13) + (4)(0.26) + (5)(0.36) + (6)(0.15)$$
$$= 4.28$$

$$\mu_Y = (1)(0.06) + (2)(0.08) + (3)(0.15) + (4)(0.25) + (5)(0.34) + (6)(0.12)$$
$$= 4.09$$

The mean course load for the fall is 4.28 courses, and the mean course load for the spring is 4.09 courses. We assume that these distributions apply to students who earned credit for courses taken in the fall and the spring semesters. The mean of the total number of courses taken for the academic year is $X + Y$. Using Rule 2, we calculate the mean of the total number of courses:

$$\mu_Z = \mu_X + \mu_Y$$
$$= 4.28 + 4.09 = 8.37$$

Note that it is not possible for a student to take 8.37 courses in an academic year. This number is the mean of the probability distribution.

EXAMPLE 4.33

What about credit hours? In the previous exercise, we examined the number of courses taken in the fall and in the spring at a small liberal arts college. Suppose that we were interested in the total number of credit hours earned for the academic year. We assume that for each course taken at this college, three credit hours are earned. Let T be the mean of the distribution of the total number of credit hours earned for the academic year. What is the mean of the distribution of T? To find the answer, we can use Rule 1 with $a = 0$ and $b = 3$. Here is the calculation:

$$\mu_T = \mu_{a+bZ}$$
$$= a + b\mu_Z$$
$$= 0 + (3)(8.37) = 25.11$$

The mean of the distribution of the total number of credit hours earned is 25.11.

USE YOUR KNOWLEDGE

4.65 Find μ_Y. The random variable X has mean $\mu_X = 12$. If $Y = 12 + 6X$, what is μ_Y?

4.66 Find μ_W. The random variable U has mean $\mu_U = 25$, and the random variable V has mean $\mu_V = 25$. If $W = 0.5U + 0.5V$, find μ_W.

The variance of a random variable

The mean is a measure of the center of a distribution. A basic numerical description requires, in addition, a measure of the spread or variability of the distribution. The variance and the standard deviation are the measures of spread that accompany the choice of the mean to measure center. Just as for the mean, we need a distinct symbol to distinguish the variance of a random variable from the variance s^2 of a data set. We write the variance of a random variable X as σ_X^2. Once again, the subscript reminds us which variable we have in mind. The definition of the variance σ_X^2 of a random variable is similar to the definition of the variance s^2 given in Chapter 1 (page 38). That is, the variance is an average value of the squared deviation $(X - \mu_X)^2$ of the variable X

from its mean μ_X. As for the mean, the average we use is a weighted average in which each outcome is weighted by its probability in order to take account of outcomes that are not equally likely. Calculating this weighted average is straightforward for discrete random variables but requires advanced mathematics in the continuous case. Here is the definition.

VARIANCE OF A DISCRETE RANDOM VARIABLE

Suppose that X is a **discrete random variable** whose distribution is

Value of X	x_1	x_2	x_3	$\cdots$
Probability	p_1	p_2	p_3	$\cdots$

and that μ_X is the mean of X. The **variance** of X is

$$\sigma_X^2 = (x_1 - \mu_X)^2 p_1 + (x_2 - \mu_X)^2 p_2 + \cdots$$
$$= \Sigma (x_i - \mu_X)^2 p_i$$

The **standard deviation** σ_X of X is the square root of the variance.

EXAMPLE 4.34

Find the mean and the variance. In Example 4.32 (pages 254–255) , we saw that the distribution of the number X of fall courses taken by students at a small liberal arts college is

Courses in the fall	1	2	3	4	5	6
Probability	0.05	0.05	0.13	0.26	0.36	0.15

We can find the mean and variance of X by arranging the calculation in the form of a table. Both μ_X and σ_X^2 are sums of columns in this table.

x_i	p_i	$x_i p_i$	$(x_i - \mu_X)^2 p_i$
1	0.05	0.05	$(1 - 4.28)^2 (0.05) = 0.53792$
2	0.05	0.10	$(2 - 4.28)^2 (0.05) = 0.25992$
3	0.13	0.39	$(3 - 4.28)^2 (0.13) = 0.21299$
4	0.26	1.04	$(4 - 4.28)^2 (0.26) = 0.02038$
5	0.36	1.80	$(5 - 4.28)^2 (0.36) = 0.18662$
6	0.15	0.90	$(6 - 4.28)^2 (0.15) = 0.44376$
		$\mu_X = 4.28$	$\sigma_X^2 = 1.662$

We see that $\sigma_X^2 = 1.662$. The standard deviation of X is $\sigma_X = \sqrt{1.662} = 1.289$. The standard deviation is a measure of the variability of the number of fall courses taken by the students at the small liberal arts college. As in the case of distributions for data, the standard deviation of a probability distribution is easiest to understand for Normal distributions.

USE YOUR KNOWLEDGE

4.67 Find the variance and the standard deviation. The random variable X has the following probability distribution:

Value of X	0	3
Probability	0.3	0.7

Find the variance σ_X^2 and the standard deviation σ_X for this random variable.

Rules for variances and standard deviations

What are the facts for variances that parallel Rules 1, 2, and 3 for means? *The mean of a sum of random variables is always the sum of their means, but this addition rule is true for variances only in special situations.* To understand why, take X to be the percent of a family's after-tax income that is spent, and take Y to be the percent that is saved. When X increases, Y decreases by the same amount. Though X and Y may vary widely from year to year, their sum $X + Y$ is always 100% and does not vary at all. It is the association between the variables X and Y that prevents their variances from adding.

If random variables are independent, this kind of association between their values is ruled out and their variances do add. Two random variables independence X and Y are **independent** if knowing that any event involving X alone did or did not occur tells us nothing about the occurrence of any event involving Y alone.

Probability models often assume independence when the random variables describe outcomes that appear unrelated to each other. You should ask in each instance whether the assumption of independence seems reasonable.

When random variables are not independent, the variance of their sum correlation depends on the **correlation** between them as well as on their individual variances. In Chapter 2, we met the correlation r between two observed variables measured on the same individuals. We defined (page 101) the correlation r as an average of the products of the standardized x and y observations. The correlation between two random variables is defined in the same way, once again using a weighted average with probabilities as weights. We won't give the details—it is enough to know that the correlation between two random variables has the same basic properties as the correlation r calculated from data. We use ρ, the Greek letter rho, for the correlation between two random variables. The correlation ρ is a number between -1 and 1 that measures the direction and strength of the linear relationship between two variables. **The correlation between two independent random variables is zero**.

Returning to family finances, if X is the percent of a family's after-tax income that is spent and Y is the percent that is saved, then $Y = 100 - X$. This is a perfect linear relationship with a negative slope, so the correlation between X and Y is $\rho = -1$. With the correlation at hand, we can state the rules for manipulating variances.

RULES FOR VARIANCES AND STANDARD DEVIATIONS OF LINEAR TRANSFORMATIONS, SUMS, AND DIFFERENCES

Rule 1. If X is a random variable and a and b are fixed numbers, then

$$\sigma^2_{a+bX} = b^2\sigma^2_X$$

Rule 2. If X and Y are independent random variables, then

$$\sigma^2_{X+Y} = \sigma^2_X + \sigma^2_Y$$
$$\sigma^2_{X-Y} = \sigma^2_X + \sigma^2_Y$$

This is the **addition rule for variances of independent random variables**.

Rule 3. If X and Y have correlation ρ, then

$$\sigma^2_{X+Y} = \sigma^2_X + \sigma^2_Y + 2\rho\sigma_X\sigma_Y$$
$$\sigma^2_{X-Y} = \sigma^2_X + \sigma^2_Y - 2\rho\sigma_X\sigma_Y$$

This is the **general addition rule for variances of random variables**.

To find the standard deviation, take the square root of the variance.

Because a variance is the average of squared deviations from the mean, multiplying X by a constant b multiplies σ^2_X by the square of the constant. Adding a constant a to a random variable changes its mean but does not change its variability. The variance of $X + a$ is, therefore, the same as the variance of X. Because the square of -1 is 1, the addition rule says that the variance of a difference between independent random variables is the *sum* of the variances. For independent random variables, the difference $X - Y$ is more variable than either X or Y alone because variations in both X and Y contribute to variation in their difference.

As with data, we prefer the standard deviation to the variance as a measure of the variability of a random variable. *Rule 2 for variances implies that standard deviations of independent random variables do not add. To combine standard deviations, use the rules for variances.* For example, the standard deviations of $2X$ and $-2X$ are both equal to $2\sigma_X$ because this is the square root of the variance $4\sigma^2_X$.

EXAMPLE 4.35

Payoff in the Tri-State Pick 3 lottery. The payoff X of a \$1 ticket in the Tri-State Pick 3 game is \$500 with probability 1/1000 and 0 the rest of the time. Here is the combined calculation of mean and variance:

x_i	p_i	x_ip_i	$(x_i - \mu_X)^2p_i$
0	0.999	0	$(0 - 0.5)^2(0.999) =$ 0.24975
500	0.001	0.5	$(500 - 0.5)^2(0.001) =$ 249.50025
	$\mu_X = 0.5$		$\sigma^2_X = 249.75$

The mean payoff is 50 cents. The standard deviation is $\sigma_X = \sqrt{249.75} = 15.80$. It is usual for games of chance to have large standard deviations because large variability makes gambling exciting.

If you buy a Pick 3 ticket, your winnings are $W = X - 1$ because the dollar you paid for the ticket must be subtracted from the payoff. Let's find the mean and variance for this random variable.

EXAMPLE 4.36

Winnings in the Tri-State Pick 3 lottery. By the rules for means, the mean amount you win is

$$\mu_W = \mu_X - 1 = -\$0.50$$

That is, you lose an average of 50 cents on a ticket. The rules for variances remind us that the variance and standard deviation of the winnings $W = X - 1$ are the same as those of X. Subtracting a fixed number changes the mean but not the variance.

Suppose now that you buy a $1 ticket on each of two different days. The payoffs X and Y on the two tickets are independent because separate drawings are held each day. Your total payoff is $X + Y$. Let's find the mean and standard deviation for this payoff.

EXAMPLE 4.37

Two tickets. The mean for the payoff for the two tickets is

$$\mu_{X+Y} = \mu_X + \mu_Y = \$0.50 + \$0.50 = \$1.00$$

Because X and Y are independent, the variance of $X + Y$ is

$$\sigma^2_{X+Y} = \sigma^2_X + \sigma^2_Y = 249.75 + 249.75 = 499.5$$

The standard deviation of the total payoff is

$$\sigma_{X+Y} = \sqrt{499.5} = \$22.35$$

This is not the same as the sum of the individual standard deviations, which is $15.80 + \$15.80 = \$31.60$. Variances of independent random variables add; standard deviations do not.

When we add random variables that are correlated, we need to use the correlation for the calculation of the variance but not for the calculation of the mean. Here is an example.

EXAMPLE 4.38

Utility bills. Consider a household where the monthly bill for natural-gas averages $125 with a standard deviation of $75, while the monthly bill for electricity averages $174 with a standard deviation of $41. The correlation between the two bills is -0.55.

Let's compute the mean and standard deviation of the sum of the natural-gas bill and the electricity bill. We let X stand for the natural-gas bill and Y stand for the electricity bill. Then the total is $X + Y$. Using the rules for means, we have

$$\mu_{X+Y} = \mu_X + \mu_Y = 125 + 174 = 299$$

To find the standard deviation, we first find the variance and then take the square root to determine the standard deviation. From the general addition rule for variances of random variables,

$$\sigma_{X+Y}^2 = \sigma_X^2 + \sigma_Y^2 + 2\rho\sigma_X\sigma_Y$$
$$= (75)^2 + (41)^2 + (2)(-0.55)(75)(41)$$
$$= 3923.5$$

Therefore, the standard deviation is

$$\sigma_{X+Y} = \sqrt{3923.5} = 63$$

The total of the natural-gas bill and the electricity bill has mean $299 and standard deviation $63.

The negative correlation in Example 4.38 is due to the fact that, in this household, natural gas is used for heating and electricity is used for air-conditioning. So, when it is warm, the electricity charges are high and the natural-gas charges are low. When it is cool, the reverse is true. This causes the standard deviation of the sum to be less than it would be if the two bills were uncorrelated (see Exercise 4.79, page 263).

There are situations where we need to combine several of our rules to find means and standard deviations. Here is an example.

EXAMPLE 4.39

Calcium intake. To get enough calcium for optimal bone health, tablets containing calcium are often recommended to supplement the calcium in the diet. One study designed to evaluate the effectiveness of a supplement followed a group of young people for seven years. Each subject was assigned to take either a tablet containing 1000 milligrams of calcium per day (mg/d) or a placebo tablet that was identical except that it had no calcium.[17] A major problem with studies like this one is compliance: subjects do not always take the treatments assigned to them.

In this study, the compliance rate declined to about 47% toward the end of the seven-year period. The standard deviation of compliance was 22%. Calcium from the diet averaged 850 mg/d with a standard deviation of 330 mg/d. The correlation between compliance and dietary intake was 0.68. Let's find the mean and standard deviation for the total calcium intake. We let S stand for the intake from the supplement and D stand for the intake from the diet.

We start with the intake from the supplement. Because the compliance is 47% and the amount in each tablet is 1000 mg, the mean for S is

$$\mu_S = 1000(0.47) = 470$$

Because the standard deviation of the compliance is 22%, the variance of S is

$$\sigma_S^2 = 1000^2(0.22)^2 = 48,400$$

The standard deviation is

$$\sigma_S = \sqrt{48,400} = 220$$

Be sure to verify which rules for means and variances are used in these calculations.

We can now find the mean and standard deviation for the total intake. The mean is

$$\mu_{S+D} = \mu_S + \mu_D = 470 + 850 = 1320$$

the variance is

$$\sigma_{S+D}^2 = \sigma_S^2 + \sigma_D^2 + 2\rho\sigma_S\sigma_D = (220)^2 + (330)^2 + 2(0.68)(220)(330) = 256,036$$

and the standard deviation is

$$\sigma_{S+D} = \sqrt{256{,}036} = 506$$

The mean of the total calcium intake is 1320 mg/d and the standard deviation is 506 mg/d.

The correlation in this example illustrates an unfortunate fact about compliance and having an adequate diet. Some of the subjects in this study have diets that provide an adequate amount of calcium while others do not. The positive correlation between compliance and dietary intake tells us that those who have relatively high dietary intakes are more likely to take the assigned supplements. On the other hand, those subjects with relatively low dietary intakes, the ones who need the supplement the most, are less likely to take the assigned supplements.

SECTION 4.4 SUMMARY

• The probability distribution of a random variable X, like a distribution of data, has a **mean** μ_X and a **standard deviation** σ_X.

• The **law of large numbers** says that the average of the values of X observed in many trials must approach μ.

• The **mean** μ is the balance point of the probability histogram or density curve. If X is **discrete** with possible values x_i having probabilities p_i, the mean is the average of the values of X, each weighted by its probability:

$$\mu_X = x_1 p_1 + x_2 p_2 + \cdots$$

• The **variance** σ_X^2 is the average squared deviation of the values of the variable from their mean. For a discrete random variable,

$$\sigma_X^2 = (x_1 - \mu_X)^2 p_1 + (x_2 - \mu_X)^2 p_2 + \cdots$$

• The **standard deviation** σ_X is the square root of the variance. The standard deviation measures the variability of the distribution about the mean. It is easiest to interpret for Normal distributions.

• The **mean and variance of a continuous random variable** can be computed from the density curve, but to do so requires more advanced mathematics.

• The means and variances of random variables obey the following rules. If a and b are fixed numbers, then

$$\mu_{a+bX} = a + b\mu_X$$
$$\sigma_{a+bX}^2 = b^2 \sigma_X^2$$

• If X and Y are any two random variables having correlation ρ, then

$$\mu_{X+Y} = \mu_X + \mu_Y$$
$$\mu_{X-Y} = \mu_X - \mu_Y$$
$$\sigma_{X+Y}^2 = \sigma_X^2 + \sigma_Y^2 + 2\rho\sigma_X\sigma_Y$$
$$\sigma_{X-Y}^2 = \sigma_X^2 + \sigma_Y^2 - 2\rho\sigma_X\sigma_Y$$

• If X and Y are **independent**, then $\rho = 0$. In this case,

$$\sigma_{X+Y}^2 = \sigma_X^2 + \sigma_Y^2$$
$$\sigma_{X-Y}^2 = \sigma_X^2 + \sigma_Y^2$$

• To find the standard deviation, take the square root of the variance.

SECTION 4.4 EXERCISES

For Exercise 4.63, see page 247; for Exercise 4.64, see page 251; for Exercises 4.65 and 4.66, see page 255; and for Exercise 4.67, see page 257.

4.68 Find the mean of the random variable. A random variable X has the following distribution.

X	−1	0	1	2
Probability	0.2	0.3	0.2	0.3

Find the mean for this random variable. Show your work.

4.69 Explain what happens when the sample size gets large. Consider the following scenarios: (1) You take a sample of two observations on a random variable and compute the sample mean, (2) you take a sample of 100 observations on the same random variable and compute the sample mean, (3) you take a sample of 1000 observations on the same random variable and compute the sample mean. Explain in simple language how close you expect the sample mean to be to the mean of the random variable as you move from Scenario 1 to Scenario 2 to Scenario 3.

4.70 Find some means. Suppose that X is a random variable with mean 30 and standard deviation 4. Also suppose that Y is a random variable with mean 50 and standard deviation 8. Find the mean of the random variable Z for each of the following cases. Be sure to show your work.

(a) $Z = 35 - 10X$.

(b) $Z = 12X - 5$.

(c) $Z = X + Y$.

(d) $Z = X - Y$.

(e) $Z = -2X + 2Y$.

4.71 Find the variance and the standard deviation. A random variable X has the following distribution.

X	−1	0	1	2
Probability	0.3	0.2	0.3	0.2

Find the variance and the standard deviation for this random variable. Show your work.

4.72 Find some variances and standard deviations. Suppose that X is a random variable with mean 30 and standard deviation 4. Also suppose that Y is a random variable with mean 50 and standard deviation 8. Assume that the correlation between X and Y is zero. Find the variance and the standard deviation of the random variable Z for each of the following cases. Be sure to show your work.

(a) $Z = 35 - 10X$.

(b) $Z = 12X - 5$.

(c) $Z = X + Y$.

(d) $Z = X - Y$.

(e) $Z = -2X + 2Y$.

4.73 What happens if the correlation is not zero? Suppose that X is a random variable with mean 30 and standard deviation 4. Also suppose that Y is a random variable with mean 50 and standard deviation 8. Assume that the correlation between X and Y is 0.5. Find the variance and the standard deviation of the random variable Z for each of the following cases. Be sure to show your work.

(a) $Z = 35 - 10X$.

(b) $Z = 12X - 5$.

(c) $Z = X + Y$.

(d) $Z = X - Y$.

(e) $Z = -2X + 2Y$.

4.74 What's wrong? In each of the following scenarios, there is something wrong. Describe what is wrong and give a reason for your answer.

(a) If you toss a fair coin three times and get heads all three times, then the probability of getting a tail on the next toss is much greater than one-half.

(b) If you multiply a random variable by 10, then the mean is multiplied by 10 and the variance is multiplied by 10.

(c) When finding the mean of the sum of two random variables, you need to know the correlation between them.

4.75 Servings of fruits and vegetables. The following table gives the distribution of the number of servings of fruits and vegetables consumed per day in a population.

Number of servings X	0	1	2	3	4	5
Probability	0.3	0.1	0.1	0.2	0.2	0.1

Find the mean for this random variable.

4.76 Mean of the distribution for the number of aces. In Exercise 4.54 (page 244) you examined the probability distribution for the number of aces when you are dealt two cards in the game of Texas hold 'em. Let X represent the number of aces in a randomly selected deal of two cards in this game. Here is the probability distribution for the random variable X:

Value of X	0	1	2
Probability	0.8507	0.1448	0.0045

Find μ_X, the mean of the probability distribution of X.

4.77 Standard deviation of the number of aces. Refer to Exercise 4.76. Find the standard deviation of the number of aces.

4.78 Standard deviation for fruits and vegetables. Refer to Exercise 4.75. Find the variance and the standard deviation for the distribution of the number of servings of fruits and vegetables.

4.79 Suppose that the correlation is zero. Refer to Example 4.38 (page 259).

(a) Recompute the standard deviation for the total of the natural-gas bill and the electricity bill, assuming that the correlation is zero.

(b) Is this standard deviation larger or smaller than the standard deviation computed in Example 4.38? Explain why.

4.80 Find the mean of the sum. Figure 4.12 (page 245) displays the density curve of the sum $Y = X_1 + X_2$ of two independent random numbers, each uniformly distributed between 0 and 1.

(a) The mean of a continuous random variable is the balance point of its density curve. Use this fact to find the mean of Y from Figure 4.12.

(b) Use the same fact to find the means of X_1 and X_2. (They have the density curve pictured in Figure 4.9, page 240.) Verify that the mean of Y is the sum of the mean of X_1 and the mean of X_2.

4.81 Calcium supplements and calcium in the diet. Refer to Example 4.39 (page 260). Suppose that people who have high intakes of calcium in their diets are more compliant than those who have low intakes. What effect would this have on the calculation of the standard deviation for the total calcium intake? Explain your answer.

4.82 Toss a four-sided die twice. Role-playing games like *Dungeons & Dragons* use many different types of dice. Suppose that a four-sided die has faces marked 1, 2, 3, and 4. The intelligence of a character is determined by rolling this die twice and adding 1 to the sum of the spots. The faces are equally likely, and the two rolls are independent. What is the average (mean) intelligence for such characters? How spread out are their intelligences, as measured by the standard deviation of the distribution?

4.83 Means and variances of sums. The rules for means and variances allow you to find the mean and variance of a sum of random variables without first finding the distribution of the sum, which is usually much harder to do.

(a) A single toss of a balanced coin has either 0 or 1 head, each with probability 1/2. What are the mean and standard deviation of the number of heads?

(b) Toss a coin four times. Use the rules for means and variances to find the mean and standard deviation of the total number of heads.

(c) Example 4.23 (page 238) finds the distribution of the number of heads in four tosses. Find the mean and standard deviation from this distribution. Your results in parts (b) and (c) should agree.

4.84 What happens when the correlation is 1? We know that variances add if the random variables involved are uncorrelated ($\rho = 0$), but not otherwise. The opposite extreme is perfect positive correlation ($\rho = 1$). Show by using the general addition rule for variances that, in this case, the standard deviations add. That is, $\sigma_{X+Y} = \sigma_X + \sigma_Y$ if $\rho_{XY} = 1$.

4.85 Will you assume independence? In which of the following games of chance would you be willing to assume independence of X and Y in making a probability model? Explain your answer in each case.

(a) In blackjack, you are dealt two cards and examine the total points X on the cards (face cards count 10 points). You can choose to be dealt another card and compete based on the total points Y on all three cards.

(b) In craps, the betting is based on successive rolls of two dice. X is the sum of the faces on the first roll, and Y the sum of the faces on the next roll.

4.86 Transform the distribution of heights from centimeters to inches. A report of the National Center for Health Statistics says that the heights of 20-year-old men have mean 176.8 centimeters (cm) and standard deviation 7.2 cm. There are 2.54 centimeters in an inch. What are the mean and standard deviation in inches?

4.87 Fire insurance. An insurance company looks at the records for millions of homeowners and sees that the mean loss from fire in a year is $\mu =$ \$300 per person. (Most of us have no loss, but a few lose their homes. The \$300 is the average loss.) The company plans to sell fire insurance for \$300 plus enough to cover its costs and profit. Explain clearly why it would be stupid to sell only 10 policies. Then explain why selling thousands of such policies is a safe business.

4.88 Mean and standard deviation for 10 and for 12 policies. In fact, the insurance company sees that in the entire population of homeowners, the mean loss from fire is $\mu =$ \$300 and the standard deviation of the loss is $\sigma =$ \$400. What are the mean and standard deviation of the average loss for 10 policies? (Losses on separate policies are independent.) What are the mean and standard deviation of the average loss for 12 policies?

4.5 General Probability Rules

When you complete this section, you will be able to:

- Apply the five rules of probability.
- Apply the general addition rule for unions of two or more events.
- Find conditional probabilities.
- Apply the multiplication rule.
- Use a tree diagram to find probabilities.
- Use Bayes's rule to find probabilities.
- Determine whether or not two events that both have positive probability are independent.

Our study of probability has concentrated on random variables and their distributions. Now we return to the laws that govern any assignment of probabilities. The purpose of learning more laws of probability is to be able to give probability models for more complex random phenomena. We have already met and used five rules.

PROBABILITY RULES

Rule 1. $0 \leq P(A) \leq 1$ for any event A

Rule 2. $P(S) = 1$

Rule 3. Addition rule: If A and B are **disjoint** events, then
$$P(A \text{ or } B) = P(A) + P(B)$$

Rule 4. Complement rule: For any event A,
$$P(A^c) = 1 - P(A)$$

Rule 5. Multiplication rule: If A and B are **independent** events, then
$$P(A \text{ and } B) = P(A)P(B)$$

General addition rules

Probability has the property that if A and B are disjoint events, then $P(A \text{ or } B) = P(A) + P(B)$. What if there are more than two events or if the events are not disjoint? These circumstances are covered by more general addition rules for probability.

UNION

The **union** of any collection of events is the event that at least one of the collection occurs.

For two events A and B, the union is the event $\{A \text{ or } B\}$ that A or B or both occur. From the addition rule for two disjoint events, we can obtain rules for more general unions. Suppose first that we have several events—say, A, B, and C—that are disjoint in pairs. That is, no two can occur simultaneously. The

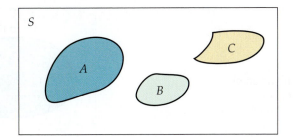

FIGURE 4.15 The addition rule for disjoint events: $P(A$ or B or $C) = P(A) + P(B) + P(C)$ when events A, B, and C are disjoint.

Venn diagram in Figure 4.15 illustrates three disjoint events. The addition rule for two disjoint events extends to the following law.

ADDITION RULE FOR DISJOINT EVENTS

If events A, B, and C are disjoint in the sense that no two have any outcomes in common, then

$$P(\text{one or more of } A,B,C) = P(A) + P(B) + P(C)$$

This rule extends to any number of disjoint events.

EXAMPLE 4.40

Probabilities as areas. Generate a random number X between 0 and 1. What is the probability that the first digit after the decimal point will be a 3, a 6, or a 9? The random number X is a continuous random variable whose density curve has constant height 1 between 0 and 1 and is 0 elsewhere. The event that the first digit of X is odd is the union of five disjoint events. These events are

$$0.30 \leq X < 0.40$$
$$0.60 \leq X < 0.70$$
$$0.90 \leq X < 1.00$$

Figure 4.16 illustrates the probabilities of these events as areas under the density curve. Each area is 0.1. Therefore, the union of the three has probability equal to the sum, or 0.3.

FIGURE 4.16 The probability that the first digit after the decimal point of a random number is a 3, a 6, or a 9 is the sum of the probabilities of the three disjoint events shown, Example 4.40.

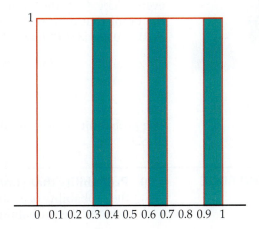

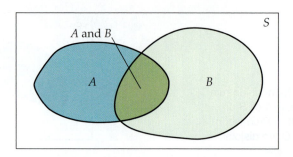

FIGURE 4.17 The union of two events that are not disjoint. The general addition rule says that $P(A \text{ or } B) = P(A) + P(B) - P(A \text{ and } B)$.

USE YOUR KNOWLEDGE

4.89 Probability that you roll a 3 or a 4 or a 5. If you roll a die, the probability of each of the six possible outcomes (1, 2, 3, 4, 5, 6) is 1/6. What is the probability that you roll a 3 or a 4 or a 5?

If events A and B are not disjoint, they can occur simultaneously. The probability of their union is then *less* than the sum of their probabilities. As Figure 4.17 suggests, the outcomes common to both are counted twice when we add probabilities, so we must subtract this probability once. Here is the addition rule for the union of any two events, disjoint or not.

GENERAL ADDITION RULE FOR UNIONS OF TWO EVENTS

For any two events A and B,

$$P(A \text{ or } B) = P(A) + P(B) - P(A \text{ and } B)$$

If A and B are disjoint, the event $\{A \text{ and } B\}$ that both occur has no outcomes in it. This *empty event* is the complement of the sample space S and must have probability 0. So the general addition rule includes Rule 3, the addition rule for disjoint events.

EXAMPLE 4.41

Adequate sleep and exercise. Suppose that 40% of adults get enough sleep and 46% exercise regularly. What is the probability that an adult gets enough sleep or exercises regularly? To find this probability, we also need to know the percent who get enough sleep and exercise. Let's assume that 24% do both.

We will use the notation of the general addition rule for unions of two events. Let A be the event that an adult gets enough sleep, and let B be the event that a person exercises regularly. We are given that $P(A) = 0.40$, $P(B) = 0.46$, and $P(A \text{ and } B) = 0.24$. Therefore,

$$P(A \text{ or } B) = P(A) + P(B) - P(A \text{ and } B)$$
$$= 0.40 + 0.46 - 0.24$$
$$= 0.62$$

The probability that an adult gets enough sleep or exercises regularly is 0.62, or 62%.

USE YOUR KNOWLEDGE

4.90 Probability that your roll is even or greater than 5. If you roll a die, the probability of each of the six possible outcomes (1, 2, 3, 4, 5, 6) is 1/6. What is the probability that your roll is even or greater than 5?

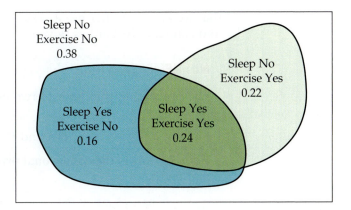

FIGURE 4.18 Venn diagram and probabilities, Example 4.41.

Venn diagrams are a great help in finding probabilities for unions because you can just think of adding and subtracting areas. Figure 4.18 shows some events and their probabilities for Example 4.41. What is the probability that an adult gets adequate sleep and does not exercise?

The Venn diagram shows that the probability that an adult gets adequate sleep minus the probability that an adult gets adequate sleep and exercises regularly is $0.40 - 0.24 = 0.16$. Similarly, the probability that an adult does not get adequate sleep and exercises regularly is $0.46 - 0.24 = 0.22$. The four probabilities that appear in the figure add to 1 because they refer to four disjoint events whose union is the entire sample space.

Conditional probability

The probability we assign to an event can change if we know that some other event has occurred. This idea is the key to many applications of probability.

EXAMPLE 4.42

Probability of being dealt a heart. Doyle is a professional poker player. He stares at the dealer, who prepares to deal. What is the probability that the card dealt to Doyle is a heart? There are 52 cards in the deck. Because the deck was carefully shuffled, the next card dealt is equally likely to be any of the cards that Doyle has not seen. Thirteen of the 52 cards are hearts. So

$$P(\text{heart}) = \frac{13}{52} = \frac{1}{4}$$

This calculation assumes that Doyle knows nothing about any cards already dealt. Suppose now that he is looking at four cards already in his hand and that they are all hearts. He knows nothing about the other 48 cards except that exactly nine $(13 - 4)$ hearts are among them. Doyle's probability of being dealt a heart *given what he knows* is now

$$P(\text{heart} \mid 4 \text{ hearts in 4 visible cards}) = \frac{9}{48} = \frac{3}{16}$$

Knowing that there are four hearts among the four cards Doyle can see changes the probability that the next card dealt is a heart.

conditional probability   The new notation $P(A \mid B)$ is a **conditional probability**. That is, it gives the probability of one event (the next card dealt is a heart) under the condition that we know another event (exactly one of the four visible cards is a heart). You can read the bar | as "given the information that."

MULTIPLICATION RULE

The probability that both of two events A and B happen together can be found by

$$P(A \text{ and } B) = P(A)P(B \mid A)$$

Here $P(B \mid A)$ is the conditional probability that B occurs, given the information that A occurs.

USE YOUR KNOWLEDGE

4.91 The probability of a heart. Refer to Example 4.42. Suppose that none of the four cards in Doyle's hand are hearts. What is the probability that the next card dealt to him is a heart?

EXAMPLE 4.43

Downloading music from the Internet. The multiplication rule is just common sense made formal. For example, suppose that 30% of Internet users download music files, and 70% of downloaders say they don't care if the music is copyrighted. So the percent of Internet users who download music (event A) *and* don't care about copyright (event B) is 70% of the 30% who download, or

$$(0.7)(0.3) = 0.21 = 21\%$$

The multiplication rule expresses this as

$$\begin{aligned} P(A \text{ and } B) &= P(A) \times P(B \mid A) \\ &= (0.3)(0.7) = 0.21 \end{aligned}$$

Here is another example that uses conditional probability.

EXAMPLE 4.44

Probability of a favorable draw. Doyle is still at the poker table. At the moment, he has two cards and they are both hearts. He has seen 24 cards and none of other players have any hearts. What is the chance that the next three cards he draws will be hearts? The full deck of 52 cards contains 13 hearts. Therefore, 11 of the unseen cards are hearts. There are 28 (52 − 24) unseen cards. To find Doyle's probability of drawing three hearts, we first calculate

$$P(\text{first card is a heart}) = \frac{11}{28}$$

$$P(\text{second card is a heart} \mid \text{first card is a heart}) = \frac{10}{27}$$

$$P(\text{third card is a heart} \mid \text{first two cards are hearts}) = \frac{9}{26}$$

Doyle finds both probabilities by counting cards. The probability that the first card drawn is a heart is 11/28 because 11 of the 28 unseen cards are hearts. If the first card is a heart, that leaves 10 hearts among the

27 remaining cards. So the *conditional* probability of another diamond is 10/27. The multiplication rule now says that

$$P(\text{next two cards are hearts}) = \frac{11}{28} \times \frac{10}{27} = 0.146$$

We again apply the multiplication rule for the third card. The probability that the next three draws are hearts is equal to the probability that the first two draws are hearts times the probability that the third card is a heart given that the first two draws are hearts. This probability is

$$P(\text{next three cards are hearts}) = \frac{11}{28} \times \frac{10}{27} \times \frac{9}{26} = 0.050$$

It is very unlikely that Doyle's next three cards will be hearts, even though his hearts are the only ones that he has seen.

USE YOUR KNOWLEDGE

4.92 **The probability that the next two cards are hearts.** In the setting of Example 4.44, suppose that Doyle's third card is a heart, so he now has three hearts, and that none of the five additional cards that he sees are hearts What is the probability that the next two cards dealt to Doyle will be hearts?

If $P(A)$ and $P(A \text{ and } B)$ are given, we can rearrange the multiplication rule to produce a *definition* of the conditional probability $P(B \mid A)$ in terms of unconditional probabilities.

DEFINITION OF CONDITIONAL PROBABILITY

When $P(A) > 0$, the **conditional probability** of B given A is

$$P(B \mid A) = \frac{P(A \text{ and } B)}{P(A)}$$

Be sure to keep in mind the distinct roles in $P(B \mid A)$ of the event B whose probability we are computing and the event A that represents the information we are given. The conditional probability $P(B \mid A)$ makes no sense if the event A can never occur, so we require that $P(A) > 0$ whenever we talk about $P(B \mid A)$.

EXAMPLE 4.45

College students. Here is the distribution of U.S. college students classified by age and full-time or part-time status:

Age (years)	Full-time	Part-time
15 to 19	0.21	0.02
20 to 24	0.32	0.07
25 to 34	0.10	0.10
30 and over	0.05	0.13

Let's compute the probability that a student is aged 20 to 24, given that the student is full-time. We know that the probability that a student is part-time *and* aged 20 to 24 is 0.32 from the table of probabilities. But what we want here is a conditional probability, given that a student is full-time. Rather than asking about age among all students, we restrict our attention to the subpopulation of students who are full-time. Let

$$A = \text{the student is between 20 and 24 years of age}$$
$$B = \text{the student is a full-time student}$$

Our formula is

$$P(A \mid B) = \frac{P(A \text{ and } B)}{P(B)}$$

We read $P(A \text{ and } B) = 0.32$ from the table as we mentioned previously. What about $P(B)$? This is the probability that a student is full-time. Notice that there are four groups of students in our table that fit this description. To find the probability needed, we add the entries:

$$P(B) = 0.21 + 0.32 + 0.10 + 0.05 = 0.68$$

We are now ready to complete the calculation of the conditional probability:

$$P(A \mid B) = \frac{P(A \text{ and } B)}{P(B)}$$
$$= \frac{0.32}{0.68}$$
$$= 0.47$$

The probability that a student is 20 to 24 years of age, given that the student is full-time, is 0.47.

Here is another way to give the information in the last sentence of this example: 47% of full-time college students are 20 to 24 years old. Which way do you prefer?

USE YOUR KNOWLEDGE

4.93 What rule did we use? In Example 4.45, we calculated $P(B)$. What rule did we use for this calculation? Explain why this rule applies in this setting.

4.94 Find the conditional probability. Refer to Example 4.45. What is the probability that a student is part-time, given that the student is 20 to 24 years old? Explain in your own words the difference between this calculation and the one that we did in Example 4.45.

General multiplication rules

The definition of conditional probability reminds us that, in principle, all probabilities—including conditional probabilities—can be found from the assignment of probabilities to events that describe random phenomena. More often, however, conditional probabilities are part of the information given to us in a probability model, and the multiplication rule is used to compute $P(A$ and $B)$. This rule extends to more than two events.

The union of a collection of events is the event that *any* of them occur. Here is the corresponding term for the event that *all* of them occur.

INTERSECTION

The **intersection** of any collection of events is the event that *all* the events occur.

To extend the multiplication rule to the probability that all of several events occur, the key is to condition each event on the occurrence of *all* the preceding events. For example, the intersection of three events A, B, and C has probability

$$P(A \text{ and } B \text{ and } C) = P(A)P(B \mid A)P(C \mid A \text{ and } B)$$

EXAMPLE 4.46

High school athletes and professional careers. Only 5% of male high school basketball, baseball, and football players go on to play at the college level. Of these, only 1.7% enter major league professional sports. About 40% of the athletes who compete in college and then reach the pros have a career of more than three years. Define these events:

$$A = \{\text{competes in college}\}$$
$$B = \{\text{competes professionally}\}$$
$$C = \{\text{pro career longer than 3 years}\}$$

What is the probability that a high school athlete competes in college and then goes on to have a pro career of more than three years? We know that

$$P(A) = 0.05$$
$$P(B \mid A) = 0.017$$
$$P(C \mid A \text{ and } B) = 0.4$$

Therefore, the probability we want is

$$P(A \text{ and } B \text{ and } C) = P(A)P(B \mid A)P(C \mid A \text{ and } B)$$
$$= 0.05 \times 0.017 \times 0.4 = 0.00034$$

Only about 3 of every 10,000 high school athletes can expect to compete in college and have a professional career of more than three years. High school students would be wise to concentrate on studies rather than on unrealistic hopes of fortune from pro sports.

Tree diagrams

Probability problems often require us to combine several of the basic rules into a more elaborate calculation. Here is an example that illustrates how to solve problems that have several stages.

EXAMPLE 4.47

tree diagram

Online chat rooms. Online chat rooms are dominated by the young. Teens are the biggest users. If we look only at adult Internet users (aged 18 and over), 47% of the 18 to 29 age group chat, as do 21% of the 30 to 49 age group and just 7% of those 50 and over. To learn what percent of all Internet users participate in chat, we also need the age breakdown of users. Here it is: 29% of adult Internet users are 18 to 29 years old (event A_1), another 47% are 30 to 49 (event A_2), and the remaining 24% are 50 and over (event A_3).

What is the probability that a randomly chosen adult user of the Internet participates in chat rooms (event C)? To find out, use the **tree diagram** in Figure 4.19 to organize your thinking. Each segment in the tree is one stage

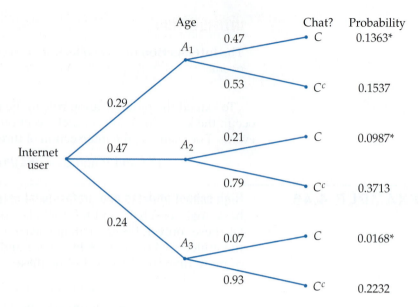

FIGURE 4.19 Tree diagram, Example 4.47. The probability $P(C)$ is the sum of the probabilities of the three branches marked with asterisks (*).

of the problem. Each complete branch shows a path through the two stages. The probability written on each segment is the conditional probability of an Internet user following that segment, given that he or she has reached the node from which it branches.

Starting at the left, an Internet user falls into one of the three age groups. The probabilities of these groups

$$P(A_1) = 0.29 \qquad P(A_2) = 0.47 \qquad P(A_3) = 0.24$$

mark the leftmost branches in the tree. Conditional on being 18 to 29 years old, the probability of participating in chat is $P(C \mid A_1) = 0.47$. So the conditional probability of *not* participating is

$$P(C^c \mid A_1) = 1 - 0.47 = 0.53$$

These conditional probabilities mark the paths branching out from the A_1 node in Figure 4.19. The other two age group nodes similarly lead to two branches marked with the conditional probabilities of chatting or not. The probabilities on the branches from any node add to 1 because they cover all possibilities, given that this node was reached.

There are three disjoint paths to C, one for each age group. By the addition rule, $P(C)$ is the sum of their probabilities. The probability of reaching C through the 18 to 29 age group is

$$P(C \text{ and } A_1) = P(A_1)P(C \mid A_1)$$
$$= 0.29 \times 0.47 = 0.1363$$

Follow the paths to C through the other two age groups. The probabilities of these paths are

$$P(C \text{ and } A_2) = P(A_2)P(C \mid A_2) = (0.47)(0.21) = 0.0987$$
$$P(C \text{ and } A_3) = P(A_3)P(C \mid A_3) = (0.24)(0.07) = 0.0168$$

The final result is

$$P(C) = 0.1363 + 0.0987 + 0.0168 = 0.2518$$

About 25% of all adult Internet users take part in chat rooms.

It takes longer to explain a tree diagram than it does to use it. Once you have understood a problem well enough to draw the tree, the rest is easy. Tree diagrams combine the addition and multiplication rules. The multiplication rule says that the probability of reaching the end of any complete branch is the product of the probabilities written on its segments. The probability of any outcome, such as the event C that an adult Internet user takes part in chat rooms, is then found by adding the probabilities of all branches that are part of that event.

USE YOUR KNOWLEDGE

4.95 Draw a tree diagram. Refer to Doyle's chances of five hearts in Example 4.44 (page 268). Draw a tree diagram to describe the outcomes for the three cards that he will be dealt. At the first stage, his draw can be a heart or a nonheart. At the second and third stages, he has the same possible outcomes but the probabilities are different.

Bayes's rule

There is another kind of probability question that we might ask in the context of thinking about online chat. What percent of adult chat room participants are aged 18 to 29?

EXAMPLE 4.48

Conditional versus unconditional probabilities. In the notation of Example 4.47, this is the conditional probability $P(A_1 \mid C)$. Start from the definition of conditional probability and then apply the results of Example 4.47:

$$P(A_1 \mid C) = \frac{P(A_1 \text{ and } C)}{P(C)}$$

$$= \frac{0.1363}{0.2518} = 0.5413$$

More than half of adult chat room participants are between 18 and 29 years old. Compare this conditional probability with the original information (unconditional) that 29% of adult Internet users are between 18 and 29 years old. Knowing that a person chats increases the probability that he or she is young.

We know the probabilities $P(A_1)$, $P(A_2)$, and $P(A_3)$ that give the age distribution of adult Internet users. We also know the conditional probabilities $P(C \mid A_1)$, $P(C \mid A_2)$, and $P(C \mid A_3)$ that a person from each age group chats. Example 4.47 shows how to use this information to calculate $P(C)$. The method can be summarized in a single expression that adds the probabilities of the three paths to C in the tree diagram:

$$P(C) = P(A_1)P(C \mid A_1) + P(A_2)P(C \mid A_2) + P(A_3)P(C \mid A_3)$$

In Example 4.48, we calculated the "reverse" conditional probability $P(A_1 \mid C)$. The denominator 0.2518 in that example came from the previous expression. Put in this general notation, we have another probability law.

BAYES'S RULE

Suppose that $A_1, A_2, \ldots, A_k$ are disjoint events whose probabilities are not 0 and add to exactly 1. That is, any outcome is in exactly one of these events. Then if C is any other event whose probability is not 0 or 1,

$$P(A_i \mid C) = \frac{P(C \mid A_i)P(A_i)}{P(C \mid A_1)P(A_1) + P(C \mid A_2)P(A_2) + \cdots + P(A_k)P(C \mid A_k)}$$

The numerator in Bayes's rule is always one of the terms in the sum that makes up the denominator. The rule is named after Thomas Bayes, who wrestled with arguing from outcomes like C back to the A_i in a book published in 1763. It is far better to think your way through problems like Examples 4.47 and 4.48 than to memorize these formal expressions.

Independence again

The conditional probability $P(B \mid A)$ is generally not equal to the unconditional probability $P(B)$. That is because the occurrence of event A generally gives us some additional information about whether or not event B occurs. If knowing that A occurs gives no additional information about B, then A and B are independent events. The formal definition of independence is expressed in terms of conditional probability.

INDEPENDENT EVENTS

Two events A and B that both have positive probability are **independent** if

$$P(B \mid A) = P(B)$$

This definition makes precise the informal description of independence given in Section 4.2 (page 229). We now see that the multiplication rule for independent events, $P(A \text{ and } B) = P(A)P(B)$, is a special case of the general multiplication rule, $P(A \text{ and } B) = P(A)P(B \mid A)$, just as the addition rule for disjoint events is a special case of the general addition rule.

SECTION 4.5 SUMMARY

• The **complement** A^c of an event A contains all outcomes that are not in A. The **union** $\{A \text{ or } B\}$ of events A and B contains all outcomes in A, in B, and in both A and B. The **intersection** $\{A \text{ and } B\}$ contains all outcomes that are in both A and B, but not outcomes in A alone or B alone.

• The **conditional probability** $P(B \mid A)$ of an event B, given an event A, is defined by

$$P(B \mid A) = \frac{P(A \text{ and } B)}{P(A)}$$

when $P(A) > 0$. In practice, conditional probabilities are most often found from directly available information.

- The essential general rules of elementary probability are

  Legitimate values: $0 \le P(A) \le 1$ for any event A

  Total probability 1: $P(S) = 1$

  Complement rule: $P(A^c) = 1 - P(A)$

  Addition rule: $P(A \text{ or } B) = P(A) + P(B) - P(A \text{ and } B)$

  Multiplication rule: $P(A \text{ and } B) = P(A)P(B \mid A)$

- If A and B are **disjoint**, then $P(A \text{ and } B) = 0$. The general addition rule for unions then becomes the special addition rule, $P(A \text{ or } B) = P(A) + P(B)$.

- A and B are **independent** when $P(B \mid A) = P(B)$. The multiplication rule for intersections then becomes $P(A \text{ and } B) = P(A)P(B)$.

- In problems with several stages, draw a **tree diagram** to organize use of the multiplication and addition rules.

SECTION 4.5 EXERCISES

For Exercise 4.89, see page 266; for Exercise 4.90, see page 266; for Exercise 4.91, see page 268; for Exercise 4.92, see page 269; for Exercises 4.93 and 4.94, see page 270; and for Exercise 4.95, see page 273.

4.96 Find and explain some probabilities.

(a) Can we have an event A that has negative probability? Explain your answer.

(b) Suppose $P(A) = 0.3$ and $P(B) = 0.5$. Explain what it means for A and B to be disjoint. Assuming that they are disjoint, find the probability that A or B occurs.

(c) Explain in your own words the meaning of the rule $P(S) = 1$.

(d) Consider an event A. What is the name for the event that A does not occur? If $P(A) = 0.4$, what is the probability that A does not occur?

(e) Suppose that A and B are independent and that $P(A) = 0.8$ and $P(B) = 0.3$. Explain the meaning of the event $\{A \text{ and } B\}$, and find its probability.

4.97 Unions.

(a) Assume that $P(A) = 0.2$, $P(B) = 0.4$, and $P(C) = 0.1$. If the events A, B, and C are disjoint, find the probability that the union of these events occurs.

(b) Draw a Venn diagram to illustrate your answer to part (a).

(c) Find the probability of the complement of the union of A, B, and C.

4.98 Conditional probabilities. Suppose that $P(A) = 0.4$, $P(B) = 0.3$, and $P(B \mid A) = 0.4$.

(a) Find the probability that both A and B occur.

(b) Use a Venn diagram to explain your calculation.

(c) What is the probability of the event that B occurs and A does not?

4.99 Find the probabilities. Suppose that the probability that A occurs is 0.5 and the probability that A and B occur is 0.2.

(a) Find the probability that B occurs given that A occurs.

(b) Illustrate your calculations in part (a) using a Venn diagram.

4.100 Why not? Suppose that $P(B) = 0.6$. Explain why $P(A \text{ and } B)$ cannot be 0.7.

4.101 Is the calcium intake adequate? In the population of young children eligible to participate in a study of whether or not their calcium intake is adequate, 52% are 5 to 10 years of age and 48% are 11 to 13 years of age. For those who are 5 to 10 years of age, 18% have inadequate calcium intake. For those who are 11 to 13 years of age, 57% have inadequate calcium intake.[18]

(a) Use letters to define the events of interest in this exercise.

(b) Convert the percents given to probabilities of the events you have defined.

(c) Use a tree diagram similar to Figure 4.19 (page 272) to calculate the probability that a randomly selected child from this population has an inadequate intake of calcium.

4.102 Use Bayes's rule. Refer to the previous exercise. Use Bayes's rule to find the probability that a child from this population who has inadequate intake is 11 to 13 years old.

4.103 Are the events independent? Refer to the previous two exercises. Are the age of the child and whether or not the child has adequate calcium intake independent? Calculate the probabilities that you need to answer this question and write a short summary of your conclusion.

4.104 What's wrong? In each of the following scenarios, there is something wrong. Describe what is wrong and give a reason for your answer.

(a) $P(A$ or $B)$ is always equal to the sum of $P(A)$ and $P(B)$.

(b) The probability of an event minus the probability of its complement is always equal to 1.

(c) Two events are disjoint if $P(B \mid A) = P(B)$.

4.105 Exercise and sleep. Suppose that 42% of adults get enough sleep, 39% get enough exercise, and 28% do both. Find the probabilities of the following events:

(a) Enough sleep and not enough exercise.

(b) Not enough sleep and enough exercise.

(c) Not enough sleep and not enough exercise.

(d) For each of parts (a), (b), and (c), state the rule that you used to find your answer.

4.106 Exercise and sleep. Refer to the previous exercise. Draw a Venn diagram showing the probabilities for exercise and sleep.

4.107 Lying to a teacher. Suppose that 53% of high school students would admit to lying at least once to a teacher during the past year and that 24% of students are male and would admit to lying at least once to a teacher during the past year.[19] Assume that 44% of the students are male. What is the probability that a randomly selected student is either male or would admit to lying to a teacher during the past year? Be sure to show your work and indicate all the rules that you use to find your answer.

4.108 Lying to a teacher. Refer to the previous exercise. Suppose that you select a student from the subpopulation of those who would admit to lying to a teacher during the past year. What is the probability that the student is female? Be sure to show your work and indicate all the rules that you use to find your answer.

4.109 Attendance at two-year and four-year colleges. In a large national population of college students, 61% attend four-year institutions and the rest attend two-year institutions. Males make up 44% of the students

in the four-year institutions and 41% of the students in the two-year institutions.

(a) Find the four probabilities for each combination of gender and type of institution in the following table. Be sure that your probabilities sum to 1.

	Men	Women
Four-year institution		
Two-year institution		

(b) Consider randomly selecting a female student from this population. What is the probability that she attends a four-year institution?

4.110 Draw a tree diagram. Refer to the previous exercise. Draw a tree diagram to illustrate the probabilities in a situation where you first identify the type of institution attended and then identify the gender of the student.

4.111 Draw a different tree diagram for the same setting. Refer to the previous two exercises. Draw a tree diagram to illustrate the probabilities in a situation where you first identify the gender of the student and then identify the type of institution attended. Explain why the probabilities in this tree diagram are different from those that you used in the previous exercise.

4.112 Education and income. Call a household prosperous if its income exceeds $100,000. Call the household educated if the householder completed college. Select an American household at random, and let A be the event that the selected household is prosperous and B the event that it is educated. According to the Current Population Survey, $P(A) = 0.138$, $P(B) = 0.261$, and the probability that a household is both prosperous and educated is $P(A$ and $B) = 0.082$. What is the probability $P(A$ or $B)$ that the household selected is either prosperous or educated?

4.113 Find a conditional probability. In the setting of the previous exercise, what is the conditional probability that a household is prosperous, given that it is educated? Explain why your result shows that events A and B are not independent.

4.114 Draw a Venn diagram. Draw a Venn diagram that shows the relation between the events A and B in Exercise 4.112. Indicate each of the following events on your diagram and use the information in Exercise 4.112 to calculate the probability of each event. Finally, describe in words what each event is.

(a) $\{A$ and $B\}$.

(b) $\{A^c$ and $B\}$.

(c) $\{A$ and $B^c\}$.

(d) $\{A^c$ and $B^c\}$.

4.115 Sales of cars and light trucks. Motor vehicles sold to individuals are classified as either cars or light trucks (including SUVs) and as either domestic or imported. In a recent year, 69% of vehicles sold were light trucks, 78% were domestic, and 55% were domestic light trucks. Let A be the event that a vehicle is a car and B the event that it is imported. Write each of the following events in set notation and give its probability.

(a) The vehicle is a light truck.

(b) The vehicle is an imported car.

4.116 Job offers. Emily is graduating from college. She has studied biology, chemistry, and computing and hopes to work as a forensic scientist applying her science background to crime investigation. Late one night she thinks about some jobs she has applied for. Let A, B, and C be the events where Emily is offered a job by

A = the Connecticut Office of the Chief Medical Examiner

B = the New Jersey Division of Criminal Justice

C = the federal Disaster Mortuary Operations Response Team

Julie writes down her personal probabilities for being offered these jobs:

$P(A) = 0.6$ $P(B) = 0.5$ $P(C) = 0.3$

$P(A \text{ and } B) = 0.3$ $P(A \text{ and } C) = 0.1$ $P(B \text{ and } C) = 0.1$

$P(A \text{ and } B \text{ and } C) = 0$

Make a Venn diagram of the events A, B, and C. As in Figure 4.18 (page 267), mark the probabilities of every intersection involving these events and their complements. Use this diagram for Exercises 4.117, 4.118, and 4.119.

4.117 Find the probability of at least one offer. What is the probability that Julie is offered at least one of the three jobs?

4.118 Find the probability of another event. What is the probability that Julie is offered both the Connecticut and New Jersey jobs, but not the federal job?

4.119 Find a conditional probability. If Julie is offered the federal job, what is the conditional probability that she is also offered the New Jersey job? If Julie is offered the New Jersey job, what is the conditional probability that she is also offered the federal job?

4.120 Conditional probabilities and independence. Using the information in Exercise 4.115, answer these questions.

(a) Given that a vehicle is imported, what is the conditional probability that it is a light truck?

(b) Are the events "vehicle is a light truck" and "vehicle is imported" independent? Justify your answer.

Genetic counseling. Conditional probabilities and Bayes's rule are a basis for counseling people who may have genetic defects that can be passed to their children. Exercises 4.121, 4.112, and 4.123 concern genetic counseling settings.

4.121 Albinism. People with albinism have little pigment in their skin, hair, and eyes. The gene that governs albinism has two forms (called alleles), which we denote by a and A. Each person has a pair of these genes, one inherited from each parent. A child inherits one of each parent's two alleles independently with probability 0.5. Albinism is a recessive trait, so a person is albino only if the inherited pair is aa.

(a) Beth's parents are not albino but she has an albino brother. This implies that both of Beth's parents have type Aa. Why?

(b) Which of the types aa, Aa, AA could a child of Beth's parents have? What is the probability of each type?

(c) Beth is not albino. What are the conditional probabilities for Beth's possible genetic types, given this fact? (Use the definition of conditional probability.)

4.122 Find some conditional probabilities. Beth knows the probabilities for her genetic types from part (c) of the previous exercise. She marries Bob, who is albino. Bob's genetic type must be aa.

(a) What is the conditional probability that a child of Beth and Bob is non-albino if Beth has type Aa? What is the conditional probability of a non-albino child if Beth has type AA?

(b) Beth and Bob's first child is non-albino. What is the conditional probability that Beth is a carrier, type Aa?

4.123 Muscular dystrophy. Muscular dystrophy is an incurable muscle-wasting disease. The most common and serious type, called DMD, is caused by a sex-linked recessive mutation. Specifically, women can be carriers but do not get the disease; a son of a carrier has probability 0.5 of having DMD; a daughter has probability 0.5 of being a carrier. As many as one-third of DMD cases, however, are due to spontaneous mutations in sons of mothers who are not carriers. Toni has one son, who has DMD.

In the absence of other information, the probability is 1/3 that the son is the victim of a spontaneous mutation and 2/3 that Toni is a carrier. There is a screening test called the CK test that is positive with probability 0.7 if a woman is a carrier and with probability 0.1 if she is not. Toni's CK test is positive. What is the probability that she is a carrier?

CHAPTER 4 EXERCISES

4.124 Repeat the experiment many times. Here is a probability distribution for a random variable X:

Value of X	−3	4
Probability	0.3	0.7

A single experiment generates a random value from this distribution. If the experiment is repeated many times, what will be the approximate proportion of times that the value is −3? Give a reason for your answer.

4.125 Repeat the experiment many times and take the mean. Here is a probability distribution for a random variable X:

Value of X	−8	5
Probability	0.6	0.4

A single experiment generates a random value from this distribution. If the experiment is repeated many times, what will be the approximate value of the mean of these random variables? Give a reason for your answer.

4.126 Work with a transformation. Here is a probability distribution for a random variable X:

Value of X	2	3
Probability	0.2	0.8

(a) Find the mean and the standard deviation of this distribution.

(b) Let $Y = 5X - 1$. Use the rules for means and variances to find the mean and the standard deviation of the distribution of Y.

(c) For part (b), give the rules that you used to find your answer.

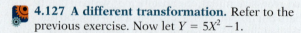 **4.127 A different transformation.** Refer to the previous exercise. Now let $Y = 5X^2 - 1$.

(a) Find the distribution of Y.

(b) Find the mean and standard deviation for the distribution of Y.

(c) Explain why the rules that you used for part (b) of the previous exercise do not work for this transformation.

4.128 Roll a pair of dice two times. Consider rolling a pair of fair dice two times. Let A be the total on the up-faces for the first roll and let B be the total on the up-faces for the second roll. For each of the following pairs of events, tell whether they are disjoint, independent, or neither.

(a) $A = \{2$ on the first roll$\}$, $B = \{8$ or more on the first roll$\}$.

(b) $A = \{2$ on the first roll$\}$, $B = \{8$ or more on the second roll$\}$.

(c) $A = \{5$ or less on the second roll$\}$, $B = \{4$ or less on the first roll$\}$.

(d) $A = \{5$ or less on the second roll$\}$, $B = \{4$ or less on the second roll$\}$.

4.129 Find the probabilities. Refer to the previous exercise. Find the probabilities for each event.

4.130 Some probability distributions. Here is a probability distribution for a random variable X:

Value of X	2	3	4
Probability	0.4	0.3	0.3

(a) Find the mean and standard deviation for this distribution.

(b) Construct a different probability distribution with the same possible values, the same mean, and a larger standard deviation. Show your work and report the standard deviation of your new distribution.

(c) Construct a different probability distribution with the same possible values, the same mean, and a smaller standard deviation. Show your work and report the standard deviation of your new distribution.

4.131 A fair bet at craps. Almost all bets made at gambling casinos favor the house. In other words, the difference between the amount bet and the mean of the distribution of the payoff is a positive number. An exception is "taking the odds" at the game of craps, a bet that a player can make under certain circumstances. The bet becomes available when a shooter throws a 4, 5, 6, 8, 9, or 10 on the initial roll. This number is called the "point"; when a point is rolled, we say that a point has been established. If a 4 is the point, an odds bet can be made that wins if a 4 is rolled before a 7 is rolled. The probability of winning this bet is 1/3, and the same payoff for a $10 bet is $20 (you keep the $10 you bet and you receive an additional $20). The same probability of winning and payoff apply for an odds bet on a 10. For an initial roll of 5 or 9, the odds bet has a winning probability of 2/5, and the payoff for a $10 bet is $15. Similarly, when the initial roll is 6 or 8, the odds bet has a winning probability of 5/11, and the payoff for a $10 bet is $12.

(a) Find the mean of the payoff distribution for each of these bets.

(b) Confirm that the bets are fair by showing that the difference between the amount bet and the mean of the distribution of the payoff is zero.

4.132 An interesting case of independence.
Independence of events is not always obvious. Toss two balanced coins independently. The four possible combinations of heads and tails in order each have probability 0.25. The events

A = head on the first toss

B = both tosses have the same outcome

may seem intuitively related. Show that $P(B \mid A) = P(B)$, so that A and B are, in fact, independent.

4.133 Wine tasters.
Two wine tasters rate each wine they taste on a scale of 1 to 5. From data on their ratings of a large number of wines, we obtain the following probabilities for both tasters' ratings of a randomly chosen wine:

		Taster 2			
Taster 1	1	2	3	4	5
1	0.03	0.02	0.01	0.00	0.00
2	0.02	0.07	0.06	0.02	0.01
3	0.01	0.05	0.25	0.05	0.01
4	0.00	0.02	0.05	0.20	0.02
5	0.00	0.01	0.01	0.02	0.06

(a) Why is this a legitimate assignment of probabilities to outcomes?

(b) What is the probability that the tasters agree when rating a wine?

(c) What is the probability that Taster 1 rates a wine higher than 3? What is the probability that Taster 2 rates a wine higher than 3?

4.134 Wine tasting.
In the setting of the previous exercise, Taster 1's rating for a wine is 3. What is the conditional probability that Taster 2's rating is higher than 3?

4.135 Lottery tickets.
Michael buys a ticket in the Tri-State Pick 3 lottery every day, always betting on 812. He will win something if the winning number contains 8, 1, and 2 in any order. Each day, Michael has probability 0.006 of winning, and he wins (or not) independently of other days because a new drawing is held each day. What is the probability that Michael's first winning ticket comes on the 10th day?

4.136 Higher education at two-year and four-year institutions.
The following table gives the counts of U.S. institutions of higher education classified as public or private and as two-year or four-year:[20]

	Public	Private
Two-year	1000	721
Four-year	2774	672

Convert the counts to probabilities and summarize the relationship between these two variables using conditional probabilities.

4.137 Odds bets at craps.
Refer to the odds bets at craps in Exercise 4.131. Suppose that whenever the shooter has an initial roll of 4, 5, 6, 8, 9, or 10, you take the odds. Here are the probabilities for these initial rolls:

Point	4	5	6	8	9	10
Probability	3/36	4/36	5/36	5/36	4/36	3/36

Draw a tree diagram with the first stage showing the point rolled and the second stage showing whether the point is again rolled before a 7 is rolled. Include a first-stage branch showing the outcome that a point is not established. In this case, the amount bet is zero and the distribution of the winnings is the special random variable that has $P(X = 0) = 1$. For the combined betting system where the player always makes a $10 odds bet when it is available, show that the game is fair.

4.138 Sample surveys for sensitive issues.
It is difficult to conduct sample surveys on sensitive issues because many people will not answer questions if the answers might embarrass them. **Randomized response** is an effective way to guarantee anonymity while collecting information on topics such as student cheating or sexual behavior. Here is the idea. To ask a sample of students whether they have plagiarized a term paper while in college, have each student toss a coin in private. If the coin lands heads *and* they have not plagiarized, they are to answer No. Otherwise, they are to give Yes as their answer. Only the student knows whether the answer reflects the truth or just the coin toss, but the researchers can use a proper random sample with follow-up for nonresponse and other good sampling practices.

Suppose that, in fact, the probability is 0.3 that a randomly chosen student has plagiarized a paper. Draw a tree diagram in which the first stage is tossing the coin and the second is the truth about plagiarism. The outcome at the end of each branch is the answer given to the randomized-response question. What is the probability of a No answer in the randomized-response poll? If the probability of plagiarism were 0.2, what would be the probability of a No response on the poll? Now suppose that you get 39% No answers in a randomized-response poll of a large sample of students at your college. What do you estimate to be the percent of the population who have plagiarized a paper?

4.139 Find some conditional probabilities. Choose a point at random in the square with sides $0 \leq x \leq 1$ and $0 \leq y \leq 1$. This means that the probability that the point falls in any region within the square is the area of that region. Let X be the x coordinate and Y the y coordinate of the point chosen. Find the conditional probability $P(Y < 1/3 \mid Y > X)$. (*Hint:* Sketch the square and the events $Y < 1/3$ and $Y > X$.)

Jose Luis Pelaez Inc/Getty Images

Sampling Distributions

Introduction

Statistical inference draws conclusions about a population or process from data. It emphasizes substantiating these conclusions via probability calculations because probability allows us to take chance variation into account. We have already examined data and arrived at conclusions many times. How do we move from summarizing a single data set to formal inference involving probability calculations?

The foundation for statistical inference is described in Section 5.1. There, we not only discuss the use of *statistics* as estimates of population *parameters*, but also describe the chance variation of a statistic when the data are produced by random sampling or randomized experimentation.

The *sampling distribution* of a statistic shows how the statistic would vary in identical repeated data collections. That is, the sampling distribution is a probability distribution that answers the question, "What would happen if we did this experiment or sampling many times?" It is these distributions that provide the necessary link between probability and the data in your sample or from your experiment. They are the key to understanding statistical inference.

The last two sections of this chapter study the sampling distributions of two common statistics: the sample mean (for quantitative data) and the sample proportion or count (for categorical data). The general

5.1 Toward Statistical Inference

5.2 The Sampling Distribution of a Sample Mean

5.3 Sampling Distributions for Counts and Proportions

281

framework for constructing a sampling distribution is the same for all statistics, so we focus on those statistics commonly used in inference. As part of this study, we revisit the Normal distributions and are introduced to two common discrete probability distributions, the binomial and Poisson distributions.

5.1 Toward Statistical Inference

When you complete this section, you will be able to:

- Identify parameters, populations, statistics, and samples and the relationships among these items.
- Use simulation to study a sampling distribution.
- Interpret and use a sampling distribution to describe a property of a statistic.
- Identify bias in a statistic by examining its sampling distribution and characterize an unbiased estimator of a parameter.
- Describe the relationship between the sample size and the variability of a statistic.
- Identify ways to reduce bias and variability of a statistic.
- Use the margin of error to describe the variability of a statistic.

A market research firm interviews a random sample of 1200 undergraduates enrolled in four-year colleges and universities throughout the United States. One result: the average number of hours spent online weekly is 19.0 hours. That's the truth about the 1200 students in the sample. What is the truth about the millions of undergraduates who make up this population?

Because the sample was chosen at random, it's reasonable to think that these 1200 students represent the entire population fairly well. So the market researchers turn the *fact* that the *sample mean* is $\bar{x} = 19.0$ hours into an *estimate* that the average time spent online weekly in the *population of undergraduates* enrolled in four-year colleges and universities is 19.0 hours.

statistical inference   That's a basic idea in statistics: use a fact about a sample to estimate the truth about the whole population. We call this **statistical inference** because we infer conclusions about the larger population from data on selected individuals.

To think about inference, we must keep straight whether a number describes a sample or a population. Here is the vocabulary we use.

PARAMETERS AND STATISTICS

A **parameter** is a number that describes the **population**. A parameter is a fixed number, but in practice, we do not know its value.

A **statistic** is a number that describes a **sample**. The value of a statistic is known when we have taken a sample, but it can change from sample to sample. We often use a statistic to estimate an unknown parameter.

EXAMPLE 5.1

Understanding the college student market. Since 1987, *Student Monitor* has published an annual market research study that provides clients with information about the college student market. The firm uses a random sample of 1200 students located throughout the United States.[1] One phase of the research focuses on computing and technology. The firm reports that undergraduates spend an average of 19.0 hours per week on the Internet and that 88% own a cell phone.

The sample mean $\bar{x} = 19.0$ hours is a *statistic*. The corresponding *parameter* is the average (call it μ) of all undergraduates enrolled in four-year colleges and universities. Similarly, the **proportion of the sample** who own a cell phone

sample proportion

$$\hat{p} = \frac{1056}{1200} = 0.88 = 88\%$$

population proportion

is a *statistic*. The corresponding *parameter* is the **proportion** (call it p) of all undergraduates at four-year colleges and universities who own a cell phone. We don't know the values of the parameters μ and p, so we use the statistics $\bar{x}$ and $\hat{p}$, respectively, to estimate them.

USE YOUR KNOWLEDGE

5.1 **Street harassment.** A large-scale survey of 16,607 women from 42 cities around the world reports that 84% of women experience their first street harassment before the age of 17.[2] Describe the statistic, population, and population parameter for this setting.

Sampling variability

If *Student Monitor* took a second random sample of 1200 students, the new sample would have different undergraduates in it. It is almost certain that the sample mean $\bar{x}$ would not again be 19.0. Likewise, we would not expect there to be exactly 1056 students who own a cell phone. In other words, the value of a statistic will vary from sample to sample. This basic fact is called **sampling variability**: the value of a statistic varies in repeated random sampling.

sampling variability

Random samples eliminate any preferences or favoritism from the act of choosing a sample, but they can still be misleading because of this *variability* that results when we choose at random. For example, what if a second random sample of 1200 undergraduates resulted in only 57% of the students owning a cell phone? Do these two results, 88% and 57%, leave you more or less confident in the value of the true population proportion? When sampling variability is too great, we can't trust the results of any one sample.

We can assess this variability by using the second advantage of random samples (the first advantage being the elimination of *bias*). Specifically, the fact that if we take lots of random samples of the same size from the same population, the variation from sample to sample will follow a predictable pattern. **All of statistical inference is based on one idea: to see how trustworthy a procedure is, ask what would happen if we repeated it many times**.

LOOK BACK

bias, p. 174

To understand why sampling variability is not fatal, we ask, "What would happen if we took many samples?" Here's how to answer that question for any statistic:

- Take a large number of random samples of size n from the same population.

- Calculate the statistic for each sample.

- Make a histogram of the values of the statistic.

- Examine the distribution displayed in the histogram for shape, center, and spread, as well as outliers or other deviations.

In practice, it is too expensive to take many samples from a large population such as all undergraduates enrolled in four-year colleges and universities. But we can imitate taking many samples by using random digits from a table **simulation** or computer software to emulate chance behavior. This is called **simulation**.

EXAMPLE 5.2

LOOK BACK

random digits,
p. 179

Simulate a random sample. Let's simulate drawing simple random samples (SRSs) of size 100 from the population of undergraduates. Suppose that, in fact, 90% of the population owns a cell phone. Then the true value of the parameter we want to estimate is $p = 0.9$. (Of course, we would not sample in practice if we already knew that $p = 0.9$. We are sampling here to understand how the statistic $\hat{p}$ behaves.)

For cell phone ownership, we can imitate the population by a table of random digits, with each entry standing for a person. Nine of the 10 digits (say, 0 to 8) stand for students who own a cell phone. The remaining digit, 9, stands for those who do not. Because all digits in a random number table are equally likely, this assignment produces a population proportion of cell phone owners equal to $p = 0.9$. We then simulate an SRS of 100 students from the population by taking 100 consecutive digits from Table B. The statistic $\hat{p}$ is the proportion of 0s to 8s in the sample.

Here are the first 100 entries in Table B with digits 0 to 8 highlighted:

19223	95034	05756	28713	96409	12531	42544	82853
73676	47150	99400	01927	27754	42648	82425	36290
45467	71709	77558	00095				

There are 90 digits between 0 and 8, so $\hat{p} = 90/100 = 0.90$. We are fortunate here that our estimate is the true population value $p = 0.9$. A second SRS based on the second 100 entries in Table B gives a different result, $\hat{p} = 0.86$. The third SRS gives the result $\hat{p} = 0.92$. All three sample results are different. That's sampling variability.

USE YOUR KNOWLEDGE

5.2 Using a random numbers table. In Example 5.2, we considered $p = 0.9$ and used each entry in Table B as a person for our simulations. Suppose instead that $p = 0.85$. How might we use Table B for simulations in this setting?

Sampling distributions

Simulation is a powerful tool for studying chance variation. Now that we see how simulation works, it is faster to abandon Table B and to use a computer to generate random numbers. This also allows us to study other statistics, such as the sample mean, when the population cannot be easily imitated by a table of random numbers. We address the sampling distribution of $\bar{x}$ in the next section.

EXAMPLE 5.3

LOOK BACK

histogram, p. 14

Take many random samples. Figure 5.1 illustrates the process of choosing many samples and finding the statistic $\hat{p}$ for each one. Follow the flow of the figure from the population at the left, to choosing an SRS and finding the $\hat{p}$ for this sample, to collecting together the $\hat{p}$'s from many samples. The histogram at the right of the figure shows the distribution of the values of $\hat{p}$ from 1000 separate SRSs of size 100 drawn from a population with $p = 0.9$.

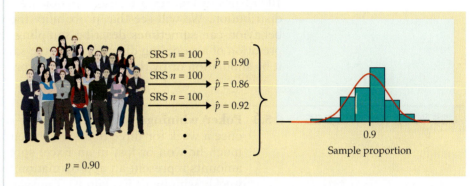

FIGURE 5.1 The results of many SRSs have a regular pattern, Example 5.3. Here we draw 1000 SRSs of size 100 from the same population. The population parameter is $p = 0.9$. The histogram shows the distribution of 1000 sample proportions.

Of course, *Student Monitor* samples 1200 students, not just 100. Figure 5.2 is parallel to Figure 5.1. It shows the process of choosing 1000 SRSs, each of size 1200, from a population in which the true proportion is $p = 0.9$. The 1000 values of $\hat{p}$ from these samples form the histogram at the right of the figure. Figures 5.1 and 5.2 are drawn on the same scale. Comparing them shows what happens when we increase the size of our samples from 100 to 1200. These histograms display the *sampling distribution* of the statistic $\hat{p}$ for two sample sizes.

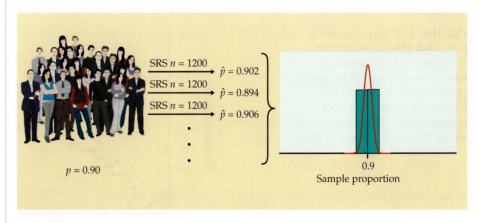

FIGURE 5.2 The distribution of the sample proportion for 1000 SRSs of size 1200 drawn from the same population as in Figure 5.1. The two histograms have the same scale. The statistic from the larger sample is less variable.

SAMPLING DISTRIBUTION

The **sampling distribution** of a statistic is the distribution of values taken by the statistic in all possible samples of the same size from the same population.

Strictly speaking, the sampling distribution is the ideal pattern that would emerge if we looked at all possible samples of size n (here, 100 or 1200) from our population. A distribution obtained from a fixed number of trials, like the 1000 trials in Figures 5.1 and 5.2, is only an approximation to the sampling distribution. We will see that probability theory, the mathematics of chance behavior, can sometimes describe sampling distributions exactly. The interpretation of a sampling distribution is the same, however, whether we obtain it by simulation or by the mathematics of probability.

USE YOUR KNOWLEDGE

5.3 Poker winnings. Doug plays poker with the same group of friends once a week for three hours. At the end of each night, he records how much he won or lost in an Excel spreadsheet. Does this collection of amounts represent an approximation to a sampling distribution of his weekly winnings? Explain your answer.

We can use the tools of data analysis to describe any distribution. Let's apply those tools to Figures 5.1 and 5.2.

• **Shape:** The histograms look Normal. Figure 5.3 is a Normal quantile plot of the values of $\hat{p}$ for our samples of size 100. It confirms that the distribution in Figure 5.1 is close to Normal. The 1000 values for samples of size 1200 in Figure 5.2 are even closer to Normal. The Normal curves drawn through the histograms describe the overall shape quite well.

• **Center:** In both cases, the values of the sample proportion $\hat{p}$ vary from sample to sample, but the values are centered at 0.9. Recall that $p = 0.9$ is the true population parameter. Some samples have a $\hat{p}$ less than 0.9 and

FIGURE 5.3 Normal quantile plot of the sample proportions in Figure 5.1. The distribution is close to Normal except for some clustering due to the fact that the sample proportions from a sample of size 100 can take only values that are a multiple of 0.01.

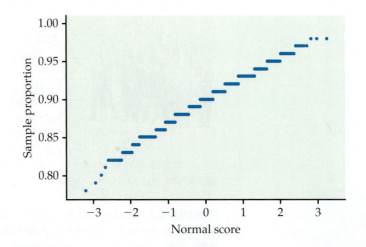

some greater, but there is no tendency to be always low or always high. That is, $\hat{p}$ has no *bias* as an estimator of p. This is true for both large and small samples. (Want the details? The mean of the 1000 values of $\hat{p}$ is 0.8985 for samples of size 100 and 0.8994 for samples of size 1200. The median value of $\hat{p}$ is exactly 0.9 for samples of both sizes.)

• **Spread:** The values of $\hat{p}$ from samples of size 1200 are much less spread out than the values from samples of size 100. In fact, the standard deviations are 0.0304 for Figure 5.1 and 0.0083 for Figure 5.2.

Although these results describe just two sets of simulations, they reflect facts that are true whenever we use random sampling.

Bias and variability

Our simulations show that a sample of size 1200 will almost always give an estimate $\hat{p}$ that is close to the truth about the population. Figure 5.2 illustrates this fact for just one value of the population proportion ($p = 0.9$), but it is true for any proportion. That is a primary reason *Student Monitor* uses a sample of size of 1200. There is more sampling variability the smaller the sample size. Samples of size 100, for example, might give an estimate of 83% or 97% when the truth is 90%.

Thinking about Figures 5.1 and 5.2 helps us restate the idea of bias when we use a statistic like $\hat{p}$ to estimate a parameter like p. It also reminds us that variability matters as much as bias.

> ### BIAS AND VARIABILITY
>
> **Bias** concerns the center of the sampling distribution. A statistic used to estimate a parameter is an **unbiased estimator** if the mean of its sampling distribution is equal to the true value of the parameter being estimated.
>
> The **variability of a statistic** is described by the spread of its sampling distribution. This spread is determined by the sampling design and the sample size n. Statistics from larger probability samples have smaller spreads.
>
> The **margin of error** is a numerical measure of the spread of a sampling distribution. It can be used to set bounds on the size of the likely error in using the statistic as an estimator of a population parameter.

We can think of the true value of the population parameter as the bull's-eye on a target and of the sample statistic as an arrow fired at the bull's-eye. Bias and variability describe what happens when an archer fires many arrows at the target. *Bias* means that the aim is off, and the arrows do not center about the bull's-eye. Large *variability* means that arrows are widely scattered about the target. In other words, there is a lack of precision, or consistency, among the arrows. Figure 5.4 shows this target illustration of the two types of error.

Notice that small variability (repeated shots are close together) can accompany large bias (the arrows are consistently away from the bull's-eye in one

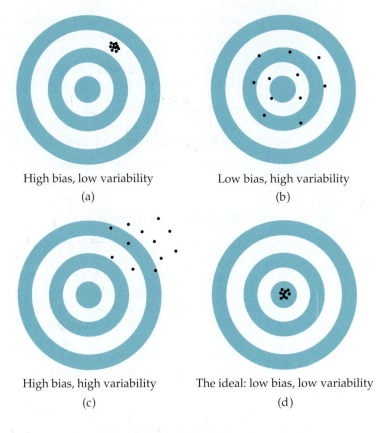

High bias, low variability
(a)

Low bias, high variability
(b)

High bias, high variability
(c)

The ideal: low bias, low variability
(d)

FIGURE 5.4 Bias and variability in shooting arrows at a target. Bias means the shots do not center around the bull's-eye. Variability means that the shots are scattered.

direction). And small bias (the arrows center on the bull's-eye) can accompany large variability (repeated shots are widely scattered). A good sampling scheme, like a good archer, must have both small bias and small variability. Here's how we do this.

MANAGING BIAS AND VARIABILITY

To reduce bias, use random sampling. When we start with a list of the entire population, simple random sampling produces unbiased estimates—the values of a statistic computed from an SRS neither consistently overestimate nor consistently underestimate the value of the population parameter.

To reduce the variability of a statistic from an SRS, use a larger sample. You can make the variability as small as you want by taking a large enough sample.

In practice, *Student Monitor* takes only one random sample. We don't know how close to the truth the estimate from this one sample is because we don't know what the true population parameter value is. But *large random samples almost always give an estimate that is close to the truth*. Looking at the pattern of many samples when $n = 1200$ shows that we can trust the result of one sample.

Similarly, the Current Population Survey's sample of about 60,000 households estimates the national unemployment rate very accurately. Of course, only probability samples carry this guarantee. Using a probability sampling design and taking care to deal with practical difficulties reduce bias in a sample.

The size of the sample then determines how close to the population truth the sample result is likely to fall. Results from a sample survey usually come with a *margin of error* that sets bounds on the size of the likely error. The margin of error directly reflects the variability of the sample statistic, so it is smaller for larger samples. We will describe the details of its calculation in later chapters.

USE YOUR KNOWLEDGE

5.4   Bigger is better? Radio talk shows often report opinion polls based on tens of thousands of listeners. These sample sizes are typically much larger than those used in opinion polls that incorporate probability sampling. Does a larger sample size mean more trustworthy results? Explain your answer.

5.5   Effect of sample size on the sampling distribution. You are planning an opinion study and are considering taking an SRS of either 200 or 600 people. Explain how the sampling distributions of the population proportion p would differ in terms of center and spread for these two scenarios.

Sampling from large populations

Student Monitor's sample of 1200 students is only about 1 out of every 90,000 undergraduate students in the United States. Does it matter whether we sample 1-in-1000 individuals in the population or 1-in-90,000?

LARGE POPULATIONS DO NOT REQUIRE LARGE SAMPLES

The variability of a statistic from a random sample depends little on the size of the population, as long as the population is at least 20 times larger than the sample.

Why does the size of the population have little influence on the behavior of statistics from random samples? To see why this is plausible, imagine sampling harvested corn by thrusting a scoop into a lot of corn kernels. The scoop doesn't know whether it is surrounded by a bag of corn or by an entire truckload. As long as the corn is well mixed (so that the scoop selects a random sample), the variability of the result depends only on the size of the scoop.

The fact that the variability of sample results is controlled by the size of the sample has important consequences for sampling design. An SRS of size 1200 from the 10.5 million undergraduates gives results as precise as an SRS of size 1200 from the roughly 156,000 inhabitants of San Francisco between the ages of 20 and 29. This is good news for designers of national samples but bad news for those who want accurate information about these citizens of San Francisco. If both use an SRS, both must use the same size sample to obtain equally trustworthy results.

Why randomize?

Why randomize? The act of randomizing guarantees that the results of analyzing our data are subject to the laws of probability. The behavior of statistics is described by a sampling distribution. The form of the distribution is known and, in many cases, is approximately Normal. Often, the center of the distribution lies at the true parameter value so that the notion that randomization eliminates bias is made more explicit. The spread of the distribution describes the variability of the statistic and can be made as small as we wish by choosing a large enough sample. In a randomized experiment, we can reduce variability by choosing larger groups of subjects for each treatment.

These facts are at the heart of formal statistical inference. The remainder of this chapter has much to say in more technical language about sampling distributions. Later chapters describe the way statistical conclusions are based on them. What any user of statistics must understand is that all the technical talk has its basis in a simple question: *What would happen if the sample or the experiment were repeated many times?* The reasoning applies not only to an SRS, but also to the complex sampling designs actually used by opinion polls and other national sample surveys. The same conclusions hold as well for randomized experimental designs. The details vary with the design, but the basic facts are true whenever randomization is used to produce data.

Remember that proper statistical design is not the only aspect of a good sample or experiment. *The sampling distribution shows only how a statistic varies due to the operation of chance in randomization. It reveals nothing about possible bias due to undercoverage or nonresponse in a sample (page 196) or to lack of realism in an experiment.* The actual error in estimating a parameter by a statistic can be much larger than the sampling distribution suggests. What is worse, there is no way to say how large the added error is. The real world is less orderly than statistics textbooks imply.

In the next two sections, we will study the sampling distributions of two common statistics, the sample mean and the sample proportion. The focus will be on the important features of these distributions so that we can quickly describe and use them in the later chapters on statistical inference. We will see that, in each case, the sampling distribution depends on **both** the population and the way we collect the data from the population.

SECTION 5.1 SUMMARY

• A number that describes a population is a **parameter**. A number that describes a sample (is computed from the sample data) is a **statistic**. The purpose of sampling or experimentation is usually **inference**: use sample statistics to make statements about unknown population parameters.

• A statistic from a probability sample or a randomized experiment has a **sampling distribution** that describes how the statistic varies in repeated data productions. The sampling distribution answers the question "What would happen if we repeated the sample or experiment many times?" Formal statistical inference is based on the sampling distributions of statistics.

• A statistic as an estimator of a parameter may suffer from **bias** or from high **variability**. Bias means that the center of the sampling distribution is not equal to the true value of the parameter. The variability of the statistic is described by the spread of its sampling distribution. Variability is usually reported by giving a **margin of error** for conclusions based on sample results.

- Properly chosen statistics from randomized data production designs have no bias resulting from the way the sample is selected or the way the experimental units are assigned to treatments. We can reduce the variability of the statistic by increasing the size of the sample or the size of the experimental groups.

SECTION 5.1 EXERCISES

For Exercise 5.1, see page 283; for Exercise 5.2, see page 284; for Exercise 5.3, see page 286; and for Exercises 5.4 and 5.5, see page 289.

5.6 Web polls. If you connect to the website **peopleschoice .com/pca/polls/polls.jsp**, you are given the opportunity to vote on various entertainment questions. Can you apply the ideas about populations and samples to these polls? Explain why or why not.

5.7 What population and sample? Thirty students from your college who are majoring in English are randomly selected to be on a committee to evaluate immediate changes in the statistics requirement for the major. There are 153 English majors at your college. The current rules say that a statistics course is one of three options for a quantitative competency requirement. The proposed change would be to require a statistics course. Each of the committee members is asked to vote Yes or No on the new requirement.

(a) Describe the population for this setting.

(b) What is the sample?

(c) Describe the statistic and how it would be calculated.

(d) What is the population parameter?

(e) Write a short summary based on your answers to parts (a) through (d) using this setting to explain population, sample, parameter, statistic, and the relationships among these items.

5.8 What's wrong? State what is wrong in each of the following scenarios.

(a) A parameter describes a sample.

(b) Bias and variability are two names for the same thing.

(c) Large samples are always better than small samples.

(d) A sampling distribution is something generated by a computer.

5.9 Describe the population and the sample. For each of the following situations, describe the population and the sample.

(a) A survey of 17,096 students in U.S. four-year colleges reported that 19.4% were binge drinkers.

(b) In a study of work stress, 100 restaurant workers were asked about the impact of work stress on their personal lives.

(c) A tract of forest has 584 longleaf pine trees. The diameters of 40 of these trees were measured.

5.10 Is it unbiased? A statistic has a sampling distribution that is somewhat skewed. The mean is 17, the median is 15, the quartiles are 13 and 19.

(a) If the population parameter is 15, is the estimator unbiased?

(b) If the population parameter is 17, is the estimator unbiased?

(c) If the population parameter is 16, is the estimator unbiased?

(d) Write a short summary of your results in parts (a), (b), and (c) and include a discussion of bias and unbiased estimators.

5.11 Constructing a sampling distribution. Refer to Example 5.1 (page 283). Suppose *Student Monitor* also reported that the median number of hours per week spent on the Internet was 12.5 hours.

(a) Explain why we'd expect the population median to be less than the population mean in this setting by drawing the distribution of times spent on the Internet for all undergraduates. This is called the *population distribution*.

(b) Using Figure 5.2 (page 285) as a guide and your distribution from part (a), describe how to approximate the sampling distribution of the sample median in this setting.

5.12 Bias and variability. Figure 5.5 shows histograms of four sampling distributions of statistics intended to estimate the same parameter. Label each distribution relative to the others as high or low bias and as high or low variability.

5.13 Constructing sampling distributions. The *Probability* applet simulates tossing a coin, with the advantage that you can choose the true long-term proportion, or probability, of a head. Suppose that we have a population in which proportion $p = 0.4$ (the

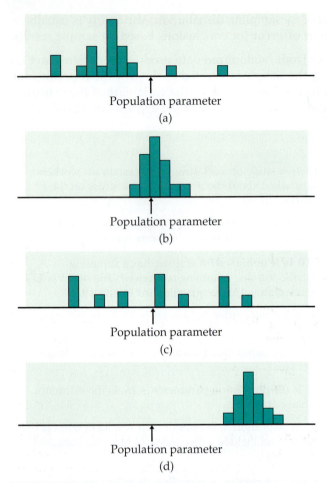

Population parameter
(a)

Population parameter
(b)

Population parameter
(c)

Population parameter
(d)

FIGURE 5.5 Determine which of these sampling distributions displays high or low bias and high or low variability, Exercise 5.12.

parameter) plan to vote in the next election. Tossing a coin with probability $p = 0.4$ of a head simulates this situation: each head is a person who plans to vote, and each tail is a person who does not. Set the "Probability of heads" in the applet to 0.4 and the number of tosses to 25. This simulates an SRS of size 25 from this population. By alternating between "Toss" and "Reset," you can take many samples quickly.

(a) Take 50 samples, recording the number of heads in each sample. Make a histogram of the 50 sample proportions (count of heads divided by 25). You are constructing the sampling distribution of this statistic.

(b) Another population contains only 20% who plan to vote in the next election. Take 50 samples of size 25 from this population, record the number in each sample who approve, and make a histogram of the 50 sample proportions.

5.14 Comparing sampling distributions. Refer to the previous exercise.

(a) How do the centers of your two histograms reflect the differing truths about the two populations?

(b) Describe any differences in the shapes of the two histograms. Is one more skewed than the other?

(c) Compare the spreads of the two histograms. For which population is there less sampling variability?

(d) Suppose instead that the population proportions were 0.6 and 0.8, respectively. Describe how the sampling distributions of $\hat{p}$ would differ from those constructed in Exercise 5.13.

5.15 Use the *Simple Random Sample* applet. The *Simple Random Sample* applet can illustrate the idea of a sampling distribution. Form a population labeled 1 to 100. We will choose an SRS of 15 of these numbers. That is, in this exercise, the numbers themselves are the population, not just labels for 100 individuals. The mean of the whole numbers 1 to 100 is 50.5. This is the parameter, the mean of the population.

(a) Use the applet to choose an SRS of size 15. Which 15 numbers were chosen? What is their mean? This is a statistic, the sample mean $\bar{x}$.

(b) Although the population and its mean 50.5 remain fixed, the sample mean changes as we take more samples. Take another SRS of size 15. (Use the "Reset" button to return to the original population before taking the second sample.) What are the 15 numbers in your sample? What is their mean? This is another value of $\bar{x}$.

(c) Take 18 more SRSs from this same population and record their means. You now have 20 values of the sample mean $\bar{x}$ from 20 SRSs of the same size from the same population. Make a histogram of the 20 values and mark the population mean 50.5 on the horizontal axis. Are your 20 sample values roughly centered at the population value? (If you kept going forever, your $\bar{x}$-values would form the sampling distribution of the sample mean; the population mean would indeed be the center of this distribution.)

5.16 Use the *Simple Random Sample* applet, continued. Refer to the previous exercise.

(a) Suppose instead that a sample size of $n = 10$ was used. Based on what you know about the effect of the sample size on the sampling distribution, which sampling distribution should have the smaller variability?

(b) Repeat the previous exercise using $n = 10$. Did your simulations confirm your answer in part (a)? Explain your answer.

(c) Write a short paragraph about the effect of the sample size on the variability of a sampling distribution using these simulations to illustrate the basic idea.

5.2 The Sampling Distribution of a Sample Mean

When you complete this section, you will be able to:

- Explain the difference between the sampling distribution of $\bar{x}$ and the population distribution.
- Determine the mean and standard deviation of $\bar{x}$ for an SRS of size n from a population with mean μ and standard deviation σ.
- Describe how much larger n has to be for an SRS to reduce the standard deviation of $\bar{x}$ by a certain factor.
- Utilize the central limit theorem to approximate the sampling distribution of $\bar{x}$ and perform probability calculations based on this approximation.

A variety of statistics are used to describe quantitative data. The sample mean, median, and standard deviation are all examples of statistics based on quantitative data. Statistical theory describes the sampling distributions of these statistics. However, the general framework for constructing a sampling distribution is the same for all statistics. In this section, we will concentrate on the sample mean. Because sample means are just averages of observations, they are among the most frequently used statistics.

Suppose that you plan to survey 1000 undergraduates enrolled in four-year U.S. universities about their sleeping habits. The sampling distribution of the average hours of sleep per night describes what this average would be if many simple random samples of 1000 students were drawn from the population of students in the United States. In other words, it gives you an idea of what you are likely to see from your survey. It tells you whether you should expect this average to be near the population mean and whether the variation of the statistic is roughly ±2 hours or ±2 minutes.

Before constructing this distribution, however, we need to consider another set of probability distributions that also plays a role in statistical inference. Any quantity that can be measured on each member of a population is described by the distribution of its values for all members of the population. This is the context in which we first met distributions, as density curves that provide models for the overall pattern of data.

LOOK BACK

density curves, p. 51

Imagine choosing one individual at random from a population and measuring a quantity. The quantities obtained from repeated draws of one individual from a population have a probability distribution that is the distribution of the population.

EXAMPLE 5.4

Total sleep time of college students. A recent survey describes the distribution of total sleep time among college students as approximately Normal with a mean of 6.78 hours and standard deviation of 1.24 hours.[3] Suppose that we select a college student at random and obtain his or her sleep time. This result is a random variable X because, prior to the random sampling, we don't know the sleep time. We do know, however, that in repeated sampling, X will have the same $N(6.78, 1.24)$ distribution that describes the pattern of sleep time in the entire population. We call $N(6.78, 1.24)$ the *population distribution*.

POPULATION DISTRIBUTION

The **population distribution** of a variable is the distribution of its values for all members of the population. The population distribution is also the probability distribution of the variable when we choose one individual at random from the population.

In this example, the population of all college students actually exists so that we can, in principle, draw an **SRS** of students from it. Sometimes, our population of interest does not actually exist. For example, suppose that we are interested in studying final-exam scores in a statistics course, and we have the scores of the 34 students who took the course last semester. For the purposes of statistical inference, we might want to consider these 34 students as part of a hypothetical population of similar students who would take this course. In this sense, these 34 students represent not only themselves, but also a larger population of similar students. The key idea is to think of the observations that you have as coming from a population with a probability distribution.

USE YOUR KNOWLEDGE

5.17 Time spent using apps on a mobile device. Nielsen has installed, with permission, Mobile Netview 3 on approximately 5000 cell phones to gather information on mobile app usage among adults in the United States. Nielsen reported that 18–24 year olds spend an average of 37 hours and 6 minutes a month using mobile apps.[4] State the population that this survey describes, the statistic, and some likely values from the population distribution.

Now that we have made the distinction between the population distributions and sampling distributions, we can proceed with an in-depth study of the sampling distribution of a sample mean $\bar{x}$.

EXAMPLE 5.5

HELP60

Sample means are approximately Normal. Figure 5.6 illustrates two striking facts about the sampling distribution of a sample mean. Figure 5.6(a) displays the distribution of student visit lengths (in minutes) to a statistics

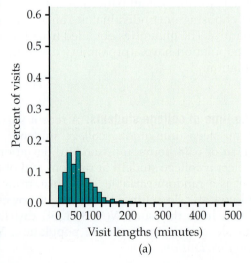

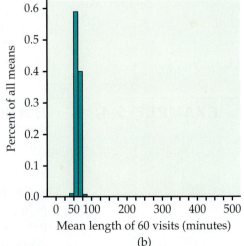

FIGURE 5.6 (a) The distribution of visit lengths to a statistics help room during the school year, Example 5.5. (b) The distribution of the sample means $\bar{x}$ for 500 random samples of size 60 from this population. The scales and histogram classes are exactly the same in both panels.

TABLE 5.1		Length (in Minutes) of 60 Visits to a Statistics Help Room							
10	14	15	16	18	20	20	20	23	25
28	30	30	30	30	30	31	33	35	35
46	48	50	50	50	50	51	54	55	55
60	60	60	60	60	60	60	65	65	65
75	77	80	80	84	85	88	98	100	100
105	105	105	115	120	135	135	136	157	210

help room at a large midwestern university. Students visiting the help room were asked to sign in upon arrival and then sign out when leaving. During the school year, there were 1838 visits to the help room but only 1264 recorded visit lengths. This is because many visiting students forgot to sign out. We also omitted a few large outliers (visits lasting more than 10 hours).[5] The distribution is strongly skewed to the right. The population mean is $\mu = 61.28$ minutes.

Table 5.1 contains the lengths of a random sample of 60 visits from this population. The mean of these 60 visits is $\bar{x} = 63.45$ minutes. If we were to take another sample of size 60, we would likely get a different value of $\bar{x}$. This is because this new sample would contain a different set of visits. To find the sampling distribution of $\bar{x}$, we take many SRSs of size 60 and calculate $\bar{x}$ for each sample. Figure 5.6(b) is the distribution of the values of $\bar{x}$ for 500 random samples. The scales and choice of classes are exactly the same as in Figure 5.6(a) so that we can make a direct comparison.

The sample means are much less spread out than the individual visit lengths. What is more, the Normal quantile plot in Figure 5.7 confirms that the distribution in Figure 5.6(b) is close to Normal.

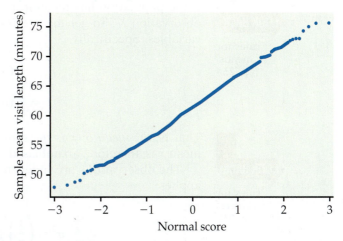

FIGURE 5.7 Normal quantile plot of the 500 sample means in Figure 5.6(b). The distribution is close to Normal.

This example illustrates two important facts about sample means that we will discuss in this section.

FACTS ABOUT SAMPLE MEANS

1. Sample means are less variable than individual observations.

2. Sample means are more Normal than individual observations.

These two facts contribute to the popularity of sample means in statistical inference.

The mean and standard deviation of $\bar{x}$

The sample mean $\bar{x}$ from a sample or an experiment is an estimate of the mean μ of the underlying population. The sampling distribution of $\bar{x}$ is determined by

- the design used to produce the data,
- the sample size n, and
- the population distribution.

Select an SRS of size n from a population, and measure a variable X on each individual in the sample. The n measurements are values of n random variables $X_1, X_2, \ldots, X_n$. A single X_i is a measurement on one individual selected at random from the population and, therefore, has the distribution of the population. If the population is large relative to the sample, we can consider $X_1, X_2, \ldots, X_n$ to be independent random variables, each having the same distribution. This is our probability model for measurements on each individual in an SRS.

The sample mean of an SRS of size n is

$$\bar{x} = \frac{1}{n}(X_1 + X_2 + \cdots + X_n)$$

LOOK BACK

rules for means, p. 254

If the population has mean μ, then μ is the mean of the distribution of each observation X_i. To get the mean of $\bar{x}$, we use the rules for means of random variables. Specifically,

$$\mu_{\bar{x}} = \frac{1}{n}(\mu_{X_1} + \mu_{X_2} + \cdots + \mu_{X_n})$$
$$= \frac{1}{n}(\mu + \mu + \cdots + \mu) = \mu$$

LOOK BACK

rules for variances, p. 258

That is, *the mean of $\bar{x}$ is the same as the mean of the population.* The sample mean $\bar{x}$ is, therefore, an unbiased estimator of the unknown population mean μ.

The observations are independent, so the addition rule for variances also applies:

$$\sigma_{\bar{x}}^2 = \left(\frac{1}{n}\right)^2 (\sigma_{X_1}^2 + \sigma_{X_2}^2 + \cdots + \sigma_{X_n}^2)$$
$$= \left(\frac{1}{n}\right)^2 (\sigma^2 + \sigma^2 + \cdots + \sigma^2)$$
$$= \frac{\sigma^2}{n}$$

With n in the denominator, the variability of $\bar{x}$ about its mean decreases as the sample size grows. Thus, a sample mean from a large sample will usually be very close to the true population mean μ. Here is a summary of these facts.

MEAN AND STANDARD DEVIATION OF A SAMPLE MEAN

Let $\bar{x}$ be the mean of an SRS of size n from a population having mean μ and standard deviation σ. The mean and standard deviation of $\bar{x}$ are

$$\mu_{\bar{x}} = \mu$$

$$\sigma_{\bar{x}} = \frac{\sigma}{\sqrt{n}}$$

How precisely does a sample mean $\bar{x}$ estimate a population mean μ? Because the values of $\bar{x}$ vary from sample to sample, we must give an answer in terms of the sampling distribution. We know that $\bar{x}$ is an unbiased estimator of μ, so its values in repeated samples are not systematically too high or too low. Most samples will give an $\bar{x}$-value close to μ if the sampling distribution is concentrated close to its mean μ. So the precision of estimation depends on the spread of the sampling distribution.

Because the standard deviation of $\bar{x}$ is $\sigma/\sqrt{n}$, the standard deviation of the statistic decreases in proportion to the square root of the sample size. This means, for example, that a sample size must be multiplied by 4 in order to divide the statistic's standard deviation in half. By comparison, a sample size must be multiplied by 100 in order to reduce the standard deviation by a factor of 10.

EXAMPLE 5.6

Standard deviations for sample means of visit lengths. The standard deviation of the population of visit lengths in Figure 5.6(a) is $\sigma = 41.84$ minutes. The length of a single visit will often be far from the population mean. If we choose an SRS of 15 visits, the standard deviation of their mean length is

$$\sigma_{\bar{x}} = \frac{41.84}{\sqrt{15}} = 10.80 \text{ minutes}$$

Averaging over more visits reduces the variability and makes it more likely that $\bar{x}$ is close to μ. Our sample size of 60 visits is 4 times 15, so the standard deviation will be half as large:

$$\sigma_{\bar{x}} = \frac{41.84}{\sqrt{60}} = 5.40 \text{ minutes}$$

USE YOUR KNOWLEDGE

5.18 Find the mean and the standard deviation of the sampling distribution. Compute the mean and standard deviation of the sampling distribution of the sample mean when you plan to take an SRS of size 64 from a population with mean 44 and standard deviation 16.

5.19 The effect of increasing the sample size. In the setting of the previous exercise, repeat the calculations for a sample size of 576. Explain the effect of the sample size increase on the mean and standard deviation of the sampling distribution.

The central limit theorem

We have described the center and spread of the probability distribution of a sample mean $\bar{x}$, but not its shape. The shape of the distribution of $\bar{x}$ depends on the shape of the population distribution. Here is one important case: if the population distribution is Normal, then so is the distribution of the sample mean.

SAMPLING DISTRIBUTION OF A SAMPLE MEAN

If a population has the $N(\mu, \sigma)$ distribution, then the sample mean $\bar{x}$ of n independent observations has the $N(\mu, \sigma/\sqrt{n})$ distribution.

This is a somewhat special result. Many population distributions are not Normal. The help room visit lengths in Figure 5.6(a), for example, are strongly skewed. Yet Figures 5.6(b) and 5.7 show that means of samples of size 60 are close to Normal.

One of the most famous facts of probability theory says that, for large sample sizes, the distribution of $\bar{x}$ is close to a Normal distribution. This is true no matter what shape the population distribution has, as long as the population has a finite standard deviation σ. This is the **central limit theorem**. It is much more useful than the fact that the distribution of $\bar{x}$ is exactly Normal if the population is exactly Normal.

central limit theorem

CENTRAL LIMIT THEOREM

Draw an SRS of size n from any population with mean μ and finite standard deviation σ. When n is large, the sampling distribution of the sample mean $\bar{x}$ is approximately Normal:

$$\bar{x} \text{ is approximately } N\left(\mu, \frac{\sigma}{\sqrt{n}}\right)$$

EXAMPLE 5.7

LOOK BACK

68–95–99.7 rule, p. 57

How close will the sample mean be to the population mean? With the Normal distribution to work with, we can better describe how precisely a random sample of 60 visits estimates the mean length of all visits to the help room. The population standard deviation for the 1264 visits in the population of Figure 5.6(a) is $\sigma = 41.84$ minutes. From Example 5.6 we know $\sigma_{\bar{x}} = 5.4$ minutes. By the 95 part of the 68–95–99.7 rule, about 95% of all samples will have mean $\bar{x}$ within two standard deviations of μ, that is, within ± 10.8 minutes of μ.

USE YOUR KNOWLEDGE

5.20 Use the 68–95–99.7 rule. You take an SRS of size 64 from a population with mean 82 and standard deviation 24. According to the central limit theorem, what is the approximate sampling distribution of the sample mean? Use the 95 part of the 68–95–99.7 rule to describe the variability of $\bar{x}$.

For the sample size of $n = 60$ in Example 5.7, the sample mean is not very precise. The population of help room visit lengths is very spread out, so the sampling distribution of $\bar{x}$ has a large standard deviation.

EXAMPLE 5.8

How can we reduce the standard deviation? In the setting of Example 5.7, if we want to reduce the standard deviation of $\bar{x}$ by a factor of 2, we must take a sample four times as large, $n = 4 \times 60$, or 240. Then

$$\sigma_{\bar{x}} = \frac{41.84}{\sqrt{240}} = 2.70 \text{ minutes}$$

For samples of size 240, about 95% of the sample means will be within twice 2.70, or 5.40 minutes, of the population mean μ.

finite population correction factor

The standard deviation computed in Example 5.8 is actually too large. This is due to the fact that the population size, $N = 1264$, is not at least 20 times larger than the sample size, $n = 240$. In these settings, it is better to adjust the standard deviation of $\bar{x}$ to reflect only the variance remaining in the population that is not in the sample. This is done by multiplying the unadjusted standard deviation by the **finite population correction factor**. This quantity is $\sqrt{\dfrac{N - n}{N - 1}}$ and moves the standard deviation of $\bar{x}$ toward 0 as n moves toward N. Applying this correction to Example 5.8, the standard deviation of $\bar{x}$ is reduced 10% to

$$\frac{41.84}{\sqrt{240}} \sqrt{\frac{1264 - 240}{1264 - 1}} = 2.43 \text{ minutes}$$

Thus, for samples of size 240, about 95% of the sample means will be within twice 2.43, or 4.86 minutes, of the population mean μ, rather than the 5.40 minutes reported in Example 5.8.

USE YOUR KNOWLEDGE

5.21 The effect of increasing the sample size. In the setting of Exercise 5.20, suppose that we increase the sample size to 2304. Use the 95 part of the 68–95–99.7 rule to describe the variability of this sample mean. Compare your results with those you found in Exercise 5.20.

Example 5.8 reminds us that if the population is very spread out, the $\sqrt{n}$ in the formula for the deviation of $\bar{x}$ implies that very large samples are needed to estimate the population mean precisely. The main point of the example, however, is that the central limit theorem allows us to use Normal probability calculations to answer questions about sample means even when the population distribution is not Normal.

How large a sample size n is needed for $\bar{x}$ to be close to Normal depends on the population distribution. More observations are required if the shape of the population distribution is far from Normal. For the very skewed visit length population, samples of size 60 are large enough. Further study would be needed to see if the distribution of $\bar{x}$ is close to Normal for smaller samples like $n = 20$ or $n = 40$. Here is a more detailed study of another skewed distribution.

EXAMPLE 5.9

exponential distribution

The central limit theorem in action. Figure 5.8 shows the central limit theorem in action for another very non-Normal population. Figure 5.8(a) displays the density curve of a single observation from the population. The distribution is strongly right-skewed, and the most probable outcomes are near 0. The mean μ of this distribution is 1, and its standard deviation σ is also 1. This particular continuous distribution is called an **exponential distribution**. Exponential distributions are used as models for how long an iPhone will function properly and for the time between snaps you receive on Snapchat.

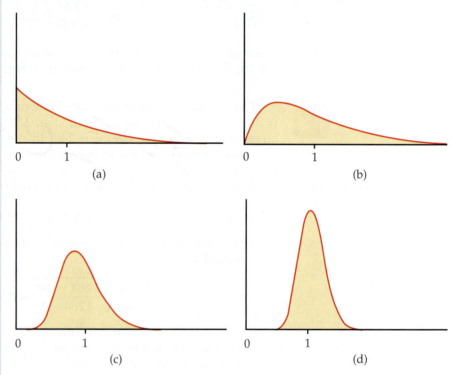

(a)   (b)

(c)   (d)

FIGURE 5.8 The central limit theorem in action: the sampling distribution of sample means from a strongly non-Normal population becomes more Normal as the sample size increases, Example 5.9. (a) The distribution of 1 observation. (b) The distribution of $\bar{x}$ for 2 observations. (c) The distribution of $\bar{x}$ for 10 observations. (d) The distribution of $\bar{x}$ for 25 observations.

Figures 5.8(b), (c), and (d) are the density curves of the sample means of 2, 10, and 25 observations from this population. As n increases, the shape becomes more Normal. The mean remains at $\mu = 1$, but the standard deviation decreases, taking the value $1/\sqrt{n}$. The density curve for 10 observations is still somewhat skewed to the right but already resembles a Normal curve having $\mu = 1$ and $\sigma = 1/\sqrt{10} = 0.32$. The density curve for $n = 25$ is yet more Normal. The contrast between the shape of the population distribution and of the distribution of the mean of 10 or 25 observations is striking.

You can also use the *Central Limit Theorem* applet to study the sampling distribution of $\bar{x}$. From one of three population distributions, 10,000 SRSs of a user-specified sample size n are generated, and a histogram of the sample means is constructed. You can then compare this estimated sampling distribution with the Normal curve that is based on the central limit theorem.

EXAMPLE 5.10

Using the *Central Limit Theorem* applet. In Example 5.9, we considered sample sizes of $n = 2, 10,$ and 25 from an exponential distribution. Figure 5.9 shows a screenshot of the *Central Limit Theorem* applet for the exponential distribution when $n = 10$. The mean and standard deviation of this sampling distribution are 1 and $1/\sqrt{10} = 0.316$, respectively. From the 10,000 SRSs, the mean is estimated to be 1.001 and the estimated standard deviation is 0.319. These are both quite close to the true values. In Figure 5.8(c), we saw that the density curve for 10 observations is still somewhat skewed to the right. We can see this same behavior in Figure 5.9 when we compare the histogram with the Normal curve based on the central limit theorem.

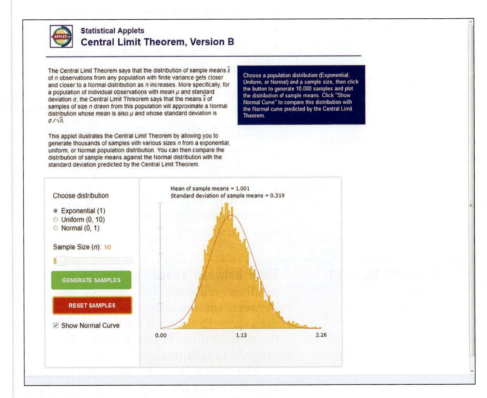

FIGURE 5.9 Screenshot of the *Central Limit Theorem* applet for the exponential distribution when $n = 10$, Example 5.10.

Try using the applet for the other sample sizes in Example 5.9. You should get histograms shaped like the density curves shown in Figure 5.8. You can also consider other sample sizes by sliding n from 1 to 100. As you increase n, the shape of the histogram moves closer to the Normal curve that is based on the central limit theorem.

USE YOUR KNOWLEDGE

5.22   Use the *Central Limit Theorem* applet. Let's consider the uniform distribution between 0 and 10. For this distribution, all intervals of the same length between 0 and 10 are equally likely. This distribution has a mean of 5 and standard deviation of 2.89.

(a) Approximate the population distribution by setting $n = 1$ and clicking the "Generate samples" button.

(b) What are your estimates of the population mean and population standard deviation based on the 10,000 SRSs? Are these population estimates close to the true values?

(c) Describe the shape of the histogram and compare it with the Normal curve.

5.23 Use the *Central Limit Theorem* applet again. Refer to the previous exercise. In the setting of Example 5.9, let's approximate the sampling distribution for samples of size $n = 2$, 10, and 25 observations.

(a) For each sample size, compute the mean and standard deviation of $\bar{x}$.

(b) For each sample size, use the applet to approximate the sampling distribution. Report the estimated mean and standard deviation. Are they close to the true values calculated in part (a)?

(c) For each sample size, compare the shape of the sampling distribution with the Normal curve based on the central limit theorem.

(d) For this population distribution, what sample size do you think is needed to make you feel comfortable using the central limit theorem to approximate the sampling distribution of $\bar{x}$? Explain your answer.

Now that we know that the sampling distribution of the sample mean $\bar{x}$ is approximately Normal for a sufficiently large n, let's consider some probability calculations.

EXAMPLE 5.11

Time between snaps. Snapchat has more than 100 million daily users sending well over 400 million snaps a day.[6] Suppose that the time X between snaps received is governed by the exponential distribution with mean $\mu = 15$ minutes and standard deviation $\sigma = 15$ minutes. You record the next 50 times between snaps. What is the probability that their average exceeds 13 minutes?

The central limit theorem says that the sample mean time $\bar{x}$ (in minutes) between snaps has approximately the Normal distribution with mean equal to the population mean $\mu = 15$ minutes and standard deviation

$$\frac{\sigma}{\sqrt{50}} = \frac{15}{\sqrt{50}} = 2.12 \text{ minutes}$$

The sampling distribution of $\bar{x}$ is, therefore, approximately $N(15, 2.12)$. Figure 5.10 shows this Normal curve (solid) and also the actual density curve of $\bar{x}$ (dashed).

The probability we want is $P(\bar{x} > 13.0)$. This is the area to the right of 13 under the solid Normal curve in Figure 5.10. A Normal distribution calculation gives

$$P(\bar{x} > 13.0) = P\left(\frac{\bar{x} - 15}{2.12} > \frac{13.0 - 15}{2.12}\right)$$

$$= P(Z > -0.94) = 0.8264$$

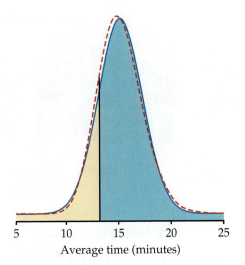

FIGURE 5.10 The exact distribution (dashed) and the Normal approximation from the central limit theorem (solid) for the average time between snaps received, Example 5.11.

The exactly correct probability is the area under the dashed density curve in the figure. It is 0.8265. The central limit theorem Normal approximation is off by only about 0.0001.

We can also use this sampling distribution to talk about the total time between the 1st and 51st snap received.

EXAMPLE 5.12

Convert the results to the total time. There are 50 time intervals between the 1st and 51st snap. According to the central limit theorem calculations in Example 5.11,

$$P(\bar{x} > 13.0) = 0.8264$$

We know that the sample mean is the total time divided by 50, so the event $\{\bar{x} > 13.0\}$ is the same as the event $\{50\bar{x} > 50(13.0)\}$. We can say that the probability is 0.8264 that the total time is $50(13.0) = 650$ minutes (10.8 hours) or greater.

USE YOUR KNOWLEDGE

5.24 Find a probability. Refer to Example 5.11. Find the probability that the mean time between snaps is less than 15 minutes. The exact probability is 0.5188. Compare your answer with the exact one.

Figure 5.11 summarizes the facts about the sampling distribution of $\bar{x}$ in a way that emphasizes the big idea of a sampling distribution. The general framework for constructing the sampling distribution of $\bar{x}$ is shown on the left.

- Take many random samples of size n from a population with mean μ and standard deviation σ.

- Find the sample mean $\bar{x}$ for each sample.

- Collect all the $\bar{x}$'s and display their distribution.

The sampling distribution of $\bar{x}$ is shown on the right. Keep this figure in mind as you go forward.

FIGURE 5.11 The sampling distribution of a sample mean $\bar{x}$ has mean μ and standard deviation $\sigma/\sqrt{n}$. The sampling distribution is Normal if the population distribution is Normal; it is approximately Normal for large samples in any case.

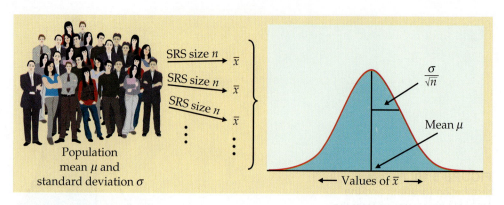

A few more facts

The central limit theorem is the big fact of probability theory in this section. Here are three additional facts related to our investigations that will be useful in describing methods of inference in later chapters.

The fact that the sample mean of an SRS from a Normal population has a Normal distribution is a special case of a more general fact: **any linear combination of independent Normal random variables is also Normally distributed.** That is, if X and Y are independent Normal random variables and a and b are any fixed numbers, $aX + bY$ is also Normally distributed, and this is true for any number of Normal random variables. In particular, the sum or difference of independent Normal random variables has a Normal distribution. The mean and standard deviation of $aX + bY$ are found as usual from the rules for means and variances. These facts are often used in statistical calculations. Here is an example.

LOOK BACK

rules for means, p. 254

rules for variances, p. 258

EXAMPLE 5.13

Getting to and from campus. You live off campus and take the shuttle, provided by your apartment complex, to and from campus. Your time on the shuttle in minutes varies from day to day. The time going to campus X has the $N(20,4)$ distribution, and the time returning from campus Y varies according to the $N(18, 8)$ distribution. If they vary independently, what is the probability that you will be on the shuttle for less time going to campus?

The difference in times $X - Y$ is Normally distributed, with mean and variance

$$\mu_{X-Y} = \mu_X - \mu_Y = 20 - 18 = 2$$
$$\sigma^2_{X-Y} = \sigma^2_X + \sigma^2_Y = 4^2 + 8^2 = 80$$

Because $\sqrt{80} = 8.94$, $X - Y$ has the $N(2, 8.94)$ distribution. Figure 5.12 illustrates the probability computation:

$$P(X < Y) = P(X - Y < 0)$$
$$= P\left(\frac{(X - Y) - 2}{8.94} < \frac{0 - 2}{8.94}\right)$$
$$= P(Z < -0.22) = 0.4129$$

Although, on average, it takes longer to go to campus than return, the trip to campus will take less time on roughly two of every five days.

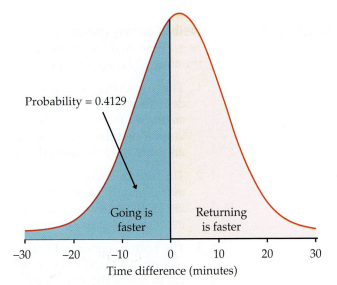

FIGURE 5.12 The Normal probability calculation, Example 5.13. The difference in times going to campus and returning from campus ($X - Y$) is Normal with mean 2 minutes and standard deviation 8.94 minutes.

The second useful fact is that **more general versions of the central limit theorem say that the distribution of a sum or average of many small random quantities is close to Normal**. This is true even if the quantities are not independent (as long as they are not too highly correlated) and even if they have different distributions (as long as no single random quantity is so large that it dominates the others). These more general versions of the central limit theorem suggest why the Normal distributions are common models for observed data. Any variable that is a sum of many small random influences will have approximately a Normal distribution.

Finally, **the central limit theorem also applies to discrete random variables**. An average of discrete random variables will never result in a continuous sampling distribution, but the Normal distribution often serves as a good approximation. In Section 5.3, we will discuss the sampling distribution and Normal approximation for counts and proportions. This Normal approximation is just an example of the central limit theorem applied to these discrete random variables.

BEYOND THE BASICS

Weibull Distributions

Our discussion of sampling distributions so far has concentrated on the Normal model to approximate the sampling distribution of the sample mean $\bar{x}$. This model is important in statistical practice because of the central limit theorem and the fact that sample means are among the most frequently used statistics. Simplicity also contributes to its popularity. The parameter μ is easy to understand, and to estimate it, we use a statistic $\bar{x}$ that is also easy to understand and compute.

There are, however, many other probability distributions that are used to model data in various circumstances. The time that a product, such as a computer hard drive, lasts before failing rarely has a Normal distribution. Earlier, we mentioned the use of the exponential distribution to model time to failure. Another class of continuous distributions, the **Weibull distributions**, is more commonly used in these situations.

Weibull distributions

EXAMPLE 5.14

Weibull density curves. Figure 5.13 shows the density curves of three members of the Weibull family. Each describes a different type of distribution for the time to failure of a product.

1. The top curve in Figure 5.13 is a model for *infant mortality*. This describes products that often fail immediately, prior to delivery to the customer. However, if the product does not fail right away, it will likely last a long time. For products like this, a manufacturer might test them and ship only the ones that do not fail immediately.

2. The middle curve in Figure 5.13 is a model for *early failure*. These products do not fail immediately, but many fail early in their lives after they are in the hands of customers. This is disastrous—the product or the process that makes it must be changed at once.

3. The bottom curve in Figure 5.13 is a model for *old-age wear-out*. Most of these products fail only when they begin to wear out, and then many fail at about the same age.

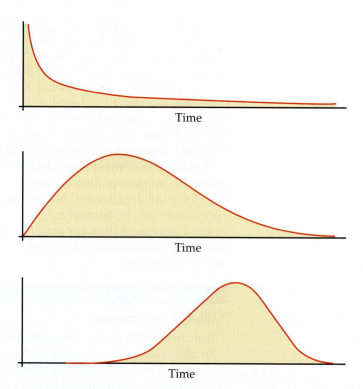

FIGURE 5.13 Density curves for three members of the Weibull family of distributions, Example 5.14.

A manufacturer certainly wants to know to which of these classes a new product belongs. To find out, engineers operate a random sample of products until they fail. From the failure time data, we can estimate the parameter (called the "shape parameter") that distinguishes among the three Weibull distributions in Figure 5.13. The shape parameter has no simple definition like that of a population proportion or mean, and it cannot be estimated by a simple statistic such as $\hat{p}$ or $\bar{x}$.

Two things save the situation. First, statistical theory provides general approaches for finding good estimates of any parameter. These general methods not only tell us how to use $\bar{x}$ in the Normal settings, but also how to estimate the Weibull shape parameter. Second, software can calculate the estimate from data even though there is no algebraic formula that we can write for the estimate. Statistical practice often relies on both mathematical theory and methods of computation more elaborate than the ones we will meet in this book. Fortunately, big ideas such as sampling distributions carry over to more complicated situations.[7]

SECTION 5.2 SUMMARY

• The **population distribution** of a variable is the distribution of its values for all members of the population.

• The **sample mean** $\bar{x}$ of an SRS of size n drawn from a large population with mean μ and standard deviation σ has a sampling distribution with mean and standard deviation

$$\mu_{\bar{x}} = \mu$$
$$\sigma_{\bar{x}} = \frac{\sigma}{\sqrt{n}}$$

The sample mean $\bar{x}$ is an unbiased estimator of the population mean μ and is less variable than a single observation. The standard deviation decreases in proportion to the square root of the sample size n. This means that to reduce the standard deviation by a factor of C, we need to increase the sample size by a factor of C^2.

• The **central limit theorem** states that, for large n, the sampling distribution of $\bar{x}$ is approximately $N(\mu, \sigma/\sqrt{n})$ for any population with mean μ and finite standard deviation σ. This allows us to approximate probability calculations of $\bar{x}$ using the Normal distribution.

• Linear combinations of independent Normal random variables have Normal distributions. In particular, if the population has a Normal distribution, so does $\bar{x}$.

SECTION 5.2 EXERCISES

For Exercise 5.17, see page 294; for Exercises 5.18 and 5.19, see page 297; for Exercise 5.20, see page 298; for Exercise 5.21, see page 299; for Exercises 5.22 and 5.23, see pages 301–302; and for Exercise 5.24, see page 303.

5.25 What is wrong? Explain what is wrong in each of the following statements.

(a) If the population standard deviation is 10, then the standard deviation of $\bar{x}$ for an SRS of 10 observations is 10/10 = 1.

(b) When taking SRSs from a population, larger sample sizes will result in larger standard deviations of $\bar{x}$.

(c) For an SRS from a population, both the mean and the standard deviation of $\bar{x}$ depend on the sample size n.

(d) The larger the population size N, the larger the sample size n needs to be for a desired standard deviation of $\bar{x}$.

5.26 What is wrong? Explain what is wrong in each of the following statements.

(a) The central limit theorem states that for large n, the population mean μ is approximately Normal.

(b) For large n, the distribution of observed values will be approximately Normal.

(c) For sufficiently large n, the 68–95–99.7 rule says that $\bar{x}$ should be within $\mu \pm 2\sigma$ about 95% of the time.

(d) As long as the sample size n is less than half the population size N, the standard deviation of $\bar{x}$ is $\sigma/\sqrt{n}$.

5.27 Generating a sampling distribution. Let's illustrate the idea of a sampling distribution in the case of a very small sample from a very small population. The population is the 10 scholarship players currently on your women's basketball team. For convenience, the 10 players have been labeled with the integers

0 to 9. For each player, the total amount of time spent (in minutes) on Twitter during the last week is recorded in the following table.

Player	0	1	2	3	4	5	6	7	8	9
Total time (min)	98	63	137	210	52	88	151	133	105	168

The parameter of interest is the average amount of time on Twitter. The sample is an SRS of size $n = 3$ drawn from this population of players. Because the players are labeled 0 to 9, a single random digit from Table B chooses one player for the sample.

(a) Find the mean for the 10 players in the population. This is the population mean μ.

(b) Use Table B to draw an SRS of size 3 from this population. (*Note:* You may sample the same player's time more than once.) Write down the three times in your sample and calculate the sample mean $\bar{x}$. This statistic is an estimate of μ.

(c) Repeat this process nine more times using different parts of Table B. Make a histogram of the 10 values of $\bar{x}$. You are approximating the sampling distribution of $\bar{x}$.

(d) Is the center of your histogram close to μ? Explain why you'd expect it to get closer to μ the more times you repeated this sampling process.

5.28 Total sleep time of college students. In Example 5.4, the total sleep time per night among college students was approximately Normally distributed with mean $\mu = 6.78$ hours and standard deviation $\sigma = 1.24$ hours. You plan to take an SRS of size $n = 120$ and compute the average total sleep time.

(a) What is the standard deviation for the average time?

(b) Use the 95 part of the 68–95–99.7 rule to describe the variability of this sample mean.

(c) What is the probability that your average will be below 6.9 hours?

5.29 Determining sample size. Refer to the previous exercise. You want to use a sample size such that about 95% of the averages fall within ±5 minutes (0.08 hour) of the true mean $\mu = 6.78$.

(a) Based on your answer to part (b) in Exercise 5.28, should the sample size be larger or smaller than 120? Explain.

(b) What standard deviation of $\bar{x}$ do you need such that approximately 95% of all samples will have a mean within 5 minutes of μ?

(c) Using the standard deviation you calculated in part (b), determine the number of students you need to sample.

5.30 Music file size on a tablet PC. A tablet PC contains 3217 music files. The distribution of file size is highly skewed with many small file sizes. Assume that the standard deviation for this population is 3.25 megabytes (MB).

(a) What is the standard deviation of the average file size when you take an SRS of 25 files from this population?

(b) How many files would you need to sample if you wanted the standard deviation of $\bar{x}$ to be no larger than 0.50 MB?

5.31 Bottling an energy drink. A bottling company uses a filling machine to fill cans with an energy drink. The cans are supposed to contain 250 milliliters (ml). The machine, however, has some variability, so the standard deviation of the volume is $\sigma = 0.4$ ml. A sample of five cans is inspected each hour for process control purposes, and records are kept of the sample mean volume. If the process mean is exactly equal to the target value, what is the mean and standard deviation of the numbers recorded?

5.32 Average file size on a tablet. Refer to Exercise 5.30. Suppose that the true mean file size of the music and video files on the tablet is 2.35 MB and you plan to take an SRS of $n = 50$ files.

(a) Explain why it may be reasonable to assume that the average $\bar{x}$ is approximately Normal even though the population distribution is highly skewed.

(b) Sketch the approximate Normal curve for the sample mean, making sure to specify the mean and standard deviation.

(c) What is the probability that your sample mean will differ from the population mean by more than 0.15 MB?

5.33 Can volumes. Averages are less variable than individual observations. It is reasonable to assume that the can volumes in Exercise 5.31 vary according to a Normal distribution. In that case, the mean $\bar{x}$ of an SRS of cans also has a Normal distribution.

(a) Make a sketch of the Normal curve for a single can. Add the Normal curve for the mean of an SRS of five cans on the same sketch.

(b) What is the probability that the volume of a single randomly chosen can differs from the target value by 0.1 ml or more?

(c) What is the probability that the mean volume of an SRS of five cans differs from the target value by 0.1 ml or more?

5.34 Number of friends on Facebook. To commemorate Facebook's 10-year milestone, Pew Research reported several facts about Facebook obtained from its Internet Project survey. One was that the average adult user of Facebook has 338 friends. This population distribution takes only integer values, so it is certainly not Normal. It is also highly skewed to the right, with a reported median of 200 friends.[8] Suppose that $\sigma = 380$ and you take an SRS of 80 adult Facebook users.

(a) For your sample, what are the mean and standard deviation of $\bar{x}$, the mean number of friends per adult user?

(b) Use the central limit theorem to find the probability that the average number of friends for 80 Facebook users is greater than 350.

(c) What are the mean and standard deviation of the total number of friends in your sample?

(d) What is the probability that the total number of friends among your sample of 80 Facebook users is greater than 28,000?

5.35 Cholesterol levels of teenagers. A study of the health of teenagers plans to measure the blood cholesterol level of an SRS of 13- to 16-year olds. The researchers will report the mean $\bar{x}$ from their sample as an estimate of the mean cholesterol level μ in this population.

(a) Explain to someone who knows no statistics what it means to say that $\bar{x}$ is an "unbiased" estimator of μ.

(b) The sample result $\bar{x}$ is an unbiased estimator of the population truth μ no matter what size SRS the study chooses. Explain to someone who knows no statistics why a large sample gives more trustworthy results than a small sample.

5.36 Grades in a math course. Indiana University posts the grade distributions for its courses online.[9] Students in one section of Math 118 in the fall semester received 18% A's, 31% B's, 26% C's, 13% D's and 12% F's.

(a) Using the common scale A = 4, B = 3, C = 2, D = 1, F = 0, take X to be the grade of a randomly chosen Math 118 student. Use the definitions of the mean (page 248) and standard deviation (page 256) for discrete random variables to find the mean μ and the standard deviation σ of grades in this course.

(b) Math 118 is a large enough course that we can take the grades of an SRS of 25 students and not worry about the finite population correction factor. If $\bar{x}$ is the average of these 25 grades, what are the mean and standard deviation of $\bar{x}$?

(c) What is the probability that a randomly chosen Math 118 student gets a B or better, $P(X \geq 3)$?

(d) What is the approximate probability that the grade point average for 25 randomly chosen Math 118 students is B or better, $P(\bar{x} \geq 3)$?

5.37 Monitoring the emerald ash borer. The emerald ash borer is a beetle that poses a serious threat to ash trees. Purple traps are often used to detect or monitor populations of this pest. In the counties of your state where the beetle is present, thousands of traps are used to monitor the population. These traps are checked periodically. The distribution of beetle counts per trap is discrete and strongly skewed. A majority of traps have no beetles, and only a few will have more than two beetles. For this exercise, assume that the mean number of beetles trapped is 0.4 with a standard deviation of 0.9.

(a) Suppose that your state does not have the resources to check all the traps, so it plans to check only an SRS of $n = 100$ traps. What are the mean and standard deviation of the average number of beetles $\bar{x}$ in 100 traps?

(b) Use the central limit theorem to find the probability that the average number of beetles in 100 traps is greater than 0.5.

(c) Do you think it is appropriate in this situation to use the central limit theorem? Explain your answer.

5.38 Risks and insurance. The idea of insurance is that we all face risks that are unlikely but carry high cost. Think of a fire destroying your home. So we form a group to share the risk: we all pay a small amount, and the insurance policy pays a large amount to those few of us whose homes burn down. An insurance company looks at the records for millions of homeowners and sees that the mean loss from fire in a year is $\mu = \$500$ per house and that the standard deviation of the loss is $\sigma = \$10,000$. (The distribution of losses is extremely right-skewed: most people have $0 loss, but a few have large losses.) The company plans to sell fire insurance for $500 plus enough to cover its costs and profit.

(a) Explain clearly why it would be unwise to sell only 100 policies. Then explain why selling many thousands of such policies is a safe business.

(b) Suppose the company sells the policies for $600. If the company sells 50,000 policies, what is the approximate probability that the average loss in a year will be greater than $600?

5.39 Weights of airline passengers. In 2005, the Federal Aviation Administration (FAA) updated its passenger weight standards to an average of 190 pounds in the summer (195 in the winter). This includes clothing and carry-on baggage. The FAA, however, did not specify a standard deviation. A reasonable standard deviation is 35 pounds. Weights are not Normally distributed, especially when the population includes both men and women, but they are not very non-Normal. A commuter plane carries 25 passengers. What is the approximate probability that, in the summer, the total weight of the passengers exceeds 5200 pounds? (*Hint*: To apply the central limit theorem, restate the problem in terms of the mean weight.)

5.40 Iron depletion without anemia and physical performance. Several studies have shown a link between iron depletion without anemia (IDNA) and physical performance. In one recent study, the physical performance of 24 female collegiate rowers with IDNA was compared with 24 female collegiate rowers with normal iron status.[10] Several different measures of physical performance were studied, but we'll focus here on training-session duration. Assume that training-session duration of female rowers with IDNA is Normally distributed, with mean 58 minutes and standard deviation 11 minutes. Training-session duration of female rowers with normal iron status is Normally distributed, with mean 69 minutes and standard deviation 18 minutes.

(a) What is the probability that the mean duration of the 24 rowers with IDNA exceeds 63 minutes?

(b) What is the probability that the mean duration of the 24 rowers with normal iron status is less than 63 minutes?

(c) What is the probability that the mean duration of the 24 rowers with IDNA is greater than the mean duration of the 24 rowers with normal iron status?

5.41 Treatment and control groups. The previous exercise illustrates a common setting for statistical inference. This exercise gives the general form of the sampling distribution needed in this setting. We have a sample of n observations from a treatment group and an independent sample of m observations from a control group. Suppose that the response to the treatment has the $N(\mu_X, \sigma_X)$ distribution and that the response of control subjects has the $N(\mu_Y, \sigma_Y)$ distribution. Inference about the difference $\mu_Y - \mu_X$ between the population means is based on the difference $\bar{y} - \bar{x}$ between the sample means in the two groups.

(a) Under the assumptions given, what is the distribution of $\bar{y}$? Of $\bar{x}$?

(b) What is the distribution of $\bar{y} - \bar{x}$?

5.42 Investments in two funds. Jennifer invests her money in a portfolio that consists of 70% Fidelity Spartan 500 Index Fund and 30% Fidelity Diversified International Fund. Suppose that, in the long run, the annual real return X on the 500 Index Fund has mean 10% and standard deviation 15%, the annual real return Y on the Diversified International Fund has mean 9% and standard deviation 19%, and the correlation between X and Y is 0.6.

(a) The return on Jennifer's portfolio is $R = 0.7X + 0.3Y$. What are the mean and standard deviation of R?

(b) The distribution of returns is typically roughly symmetric but with more extreme high and low observations than a Normal distribution. The average return over a number of years, however, is close to Normal. If Jennifer holds her portfolio for 20 years, what is the approximate probability that her average return is less than 5%?

(c) The calculation you just made is not overly helpful because Jennifer isn't really concerned about the mean return $\bar{R}$. To see why, suppose that her portfolio returns 12% this year and 6% next year. The mean return for the two years is 9%. If Jennifer starts with $1000, how much does she have at the end of the first year? At the end of the second year? How does this amount compare with what she would have if both years had the mean return, 9%? Over 20 years, there may be a large difference between the ordinary mean $\bar{R}$ and the *geometric mean*, which reflects the fact that returns in successive years multiply rather than add.

5.3 Sampling Distributions for Counts and Proportions

When you complete this section, you will be able to:

- Determine when a count X can be modeled using the binomial distribution.
- Determine when the sampling distribution of a count can be modeled using the binomial distribution.
- Calculate the mean and standard deviation of X when it has the $B(n, p)$ distribution.
- Explain the difference between the sampling distribution of a count X and the sampling distribution of the sample proportion $\hat{p} = X/n$.
- Determine when one can approximate the sampling distribution of a count using the Normal distribution.
- Determine when one can approximate the sampling distribution of the sample proportion using the Normal distribution.
- Use the Normal approximation for counts and proportions to perform probability calculations about the statistics.

LOOK BACK

categorical
variable,
p. 3

In the previous section, we discussed the probability distribution of the sample mean, which meant a focus on population values that were quantitative. We will now shift our focus to population values that are categorical. Counts and proportions are common discrete statistics that describe categorical data.

In Section 5.1 (pages 283–287), we discussed the use of simulation to study the sampling distribution of the sample proportion. In this section, we will use probability theory to more precisely describe the sampling distributions of the sample count and proportion. Let's start with an example.

EXAMPLE 5.15

Istockphoto/Thinkstock

Work hours make it difficult to spend time with children. A sample survey asks 1006 British parents whether they think long working hours are making it difficult to spend enough time with their children.[11] We would like to view the responses of these parents as representative of a larger population of British parents who hold similar beliefs. That is, we will view the responses of the sampled parents as an SRS from a population.

When there are only two possible outcomes for a random variable, we can summarize the results by giving the count for one of the possible outcomes. We let n represent the sample size, and we use X to represent the random variable that gives the count for the outcome of interest.

EXAMPLE 5.16

The random variable of interest. In this sample survey of British parents, $n = 1006$. The parents in the sample were asked if they agree with the statement "These days, long working hours make it difficult for parents to spend enough time with their children." The variable X is the number of parents who agreed with the statement. In this case, $X = 755$.

In our example, we chose the random variable X to be the number of parents who think that long working hours make it difficult to spend enough time with their children. We could have chosen X to be the number of parents who do not think that long working hours make it difficult to spend enough time with their children. The choice is yours. Often, we make the choice based on how we would like to describe the results in a summary. Which choice do you prefer in this case?

LOOK BACK

sample
proportion,
p. 283

When a random variable has only two possible outcomes, it is more common to use the sample proportion $\hat{p} = X/n$ as the summary rather than the count X.

EXAMPLE 5.17

The sample proportion. The sample proportion of parents surveyed who think that long working hours make it difficult to spend enough time with their children is

$$\hat{p} = \frac{755}{1006} = 0.75$$

Notice that this summary takes into account the sample size n. We need to know n in order to properly interpret the meaning of the random variable X. For example, the conclusion we would draw about parent opinions in this survey would be quite different if we had observed $X = 755$ from a sample twice as large, $n = 2012$.

5.43 Sexual harassment in middle school. A survey of 1391 students in grades 5 to 8 reports that 26% of the students say they have encountered some type of sexual harassment while at school.[12] Give the sample size n, the count X, and the sample proportion $\hat{p}$ for this survey.

5.44 High school graduates who took a statistics course. In a random sample of $n = 4012$ high school graduates, 10.8% reported that they had taken a statistics course.[13] Give the sample size n, the count X, and the sample proportion $\hat{p}$ for this setting.

5.45 Use of the Internet to find a place to live. A poll of 1500 college students asked whether or not they have used the Internet to find a place to live sometime within the past year. There were 1234 students who answered Yes; the other 266 answered No.

(a) What is the sample size n?

(b) Choose one of the two possible outcomes to define the random variable, X. Give a reason for your choice.

(c) What is the value of the count X?

(d) Find the sample proportion, $\hat{p}$.

Just like the sample mean, sample counts and sample proportions are commonly used statistics, and understanding their sampling distributions is important for statistical inference. These statistics, however, are discrete random variables, so their sampling distributions introduce us to a new family of probability distributions.

The binomial distributions for sample counts

The distribution of a count X depends on how the data are produced. Here is a simple but common situation.

THE BINOMIAL SETTING

1. There is a fixed number of observations n.

2. The n observations are all independent.

3. Each observation falls into one of just two categories, which for convenience we call "success" and "failure."

4. The probability of a success, call it p, is the same for each observation.

Think of tossing a coin n times as an example of the binomial setting. Each toss gives either heads or tails, and the outcomes of successive tosses are independent. If we call heads a success, then p is the probability of a head and remains the same as long as we toss the same coin. The number of heads we count is a random variable X. The distribution of X (and, more generally, the distribution of the count of successes in any binomial setting) is completely determined by the number of observations n and the success probability p.

BINOMIAL DISTRIBUTIONS

The distribution of the count X of successes in the binomial setting is called the **binomial distribution** with parameters n and p. The parameter n is the number of observations, and p is the probability of a success on any one observation. The possible values of X are the whole numbers from 0 to n. As an abbreviation, we say that the distribution of X is $B(n, p)$.

The binomial distributions are an important class of discrete probability distributions. Later in this section, we will learn how to assign probabilities to outcomes and how to find the mean and standard deviation of binomial distributions. *The most important skill for using binomial distributions is the ability to recognize situations to which they do and do not apply.* This can be done by checking all the facets of the binomial setting.

EXAMPLE 5.18

Binomial examples? **(a)** Genetics says that children receive genes from their parents independently. Each child of a particular pair of parents has probability 0.25 of having type O blood. If these parents have three children, the number who have type O blood is the count X of successes in three independent trials with probability 0.25 of a success on each trial. So X has the $B(3, 0.25)$ distribution.

(b) Engineers define reliability as the probability that an item will perform its function under specific conditions for a specific period of time. Replacement heart valves made of animal tissue, for example, have probability 0.77 of performing well for 15 years.[14] The probability of failure within 15 years is, therefore, 0.23. It is reasonable to assume that valves in different patients fail (or not) independently of each other. The number of patients in a group of 500 who will need another valve replacement within 15 years has the $B(500, 0.23)$ distribution.

(c) A multicenter trial is designed to assess a new surgical procedure. A total of 540 patients will undergo the procedure, and the count of patients X who suffer a major adverse cardiac event (MACE) within 30 days of surgery will be recorded. Because these patients will receive this procedure from different surgeons at different hospitals, it may not be true that the probability of a MACE is the same for each patient. Thus, X may not have the binomial distribution.

USE YOUR KNOWLEDGE

5.46 Genetics and blood types. Genetics says that children receive genes from each of their parents independently. Suppose that each child of a particular pair of parents has probability 0.375 of having type AB blood. If these parents have three children, what is the distribution of the number who have type AB blood? Explain your answer.

5.47 Tossing a coin. Suppose you plan to toss a coin 20 times and record X, the number of heads that you observe. If the coin is fair ($p = 0.5$), what is the distribution of X? Also, explain why this distribution is also the sampling distribution of X.

Binomial distributions in statistical sampling

The binomial distributions are important in statistics when we wish to make inferences about the proportion p of "successes" in a population. Here is a typical example.

EXAMPLE 5.19

Audits of financial records. The financial records of businesses are often audited by state tax authorities to test compliance with tax laws. Suppose that for one retail business, 800 of the 10,000 sales are incorrectly classified as subject to state sales tax. It would be too time-consuming for authorities to examine all these sales. Instead, an auditor examines an SRS of sales records. Is the count X of misclassified records in an SRS of 150 records a binomial random variable?

Choosing an SRS from a population is not quite a binomial setting. Removing one record in Example 5.19 changes the proportion of bad records in the remaining population, so the state of the second record chosen is not independent of the first. Because the population is large, however, removing a few items has a very small effect on the composition of the remaining population. Successive inspection results are very nearly independent. The population proportion of misclassified records is

$$p = \frac{800}{10,000} = 0.08$$

If the first record chosen is bad, the proportion of bad records remaining is $799/9999 = 0.079908$. If the first record is good, the proportion of bad records left is $800/9999 = 0.080008$. These proportions are so close to 0.08 that, for practical purposes, we can act as if removing one record has no effect on the proportion of misclassified records remaining. We act as if the count X of misclassified sales records in the audit sample has the binomial distribution $B(150, 0.08)$.

Populations like the one described in Example 5.19 often contain a relatively small number of items with very large values. For this example, these values would be very large sale amounts and likely represent an important group of items to the auditor. An SRS taken from such a population will likely include very few items of this type. Therefore, it is common to use a stratified sample in settings like this. Strata are defined based on dollar value of the sale, and within each stratum, an SRS is taken. The results are then combined to obtain an estimate for the entire population.

LOOK BACK

stratified random
sample,
p. 194

SAMPLING DISTRIBUTION OF A COUNT

A population contains proportion p of successes. If the population is much larger than the sample, the count X of successes in an SRS of size n has approximately the binomial distribution $B(n, p)$.

The accuracy of this approximation improves as the size of the population increases relative to the size of the sample. As a rule of thumb, we will use the binomial sampling distribution for counts when the population is at least 20 times as large as the sample.

Finding binomial probabilities

We will later give a formula for the probability that a binomial random variable takes any of its values. In practice, you will rarely have to use this formula for calculations because some calculators and most statistical software packages will calculate binomial probabilities for you.

EXAMPLE 5.20

Probabilities for misclassified sales records. In the audit setting of Example 5.19, what is the probability that the audit finds exactly 10 misclassified sales records? What is the probability that the audit finds no more than 10 misclassified records? Figure 5.14 shows the output from one statistical software system. You see that if the count X has the $B(150, 0.08)$ distribution,

$$P(X = 10) = 0.106959$$
$$P(X \leq 10) = 0.338427$$

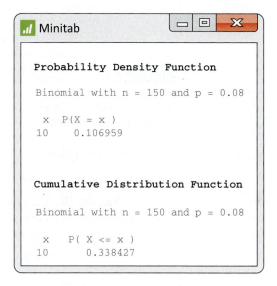

FIGURE 5.14 Binomial probabilities, Example 5.20: output from the Minitab statistical software.

It was easy to request these calculations in the software's menus. For the TI-83/84 calculator, the functions **binompdf** and **binomcdf** would be used. In R, the functions **dbinom** and **pbinom** would be used. Typically, the output supplies more decimal places than we need and uses labels that may not be helpful (for example, "Probability Density Function" when the distribution is discrete, not continuous). But, as usual with software, we can ignore distractions and find the results we need.

If you do not have suitable computing facilities, you can still shorten the work of calculating binomial probabilities for some values of n and p by looking up probabilities in Table C in the back of this book. The entries in the table are the probabilities $P(X = k)$ of individual outcomes for a binomial random variable X.

EXAMPLE 5.21

The probability histogram. Suppose that the audit in Example 5.19 chose just 15 sales records. What is the probability that no more than one of the 15 is misclassified? The count X of misclassified records in the sample has approximately the $B(15, 0.08)$ distribution. Figure 5.15 is a probability histogram for this distribution. The distribution is strongly skewed. Although X can take any whole-number value from 0 to 15, the probabilities of values larger than 5 are so small that they do not appear in the histogram.

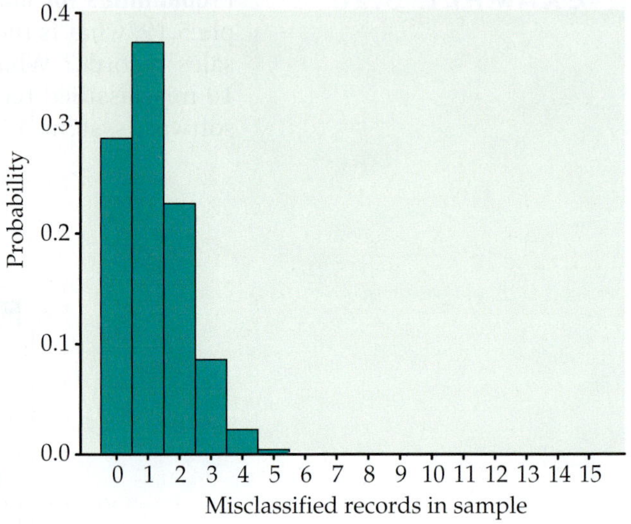

FIGURE 5.15 Probability histogram for the binomial distribution with $n = 15$ and $p = 0.08$, Example 5.21.

		p
n	k	.08
15	0	.2863
	1	.3734
	2	.2273
	3	.0857
	4	.0223
	5	.0043
	6	.0006
	7	.0001
	8	
	9	

We want to calculate

$$P(X \leq 1) = P(X = 0) + P(X = 1)$$

when X has the $B(15, 0.08)$ distribution. To use Table C for this calculation, look opposite $n = 15$ and under $p = 0.08$. The entries in the rows for each k are $P(X = k)$. Blank cells in the table are 0 to four decimal places. You see that

$$P(X \leq 1) = P(X = 0) + P(X = 1)$$
$$= 0.2863 + 0.3734 = 0.6597$$

About two-thirds of all samples will contain no more than one bad record. In fact, almost 29% of the samples will contain no bad records. The sample of size 15 cannot be trusted to provide adequate evidence about misclassified sales records. A larger number of observations is needed.

The values of p that appear in Table C are all 0.5 or smaller. When the probability of a success is greater than 0.5, restate the problem in terms of the number of failures. The probability of a failure is less than 0.5 when the probability of a success exceeds 0.5. When using the table, always stop to ask whether you must count successes or failures.

EXAMPLE 5.22

Falling asleep in class. In the survey of 4513 college students described in Example 5.4, 46% of the respondents reported falling asleep in class due to poor sleep. You randomly sample 10 students in your dormitory, and eight state that they fell asleep in class during the last week due to poor sleep. Relative to the survey results, is this an unusually high number of students?

To answer this question, assume that the students' actions (falling asleep or not) are independent, with the probability of falling asleep equal to 0.46. This independence assumption may not be reasonable if the students study and socialize together or if there is a loud student in the dormitory who keeps everyone up. We'll assume this is not an issue here, so the number X of students who fell asleep in class out of 10 students has the $B(10, 0.46)$ distribution.

We want the probability of classifying at least eight students as having fallen asleep in class. Using software, we find

$$P(X \geq 8) = P(X = 8) + P(X = 9) + P(X = 10)$$
$$= 0.0263 + 0.0050 + 0.0004 = 0.0317$$

We would expect to find eight or more students falling asleep in class about 3% of the time or in fewer than one of every 30 surveys of 10 students. This is a pretty rare outcome and falls outside the range of the usual chance variation due to random sampling.

USE YOUR KNOWLEDGE

5.48   Free-throw shooting. April is a college basketball player who makes 80% of her free throws. In a recent game, she had 10 free throws and missed three of them. How unusual is this outcome? Using software, calculator, or Table C, compute $1 - P(X \leq 2)$, where X is the number of free throws missed in 10 shots. Explain your answer.

5.49   Find the probabilities.

(a) Suppose that X has the $B(8, 0.3)$ distribution. Use software, calculator, or Table C to find $P(X = 0)$ and $P(X \geq 6)$.

(b) Suppose that X has the $B(8, 0.7)$ distribution. Use software, calculator, or Table C to find $P(X = 8)$ and $P(X \leq 2)$.

(c) Explain the relationship between your answers to parts (a) and (b) of this exercise.

Binomial mean and standard deviation

If a count X has the $B(n, p)$ distribution, what are the mean μ_X and the standard deviation σ_X? We can guess the mean. If we expect 46% of the students to have fallen asleep in class due to poor sleep, the mean number in 10 students should be 46% of 10, or 4.6. That's μ_X when X has the $B(10, 0.46)$ distribution.

Intuition suggests more generally that the mean of the $B(n, p)$ distribution should be np. Can we show that this is correct and also obtain a short formula for the standard deviation? Because binomial distributions are discrete probability distributions, we could find the mean and variance by using the definitions in Section 4.4. Here is an easier way.

LOOK BACK

mean and variance of a discrete random variable, pp. 248, 256

A binomial random variable X is the count of successes in n independent observations that each have the same probability p of success. Let the random variable S_i indicate whether the ith observation is a success or failure by taking the values $S_i = 1$ if a success occurs and $S_i = 0$ if the outcome is a failure. The S_i are independent because the observations are, and each S_i has the same simple distribution:

Outcome	1	0
Probability	p	$1 - p$

From the definition of the mean of a discrete random variable, we know that the mean of each S_i is

$$\mu_S = (1)(p) + (0)(1 - p) = p$$

Similarly, the definition of the variance shows that $\sigma_S^2 = p(1 - p)$. Because each S_i is 1 for a success and 0 for a failure, to find the total number of successes X we add the S_i's:

$$X = S_1 + S_2 + \cdots + S_n$$

LOOK BACK

addition rules for means and variances, pp. 254, 258

Apply the addition rules for means and variances to this sum. To find the mean of X we add the means of the S_i's:

$$\mu_X = \mu_{S1} + \mu_{S2} + \cdots + \mu_{Sn}$$
$$= n\mu_S = np$$

Similarly, the variance is n times the variance of a single S, so that $\sigma_X^2 = np(1 - p)$. The standard deviation σ_X is the square root of the variance. Here is the result.

BINOMIAL MEAN AND STANDARD DEVIATION

If a count X has the binomial distribution $B(n, p)$, then
$$\mu_X = np$$
$$\sigma_X = \sqrt{np(1 - p)}$$

EXAMPLE 5.23

The Helsinki Heart Study. The Helsinki Heart Study asked whether the anticholesterol drug gemfibrozil reduces heart attacks. In planning such an experiment, the researchers must be confident that the sample sizes are large enough to enable them to observe enough heart attacks. The Helsinki study planned to give gemfibrozil to about 2000 men aged 40 to 55 and a placebo to another 2000. The probability of a heart attack during the five-year period of the study for men this age is about 0.04. What are the mean and standard deviation of the number of heart attacks that will be observed in one group if the treatment does not change this probability?

There are 2000 independent observations, each having probability $p = 0.04$ of a heart attack. The count X of heart attacks has the $B(2000, 0.04)$ distribution, so that

$$\mu_X = np = (2000)(0.04) = 80$$
$$\sigma_X = \sqrt{np(1 - p)} = \sqrt{(2000)(0.04)(0.96)} = 8.76$$

The expected number of heart attacks is large enough to permit conclusions about the effectiveness of the drug. In fact, there were 84 heart attacks among the 2035 men actually assigned to the placebo, quite close to the mean. The gemfibrozil group of 2046 men suffered only 56 heart attacks. This is evidence that the drug reduces the chance of a heart attack. In a later chapter, we will learn how to determine if this is strong enough evidence to conclude the drug is effective.

USE YOUR KNOWLEDGE

5.50 **Free-throw shooting.** Refer to Exercise 5.48 (page 317). If April takes 85 free throws in the upcoming season, what are the mean and standard deviation of the number of free throws made?

5.51 **Find the mean and standard deviation**

(a) Suppose that X has the $B(8, 0.3)$ distribution. Compute the mean and standard deviation of X.

(b) Suppose that X has the $B(8, 0.7)$ distribution. Compute the mean and standard deviation of X.

(c) Explain the relationship between your answers to parts (a) and (b) of this exercise.

Sample proportions

LOOK BACK

population proportion, p. 283

What proportion of a company's sales records have an incorrect sales tax classification? What percent of adults favor stronger laws restricting firearms? In statistical sampling, we often want to estimate the proportion p of "successes" in a population. Our estimator is the sample proportion of successes:

$$\hat{p} = \frac{\text{count of successes in sample}}{\text{size of sample}}$$

$$= \frac{X}{n}$$

Be sure to distinguish between the proportion $\hat{p}$ and the count X. The count takes whole-number values between 0 and n, but a proportion is always a number between 0 and 1. In the binomial setting, the count X has a binomial distribution. The proportion $\hat{p}$ does *not* have a binomial distribution. We can, however, do probability calculations about $\hat{p}$ by restating them in terms of the count X and using binomial methods. In Example 5.12 (page 303), we took a similar approach for the sum, restating the problem in terms of the sample mean and then using the Normal distribution to calculate the probability.

EXAMPLE 5.24

Shopping online. A survey by the Consumer Reports National Research Center revealed that 84% of all respondents were very satisfied with their online shopping experience.[15] It was also reported, however, that people over the age of 40 were generally more satisfied than younger respondents. You decide to take a nationwide random sample of 2500 college students and ask if they agree or disagree that "I am very satisfied with my online shopping experience." Suppose that 60% of all college students would agree if asked this question. What is the probability that the sample proportion who agree is at least 58%?

The count X who agree has the binomial distribution $B(2500, 0.6)$. The sample proportion $\hat{p} = X/2500$ does *not* have a binomial distribution because it is not a count. But we can translate any question about a sample proportion $\hat{p}$ into a question about the count X. Because 58% of 2500 is 1450,

$$P(\hat{p} \geq 0.58) = P(X \geq 1450)$$

$$= P(X = 1450) + P(X = 1451) + \cdots + P(X = 2500)$$

This is a rather elaborate calculation. We must add more than 1000 binomial probabilities. Software tells us that $P(\hat{p} \geq 0.58) = 0.9802$. But what do we do if we don't have access to software?

LOOK BACK

rules for means and variances, pp. 254, 258

As a first step, find the mean and standard deviation of a sample proportion. We know the mean and standard deviation of a sample count, so apply the rules from Section 4.4 for the mean and variance of a constant times a random variable. Here is the result.

MEAN AND STANDARD DEVIATION OF A SAMPLE PROPORTION

Let $\hat{p}$ be the sample proportion of successes in an SRS of size n drawn from a large population having population proportion p of successes. The mean and standard deviation of $\hat{p}$ are

$$\mu_{\hat{p}} = p$$

$$\sigma_{\hat{p}} = \sqrt{\frac{p(1-p)}{n}}$$

The formula for $\sigma_{\hat{p}}$ is exactly correct in the binomial setting. It is approximately correct for an SRS from a large population. We will use it when the population is at least 20 times as large as the sample.

Let's now use these formulas to calculate the mean and standard deviation for Example 5.24.

EXAMPLE 5.25

The mean and the standard deviation. The mean and standard deviation of the proportion of the survey respondents in Example 5.24 who are satisfied with their online shopping experience are

$$\mu_{\hat{p}} = p = 0.6$$

$$\sigma_{\hat{p}} = \sqrt{\frac{p(1-p)}{n}} = \sqrt{\frac{(0.6)(0.4)}{2500}} = 0.0098$$

USE YOUR KNOWLEDGE

5.52 Find the mean and the standard deviation. If we toss a fair coin 150 times, the number of heads is a random variable that is binomial.

(a) Find the mean and the standard deviation of the sample proportion of heads.

(b) Is your answer to part (a) the same as the mean and the standard deviation of the sample count of heads in 150 throws? Explain your answer.

The fact that the mean of $\hat{p}$ is p states in statistical language that the sample proportion $\hat{p}$ in an SRS is an *unbiased estimator* of the population proportion p. When a sample is drawn from a new population having a different value of the population proportion p, the sampling distribution of the unbiased estimator $\hat{p}$ changes so that its mean moves to the new value of p. We observed this fact empirically in Section 5.1 and have now verified it from the laws of probability.

The variability of $\hat{p}$ about its mean, as described by the variance or standard deviation, gets smaller as the sample size increases. So a sample proportion from a large sample will usually lie quite close to the population proportion p. We observed this in the simulation experiment on page 285 in Section 5.1. Now we have discovered exactly how the variability decreases: the standard deviation is $\sqrt{p(1-p)/n}$. Similar to what we observed in the previous section, the $\sqrt{n}$ in the denominator means that the sample size must be multiplied by 4 if we wish to divide the standard deviation in half.

Normal approximation for counts and proportions

Using simulation, we discovered in Section 5.1 that the sampling distribution of a sample proportion $\hat{p}$ is close to Normal. Now we know that the distribution of $\hat{p}$ is that of a binomial count divided by the sample size n. This seems at first to be a contradiction. To clear up the matter, look at Figure 5.16. This is a probability histogram of the exact distribution of the proportion of very satisfied shoppers $\hat{p}$, based on the binomial distribution $B(2500, 0.6)$. There are hundreds of narrow bars, one for each of the 2501 possible values of $\hat{p}$. Most have probabilities too small to show in a graph. *The probability histogram looks very Normal!* In fact, both the count X and the sample proportion $\hat{p}$ are approximately Normal in large samples.

We also know this to be true as a result of the central limit theorem discussed in the previous section (page 298). Recall that we can consider the count X as a sum

$$X = S_1 + S_2 + \cdots + S_n$$

of independent random variables S_i that take the value 1 if a success occurs on the ith trial and the value 0 otherwise. The proportion of successes $\hat{p} = X/n$

FIGURE 5.16 Probability histogram of the sample proportion $\hat{p}$ based on a binomial count with $n = 2500$ and $p = 0.6$. The distribution is very close to Normal.

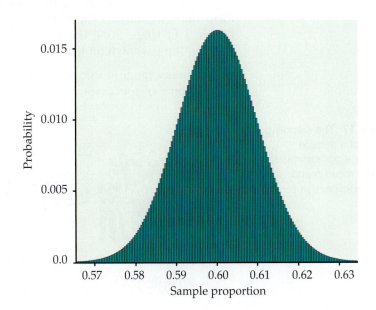

can then be thought of as the sample mean of the S_i and, like all sample means, is approximately Normal when n is large. Given that $\hat{p}$ is approximately Normal, the count will also be approximately Normal because it is just a constant n times $\hat{p}$, an approximately Normal random variable.

NORMAL APPROXIMATION FOR COUNTS AND PROPORTIONS

Draw an SRS of size n from a large population having population proportion p of successes. Let X be the count of successes in the sample and $\hat{p} = X/n$ be the sample proportion of successes. When n is large, the sampling distributions of these statistics are approximately Normal:

$$X \text{ is approximately } N\left(np, \sqrt{np(1-p)}\right)$$

$$\hat{p} \text{ is approximately } N\left(p, \sqrt{\frac{p(1-p)}{n}}\right)$$

As a rule of thumb, we will use this approximation for values of n and p that satisfy $np \geq 10$ and $n(1-p) \geq 10$.

These Normal approximations are easy to remember because they say that $\hat{p}$ and X are Normal, with their usual means and standard deviations. Whether or not you use the Normal approximations should depend on how accurate your calculations need to be. For most statistical purposes, great accuracy is not required. Our "rule of thumb" for use of the Normal approximations reflects this judgment.

The accuracy of the Normal approximations improves as the sample size n increases. They are most accurate for any fixed n when p is close to 0.5, and least accurate when p is near 0 or 1. You can compare binomial distributions with their Normal approximations by using the *Normal Approximation to Binomial* applet. This applet allows you to change n or p while watching the effect on the binomial probability histogram and the Normal curve that approximates it.

Figure 5.17 summarizes the distribution of a sample proportion in a form that emphasizes the big idea of a sampling distribution. Just as with Figure 5.11 (page 304), the general framework for constructing a sampling distribution is shown on the left.

• Take many random samples of size n from a population that contains proportion p of successes.

FIGURE 5.17 The sampling distribution of a sample proportion $\hat{p}$ is approximately Normal with mean p and standard deviation $\sqrt{p(1-p)/n}$.

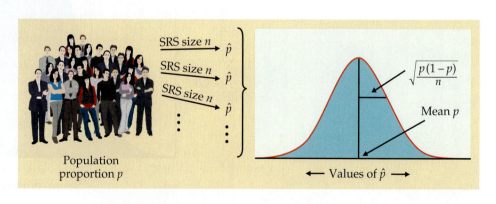

- Find the sample proportion $\hat{p}$ for each sample.

- Collect all the $\hat{p}$'s and display their distribution.

The sampling distribution of $\hat{p}$ is shown on the right. Keep this figure in mind as you move toward statistical inference.

EXAMPLE 5.26

Compare the Normal approximation with the exact calculation. Let's compare the Normal approximation for the calculation of Example 5.24 with the exact calculation from software. We want to calculate $P(\hat{p} \geq 0.58)$ when the sample size is $n = 2500$ and the population proportion is $p = 0.6$. Example 5.25 shows that

$$\mu_{\hat{p}} = p = 0.6$$

$$\sigma_{\hat{p}} = \sqrt{\frac{p(1-p)}{n}} = 0.0098$$

Act as if $\hat{p}$ were Normal with mean 0.6 and standard deviation 0.0098. The approximate probability, as illustrated in Figure 5.18, is

$$P(\hat{p} \geq 0.58) = P\left(\frac{\hat{p} - 0.6}{0.0098} \geq \frac{0.58 - 0.6}{0.0098}\right)$$

$$\doteq P(Z \geq -2.04) = 0.9793$$

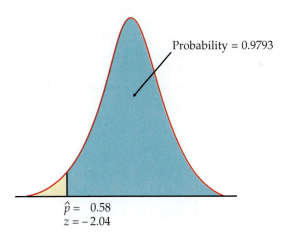

Probability = 0.9793

$\hat{p} = 0.58$
$z = -2.04$

FIGURE 5.18 The Normal probability calculation, Example 5.26.

That is, about 98% of all samples have a sample proportion that is at least 0.58. Because the sample was large, this Normal approximation is quite accurate. It misses the software value 0.9802 by only 0.0009.

EXAMPLE 5.27

Using the Normal approximation. The audit described in Example 5.19 examined an SRS of 150 sales records for compliance with sales tax laws. In fact, 8% of all the company's sales records have an incorrect sales tax classification. The count X of bad records in the sample has approximately the $B(150, 0.08)$ distribution.

According to the Normal approximation to the binomial distributions, the count X is approximately Normal with mean and standard deviation

$$\mu_X = np = (150)(0.08) = 12$$

$$\sigma_X = \sqrt{np(1-p)} = \sqrt{(150)(0.08)(0.92)} = 3.3226$$

The Normal approximation for the probability of no more than 10 misclassified records is the area to the left of $X = 10$ under the Normal curve. Using Table A,

$$P(X \le 10) = P\left(\frac{X-12}{3.3226} \le \frac{10-12}{3.3226}\right)$$

$$\doteq P(Z \le -0.60) = 0.2743$$

Software tells us that the actual binomial probability that no more than 10 of the records in the sample are misclassified is $P(X \le 10) = 0.3384$. The Normal approximation is only roughly accurate. Because $np = 12$, this combination of n and p is close to the border of the values for which we are willing to use the approximation.

The distribution of the count of bad records in a sample of 15 is distinctly non-Normal, as Figure 5.15 (page 316) showed. When we increase the sample size to 150, however, the shape of the binomial distribution becomes roughly Normal. Figure 5.19 displays the probability histogram of the binomial distribution with the density curve of the approximating Normal distribution superimposed. Both distributions have the same mean and standard deviation, and both the area under the histogram and the area under the curve are 1. The Normal curve fits the histogram reasonably well. Look closely: the histogram is slightly skewed to the right, a property that the symmetric Normal curve can't match.

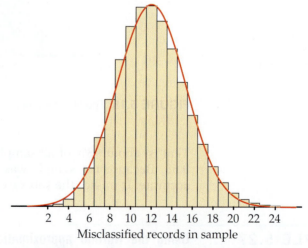

FIGURE 5.19 Probability histogram and Normal approximation for the binomial distribution with $n = 150$ and $p = 0.08$, Example 5.27.

5.53 Use the Normal approximation. Suppose that we toss a fair coin 150 times. Use the Normal approximation to find the probability that the sample proportion of heads is

(a) between 0.4 and 0.6.

(b) between 0.45 and 0.55.

The continuity correction

Figure 5.20 illustrates an idea that greatly improves the accuracy of the Normal approximation to binomial probabilities. The binomial probability $P(X \leq 10)$ is the area of the histogram bars for values 0 to 10. The bar for $X = 10$ actually extends from 9.5 to 10.5. Because the discrete binomial distribution puts probability only on whole numbers, the probabilities $P(X \leq 10)$ and $P(X \leq 10.5)$ are the same. The Normal distribution spreads probability continuously, so these two Normal probabilities are different. The Normal approximation is more accurate if we consider $X = 10$ to extend from 9.5 to 10.5, matching the bar in the probability histogram.

The event $\{X \leq 10\}$ includes the outcome $X = 10$. Figure 5.20 shades the area under the Normal curve that matches all the histogram bars for outcomes 0 to 10, bounded on the right not by 10, but by 10.5. So $P(X \leq 10)$ is calculated as $P(X \leq 10.5)$. On the other hand, $P(X < 10)$ excludes the outcome $X = 10$, so we exclude the entire interval from 9.5 to 10.5 and calculate $P(X \leq 9.5)$ from the Normal table. Here is the result of the Normal calculation in Example 5.27 improved in this way:

$$P(X \leq 10) = P(X \leq 10.5)$$

$$= P\left(\frac{X - 12}{3.3226} \leq \frac{10.5 - 12}{3.3226}\right)$$

$$\doteq P(Z \leq -0.45) = 0.3264$$

The improved approximation misses the binomial probability by only 0.012. Acting as though a whole number occupies the interval from 0.5 below to 0.5 above the number is called the **continuity correction** to the Normal

continuity correction

FIGURE 5.20 Area under the Normal approximation curve for the probability in Example 5.27.

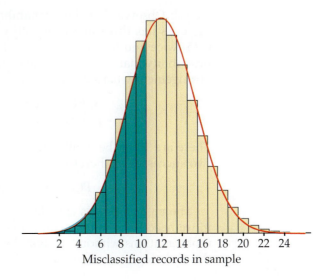

Misclassified records in sample

approximation. If you need accurate values for binomial probabilities, try to use software to do exact calculations. If no software is available, use the continuity correction unless n is very large. Because most statistical purposes do not require extremely accurate probability calculations, we do not emphasize use of the continuity correction.

Binomial formula

We can find a formula for the probability that a binomial random variable takes any value by adding probabilities for the different ways of getting exactly that many successes in n observations. Here is the example we will use to show the idea.

EXAMPLE 5.28

Blood types of children. Each child born to a particular set of parents has probability 0.25 of having blood type O. If these parents have five children, what is the probability that exactly two of them have type O blood?

The count of children with type O blood is a binomial random variable X with $n = 5$ tries and probability $p = 0.25$ of a success on each try. We want $P(X = 2)$.

Because the method doesn't depend on the specific example, we will use "S" for success and "F" for failure. In Example 5.28, "S" would stand for type O blood. Do the work in two steps.

Step 1: Find the probability that a specific two of the five tries give successes—say, the first and the third. This is the outcome SFSFF. The multiplication rule for independent events tells us that

$$P(\text{SFSFF}) = P(S)P(F)P(S)P(F)P(F)$$
$$= (0.25)(0.75)(0.25)(0.75)(0.75)$$
$$= (0.25)^2(0.75)^3$$

Step 2: Observe that the probability of *any one* arrangement of two S's and three F's has this same probability. That's true because we multiply together 0.25 twice and 0.75 three times whenever we have two S's and three F's. The probability that $X = 2$ is the probability of getting two S's and three F's in any arrangement whatsoever. Here are all the possible arrangements:

SSFFF   SFSFF   SFFSF   SFFFS   FSSFF

FSFSF   FSFFS   FFSSF   FFSFS   FFFSS

There are 10 of them, all with the same probability. The overall probability of two successes is, therefore,

$$P(X = 2) = 10(0.25)^2(0.75)^3 = 0.2637$$

The pattern of this calculation works for any binomial probability. To use it, we need to be able to count the number of arrangements of k successes in n observations without actually listing them. We use the following fact to do the counting.

BINOMIAL COEFFICIENT

The number of ways of arranging k successes among n observations is given by the **binomial coefficient**

$$\binom{n}{k} = \frac{n!}{k!(n-k)!}$$

for $k = 0, 1, 2, \ldots, n$.

factorial   The formula for binomial coefficients uses the **factorial** notation. The factorial $n!$ for any positive whole number n is

$$n! = n \times (n-1) \times (n-2) \times \cdots \times 3 \times 2 \times 1$$

Also, $0! = 1$. Notice that the larger of the two factorials in the denominator of a binomial coefficient will cancel much of the $n!$ in the numerator. For example, the binomial coefficient we need for Example 5.28 is

$$\binom{5}{2} = \frac{5!}{2!3!}$$
$$= \frac{(5)(4)(3)(2)(1)}{(2)(1) \times (3)(2)(1)}$$
$$= \frac{(5)(4)}{(2)(1)} = \frac{20}{2} = 10$$

This agrees with our previous calculation.

The notation $\binom{n}{k}$ *is not related to the fraction* $\frac{n}{k}$. A helpful way to remember its meaning is to read it as "binomial coefficient n choose k." Binomial coefficients have many uses in mathematics, but we are interested in them only as an aid to finding binomial probabilities. The binomial coefficient $\binom{n}{k}$ counts the number of ways in which k successes can be distributed among n observations. The binomial probability $P(X = k)$ is this count multiplied by the probability of any specific arrangement of the k successes. Here is the formula we seek.

BINOMIAL PROBABILITY

If X has the binomial distribution $B(n, p)$ with n observations and probability p of success on each observation, the possible values of X are 0, 1, 2, \ldots, n. If k is any one of these values, the **binomial probability** is

$$P(X = k) = \binom{n}{k}p^k(1-p)^{n-k}$$

Here is an example of the use of the binomial probability formula.

EXAMPLE 5.29

Using the binomial probability formula. The number X of misclassified sales records in the auditor's sample in Example 5.21 (page 316) has the $B(15, 0.08)$ distribution. The probability of finding no more than one misclassified record is

$$P(X \leq 1) = P(X = 0) + P(X = 1)$$

$$= \binom{15}{0}(0.08)^0(0.92)^{15} + \binom{15}{1}(0.08)^1(0.92)^{14}$$

$$= \frac{15!}{0!15!}(1)(0.2863) + \frac{15!}{1!14!}(0.08)(0.3112)$$

$$= (1)(1)(0.2863) + (15)(0.08)(0.3112)$$

$$= 0.2863 + 0.3734 = 0.6597$$

The calculation used the facts that $0! = 1$ and that $a^0 = 1$ for any number $a \neq 0$. The result agrees with that obtained from Table C in Example 5.21.

USE YOUR KNOWLEDGE

5.54 An unfair coin. A coin is slightly bent, and as a result, the probability of a head is 0.53. Suppose that you toss the coin five times.

(a) Use the binomial formula to find the probability of three or more heads.

(b) Compare your answer with the one that you would obtain if the coin were fair.

The Poisson distributions

A count X has a binomial distribution when it is produced under the binomial setting. If one or more facets of this setting do not hold, the count X will have a different distribution. In this subsection, we discuss one of these distributions.

Frequently, we meet counts that are open-ended; that is, they are not based on a fixed number of n observations: the number of customers at a popular café between 12:00 P.M. and 1:00 P.M.; the number of dings on your car door; the number of reported pedestrian/bicyclist collisions on campus during the academic year. These are all counts that could be 0, 1, 2, 3, and so on indefinitely.

The Poisson distribution is another model for a count and can often be used in these open-ended situations. The count represents the number of events (call them "successes") that occur in some fixed unit of measure such as a period of time or region of space. The Poisson distribution is appropriate under the following conditions.

THE POISSON SETTING

1. The number of successes that occur in two nonoverlapping units of measure are **independent**.

2. The probability that a success will occur in a unit of measure is the same for all units of equal size and is proportional to the size of the unit.

3. The probability that more than one event occurs in a unit of measure is negligible for very small-sized units. In other words, the events occur one at a time.

For binomial distributions, the important quantities were n, the fixed number of observations, and p, the probability of success on any given observation. For Poisson distributions, the only important quantity is the mean number of successes μ occurring per unit of measure.

POISSON DISTRIBUTION

The distribution of the count X of successes in the Poisson setting is the **Poisson distribution** with **mean** μ. The parameter μ is the mean number of successes per unit of measure. The possible values of X are the whole numbers 0, 1, 2, 3, ... If k is any whole number, then*

$$P(X = k) = \frac{e^{-\mu}\mu^k}{k!}$$

The **standard deviation** of the distribution is $\sqrt{\mu}$.

EXAMPLE 5.30

Number of dropped calls. Suppose that the number of dropped calls on your cell phone varies, with an average of 2.1 calls per day. If we assume that the Poisson setting is reasonable for this situation, we can model the daily count of dropped calls X using the Poisson distribution with $\mu = 2.1$. What is the probability of having no more than two dropped calls tomorrow?

We can calculate $P(X \leq 2)$ either using software or the Poisson probability formula. Using the probability formula:

$$P(X \leq 2) = P(X = 0) + P(X = 1) + P(X = 2)$$
$$= \frac{e^{-2.1}(2.1)^0}{0!} + \frac{e^{-2.1}(2.1)^1}{1!} + \frac{e^{-2.1}(2.1)^2}{2!}$$
$$= 0.1225 + 0.2572 + 0.2700$$
$$= 0.6497$$

Using the R software, the probability is

dpois(0,2.1) + dpois(1,2.1) + dpois(2,2.1)

[1] 0.6496314

These two answers differ slightly due to roundoff error in the hand calculation. There is roughly a 65% chance that you will have no more than two dropped calls tomorrow.

Similar to the binomial, Poisson probability calculations are rarely done by hand if the event includes numerous possible values for X. Most software provides functions to calculate $P(X = k)$ and the cumulative probabilities of the form $P(X \leq k)$. These cumulative probability calculations make solving many problems less tedious. Here's an example.

---

*The e in the Poisson probability formula is a mathematical constant equal to 2.71828 to six decimal places. Many calculators have an e^x function.

EXAMPLE 5.31

Counting software remote users. Your university supplies online remote access to various software programs used in courses. Suppose that the number of students remotely accessing these programs in any given hour can be modeled by a Poisson distribution with $\mu = 17.2$. What is the probability that more than 25 students will remotely access these programs in the next hour?

Calculating this probability requires two steps.

1. Write $P(X > 25)$ as an expression involving a cumulative probability:

$$P(X > 25) = 1 - P(X \leq 25)$$

2. Obtain $P(X \leq 25)$ and subtract the value from 1. Again using R,

1-ppois(25,17.2)

[1] 0.02847261

The probability that more than 25 students will use this remote access in the next hour is only 0.028. Relying on software to get the cumulative probability is much quicker and less prone to error than the method of Example 5.30. For this case, that method would involve determining 26 probabilities and then summing their values.

Under the Poisson setting, this probability of 0.028 applies not only to the next hour, but also to any other hour in the future. The probability does not change because the units of measure are the same size and nonoverlapping.

USE YOUR KNOWLEDGE

5.55 Number of aphids. The milkweed aphid is a common pest to many ornamental plants. Suppose that the number of aphids on a shoot of a Mexican butterfly weed follows a Poisson distribution with $\mu = 4.4$ aphids.

(a) What is the probability of observing exactly five aphids on a shoot?

(b) What is the probability of observing five or fewer aphids on a shoot?

5.56 Number of aphids, continued. Refer to the previous exercise.

(a) What proportion of shoots would you expect to have no aphids present?

(b) If you do not observe any aphids on a shoot, is the probability that a nearby shoot has no aphids smaller than, equal to, or larger than your answer in part (a)? Explain your reasoning.

If we add counts from successive nonoverlapping areas of equal size, we are just counting the successes in a larger area. That count still meets the conditions of the Poisson setting. However, because our unit of measure has doubled, the mean of this new count is twice as large. Put more formally, if X is a Poisson random variable with mean μ_X and Y is a Poisson random variable with mean μ_Y and Y is independent of X, then $X + Y$ is a Poisson random variable with mean $\mu_X + \mu_Y$. This fact means that we can combine areas or look at a portion of an area and still use Poisson distributions to model the count.

EXAMPLE 5.32

Number of potholes. The Automobile Association (AA) in Britain had member volunteers make a 60-minute, two-mile walk around their neighborhoods and survey the condition of their roads and sidewalks. One outcome was the number of potholes, defined as being at least 2 inches deep and at least 6 inches in diameter, in their roads.[16] It was reported that Scotland averages 8.9 potholes per mile of road and London averages 4.9 potholes per mile of road. Suppose that the number of potholes per mile in each of these two regions follow the Poisson distribution. Then

- The number of potholes per 20 miles of road in Scotland is a Poisson random variable with mean $20 \times 8.9 = 178$.

- The number of potholes per half mile of road in London is a Poisson random variable with mean $0.5 \times 4.9 = 2.45$.

- The number of potholes per 500 miles of road in Scotland is a Poisson random variable with mean $500 \times 8.9 = 4450$.

- If we examined 2 miles of road in Scotland and 5 miles of road in London, the total number of potholes would be a Poisson random variable with mean $2 \times 8.9 + 5 \times 4.9 = 42.3$.

When the mean of the Poisson distribution is large, it may be difficult to calculate Poisson probabilities using a calculator or software. Fortunately, when μ is large, Poisson probabilities can be approximated using the Normal distribution with mean μ and standard deviation $\sqrt{\mu}$. Here is an example.

EXAMPLE 5.33

Number of snaps received. In Example 5.11, it was reported that Snapchat has more than 100 million daily users who send over 400 million snaps a day. Suppose that the number of snaps you receive per day follows a Poisson distribution with mean 12. What is the probability that, over a week, you would receive more than 100 snaps?

To answer this using software, we first compute the mean number of snaps sent per week. Because there are seven days in a week, the mean is $7 \times 12 = 84$. Plugging this into R tells us that there is slightly less than an 4% chance of receiving this many snaps:

$$1\text{-ppois}(100,84)$$

$$[1]\ 0.03891883$$

For the Normal approximation we compute

$$P(X > 100) = P\left(\frac{X - 84}{\sqrt{84}} > \frac{100 - 84}{\sqrt{84}}\right)$$

$$= P(Z > 1.75)$$

$$= 1 - P(Z < 1.75)$$

$$= 1 - 0.9599 = 0.0401$$

The approximation is quite accurate, differing from the actual probability by only 0.0012.

While the Normal approximation is adequate for many practical purposes, we recommend using statistical software when possible so you can get exact Poisson probabilities.

There is one other approximation associated with the Poisson distribution that is worth mentioning. It is related to the binomial distribution. Previously, we recommended using the Normal distribution to approximate the binomial distribution when n and p satisfy $np \geq 10$ and $n(1 - p) \geq 10$. In cases where n is large but p is so small that $np < 10$, the Poisson distribution with $\mu = np$ yields more accurate results. For example, suppose that you wanted to calculate $P(X \leq 2)$ when X has the $B(1000, .001)$ distribution. Using R, the actual binomial probability and the Poisson approximation are

pbinom(2,1000,.001)   ppois(2,1)

[1] 0.9197907   [1] 0.9196986

The Poisson approximation gives a very accurate probability calculation for the binomial distribution in this case.

SECTION 5.3 SUMMARY

• A **count** X of successes has the **binomial distribution** $B(n, p)$ in the **binomial setting**: there are n trials, all independent, each resulting in a success or a failure, and each having the same probability p of a success.

• The binomial distribution $B(n, p)$ is a good approximation to the **sampling distribution of the count of successes** in an SRS of size n from a large population containing proportion p of successes. We will use this approximation when the population is at least 20 times larger than the sample.

• The **sample proportion** of successes $\hat{p} = X/n$ is an estimator of the population proportion p. It does not have a binomial distribution, but we can do probability calculations about $\hat{p}$ by restating them in terms of X.

• **Binomial probabilities** are most easily found by software. There is an exact formula that is practical for calculations when n is small. Table C contains binomial probabilities for some values of n and p. For large n, you can use the Normal approximation.

• The mean and standard deviation of a **binomial count** X and a **sample proportion** $\hat{p} = X/n$ are

$$\mu_X = np \qquad\qquad \mu_{\hat{p}} = p$$

$$\sigma_X = \sqrt{np(1 - p)} \qquad \sigma_{\hat{p}} = \sqrt{\frac{p(1 - p)}{n}}$$

The sample proportion $\hat{p}$ is, therefore, an unbiased estimator of the population proportion p.

• The **Normal approximation** to the binomial distribution says that if X is a count having the $B(n,p)$ distribution, then when n is large,

$$X \text{ is approximately } N\left(np, \sqrt{np(1 - p)}\right)$$

$$\hat{p} \text{ is approximately } N\left(p, \sqrt{\frac{p(1 - p)}{n}}\right)$$

We will use this approximation when $np \geq 10$ and $n(1 - p) \geq 10$. It allows us to approximate probability calculations about X and $\hat{p}$ using the Normal distribution.

• The **continuity correction** improves the accuracy of the Normal approximations.

• The exact **binomial probability formula** is

$$P(X = k) = \binom{n}{k} p^k (1 - p)^{n-k}$$

where the possible values of X are $k = 0, 1, \ldots, n$. The binomial probability formula uses the **binomial coefficient**

$$\binom{n}{k} = \frac{n!}{k!(n - k)!}$$

• Here the **factorial** $n!$ is

$$n! = n \times (n - 1) \times (n - 2) \times \cdots \times 3 \times 2 \times 1$$

for positive whole numbers n and $0! = 1$. The binomial coefficient counts the number of ways of distributing k successes among n trials.

• A count X of successes has a **Poisson distribution** in the **Poisson setting**: the number of successes that occur in two nonoverlapping units of measure are independent; the probability that a success will occur in a unit of measure is the same for all units of equal size and is proportional to the size of the unit; the probability that more than one event occurs in a unit of measure is negligible for very small-sized units. In other words, the events occur one at a time.

• If X has the Poisson distribution with mean μ, then the standard deviation of X is $\sqrt{\mu}$, and the possible values of X are the whole numbers 0, 1, 2, 3, and so on.

• The **Poisson probability** that X takes any of these values is

$$P(X = k) = \frac{e^{-\mu}\mu^k}{k!} \quad k = 0, 1, 2, 3, \ldots$$

Sums of independent Poisson random variables also have the Poisson distribution. For example, in a Poisson model with mean μ per unit of measure, the count of successes in a units is a Poisson random variable with mean $a\mu$.

SECTION 5.3 EXERCISES

For Exercises 5.43, 5.44, and 5.45, see page 312; for Exercises 5.46 and 5.47, see page 313; for Exercises 5.48 and 5.49, see page 317; for Exercises 5.50 and 5.51, see page 319; for Exercise 5.52, see page 320; for Exercise 5.53, see page 325; for Exercise 5.54, see page 328; and for Exercises 5.55 and 5.56, see page 330.

Most binomial probability calculations required in these exercises can be done by using Table C or the Normal approximation. Your instructor may request that you use the binomial probability formula or software. In exercises requiring the Normal

approximation, you should use the continuity correction if you studied that topic.

5.57 What is wrong? Explain what is wrong in each of the following scenarios.

(a) If you toss a fair coin four times and a head appears each time, then the next toss is more likely to be a tail than a head.

(b) If you toss a fair coin four times and observe the pattern HTHT, then the next toss is more likely to be a head than a tail.

(c) The quantity $\hat{p}$ is one of the parameters for a binomial distribution.

(d) The binomial distribution can be used to model the daily number of pedestrian/cyclist near-crash events on campus.

5.58 What is wrong? Explain what is wrong in each of the following scenarios.

(a) In the binomial setting, X is a proportion.

(b) The variance for a binomial count is $\sqrt{p(1-p)/n}$.

(c) The Normal approximation to the binomial distribution is always accurate when n is greater than 1000.

(d) We can use the binomial distribution to approximate the sampling distribution of $\hat{p}$ when we draw an SRS of size $n = 50$ students from a population of 500 students.

5.59 Should you use the binomial distribution? In each of the following situations, is it reasonable to use a binomial distribution for the random variable X? Give reasons for your answer in each case. If a binomial distribution applies, give the values of n and p.

(a) A poll of 200 college students asks whether or not you usually feel irritable in the morning. X is the number who reply that they do usually feel irritable in the morning.

(b) You toss a fair coin until a head appears. X is the count of the number of tosses that you make.

(c) Most calls made at random by sample surveys don't succeed in talking with a person. Of calls to New York City, only one-twelfth succeed. A survey calls 500 randomly selected numbers in New York City. X is the number of times that a person is reached.

(d) You deal 10 cards from a shuffled deck of standard playing cards and count the number X of black cards.

5.60 Should you use the binomial distribution? In each of the following situations, is it reasonable to use a binomial distribution for the random variable X? Give reasons for your answer in each case.

(a) In a random sample of students in a fitness study, X is the mean daily exercise time of the sample.

(b) A manufacturer of running shoes picks a random sample of 20 shoes from the production of shoes each day for a detailed inspection. X is the number of pairs of shoes with a defect.

(c) A nutrition study chooses an SRS of college students. They are asked whether or not they usually eat at least five servings of fruits or vegetables per day. X is the number who say that they do.

(d) X is the number of days during the school year when you skip a class.

5.61 Stealing from a store. A survey of more than 20,000 U.S. high school students revealed that 20% of the students say that they stole something from a store in the past year.[17] This is down 7% from the last survey, which was performed two years earlier. You decide to take a random sample of 10 high school students from your city and ask them this question.

(a) If the high school students in your city match this 20% rate, what is the distribution of the number of students who say that they stole something from a store in the past year? What is the distribution of the number of students who do not say that they stole something from a store in the past year?

(b) What is the probability that four or more of the 10 students in your sample say that they stole something from a store in the past year?

5.62 Illegal downloading. New regulations in Canada require all Internet service providers (ISPs) to send a notice to subscribers who are downloading files illegally asking them to stop. This "notice and notice" system was already in place with Rogers Cable. That company says that prior to these new regulations, 67% of its subscribers who received a notice did not reoffend.[18] Consider a random sample of 50 of these Rogers subscribers who received a first notice.

(a) What is the distribution of the number X of subscribers who reoffend? Explain your answer.

(b) What is the probability that at least 18 of the 50 subscribers in your sample reoffend?

5.63 Stealing from a store, continued. Refer to Exercise 5.61.

(a) What is the expected number of students in your sample who say that they stole something from a store in the past year? What is the expected number of students who do not say that they stole? You should see that these two means add to 10, the total number of students.

(b) What is the standard deviation σ of the number of students in your sample who say that they stole something?

(c) Suppose that you live in a city where only 10% of the high school students say that they stole something from a store in the past year. What is σ in this case? What is σ if $p = 0.01$? What happens to the standard deviation of a binomial distribution as the probability of a success gets close to 0?

5.64 Illegal downloading, continued. Refer to Exercise 5.62. Given the new regulations, suppose that 75% of the Canadian ISP subscribers will not reoffend after receiving a notice.

(a) If you choose at random 15 subscribers who received a notice, what is the mean of the count X who will not reoffend? What is the mean of the proportion $\hat{p}$ in your sample who will not reoffend?

(b) Repeat the calculations in part (a) for samples of size 150 and 1500. What happens to the mean count of successes as the sample size increases? What happens to the mean proportion of successes?

 5.65 More on illegal downloading. Consider the settings of Exercises 5.62 and 5.64.

(a) Using the 67% rate of Rogers subscribers prior to the new regulations, what is the smallest number m out of $n = 15$ Canadian ISP subscribers who receive a notice such that $P(X \geq m)$ is no larger than 0.05? You might consider m or more subscribers as evidence that the rate in your sample is larger than 67%.

(b) Now using the 75% rate of Canadian ISP subscribers after the new regulations and your answer to part (a), what is $P(X \geq m)$? This represents the chance of obtaining enough evidence given that the rate is 75%.

(c) If you were to increase the sample size from $n = 15$ to $n = 100$ and repeat parts (a) and (b), would you expect the probability in part (b) to increase or decrease? Explain your answer.

5.66 Attitudes toward drinking and studies of behavior. Some of the methods in this section are approximations rather than exact probability results. We have given rules of thumb for safe use of these approximations.

(a) You are interested in attitudes toward drinking among the 75 members of a fraternity. You choose 30 members at random to interview. One question is "Have you had five or more drinks at one time during the last week?" Suppose that, in fact, 30% of the 75 members would say Yes. Explain why you *cannot* safely use the $B(30, 0.3)$ distribution for the count X in your sample who say Yes.

(b) The National AIDS Behavioral Surveys found that 0.2% (that's 0.002 as a decimal fraction) of adult heterosexuals had both received a blood transfusion and had a sexual partner from a group at high risk of AIDS. Suppose that this national proportion holds for your region. Explain why you *cannot* safely use the Normal approximation for the sample proportion who fall in this group when you interview an SRS of 1000 adults.

5.67 Random digits. Each entry in a table of random digits like Table B has probability 0.1 of being any given digit, and digits are independent of each other.

(a) What is the probability that a group of six digits from the table will contain at least one digit greater than 5?

(b) What is the mean number of digits greater than 5 in lines 40 digits long?

5.68 Use the *Probability* applet. The *Probability* applet simulates tosses of a coin. You can choose the number of tosses n and the probability p of a head. You can therefore use the applet to simulate binomial random variables.

The count of misclassified sales records in Example 5.21 has the binomial distribution with $n = 15$ and $p = 0.08$.

Set these values for the number of tosses and probability of heads in the applet. Table C shows that the probability of getting a sample with exactly 0 misclassified records is 0.2863. This is the long-run proportion of samples with no bad records. Click "Toss" and "Reset" repeatedly to simulate 25 samples of 15 tosses. Record the number of bad records (the count of heads) in each of the 25 samples.

(a) What proportion of the 25 samples had exactly 0 bad records? Do you think this sample proportion is close to the probability?

(b) Remember that this probability of 0.2863 tells us only what happens in the long run. Here we're considering only 25 samples. If X is the number of samples out of 25 with exactly 0 misclassified records, what is the distribution of X?

(c) Explain how to use the distribution in part (b) to describe the sampling distribution of $\hat{p}$ in part (a).

5.69 Cyberbullying. An online survey, in partnership with Habbo, was conducted to study cyberbullying among 13- to 25-year-olds in the United Kingdom. It was reported that 62% of the young people had received nasty private messages on a smartphone social network app.[19] You randomly sample four young people from the United Kingdom and ask them if they've received nasty messages. Let X be the number who say Yes.

(a) What are n and p in the binomial distribution of X?

(b) Find the probability of each possible value of X, and draw a probability histogram for this distribution.

(c) Find the mean number of positive responders and mark the location of this value on your histogram.

5.70 The ideal number of children. "What do you think is the ideal number of children for a family to have?" A Gallup Poll asked this question of 1020 randomly chosen adults. Slightly less than half (48%) thought that a total of two children was ideal.[20] Suppose that $p = 0.48$ is exactly true for the population of all adults. Gallup announced a margin of error of ±4 percentage points for this poll. What is the probability that the sample proportion $\hat{p}$ for an SRS of size $n = 1020$ falls between 0.44 and 0.52? You see that it is likely, but not certain, that polls like this give results that are correct within their margin of error. We say more about margins of error in Chapter 6.

5.71 Cyberbullying, continued. Refer to Exercise 5.69. Assume instead that that you sample $n = 500$ young people from the United Kingdom.

(a) What is the probability that the sample proportion $\hat{p}$ of those who received nasty messages is between 0.59 and 0.65 if the population proportion is $p = 0.62$?

(b) What is the probability that the sample proportion $\hat{p}$ is between 0.87 and 0.93 if the population proportion is $p = 0.90$?

(c) Using the results from parts (a) and (b), how does the probability that $\hat{p}$ falls within ± 0.03 of the true p change as p gets closer to 1?

5.72 How do the results depend on the sample size? Return to the Gallup Poll setting of Exercise 5.70. We are supposing that the proportion of all adults who think that having two children is ideal is $p = 0.48$. What is the probability that a sample proportion $\hat{p}$ falls between 0.44 and 0.52 (that is, within ± 4 percentage points of the true p) if the sample is an SRS of size $n = 300$? Of size $n = 5000$? Combine these results with your work in Exercise 5.70 to make a general statement about the effect of larger samples in a sample survey.

5.73 Shooting free throws. Since the mid-1960s, the overall free-throw percent at all college levels, for both men and women, has remained pretty consistent. For men, players have been successful on roughly 69% of these free throws, with the season percent never falling below 67% or above 70%.[21] Assume that 300,000 free throws will be attempted in the upcoming season.

(a) What are the mean and standard deviation of $\hat{p}$ if the population proportion is $p = 0.69$?

(b) Using the 68–95–99.7 rule, we expect $\hat{p}$ to fall between what two percents about 95% of the time?

(c) Given the width of the interval in part (b) and the range of season percents, do you think that it is reasonable to assume that the population proportion has been the same over the last 50 seasons? Explain your answer.

5.74 Online learning. The U.S. Department of Education released a report on online learning stating that blended instruction, a combination of conventional face-to-face and online instruction, appears more effective in terms of student performance than conventional teaching.[22] You decide to poll incoming students at your institution to see if they prefer courses that blend face-to-face instruction with online components. In an SRS of 400 incoming students, you find that 373 prefer this type of course.

(a) What is the sample proportion of incoming students at your school who prefer this type of blended instruction?

(b) Assume the population proportion for all students nationwide is 85%. Assuming this is true for your insitution too, what is the standard deviation of $\hat{p}$?

(c) Using the 68–95–99.7 rule, you would expect $\hat{p}$ to fall between what two percents about 95% of the time?

(d) Based on your result in part (a), do you think that the incoming students at your institution prefer this type of instruction more, less, or about the same as students nationally? Explain your answer.

5.75 Binge drinking. The Centers for Disease Control and Prevention finds that 28% of people

aged 18 to 24 years binge drank. Those who binge drank averaged 9.3 drinks per episode and 4.2 episodes per month. The study took a sample of over 18,000 people aged 18 to 24 years, so the population proportion of people who binge drank is very close to $p = 0.28$.[23] The administration of your college surveys an SRS of 200 students and finds that 56 binge drink.

(a) What is the sample proportion of students at your college who binge drink?

(b) If, in fact, the proportion of all students on your campus who binge drink is the same as the national 28%, what is the probability that the proportion in an SRS of 200 students is as large or larger than the result of the administration's sample?

(c) A writer for the student paper says that the percent of students who binge brink is higher on your campus than nationally. Write a short letter to the editor explaining why the survey does not support this conclusion.

5.76 How large a sample is needed? The changing probabilities you found in Exercises 5.70 and 5.72 are due to the fact that the standard deviation of the sample proportion $\hat{p}$ gets smaller as the sample size n increases. If the population proportion is $p = 0.48$, how large a sample is needed to reduce the standard deviation of $\hat{p}$ to $\sigma_{\hat{p}} = 0.005$? (The 68–95–99.7 rule then says that about 95% of all samples will have $\hat{p}$ within 0.01 of the true p.)

5.77 A test for ESP. In a test for ESP (extrasensory perception), the experimenter looks at cards that are hidden from the subject. Each card contains either a star, a circle, a wave, or a square. As the experimenter looks at each of 20 cards in turn, the subject names the shape on the card.

(a) If a subject simply guesses the shape on each card, what is the probability of a successful guess on a single card? Because the cards are independent, the count of successes in 20 cards has a binomial distribution.

(b) What is the probability that a subject correctly guesses at least 10 of the 20 shapes?

(c) In many repetitions of this experiment with a subject who is guessing, how many cards will the subject guess correctly on the average? What is the standard deviation of the number of correct guesses?

(d) A standard ESP deck actually contains 25 cards. There are five different shapes, each of which appears on five cards. The subject knows that the deck has this makeup. Is a binomial model still appropriate for the count of correct guesses in one pass through this deck? If so, what are n and p? If not, why not?

5.78 Admitting students to college. A selective college would like to have an entering class of 1000 students.

Because not all students who are offered admission accept, the college admits more than 1000 students. Past experience shows that about 83% of the students admitted will accept. The college decides to admit 1200 students. Assuming that students make their decisions independently, the number who accept has the $B(1200, 0.83)$ distribution. If this number is less than 1000, the college will admit students from its waiting list.

(a) What are the mean and the standard deviation of the number X of students who accept?

(b) Use the Normal approximation to find the probability that at least 800 students accept.

(c) The college does not want more than 1000 students. What is the probability that more than 1000 will accept?

(d) If the college decides to decrease the number of admission offers to 1150, what is the probability that more than 1000 will accept?

5.79 Is the ESP result better than guessing? When the ESP study of Exercise 5.77 discovers a subject whose performance appears to be better than guessing, the study continues at greater length. The experimenter looks at many cards bearing one of five shapes (star, square, circle, wave, and cross) in an order determined by random numbers. The subject cannot see the experimenter as the experimenter looks at each card in turn, in order to avoid any possible nonverbal clues. The answers of a subject who does not have ESP should be independent observations, each with probability 1/5 of success. We record 900 attempts.

(a) What are the mean and the standard deviation of the count of successes?

(b) What are the mean and the standard deviation of the proportion of successes among the 900 attempts?

(c) What is the probability that a subject without ESP will be successful in at least 24% of 900 attempts?

(d) The researcher considers evidence of ESP to be a proportion of successes so large that there is only probability 0.01 that a subject could do this well or better by guessing. What proportion of successes must a subject have to meet this standard? (Example 1.45, on pages 65–66, shows how to do an inverse calculation for the Normal distribution that is similar to the type required here.)

5.80 Show that these facts are true. Use the definition of binomial coefficients to show that each of the following facts is true. Then restate each fact in words in terms of the number of ways that k successes can be distributed among n observations.

(a) $\binom{n}{n} = 1$ for any whole number $n \geq 1$.

(b) $\binom{n}{n-1} = n$ for any whole number $n \geq 1$.

(c) $\binom{n}{k} = \binom{n}{n-k}$ for any n and k with $k \leq n$.

5.81 English Premier League Goals. The total number of goals scored per soccer match in the English Premier League (EPL) often follows the Poisson distribution. In one recent season, the average number of goals scored per match (over 380 games played) was 2.768. Compute the following probabilities.

(a) What is the probability that three or more goals are scored in a game?

(b) What is the probability that a game will end in a 0–0 tie?

(c) Explain why you cannot compute the probability that a game will end in a 1–1 tie but can provide an upper bound on this probability.

5.82 Number of colony-forming units. In microbiology, colony-forming units (CFUs) are used to measure the number of microorganisms present in a sample. To determine the number of CFUs, the sample is prepared, spread uniformly on an agar plate, and then incubated at some suitable temperature. Suppose that the number of CFUs that appear after incubation follows a Poisson distribution with $\mu = 15$.

(a) If the area of the agar plate is 75 square centimeters (cm^2), what is the probability of observing fewer than 4 CFUs in a 25 cm^2 area of the plate?

(b) If you were to count the total number of CFUs in five plates, what is the probability you would observe more than 90 CFUs? Use the Poisson distribution to obtain this probability.

(c) Repeat the probability calculation in part (b), but now use the Normal approximation. How close is your answer to your answer in part (b)?

5.83 Metal fatigue. Metal fatigue refers to the gradual weakening and eventual failure of metal that undergoes cyclic loads. The wings of an aircraft, for example, are subject to cyclic loads when in the air, and cracks can form. It is thought that these cracks start at large particles found in the metal. Suppose that the number of particles large enough to initiate a crack follows a Poisson distribution with mean $\mu = 0.5$ per square centimeter (cm^2).

(a) What is the mean of the Poisson distribution if we consider a 100 cm^2 area?

(b) Using the Normal approximation, what is the probability that this section has more than 60 of these large particles?

CHAPTER 5 EXERCISES

5.84 The cost of Internet access. In Canada, households spent an average of $80.63 CDN monthly for high-speed broadband access.[24] Assume that the standard deviation is $27.32. If you ask an SRS of 500 Canadian households with high-speed broadband access how much they pay, what is the probability that the average amount will exceed $85?

5.85 Dust in coal mines. A laboratory weighs filters from a coal mine to measure the amount of dust in the mine atmosphere. Repeated measurements of the weight of dust on the same filter vary Normally with standard deviation $\sigma = 0.09$ milligram (mg) because the weighing is not perfectly precise. The dust on a particular filter actually weighs 137 mg.

(a) The laboratory reports the mean of three weighings of this filter. What is the distribution of this mean?

(b) What is the probability that the laboratory reports a weight of 140 mg or higher for this filter?

5.86 The effect of sample size on the standard deviation. Assume that the standard deviation in a very large population is 100.

(a) Calculate the standard deviation for the sample mean for samples of size 1, 4, 25, 100, 250, 500, 1000, and 5000.

(b) Graph your results with the sample size on the x axis and the standard deviation on the y axis.

(c) Summarize the relationship between the sample size and the standard deviation that your graph shows.

5.87 Marks per round in cricket. Cricket is a dart game that uses the numbers 15 to 20 and the bull's-eye. Each time you hit one of these regions, you score either 0, 1, 2 or 3 marks. Thus, in a round of three throws, a person can score 0 to 9 marks. Lex plans to play 20 games. Her distribution of marks per round is discrete and strongly skewed. A majority of her rounds result in 0, 1, or 2 marks and only a few are more than 4 marks. Assume that her mean is 2.07 marks per round with a standard deviation of 2.11.

(a) Her 20 games involve 140 rounds of three throws each. What are the mean and standard deviation of the average number of marks $\bar{x}$ in 140 rounds?

(b) Using the central limit theorem, what is the probability that she averages fewer than 2 marks per round?

(c) Do you think that the central limit theorem can be used in this setting? Explain your answer.

5.88 Common last names. The U.S. Census Bureau says that the 10 most common names in the United States are (in order) Smith, Johnson, Williams, Brown, Jones, Miller,

Davis, Garcia, Rodriguez, and Wilson.[25] These names account for 4.9% of all U.S. residents. Out of curiosity, you look at the authors of the textbooks for your current courses. There are 12 authors in all. Would you be surprised if none of the names of these authors were among the 10 most common? Give a probability to support your answer and explain the reasoning behind your calculation.

5.89 Benford's law. It is a striking fact that the first digits of numbers in legitimate records often follow Benford's law (see Example 4.12, page 226). Here it is:

First digit	1	2	3	4	5	6	7	8	9
Proportion	0.301	0.176	0.125	0.097	0.079	0.067	0.058	0.051	0.046

Fake records usually have fewer first digits 1, 2, and 3. What is the approximate probability, if Benford's law holds, that among 1000 randomly chosen invoices there are 575 or fewer in amounts with first digit 1, 2, or 3?

5.90 Genetics of peas. According to genetic theory, the blossom color in the second generation of a certain cross of sweet peas should be red or white in a 3:1 ratio. That is, each plant has probability 3/4 of having red blossoms, and the blossom colors of separate plants are independent.

(a) What is the probability that exactly 8 out of 10 of these plants have red blossoms?

(b) What is the mean number of red-blossomed plants when 130 plants of this type are grown from seeds?

(c) What is the probability of obtaining at least 90 red-blossomed plants when 130 plants are grown from seeds?

5.91 Leaking gas tanks. Leakage from underground gasoline tanks at service stations can damage the environment. It is estimated that 25% of these tanks leak. You examine 15 tanks chosen at random, independently of each other.

(a) What is the mean number of leaking tanks in such samples of 15?

(b) What is the probability that 10 or more of the 15 tanks leak?

(c) Now you do a larger study, examining a random sample of 2000 tanks nationally. What is the probability that at least 540 of these tanks are leaking?

5.92 A roulette payoff. A $1 bet on a single number on a casino's roulette wheel pays $35 if the ball ends up in the number slot you choose. Here is the distribution of the payoff X:

Payoff X	$0	$35
Probability	0.974	0.026

Each spin of the roulette wheel is independent of other spins.

(a) What are the mean and standard deviation of X?

(b) Sam comes to the casino weekly and bets on 10 spins of the roulette wheel. What does the law of large numbers say about the average payoff Sam receives from his bets each visit?

(c) What does the central limit theorem say about the distribution of Sam's average payoff after betting on 520 spins in a year?

(d) Sam comes out ahead for the year if his average payoff is greater than $1 (the amount he bet on each spin). What is the probability that Sam ends the year ahead? The true probability is 0.396. Does using the central limit theorem provide a reasonable approximation?

5.93 A roulette payoff revisited. Refer to the previous exercise. In part (d), the central limit theorem was used to approximate the probability that Sam ends the year ahead. The estimate was about 0.10 too large. Let's see if we can get closer using the Normal approximation to the binomial with the continuity correction.

(a) If Sam plans to bet on 520 roulette spins, he needs to win at least $520 to break even. If each win gives him $35, what is the minimum number of wins m he must have?

(b) Given $p = 1/38 = 0.026$, what are the mean and standard deviation of X, the number of wins in 520 roulette spins?

(c) Use the information in the previous two parts to compute $P(X \geq m)$ with the continuity correction. Does your answer get closer to the exact probability 0.396?

5.94 Learning a foreign language. Does delaying oral practice hinder learning a foreign language? Researchers randomly assigned 25 beginning students of Russian to begin speaking practice immediately and another 25 to delay speaking for four weeks. At the end of the semester both groups took a standard test of comprehension of spoken Russian. Suppose that in the population of all beginning students, the test scores for early speaking vary according to the $N(32, 6)$ distribution and scores for delayed speaking have the $N(29, 5)$ distribution.

(a) What is the sampling distribution of the mean score $\bar{x}$ in the early-speaking group in many repetitions of the experiment? What is the sampling distribution of the mean score $\bar{y}$ in the delayed-speaking group?

(b) If the experiment were repeated many times, what would be the sampling distribution of the difference $\bar{y} - \bar{x}$ between the mean scores in the two groups?

(c) What is the probability that the experiment will find (misleadingly) that the mean score for delayed speaking is at least as large as that for early speaking?

5.95 Summer employment of college students. Suppose (as is roughly true) that 88% of college men and 82% of college women were employed last summer. A sample survey interviews SRSs of 400 college men and 400 college women. The two samples are of course independent.

(a) What is the approximate distribution of the proportion $\hat{p}_F$ of women who worked last summer? What is the approximate distribution of the proportion $\hat{p}_M$ of men who worked?

(b) The survey wants to compare men and women. What is the approximate distribution of the difference in the proportions who worked, $\hat{p}_M - \hat{p}_F$? Explain the reasoning behind your answer.

(c) What is the probability that in the sample a higher proportion of women than men worked last summer?

5.96 Income of working couples. A study of working couples measures the income X of the husband and the income Y of the wife in a large number of couples in which both partners are employed. Suppose that you knew the means μ_X and μ_Y and the variances σ_X^2 and σ_Y^2 of both variables in the population.

(a) Is it reasonable to take the mean of the total income $X + Y$ to be $\mu_X + \mu_Y$? Explain your answer.

(b) Is it reasonable to take the variance of the total income to be $\sigma_X^2 + \sigma_Y^2$? Explain your answer.

5.97 A random walk. A particle moves along the line in a random walk. That is, the particle starts at the origin (position 0) and moves either right or left in independent steps of length 1. If the particle moves to the right with probability 0.6, its movement at the ith step is a random variable X_i with distribution

$$P(X_i = 1) = 0.6$$
$$P(X_i = -1) = 0.4$$

The position of the particle after k steps is the sum of these random movements,

$$Y = X_1 + X_2 + \cdots + X_k$$

Use the central limit theorem to find the approximate probability that the position of the particle after 500 steps is at least 200 to the right.

5.98 A lottery payoff. A $1 bet in a state lottery's Pick 3 game pays $500 if the three-digit number you choose exactly matches the winning number, which is drawn at random. Here is the distribution of the payoff X:

Payoff X	$0	$500
Probability	0.999	0.001

Each day's drawing is independent of other drawings.

(a) Joe buys a Pick 3 ticket twice a week. The number of times he wins follows a $B(104, 0.001)$ distribution. Using the Poisson approximation to the binomial, what is the probability that he wins at least once?

(b) The exact binomial probability is 0.0988. How accurate is the Poisson approximation here?

(c) If Joe pays $5 a ticket, he needs to win at least twice a year to come out ahead. Using the Poisson approximation, what is the probability that Joe comes out ahead?

5.99 Poisson distribution? Suppose you find in your spam folder an average of two spam emails every 10 minutes. Furthermore, you find that the rate of spam mail from midnight to 6 A.M. is twice the rate during other parts of the day. Explain whether or not the Poisson distribution is an appropriate model for the spam process.

5.100 Tossing a die. You are tossing a balanced die that has probability 1/6 of coming up 1 on each toss. Tosses are independent. We are interested in how long we must wait to get the first 1.

(a) The probability of a 1 on the first toss is 1/6. What is the probability that the first toss is not a 1 and the second toss is a 1?

(b) What is the probability that the first two tosses are not 1s and the third toss is a 1? This is the probability that the first 1 occurs on the third toss.

(c) Now you see the pattern. What is the probability that the first 1 occurs on the fourth toss? On the fifth toss?

5.101 The geometric distribution. Generalize your work in Exercise 5.100. You have independent trials, each resulting in a success or a failure. The probability of a success is p on each trial. The binomial distribution describes the count of successes in a fixed number of trials. Now the number of trials is not fixed; instead, continue until you get a success. The random variable Y is the number of the trial on which the first success occurs. What are the possible values of Y? What is the probability $P(Y = k)$ for any of these values? (*Comment:* The distribution of the number of trials to the first success is called a **geometric distribution**.)

5.102 Wi-fi interruptions. Suppose that the number of wi-fi interruptions on your home network follows the Poisson distribution with an average of 0.9 wi-fi interruptions per day.

(a) Show that the probability of no interruptions on a given day is 0.4066.

(b) Treating each day as a trial in a binomial setting, use the binomial formula to compute the probability of no interruptions in a week.

(c) Now, instead of using the binomial model, let's use the Poisson distribution exclusively. What is the mean number of wi-fi interruptions during a week?

(d) Based on the Poisson mean of part (c), use the Poisson distribution to compute the probability of no interruptions in a week. Confirm that this probability is the same as found part (b). Explain in words why the two ways of computing no interruptions in a week give the same result.

(e) Explain why using the binomial distribution to compute the probability that only one day in the week will not be interruption free would not give the same probability had we used the Poisson distribution to compute that only one interruption occurs during the week.

Danita Delimont/Getty Images

Introduction to Inference

Introduction

Statistical inference draws conclusions about a population or process from sample data. It also provides a statement of how much confidence we can place in our conclusions. Although there are numerous methods for inference, there are only a few general types of statistical inference. This chapter introduces the two most common types: *confidence intervals* and *tests of significance*.

Because the underlying reasoning for these two types of inference remains the same across different settings, this chapter considers just one simple setting that is closely related to our study of the sampling distributions of $\bar{x}$ in Section 5.2 (page 293): inference about the mean of a large population whose standard deviation is known. This setting, although unrealistic, allows us to focus on the underlying rationale of statistical inference rather than the calculations.

Later chapters present inference methods to use in most of the settings we met in learning to explore data. In fact, there are libraries—both of books and of computer software—full of more elaborate statistical techniques. Informed use of any of these methods, however, requires a firm understanding of the underlying reasoning. That is the goal of this chapter. A computer or calculator will do the arithmetic, but *you must exercise sound judgment based on understanding*.

6.1 Estimating with Confidence

6.2 Tests of Significance

6.3 Use and Abuse of Tests

6.4 Power and Inference as a Decision

Overview of Inference

The purpose of statistical inference is to draw conclusions from data. Formal inference emphasizes substantiating our conclusions via probability calculations. Probability allows us to take chance variation into account. Here is an example.

EXAMPLE 6.1

WADE

Clustering of trees in a forest. The Wade Tract in Thomas County, Georgia, is an old-growth forest of longleaf pine trees (*Pinus palustris*) that has survived in a relatively undisturbed state since before the settlement of the area by Europeans. Foresters who study these trees are interested in how the trees are distributed in the forest. Is there some sort of clustering, resulting in regions of the forest with more trees than others? Or are the tree locations random, resulting in no particular patterns? Figure 6.1 gives a plot of the locations of all 584 longleaf pine trees in a 200-meter by 200-meter region in the Wade Tract.[1]

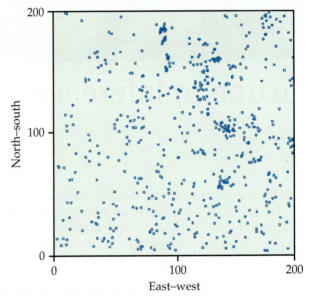

FIGURE 6.1 The distribution of longleaf pine trees, Example 6.1.

Do the locations appear to be random, or do there appear to be clusters of trees? One approach to the analysis of these data indicates that a pattern as clustered as, or more clustered than, the one in Figure 6.1 would occur only 4% of the time if, in fact, the locations of longleaf pine trees in the Wade Tract are random. Because this chance is fairly small, we conclude that there is some clustering of these trees.

This probability calculation helps us to distinguish between patterns that are consistent or inconsistent with the random location scenario. Here is an example assessing a new oral antibiotic for acne—with a different conclusion.

EXAMPLE 6.2

Effectiveness of a new oral antibiotic. Researchers want to know if a new oral antibiotic is more effective in relieving acne than a popular topical (on the skin) antibiotic. Twenty patients are randomly assigned to receive the oral medication, and another 20 receive the topical medication. Fifteen (75%) of those taking the oral medication find satisfactory symptom relief versus only 11 (55%) of the topical medication patients.

Our unaided judgment suggests that the oral medication is better, 75% to 55%. However, probability calculations tell us that a difference this large or larger between the results in the two groups of 20 patients would occur about one time in five simply because of chance variation. In this case, it is better to conclude that the data fail to establish a real difference between the two treatments. This probability (nearly 0.19) is too large to ignore.

In this chapter, we introduce the two most frequently used types of statistical inference. Section 6.1 concerns *confidence intervals* for estimating the value of a population parameter. Section 6.2 presents *tests of significance*, which assess the evidence for a claim, such as those in Examples 6.1 and 6.2.

LOOK BACK

sampling distribution, p. 286

Both types of inference are based on the sampling distributions of statistics. That is, both report probabilities that state *what would happen if we used the inference method many times.* This kind of probability statement is characteristic of standard statistical inference. Users of statistics must understand the nature of this reasoning and the meaning of the probability statements that appear, for example, online and in journal articles and statistical software output.

Because the methods of formal inference are based on sampling distributions, they require a probability model for the data. Trustworthy probability models can arise in many ways, but the model is most secure and inference is most reliable when the data are produced by a properly randomized design.

When you use statistical inference, you are acting as if the data come from a random sample or a randomized experiment. If this is not true, your conclusions may be open to challenge. Do not be overly impressed by the complex details of formal inference. This elaborate machinery cannot remedy basic flaws in producing the data such as voluntary response samples and confounded experiments. Use the common sense developed in your study of the first three chapters of this book, and proceed to detailed formal inference only when you are satisfied that the data deserve such analysis.

6.1 Estimating with Confidence

When you complete this section, you will be able to:

- Describe a level *C* confidence interval for a population parameter in terms of an estimate and its margin of error.

- Construct a level *C* confidence interval for μ from a simple random sample (SRS) of size *n* from a large population having known standard deviation σ.

- Explain how the margin of error changes with a change in the confidence level *C*.

- Determine the sample size needed to obtain a specified margin of error for a level *C* confidence interval for μ.

- Identify situations where inference about μ based on the confidence interval $\bar{x} \pm z^*\sigma/\sqrt{n}$ may be suspect.

LOOK BACK

linear
transformations,
p. 44

The SAT is a widely used measure of readiness for college study. It consists of two sections, one for mathematical reasoning ability (SATM), one for reading and writing ability (SATV). Possible scores on each section range from 200 to 800, for a total range of 400 to 1600. Since 1995, section scores have been *recentered* so that the mean is approximately 500 with a standard deviation of 100 in a large "standardized group." This scale has been maintained so that scores have a constant interpretation.

EXAMPLE 6.3

Peter Cade/The Image Bank/Getty Images

Estimating the mean SATM score for seniors in California. Suppose that you want to estimate the mean SATM score for the 485,264 high school seniors in California.[2] You know better than to trust data from the students who choose to take the SAT. Only about 38% of California students typically take the SAT. These self-selected students are planning to attend college and are not representative of all California seniors. At considerable effort and expense, you give the test to an SRS of 500 California high school seniors. The mean score for your sample is $\bar{x} = 495$. What can you say about the mean score μ in the population of all 485,264 seniors?

LOOK BACK

unbiased
estimator,
p. 287
law of large
numbers,
p. 250

The sample mean $\bar{x}$ is the natural estimator of the unknown population mean μ. We know that $\bar{x}$ is an unbiased estimator of μ. More important, the law of large numbers says that the sample mean must approach the population mean as the size of the sample grows. The value $\bar{x} = 495$, therefore, appears to be a reasonable estimate of the mean score μ that all 485,264 students would achieve if they took the test.

But how reliable is this estimate? A second sample of 500 students would surely not give a sample mean of 495 again. Unbiasedness says only that there is no systematic tendency to underestimate or overestimate the truth. Could we plausibly get a sample mean of 485 and a sample mean of 520 in repeated samples? *An estimate without an indication of its variability is of little value.*

Statistical confidence

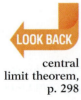

LOOK BACK

central
limit theorem,
p. 298

The unbiasedness of an estimator concerns the center of its sampling distribution, but questions about variation are answered by looking at its spread. The central limit theorem says that if the entire population of SATM scores has mean μ and standard deviation σ, then in repeated SRSs of size 500, the sample mean $\bar{x}$ is approximately $N(\mu, \sigma/\sqrt{500})$. Let us suppose that we know that the standard deviation σ of SATM scores in our California population is $\sigma = 100$. (We will see in the next chapter how to proceed when σ is not known. For now, we are more interested in statistical reasoning than in details of realistic methods.) This means that in repeated sampling the sample mean $\bar{x}$ has an approximately Normal distribution centered at the unknown population mean μ and a standard deviation of

$$\sigma_{\bar{x}} = \frac{100}{\sqrt{500}} = 4.5$$

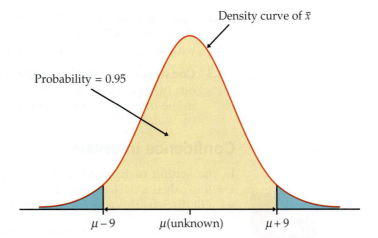

FIGURE 6.2 Distribution of the sample mean, Example 6.3. $\bar{x}$ lies within ±9 points of μ in 95% of all samples. This also means that μ is within ±9 points of $\bar{x}$ in those samples.

Now we are ready to proceed. Consider this line of thought, which is illustrated in Figure 6.2:

- The 68–95–99.7 rule says that the probability is about 0.95 that $\bar{x}$ will be within 9 points (that is, two standard deviations of $\bar{x}$) of the population mean score μ.

- To say that $\bar{x}$ lies within 9 points of μ is the same as saying that μ is within 9 points of $\bar{x}$.

- So about 95% of all samples will contain the true μ in the interval from $\bar{x} - 9$ to $\bar{x} + 9$.

We have simply restated a fact about the sampling distribution of $\bar{x}$. *The language of statistical inference uses this fact about what would happen in the long run to express our confidence in the results of any one sample.* Our sample gave $\bar{x} = 495$. We say that we are *95% confident* that the unknown mean score for all California seniors lies between

$$\bar{x} - 9 = 495 - 9 = 486$$

and

$$\bar{x} + 9 = 495 + 9 = 504$$

Be sure you understand the grounds for our confidence. There are only two possibilities for our SRS:

1. The interval between 486 and 504 contains the true μ.

2. The interval between 486 and 504 does not contain the true μ.

We cannot know whether our sample is one of the 95% for which the interval $\bar{x} \pm 9$ contains μ or one of the unlucky 5% for which it does not contain μ. The statement that we are 95% confident is shorthand for saying, "We arrived at these numbers by a method that gives correct results 95% of the time."

USE YOUR KNOWLEDGE

6.1 How much do you spend on lunch? The average amount you spend on a lunch during the week is not known. Based on past experience, you are willing to assume that the standard deviation is $2.10. If you take a random sample of 28 lunches, what is the value of the standard deviation of $\bar{x}$?

6.2 Applying the 68–95–99.7 rule. In the setting of the previous exercise, the 68–95–99.7 rule says that the probability is about 0.95 that $\bar{x}$ is within \$_____ of the population mean μ. Fill in the blank.

6.3 Constructing a 95% confidence interval. In the setting of the previous two exercises, about 95% of all samples will capture the true mean in the interval $\bar{x}$ plus or minus \$_____. Fill in the blank.

Confidence intervals

In the setting of Example 6.3, the interval of numbers between the values $\bar{x} \pm 9$ is called a *95% confidence interval* for μ. Like most confidence intervals we will discuss, this one has the form

LOOK BACK

margin
of error,
p. 287

$$\text{estimate} \pm \text{margin of error}$$

The estimate ($\bar{x} = 495$ in this case) is our guess for the value of the unknown parameter. The margin of error (9 here) reflects how accurate we believe our guess is, based on the variability of the estimate, and how confident we are that the procedure will produce an interval that will contain the true population mean μ.

Figure 6.3 illustrates the behavior of 95% confidence intervals in repeated sampling from a Normal distribution with mean μ. The center of each interval (marked by a dot) is at $\bar{x}$ and varies from sample to sample. The sampling distribution of $\bar{x}$ (also Normal) appears at the top of the figure to show the long-term pattern of this variation.

The 95% confidence intervals, $\bar{x} \pm$ margin of error, from 25 SRSs appear below the sampling distribution. The arrows on either side of the dot ($\bar{x}$) span

FIGURE 6.3 Twenty-five samples from the same population gave these 95% confidence intervals. In the long run, 95% of all samples give an interval that covers μ. The sampling distribution of $\bar{x}$ is shown at the top.

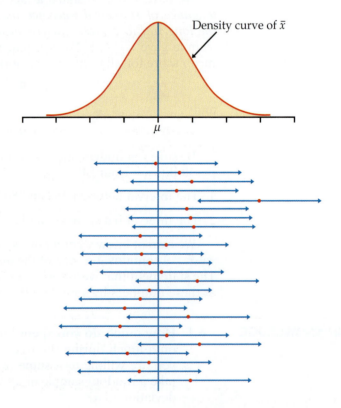

the confidence interval. All except one of the 25 intervals contain the true value of μ. In those intervals that contain μ, sometimes μ is near the middle of the interval and sometimes it is closer to one of the ends. This again reflects the variation of $\bar{x}$. In practice, we don't know the value of μ, but we have a method such that, in a very large number of samples, 95% of the confidence intervals will contain μ.

We can construct confidence intervals for many different parameters based on a variety of designs for data collection. We will learn the details of a number of these in later chapters. Two important things about a confidence interval are common to all settings:

1. It is an interval of the form (a, b), where a and b are numbers computed from the sample data.

confidence level 2. It has a property called a **confidence level** that gives the probability of producing an interval that contains the unknown parameter.

Users can choose the confidence level, but 95% is the standard for most situations. Occasionally, 90% or 99% is used. We use C to stand for the confidence level in decimal form. For example, a 95% confidence level corresponds to $C = 0.95$.

CONFIDENCE INTERVAL

A level C **confidence interval** for a parameter is an interval computed from sample data by a method that has probability C of producing an interval containing the true value of the parameter.

With the *Confidence Interval* applet, you can construct diagrams similar to the one displayed in Figure 6.3. The only difference is that the applet displays the Normal population distribution at the top rather than the Normal sampling distribution of $\bar{x}$. You choose the confidence level C, the sample size n, and whether you want to generate 1 or 25 samples at a time. A running total (and percent) of the number of intervals that contain μ is displayed so you can consider a larger number of samples.

When generating single samples, the data for the latest SRS are shown below the confidence interval. The spread in these data reflects the spread of the population distribution. This spread is assumed known, and it does not change with sample size. What does change, as you vary n, is the margin of error, since it reflects the uncertainty in the estimate of μ. As you increase n, you'll find that the span of the interval gets smaller.

USE YOUR KNOWLEDGE

6.4 Generating a single confidence interval. Using the default settings in the *Confidence Interval* applet (95% confidence level and $n = 20$), click "Sample" to choose an SRS and display its confidence interval.

(a) Is the spread in the data, shown as yellow dots below the confidence interval, larger than the span of the confidence interval? Explain why this would typically be the case.

(b) For the same data set, you can compare the span of the confidence interval for different values of C by sliding the confidence level to a new value. For the SRS you generated in part (a), what happens to the span of the interval when you move C to 99%? What about 90%? Describe the relationship you find between the confidence level C and the span of the confidence interval.

6.5 80% confidence intervals. The idea of an 80% confidence interval is that the interval captures the true parameter value in 80% of all samples. That's not high enough confidence for practical use, but 80% hits and 20% misses make it easy to see how a confidence interval behaves in repeated samples from the same population.

(a) Set the confidence level in the *Confidence Interval* applet to 80%. Click "Sample 25" to choose 25 SRSs and display their confidence intervals. How many of the 25 intervals contain the true mean μ? What proportion contain the true mean?

(b) We can't determine whether a new SRS will result in an interval that contains μ or not. The confidence level only tells us what percent will contain μ in the long run. Click "Sample 25" again to get the confidence intervals from 50 SRSs. What proportion hit? Keep clicking "Sample 25" and record the proportion of hits among 100, 200, 300, 400, and 500 SRSs. As the number of samples increases, we expect the percent of captures to get closer to the confidence level, 80%. Do you find this pattern in your results?

Confidence interval for a population mean

LOOK BACK

central limit theorem, p. 298

We now construct a level C confidence interval for the mean μ of a population when the data are an SRS of size n. The construction is based on the sampling distribution of the sample mean $\bar{x}$. This distribution is exactly $N(\mu, \sigma/\sqrt{n})$ when the population has the $N(\mu, \sigma)$ distribution. The central limit theorem says that this same sampling distribution is approximately correct for large samples whenever the population mean and standard deviation are μ and σ. For now, we will assume we are in one of these two situations. We discuss what we mean by "large sample" after we briefly study these intervals.

Our construction of a 95% confidence interval for the mean SATM score began by noting that any Normal distribution has probability about 0.95 within ± 2 standard deviations of its mean. To construct a level C confidence interval we first catch the central C area under a Normal curve. That is, we must find the number z^* such that any Normal distribution has probability C within $\pm z^*$ standard deviations of its mean.

Because all Normal distributions have the same standardized form, we can obtain everything we need from the standard Normal curve. Figure 6.4 shows how C and z^* are related. Values of z^* for many choices of C appear in the row labeled z^* at the bottom of Table D. Here are the most important entries from that row:

z^*	1.645	1.960	2.576
C	90%	95%	99%

Notice that for 95% confidence the value 2 obtained from the 68–95–99.7 rule is replaced with the more precise 1.96.

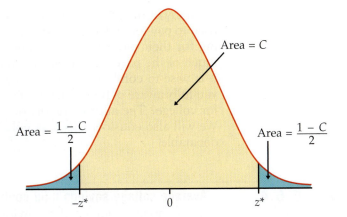

Area = C

Area = $\dfrac{1-C}{2}$

Area = $\dfrac{1-C}{2}$

$-z^*$ 0 z^*

FIGURE 6.4 To construct a level C confidence interval, we must find the number z^*. The area between $-z^*$ and z^* under the standard Normal curve is C.

As Figure 6.4 reminds us, any Normal curve has probability C between the point z^* standard deviations below the mean and the point z^* standard deviations above the mean. The sample mean $\bar{x}$ has the Normal distribution with mean μ and standard deviation $\sigma/\sqrt{n}$, so there is probability C that $\bar{x}$ lies between

$$\mu - z^*\frac{\sigma}{\sqrt{n}} \quad \text{and} \quad \mu + z^*\frac{\sigma}{\sqrt{n}}$$

This is exactly the same as saying that the unknown population mean μ lies between

$$\bar{x} - z^*\frac{\sigma}{\sqrt{n}} \quad \text{and} \quad \bar{x} + z^*\frac{\sigma}{\sqrt{n}}$$

That is, there is probability C that the interval $\bar{x} \pm z^*\sigma/\sqrt{n}$ contains μ. This is our confidence interval. The estimate of the unknown μ is $\bar{x}$, and the margin of error is $z^*\sigma/\sqrt{n}$.

CONFIDENCE INTERVAL FOR A POPULATION MEAN

Choose an SRS of size n from a population having unknown mean μ and known standard deviation σ. The **margin of error** for a level C confidence interval for μ is

$$m = z^*\frac{\sigma}{\sqrt{n}}$$

Here, z^* is the value on the standard Normal curve with area C between the critical points $-z^*$ and z^*. The level C **confidence interval** for μ is

$$\bar{x} \pm m$$

The confidence level of this interval is exactly C when the population distribution is Normal and is approximately C when n is large in other cases.

Starting in 2008, Sallie Mae, a major provider of education loans and savings programs, has conducted an annual study titled "How America Pays for College." In the 2015 survey, 1600 randomly selected individuals (800 parents of undergraduate students and 800 undergraduate students) were surveyed by telephone.[3]

Many of the survey questions focus on the composition of funding sources used to pay for college, so the undergraduates in the survey are often responding for their parents. For example, each participant is asked to report how much of the parent's current income is used to pay for college. Do you think it is wise to combine responses across the parents and undergraduates? Are you fully aware of how much money your parents are spending and borrowing for college? The authors report overall averages and percents in their report. We will also consider this a sample from one population but this is certainly debatable.

EXAMPLE 6.4

Average college savings fund contribution. One survey question asked how much money from a college savings fund, such as a 529 plan, is used to pay for college. Of the 1600 who were surveyed, $n = 1593$ provided an answer. *Nonresponse should always be considered as a source of bias.* In this case, the nonresponse is very low, so we'll proceed by treating the $n = 1593$ sample as if it were an unbiased sample.

The average amount is $1768. It's very likely that this distribution is highly skewed to the right with many small amounts and a few very large amounts. Nevertheless, because the sample size is quite large, we can rely on the central limit theorem to assure us that the confidence interval based on the Normal distribution will be a good approximation.

Let's compute an approximate 95% confidence interval for the true mean amount contributed from a college savings fund among all undergraduates. We'll assume that the standard deviation for the population of college savings fund contributions is $1483. For 95% confidence, we see from Table D that $z^* = 1.960$. The margin of error for the 95% confidence interval for μ is, therefore,

$$m = z^* \frac{\sigma}{\sqrt{n}}$$

$$= 1.960 \frac{1483}{\sqrt{1593}}$$

$$= 37.16$$

We have computed the margin of error with more digits than we really need. Our mean is rounded to the nearest $1, so we will do the same for the margin of error. Keeping additional digits would provide no additional useful information. Therefore, we will use $m = 37$. The approximate 95% confidence interval is

$$\bar{x} \pm m = 1768 \pm 37$$

$$= (1731, 1805)$$

We are 95% confident that the mean amount contributed from a college savings fund among all undergraduates is between $1731 and $1805.

Suppose that the researchers who designed this study had used a different sample size. How would this affect the confidence interval? We can answer this question by changing the sample size in our calculations and assuming that the sample mean is the same.

EXAMPLE 6.5

How sample size affects the confidence interval. As in Example 6.4, the sample mean of the college savings fund contribution is $1768 and the population standard deviation is $1483. Suppose that the sample size is only 177 but still large enough for us to rely on the central limit theorem. In this case, the margin of error for 95% confidence is

$$m = z^* \frac{\sigma}{\sqrt{n}}$$

$$= 1.960 \frac{1483}{\sqrt{177}}$$

$$= 111.47$$

and the approximate 95% confidence interval is

$$\bar{x} \pm m = 1768 \pm 111$$
$$= (1657, 1879)$$

Notice that the margin of error for this example is three times as large as the margin of error that we computed in Example 6.4. The only change that we made was to assume that the sample size is 177 rather than 1593. This sample size is one-ninth of the original 1593. Thus, we triple the margin of error when we reduce the sample size to one-ninth of the original value. Figure 6.5 illustrates the effect in terms of the intervals.

FIGURE 6.5 Confidence intervals for $n = 1593$ and $n = 177$, Examples 6.4 and 6.5. A sample size nine times as large results in a confidence interval that is one-third as wide.

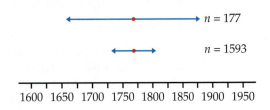

USE YOUR KNOWLEDGE

6.6 **Average amount paid for college.** Refer to Example 6.4. The average annual amount the $n = 1593$ families paid for college was $24,164.[4] If the population standard deviation is $8500, give the 95% confidence interval for μ, the average annual amount a family pays for a college undergraduate.

6.7 **Changing the sample size.** In the setting of the previous exercise, would the margin of error for 95% confidence be roughly doubled or halved if the sample size were raised to $n = 6375$? Verify your answer by performing the calculations.

6.8 **Changing the confidence level.** In the setting of Exercise 6.7, would the margin of error for 99% confidence be larger or smaller? Verify your answer by performing the calculations.

The argument leading to the form of confidence intervals for the population mean μ rested on the fact that the statistic $\bar{x}$ used to estimate μ has a Normal distribution. Because many sample estimates have Normal distributions (at least approximately), it is useful to notice that the confidence interval has the form

$$\text{estimate} \pm z^* \sigma_{\text{estimate}}$$

The estimate based on the sample is the center of the confidence interval. The margin of error is $z^* \sigma_{\text{estimate}}$. The desired confidence level determines z^* from Table D. The standard deviation of the estimate is found from knowledge of the sampling distribution in a particular case. When the estimate is $\bar{x}$ from an SRS, the standard deviation of the estimate is $\sigma_{\text{estimate}} = \sigma/\sqrt{n}$. We return to this general form numerous times in the following chapters.

How confidence intervals behave

The margin of error $z^*\sigma/\sqrt{n}$ for the mean of a Normal population illustrates several important properties that are shared by all confidence intervals in common use. The user chooses the confidence level, and the margin of error follows from this choice.

Both high confidence and a small margin of error are desirable characteristics of a confidence interval. High confidence says that our method almost always gives correct answers. A small margin of error says that we have pinned down the parameter quite precisely.

Suppose that in planning a study you calculate the margin of error and decide that it is too large. Here are your choices to reduce it:

- Use a lower level of confidence (smaller C).

- Choose a larger sample size (larger n).

- Reduce σ.

For most problems, you would choose a confidence level of 90%, 95%, or 99%, so z^* will be 1.645, 1.960, or 2.576, respectively. Figure 6.4 (page 349) shows that z^* will be smaller for lower confidence (smaller C). The bottom row of Table D also shows this. If n and σ are unchanged, a smaller z^* leads to a smaller margin of error.

EXAMPLE 6.6

How the confidence level affects the confidence interval. Suppose that for the college saving fund contribution data in Example 6.4 (page 350), we wanted 99% confidence. Table D tells us that for 99% confidence, $z^* = 2.576$. The margin of error for 99% confidence based on 1593 observations is

$$m = z^* \frac{\sigma}{\sqrt{n}}$$

$$= 2.576 \frac{1483}{\sqrt{1593}}$$

$$= 95.71$$

and the 99% confidence interval is

$$\bar{x} \pm m = 1768 \pm 96$$

$$= (1672, 1864)$$

Requiring 99%, rather than 95%, confidence has increased the margin of error from 37 to 96. Figure 6.6 compares the two intervals.

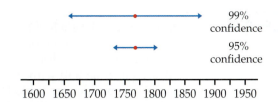

FIGURE 6.6 Confidence intervals, Examples 6.4 and 6.6. The larger the value of C, the wider the interval.

Similarly, choosing a larger sample size n reduces the margin of error for any fixed confidence level. The square root in the formula implies that we must multiply the number of observations by 4 in order to cut the margin of error in half. Likewise, if we want to reduce the standard deviation of $\bar{x}$ by a factor of 4, we must take a sample 16 times as large.

The standard deviation σ measures the variation in the population. You can think of the variation among individuals in the population as noise that obscures the average value μ. It is harder to pin down the mean μ of a highly variable population; that is why the margin of error of a confidence interval increases with σ.

In practice, we can sometimes reduce σ by carefully controlling the measurement process. We also might change the mean of interest by restricting our attention to only part of a large population. Focusing on a subpopulation will often result in a smaller σ. This is why many medical studies only use healthy male subjects. The tradeoff, however, is less generalizable results.

Choosing the sample size

A wise user of statistics never plans data collection without, at the same time, planning the inference. You can arrange to have both high confidence and a small margin of error. The margin of error of the confidence interval for a population mean is

$$m = z^* \frac{\sigma}{\sqrt{n}}$$

Notice once again that it is the size of the *sample* that determines the margin of error. The size of the *population* (as long as the population is much larger than the sample) does not influence the sample size we need.

To obtain a desired margin of error m, plug in the value of σ and the value of z^* for your desired confidence level, and solve for the sample size n. Here is the result.

SAMPLE SIZE FOR DESIRED MARGIN OF ERROR

The confidence interval for a population mean will have a specified margin of error m when the sample size is

$$n = \left(\frac{z^* \sigma}{m} \right)^2$$

This formula does not account for collection costs. In practice, taking observations costs time and money. The required sample size may be impossibly expensive. In those situations, you might consider a larger margin of error and/or a lower confidence level to find a workable sample size.

EXAMPLE 6.7

How many undergraduates should we survey? Suppose that we are planning a survey similar to the one described in Example 6.4 (page 350). If we want the margin of error for the average amount contributed from a college savings plan to be $30 with 95% confidence, what sample size n do we need? For 95% confidence, Table D gives $z^* = 1.960$. For σ we will use the value from the previous study, $1483. If the margin of error is $30, we have

$$n = \left(\frac{z^*\sigma}{m}\right)^2 = \left(\frac{1.96 \times 1483}{30}\right)^2 = 9387.54$$

Because 9387 measurements will give a slightly wider interval than desired and 9388 measurements a slightly narrower interval, we should choose $n = 9388$. We need information from 9388 undergraduates to determine an estimate of mean college savings fund contribution with the desired margin of error.

It is always safe to round *up* to the next higher whole number when finding n because this will give us a smaller margin of error. The purpose of this calculation is to determine a sample size that is sufficient to provide useful results, but the determination of what is useful is a matter of judgment.

Would we need a much larger sample size to obtain a margin of error of $25? Here is the calculation:

$$n = \left(\frac{z^*\sigma}{m}\right)^2 = \left(\frac{1.96 \times 1483}{25}\right)^2 = 13,518.06$$

A sample of $n = 13,519$ is much larger, and the costs of such a large sample may be prohibitive.

Unfortunately, the actual number of usable observations is often less than what we plan at the beginning of a study. This is particularly true of data collected in surveys but is an important consideration in most studies. Careful study designers often assume a nonresponse rate or dropout rate that specifies what proportion of the originally planned sample will fail to provide data. We use this information to calculate the sample size to be used at the start of the study.

For example, if in Example 6.7 we expect only 50% of those contacted to respond, we would need to start with a sample size of $2 \times 9388 = 18,776$ to obtain usable information from 9388 undergraduates and parents of undergraduates.

USE YOUR KNOWLEDGE

6.9 **Starting salaries.** You are planning a survey of starting salaries for recent computer science majors. In the latest survey by the National Association of Colleges and Employers, the average starting salary was reported to be $61,287.[5] If you assume that the standard deviation is $3850, what sample size do you need to have a margin of error equal to $500 with 95% confidence?

6.10 **Changes in sample size.** Suppose that in the setting of the previous exercise you have the resources to contact 300 recent graduates. If all respond, will your margin of error be larger or smaller than $500? What if only 50% respond? Verify your answers by performing the calculations.

Some cautions

We have already seen that small margins of error and high confidence can require large numbers of observations. You should also be keenly aware that *any formula for inference is correct only in specific circumstances*. If the government required statistical procedures to carry warning labels like those on drugs, most inference methods would have long labels. Our formula $\bar{x} \pm z^*\sigma/\sqrt{n}$ for estimating a population mean comes with the following list of warnings for the user:

• The data should be an SRS from the population. We are completely safe if we actually did a randomization and drew an SRS. We are not in great danger if the data can plausibly be thought of as independent observations from a population. That is the case in Examples 6.4 through 6.7, provided the undergraduates and parents can be considered one population.

• The formula is not correct for probability sampling designs more complex than an SRS. Correct methods for other designs are available. We will not discuss confidence intervals based on multistage or stratified samples (page 195). If you plan such samples, be sure that you (or your statistical consultant) know how to carry out the inference you desire.

• There is no correct method for inference from data haphazardly collected with bias of unknown size. Fancy formulas cannot rescue badly produced data.

resistant
measure,
p. 30

• Because $\bar{x}$ is not a resistant measure, outliers can have a large effect on the confidence interval. *You should search for outliers and try to correct them or justify their removal before computing the interval*. If the outliers cannot be removed, ask your statistical consultant about procedures that are not sensitive to outliers.

• If the sample size is small and the population is not Normal, the true confidence level will be different from the value C used in computing the interval. *Prior to any calculations, examine your data carefully for skewness and other signs of non-Normality*. Remember though that the interval relies only on the distribution of $\bar{x}$, which even for quite small sample sizes is much closer to Normal than is the distribution of the individual observations. When $n \geq 15$, the confidence level is not greatly disturbed by non-Normal populations unless extreme outliers or quite strong skewness are present. Our college fund contribution data in Example 6.4 are very likely skewed, but because of the large sample size, we are confident that the distribution of the sample mean will be approximately Normal.

standard
deviation s,
p. 38

• The interval $\bar{x} \pm z^*\sigma/\sqrt{n}$ assumes that the standard deviation σ of the population is known. This unrealistic requirement renders the interval of little use in statistical practice. We will learn in the next chapter what to do when σ is unknown. If, however, the sample is large, the sample standard deviation s will be close to the unknown σ. The interval $\bar{x} \pm z^*s/\sqrt{n}$ is then an approximate confidence interval for μ.

The most important caution concerning confidence intervals is a consequence of the first of these warnings. *The margin of error in a confidence interval covers only random sampling errors*. The margin of error is obtained from the sampling distribution and indicates how much error can be expected because of chance variation in randomized data production.

Practical difficulties such as undercoverage and nonresponse in a sample survey cause additional errors. These errors can be larger than the random sampling error. This often happens when the sample size is large (so that $\sigma/\sqrt{n}$ is small). Remember this unpleasant fact when reading the results of an opinion poll or other sample survey. The practical conduct of the survey influences the trustworthiness of its results in ways that are not included in the announced margin of error.

Every inference procedure that we will meet has its own list of warnings. Because many of the warnings are similar to those we have mentioned, we will not print the full warning label each time. It is easy to state (from the mathematics of probability) conditions under which a method of inference is exactly correct. These conditions are *never* fully met in practice.

For example, no population is exactly Normal. *Deciding when a statistical procedure should be used in practice often requires judgment assisted by exploratory analysis of the data.* Mathematical facts are, therefore, only a part of statistics. The difference between statistics and mathematics can be stated thusly: mathematical theorems are true; statistical methods are often effective when used with skill.

Finally, you should understand what statistical confidence does not say. Based on our SRS in Example 6.3, we are 95% confident that the mean SATM score for the California students lies between 486 and 504. This says that this interval was calculated by a method that gives correct results in 95% of all possible samples. It does *not* say that the probability is 0.95 that the true mean falls between 486 and 504. *No randomness remains after we draw a particular sample and compute the interval.* The true mean either is or is not between 486 and 504. The probability calculations of standard statistical inference describe how often the *method,* not a particular sample, gives correct answers.

USE YOUR KNOWLEDGE

6.11 Nonresponse in a survey. In earlier versions of the Sallie Mae survey of Example 6.4 (page 350), participants were asked to report the undergraduate's outstanding credit card balance. Only about a third reported this amount. Provide a couple of reasons why a survey respondent might not provide an amount. Based on these reasons, do you think the sample mean using just the reported amounts is biased? Is the margin of error based just on the reported amounts a good measure of precision? Explain your answers.

SECTION 6.1 SUMMARY

• The purpose of a **confidence interval** is to estimate an unknown parameter with an indication of how accurate the estimate is and of how confident we are that the result is correct.

• Any confidence interval has two parts: an interval computed from the data and a confidence level. The interval often has the form

$$\text{estimate} \pm \text{margin of error}$$

• The **confidence level** states the probability that the method will give a correct answer. That is, if you use 95% confidence intervals, in the long run 95% of your intervals will contain the true parameter value. When you apply the method once (that is, to a single sample), you do not know if your interval gave a correct answer (this happens 95% of the time) or not (this happens 5% of the time).

- The **margin of error** for a level C confidence interval for the mean μ of a Normal population with known standard deviation σ, based on an SRS of size n, is given by

$$m = z^* \frac{\sigma}{\sqrt{n}}$$

Here z^* is obtained from the row labeled z^* at the bottom of Table D. The probability is C that a standard Normal random variable takes a value between $-z^*$ and z^*. The confidence interval is

$$\bar{x} \pm m$$

If the population is not Normal and n is large, the confidence level of this interval is approximately correct.

- Other things being equal, the margin of error of a confidence interval decreases as

  - the confidence level C decreases,

  - the sample size n increases, and

  - the population standard deviation σ decreases.

- The sample size n required to obtain a confidence interval of specified margin of error m for a population mean is

$$n = \left(\frac{z^* \sigma}{m} \right)^2$$

where z^* is the critical point for the desired level of confidence.

- A specific confidence interval formula is correct only under specific conditions. The most important conditions concern the method used to produce the data. Other factors such as the form of the population distribution may also be important. These conditions should be investigated *prior* to any calculations.

SECTION 6.1 EXERCISES

For Exercises 6.1 through 6.3, see pages 345–346; for Exercises 6.4 and 6.5, see pages 347–348; for Exercises 6.6 through 6.8, see page 351; for Exercises 6.9 and 6.10, see page 354; and for Exercise 6.11, see page 356.

6.12 Margin of error and the confidence interval. A study of stress on the campus of your university reported a mean stress level of 78 (on a 0 to 100 scale with a higher score indicating more stress) with a margin of error of 5 for 95% confidence. The study was based on a random sample of 64 undergraduates.

(a) Give the 95% confidence interval.

(b) If you wanted 99% confidence for the same study, would your margin of error be greater than, equal to, or less than 5? Explain your answer.

6.13 Changing the sample size. Consider the setting of the previous exercise. Suppose that the sample mean is again 78 and the population standard deviation is 20. Make a diagram similar to Figure 6.5 (page 351) that illustrates the effect of sample size on the width of a 95% interval. Use the following sample sizes: 9, 25, 81, and 100. Summarize what the diagram shows.

6.14 Changing the confidence level. Consider the setting of the previous two exercises. Suppose that the sample mean is still 78, the sample size is 64, and the population standard deviation is 20. Make a diagram similar to Figure 6.6 (page 353) that illustrates the effect of the confidence level on the width of the interval. Use 80%, 90%, 95%, and 99%. Summarize what the diagram shows.

6.15 Confidence interval mistakes and misunderstandings. Suppose that 500 randomly selected alumni of the University of Okoboji were asked to rate the university's academic advising services on a 1 to 10 scale. The sample mean $\bar{x}$ was found to be 8.6. Assume that the population standard deviation is known to be $\sigma = 2.2$.

(a) Ima Bitlost computes the 95% confidence interval for the average satisfaction score as $8.6 \pm 1.96(2.2)$. What is her mistake?

(b) After correcting her mistake in part (a), she states, "I am 95% confident that the sample mean falls between 8.4 and 8.8." What is wrong with this statement?

(c) She quickly realizes her mistake in part (b) and instead states, "The probability that the true mean is between 8.4 and 8.8 is 0.95." What misinterpretation is she making now?

(d) Finally, in her defense for using the Normal distribution to determine the confidence interval she says, "Because the sample size is quite large, the population of alumni ratings will be approximately Normal." Explain to Ima her misunderstanding and correct this statement.

6.16 More confidence interval mistakes and misunderstandings. Suppose that 100 randomly selected members of the Karaoke Channel were asked how much time they typically spend on the site during the week.[6] The sample mean $\bar{x}$ was found to be 3.8 hours. Assume that the population standard deviation is known to be $\sigma = 2.9$.

(a) Cary Oakey computes the 95% confidence interval for the average time on the site as $3.8 \pm 1.96(2.9/100)$. What is his mistake?

(b) He corrects this mistake and then states that "95% of the members spend between 3.23 and 4.37 hours a week on the site." What is wrong with his interpretation of this interval?

(c) The margin of error is slightly larger than half an hour. To reduce this to roughly 15 minutes, Cary says that the sample size needs to be doubled to 200. What is wrong with this statement?

6.17 The state of stress in the United States. Since 2007, the American Psychological Association has supported an annual nationwide survey to examine stress across the United States.[7] This year, a total of 720 millennials (18- to 33-year-olds) were asked to indicate their average stress level (on a 10-point scale) during the past month. The mean score was 5.5. Assume that the population standard deviation is 2.8.

(a) Give the margin of error and find the 95% confidence interval for this sample.

(b) Repeat these calculations for a 99% confidence interval. How do the results compare with those in part (a)?

6.18 Inference based on integer values. Refer to Exercise 6.17. The data for this study are integer values between 1 and 10. Explain why the confidence interval based on the Normal distribution should be a good approximation.

6.19 Mean TRAP in young women. For many important processes that occur in the body, direct measurement of characteristics of the process is not possible. In many cases, however, we can measure a *biomarker*, a biochemical substance that is relatively easy to measure and is associated with the process of interest. Bone turnover is the net effect of two processes: the breaking down of old bone, called resorption, and the building of new bone, called formation. One biochemical measure of bone resorption is tartrate-resistant acid phosphatase (TRAP), which can be measured in blood. In a study of bone turnover in young women, serum TRAP was measured in 31 subjects.[8] The mean was 13.2 units per liter (U/l). Assume that the standard deviation is known to be 6.5 U/l. Give the margin of error and find a 95% confidence interval for the mean TRAP amount in young women represented by this sample.

6.20 Mean OC in young women. Refer to the previous exercise. A biomarker for bone formation measured in the same study was osteocalcin (OC), measured in the blood. For the 31 subjects in the study, the mean was 33.4 nanograms per milliliter (ng/ml). Assume that the standard deviation is known to be 19.6 ng/ml. Report the 95% confidence interval.

6.21 Populations sampled and margins of error. Consider the following two scenarios. (A) Take a simple random sample of 200 freshman students at your college or university. (B) Take a simple random sample of 200 students at your college or university. For each of these samples, you will record the amount spent on textbooks used for classes during the fall semester. Which sample should have the smaller margin of error? Explain your answer.

6.22 Average starting salary. The National Association of Colleges and Employers (NACE) Spring Salary Survey shows that the current class of college graduates received an average starting-salary offer of $48,127.[9] Your institution collected an SRS ($n = 300$) of its recent graduates and obtained a 95% confidence interval of ($46,382, $48,008). What can we conclude about the *difference* between the average starting salary of recent graduates at your institution and the overall NACE average? Write a short summary.

6.23 Consumption of sweet snacks. A recent study reported that the U.S. per capita consumption of sweet snacks among healthy weight children aged 12 to 19 years is 251.2 kilocalories per day (kcal/d).[10] This was based on 24-hour dietary recall records of $n = 2265$ adolescents.

(a) Suppose that the population distribution is heavily skewed, with a standard deviation equal to 540 kcal/d. What is the margin of error for a 95% confidence interval of the per capita consumption of sweet snacks?

(b) A future study is being planned and the goal is to have the margin of error no more than 15 kcal/d. Based on your answer to part (a), will this study require an examination of more or fewer recall records? Explain your answer without calculations.

(c) Compute the sample size necessary for the study described in part (b).

6.24 Total sleep time of college students. In Example 5.4 (page 293), the total sleep time per night among college students was approximately Normally distributed with mean $\mu = 6.78$ hours and standard deviation $\sigma = 1.24$ hours. You initially plan to take an SRS of size $n = 175$ and compute the average total sleep time.

(a) What is the standard deviation for the average time in hours? in minutes?

(b) Use the 95 part of the 68–95–99.7 rule to describe the variability of this sample mean.

(c) What is the probability that your average will be below 6.9 hours?

6.25 Determining sample size. Refer to the previous exercise. You really want to use a sample size such that about 95% of the averages fall within ±5 minutes of the true mean $\mu = 6.78$.

(a) Based on your answer to part (b) in Exercise 6.24, should the sample size be larger or smaller than 175? Explain.

(b) What standard deviation of $\bar{x}$ do you need such that 95% of all samples will have a mean within 5 minutes of μ?

(c) Using the standard deviation you calculated in part (b), determine the number of students you need to sample.

6.26 Inference based on skewed data. The mean OC for the 31 subjects in Exercise 6.20 was 33.4 ng/ml. In our calculations, we assumed that the standard deviation was known to be 19.6 ng/ml. Use the 68–95–99.7 rule from Chapter 1 (page 57) to find the approximate bounds on the values of OC that would include these percents of the population. If the assumed standard deviation is correct, this distribution may be highly skewed. Why? (*Hint:* The measured values for a variable such as this are all positive.) Do you think that this skewness will invalidate the use of the Normal confidence interval in this case? Explain your answer.

6.27 Average hours per week listening to the radio. The *Student Monitor* surveys 1200 undergraduates from four-year colleges and universities throughout the United States semiannually to understand trends among college students.[11] Recently, the *Student Monitor* reported that the average amount of time listening to the radio per week was 11.5 hours. Of the 1200 students surveyed, 83% said that they listened to the radio, so this collection of listening times has around 204 (17% × 1200) zeros. Assume that the standard deviation is 8.3 hours.

(a) Give a 95% confidence interval for the mean time spent per week listening to the radio.

(b) Is it true that 95% of the 1200 students reported weekly times that lie in the interval you found in part (a)? Explain your answer.

(c) It appears that the population distribution has many zeros and is skewed to the right. Explain why the confidence interval based on the Normal distribution should nevertheless be a good approximation.

6.28 Average minutes per week listening to the radio. Refer to the previous exercise.

(a) Give the mean and standard deviation in minutes.

(b) Calculate the 95% confidence interval in minutes from your answer to part (a).

(c) Explain how you could have directly calculated this interval from the 95% interval that you calculated in the previous exercise.

6.29 Outlook on life. Since 2008, the Gallup-Healthways Well-Being Index tracks how people feel about their daily lives. In 2014, 54.1% of the respondents were classified as "thriving." This classification is based on how a respondent rates his or her current and future lives. This is the highest percent of respondents in this category since the index started. Material provided with the results noted:

Results are based on telephone interviews . . . with a random sample of 176,903 adults, living in all 50 U.S. states and the District of Columbia. For results based on the total sample of national adults, the margin of sampling error is ±1 percentage points at the 95% confidence level.[12]

The poll uses a complex multistage sample design, but the sample percent has approximately a Normal sampling distribution.

(a) The announced poll result was 54.1% ± 1%. Can we be certain that the true population percent falls in this interval? Explain your answer.

(b) Explain to someone who knows no statistics what the announced result 54.1% ± 1% means.

(c) This confidence interval has the same form we have met earlier:

$$\text{estimate} \pm z^* \, \sigma_{\text{estimate}}$$

What is the standard deviation σ_{estimate} of the estimated percent?

(d) Does the announced margin of error include errors due to practical problems such as nonresponse? Explain your answer.

6.30 Fuel efficiency. Computers in some vehicles calculate various quantities related to performance. One of these is the fuel efficiency, or gas mileage, usually expressed as miles per gallon (mpg). For one vehicle equipped in this way, the miles per gallon were recorded each time the gas tank was filled, and the computer was then reset.[13] Here are the mpg values for a random sample of 20 of these records: **MPG**

| 41.5 | 50.7 | 36.6 | 37.3 | 34.2 | 45.0 | 48.0 | 43.2 | 47.7 | 42.2 |
| 43.2 | 44.6 | 48.4 | 46.4 | 46.8 | 39.2 | 37.3 | 43.5 | 44.3 | 43.3 |

Suppose that the standard deviation is known to be $\sigma = 3.5$ mpg.

(a) What is $\sigma_{\bar{x}}$, the standard deviation of $\bar{x}$?

(b) Examine the data for skewness and other signs of non-Normality. Show your plots and numerical summaries. Do you think it is reasonable to construct a confidence interval based on the Normal distribution? Explain your answer.

(c) Give a 95% confidence interval for μ, the mean miles per gallon for this vehicle.

6.31 Fuel efficiency in metric units. In the previous exercise, you found an estimate with a margin of error for the average miles per gallon. Convert your estimate and margin of error to the metric units kilometers per liter (kpl). To change mpg to kpl, use the fact that 1 mile = 1.609 kilometers and 1 gallon = 3.785 liters.

6.32 How many "hits"? The *Confidence Interval* applet lets you simulate large numbers of confidence intervals quickly. Select 95% confidence and then sample 50 intervals. Record the number of intervals that cover the true value (this appears in the "Hit" box in the applet). Press the "Reset" button and repeat 30 times. Make a stemplot of the results and find the mean. Describe the results. If you repeated this experiment very many times, what would you expect the average number of hits to be?

6.33 Required sample size for specified margin of error. A new bone study is being planned that will measure the biomarker TRAP described in Exercise 6.19. Using the value of σ given there, 6.5 U/l, find the sample size required to provide an estimate of the mean TRAP with a margin of error of 1.5 U/l for 95% confidence.

6.34 Adjusting required sample size for dropouts. Refer to the previous exercise. In similar previous studies, about 20% of the subjects drop out before the study is completed. Adjust your sample size requirement so that you will have enough subjects at the end of the study to meet the margin of error criterion.

6.35 Radio poll. A national public radio (NPR) station invites listeners to enter a dispute about a proposed "pay as you throw" waste collection program. The station asks listeners to call in and state how much each 10 gallon bag of trash should cost. A total of 179 listeners call in. The station calculates the 95% confidence interval for the average fee to be $0.53 to $1.39. Is this result trustworthy? Explain your answer.

6.36 Accuracy of a laboratory scale. To assess the accuracy of a laboratory scale, a standard weight known to weigh 10 grams is weighed repeatedly. The scale readings are Normally distributed with unknown mean (this mean is 10 grams if the scale has no bias). The standard deviation of the scale readings is known to be 0.0002 gram.

(a) The weight is measured six times. The mean result is 10.0023 grams. Give a 99% confidence interval for the mean of repeated measurements of the weight.

(b) Based on the interval in part (a), do you think the scale is accurate? Explain your answer.

(c) How many measurements must be averaged to get a margin of error of ± 0.0001 with 99% confidence?

6.37 More than one confidence interval. As we prepare to take a sample and compute a 95% confidence interval, we know that the probability that the interval we compute will cover the parameter is 0.95. That's the meaning of 95% confidence. If we plan to use several such intervals, however, our confidence that *all* of them will give correct results is less than 95%. Suppose that we plan to take independent samples each month for five months and report a 95% confidence interval for each set of data.

(a) What is the probability that all five intervals will cover the true means? This probability (expressed as a percent) is our overall confidence level for the five simultaneous statements.

(b) Suppose we instead considered individual 99% confidence intervals. Now, what is the overall confidence level for the five simultaneous statements?

(c) Based on the results of parts (a) and (b), how could you keep the overall confidence level near 95% if you were considering 10 simultaneous intervals?

6.2 Tests of Significance

- Outline the four steps common to all tests of significance.
- Formulate the null and alternative hypotheses of a significance test.
- Describe a common form for the test statistic in terms of the parameter estimate, its standard deviation, and the hypothesized value.
- Define what a *P*-value is and explain whether a small *P*-value provides evidence for or against the null hypothesis.
- Draw a conclusion from a test of significance based on the test's *P*-value and significance level α.
- Describe the relationship between a level α two-sided significance test for μ and the $1 - \alpha$ confidence interval.

The confidence interval is appropriate when our goal is to estimate population parameters. The second common type of inference is directed at a quite different goal: to assess the evidence provided by the data in favor of some claim about the population parameters.

The reasoning of significance tests

A significance test is a formal procedure for comparing observed data with a hypothesis whose truth we want to assess. The hypothesis is a statement about the population parameters. The results of a test are expressed in terms of a probability that measures how well the data and the hypothesis agree. We use the following examples to illustrate these concepts.

EXAMPLE 6.8

Scholarship amount by borrower status. One purpose of Sallie Mae's annual study described in Example 6.4 (page 350) is to allow comparisons of different subgroups. For example, in the latest report, 980 of the 1593 participants (61.5%) did not borrow any money to pay for college. The average scholarship amount among these participants was $3925. The average scholarship among those who did borrow was $4350. The difference of $425 is fairly large, but we know that these numbers are estimates of the population means. If we took different samples, we would get different estimates.

Can we conclude from these data that the average scholarship amounts in these two groups are different? One way to answer this question is to compute the probability of obtaining a difference as large or larger than the observed $425 assuming that, in fact, there is no difference in the population means. This probability is 0.23. Because this probability is not particularly small, we conclude that observing a difference of $425 is not very surprising when the population means are equal. The data do not provide enough evidence for us to conclude that the average scholarship amount for borrowers and non-borrowers differ.

Here is an example with a different conclusion.

EXAMPLE 6.9

Parent income contribution by school type. Sallie Mae's study also reports that the parents' current income contribution among undergraduates going to a four-year public or four-year private college. The parents' contribution averages $4444 among undergraduates at public colleges, while it is $6083 among undergraduates at private schools. Do parents pay more of their current income for undergraduates going to private schools? The observed difference is $1639, but as we learned in the previous example, an observed difference in means is not necessarily sufficient for us to conclude that the population means are different.

Again, we answer this question with a probability calculated under the assumption that there is *no difference in the population means*. The probability is 0.0001 of observing a difference in mean contributions that is $1639 or more when there really is no difference. Because this probability is so small, we have sufficient evidence in the data to conclude that the average current income contribution of parents is higher for undergraduates going to a private school than undergraduates going to a public school.

What are the key steps in these examples?

- We started each with a question about the difference between two means. In Example 6.8, we compare borrowers with nonborrowers. In Example 6.9, we compare undergraduates attending private and public four-year colleges. In both cases, we ask whether or not the data are compatible with "no difference," that is, a difference of $0.

- Next we compared the difference given by the data, $425 in the first case and $1639 in the second, with the value assumed in the question, $0.

- The results of the comparisons are probabilities, 0.23 in the first case and 0.0001 in the second.

The 0.23 probability is not particularly small, so we have limited evidence to question the possibility that the true difference is zero. In the second case, however, the probability is very small. Something that happens with probability 0.0001 occurs only about 1 time out of 10,000. In this case we have two possible explanations:

1. We have observed something that is very unusual.

2. The assumption that underlies the calculation, no difference in mean balance, is not true.

Because this probability is so small, we prefer the second conclusion: the average current income contribution from parents for undergraduates attending public colleges and for undergraduates attending private colleges is different, with the private school group contribution higher than that of the public school group.

The probabilities in Examples 6.8 and 6.9 are measures of the compatibility of the data (a difference in means of $425 and $1639) with the *null hypothesis* that there is no difference in the true means. Figures 6.7 and 6.8 compare the two results graphically. For each, a Normal curve centered at 0 is the sampling distribution. You can see from Figure 6.7 that we should not be particularly surprised to observe the difference $425, but the difference $1639 in Figure 6.8 is clearly an unusual observation. We will now consider some of the formal aspects of significance testing.

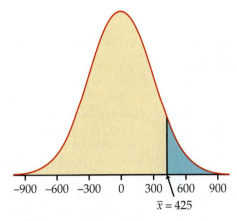

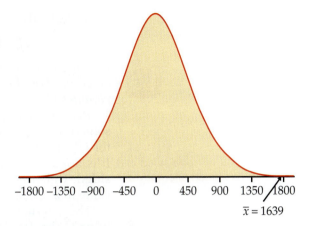

FIGURE 6.7 Comparison of the sample mean in Example 6.8 with the null hypothesized value 0.

FIGURE 6.8 Comparison of the sample mean in Example 6.9 with the null hypothesized value 0.

Stating hypotheses

In Examples 6.8 and 6.9, we asked whether the difference in the observed means is reasonable if, in fact, there is no difference in the population means. To answer this, we begin by supposing that the statement following the "if" in the previous sentence is true. In other words, we suppose that the true difference is $0. We then ask whether the data provide evidence against the supposition we have made. If so, we have evidence in favor of an effect (the means are different) we are seeking. Often, the first step in a test of significance is to state a claim that we will try to find evidence *against*.

> ### NULL HYPOTHESIS
>
> The statement being tested in a test of significance is called the **null hypothesis**. The test of significance is designed to assess the strength of the evidence against the null hypothesis. Usually, the null hypothesis is a statement of "no effect" or "no difference."

We abbreviate "null hypothesis" as H_0. A null hypothesis is a statement about the population parameters. For example, our null hypothesis for Example 6.8 is

H_0: there is no difference in the population means

or equivalently,

H_0: the difference in population means is zero

Note that the null hypothesis refers to the *population* means for all undergraduates, including those for whom we do not have data.

It is convenient also to give a name to the statement we hope or suspect is true instead of H_0. This is called the **alternative hypothesis** and is abbreviated as H_a. In Example 6.8, the alternative hypothesis states that the means are different. We write this as

alternative hypothesis

H_a: the population means are not the same

or equivalently,

$$H_a: \text{the difference in population means is not zero}$$

Hypotheses always refer to some populations or a model, not to a particular outcome. For this reason, we must state H_0 and H_a in terms of population parameters.

Because H_a expresses the effect that we hope to find evidence *for*, we will sometimes begin with H_a and then set up H_0 as the statement that the hoped-for effect is not present. Stating H_a, however, is often the more difficult task. **one-sided or two-sided alternatives** It is not always clear, in particular, whether H_a should be **one-sided** or **two-sided**, which refers to whether a parameter differs from its null hypothesis value in a specific direction or in either direction.

The alternative hypothesis should express the hopes or suspicions we bring to the data. *It is cheating to first look at the data and then frame H_a to fit what the data show.* If you do not have a specific direction firmly in mind in advance, you must use a two-sided alternative. Moreover, some users of statistics argue that we should always use a two-sided alternative.

USE YOUR KNOWLEDGE

6.38 Dining court survey. The dining court closest to your university residence has been redesigned. A survey is planned to assess whether or not students think that the new design is an improvement. It will contain eight questions; a seven-point scale will be used for the answers, with scores less than 4 favoring the previous design and scores greater than 4 favoring the new design (to varying degrees). The average of these eight questions will be used as the student's response. State the null and alternative hypotheses you would use for examining whether or not the new design is viewed more favorably.

6.39 DXA scanners. A dual-energy X-ray absorptiometry (DXA) scanner is used to measure bone mineral density for people who may be at risk for osteoporosis. One researcher believes that her scanner is not giving accurate readings. To assess this, the researcher uses an object called a "phantom" that has known mineral density $\mu = 1.4$ grams per square centimeter. The researcher scans the phantom 10 times and compares the sample mean reading $\bar{x}$ with the theoretical mean μ using a significance test. State the null and alternative hypotheses for this test.

Test statistics

We will learn the form of significance tests in a number of common situations. Here are some principles that apply to most tests and that help in understanding these tests:

- The test is based on a statistic that estimates the parameter that appears in the hypotheses. Usually, this is the same estimate we would use in a confidence interval for the parameter. When H_0 is true, we expect the estimate to take a value near the parameter value specified by H_0. We call this specified value the hypothesized value.

- Values of the estimate far from the hypothesized value give evidence against H_0. The alternative hypothesis determines which directions count against H_0.

• To assess how far the estimate is from the hypothesized value, standardize the estimate. In many common situations the test statistic has the form

$$z = \frac{\text{estimate} - \text{hypothesized value}}{\text{standard deviation of the estimate}}$$

test statistic   A **test statistic** measures compatibility between the null hypothesis and the data. We use it for the probability calculation that we need for our test of significance. It is a random variable with a distribution that we know.

Let's return to our comparison of the scholarship amount among borrowers and nonborrowers and specify the hypotheses as well as calculate the test statistic.

EXAMPLE 6.10

Average scholarship amount of borrowers and nonborrowers: The hypotheses.
In Example 6.8, the hypotheses are stated in terms of the difference in the average scholarship amount between borrowers and nonborrowers:

H_0: there is no difference in the population means

H_a: there is a difference in the population means

Because H_a is two-sided, large values of both positive and negative differences count as evidence against the null hypothesis.

We can also state the null hypothesis as H_0: the true mean difference is 0. This statement makes it more clear that the hypothesized value for this comparison of average scholarship amounts is 0.

EXAMPLE 6.11

Average scholarship amount of borrowers and nonborrowers: The test statistic.
In Example 6.8, the estimate of the difference is $425. Using methods that we will discuss in detail later, we can determine that the standard deviation of the estimate is $353. For this problem the test statistic is

$$z = \frac{\text{estimate} - \text{hypothesized value}}{\text{standard deviation of the estimate}}$$

For our data,

$$z = \frac{425 - 0}{353} = 1.20$$

We have observed a sample estimate that is one and one-fifth standard deviations away from the hypothesized value of the parameter.

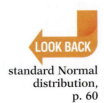

LOOK BACK

standard Normal distribution, p. 60

Because the sample sizes are sufficiently large for us to conclude that the distribution of the sample estimate is approximately Normal, the standardized test statistic z will have approximately the $N(0, 1)$ distribution. We will use facts about the Normal distribution in what follows.

P-values

If all test statistics were Normal, we could base our conclusions on the value of the z test statistic. In fact, the Supreme Court of the United States has said that "two or three standard deviations" ($z = 2$ or 3) is its criterion for rejecting H_0 (see Exercise 6.44 on page 370), and this is the criterion used in most applications involving the law. But because not all test statistics are Normal, we use the language of probability to express the meaning of a test statistic.

A test of significance finds the probability of getting an outcome *as extreme or more extreme than the actually observed outcome*. "Extreme" means "far from what we would expect if H_0 were true." The direction or directions that count as "far from what we would expect" are determined by H_a and H_0.

> **P-VALUE**
>
> The probability, assuming H_0 is true, that the test statistic would take a value as extreme or more extreme than that actually observed is called the **P-value** of the test. The smaller the P-value, the stronger the evidence against H_0 provided by the data.

The key to calculating the P-value is the sampling distribution of the test statistic. For the problems we consider in this chapter, we need only the standard Normal distribution for the test statistic z.

In Example 6.8, we want to know if the average scholarship amount for borrowers differs from the average scholarship amount for non-borrowers. The difference we calculated based on our sample is $425, which corresponds to 1.20 standard deviations away from zero—that is, $z = 1.20$. Because we are using a two-sided alternative for this problem, the evidence against H_0 is measured by the probability that we observe a value of Z as extreme or more extreme than 1.20.

EXAMPLE 6.12

Average scholarship amount of borrowers and nonborrowers: The *P*-value. In Example 6.11, we found that the test statistic for testing

$$H_0\text{: the true mean difference is 0}$$

versus

$$H_a\text{: there is a difference in the population means}$$

is

$$z = \frac{425 - 0}{353} = 1.20$$

If H_0 is true, then z is a single observation from the standard Normal, $N(0, 1)$, distribution. Figure 6.9 illustrates this calculation. The P-value is the probability of observing a value of Z at least as extreme as the one that we observed, $z = 1.20$. From Table A, our table of standard Normal probabilities, we find

$$P(Z \geq 1.20) = 1 - 0.8849 = 0.1151$$

The probability for being extreme in the negative direction is the same:

$$P(Z \leq -1.20) = 0.1151$$

So the P-value is

$$P = 2P(Z \geq 1.20) = 2(0.1151) = 0.2302$$

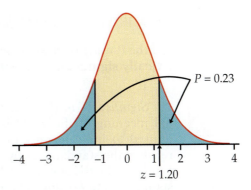

FIGURE 6.9 The *P*-value, Example 6.12. The *P*-value is the probability (when H_0 is true) that $\bar{x}$ takes a value as extreme or more extreme than the actual observed value, $z = 1.20$. Because the alternative hypothesis is two-sided, we use both tails of the distribution.

This is the value that we reported on page 361. There is a 23% chance of observing a difference as extreme as the $425 in our sample if the true population difference is zero. This *P*-value tells us that our outcome is not particularly extreme. In other words, the data do not provide substantial evidence for us to doubt the validity of the null hypothesis.

USE YOUR KNOWLEDGE

6.40 Normal curve and the *P*-value. A test statistic for a two-sided significance test for a population mean is $z = 2.47$. Sketch a standard Normal curve and mark this value of z on it. Find the *P*-value and shade the appropriate areas under the curve to illustrate your calculations.

6.41 More on the Normal curve and the *P*-value. A test statistic for a two-sided significance test for a population mean is $z = -1.57$. Sketch a standard Normal curve and mark this value of z on it. Find the *P*-value and shade the appropriate areas under the curve to illustrate your calculations.

Statistical significance

We started our discussion of the reasoning of significance tests with the statement of null and alternative hypotheses. We then learned that a test statistic is the tool used to examine the compatibility of the observed data with the null hypothesis. Finally, we translated the test statistic into a *P*-value to quantify the evidence against H_0. One important final step is needed: to state our conclusion.

We can compare the *P*-value we calculated with a fixed value that we regard as decisive. This amounts to announcing in advance how much evidence against H_0 we will require to reject H_0. The decisive value is called the **significance level**. It is commonly denoted by α (the Greek letter alpha). If we choose $\alpha = 0.05$, we are requiring that the data give evidence against H_0 so strong that it would happen no more than 5% of the time (1 time in 20) when H_0 is true. If we choose $\alpha = 0.01$, we are insisting on stronger evidence against H_0, evidence so strong that it would appear only 1% of the time (1 time in 100) if H_0 is in fact true.

significance level

STATISTICAL SIGNIFICANCE

If the *P*-value is as small or smaller than α, we say that the data are **statistically significant at level** α.

"Significant" in the statistical sense does not mean *"important."* The original meaning of the word is "signifying something." In statistics, the term is used to indicate only that the evidence against the null hypothesis has reached the standard set by α. For example, significance at level 0.01 is often expressed by the statement "The results were significant ($P < 0.01$)." Here, *P* stands for the *P*-value. The *P*-value is more informative than a statement of significance because we can then assess significance at any level we choose. For example, a result with $P = 0.03$ is significant at the $\alpha = 0.05$ level but is not significant at the $\alpha = 0.01$ level. We discuss this in more detail at the end of this section.

EXAMPLE 6.13

Average scholarship amount of borrowers and nonborrowers: The conclusion. In Example 6.12, we found that the *P*-value is

$$P = 2P(Z \geq 1.20) = 2(0.1151) = 0.2302$$

There is an 23% chance of observing a difference as extreme as the $425 in our sample if the true population difference is zero. Because this *P*-value is larger than the $\alpha = 0.05$ significance level, we conclude that our test result is not significant. We could report the result as "the data fail to provide evidence that would cause us to conclude that there is a difference in average scholarship amount between borrowers and nonborrowers ($z = 1.20$, $P = 0.23$)."

This statement does not mean that we conclude that the null hypothesis is true, only that the level of evidence we require to reject the null hypothesis is not met. Our criminal court system follows a similar procedure in which a defendant is presumed innocent (H_0) until proven guilty. If the level of evidence presented is not strong enough for the jury to find the defendant guilty beyond a reasonable doubt, the defendant is acquitted. Acquittal does not imply innocence, only that the degree of evidence was not strong enough to prove guilt.

If the *P*-value is small, we reject the null hypothesis. Here is the conclusion for our second example.

EXAMPLE 6.14

Parent income contribution by school type: The conclusion. In Example 6.9, we found that the difference in the average parent current income contribution between undergraduates going to a private college versus public college was $1639. Because the cost of tuition at a private college is typically higher than the cost at a public college,[14] we had a prior expectation that the parental current income contribution would be higher for undergraduates going to a private college. It is appropriate to use a one-sided alternative in this situation. So, our hypotheses are

$$H_0: \text{the true mean difference is } 0$$

versus

H_a: the difference between the average parent income contribution of undergraduates at a private college and public college is positive

The standard deviation is \$428 (again, we defer details regarding this calculation), and the test statistic is

$$z = \frac{\text{estimate} - \text{hypothesized value}}{\text{standard deviation of the estimate}}$$

$$z = \frac{1639 - 0}{430}$$

$$= 3.81$$

Because only positive differences in parental contributions count against the null hypothesis, the one-sided alternative leads to the calculation of the *P*-value using the upper tail of the Normal distribution. In Table A, the largest *z* is 3.49. This means that for $z = 3.81$, $P < 0.0002$. Using software, we can be more precise. The *P*-value is

$$P = P(Z \geq 3.81)$$

$$= 0.0001$$

The calculation is illustrated in Figure 6.10. There is about a 1-in-10,000 chance of observing a difference as large or larger than the \$1639 in our sample if the true population difference is zero. This *P*-value tells us that our outcome is extremely rare. We conclude that the null hypothesis must be false. Because the observed difference is positive, here is one way to report the result: "The data clearly show that the average parent income contribution for undergraduates at a private college is larger than the average parent income contribution for undergraduates at a public college ($z = 3.81$, $P = 0.0001$)."

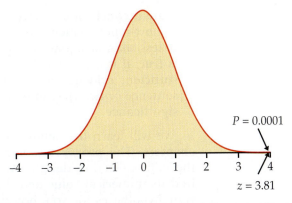

FIGURE 6.10 The *P*-value, Example 6.14. The *P*-value is the probability (when H_0 is true) that $\bar{x}$ takes a value as extreme or more extreme than the actual observed value, $z = 3.81$. We look at only the right tail because we are considering the one-sided (>) alternative.

6.42 Finding significant z-scores. Consider a two-sided significance test for a population mean.

(a) Sketch a Normal curve similar to that shown in Figure 6.9 (page 367), but find the value z such that $P = 0.05$.

(b) Based on your curve from part (a), what values of the z statistic are statistically significant at the $\alpha = 0.05$ level?

6.43 More on finding significant z-scores. Consider a one-sided significance test for a population mean, where the alternative is "greater than."

(a) Sketch a Normal curve similar to that shown in Figure 6.10, but find the value z such that $P = 0.05$.

(b) Based on your curve from part (a), what values of the z statistic are statistically significant at the $\alpha = 0.05$ level?

6.44 The Supreme Court speaks. The Supreme Court has said that z-scores beyond 2 or 3 are generally convincing statistical evidence. For a two-sided test, what significance level corresponds to $z = 2$? To $z = 3$?

A test of significance is a process for assessing the significance of the evidence provided by data against a null hypothesis. **The four steps common to all tests of significance are as follows:**

1. State the *null hypothesis* H_0 and the *alternative hypothesis* H_a. The test is designed to assess the strength of the evidence against H_0; H_a is the statement that we will accept if the evidence enables us to reject H_0.

2. Calculate the value of the *test statistic* on which the test will be based. This statistic usually measures how far the data are from H_0.

3. Find the *P-value* for the observed data. This is the probability, calculated assuming that H_0 is true, that the test statistic will weigh against H_0 at least as strongly as it does for these data.

4. State a conclusion. One way to do this is to choose a *significance level* α, how much evidence against H_0 you regard as decisive. If the P-value is less than or equal to α, you conclude that the alternative hypothesis is true; if it is greater than α, you conclude that the data do not provide sufficient evidence to reject the null hypothesis. Your conclusion is a sentence or two that summarizes what you have found by using a test of significance.

We will learn the details of many tests of significance in the following chapters. The proper test statistic is determined by the hypotheses and the data collection design. We use computer software or a calculator to find its numerical value and the P-value. The computer will not formulate your hypotheses for you, however. Nor will it decide if significance testing is appropriate or help you to interpret the P-value that it presents to you. These steps require judgment based on a sound understanding of this type of inference.

Tests for a population mean

Our discussion has focused on the reasoning of statistical tests, and we have outlined the key ideas for one type of procedure. Our examples focused on the comparison of two population means. Here is a summary for a test about one population mean.

We want to test the hypothesis that a parameter has a specified value. This is the null hypothesis. For a test of a population mean μ, the null hypothesis is

$$H_0: \text{the true population mean is equal to } \mu_0$$

which often is expressed as

$$H_0: \mu = \mu_0$$

where μ_0 is the hypothesized value of μ that we would like to examine.

The test is based on data summarized as an estimate of the parameter. For a population mean this is the sample mean $\bar{x}$. Our test statistic measures the difference between the sample estimate and the hypothesized parameter in terms of standard deviations of the test statistic:

$$z = \frac{\text{estimate} - \text{hypothesized value}}{\text{standard deviation of the estimate}}$$

LOOK BACK

distribution of
sample mean,
p. 298

Recall from Chapter 5 that the standard deviation of $\bar{x}$ is $\sigma/\sqrt{n}$. Therefore, the test statistic is

$$z = \frac{\bar{x} - \mu_0}{\sigma/\sqrt{n}}$$

LOOK BACK

central limit
theorem,
p. 298

Again recall from Chapter 5 that, if the population is Normal, then $\bar{x}$ will be Normal and z will have the standard Normal distribution when H_0 is true. By the central limit theorem, both distributions will be approximately Normal when the sample size is large even if the population is not Normal. We'll assume that we're in one of these two settings for now.

Suppose that we have calculated a test statistic $z = 1.7$. If the alternative is one-sided on the high side, then the P-value is the probability that a standard Normal random variable Z takes a value as large or larger than the observed 1.7. That is,

$$P = P(Z \geq 1.7)$$
$$= 1 - P(Z < 1.7)$$
$$= 1 - 0.9554$$
$$= 0.0446$$

Similar reasoning applies when the alternative hypothesis states that the true μ lies below the hypothesized μ_0 (one-sided). When H_a states that μ is simply unequal to μ_0 (two-sided), values of z away from zero in either direction count against the null hypothesis. The P-value is the probability that a standard Normal Z is at least as far from zero as the observed z. Again, if the test statistic is $z = 1.7$, the two-sided P-value is the probability that $Z \leq -1.7$ or $Z \geq 1.7$. Because the standard Normal distribution is symmetric, we calculate this probability by finding $P(Z \geq 1.7)$ and *doubling* it:

$$P(Z \leq -1.7 \text{ or } Z \geq 1.7) = 2P(Z \geq 1.7)$$
$$= 2(1 - 0.9554) = 0.0892$$

We would make exactly the same calculation if we observed $z = -1.7$. It is the absolute value $|z|$ that matters, not whether z is positive or negative. Here is a statement of the test in general terms.

z TEST FOR A POPULATION MEAN

To test the hypothesis $H_0: \mu = \mu_0$ based on an SRS of size n from a population with unknown mean μ and known standard deviation σ, compute the **test statistic**

$$z = \frac{\bar{x} - \mu_0}{\sigma/\sqrt{n}}$$

In terms of a standard Normal random variable Z, the P-value for a test of H_0 against

$H_a: \mu > \mu_0$ is $P(Z \geq z)$

$H_a: \mu < \mu_0$ is $P(Z \leq z)$

$H_a: \mu \neq \mu_0$ is $2P(Z \geq |z|)$

These P-values are exact if the population distribution is Normal and are approximately correct for large n in other cases.

EXAMPLE 6.15

Joe Raedle/Getty Images

Energy intake from sugar-sweetened beverages. Consumption of sugar-sweetened beverages (SSBs) has been positively associated with weight gain and obesity and negatively associated with the intake of important micronutrients. One study used data from the National Health and Nutrition Examination Survey (NHANES) to estimate SSB consumption among adolescents (aged 12 to 19 years). More than 2400 individuals provided data for this study.[15] The mean consumption was 298 calories per day.

You survey 100 students at your large university and find the average consumption of SSBs per day to be 262 calories. Is there evidence that the average calories per day from SSBs at your university differs from this large U.S. survey average?

The null hypothesis is "no difference" from the published mean $\mu_0 = 298$. The alternative is two-sided because you did not have a particular direction in mind before examining the data. So the hypotheses about the unknown mean μ of the students at your university are

$$H_0: \mu = 298$$
$$H_a: \mu \neq 298$$

As usual in this chapter, we make the unrealistic assumption that the population standard deviation is known. In this case, we'll use the standard deviation from the large national study, $\sigma = 435$ calories.

We compute the test statistic:

$$z = \frac{\bar{x} - \mu_0}{\sigma/\sqrt{n}} = \frac{262 - 298}{435/\sqrt{100}}$$
$$= -0.83$$

Figure 6.11 illustrates the *P*-value, which is the probability that a standard Normal variable *Z* takes a value at least 0.83 away from zero. From Table A, we find that this probability is

$$P = 2P(Z \geq 0.83) = 2(1 - 0.7967) = 0.4066$$

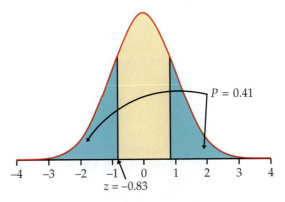

FIGURE 6.11 Sketch of the *P*-value calculation for the two-sided test, Example 6.15. The test statistic is $z = -0.83$.

That is, if the population mean were 298, more than 40% of the time an SRS of size 100 from the students at your university would have a mean consumption from SSBs at least as far from 298 as that of this sample. The observed $\bar{x} = 262$ is, therefore, not strong evidence that the student population mean at your university differs from that of the large population of adolescents.

This *z* test requires that the 100 students in the sample are an SRS from the population of students at your university. We will assume that the students in the sample were selected in a proper random manner. We'll also assume that $n = 100$ is sufficiently large that we can rely on the central limit theorem to assure us that the *P*-value based on the Normal distribution will be a good approximation.

The data in Example 6.15 do *not* establish that the mean consumption μ for the students at your university is 298 calories. We sought evidence that μ differed from 298 and failed to find convincing evidence. That is all we can say. No doubt the mean amount at your university is not exactly equal to 298 calories. A large enough sample would give evidence of the difference, even if it is very small.

Tests of significance assess the evidence *against* H_0. If the evidence is strong, we can confidently reject H_0 in favor of the alternative. *Failing to find evidence against H_0 means only that the data are consistent with H_0, not that we have clear evidence that H_0 is true.*

EXAMPLE 6.16

Significance test of the mean SATM score. In a discussion of SAT Mathematics (SATM) scores, someone comments: "Because only a select minority of California high school students take the test, the scores overestimate the ability of typical high school seniors. I think that if all seniors took the test, the mean score would be no more than 485." You do not agree with this claim and decide to use the SRS of 500 seniors from Example 6.3 (page 344) to assess the degree of evidence against it. Those 500 seniors had a mean SATM score of $\bar{x} = 495$. Is this strong enough evidence to conclude that this person's claim is wrong?

Because the claim states that the mean is "no more than 485," the alternative hypothesis is one-sided. The hypotheses are

$$H_0: \mu = 485$$
$$H_a: \mu > 485$$

As we did in the discussion following Example 6.3, we assume that $\sigma = 100$. The z statistic is

$$z = \frac{\bar{x} - \mu_0}{\sigma/\sqrt{n}} = \frac{495 - 485}{100/\sqrt{500}}$$
$$= 2.24$$

Because H_a is one-sided on the high side, large values of z count against H_0. From Table A, we find that the P-value is

$$P = P(Z \geq 2.24) = 1 - 0.9875 = 0.0125$$

Figure 6.12 illustrates this P-value. A mean score as large as that observed would occur roughly 12 times in 1000 samples if the population mean were 485. This is convincing evidence that the mean SATM score for all California high school seniors is higher than 485. You can confidently tell this person that his or her claim is incorrect.

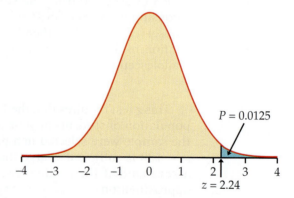

FIGURE 6.12 Sketch of the P-value calculation for the one-sided test, Example 6.16. The test statistic is $z = 2.24$.

USE YOUR KNOWLEDGE

6.45 Computing the test statistic and P-value. You will perform a significance test of $H_0: \mu = 30$ based on an SRS of $n = 49$. Assume that $\sigma = 14$.

(a) If $\bar{x} = 33.5$, what is the test statistic z?

(b) What is the P-value if $H_a: \mu > 30$?

(c) What is the P-value if $H_a: \mu \neq 30$?

6.46 Testing a random number generator. Statistical software often has a "random number generator" that is supposed to produce numbers uniformly distributed between 0 and 1. If this is true, the numbers generated come from a population with $\mu = 0.5$. A command to generate 100 random numbers gives outcomes with mean $\bar{x} = 0.469$ and $s = 0.286$. Because the sample is reasonably large, take the population standard deviation also to be $\sigma = 0.286$. Do we have evidence that the mean of all numbers produced by this software is not 0.5?

Two-sided significance tests and confidence intervals

Recall the basic idea of a confidence interval, discussed in Section 6.1. We constructed an interval that would include the true value of μ with a specified probability C. Suppose that we use a 95% confidence interval ($C = 0.95$). Then the values of μ_0 that are not in our interval would seem to be incompatible with the data. This sounds like a significance test with $\alpha = 0.05$ (or 5%) as our standard for drawing a conclusion. The following examples demonstrate that this is correct.

EXAMPLE 6.17

Voisin/ Phanie/Science Source

PBTEST

Water quality testing. The Deely Laboratory is a drinking-water testing and analysis service. One of the common contaminants it tests for is lead. Lead enters drinking water through corrosion of plumbing materials, such as lead pipes, fixtures, and solder. The service knows that their analysis procedure is unbiased but not perfectly precise, so the laboratory analyzes each water sample three times and reports the mean result. The repeated measurements follow a Normal distribution quite closely. The standard deviation of this distribution is a property of the analytic procedure and is known to be $\sigma = 0.25$ parts per billion (ppb).

The Deely Laboratory has been asked by a university to evaluate a claim that the drinking water in the Student Union has a lead concentration above the Environmental Protection Agency's (EPA) action level of 15 ppb. Because the true concentration of the sample is the mean μ of the population of repeated analyses, the hypotheses are

$$H_0: \mu = 15$$
$$H_a: \mu \neq 15$$

We use the two-sided alternative here because there is no prior evidence to substantiate a one-sided alternative. The lab chooses the 1% level of significance, $\alpha = 0.01$.

Three analyses of one specimen give concentrations

$$15.84 \quad 15.33 \quad 15.58$$

The sample mean of these readings is

$$\bar{x} = \frac{15.84 + 15.33 + 15.58}{3} = 15.58$$

The test statistic is

$$z = \frac{\bar{x} - \mu_0}{\sigma/\sqrt{n}} = \frac{15.58 - 15.00}{0.25/\sqrt{3}} = 4.02$$

Because the alternative is two-sided, the P-value is

$$P = 2P(Z \geq 4.02)$$

We cannot find this probability in Table A. The largest value of z in that table is 3.49. All that we can say from Table A is that P is less than $2P(Z \geq 3.49) = 2(1-0.9998) = 0.0004$. Software or a calculator could be used to give an accurate value of the P-value. However, because the P-value is clearly less than the lab's standard of 1%, we reject H_0. Because $\bar{x}$ is larger than 15.00, we can conclude that the true concentration level of lead in this one specimen is higher than the EPA's action level.

We can compute a 99% confidence interval for the same data to get a likely range for the actual mean concentration μ of this specimen.

EXAMPLE 6.18

PBTEST

99% confidence interval for the mean concentration. The 99% confidence interval for μ in Example 6.17 is

$$\bar{x} \pm z^* \frac{\sigma}{\sqrt{n}} = 15.58 \pm 2.576(0.25/\sqrt{3})$$

$$= 15.58 \pm 0.37$$

$$= (15.21, 15.95)$$

The hypothesized value $\mu_0 = 15.00$ in Example 6.17 falls outside the confidence interval we computed in Example 6.18. In other words, it is in the region we are 99% confident that μ is *not* in. Thus, we can reject

$$H_0: \mu = 15.00$$

at the 1% significance level. On the other hand, we cannot reject

$$H_0: \mu = 15.30$$

at the 1% level in favor of the two-sided alternative $H_a: \mu \neq 15.30$, because 15.30 lies inside the 99% confidence interval for μ. Figure 6.13 illustrates both cases.

FIGURE 6.13 The link between two-sided significance tests and confidence intervals. For the study described in Examples 6.17 and 6.18, values of μ falling outside a 99% confidence interval can be rejected at the 1% significance level; values falling inside the interval cannot be rejected. This holds for any significance level α and $1 - \alpha$ confidence interval.

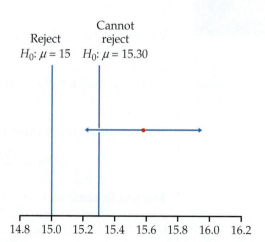

The calculation in Example 6.17 for a 1% significance test is very similar to the calculation for a 99% confidence interval. In fact, a two-sided test at significance level α can be carried out directly from a confidence interval with confidence level $C = 1 - \alpha$.

TWO-SIDED SIGNIFICANCE TESTS AND CONFIDENCE INTERVALS

A level α two-sided significance test rejects a hypothesis H_0: $\mu = \mu_0$ exactly when the value μ_0 falls outside a level $1 - \alpha$ confidence interval for μ.

USE YOUR KNOWLEDGE

6.47 Two-sided significance tests and confidence intervals. The P-value for a two-sided test of the null hypothesis H_0: $\mu = 30$ is 0.037.

(a) Does the 95% confidence interval include the value 30? Explain.

(b) Does the 99% confidence interval include the value 30? Explain.

6.48 More on two-sided tests and confidence intervals. A 95% confidence interval for a population mean is (29, 58).

(a) Can you reject the null hypothesis that $\mu = 50$ against the two-sided alternative at the 5% significance level? Explain.

(b) Can you reject the null hypothesis that $\mu = 60$ against the two-sided alternative at the 5% significance level? Explain.

The *P*-value versus a statement of significance

The observed result in Example 6.17 was $z = 4.02$. The conclusion that this result is significant at the 1% level does not tell the whole story. The observed z is far beyond the z corresponding to 1%, and the evidence against H_0 is far stronger than 1% significance suggests. The actual P-value

$$2P(Z \geq 4.02) = 0.000058$$

gives a better sense of how strong the evidence is. *The P-value is the smallest level α at which the data are significant.* Knowing the P-value allows us to assess significance at any level.

EXAMPLE 6.19

Test of the mean SATM score: Significance. In Example 6.16, we tested the hypotheses

$$H_0: \mu = 485$$
$$H_a: \mu > 485$$

concerning the mean SAT Mathematics score μ of California high school seniors. The test had the P-value $P = 0.0125$. This result is significant at the $\alpha = 0.05$ level because $0.0125 \leq 0.05$. It is not significant at the $\alpha = 0.01$ level, because the P-value is larger than 0.01. See Figure 6.14.

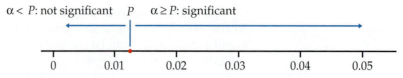

FIGURE 6.14 Link between the *P*-value and the significance level α. An outcome with *P*-value *P* is significant at all levels α at or above *P* and is not significant at smaller levels α.

A *P*-value is more informative than a reject-or-not finding at a fixed significance level. But assessing significance at a fixed level α is easier because no probability calculation is required. You need only look up a number in a table. A value z^* with a specified area to its right under the standard Normal **critical value**   curve is called a **critical value** of the standard Normal distribution. Because the practice of statistics almost always employs computer software or a calculator that calculates *P*-values automatically, the use of tables of critical values is becoming outdated. We include the usual tables of critical values (such as Table D) at the end of the book for learning purposes and to rescue students without good computing facilities. The tables can be used directly to carry out fixed α tests. They also allow us to approximate *P*-values quickly without a probability calculation. The following example illustrates the use of Table D to find an approximate *P*-value.

EXAMPLE 6.20

Average scholarship amount of borrowers and nonborrowers: Assessing significance. In Example 6.11 (page 365), we found the test statistic $z = 1.20$ for testing the null hypothesis that there was no difference in the mean scholarship amount between borrowers and nonborrowers. The alternative was two-sided. Under the null hypothesis, z has a standard Normal distribution, and from the last row in Table D, we can see that there is a 95% chance that z is between ± 1.96. Therefore, we reject H_0 in favor of H_a whenever z is outside this range. Because our calculated value is 1.20, we are within the range and we do not reject the null hypothesis at the 5% level of significance.

USE YOUR KNOWLEDGE

6.49 *P*-value and the significance level. The *P*-value for a significance test is 0.033.

(a) Do you reject the null hypothesis at level $\alpha = 0.05$?

(b) Do you reject the null hypothesis at level $\alpha = 0.01$?

(c) Explain how you determined your answers to parts (a) and (b).

6.50 More on *P*-value and the significance level. The *P*-value for a significance test is 0.069.

(a) Do you reject the null hypothesis at level $\alpha = 0.05$?

(b) Do you reject the null hypothesis at level $\alpha = 0.01$?

(c) Explain how you determined your answers to parts (a) and (b).

6.51 One-sided and two-sided P-values. The *P*-value for a two-sided significance test is 0.076.

(a) State the *P*-values for the two one-sided tests.

(b) What additional information do you need to properly assign these *P*-values to the $>$ and $<$ (one-sided) alternatives?

SECTION 6.2 SUMMARY

• A **test of significance** is intended to assess the evidence provided by data against a **null hypothesis H_0** in favor of an **alternative hypothesis H_a**.

• The hypotheses are stated in terms of population parameters. Usually, H_0 is a statement that no effect or no difference is present, and H_a says that there is an effect or difference in a specific direction (**one-sided alternative**) or in either direction (**two-sided alternative**).

• The test is based on a **test statistic**. The **P-value** is the probability, computed assuming that H_0 is true, that the test statistic will take a value at least as extreme as that actually observed. Small *P*-values indicate strong evidence against H_0. Calculating *P*-values requires knowledge of the sampling distribution of the test statistic when H_0 is true.

• If the *P*-value is as small or smaller than a specified value α, the data are **statistically significant** at significance level α.

• Significance tests for the hypothesis $H_0: \mu = \mu_0$ concerning the unknown mean μ of a population are based on the **z statistic**:

$$z = \frac{\bar{x} - \mu_0}{\sigma/\sqrt{n}}$$

The *z* test assumes an SRS of size n, known population standard deviation σ, and either a Normal population or a large sample. *P*-values are computed from the Normal distribution (Table A). Fixed α tests use the table of **standard Normal critical values** (Table D).

SECTION 6.2 EXERCISES

For Exercises 6.38 and 6.39, see page 364; for Exercises 6.40 and 6.41, see page 367; for Exercises 6.42 through 6.44, see page 370; for Exercises 6.45 and 6.46, see pages 374–375; for Exercises 6.47 and 6.48, see page 377; and for Exercises 6.49 through 6.51, see pages 378–379.

6.52 What's wrong? Here are several situations where there is an incorrect application of the ideas presented in this section. Write a short paragraph explaining what is wrong in each situation and why it is wrong.

(a) A researcher tests the following null hypothesis: H_0: $\bar{x} = 23$.

(b) A random sample of size 30 is taken from a population that is assumed to have a standard deviation of 5. The standard deviation of the sample mean is 5/30.

(c) A study with $\bar{x} = 45$ reports statistical significance for $H_a: \mu > 50$.

(d) A researcher tests the hypothesis H_0: $\mu = 350$ and concludes that the population mean is equal to 350.

6.53 What's wrong? Here are several situations where there is an incorrect application of the ideas presented in this section. Write a short paragraph explaining what is wrong in each situation and why it is wrong.

(a) A significance test rejected the null hypothesis that the sample mean is equal to 500.

(b) A test preparation company wants to test that the average score of its students on the ACT is better than the national average score of 21.2. The company states its null hypothesis to be $H_0: \mu > 21.2$.

(c) A study summary says that the results are statistically significant and the P-value is 0.98.

(d) The z test statistic is equal to 0.018. Because this is less than $\alpha = 0.05$, the null hypothesis was rejected.

6.54 Determining hypotheses. State the appropriate null hypothesis H_0 and alternative hypothesis H_a in each of the following cases.

(a) A 2015 study reported that 96% of students owned a cell phone. You plan to take an SRS of students to see if the percent has increased.

(b) The examinations in a large freshman chemistry class are scaled after grading so that the mean score is 75. The professor thinks that students who attend early-morning recitation sections will have a higher mean score than the class as a whole. Her students in these sections this semester can be considered a sample from the population of all students who might attend an early-morning section, so she compares their mean score with 75.

(c) The student newspaper at your college recently changed the format of its opinion page. You want to test whether students find the change an improvement. You take a random sample of students and select those who regularly read the newspaper. They are asked to indicate their opinions on the changes using a five-point scale: −2 if the new format is much worse than the old, −1 if the new format is somewhat worse than the old, 0 if the new format is the same as the old, +1 if the new format is somewhat better than the old, and +2 if the new format is much better than the old.

6.55 More on determining hypotheses. State the null hypothesis H_0 and the alternative hypothesis H_a in each case. Be sure to identify the parameters that you use to state the hypotheses.

(a) A university gives credit in first-year calculus to students who pass a placement test. The mathematics department wants to know if students who get credit in this way differ in their success with second-year calculus. Scores in second-year calculus are scaled so the average each year is equivalent to a 77. This year, 21 students who took second-year calculus passed the placement test.

(b) Experiments on learning in animals sometimes measure how long it takes a mouse to find its way through a maze. The mean time is 20 seconds for one particular maze. A researcher thinks that playing rap music will cause the mice to complete the maze more slowly. She measures how long each of 12 mice takes with the rap music as a stimulus.

(c) The average square footage of one-bedroom apartments in a new student-housing development is advertised to be 880 square feet. A student group thinks that the apartments are smaller than advertised. They hire an engineer to measure a sample of apartments to test their suspicion.

6.56 Even more on determining hypotheses. In each of the following situations, state an appropriate null hypothesis H_0 and alternative hypothesis H_a. Be sure to identify the parameters that you use to state the hypotheses. (We have not yet learned how to test these hypotheses.)

(a) A sociologist asks a large sample of high school students which television channel they like best. She suspects that a higher percent of males than of females will name MTV as their favorite channel.

(b) An education researcher randomly divides sixth-grade students into two groups for physical education class. He teaches both groups basketball skills, using the same methods of instruction in both classes. He encourages Group A with compliments and other positive behavior but acts cool and neutral toward Group B. He hopes to show that positive teacher attitudes result in a higher mean score on a test of basketball skills than do neutral attitudes.

(c) An education researcher believes that, among college students, there is a negative correlation between time spent at social network sites and self-esteem, measured on a 0 to 100 scale. To test this, she gathers social-networking information and self-esteem data from a sample of students at your college.

6.57 Translating research questions into hypotheses. Translate each of the following research questions into appropriate H_0 and H_a.

(a) U.S. Census Bureau data show that the mean household income in the area served by a shopping mall is $42,800 per year. A market research firm questions shoppers at the mall to find out whether the mean household income of mall shoppers is higher than that of the general population.

(b) Last year, your online registration technicians took an average of 0.4 hour to respond to trouble calls from students trying to register. Do this year's data show a different average response time?

6.58 Computing the *P*-value. A test of the null hypothesis $H_0: \mu = \mu_0$ gives test statistic $z = 1.89$.

(a) What is the *P*-value if the alternative is $H_a: \mu > \mu_0$?

(b) What is the *P*-value if the alternative is $H_a: \mu < \mu_0$?

(c) What is the *P*-value if the alternative is $H_a: \mu \neq \mu_0$?

6.59 More on computing the *P*-value. A test of the null hypothesis $H_0: \mu = \mu_0$ gives test statistic $z = -1.33$.

(a) What is the *P*-value if the alternative is $H_a: \mu > \mu_0$?

(b) What is the *P*-value if the alternative is $H_a: \mu < \mu_0$?

(c) What is the *P*-value if the alternative is $H_a: \mu \neq \mu_0$?

6.60 Timing of food intake and weight loss. A study found that a large group of late lunch eaters lost less weight over a 20-week observation period than a large group of early lunch eaters ($P = 0.002$).[16] Explain what this $P = 0.002$ means in a way that could be understood by someone who has not studied statistics.

6.61 Average starting salary. Refer to Exercise 6.22 (page 358). Use the information presented in the exercise to test that the average income of graduates from your institution is different from the national average ($\alpha = 0.01$). Write a short paragraph summarizing your conclusions.

6.62 Change in consumption of sweet snacks? Refer to Exercise 6.23 (page 358). A similar study performed four years earlier reported the average consumption of sweet snacks among healthy weight children aged 12 to 19 years to be 369.4 kilocalories per day (kcal/d). Does this current study suggest a change in the average consumption? Perform a significance test using the 5% significance level. Write a short paragraph summarizing the results.

6.63 Peer pressure and choice of major. A study followed a cohort of students entering a business/economics program.[17] All students followed a common track during the first three semesters and then chose to specialize in either business or economics. Through a series of surveys, the researchers were able to classify roughly 50% of the students as either peer driven (ignored abilities and chose major to follow peers) or ability driven (ignored peers and chose major based on ability). When looking at entry wages after graduation, the researchers conclude that a peer-driven student can expect an average wage that is 13% less than that of an ability-driven student. The report states that the significance level is $P = 0.09$. Can you be confident of

the researchers' conclusion statement regarding the wage decrease? Explain your answer.

6.64 Symbol of wealth in ancient China? Every society has its own symbols of wealth and prestige. In ancient China, it appears that owning pigs was such a symbol. Evidence comes from examining burial sites. If the skulls of sacrificed pigs tend to appear along with expensive ornaments, that suggests that the pigs, like the ornaments, signal the wealth and prestige of the person buried. A study of burials from around 3500 B.C. concluded that "there are striking differences in grave goods between burials with pig skulls and burials without them... A test indicates that the two samples of total artifacts are significantly different at the 0.01 level."[18] Explain clearly why "significantly different at the 0.01 level" gives good reason to think that there really is a systematic difference between burials that contain pig skulls and those that lack them.

6.65 Alcohol awareness among college students. A study of alcohol awareness among college students reported a higher awareness for students enrolled in a health and safety class than for those enrolled in a statistics class.[19] The difference is described as being statistically significant. Explain what this means in simple terms and offer an explanation for why the health and safety students had a higher mean score.

6.66 Change in eighth-grade average mathematics score. A report based on the 2015 National Assessment of Educational Progress (NAEP)[20] states that the average score on their mathematics test for eighth-grade students attending public schools is significantly higher than in 2011. The report also states that the average score for eighth-grade students attending private schools is not significantly different from the average score in 2011. A footnote states that comparisons are determined by two-sided statistical tests with 0.05 as the level of significance. Explain what this footnote means in language understandable to someone who knows no statistics. Do not use the word "significance" in your answer.

6.67 More on change in eighth-grade average mathematics score. Refer to the previous exercise. On the basis of the NAEP study, a friend who works for the school newspaper wants to report that between 2011 and 2013 the average mathematics score improved for students attending public schools but stayed the same for students attending private schools. Do you agree with this statement? Explain your answer.

6.68 Background television in homes of U.S. children. In one study, U.S. parents were surveyed to determine the amount of background television their children were exposed to. A total of $n = 1454$ families with one child between the ages of 8 months and 8 years participated.[21] For those families in which the caregiver had a high school degree or less, the child was exposed to an average of 313.0 minutes of background television per day. For those families in which the caregiver had some college or a college degree, the child was exposed to an average of 218.8 minutes per day. These average times were reported to be significantly different with $P < 0.05$. The actual P-value is 0.003. Explain why the actual P-value is more informative than the statement of significance at the 0.05 level.

6.69 Sleep quality and elevated blood pressure. A study looked at $n = 238$ adolescents, all free of severe illness.[22] Subjects wore a wrist actigraph, which allowed the researchers to estimate sleep patterns. Those subjects classified as having low sleep efficiency had an average systolic blood pressure that was 5.8 millimeters of mercury (mm Hg) higher than that of other adolescents. The standard deviation of this difference is 1.4 mm Hg. Based on these results, test whether this difference is significant at the 0.01 level.

6.70 Are the pine trees randomly distributed from north to south? In Example 6.1 (page 342), we looked at the distribution of longleaf pine trees in the Wade Tract. One way to formulate hypotheses about whether or not the trees are randomly distributed in the tract is to examine the average location in the north–south direction. The values range from 0 to 200, so if the trees are uniformly distributed in this direction, any difference from the middle value (100) should be due to chance variation. The sample mean for the 584 trees in the tract is 99.74. A theoretical calculation based on the assumption that the trees are uniformly distributed gives a standard deviation of 58. Carefully state the null and alternative hypotheses in terms of this variable. Note that this requires that you translate the research question about the random distribution of the trees into specific statements about the mean of a probability distribution. Test your hypotheses, report your results, and write a short summary of what you have found.

6.71 Are the pine trees randomly distributed from east to west? Answer the questions in the previous exercise for the east–west direction, for which the sample mean is 113.8.

6.72 Who is the author? Statistics can help decide the authorship of literary works. Sonnets by a certain Elizabethan poet are known to contain an average of $\mu = 8.9$ new words (words not used in the poet's other works). The standard deviation of the number of new

words is $\sigma = 2.5$. Now a manuscript with six new sonnets has come to light, and scholars are debating whether it is the poet's work. The new sonnets contain an average of $\bar{x} = 10.2$ words not used in the poet's known works. We expect poems by another author to contain more new words, so to see if we have evidence that the new sonnets are not by our poet we test

$$H_0: \mu = 8.9$$
$$H_a: \mu > 8.9$$

Give the z test statistic and its P-value. What do you conclude about the authorship of the new poems?

6.73 Attitudes toward school. The Survey of Study Habits and Attitudes (SSHA) is a psychological test that measures the motivation, attitude toward school, and study habits of students. Scores range from 0 to 200. The mean score for U.S. college students is about 95, and the standard deviation is about 20. A teacher who suspects that older students have better attitudes toward school gives the SSHA to 25 students who are at least 30 years of age. Their mean score is $\bar{x} = 103.3$.

(a) Assuming that $\sigma = 30$ for the population of older students, carry out a test of

$$H_0: \mu = 95$$
$$H_a: \mu > 95$$

Report the P-value of your test, and state your conclusion clearly.

(b) Your test in part (a) required two important assumptions in addition to the assumption that the value of σ is known. What are they? Which of these assumptions is most important to the validity of your conclusion in part (a)?

6.74 Nutritional intake among Canadian high-performance athletes. Since previous studies have reported that elite athletes are often deficient in their nutritional intake (for example, total calories, carbohydrates, protein), a group of researchers decided to evaluate Canadian high-performance athletes.[23] A total of $n = 324$ athletes from eight Canadian sports centers participated in the study. One reported finding was that the average caloric intake among the $n = 201$ women was 2403.7 kilocalories per day (kcal/d). The recommended amount is 2811.5 kcal/d. Is there evidence that female Canadian athletes are deficient in caloric intake?

(a) State the appropriate H_0 and H_a to test this.

(b) Assuming a standard deviation of 880 kcal/d, carry out the test. Give the P-value, and then interpret the result in plain language.

6.75 Are the measurements similar? Refer to Exercise 6.30 (page 360). In addition to the computer's calculations of miles per gallon, the driver also recorded the miles per gallon by dividing the miles driven by the number of gallons at each fill-up. The following data are the differences between the computer's and the driver's calculations for that random sample of 20 records. The driver wants to determine if these calculations are different. Assume that the standard deviation of a difference is $\sigma = 3.0$. MPGDIFF

5.0	6.5	−0.6	1.7	3.7	4.5	8.0	2.2	4.9	3.0
4.4	0.1	3.0	1.1	1.1	5.0	2.1	3.7	−0.6	−4.2

(a) State the appropriate H_0 and H_a to test this suspicion.

(b) Carry out the test. Give the P-value, and then interpret the result in plain language.

6.76 Impact of $\bar{x}$ on significance. The *Statistical Significance* applet illustrates statistical tests with a fixed level of significance for Normally distributed data with known standard deviation. Open the applet and keep the default settings for the null ($\mu = 0$) and the alternative ($\mu > 0$) hypotheses, the sample size ($n = 10$), the standard deviation ($\sigma = 1$), and the significance level ($\alpha = 0.05$). In the "I have data, and the observed $\bar{x}$ is $\bar{x} =$" box, enter the value 1. Is the difference between $\bar{x}$ and μ_0 significant at the 5% level? Repeat for $\bar{x}$ equal to 0.1, 0.2, 0.3, 0.4, 0.5, 0.6, 0.7, 0.8, 0.9. Make a table giving $\bar{x}$ and the results of the significance tests. What do you conclude?

6.77 Effect of changing α on significance. Repeat the previous exercise with significance level $\alpha = 0.01$. How does the choice of α affect which values of $\bar{x}$ are far enough away from μ_0 to be statistically significant?

6.78 Changing to a two-sided alternative. Repeat the previous exercise but with the two-sided alternative hypothesis. How does this change affect which values of $\bar{x}$ are far enough away from μ_0 to be statistically significant at the 0.01 level?

6.79 Changing the sample size. Refer to Exercise 6.76. Suppose that you increase the sample size n from 10 to 50. Again, make a table giving $\bar{x}$ and the results of the significance tests at the 0.05 significance level. What do you conclude?

6.80 Impact of $\bar{x}$ on the P-value. We can also study the P-value using the *Statistical Significance* applet. Reset the applet to the default settings for the null ($\mu = 0$) and the alternative ($\mu > 0$) hypotheses, the sample size ($n = 10$), the standard deviation ($\sigma = 1$), and the

significance level ($\alpha = 0.05$). In the "I have data, and the observed $\bar{x}$ is $\bar{x} =$" box, enter the value 1. What is the P-value? It is shown at the top of the blue vertical line. Repeat for $\bar{x}$ equal to 0.1, 0.2, 0.3, 0.4, 0.5, 0.6, 0.7, 0.8, 0.9. Make a table giving $\bar{x}$ and P-values. How does the P-value change as $\bar{x}$ moves farther away from μ_0?

6.81 Changing to a two-sided alternative, continued. Repeat the previous exercise but with the two-sided alternative hypothesis. How does this change affect the P-values associated with each $\bar{x}$? Explain why the P-values change in this way.

6.82 Other changes and the P-value. Refer to the previous exercise.

(a) What happens to the P-values when you change the significance level α to 0.01? Explain the result.

(b) What happens to the P-values when you change the sample size n from 10 to 50? Explain the result.

6.83 Understanding levels of significance. Explain in plain language why a significance test that is significant at the 1% level must always be significant at the 5% level.

6.84 More on understanding levels of significance. You are told that a significance test is significant at the 5% level. From this information, can you determine whether or not it is significant at the 1% level? Explain your answer.

6.85 Test statistic and levels of significance. Consider a significance test for a null hypothesis versus a two-sided alternative. Give a value of z that will give a result significant at the 1% level but not at the 0.5% level.

6.86 Using Table D to find a P-value. You have performed a two-sided test of significance and obtained a value of $z = 2.08$. Use Table D to find the approximate P-value for this test.

6.87 More on using Table D to find a P-value. You have performed a one-sided test of significance and obtained a value of $z = 1.03$. Use Table D to find the approximate P-value for this test when the alternative is greater than.

6.88 Using Table A and Table D to find a P-value. Consider a significance test for a null hypothesis versus a two-sided alternative. Between what values from Table D does the P-value for an outcome $z = 1.88$ lie? Calculate the P-value using Table A and verify that it lies between the values you found from Table D.

6.89 More on using Table A and Table D to find a P-value. Refer to the previous exercise. Find the P-value for $z = -1.88$.

6.3 Use and Abuse of Tests

When you complete this section, you will be able to:

- Explain why it is important to report the *P*-value and not just report whether the result is statistically significant or not.
- Discriminate between practical (or scientific) significance and statistical significance.
- Identify poorly designed studies where formal statistical inference is suspect.
- Understand the consequences of searching solely for statistical significance, whether through the investigation of multiple tests or by identifying and testing using the same data set.

Carrying out a test of significance is often quite simple, especially if the *P*-value is given effortlessly by a computer. Using tests wisely is not so simple. Each test is valid only in certain circumstances, with properly produced data being particularly important.

The *z* test, for example, should bear the same warning label that was attached in Section 6.1 to the corresponding confidence interval (page 355). Similar warnings accompany the other tests that we will learn. There are additional caveats that concern tests more than confidence intervals, enough to warrant this separate section. Some hesitation about the unthinking use of significance tests is a sign of statistical maturity.

The reasoning of significance tests has appealed to researchers in many fields, so that tests are widely used to report research results. In this setting H_a is a "research hypothesis" asserting that some effect or difference is present. The null hypothesis H_0 says that there is no effect or no difference. A low *P*-value represents good evidence that the research hypothesis is true. Here are some comments on the use of significance tests, with emphasis on their use in reporting scientific research.

Choosing a level of significance

The intention of a test of significance is to give a clear statement of the degree of evidence provided by the sample against the null hypothesis. The *P*-value does this. It is common practice to report *P*-values and to describe results as statistically significant whenever $P \le 0.05$. *However, there is no sharp border between "significant" and "not significant," only increasingly strong evidence as the P-value decreases.* Having the *P*-value with a description of the effect that we have found allows us to draw better conclusions from our data.

EXAMPLE 6.21

Information provided by the *P*-value. Suppose that the test statistic for a two-sided significance test for a population mean is $z = 1.95$. From Table A we can calculate the *P*-value. It is

$$P = 2[1 - P(Z \le 1.95)] = 2(1 - 0.9744) = 0.0512$$

We have failed to meet the standard of evidence for $\alpha = 0.05$. However, with the information provided by the *P*-value, we can see that the result just barely missed the standard. If the effect in question is interesting and potentially important, we might want to design another study with a larger sample to investigate it further.

Here is another example where the *P*-value provides useful information beyond that provided by the statement that we reject or fail to reject the null hypothesis.

EXAMPLE 6.22

More on information provided by the *P*-value. We have a test statistic of $z = -4.66$ for a two-sided significance test on a population mean. Software tells us that the *P*-value is 0.000003. This means that there are 3 chances in 1,000,000 of observing a sample mean this far or farther away from the null hypothesized value of μ. This kind of event is virtually impossible if the null hypothesis is true. There is no ambiguity in the result; we can clearly reject the null hypothesis.

We frequently report small *P*-values such as that in the previous example as $P < 0.001$. This corresponds to a chance of 1 in 1000 and is sufficiently small to lead us to a clear rejection of the null hypothesis.

One reason for the common use of $\alpha = 0.05$ is the great influence of Sir R. A. Fisher, the inventor of formal statistical methods for analyzing experimental data. Here is his opinion on choosing a level of significance: "A scientific fact should be regarded as experimentally established only if a properly designed experiment *rarely fails* to give this level of significance."[24]

What statistical significance does not mean

When a null hypothesis ("no effect" or "no difference") can be rejected at the usual level $\alpha = 0.05$, there is good evidence that an effect is present. That effect, however, can be extremely small. *When large samples are available, even tiny deviations from the null hypothesis will be statistically significant.*

EXAMPLE 6.23

It's significant but is it important? Suppose that we are testing the null hypothesis of no correlation between two variables. With 400 observations, an observed correlation of only $r = 0.1$ is significant evidence at the $\alpha = 0.05$ level that the correlation in the population is not zero. Figure 6.15 is an

FIGURE 6.15 Scatterplot of $n = 400$ observations with an observed correlation of 0.10, Example 6.23. There is not a strong association between the two variables even though there is significant evidence ($P < 0.05$) that the population correlation is not zero.

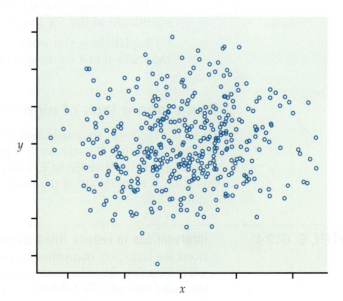

example of 400 (x, y) pairs that have an observed correlation of 0.10. The low significance level does *not* mean that there is a strong association, only that there is strong evidence of some association. The proportion of the variability in one of the variables explained by the other is $r^2 = 0.01$, or 1%.

For practical purposes, we might well decide to ignore this association. *Statistical significance is not the same as practical significance.* Statistical significance rarely tells us about the importance of the experimental results. This depends on the context of the experiment.

The remedy for attaching too much importance to statistical significance is to pay attention to the actual experimental results as well as to the *P*-value. Plot your data and examine them carefully. Beware of outliers. *The user of statistics who feeds the data to a computer without exploratory analysis will often be embarrassed.* It is usually wise to give a confidence interval for the parameter in which you are interested. Confidence intervals are not used as often as they should be, while tests of significance are overused.

USE YOUR KNOWLEDGE

6.90 Is it significant? More than 200,000 people worldwide take the GMAT examination each year when they apply for MBA programs. Their scores vary Normally with mean $\mu = 540$ and standard deviation $\sigma = 100$. One hundred students go through a rigorous training program designed to raise their GMAT scores. Test the following hypotheses about the training program

$$H_0: \mu = 540$$
$$H_a: \mu > 540$$

in each of the following situations.

(a) The students' average score is $\bar{x} = 556.4$. Is this result significant at the 5% level?

(b) Now suppose that the average score is $\bar{x} = 556.5$. Is this result significant at the 5% level?

(c) Explain how you would reconcile this difference in significance, especially if any increase greater than 15 points is considered a success.

Don't ignore lack of significance

There is a tendency to conclude that there is no effect whenever a *P*-value fails to attain the usual 5% standard. A provocative editorial in the *British Medical Journal* entitled "Absence of Evidence Is Not Evidence of Absence" deals with this issue.[25] Here is one of the examples they cite.

EXAMPLE 6.24

Interventions to reduce HIV-1 transmission. A randomized trial of interventions for reducing transmission of HIV-1 reported an incident rate ratio of 1.00, meaning that the intervention group and the control group both had the same rate of HIV-1 infection. The 95% confidence interval was reported as 0.63 to 1.58.[26] The editorial notes that a summary of these results that says the intervention has no effect on HIV-1 infection is misleading. The confidence interval indicates that the intervention may be capable of achieving a 37% decrease in infection; it might also be harmful and produce a 58% increase in infection. Clearly, more data are needed to distinguish between these possibilities.

The situation can be worse. Research in some fields has rarely been published unless significance at the 0.05 level is attained.

EXAMPLE 6.25

Journal survey of reported significance results. A survey of four journals published by the American Psychological Association showed that of 294 articles using statistical tests, only eight reported results that did not attain the 5% significance level.[27] It is very unlikely that these were the only eight studies of scientific merit that did not attain significance at the 0.05 level. Manuscripts describing other studies were likely rejected because of a lack of statistical significance or never submitted in the first place due to the expectation of rejection.

In some areas of research, small effects that are detectable only with large sample sizes can be of great practical significance. Data accumulated from a large number of patients taking a new drug may be needed before we can conclude that there are life-threatening consequences for a small number of people.

On the other hand, sometimes a meaningful result is not found significant.

EXAMPLE 6.26

A meaningful but statistically insignificant result. A sample of size 10 gave a correlation of $r = 0.5$ between two variables. The P-value is 0.102 for a two-sided significance test. In many situations, a correlation this large would be interesting and worthy of additional study. When it takes a lot of effort (say, in terms of time or money) to obtain samples, researchers often use small studies like these as pilot projects to gain interest from various funding sources. With financial support, a larger, more powerful study can then be run.

Another important aspect of planning a study is to verify that the test you plan to use does have high probability of detecting an effect of the size you hope to find. This probability is the *power* of the test. Power calculations are discussed in Section 6.4.

design of experiments, p. 171

Statistical inference is not valid for all sets of data

In Chapter 3, we learned that badly designed surveys or experiments often produce invalid results. *Formal statistical inference cannot correct basic flaws in the design.*

EXAMPLE 6.27

LOOK BACK
confounding,
p. 150

English vocabulary and studying a foreign language. There is no doubt that there is a significant difference in English vocabulary scores between high school seniors who have studied a foreign language and those who have not. But because the effect of actually studying a language is confounded with the differences between students who choose language study and those who do not, this statistical significance is hard to interpret. The most plausible explanation is that students who were already good at English chose to study another language. A randomized comparative experiment would isolate the actual effect of language study and so make significance meaningful. Do you think it would be ethical to do such a study?

Tests of significance and confidence intervals are based on the laws of probability. Randomization in sampling or experimentation ensures that these laws apply. But we must often analyze data that do not arise from randomized samples or experiments. *To apply statistical inference to such data, we must have confidence in a probability model for the data.* The diameters of successive holes bored in auto engine blocks during production, for example, may behave like independent observations from a Normal distribution. We can check this probability model by examining the data. If the Normal distribution model appears approximately correct, we can apply the methods of this chapter to do inference about the process mean diameter μ.

USE YOUR KNOWLEDGE

6.91 Home security systems. A recent TV advertisement for home security systems said that homes without an alarm system are three times more likely to be broken into. Suppose that this conclusion was obtained by examining an SRS of police records of break-ins and determining whether the percent of homes with alarm systems was significantly smaller than 50%. Explain why the significance of this study is suspect and propose an alternative study that would help clarify the importance of an alarm system.

Beware of searching for significance

Statistical significance is an outcome much desired by researchers. It means (or ought to mean) that you have found an effect that you were looking for. *The reasoning behind statistical significance works well if you decide what effect you are seeking, design an experiment or sample to search for it, and use a test of significance to weigh the evidence you get.* But because a successful search for a new scientific phenomenon often ends with statistical significance, it is all too tempting to make significance itself the object of the search. There are several ways to do this, none of them acceptable in polite scientific society.

EXAMPLE 6.28

Genomics studies. In genomics experiments, it is common to assess the differences in expression for tens of thousands of genes. If each of these genes was examined separately and statistical significance declared for all that had *P*-values that pass the 0.05 standard, we would have quite a mess. In the absence of any real biological effects, we would expect that, by chance alone, approximately 5% of these tests will show statistical significance. Much research in genomics is directed toward appropriate ways to deal with this situation.[28]

We do not mean that searching data for suggestive patterns is not proper scientific work. It certainly is. Many important discoveries have been made by accident rather than by design. Exploratory analysis of data is an essential part of statistics. We do mean that the usual reasoning of statistical inference does not apply when the search for a pattern is successful. *You cannot legitimately test a hypothesis on the same data that first suggested that hypothesis.* The remedy is clear. Once you have a hypothesis, design a study to search specifically for the effect you now think is there. If the result of this study is statistically significant, you have real evidence.

SECTION 6.3 SUMMARY

- *P*-values are more informative than the reject-or-not result of a level α test. Beware of placing too much weight on traditional values of α, such as $\alpha = 0.05$.

- Very small effects can be highly significant (small *P*), especially when a test is based on a large sample. A statistically significant effect need not be practically important. Plot the data to display the effect you are seeking, and use confidence intervals to estimate the actual values of parameters.

- On the other hand, lack of significance does not imply that H_0 is true, especially when the test has a low probability of detecting an effect.

- Significance tests are not always valid. Faulty data collection, outliers in the data, and testing a hypothesis on the same data that suggested the hypothesis can invalidate a test. Many tests run at once will probably produce some significant results by chance alone, even if all the null hypotheses are true.

SECTION 6.3 EXERCISES

For Exercise 6.90, see page 386; and for Exercise 6.91, see page 388.

6.92 A role as a statistical consultant. You are the statistical expert for a graduate student planning her PhD research. After you carefully present the mechanics of significance testing, she suggests using $\alpha = 0.20$ for the study because she would be more likely to obtain statistically significant results and she *really* needs significant results to graduate. Explain in simple terms why this would not be a good use of statistical methods.

6.93 What do you know? A research report described two results that both achieved statistical significance at the 5% level. The *P*-value for the first is 0.048; for the second it is 0.0002. Do the *P*-values add any useful information beyond that conveyed by the statement that both results are statistically significant? Write a short paragraph explaining your views on this question.

6.94 Selective publication based on results. In addition to statistical significance, selective publication can also be due to the observed outcome. A recent review of 74 studies of antidepressant agents found 38 studies with positive results and 36 studies with negative or questionable results. All but one of the 38 positive studies were published. Of the remaining 36, 22 were not published and 11 were published in such a way as to convey a positive outcome.[29] Describe how this selective reporting can have adverse consequences on health care.

6.95 What a test of significance can answer. Explain whether a test of significance can answer each of the following questions.

(a) Is the sample or experiment properly designed?

(b) Is the observed effect compatible with the null hypothesis?

(c) Is the observed effect important?

6.96 Vitamin C and colds. In a study to investigate whether vitamin C will prevent colds, 400 subjects are assigned at random to one of two groups. The experimental group takes a vitamin C tablet daily, while the control group takes a placebo. At the end of the experiment, the researchers calculate the difference between the percents of subjects in the two groups who were free of colds. This difference is statistically significant ($P = 0.03$) in favor of the vitamin C group. Can we conclude that vitamin C has a strong effect in preventing colds? Explain your answer.

6.97 How far do rich parents take us? How much education children get is strongly associated with the wealth and social status of their parents, termed "socioeconomic status," or SES. The SES of parents, however, has little influence on whether children who have graduated from college continue their education. One study looked at whether college graduates took the graduate admissions tests for business, law, and other graduate programs. The effects of the parents' SES on taking the LSAT test for law school were "both statistically insignificant and small."

(a) What does "statistically insignificant" mean?

(b) Why is it important that the effects were small in size as well as statistically insignificant?

6.98 Do you agree? State whether or not you agree with each of the following statements and provide a short summary of the reasons for your answers.

(a) If the P-value is larger than 0.05, the null hypothesis is true.

(b) Practical significance is not the same as statistical significance.

(c) We can perform a statistical analysis using any set of data.

(d) If you find an interesting pattern in a set of data, it is appropriate to then use a significance test to determine its significance.

(e) It's always better to use a significance level of $\alpha = 0.05$ than to use $\alpha = 0.01$ because it is easier to find statistical significance.

6.99 Practical significance and sample size. Every user of statistics should understand the distinction between statistical significance and practical importance. A sufficiently large sample will declare very small effects statistically significant. Consider the study of elite female Canadian athletes in Exercise 6.74 (page 382). Female athletes were consuming an average of 2403.7 kcal/d with a standard deviation of 880 kcal/d. Suppose that a nutritionist is brought in to implement a new health program for these athletes. This program should increase mean caloric intake but not change the standard deviation. Given the standard deviation and how calorie deficient these athletes are, a change in the mean of 50 kcal/d to 2453.7 is of little importance. However, with

a large enough sample, this change can be significant. To see this, calculate the P-value for the test of

$$H_0: \mu = 2403.7$$
$$H_a: \mu > 2403.7$$

in each of the following situations:

(a) A sample of 100 athletes; their average caloric intake is $\bar{x} = 2453.7$.

(b) A sample of 500 athletes; their average caloric intake is $\bar{x} = 2453.7$.

(c) A sample of 2500 athletes; their average caloric intake is $\bar{x} = 2453.7$.

6.100 Statistical versus practical significance. A study with 7500 subjects reported a result that was statistically significant at the 5% level. Explain why this result might not be particularly important.

6.101 More on statistical versus practical significance. A study with 14 subjects reported a result that failed to achieve statistical significance at the 5% level. The P-value was 0.051. Write a short summary of how you would interpret these findings.

6.102 Find journal articles. Find two journal articles that report results with statistical analyses. For each article, summarize how the results are reported and write a critique of the presentation. Be sure to include details regarding use of significance testing at a particular level of significance, P-values, and confidence intervals.

6.103 Create an example of your own. For each of the following cases, provide an example and an explanation as to why it is appropriate.

(a) A set of data or an experiment for which statistical inference is not valid.

(b) A set of data or an experiment for which statistical inference is valid.

6.104 Predicting success of trainees. What distinguishes managerial trainees who eventually become executives from those who, after expensive training, don't succeed and leave the company? We have abundant data on past trainees—data on their personalities and goals, their college preparation and performance, even their family backgrounds and their hobbies. Statistical software makes it easy to perform dozens of significance tests on these dozens of variables to see which ones best predict later success. We find that future executives are significantly more likely than washouts to have an urban or suburban upbringing and an undergraduate degree in a technical field.

Explain clearly why using these "significant" variables to select future trainees is not wise. Then suggest a follow-up study using this year's trainees as subjects that should clarify the importance of the variables identified by the first study.

6.105 Searching for significance. Give an example of a situation where searching for significance would lead to misleading conclusions.

6.106 More on searching for significance. You perform 1000 significance tests using $\alpha = 0.05$. Assuming that all null hypotheses are true, about how many of the test results would you expect to be statistically significant? Explain how you obtained your answer.

6.107 Interpreting a very small *P*-value. Assume that you are performing a large number of significance tests. Let n be the number of these tests. How large would n need to be for you to expect about one *P*-value to be 0.00001 or smaller? Use this information to write an explanation of how to interpret a result that has $P = 0.00001$ in this setting.

 6.108 An adjustment for multiple tests. One way to deal with the problem of misleading *P*-values when performing more than one significance test is to adjust the criterion you use for statistical significance. The **Bonferroni procedure** does this in a simple way. If you perform two tests and want to use the $\alpha = 5\%$ significance level, you would require a *P*-value of $0.05/2 = 0.025$ to declare either one of the tests significant. In general, if you perform k tests and want protection at level α, use α/k as your cutoff for statistical significance. You perform six tests and obtain individual *P*-values of 0.075, 0.021, 0.285, 0.002, 0.015, and <0.001. Which of these are statistically significant using the Bonferroni procedure with $\alpha = 0.05$?

6.109 Significance using the Bonferroni procedure. Refer to the previous exercise. A researcher has performed 12 tests of significance and wants to apply the Bonferroni procedure with $\alpha = 0.05$. The calculated *P*-values are 0.141, 0.519, 0.186, 0.753, 0.001, 0.008, 0.646, 0.038, 0.898, 0.013, <0.002, and 0.538. Which of these tests reject their null hypotheses with this procedure?

6.4 Power and Inference as a Decision

When you complete this section, you will be able to:	• Define what is meant by the power of a test.
	• Determine the power of a test to detect an alternative for a given sample size n.
	• Describe the two types of possible errors when performing a test that focuses on deciding between two hypotheses.
	• Relate the two errors to the significance level and power of the test.

Although we prefer to use *P*-values rather than the reject-or-not view of the level α significance test, the latter view is very important for planning studies and for understanding statistical decision theory. We will discuss these two topics in this section.

Power

Level α significance tests are closely related to confidence intervals—in fact, we saw that a two-sided test can be carried out directly from a confidence interval (pages 353–354). The significance level, like the confidence level, says how reliable the method is in repeated use. If we use 5% significance tests repeatedly when H_0 is, in fact, true, we will be wrong (the test will reject H_0) 5% of the time and right (the test will fail to reject H_0) 95% of the time.

The ability of a test to detect that H_0 is false is measured by the probability that the test will reject H_0 when an alternative is true. The higher this probability is, the more sensitive the test is.

POWER

The probability that a level α significance test will reject H_0 when a particular alternative value of the parameter is true is called the **power** of the test to detect that alternative.

EXAMPLE 6.29

The power of a TBBMC significance test. Can a six-month exercise program increase the total body bone mineral content (TBBMC) of young women? A team of researchers is planning a study to examine this question. Based on the results of a previous study, they are willing to assume that $\sigma = 2$ for the percent change in TBBMC over the six-month period. They also believe that a change in TBBMC of 1% is important, so they would like to have a reasonable chance of detecting a change this large or larger. Is 25 subjects a large enough sample for this project?

We will answer this question by calculating the power of the significance test that will be used to evaluate the data to be collected. The calculation consists of three steps:

1. State H_0, H_a (the particular alternative we want to detect), and the significance level α.

2. Find the values of $\bar{x}$ that will lead us to reject H_0.

3. Calculate the probability of observing these values of $\bar{x}$ when the alternative is true.

Step 1. The null hypothesis is that the exercise program has no effect on TBBMC. In other words, the mean percent change is zero. The alternative is that exercise is beneficial; that is, the mean change is positive. Formally, we have

$$H_0: \mu = 0$$
$$H_a: \mu > 0$$

The alternative of interest is $\mu = 1\%$ increase in TBBMC. A 5% test of significance will be used.

Step 2. The z test rejects H_0 at the $\alpha = 0.05$ level whenever

$$z = \frac{\bar{x} - \mu_0}{\sigma/\sqrt{n}} = \frac{\bar{x} - 0}{2/\sqrt{25}} \geq 1.645$$

Be sure you understand why we use 1.645. Rewrite this in terms of $\bar{x}$:

$$\bar{x} \geq 1.645 \frac{2}{\sqrt{25}}$$

$$\bar{x} \geq 0.658$$

Because the significance level is $\alpha = 0.05$, this event has probability 0.05 of occurring *when the population mean μ is 0*.

Step 3. The power to detect the alternative $\mu = 1\%$ is the probability that H_0 will be rejected *when in fact $\mu = 1\%$*. We calculate this probability

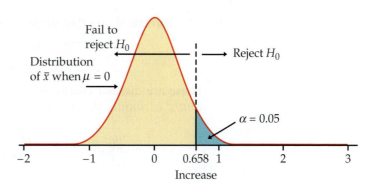

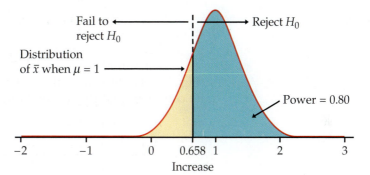

FIGURE 6.16 The sampling distributions of $\bar{x}$ when $\mu = 0$ and when $\mu = 1$, Example 6.29. The power is the probability that the test rejects H_0 when the alternative is true.

by standardizing $\bar{x}$, using the value $\mu = 1$, the population standard deviation $\sigma = 2$, and the sample size $n = 25$. The power is

$$P(\bar{x} \geq 0.658 \text{ when } \mu = 1) = P\left(\frac{\bar{x} - \mu}{\sigma/\sqrt{n}} \geq \frac{0.658 - 1}{2/\sqrt{25}}\right)$$

$$= P(Z \geq -0.855) = 0.80$$

Figure 6.16 illustrates the power with the sampling distribution of $\bar{x}$ when $\mu = 1$. This significance test rejects the null hypothesis that exercise has no effect on TBBMC 80% of the time if the true effect of exercise is a 1% increase in TBBMC. If the true effect of exercise is a greater percent increase, the test will have greater power; it will reject with a higher probability.

Here is another example of a power calculation, this time for a two-sided z test.

EXAMPLE 6.30

Power of the lead concentration test. Example 6.17 (page 375) presented a test of

$$H_0: \mu = 15.00$$
$$H_a: \mu \neq 15.00$$

at the 1% level of significance. What is the power of this test against the specific alternative $\mu = 15.50$?

The test rejects H_0 when $|z| \geq 2.576$. The test statistic is

$$z = \frac{\bar{x} - 15.00}{0.25/\sqrt{3}}$$

Some arithmetic shows that the test rejects when either of the following is true:

$$z \geq 2.576 \quad \text{(in other words, } \bar{x} \geq 15.37\text{)}$$
$$z \leq -2.576 \quad \text{(in other words, } \bar{x} \leq 14.63\text{)}$$

These are disjoint events, so the power is the sum of their probabilities, *computed assuming that the alternative $\mu = 15.50$ is true.* We find that

$$P(\bar{x} \geq 15.37) = P\left(\frac{\bar{x} - \mu}{\sigma/\sqrt{n}} \geq \frac{15.37 - 15.50}{0.25/\sqrt{3}}\right)$$

$$= P(Z \geq -0.90) = 0.8159$$

$$P(\bar{x} \leq 14.63) = P\left(\frac{\bar{x} - \mu}{\sigma/\sqrt{n}} \leq \frac{14.63 - 15.50}{0.25/\sqrt{3}}\right)$$

$$= P(Z \leq -6.03) \doteq 0$$

Figure 6.17 illustrates this calculation. A power of about 0.82, we are quite confident that the test will reject H_0 when this alternative is true.

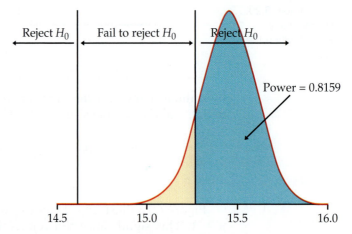

FIGURE 6.17 The power, Example 6.30. Unlike Figure 6.16, only the sampling distribution under the alternative is shown.

High power is desirable. Along with 95% confidence intervals and 5% significance tests, 80% power is becoming a standard. Many U.S. government agencies that provide research funds require that the sample size for the funded studies be sufficient to detect important results 80% of the time using a 5% test of significance.

EXAMPLE 6.31

Constructing a power curve. Example 6.30 considered one specific alternative, $\mu = 15.50$. Often, it is helpful to consider the power for a range of alternatives. Fortunately, most statistical software saves us from having to do these calculations manually. Figure 6.18 shows Minitab output for the power over the range 15.00 ppm to 15.80 ppm. The power calculation of Example 6.30 is represented by a dot on the curve at a difference of $15.50 - 15.00 = 0.50$. This curve is very informative. We see that with a sample size of three, the power is greater than 80% only for differences larger than about 0.48. If it is important to detect differences less than this, the Deely Laboratory needs to consider ways to increase the power.

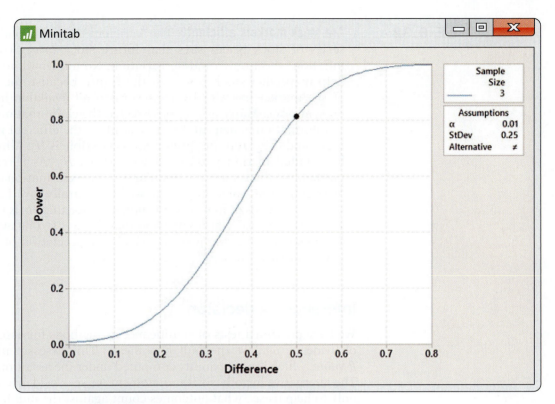

FIGURE 6.18 Minitab output (a power curve) for the one-sample power calculation, Example 6.31.

Increasing the power

Suppose that you have performed a power calculation and found that the power is too small. What can you do to increase it? Here are four ways. Note the similarity between these and the choices to reduce the margin of error (page 352).

- Increase α. A 5% test of significance will have a greater chance of rejecting the alternative than a 1% test because the strength of evidence required for rejection is less.

- Consider a particular alternative that is farther away from μ_0. Values of μ that are in H_a but lie close to the hypothesized value μ_0 are harder to detect (lower power) than values of μ that are far from μ_0.

- Increase the sample size. More data will provide more information about $\bar{x}$ so we have a better chance of distinguishing values of μ.

- Decrease σ. This has the same effect as increasing the sample size: more information about μ. Improving the measurement process and restricting attention to a subpopulation are possible ways to decrease σ.

Power calculations are important in planning studies. Using a significance test with low power makes it unlikely that you will find a significant effect even if the truth is far from the null hypothesis. A null hypothesis that is, in fact, false can become widely believed if repeated attempts to find evidence against it fail because of low power. The following example illustrates this point.

EXAMPLE 6.32

Are stock markets efficient? The "efficient market hypothesis" for the time series of stock prices says that future stock prices (when adjusted for inflation) show only random variation. No information available now will help us predict stock prices in the future because the efficient working of the market has already incorporated all available information in the present price. Many studies have tested the claim that one or another kind of information is helpful. In these studies, the efficient market hypothesis is H_0, and the claim that prediction is possible is H_a. Almost all the studies have failed to find good evidence against H_0. As a result, the efficient market hypothesis is quite popular. But an examination of the significance tests employed finds that the power is generally low. Failure to reject H_0 when using tests of low power is not evidence that H_0 is true. As one expert says, "The widespread impression that there is strong evidence for market efficiency may be due just to a lack of appreciation of the low power of many statistical tests."[30]

Inference as decision

We have presented tests of significance as methods for assessing the strength of evidence against the null hypothesis. This assessment is made by the P-value, which is a probability computed under the assumption that H_0 is true. The alternative hypothesis (the statement we seek evidence for) enters the test only to help us see what outcomes count against the null hypothesis.

There is another way to think about these issues. Sometimes, we are really concerned about making a decision or choosing an action based on our evaluation of the data. **Acceptance sampling** is one such circumstance. A producer of bearings and a skateboard manufacturer agree that each carload lot of bearings shall meet certain quality standards. When a carload arrives, the manufacturer chooses a sample of bearings to be inspected. On the basis of the sample outcome, the manufacturer will either accept or reject the carload. Let's examine how the idea of inference as a decision changes the reasoning used in tests of significance.

acceptance sampling

Two types of error

Tests of significance concentrate on H_0, the null hypothesis. If a decision is called for, however, there is no reason to single out H_0. There are simply two hypotheses, and we must accept one and reject the other. It is convenient to call the two hypotheses H_0 and H_a, but H_0 no longer has the special status (the statement we try to find evidence against) that it had in tests of significance. In the acceptance sampling problem, we must decide between

H_0: the lot of bearings meets standards

H_a: the lot does not meet standards

on the basis of a sample of bearings.

We hope that our decision will be correct, but sometimes it will be wrong. There are two types of incorrect decisions. We can accept a bad lot of bearings, or we can reject a good lot. Accepting a bad lot injures the consumer, while rejecting a good lot hurts the producer. To help distinguish these two types of error, we give them specific names.

		Truth about the population	
		H_0 true	H_a true
Decision based on sample	Reject H_0	Type I error	Correct decision
	Accept H_0	Correct decision	Type II error

FIGURE 6.19 The two types of error in testing hypotheses.

		Truth about the lot	
		Does meet standards	Does not meet standards
Decision based on sample	Reject the lot	Type I error	Correct decision
	Accept the lot	Correct decision	Type II error

FIGURE 6.20 The two types of error in the acceptance sampling setting.

TYPE I AND TYPE II ERRORS

If we reject H_0 (accept H_a) when in fact H_0 is true, this is a **Type I error**.
If we accept H_0 (reject H_a) when in fact H_a is true, this is a **Type II error**.

The possibilities are summed up in Figure 6.19. If H_0 is true, our decision either is correct (if we accept H_0) or is a Type I error. If H_a is true, our decision either is correct or is a Type II error. Only one error is possible at one time. Figure 6.20 applies these ideas to the acceptance sampling example.

Error probabilities

Any rule for making decisions is assessed in terms of the probabilities of the two types of error. This is in keeping with the idea that statistical inference is based on probability. We cannot (short of inspecting the whole lot) guarantee that good lots of bearings will never be rejected and bad lots never be accepted. But by random sampling and the laws of probability, we can say what the probabilities of both kinds of error are.

Significance tests with fixed level α give a rule for making decisions because the test either rejects H_0 or fails to reject it. If we adopt the decision-making way of thought, failing to reject H_0 means deciding that H_0 is true. We can then describe the performance of a test by the probabilities of Type I and Type II errors.

EXAMPLE 6.33

Outer diameter of a skateboard bearing. The mean outer diameter of a skateboard bearing is supposed to be 22.000 millimeters (mm). The outer diameters vary Normally with standard deviation $\sigma = 0.010$ mm. When a lot of the bearings arrives, the skateboard manufacturer takes an SRS of five bearings from the lot and measures their outer diameters. The manufacturer rejects the bearings if the sample mean diameter is significantly different from 22 mm at the 5% significance level.

This is a test of the hypotheses

$$H_0: \mu = 22$$
$$H_a: \mu \neq 22$$

To carry out the test, the manufacturer computes the z statistic:

$$z = \frac{\bar{x} - 22}{0.01/\sqrt{5}}$$

and rejects H_0 if

$$z < -1.96 \quad \text{or} \quad z > 1.96$$

A Type I error is to reject H_0 when in fact $\mu = 22$.

What about Type II errors? Because there are many values of μ in H_a, we will concentrate on one value. The producer and the manufacturer agree that a lot of bearings with mean 0.015 mm away from the desired mean 22.000 should be rejected. So a particular Type II error is to accept H_0 when in fact $\mu = 22.015$.

Figure 6.21 shows how the two probabilities of error are obtained from the two sampling distributions of $\bar{x}$, for $\mu = 22$ and for $\mu = 22.015$. When $\mu = 22$, H_0 is true and to reject H_0 is a Type I error. When $\mu = 22.015$, accepting H_0 is a Type II error. We will now calculate these error probabilities.

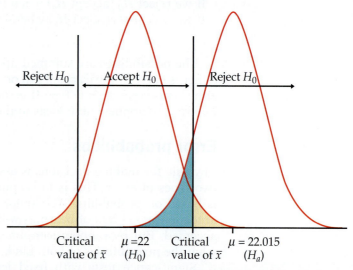

FIGURE 6.21 The two error probabilities, Example 6.33. The probability of a Type I error (yellow area) is the probability of rejecting H_0: $\mu = 22$ when, in fact, $\mu = 22$. The probability of a Type II error (blue area) is the probability of accepting H_0 when, in fact, $\mu = 22.015$.

The probability of a Type I error is the probability of rejecting H_0 when it is really true. In Example 6.33, this is the probability that $|z| \geq 1.96$ when $\mu = 22$. But this is exactly the significance level of the test. The critical value 1.96 was chosen to make this probability 0.05, so we do not have to compute it again. The definition of "significant at level 0.05" is that sample outcomes this extreme will occur with probability 0.05 when H_0 is true.

SIGNIFICANCE AND TYPE I ERROR

The significance level α of any fixed level test is the probability of a Type I error. That is, α is the probability that the test will reject the null hypothesis H_0 when H_0 is in fact true.

The probability of a Type II error for the particular alternative $\mu = 22.015$ in Example 6.33 is the probability that the test will fail to reject H_0 when μ has this alternative value. The *power* of the test to detect the alternative $\mu = 22.015$ is just the probability that the test *does* reject H_0. By following the method of Example 6.30, we can calculate that the power is about 0.92. The probability of a Type II error is therefore $1 - 0.92$, or 0.08.

POWER AND TYPE II ERROR

The power of a fixed level test to detect a particular alternative is 1 minus the probability of a Type II error for that alternative.

The two types of error and their probabilities give another interpretation of the significance level and power of a test. The distinction between tests of significance and tests as rules for deciding between two hypotheses does not lie in the calculations *but in the reasoning that motivates the calculations*. In a test of significance, we focus on a single hypothesis (H_0) and a single probability (the P-value). The goal is to measure the strength of the sample evidence against H_0. Calculations of power are done to check the sensitivity of the test. If we cannot reject H_0, we conclude only that there is not sufficient evidence against H_0, not that H_0 is actually true.

If the same inference problem is thought of as a decision problem, we focus on two hypotheses and give a rule for deciding between them based on the sample evidence. Therefore, we must focus equally on two probabilities, the probabilities of the two types of error. We must choose one hypothesis and cannot abstain on grounds of insufficient evidence.

The common practice of testing hypotheses

Such a clear distinction between the two ways of thinking is helpful for understanding. In practice, the two approaches often merge. We continued to call one of the hypotheses in a decision problem H_0. The common practice of *testing hypotheses* mixes the reasoning of significance tests and decision rules as follows:

1. State H_0 and H_a just as in a test of significance.

2. Think of the problem as a decision problem, so that the probabilities of Type I and Type II errors are relevant.

3. Because of Step 1, Type I errors are more serious. So choose an α (significance level) and consider only tests with probability of a Type I error no greater than α.

4. Among these tests, select one that makes the probability of a Type II error as small as possible (that is, power as large as possible). If this probability is too large, you will have to take a larger sample to reduce the chance of an error.

Testing hypotheses may seem to be a hybrid approach. It was, historically, the effective beginning of decision-oriented ideas in statistics. An impressive mathematical theory of hypothesis testing was developed between 1928 and 1938 by Jerzy Neyman and Egon Pearson. The decision-making approach came later (1940s). Because decision theory in its pure form leaves you with two error probabilities and no simple rule on how to balance them, it has been used less often than either tests of significance or tests of hypotheses. Decision ideas have been applied in testing problems mainly by way of the Neyman-Pearson hypothesis-testing theory. That theory asks you first to choose α, and the influence of Fisher has often led users of hypothesis testing comfortably back to $\alpha = 0.05$ or $\alpha = 0.01$. Fisher, who was exceedingly argumentative, violently attacked the Neyman-Pearson decision-oriented ideas, and the argument still continues.

SECTION 6.4 SUMMARY

• The **power** of a significance test measures its ability to detect an alternative hypothesis. The power to detect a specific alternative is calculated as the probability that the test will reject H_0 when that alternative is true. This calculation requires knowledge of the sampling distribution of the test statistic under the alternative hypothesis. Increasing the size of the sample increases the power when the significance level remains fixed.

• An alternative to significance testing regards H_0 and H_a as two statements of equal status that we must decide between. This **decision theory** point of view regards statistical inference in general as giving rules for making decisions in the presence of uncertainty.

• In the case of testing H_0 versus H_a, decision analysis chooses a decision rule on the basis of the probabilities of two types of error. A **Type I error** occurs if H_0 is rejected when it is in fact true. A **Type II error** occurs if H_0 is accepted when in fact H_a is true.

• In a fixed level α significance test, the significance level α is the probability of a Type I error, and the power to detect a specific alternative is 1 minus the probability of a Type II error for that alternative.

SECTION 6.4 EXERCISES

6.110 Make a recommendation. Your manager has asked you to review a research proposal that includes a section on sample size justification. A careful reading of this section indicates that the power is 18% for detecting an effect that would be considered important. Write a short report for your manager explaining what this means and make a recommendation on whether or not this study should be run.

6.111 Explain power and sample size. Two studies are identical in all respects except for the sample sizes. Consider the power versus a particular sample size. Will the study with the larger sample size have more power or less power than the one with the smaller sample size? Explain your answer in terms that could be understood by someone with very little knowledge of statistics.

6.112 Power for a different alternative. The power for a two-sided test of the null hypothesis $\mu = 0$ versus the alternative $\mu = 6$ is 0.83. What is the power versus the alternative $\mu = -6$? Explain your answer.

6.113 More on the power for a different alternative. A one-sided test of the null hypothesis $\mu = 20$ versus the alternative $\mu = 30$ has power equal to 0.73. Will the power for the alternative $\mu = 35$ be higher or lower than 0.73? Draw a picture and use this to explain your answer.

6.114 Effect of changing the alternative μ on power. The *Statistical Power* applet illustrates the power calculation similar to that in Figure 6.16 (page 393). Open the applet and keep the default settings for the null ($\mu = 0$) and the alternative ($\mu > 0$) hypotheses, the sample size ($n = 10$), the standard deviation ($\sigma = 1$), and the significance level ($\alpha = 0.05$). In the "alt $\mu =$" box, enter the value 1. What is the power? Repeat for alternative μ equal to 0.1, 0.2, 0.3, 0.4, 0.5, 0.6, 0.7, 0.8, 0.9. Make a table giving μ and the power. What do you conclude?

6.115 Other changes and the effect on power. Refer to the previous exercise. For each of the following changes, explain what happens to the power for each alternative μ in the table.

(a) Change to the two-sided alternative.

(b) Decrease σ to 0.5.

(c) Increase n from 10 to 30.

6.116 Power of the random north–south distribution of trees test. In Exercise 6.70 (page 382), you performed a two-sided significance test of the null hypothesis that the average north–south location of the longleaf pine trees sampled in the Wade Tract was $\mu = 100$. There were 584 trees in the sample and the standard deviation was assumed to be 58. The sample mean in that analysis was $\bar{x} = 99.74$. Use the *Statistical Power* applet to compute the power for the alternative $\mu = 99$ using a two-sided test at the 5% level of significance.

6.117 Power of the random east–west distribution of trees test. Refer to the previous exercise. Note that in the east–west direction, the average location was 113.8. Use the *Statistical Power* applet to find the power for the alternative $\mu = 110$.

6.118 Planning another test to compare consumption. Example 6.15 (page 372) gives a test of a hypothesis about the mean consumption of sugar-sweetened beverages at your university based on a sample of size $n = 100$. The hypotheses are

$$H_0: \mu = 286$$
$$H_a: \mu \neq 286$$

While the result was not statistically significant, it did provide some evidence that the mean was smaller than 286. Thus, you plan to recruit another sample of students from your university, but this time use a one-sided alternative. You were thinking of surveying $n = 100$ students but now wonder if this sample size gives adequate power to detect a decrease of 15 calories per day to $\mu = 271$.

(a) Given $\alpha = 0.05$, for what values of z will you reject the null hypothesis?

(b) Using $\sigma = 155$ and $\mu = 286$, for what values of $\bar{x}$ will you reject H_0?

(c) Using $\sigma = 155$ and $\mu = 271$, what is the probability that $\bar{x}$ will fall in the region defined in part (b)?

(d) Will a sample size of $n = 100$ give you adequate power? Or do you need to find ways to increase the power? Explain your answer.

(e) Use the *Statistical Power* applet or other statistical software to determine the sample size n that gives you power near 0.80.

6.119 Planning the dining court survey. Exercise 6.38 (page 364) describes a survey to assess whether a newly designed dining court is viewed more favorably than the old design. The organizers are considering randomly surveying $n = 100$ student patrons but would like some statistical advice. The hypotheses are

$$H_0: \mu = 4$$
$$H_a: \mu > 4$$

and they've decided they want adequate power to detect a mean of at least 4.25.

(a) The organizers have no idea of σ. You suggest a small pilot study, which gives $s = 1.73$. Based on this result, you decide to use $\sigma = 2$. Provide an explanation for this choice to the organizers.

(b) Given $\alpha = 0.05$, for what values of $\bar{x}$ will you reject H_0?

(c) Using $\mu = 4.25$, what is the probability that $\bar{x}$ will fall in the region defined in part (b)?

(d) Will a sample size of $n = 100$ give you adequate power? Explain your answer.

(e) Use the *Statistical Power* applet or statistical software to determine the sample size n that gives you power near 0.80.

6.120 Choose the appropriate distribution. You must decide which of two discrete distributions a random variable X has. We will call the distributions p_0 and p_1. Here are the probabilities they assign to the values x of X:

x	0	1	2	3	4	5	6
p_0	0.1	0.1	0.2	0.3	0.1	0.1	0.1
p_1	0.1	0.3	0.2	0.1	0.1	0.1	0.1

You have a single observation on X and wish to test

$$H_0: p_0 \text{ is correct}$$
$$H_a: p_1 \text{ is correct}$$

One possible decision procedure is to reject H_0 only if $X \leq 1$.

(a) Find the probability of a Type I error, that is, the probability that you reject H_0 when p_0 is the correct distribution.

(b) Find the probability of a Type II error.

6.121 Power of the mean SATM score test.
Example 6.16 (page 374) gives a test of a hypothesis about the SATM scores of California high school students based on an SRS of 500 students. The hypotheses are

$$H_0: \mu = 485$$
$$H_a: \mu > 485$$

Assume that the population standard deviation is $\sigma = 100$. The test rejects H_0 at the 1% level of significance when $z \geq 2.326$, where

$$z = \frac{\bar{x} - 485}{100/\sqrt{500}}$$

Is this test sufficiently sensitive to usually detect an increase of 14 points in the population mean SATM score? Answer this question by calculating the power of the test to detect the alternative $\mu = 499$.

6.122 More on choosing the appropriate distribution. Refer to Exercise 6.120. Suppose that instead of a single observation X, you obtained two observations and use the decision rule to reject when $\bar{x} \leq 1$.

(a) Under this scenario, would you expect the probabilities of a Type I and Type II errors to increase, decrease, or stay at the same values of Exercise 6.120? Explain your answer.

(b) Verify your answer to part (a) by computing the probabilities of a Type I and Type II error.

6.123 Computer-assisted career guidance systems.
A wide variety of computer-assisted career guidance systems have been developed over the last decade. These programs use factors such as student interests, aptitude, skills, personality, and family history to recommend a career path. For simplicity, suppose that a program recommends a high school graduate either to go to college or to join the workforce.

(a) What are the two hypotheses and the two types of error that the program can make?

(b) The program can be adjusted to decrease one error probability at the cost of an increase in the other error probability. Which error probability would you choose to make smaller, and why? (This is a matter of judgment. There is no single correct answer.)

CHAPTER 6 EXERCISES

6.124 Telemarketing wages. An advertisement in the student newspaper asks you to consider working for a telemarketing company. The ad states, "Earn between $500 and $1000 per week." Do you think that the ad is describing a confidence interval? Explain your answer.

6.125 Exercise and statistics exams. A study examined whether light exercise performed an hour before the final exam in statistics affects how students perform on the exam. The P-value was given as 0.13.

(a) State null and alternative hypotheses that could be used for this study. (*Note:* There is more than one correct answer.)

(b) Do you reject the null hypothesis? State your conclusion in plain language.

(c) What other facts about the study would you like to know for a proper interpretation of the results?

6.126 Roulette. A roulette wheel has 18 red slots among its 38 slots. You observe many spins and record the number of times that red occurs. Now you want to use these data to test whether the probability of a red has the

value that is correct for a fair roulette wheel. State the hypotheses H_0 and H_a that you will test.

6.127 Food selection by children in school cafeterias.
A group of researchers examined whether children's food selection in a school cafeteria met the standards set by the School Meals Initiative. They measured food selection and food intake of 2049 fourth- through sixth-grade students in 33 schools over a three-day period using digital photography. The following table summarizes some of the food intake measurements.[31]

Food intake	Boys n = 852 Mean	St. Dev.	Girls n = 1197 Mean	St. Dev.
Energy (kilojoules)	2448	717	2170	693
Protein (g)	24.5	7.5	22.1	7.7
Calcium (mg)	324.1	130.6	265.0	128.9

Given the large sample sizes, we can assume that the sample standard deviations are the population standard deviations.

(a) Compute 95% confidence intervals for all three intake measures for the boys.

(b) Compute 95% confidence intervals for all three intake measures for the girls.

(c) In the next chapter, we will describe the confidence interval for the difference between two means. For now, let's compare the boy and girl confidence intervals for each food intake measure. Do you think these pairs of intervals provide strong evidence against the null hypothesis that the boys and girls consume, on average, the same amount? Explain your answer.

6.128 Coverage percent of 95% confidence interval. For this exercise, you will use the *Confidence Interval* applet. Set the confidence level at 95% and click the "Sample" button 10 times to simulate 10 confidence intervals. Record the percent hit. Simulate another 10 intervals by clicking another 10 times (do not click the "Reset" button). Record the percent hit for your 20 intervals. Repeat the process of simulating 10 additional intervals and recording the results until you have a total of 200 intervals. Create a time plot of your results and write a summary of what you have found.

6.129 Coverage percent of 90% confidence interval. Refer to the previous exercise. Do the simulations and report the results for 90% confidence.

6.130 Effect of sample size on significance. You are testing the null hypothesis that $\mu = 0$ versus the alternative $\mu > 0$ using $\alpha = 0.05$. Assume that $\sigma = 16$. Suppose that $\bar{x} = 8$ and $n = 10$. Calculate the test statistic and its P-value. Repeat assuming the same value of $\bar{x}$ but with $n = 20$. Do the same for sample sizes of 30, 40, and 50. Plot the values of the test statistic versus the sample size. Do the same for the P-values. Summarize what this demonstration shows about the effect of the sample size on significance testing.

6.131 Survey response and margin of error. Suppose that a business conducts a marketing survey. As is often done, the survey is conducted by telephone. As it turns out, the business was only able to elicit responses from less than 10% of the randomly chosen customers. The low response rate is attributable to many factors, including caller ID screening. Undaunted, the marketing manager was pleased with the sample results because the margin of error was quite small, and thus the manager felt that the business had a good sense of the customers perceptions on various issues. Do you think the small margin of error is a good measure of the accuracy of the surveys results? Explain.

6.132 Reporting margins of error. A *U.S. News & World Report* article from July 17, 2014, reported Commerce Department *estimates* of changes in the construction industry:

> Construction fell 9.3 percent last month to a seasonally adjusted annual rate of 893,000 homes.

If we turn to the original Commerce Department report (released the same day), it states:

> Privately owned housing starts in June were at a seasonally adjusted annual rate of 893,000. This is 9.3 percent (10.3%) below the revised May estimate of 985,000.

(a) The 10.3% figure is the margin of error based on a 90% level of confidence. Given that fact, what is the 90% confidence interval for the percent change in housing starts from May to June?

(b) Explain why a credible media report should state:

> The Commerce Department has no evidence that privately owned housing starts rose or fell in June from the previous month.

6.133 Blood phosphorus level in dialysis patients. Patients with chronic kidney failure may be treated by dialysis, in which a machine removes toxic wastes from the blood, a function normally performed by the kidneys. Kidney failure and dialysis can cause other changes, such as retention of phosphorus, that must be corrected by changes in diet. A study of the nutrition of dialysis patients measured the level of phosphorus in the blood of several patients on six occasions. Here are the data for one patient (in milligrams of phosphorus per deciliter of blood):[32]

$$5.4 \quad 5.2 \quad 4.5 \quad 4.9 \quad 5.7 \quad 6.3$$

The measurements are separated in time and can be considered an SRS of the patient's blood phosphorus level. Assume that this level varies Normally with $\sigma = 0.9$ mg/dl. PMGDL

(a) Give a 95% confidence interval for the mean blood phosphorus level.

(b) The normal range of phosphorus in the blood is considered to be 2.6 to 4.8 mg/dl. Is there strong evidence that this patient has a mean phosphorus level that exceeds 4.8?

6.134 Cellulose content in alfalfa hay. An agronomist examines the cellulose content of a variety of alfalfa hay. Suppose that the cellulose content in the population has standard deviation $\sigma = 8$ milligrams per gram (mg/g). A sample of 15 cuttings has mean cellulose content $\bar{x} = 145$ mg/g.

(a) Give a 90% confidence interval for the mean cellulose content in the population.

(b) A previous study claimed that the mean cellulose content was $\mu = 140$ mg/g, but the agronomist believes that the mean is higher than that figure. State H_0 and H_a and carry out a significance test to see if the new data support this belief.

(c) The statistical procedures used in parts (a) and (b) are valid when several assumptions are met. What are these assumptions?

6.135 Odor threshold of future wine experts. Many food products contain small quantities of substances that would give an undesirable taste or smell if they are present in large amounts. An example is the "off-odors" caused by sulfur compounds in wine. Oenologists (wine experts) have determined the odor threshold, the lowest concentration of a compound that the human nose can detect. For example, the odor threshold for dimethyl sulfide (DMS) is given in the oenology literature as 25 micrograms per liter of wine (μg/l). Untrained noses may be less sensitive, however. Here are the DMS odor thresholds for 10 beginning students of oenology:

31   31   43   36   23   34   32   30   20   24

Assume (this is not realistic) that the standard deviation of the odor threshold for untrained noses is known to be $\sigma = 7$ μg/l. ODOR

(a) Make a stemplot to verify that the distribution is roughly symmetric with no outliers. (A Normal quantile plot confirms that there are no systematic departures from Normality.)

(b) Give a 95% confidence interval for the mean DMS odor threshold among all beginning oenology students.

(c) Are you convinced that the mean odor threshold for beginning students is higher than the published threshold, 25 μg/l? Carry out a significance test to justify your answer.

6.136 Where do you buy? Consumers can purchase nonprescription medications at food stores, mass merchandise stores such as Target and Walmart, or pharmacies. About 45% of consumers make such purchases at pharmacies. What accounts for the popularity of pharmacies, which often charge higher prices?

A study examined consumers' perceptions of overall performance of the three types of stores, using a long questionnaire that asked about such things as "neat and attractive store," "knowledgeable staff," and "assistance in choosing among various types of nonprescription medication." A performance score was based on 27 such questions. The subjects were 201 people chosen at random from the Indianapolis telephone directory. Here are the means and standard deviations of the performance scores for the sample:[33]

Store type	$\bar{x}$	s
Food stores	18.67	24.95
Mass merchandisers	32.38	33.37
Pharmacies	48.60	35.62

We do not know the population standard deviations, but a sample standard deviation s from so large a sample is usually close to σ. Use s in place of the unknown σ in this exercise.

(a) What population do you think the authors of the study want to draw conclusions about? What population are you certain they can draw conclusions about?

(b) Give 95% confidence intervals for the mean performance for each type of store.

(c) Based on these confidence intervals, are you convinced that consumers think that pharmacies offer higher performance than the other types of stores? (In Chapter 12, we will study a statistical method for comparing the means of several groups.)

6.137 CEO pay. A study of the pay of corporate chief executive officers (CEOs) examined the increase in cash compensation of the CEOs of 104 companies, adjusted for inflation, in a recent year. The mean increase in real compensation was $\bar{x} = 6.9$ %, and the standard deviation of the increases was $s = 55$ %. Is this good evidence that the mean real compensation μ of all CEOs increased that year? The hypotheses are

$$H_0: \mu = 0 \text{ (no increase)}$$
$$H_a: \mu > 0 \text{ (an increase)}$$

Because the sample size is large, the sample s is close to the population σ, so take $\sigma = 55$ %.

(a) Sketch the Normal curve for the sampling distribution of $\bar{x}$ when H_0 is true. Shade the area that represents the P-value for the observed outcome $\bar{x} = 6.9$ %.

(b) Calculate the P-value.

(c) Is the result significant at the $\alpha = 0.05$ level? Do you think the study gives strong evidence that the mean compensation of all CEOs went up?

6.138 Meaning of "statistically significant." When asked to explain the meaning of "statistically significant at the $\alpha = 0.01$ level," a student says, "This means there is only probability 0.01 that the null hypothesis is true." Is this an essentially correct explanation of statistical significance? Explain your answer.

6.139 More on the meaning of "statistically significant." Another student, when asked why statistical significance appears so often in research reports, says, "Because saying that results are significant tells us that they cannot easily be explained by chance variation alone." Do you think that this statement is essentially correct? Explain your answer.

6.140 Increasing the power. Refer to Example 6.17 (page 375). Suppose the Deely Laboratory wants to make sure the power is at least 80% for $\mu = 15.30$ ppm. It cannot reduce σ, so the options are changing the significance level α and/or sample size n.

(a) With $\alpha = 0.01$, what sample size n is needed to have at least 80% power?

(b) With $\alpha = 0.05$, what sample size n is needed to have at least 80% power?

(c) Which of these options do you think the Deely Laboratory should choose? Explain your reasoning.

6.141 Simulation study of the confidence interval. Use a computer to generate $n = 15$ observations from a Normal distribution with mean 20 and standard deviation 5: $N(20, 5)$. Find the 95% confidence interval for μ. Repeat this process 100 times and then count the number of times that the confidence interval includes the value $\mu = 20$. Explain your results.

6.142 Simulation study of a test of significance. Use a computer to generate $n = 15$ observations from a Normal distribution with mean 20 and standard deviation 5: $N(20, 5)$. Test the null hypothesis that $\mu = 20$ using a two-sided significance test. Repeat this process 100 times and then count the number of times that you reject H_0. Explain your results.

6.143 Another simulation study of a test of significance. Use the same procedure for generating data as in the previous exercise. Now test the null hypothesis that $\mu = 24$. Explain your results.

6.144 Simulation study of power. Refer to the previous two exercises. What is the power of detecting a difference of four units (H_0: $\mu = 24$ versus $\mu = 20$) in this setting? Compare this power with the proportion of times you rejected H_0 in the previous exercise. Explain your results.

6.145 Find published studies with confidence intervals. Search the Internet or some journals that report research in your field and find two reports that provide an estimate with a margin of error or a confidence interval. For each report,

(a) describe the method used to collect the data.

(b) describe the variable being studied.

(c) give the estimate and the confidence interval.

(d) describe any practical difficulties that may have led to errors in addition to the sampling errors quantified by the margin of error.

piranka/Getty Images

Inference for Means

Introduction

We began our study of data analysis in Chapter 1 by learning graphical and numerical tools for describing the distribution of a single variable and for comparing several distributions. Our study of the practice of statistical inference begins in the same way, with inference about a single distribution and comparison of two distributions. Comparing more than two distributions requires more elaborate methods, which are presented in Chapters 12 and 13.

Two important aspects of any distribution are its center and spread. If the distribution is Normal, we describe its center by the mean μ and its spread by the standard deviation σ.

In this chapter, we will meet confidence intervals and significance tests for inference about a population mean μ and the difference between two population means $\mu_1 - \mu_2$. Chapter 6 emphasized the reasoning of significance tests and confidence intervals; now we emphasize statistical practice and no longer assume that population standard deviations are known. As a result, we move away from the standard Normal sampling distribution to a new family of t distributions. The t procedures for inference about means are among the most commonly used statistical methods.

7.1 Inference for the Mean of a Population

7.2 Comparing Two Means

7.3 Additional Topics on Inference

7.1 Inference for the Mean of a Population

When you complete this section, you will be able to:	• Distinguish the standard deviation of the sample mean from the standard error of the sample mean.

- Distinguish the standard deviation of the sample mean from the standard error of the sample mean.
- Describe a level C confidence interval for the population mean in terms of an estimate and its margin of error.
- Construct a level C confidence interval for μ from a simple random sample (SRS) of size n from a large population.
- Perform a one-sample t significance test and summarize the results.
- Identify when the matched pairs t procedures should be used instead of two-sample t procedures.
- Explain when t procedures can be useful for non-Normal data.

Both confidence intervals and tests of significance for the mean μ of a Normal population are based on the sample mean $\bar{x}$, which estimates the unknown μ. The sampling distribution of $\bar{x}$ depends on σ. This fact causes no difficulty when σ is known. When σ is unknown, however, we must estimate σ even though we are primarily interested in μ.

In this section, we meet the sampling distribution of the standardized sample mean when we use the sample standard deviation s to estimate the population standard deviation σ. This sampling distribution is then used to produce both confidence intervals and significance tests about the mean μ.

The t distributions

LOOK BACK

sampling distribution of $\bar{x}$, p. 298

Suppose that we have a simple random sample (SRS) of size n from a Normally distributed population with mean μ and standard deviation σ. The sample mean $\bar{x}$ is then Normally distributed with mean μ and standard deviation $\sigma/\sqrt{n}$. When σ is not known, we estimate it with the sample standard deviation s, and then we estimate the standard deviation of $\bar{x}$ by $s/\sqrt{n}$. This quantity is called the *standard error* of the sample mean $\bar{x}$, and we denote it by $\mathrm{SE}_{\bar{x}}$.

STANDARD ERROR

When the standard deviation of a statistic is estimated from the data, the result is called the **standard error** of the statistic. The standard error of the sample mean is

$$\mathrm{SE}_{\bar{x}} = \frac{s}{\sqrt{n}}$$

The term "standard error" is sometimes used for the actual standard deviation of a statistic. The estimated value is then called the "estimated standard error." In this book, we will use the term "standard error" only when the standard deviation of a statistic is estimated from the data. The term has this meaning in the output of many statistical computer packages and in research reports that apply statistical methods.

In the previous chapter, the standardized sample mean, or one-sample z statistic,

$$z = \frac{\bar{x} - \mu}{\sigma/\sqrt{n}}$$

is the basis for inference about μ when σ is known. This statistic has the standard Normal distribution $N(0, 1)$. However, when we substitute the standard error $s/\sqrt{n}$ for the standard deviation of $\bar{x}$, the statistic does *not* have a Normal distribution. It has a distribution that is new to us, called a *t distribution*.

THE *t* DISTRIBUTIONS

Suppose that an SRS of size n is drawn from an $N(\mu, \sigma)$ population. Then the **one-sample *t* statistic**

$$t = \frac{\bar{x} - \mu}{s/\sqrt{n}}$$

has the ***t* distribution** with $n - 1$ **degrees of freedom**.

LOOK BACK

degrees of freedom, p. 40

A particular t distribution is specified by giving the *degrees of freedom*. We use $t(k)$ to stand for the t distribution with k degrees of freedom. The degrees of freedom for this t statistic come from the sample standard deviation s in the denominator of t. We showed earlier that s has $n - 1$ degrees of freedom. Thus, there is a different t distribution for each sample size. There are also other t statistics with different degrees of freedom, some of which we will meet later in this chapter.

The t distributions were discovered in 1908 by William S. Gosset. Gosset was a statistician employed by the Guinness brewing company, which prohibited its employees from publishing their discoveries that were brewing related. In this case, the company let him publish under the pen name "Student" using an example that did not involve brewing. The t distribution is often called "Student's t" in his honor.

The density curves of the $t(k)$ distributions are similar in shape to the standard Normal curve. That is, they are symmetric about 0 and are bell-shaped. Figure 7.1 compares the density curves of the standard Normal distribution and the t distributions with 5 and 10 degrees of freedom. The similarity in shape is apparent, as is the fact that the t distributions have more probability in the tails and less in the center.

In reference to the standardized sample mean, this greater spread is due to the extra variability caused by substituting the random variable s for the fixed parameter σ. In Figure 7.1, we see that as the degrees of freedom k increase, the $t(k)$ density gets closer to the $N(0, 1)$ curve. This reflects the fact that s will be closer to σ (more precise) as the sample size increases.

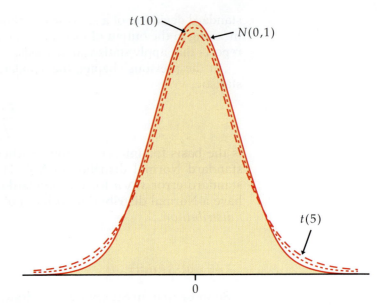

FIGURE 7.1 Density curves for the standard Normal, $t(10)$, and $t(5)$ distributions. All are symmetric with center 0. The t distributions have more probability in the tails than the standard Normal distribution.

USE YOUR KNOWLEDGE

7.1 One-bedroom apartment rates. You randomly choose 16 unfurnished one-bedroom apartments from a large number of advertisements in your local newspaper. You calculate that their mean monthly rent is $766 and their standard deviation is $180.

(a) What is the standard error of the mean?

(b) What are the degrees of freedom for a one-sample t statistic?

7.2 Changing the sample size. Refer to the previous exercise. Suppose that instead of an SRS of 16, you sampled 25 advertisements.

(a) Would you expect the standard error of the mean to be larger or smaller in this case? Explain your answer.

(b) State why you can't be certain that the standard error for this new SRS will be larger or smaller.

With the t distributions to help us, we can now analyze a sample from a Normal population with unknown σ or a large sample from a non-Normal population with unknown σ. Table D in the back of the book gives critical values t^* for the t distributions. For convenience, we have labeled the table entries both by the value of p needed for significance tests and by the confidence level C (in percent) required for confidence intervals. The standard Normal critical values are in the bottom row of entries and labeled z^*. As in the case of the Normal table (Table A), computer software often makes Table D unnecessary.

The one-sample t confidence interval

The one-sample t confidence interval is similar in both reasoning and computational detail to the z confidence interval of Chapter 6. There, the margin of error for the population mean was $z^*\sigma/\sqrt{n}$. When σ is unknown, we replace it with its estimate s and switch from z^* to t^*. This means that the margin of error for the population mean when we use the data to estimate σ is $t^*s/\sqrt{n}$.

LOOK BACK

z confidence interval, p. 349

THE ONE-SAMPLE t CONFIDENCE INTERVAL

Suppose that an SRS of size n is drawn from a population having unknown mean μ. A level C **confidence interval** for μ is

$$\bar{x} \pm t^* \frac{s}{\sqrt{n}}$$

where t^* is the value for the $t(n-1)$ density curve with area C between $-t^*$ and t^*. The quantity

$$t^* \frac{s}{\sqrt{n}}$$

is the **margin of error**. The confidence level is exactly C when the population distribution is Normal and is approximately correct for large n in other cases.

EXAMPLE 7.1

TVTIME

Watching traditional television. The Nielsen Company is a global information and media company and one of the leading suppliers of media information. In their annual Total Audience Report, the Nielsen Company states that adults age 18 to 24 years old average 18.5 hours per week watching traditional television.[1] Does this average seem reasonable for college students? They tend to watch a lot of television, but given their unusual schedules, they may be more likely to binge-watch or stream episodes after they air. To investigate, let's construct a 95% confidence interval for the average time (hours per week) spent watching traditional television among full-time U.S. college students. We draw the following SRS of size 8 from this population:

3.0 16.5 10.5 40.5 5.5 33.5 0.0 6.5

The sample mean is

$$\bar{x} = \frac{3.0 + 16.5 + \cdots + 6.5}{8} = 14.5$$

and the standard deviation is

$$s = \sqrt{\frac{(3.0 - 14.5)^2 + (16.5 - 14.5)^2 + \cdots + (6.5 - 14.5)^2}{8 - 1}} = 14.854$$

with degrees of freedom $n - 1 = 7$. The standard error is

$$\text{SE}_{\bar{x}} = s/\sqrt{n} = 14.854/\sqrt{8} = 5.252$$

From Table D, we find $t^* = 2.365$. The 95% confidence interval is

df = 7

t^*	1.895	2.365	2.517
C	0.90	0.95	0.96

$$\bar{x} \pm t^* \frac{s}{\sqrt{n}} = 14.5 \pm 2.365 \frac{14.854}{\sqrt{8}}$$

$$= 14.5 \pm (2.365)(5.252)$$

$$= 14.5 \pm 12.421$$

$$= (2.08, 26.92)$$

We are 95% confident that among U.S. college students the average time spent watching traditional television is between 2.1 and 26.9 hours per week.

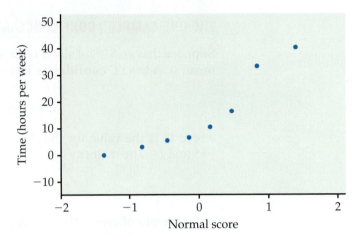

FIGURE 7.2 Normal quantile plot of data, Example 7.1.

In this example, we gave the interval (2.1, 26.9) hours per week as our answer. Sometimes, we prefer to report the mean and margin of error: the mean time is 14.5 hours per week with a margin of error of 12.4 hours per week. This is a large margin of error in relation to the estimated mean. In Section 7.3, we will return to this example and discuss determining an appropriate sample size for a desired margin of error such as ±5 hours a week.

Valid interpretation of the t confidence interval in Example 7.1 rests on assumptions that appear reasonable here. First, we assume that our random sample is an SRS from the U.S. population of college students. Second, we assume that the distribution of watching times is Normal. Figure 7.2 shows the Normal quantile plot. With only eight observations, this assumption cannot be effectively checked. In fact, because a watching time cannot be negative, we might expect this distribution to be skewed to the right. With these data, however, there are no extreme outliers to suggest a severe departure from Normality.

USE YOUR KNOWLEDGE

7.3 More on apartment rents. Recall Exercise 7.1 (page 410). Construct a 95% confidence interval for the mean monthly rent of all advertised one-bedroom apartments.

7.4 Finding critical t^*-values. What critical value t^* from Table D should be used to construct

(a) a 95% confidence interval when $n = 25$?

(b) a 99% confidence interval when $n = 11$?

(c) a 90% confidence interval when $n = 61$?

four steps of
significance test,
p. 370

The one-sample t test

Significance tests using the standard error are also very similar to the z test that we studied in the last chapter. We still carry out the four steps common to all significance tests, but because we use s in place of σ, we use a t distribution to find the P-value.

THE ONE-SAMPLE *t* TEST

Suppose that an SRS of size n is drawn from a population having unknown mean μ. To test the hypothesis $H_0: \mu = \mu_0$ based on an SRS of size n, compute the **one-sample *t* statistic**

$$t = \frac{\bar{x} - \mu_0}{s/\sqrt{n}}$$

In terms of a random variable T having the $t(n-1)$ distribution, the *P*-value for a test of H_0 against

$H_a: \mu > \mu_0$ is $P(T \geq t)$

$H_a: \mu < \mu_0$ is $P(T \leq t)$

$H_a: \mu \neq \mu_0$ is $2P(T \geq |t|)$

These *P*-values are exact if the population distribution is Normal and are approximately correct for large n in other cases.

EXAMPLE 7.2

TVTIME

Significance test for watching traditional television. We want to test whether the average time that U.S. college students spend watching traditional television differs from the reported overall U.S. average of 18- to 24-year-olds at the 0.05 significance level. Specifically, we want to test

$$H_0: \mu = 18.5$$
$$H_a: \mu \neq 18.5$$

Recall that $n = 8$, $\bar{x} = 14.5$, and $s = 14.854$. The *t* test statistic is

$$t = \frac{\bar{x} - \mu_0}{s/\sqrt{n}} = \frac{14.5 - 18.5}{14.854/\sqrt{8}}$$
$$= -0.762$$

This means that the sample mean $\bar{x} = 14.5$ is slightly more than 0.75 standard deviations below the null hypothesized value $\mu = 18.5$. Because the degrees of freedom are $n - 1 = 7$, this *t* statistic has the $t(7)$ distribution. Figure 7.3 shows that the *P*-value is $2P(T \geq 0.762)$, where T has the $t(7)$ distribution. From Table D, we see that $P(T \geq 0.711) = 0.25$ and $P(T \geq 0.896) = 0.20$.

Therefore, we conclude that the *P*-value is between $2 \times 0.20 = 0.40$ and $2 \times 0.25 = 0.50$. Software gives the exact value as $P = 0.4711$. These data are compatible with a mean of 18.5 hours per week. Under H_0, a difference this large or larger would occur about half the time simply due to chance. There is not enough evidence to reject the null hypothesis at the 0.05 level.

df = 7

p	0.25	0.20
t^*	0.711	0.896

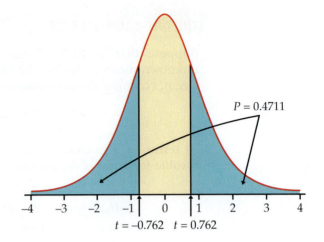

FIGURE 7.3 Sketch of the
P-value calculation, Example 7.2.

In this example, we tested the null hypothesis $\mu = 18.5$ hours per week against the two-sided alternative $\mu \neq 18.5$ hours per week because we had no prior suspicion that the average among college students would be larger or smaller. If we had suspected that the average would be smaller (for example, expected more streaming of shows), we would have used a one-sided test.

EXAMPLE 7.3

TVTIME

One-sided test for watching traditional television. For the problem described in the previous example, we want to test whether the U.S. college student average is smaller than the overall U.S. population average. Here we test

$$H_0: \mu = 18.5$$

versus

$$H_a: \mu < 18.5$$

The t test statistic does not change: $t = -0.762$. As Figure 7.4 illustrates, however, the P-value is now $P(T \leq -0.762)$, half of the value in the previous example. From Table D, we can determine that $0.20 < P < 0.25$; software gives the exact value as $P = 0.2356$. Again, there is not enough evidence to reject the null hypothesis in favor of the alternative at the 0.05 significance level.

FIGURE 7.4 Sketch of the
P-value calculation, Example 7.3.

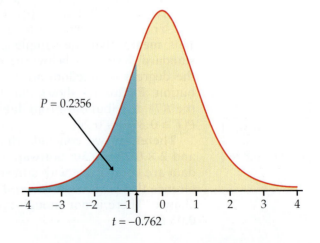

For the watching-television study our conclusion did not depend on the choice between a one-sided and a two-sided test. Sometimes, however, this choice *will* affect the conclusion, so this choice needs to be made prior to analysis. If in doubt, always use a two-sided test. *It is wrong to examine the data first and then decide to do a one-sided test in the direction indicated by the data.* Often, a significant result for a two-sided test can be used to justify a one-sided test for *another* sample from the same population.

USE YOUR KNOWLEDGE

7.5 Significance test using the *t* distribution. A test of a null hypothesis versus a two-sided alternative gives $t = 2.148$.

(a) The sample size is 23. Is the test result significant at the 5% level? Explain how you obtained your answer.

(b) The sample size is 9. Is the test result significant at the 5% level? Explain how you obtained your answer.

(c) Sketch the two *t* distributions to illustrate your answers.

7.6 Significance test for apartment rents. Refer to Exercise 7.1 (page 410). Does this SRS give good reason to believe that the mean rent of all advertised one-bedroom apartments is greater than $700? State the hypotheses, find the *t* statistic and its *P*-value, and state your conclusion.

For small data sets, such as the one in Example 7.1 (page 411), it is easy to perform the computations for confidence intervals and significance tests with an ordinary calculator. For larger data sets, however, we prefer to use software or a statistical calculator.

EXAMPLE 7.4

STOCK

Stock portfolio diversification? An investor with a stock portfolio worth several hundred thousand dollars sued his broker and brokerage firm because lack of diversification in his portfolio led to poor performance. Table 7.1 gives the rates of return for the 39 months that the account was managed by the broker.[2]

Figure 7.5 gives a histogram for these data, and Figure 7.6 gives the Normal quantile plot. There are no outliers and the distribution shows no strong skewness. We are reasonably confident that the distribution of

TABLE 7.1	Monthly Rates of Return on a Portfolio (%)						
−8.36	1.63	−2.27	−2.93	−2.70	−2.93	−9.14	−2.64
6.82	−2.35	−3.58	6.13	7.00	−15.25	−8.66	−1.03
−9.16	−1.25	−1.22	−10.27	−5.11	−0.80	−1.44	1.28
−0.65	4.34	12.22	−7.21	−0.09	7.34	5.04	−7.24
−2.14	−1.01	−1.41	12.03	−2.56	4.33	2.35	

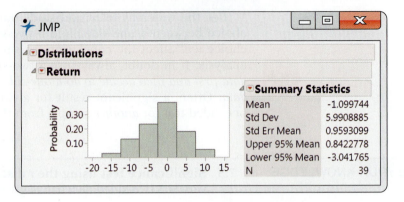

FIGURE 7.5 Histogram of monthly rates of return for a stock portfolio, Example 7.4.

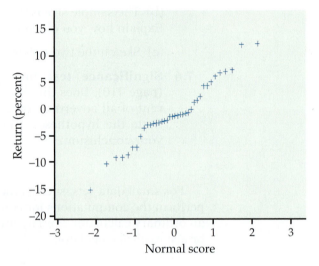

FIGURE 7.6 Normal quantile plot, Example 7.4.

$\bar{x}$ is approximately Normal, and we proceed with our inference based on Normal theory.

The arbitration panel compared these returns with the average of the Standard & Poor's 500 stock index for the same period. Consider the 39 monthly returns as a random sample from the population of monthly returns the brokerage firm would generate if it managed the account forever. Are these returns compatible with a population mean of $\mu = 0.95\%$, the S&P 500 average? Our hypotheses are

$$H_0: \mu = 0.95$$
$$H_a: \mu \neq 0.95$$

Minitab and SPSS outputs appear in Figure 7.7. Output from other software will be similar.

Here is one way to report the conclusion: the mean monthly return on investment for this client's account was $\bar{x} = -1.1\%$. This is significantly worse than the performance of the S&P 500 stock index for the same period ($t = -2.14$, df = 38, $P = 0.039$).

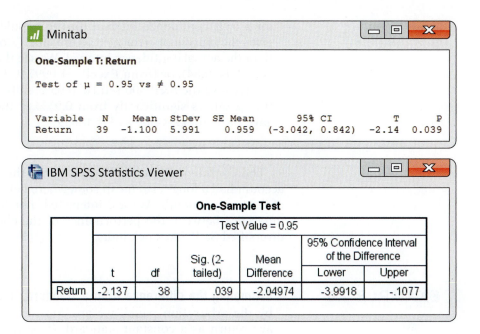

FIGURE 7.7 Minitab and SPSS outputs, Example 7.4.

The hypothesis test in Example 7.4 leads us to conclude that the mean return on the client's account differs from that of the S&P 500 stock index. Now let's assess the return on the client's account with a confidence interval.

EXAMPLE 7.5

Estimating the mean monthly return. The mean monthly return on the client's portfolio was $\bar{x} = -1.1\%$, and the standard deviation was $s = 5.99\%$. Figure 7.7 gives Minitab output, and Figure 7.8 gives Excel and JMP outputs

FIGURE 7.8 Excel and JMP outputs, Example 7.5.

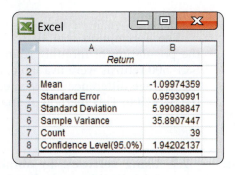

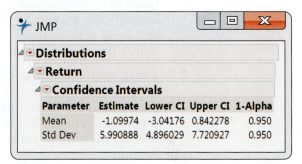

for a 95% confidence interval for the population mean μ. Note that Excel gives the margin of error next to the label "Confidence Level(95.0%)" rather than the actual confidence interval. We see that the 95% confidence interval is $(-3.04, 0.84)$, or (from Excel) -1.0997 ± 1.9420.

Because the S&P 500 return, 0.95%, falls outside this interval, we know that μ differs significantly from 0.95% at the $\alpha = 0.05$ level. Example 7.4 gave the actual P-value as $P = 0.039$.

The confidence interval suggests that the broker's management of this account had a long-term mean somewhere between a loss of 3.04% and a gain of 0.84% per month. We are interested, not in the actual mean, but in the difference between the performance of the client's portfolio and that of the diversified S&P 500 stock index.

EXAMPLE 7.6

Estimating the difference from a standard. Following the analysis accepted by the arbitration panel, we are considering the S&P 500 monthly average return as a constant standard. (It is easy to envision scenarios where we would want to treat this type of quantity as random.) The difference between the mean of the investor's account and the S&P 500 is $\bar{x} - \mu = -1.10 - 0.95 = -2.05\%$. In Example 7.5, we found that the 95% confidence interval for the investor's account was $(-3.04, 0.84)$.

To obtain the corresponding interval for the difference, subtract 0.95 from each of the endpoints. The resulting interval is $(-3.04 - 0.95, 0.84 - 0.95)$, or $(-3.99, -0.11)$. We conclude with 95% confidence that the underperformance was between -3.99 % and -0.11%. This interval is presented in the SPSS output of Figure 7.7. This estimate helps to set the compensation owed the investor.

The assumption that these 39 monthly returns represent an SRS from the population of monthly returns is certainly questionable. If the monthly S&P 500 returns were available, an alternative analysis would be to compare the average difference between each monthly return for this account and for the S&P 500. This method of analysis is discussed next.

USE YOUR KNOWLEDGE

7.7 Using software to obtain a confidence interval. In Example 7.1 (page 411), we calculated the 95% confidence interval for the U.S. college student average of hours per month spent watching traditional television. Use software to compute this interval and verify that you obtain the same interval.

7.8 Using software to perform a significance test. In Example 7.2 (page 413), we tested whether the average time that U.S. college students spend watching traditional television differs from the reported overall U.S. average of 18- to 24-year-olds at the 0.05 significance level. Use software to perform this test and obtain the exact P-value.

Matched pairs *t* procedures

confounding,
p. 150
matched pairs
design,
p. 182

The watching-television study of Example 7.1 (page 411) concerns only a single population. We know that comparative studies are usually preferred to single-sample investigations because of the protection they offer against confounding. For that reason, inference about a parameter of a single distribution is less common than comparative inference.

One common comparative design, however, makes use of single-sample procedures. In a matched pairs study, subjects are matched in pairs, and their outcomes are compared within each matched pair. For example, an experiment to compare two smartphone packages might use pairs of subjects who are the same age, sex, and income level. The experimenter could toss a coin to assign the two packages to the two subjects in each pair. The idea is that matched subjects are more similar than unmatched subjects, so comparing outcomes within each pair is more efficient (smaller σ).

Matched pairs are also common when randomization is not possible. For example, one situation calling for matched pairs is when observations are taken on the same subjects under two different conditions or before and after some intervention. Here is an example.

EXAMPLE 7.7

GEPARTS

The effect of altering a software parameter. The MeasureMind® 3D MultiSensor metrology software is used by various companies to measure complex machine parts. As part of a technical review of the software, researchers at GE Healthcare discovered that unchecking one software option reduced measurement time by 10%. This time reduction would help the company's productivity provided the option has no impact on the measurement outcome. To investigate this, the researchers measured 51 parts using the software both with and without this option checked.[3] The experimenters tossed a fair coin to decide which measurement (with or without the option) to take first.

Table 7.2 gives the measurements (in microns) for the first 20 parts. For analysis, we subtract the measurement with the option on from the measurement with the option off. These differences form a single sample and appear in the "Diff" columns for each part.

TABLE 7.2	Parts Measurements Using Optical Software						
Part	OptionOn	OptionOff	Diff	Part	OptionOn	OptionOff	Diff
1	118.63	119.01	0.38	11	119.03	118.66	−0.37
2	117.34	118.51	1.17	12	118.74	118.88	0.14
3	119.30	119.50	0.20	13	117.96	118.23	0.27
4	119.46	118.65	−0.81	14	118.40	118.96	0.56
5	118.12	118.06	−0.06	15	118.06	118.28	0.22
6	117.78	118.04	0.26	16	118.69	117.46	−1.23
7	119.29	119.25	−0.04	17	118.20	118.25	0.05
8	120.26	118.84	−1.42	18	119.54	120.26	0.72
9	118.42	117.78	−0.64	19	118.28	120.26	1.98
10	119.49	119.66	0.17	20	119.13	119.15	0.02

To assess whether there is a difference between the measurements with and without this option, we test

$$H_0: \mu = 0$$
$$H_a: \mu \neq 0$$

Here, μ is the mean difference for the entire population of parts. The null hypothesis says that there is no difference, and H_a says that there is a difference, but does not specify a direction.

The 51 differences have

$$\bar{x} = 0.0504 \quad \text{and} \quad s = 0.6943$$

Figure 7.9 shows a histogram of the differences. It is reasonably symmetric with no outliers, so we can comfortably use the one-sample t procedures. *Remember to always check assumptions before proceeding with statistical inference.*

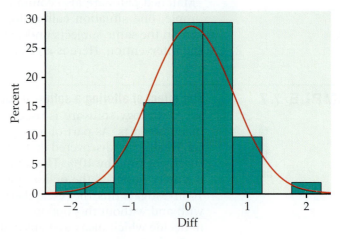

FIGURE 7.9 Histogram of differences in times, Example 7.7.

The one-sample t statistic is

$$t = \frac{\bar{x} - 0}{s/\sqrt{n}} = \frac{0.0504}{0.6943/\sqrt{51}}$$
$$= 0.52$$

The P-value is found from the $t(50)$ distribution. Remember that the degrees of freedom are 1 less than the sample size.

Table D shows that 0.52 lies to the left of the first column entry. This means the P-value is greater than $2(0.25) = 0.50$. Software gives the exact value $P = 0.6054$. There is little evidence to suggest this option has an impact on the measurements. When reporting results, it is usual to omit the details of routine statistical procedures; our test would be reported in the form: "The difference in measurements was not statistically significant ($t = 0.52$, df = 50, $P = 0.61$)."

This result, however, does not fully address the goal of this study. *A lack of statistical significance does not prove the null hypothesis is true.* If that were the case, we would simply design poor experiments whenever we wanted to prove equivalence testing the null hypothesis. The more appropriate method of inference in this setting is to consider **equivalence testing**. With this approach, we try to prove that the mean difference is within some acceptable region around 0. We can actually perform this test using a confidence interval.

EXAMPLE 7.8

GEPARTS

df = 50

t^*	1.676	2.009
C	90%	95%

Are the two means equivalent? Suppose the GE Healthcare researchers state that a mean difference less than 0.20 micron is not important. To see if the data support a mean difference within 0.00 ± 0.20 micron, we construct a 90% confidence interval for the mean difference.

The standard error is

$$\text{SE}_{\bar{x}} = \frac{s}{\sqrt{n}} = \frac{0.6943}{\sqrt{51}} = 0.0972$$

so the margin of error is

$$m = t^* \times \text{SE}_{\bar{x}} = (1.676)(0.0972) = 0.1629$$

where the critical value $t^* = 1.676$ comes from Table D using 50 degrees of freedom. The confidence interval is

$$\bar{x} \pm m = 0.0504 \pm 0.1629$$
$$= (-0.112, 0.2133)$$

This interval is *not* entirely within the 0.00 ± 0.20 micron region that the researchers state is not important. Thus, we *cannot* conclude at the 5% signficance level that the two means are equivalent. Because the observed mean difference is close to zero and well within the "equivalent region," the company may want to consider a larger study to improve precision.

ONE SAMPLE TEST OF EQUIVALENCE

Suppose that an SRS of size n is drawn from a population having unknown mean μ. To test, at significance level α, if μ is within a range of equivalency to μ_0, specified by the interval $\mu_0 \pm \delta$:

1. Compute the confidence interval with $C = 1 - 2\alpha$.

2. Compare this interval with the range of equivalency.

If the confidence interval falls entirely within $\mu_0 \pm \delta$, conclude that μ is equivalent to μ_0. If the confidence interval is outside the equivalency range or contains values both within and outside the range, conclude the μ is not equivalent to μ_0.

One can also use statistical software to perform an equivalence test. Figure 7.10 shows the Minitab output for Example 7.8. It is common to visually present the test using the confidence interval and the user-specified upper and lower equivalence limits.

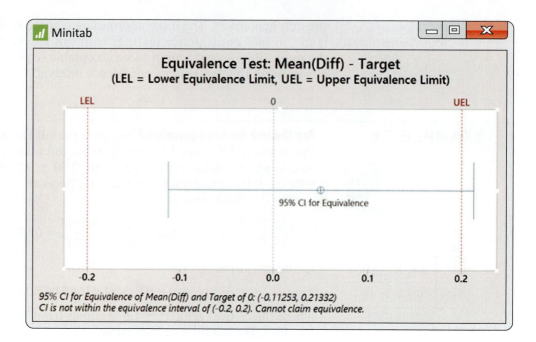

FIGURE 7.10 Minitab output of the equivalence test, Example 7.8.

Inside the Minitab window:

Equivalence Test: Mean(Diff) - Target
(LEL = Lower Equivalence Limit, UEL = Upper Equivalence Limit)

95% CI for Equivalence

95% CI for Equivalence of Mean(Diff) and Target of 0: (-0.11253, 0.21332)
CI is not within the equivalence interval of (-0.2, 0.2). Cannot claim equivalence.

USE YOUR KNOWLEDGE

7.9   Female wolf spider mate preferences. As part of a study on factors affecting mate choice, researchers exposed 18 premature female wolf spiders twice a day until maturity to iPod videos of three courting males with average size tufts. Once mature, each female spider was exposed to two videos, one involving a male with large tufts and the other involving a male with small tufts. The number of receptivity displays by the female toward each male was recorded.[4] Explain why a paired t-test is appropriate in this setting.

7.10   Oil-free deep fryer. Researchers at Purdue University are developing an oil-free deep fryer that will produce fried food faster, healthier, and safer than hot oil.[5] As part of this development, they ask food experts to compare foods made with hot oil and their oil-free fryer. Consider the following table comparing the taste of hash browns. Each hash brown was rated on a 0 to 100 scale, with 100 being the highest rating. For each expert, a coin was tossed to see which type of hash brown was tasted first.

	Expert				
	1	2	3	4	5
Hot oil:	78	84	62	73	63
Oil free:	75	85	67	75	66

Is there a difference in taste? State the appropriate hypotheses, and carry out a matched pairs t test using $\alpha = 0.05$.

7.11   95% confidence interval for the difference in taste. To a restaurant owner, the real question is how much difference there is in taste. Use the data to give a 95% confidence interval for the mean difference in taste scores between oil-free and hot-oil frying.

Robustness of the *t* procedures

The matched pairs *t* procedures and test of equivalence use one-sample *t* confidence intervals and significance tests for differences. They are, therefore, based on an assumption that the population of differences has a Normal distribution. For the histogram of the 51 differences in Example 7.7 shown in Figure 7.9 (page 420), the data appear to be slightly skewed. Does this slight non-Normality suggest that we should not use the *t* procedures for these data?

All inference procedures are based on some conditions, such as Normality. Procedures that are not strongly affected by violations of a condition are called *robust*. Robust procedures are very useful in statistical practice because they can be used over a wide range of conditions with good performance.

> ### ROBUST PROCEDURES
>
> A statistical inference procedure is called **robust** if the required probability calculations are insensitive to violations of the assumptions made.

LOOK BACK

resistant measure, p. 30

The assumption that the population is Normal rules out outliers, so the presence of outliers shows that this assumption is not valid. The *t* procedures are not robust against outliers because $\bar{x}$ and s are not resistant to outliers.

Fortunately, the *t* procedures are quite robust against non-Normality of the population except in the case of outliers or strong skewness. Larger samples improve the accuracy of *P*-values and critical values from the *t* distributions when the population is not Normal. This is true for two reasons:

LOOK BACK

central limit theorem, p. 298

law of large numbers, p. 250

1. The sampling distribution of the sample mean $\bar{x}$ from a large sample is close to Normal (that's the central limit theorem). Normality of the individual observations is of little concern when the sample is large.

2. As the sample size n grows, the sample standard deviation s will be an accurate estimate of σ whether or not the population has a Normal distribution. This fact is closely related to the law of large numbers.

To convince yourself of this fact, use the *t Statistic* applet to study the sampling distribution of the one-sample *t* statistic. From one of three population distributions, 10,000 SRSs of a user-specified sample size n are generated, and a histogram of the *t* statistics is constructed. You have the option to compare this estimated sampling distribution with the $t(n-1)$ distribution. When the population distribution is Normal, the sampling distribution of the *t* statistic is always *t* distributed. For the other two population distributions, you should see that as n increases, the histogram of *t* statistics looks more like the $t(n-1)$ distribution.

To assess whether the *t* procedures can be used in practice, a Normal quantile plot, stemplot, or boxplot is a good tool to check for skewness and outliers. For most purposes, the one-sample *t* procedures can be safely used when $n \geq 15$ unless an outlier or clearly marked skewness is present.

Except in the case of small samples, the assumption that the data are an SRS from the population of interest is more crucial than the assumption that the population distribution is Normal. Here are practical guidelines for inference on a single mean:[6]

• *Sample size less than 15:* Use *t* procedures if the data are close to Normal. If the data are clearly non-Normal or if outliers are present, do not use *t*.

• *Sample size at least 15 and less than 40:* The *t* procedures can be used except in the presence of outliers or strong skewness.

• *Large samples:* The *t* procedures can be used even for clearly skewed distributions when the sample is large, roughly $n \geq 40$.

For the measurement data in Example 7.7 (page 419), there is only slight skewness and no outliers. With $n = 51$ observations, we should feel comfortable that the *t* procedures give approximately correct results.

USE YOUR KNOWLEDGE

7.12 *t* procedures for time to start a business? Consider the data from Exercise 1.43 (page 29) but with Suriname removed. Would you be comfortable applying the *t* procedures in this case? Explain your answer.

7.13 *t* procedures for ticket prices? Consider the data on StubHub! ticket prices presented in Figure 1.32 (page 69). Would you be comfortable applying the *t* procedures in this case? In explaining your answer, recall that these *t* procedures focus on the mean μ.

BEYOND THE BASICS

The Bootstrap

Confidence intervals are based on sampling distributions. In this section, we have used the fact that the sampling distribution of $\bar{x}$ is $N(\mu, \sigma/\sqrt{n})$ when the data are an SRS from an $N(\mu, \sigma)$ population. If the data are not Normal, the central limit theorem tells us that this sampling distribution is still a reasonable approximation as long as the distribution of the data is not strongly skewed and there are no outliers. Even a fair amount of skewness can be tolerated when the sample size is large.

What if the population does not appear to be Normal and we have only a small sample? Then we do not know what the sampling distribution of *bootstrap* $\bar{x}$ looks like. The **bootstrap** is a procedure for approximating sampling distributions when theory cannot tell us their shape.[7]

The basic idea is to act as if our sample were the population. We take *resample* many samples from it. Each of these is called a **resample**. We calculate the mean $\bar{x}$ for each resample. We get different results from different resamples because we sample *with replacement*. Thus, an observation in the original sample can appear more than once in a resample. We treat the resulting distribution of $\bar{x}$'s as if it were the sampling distribution and use it to perform inference. If we want a 95% confidence interval, for example, we could use the middle 95% of this resample distribution.

EXAMPLE 7.9

TVTIME

A bootstrap confidence interval. Consider the eight time measurements (in hours per week) spent watching traditional television in Example 7.1:

3.0   16.5   10.5   40.5   5.5   33.5   0.0   6.5

We defended the use of the one-sided t confidence interval for an earlier analysis. Let's now compare those results with the confidence interval constructed using the bootstrap.

We decide to collect the $\bar{x}$'s from 1000 resamples of size $n = 8$. We use software to do this very quickly. One resample was

5.5   6.5   5.5   40.5   16.5   33.5   10.5   6.5

with $\bar{x} = 15.6251$. The middle 95% of our 1000 $\bar{x}$'s runs from 7.0 to 25.0. We repeat the procedure and get the interval (6.6, 25.1).

The two bootstrap intervals are relatively close to each other and are more narrow than the one-sample t confidence interval (2.1, 26.9). This suggests that the standard t interval is likely a little wider than it needs to be for this data set.

The bootstrap is practical only when you can use a computer to take 1000 or more resamples quickly. It is an example of how the use of fast and easy computing is changing the way we do statistics. More details about the bootstrap can be found in Chapter 16.

SECTION 7.1 SUMMARY

• Significance tests and confidence intervals for the mean μ of a Normal population are based on the sample mean $\bar{x}$ of an SRS. Because of the central limit theorem, the resulting procedures are approximately correct for other population distributions when the sample is large.

• The **standard error** of the sample mean is

$$\text{SE}_{\bar{x}} = \frac{s}{\sqrt{n}}$$

• The standardized sample mean, or **one-sample z statistic**,

$$z = \frac{\bar{x} - \mu}{\sigma/\sqrt{n}}$$

has the $N(0, 1)$ distribution. If the standard deviation $\sigma/\sqrt{n}$ of $\bar{x}$ is replaced by the **standard error** $s/\sqrt{n}$, the **one-sample t statistic**

$$t = \frac{\bar{x} - \mu}{s/\sqrt{n}}$$

has the **t distribution** with $n - 1$ degrees of freedom.

• There is a t distribution for every positive **degrees of freedom k**. All are symmetric distributions similar in shape to Normal distributions. The $t(k)$ distribution approaches the $N(0, 1)$ distribution as k increases.

• A level C **confidence interval for the mean** μ of a Normal population is

$$\bar{x} \pm t^* \frac{s}{\sqrt{n}}$$

where t^* is the value for the $t(n-1)$ density curve with area C between $-t^*$ and t^*. The quantity

$$t^* \frac{s}{\sqrt{n}}$$

is the **margin of error**.

- Significance tests for $H_0: \mu = \mu_0$ are based on the t statistic. P-values or fixed significance levels are computed from the $t(n-1)$ distribution.

- A matched pairs analysis is needed when subjects or experimental units are matched in pairs or when there are two measurements on each individual or experimental unit and the question of interest concerns the difference between the two measurements.

- The one-sample procedures are used to analyze **matched pairs** data by first taking the differences within the matched pairs to produce a single sample.

- One-sample **equivalence testing** assesses whether a population mean μ is practically different from a hypothesized mean μ_0. This test requires a threshold δ, which represents the largest difference between μ and μ_0 such that the means are considered equivalent.

- The t procedures are relatively **robust** against non-Normal populations. The t procedures are useful for non-Normal data when $15 \leq n < 40$ unless the data show outliers or strong skewness. When $n \geq 40$, the t procedures can be used even for clearly skewed distributions.

SECTION 7.1 EXERCISES

For Exercises 7.1 and 7.2, see page 410; for Exercises 7.3 and 7.4, see page 412; for Exercises 7.5 and 7.6, see page 415; for Exercises 7.7 and 7.8, see page 418; for Exercises 7.9 through 7.11, see pages 422–423; and for Exercises 7.12 and 7.13, see page 424.

7.14 What is wrong? In each of the following situations, identify what is wrong and then either explain why it is wrong or change the wording of the statement to make it true.

(a) As the degrees of freedom k decrease, the t distribution density curve gets closer to the $N(0,1)$ curve.

(b) The standard error of the sample mean is s^2/n.

(c) A researcher wants to test $H_0: \bar{x} = 30$ versus the one-sided alternative $H_a: \bar{x} < 30$.

(d) The 95% margin of error for the mean μ of a Normal population with unknown σ is the same for all SRS of size n.

7.15 Finding the critical value t^*. What critical value t^* from Table D should be used to calculate the margin of error for a confidence interval for the mean of the population in each of the following situations?

(a) A 95% confidence interval based on $n = 15$ observations.

(b) A 95% confidence interval from an SRS of 28 observations.

(c) A 90% confidence interval from a sample of size 28.

(d) These cases illustrate how the size of the margin of error depends upon the confidence level and the sample size. Summarize these relationships.

7.16 Distribution of the t statistic. Assume a sample size of $n = 24$. Draw a picture of the distribution of the t statistic under the null hypothesis. Use Table D and your picture to illustrate the values of the test statistic that would lead to rejection of the null hypothesis at the 5% level for a two-sided alternative.

7.17 More on the distribution of the t statistic. Repeat the previous exercise for the two situations where the alternative is one-sided.

7.18 One-sided versus two-sided P-values. Computer software reports $\bar{x} = 11.2$ and $P = 0.075$ for a t test of $H_0: \mu = 0$ versus $H_a: \mu \neq 0$. Based on prior knowledge, you justified testing the alternative $H_a: \mu > 0$. What is the P-value for your significance test?

7.19 More on one-sided versus two-sided P-values. Suppose that computer software reports $\bar{x} = -11.2$ and $P = 0.075$ for a t test of H_0: $\mu = 0$ versus H_a: $\mu \neq 0$. Would this change your P-value for the alternative hypothesis in the previous exercise? Use a sketch of the distribution of the test statistic under the null hypothesis to illustrate and explain your answer.

7.20 A one-sample t test. The one-sample t statistic for testing

$$H_0: \mu = 8$$
$$H_a: \mu > 8$$

from a sample of $n = 22$ observations has the value $t = 2.24$.

(a) What are the degrees of freedom for this statistic?

(b) Give the two critical values t^* from Table D that bracket t.

(c) Between what two values does the P-value of the test fall?

(d) Is the value $t = 2.24$ significant at the 5% level? Is it significant at the 1% level?

(e) If you have software available, find the exact P-value.

7.21 Another one-sample t test. The one-sample t statistic for testing

$$H_0: \mu = 40$$
$$H_a: \mu \neq 40$$

from a sample of $n = 13$ observations has the value $t = 2.78$.

(a) What are the degrees of freedom for t?

(b) Locate the two critical values t^* from Table D that bracket t.

(c) Between what two values does the P-value of the test fall?

(d) Is the value $t = 2.78$ statistically significant at the 5% level? At the 1% level?

(e) If you have software available, find the exact P-value.

7.22 A final one-sample t test. The one-sample t statistic for testing

$$H_0: \mu = 20$$
$$H_a: \mu < 20$$

based on $n = 9$ observations has the value $t = -1.85$.

(a) What are the degrees of freedom for this statistic?

(b) Between what two values does the P-value of the test fall?

(c) If you have software available, find the exact P-value.

7.23 Two-sided to one-sided P-value. Most software gives P-values for two-sided alternatives. Explain why you cannot always divide these P-values by 2 to obtain P-values for one-sided alternatives.

7.24 Business bankruptcies in Canada. Business bankruptcies in Canada are monitored by the Office of the Superintendent of Bankruptcy Canada (OSB).[8] Included in each report are the assets and liabilities the company declared at the time of the bankruptcy filing. A study is based on a random sample of 75 reports from the current year. The average debt (liabilities minus assets) is \$92,172 with a standard deviation of \$111,538.

(a) Construct a 95% one-sample t confidence interval for the average debt of these companies at the time of filing.

(b) Because the sample standard deviation is larger than the sample mean, this debt distribution is skewed. Provide a defense for using the t confidence interval in this case.

7.25 Fuel economy. Although the Environmental Protection Agency (EPA) establishes the tests to determine the fuel economy of new cars, it often does not perform them. Instead, the test protocols are given to the car companies, and the companies perform the tests themselves. To keep the industry honest, the EPA runs some spot checks each year. Recently, the EPA announced that Hyundai and Kia must lower their fuel economy estimates for many of their models.[9] Here are some city miles per gallon (mpg) values for one of the models the EPA investigated: 📊 MILEAGE

| 28.0 | 25.7 | 25.8 | 28.0 | 28.5 | 29.8 | 30.2 | 30.4 |
| 26.9 | 28.3 | 29.8 | 27.2 | 26.7 | 27.7 | 29.5 | 28.0 |

Give a 95% confidence interval for μ, the mean city mpg for this model.

7.26 Testing the sticker information. Refer to the previous exercise. The vehicle sticker information for this model stated a city average of 30 mpg. Are these mpg values consistent with the vehicle sticker? Perform a significance test using the 0.05 significance level. Be sure to specify the hypotheses, the test statistic, the P-value, and your conclusion. 📊 MILEAGE

7.27 UberX driver earnings. On its blog, Uber posted a scatterplot using a sample of several thousand drivers in New York City. The plot shows each driver's average net earnings per hour versus the number of hours worked.[10] Here is a sample of earnings (dollars) for 27 drivers working 40 hours a week. 📊 UBERX

26.25	33.51	43.91	31.91	31.78	43.37	36.66	31.69	31.25
46.86	35.44	40.30	30.93	37.80	42.44	43.80	49.64	36.79
34.10	37.54	30.93	38.40	37.83	21.73	41.62	26.25	33.51

(a) Do you think it is appropriate to use the t methods of this section to compute a 95% confidence interval for the average earnings per hour of New York City UberX drivers working 40 hours a week? Generate a plot to support your answer.

(b) Report the 95% confidence interval for μ, the average earnings per hour of New York City UberX drivers working 40 hours a week, as an estimate and margin of error.

(c) Report the 95% confidence interval for the average annual earnings of New York City UberX drivers working 40 hours a week.

(d) According to Uber, the median annual wage of an UberX driver working at least 40 hours in New York City is $90,766. Can these data be used to assess this claim? Explain your answer.

7.28 Number of friends on Facebook. To mark Facebook's 10th birthday, Pew Research surveyed people using Facebook to see what they like and dislike about the site. The survey found that among adult Facebook users, the average number of friends is 338. This distribution takes only integer values, so it is certainly not Normal. It is also highly skewed to the right with a median of 200 friends.[11] Consider the following SRS of $n = 30$ Facebook users from your large university. **FACEFR**

107	246	289	177	155	101	80	461	336	78
463	264	827	180	221	1065	79	691	70	921
126	672	296	60	11	227	84	787	18	82

(a) Are these data also heavily skewed? Use graphical methods to examine the distribution. Write a short summary of your findings.

(b) Do you think it is appropriate to use the t methods of this section to compute a 95% confidence interval for the mean number of friends that Facebook users at your large university have? Explain why or why not.

(c) Compute the sample mean and standard deviation, the standard error of the mean, and the margin of error for 95% confidence.

(d) Report the 95% confidence interval for μ, the average number of friends for Facebook users at your large university.

7.29 Alcohol content in beer. In February 2013, two California residents filed a class-action lawsuit against Anheuser-Busch, alleging the company was watering down beers to boost profits.[12] They argued that because water was being added, the true alcohol content of the beer by volume is less than the advertised amount. For example, they alleged that Budweiser beer has an alcohol

content by volume of 4.7% instead of the stated 5%. CNN, NPR, and a local St. Louis news team picked up on this suit and hired independent labs to test samples of Budweiser beer and find the alcohol content. Below is a summary of these tests each done on a single can. **BUD**

$$4.94 \quad 5.00 \quad 4.99$$

(a) Even though we have a very small sample, test the null hypothesis that the alcohol content is 4.7% by volume. Do the data provide evidence against the claim of the two residents?

(b) Construct a 95% confidence interval for the true alcohol content in Budweiser.

(c) U.S. government standards require that the true alcohol content in all cans and bottles be within ±0.3% of the advertised level. Do these tests provide strong evidence that this is the case for Budweiser beer? Explain your answer.

7.30 Using the Internet on a computer. The Nielsen Company reported that U.S. residents aged 18 to 24 years spend an average of 32.5 hours per month using the Internet on a computer.[13] You wonder if this it true for students at your large university because so many students use their smartphones to access the Internet. You collect an SRS of $n = 75$ students and obtain $\bar{x} = 28.5$ hours with $s = 23.1$ hours.

(a) Report the 95% confidence interval for μ, the average number of hours per month that students at your university use the Internet on a computer.

(b) Use this interval to test whether the average time for students at your university is different from the average reported by Nielsen. Use the 5% significance level. Summarize your results.

7.31 Rudeness and its effect on onlookers. Many believe that an uncivil environment has a negative effect on people. A pair of researchers performed a series of experiments to test whether witnessing rudeness and disrespect affects task performance.[14] In one study, 34 participants met in small groups and witnessed the group organizer being rude to a "participant" who showed up late for the group meeting. After the exchange, each participant performed an individual brainstorming task in which he or she was asked to produce as many uses for a brick as possible in five minutes. The mean number of uses was 7.88 with a standard deviation of 2.35.

(a) Suppose that prior research has shown that the average number of uses a person can produce in five minutes under normal conditions is 10. Given that the researchers hypothesize that witnessing this rudeness will decrease performance, state the appropriate null and alternative hypotheses.

(b) Carry out the significance test using a significance level of 0.05. Give the *P*-value and state your conclusion.

7.32 Fuel efficiency *t* test. Computers in some vehicles calculate various quantities related to performance. One of these is the fuel efficiency, or gas mileage, usually expressed as miles per gallon (mpg). For one vehicle equipped in this way, the miles per gallon were recorded each time the gas tank was filled, and the computer was then reset.[15] Here are the mpg values for a random sample of 20 of these records: [ıl.] MPG

| 41.5 | 50.7 | 36.6 | 37.3 | 34.2 | 45.0 | 48.0 | 43.2 | 47.7 | 42.2 |
| 43.2 | 44.6 | 48.4 | 46.4 | 46.8 | 39.2 | 37.3 | 43.5 | 44.3 | 43.3 |

(a) Describe the distribution using graphical methods. Is it appropriate to analyze these data using methods based on Normal distributions? Explain why or why not.

(b) Find the mean, standard deviation, standard error, and margin of error for 95% confidence.

(c) Report the 95% confidence interval for μ, the mean miles per gallon for this vehicle based on these data.

7.33 Tree diameter confidence interval. A study of 584 longleaf pine trees in the Wade Tract in Thomas County, Georgia, is described in Example 6.1 (page 342). For each tree in the tract, the researchers measured the diameter at breast height (DBH). This is the diameter of the tree at a height of 4.5 feet, and the units are centimeters (cm). Only trees with DBH greater than 1.5 cm were sampled. Here are the diameters of a random sample of 40 of these trees: [ıl.] PINES

10.5	13.3	26.0	18.3	52.2	9.2	26.1	17.6	40.5	31.8
47.2	11.4	2.7	69.3	44.4	16.9	35.7	5.4	44.2	2.2
4.3	7.8	38.1	2.2	11.4	51.5	4.9	39.7	32.6	51.8
43.6	2.3	44.6	31.5	40.3	22.3	43.3	37.5	29.1	27.9

(a) Use a histogram or stemplot and a boxplot to examine the distribution of DBHs. Include a Normal quantile plot if you have the necessary software. Write a careful description of the distribution.

(b) Is it appropriate to use the methods of this section to find a 95% confidence interval for the mean DBH of all trees in the Wade Tract? Explain why or why not.

(c) Report the mean with the margin of error and the confidence interval. Write a short summary describing the meaning of the confidence interval.

(d) Do you think these results would apply to other similar trees in the same area? Give reasons for your answer.

7.34 Nutritional intake among Canadian high-performance male athletes. Recall Exercise 6.74 (page 382). For one part of the study, $n = 114$ male athletes from eight Canadian sports centers were surveyed. Their average caloric intake was 3077.0 kilocalories per day (kcal/d) with a standard deviation of 987.0. The recommended amount is 3421.7. Is there evidence that Canadian high-performance male athletes are deficient in their caloric intake?

(a) State the appropriate H_0 and H_a to test this.

(b) Carry out the test, give the *P*-value, and state your conclusion.

(c) Construct a 95% confidence interval for the average deficiency in caloric intake.

7.35 Average number of Instagram posts. LocoWise provides social media analytics to companies and marketing agencies through a variety of online tools. One tool is the Instagram Analyzer, which allows a user to compare a profile with 2500 other Instagram profiles. Recently, it reported that the 2500 profiles it monitors averaged 2.55 posts per day, with a minimum value of 0 posts and a maximum value of 95 posts.[16]

(a) A common estimator of the standard deviation when provided the range R is $s = R/6$. Compute this estimate of s for these data.

(b) Construct the 95% confidence interval for the average number of Instagram posts per day.

(c) These data are clearly skewed and possibly have a few outliers. Do you think it is appropriate to use the *t* procedures? Explain your answer.

7.36 Stress levels in parents of children with ADHD. In a study of parents who have children with attention-deficit/hyperactivity disorder (ADHD), parents were asked to rate their overall stress level using the Parental Stress Scale (PSS).[17] This scale has 18 items that contain statements regarding both positive and negative aspects of parenthood. Respondents are asked to rate their agreement with each statement using a five-point scale (1 = strongly disagree to 5 = strongly agree). The scores are summed such that a higher score indicates greater stress. The mean rating for the 50 parents in the study was reported as 52.98 with a standard deviation of 10.34.

(a) Do you think that these data are approximately Normally distributed? Explain why or why not.

(b) Is it appropriate to use the methods of this section to compute a 90% confidence interval? Explain why or why not.

(c) Find the 90% margin of error and the corresponding confidence interval. Write a sentence explaining the interval and the meaning of the 90% confidence level.

(d) To recruit parents for the study, the researchers visited a psychiatric outpatient service in Rohtak, India, and selected 50 consecutive families who met the inclusion and exclusion criteria. To what extent do you think the results can be generalized to all parents with children who have ADHD in India or in other locations around the world?

7.37 Are the parents feeling extreme stress? Refer to the previous exercise. The researchers considered a score greater than 45 to represent extreme stress. Is there evidence that the average stress level for the parents in this study is above this level? Perform a test of significance using $\alpha = 0.10$ and summarize your results.

7.38 Food intake and weight gain. If we increase our food intake, we generally gain weight. Nutrition scientists can calculate the amount of weight gain that would be associated with a given increase in calories. In one study, 16 nonobese adults, aged 25 to 36 years, were fed 1000 calories per day in excess of the calories needed to maintain a stable body weight. The subjects maintained this diet for eight weeks, so they consumed a total of 56,000 extra calories.[18] According to theory, 3500 extra calories will translate into a weight gain of 1 pound. Therefore, we expect each of these subjects to gain 56,000/3500 = 16 pounds (lb). Here are the weights before and after the eight-week period, expressed in kilograms (kg): WTGAIN

Subject	1	2	3	4	5	6	7	8
Weight before	55.7	54.9	59.6	62.3	74.2	75.6	70.7	53.3
Weight after	61.7	58.8	66.0	66.2	79.0	82.3	74.3	59.3

Subject	9	10	11	12	13	14	15	16
Weight before	73.3	63.4	68.1	73.7	91.7	55.9	61.7	57.8
Weight after	79.1	66.0	73.4	76.9	93.1	63.0	68.2	60.3

(a) For each subject, subtract the weight before from the weight after to determine the weight change.

(b) Find the mean and the standard deviation for the weight change.

(c) Calculate the standard error and the margin of error for 95% confidence. Report the 95% confidence interval for weight change in a sentence that explains the meaning of the 95%.

(d) Convert the mean weight gain in kilograms to mean weight gain in pounds. Because there are 2.2 kg per pound, multiply the value in kilograms by 2.2 to obtain pounds. Do the same for the standard deviation and the confidence interval.

(e) Test the null hypothesis that the mean weight gain is 16 lb. Be sure to specify the null and alternative

hypotheses, the test statistic with degrees of freedom, and the P-value. What do you conclude?

(f) Write a short paragraph explaining your results.

7.39 Food intake and NEAT. Nonexercise activity thermogenesis (NEAT) provides a partial explanation for the results you found in the previous analysis. NEAT is energy burned by fidgeting, maintenance of posture, spontaneous muscle contraction, and other activities of daily living. In the study of the previous exercise, the 16 subjects increased their NEAT by 328 calories per day, on average, in response to the additional food intake. The standard deviation was 256.

(a) Test the null hypothesis that there was no change in NEAT versus the two-sided alternative. Summarize the results of the test and give your conclusion.

(b) Find a 95% confidence interval for the change in NEAT. Discuss the additional information provided by the confidence interval that is not evident from the results of the significance test.

7.40 Potential insurance fraud? Insurance adjusters are concerned about the high estimates they are receiving from Jocko's Garage. To see if the estimates are unreasonably high, each of 10 damaged cars was taken to Jocko's and to another garage and the estimates (in dollars) were recorded. Here are the results: JOCKO

Car	1	2	3	4	5
Jocko's	1410	1550	1250	1300	900
Other	1250	1300	1250	1200	950

Car	6	7	8	9	10
Jocko's	1520	1750	3600	2250	2840
Other	1575	1600	3380	2125	2600

(a) For each car, subtract the estimate of the other garage from Jocko's estimate. Find the mean and the standard deviation for this difference.

(b) Test the null hypothesis that there is no difference between the estimates of the two garages. Be sure to specify the null and alternative hypotheses, the test statistic with degrees of freedom, and the P-value. What do you conclude using the 0.05 significance level?

(c) Construct a 95% confidence interval for the difference in estimates.

(d) The insurance company is considering seeking repayment from 1000 claims filed with Jocko's last year. Using your answer to part (c), what repayment would you recommend the insurance company seek? Explain your answer.

7.41 Fuel efficiency comparison t test. Refer to Exercise 7.32. In addition to the computer calculating miles per gallon, the driver also recorded the miles per gallon by dividing the miles driven by the number of gallons at fill-up. The driver wants to determine if these calculations are different. **MPGDIFF**

Fill-up	1	2	3	4	5	6	7	8	9	10
Computer	41.5	50.7	36.6	37.3	34.2	45.0	48.0	43.2	47.7	42.2
Driver	36.5	44.2	37.2	35.6	30.5	40.5	40.0	41.0	42.8	39.2

Fill-up	11	12	13	14	15	16	17	18	19	20
Computer	43.2	44.6	48.4	46.4	46.8	39.2	37.3	43.5	44.3	43.3
Driver	38.8	44.5	45.4	45.3	45.7	34.2	35.2	39.8	44.9	47.5

(a) State the appropriate H_0 and H_a.

(b) Carry out the test using a significance level of 0.05. Give the P-value, and then interpret the result.

7.42 Counts of picks in a one-pound bag. A guitar supply company must maintain strict oversight on the number of picks they package for sale to customers. Their current advertisement specifies between 900 and 1000 picks in every bag. An SRS of 36 one-pound bags of picks was collected as part of a quality improvement effort within the company. The number of picks in each bag is shown in the following table. **PICKS**

924	925	967	909	959	937	970	936	952
919	965	921	913	886	956	962	916	945
957	912	961	950	923	935	969	916	952
917	977	940	924	957	920	986	895	923

(a) Create (i) a histogram or stemplot, (ii) a boxplot, and (iii) a Normal quantile plot of these counts. Write a careful description of the distribution. Make sure to note any outliers, and comment on the skewness and Normality of the data.

(b) Based on your observations in part (a), is it appropriate to analyze these data using the t procedures? Briefly explain your response.

(c) Find the mean, the standard deviation, and the standard error of the mean for this sample.

(d) Calculate the 90% confidence interval for the mean number of picks in a one-pound bag.

7.43 Significance test for the average number of picks. Refer to the previous exercise. **PICKS**

(a) Do these data provide evidence that the average number of picks in a one-pound bag is greater than 925? Using a significance level of 5%, state your hypotheses, the P-value, and your conclusions.

(b) Do these data provide evidence that the average number of picks in a one-pound bag is greater than 935?

Using a significance level of 5%, state your hypotheses, the P-value, and your conclusion.

(c) Explain the relationship between your conclusions in parts (a) and (b) and the 90% confidence interval calculated in the previous problem.

7.44 A customer satisfaction survey. Many organizations are doing surveys to determine the satisfaction of their customers. Attitudes toward various aspects of campus life were the subject of one such study conducted at Purdue University. Each item was rated on a 1 to 5 scale, with 5 being the highest rating. The average response of 1568 first-year students to "Feeling welcomed at Purdue" was 3.83 with a standard deviation of 1.10. Assuming that the respondents are an SRS, give a 90% confidence interval for the mean of all first-year students.

7.45 Comparing operators of a DXA machine. Dual-energy X-ray absorptiometry (DXA) is a technique for measuring bone health. One of the most common measures is total body bone mineral content (TBBMC). A highly skilled operator is required to take the measurements. Recently, a new DXA machine was purchased by a research lab, and two operators were trained to take the measurements. TBBMC for eight subjects was measured by both operators.[19] The units are grams (g). A comparison of the means for the two operators provides a check on the training they received and allows us to determine if one of the operators is producing measurements that are consistently higher than the other. Here are the data: **TBBMC**

	Subject							
Operator	1	2	3	4	5	6	7	8
1	1.328	1.342	1.075	1.228	0.939	1.004	1.178	1.286
2	1.323	1.322	1.073	1.233	0.934	1.019	1.184	1.304

(a) Take the difference between the TBBMC recorded for Operator 1 and the TBBMC for Operator 2. Describe the distribution of these differences. Is it appropriate to analyze these data using the t methods? Explain why or why not.

(b) Use a significance test to examine the null hypothesis that the two operators have the same mean. Be sure to give the test statistic with its degrees of freedom, the P-value, and your conclusion.

(c) The sample here is rather small, so we may not have much power to detect differences of interest. Use a 95% confidence interval to provide a range of differences that are compatible with these data.

(d) The eight subjects used for this comparison were not a random sample. In fact, they were friends of the researchers whose ages and weights were similar to those of the types of people who would be measured with this DXA machine. Comment on the appropriateness of this procedure for selecting a sample, and discuss any consequences regarding the interpretation of the significance-testing and confidence interval results.

7.46 Equivalence of paper and computer-based questionnaires. Computers are commonly being used to complete questionnaires because of the increased efficiency of data collection and reduction in coding errors. Studies, however, have shown that questionnaire format can influence responses, especially for items of a sensitive nature.[20] Consider the small study below comparing paper and computer survey formats of a self-report measure of mental health. Each participant completed both forms on adjacent days with the order determined by a flip of a coin. 🔢 EQUIV

Subject	Paper	Computer	Diff	Subject	Paper	Computer	Diff
1	5	2	3	11	6	5	1
2	4	3	1	12	5	5	0
3	4	4	0	13	3	7	-4
4	7	8	-1	14	3	6	-3
5	4	5	-1	15	4	4	0
6	6	7	-1	16	2	3	-1
7	4	3	1	17	7	10	-3
8	6	8	-2	18	8	7	1
9	6	5	1	19	4	6	-2
10	2	3	-1	20	6	8	-2

(a) Explain to someone unfamiliar with statistics why this experiment is a matched pairs design.

(b) The measure involves 10 items and produces a whole number score ranging between 0 and 20. Do you think it is appropriate to use the t procedures on the difference in survey scores? Explain your answer.

(c) Perform an equivalency test at the 0.05 level using the limits ± 0.5 and state your conclusion.

7.47 Assessment of a foreign-language institute. The National Endowment for the Humanities sponsors summer institutes to improve the skills of high

school teachers of foreign languages. One such institute hosted 20 French teachers for four weeks. At the beginning of the period, the teachers were given the Modern Language Association's listening test of understanding of spoken French. After four weeks of immersion in French in and out of class, the listening test was given again. (The actual French spoken in the two tests was different, so that simply taking the first test should not improve the score on the second test.) The maximum possible score on the test is 36.[21] Here are the data: 🔢 SUMLANG

Teacher	Pretest	Posttest	Gain	Teacher	Pretest	Posttest	Gain
1	32	34	2	11	30	36	6
2	31	31	0	12	20	26	6
3	29	35	6	13	24	27	3
4	10	16	6	14	24	24	0
5	30	33	3	15	31	32	1
6	33	36	3	16	30	31	1
7	22	24	2	17	15	15	0
8	25	28	3	18	32	34	2
9	32	26	-6	19	23	26	3
10	20	26	6	20	23	26	3

To analyze these data, we first subtract the pretest score from the posttest score to obtain the improvement for each teacher. These 20 differences form a single sample. They appear in the "Gain" columns. The first teacher, for example, improved from 32 to 34, so the gain is $34 - 32 = 2$.

(a) State appropriate null and alternative hypotheses for examining the question of whether or not the course improves French spoken-language skills.

(b) Describe the gain data. Use numerical and graphical summaries.

(c) Perform the significance test. Give the test statistic, the degrees of freedom, and the P-value. Summarize your conclusion.

(d) Give a 95% confidence interval for the mean improvement.

7.2 Comparing Two Means

When you complete this section, you will be able to:

- Describe a level C confidence interval for the difference between two population means in terms of an estimate and its margin of error.
- Construct a level C confidence interval for the difference between two population means $\mu_1 - \mu_2$ from two SRSs of size n_1 and n_2, respectively.
- Perform a two-sample t significance test and summarize the results.
- Explain when the t procedures can be useful for non-Normal data.

A psychologist wants to compare male and female college students' impressions of personality based on selected Facebook pages. A nutritionist is interested in the effect of increased calcium on blood pressure. A bank wants to know which of two incentive plans will most increase the use of its debit cards. Two-sample problems such as these are among the most common situations encountered in statistical practice.

TWO-SAMPLE PROBLEMS

- The goal of inference is to compare the means of the response variable in two groups.

- Each group is considered to be a sample from a distinct population.

- The responses in each group are independent of those in the other group.

LOOK BACK

randomized
comparative
experiment,
p. 177

side-by-side
boxplots,
p. 37

A two-sample problem can arise from a randomized comparative experiment that randomly divides the subjects into two groups and exposes each group to a different treatment. A two-sample problem can also arise when comparing random samples separately selected from two populations. Unlike the matched pairs designs studied earlier, there is no matching of the units in the two samples, and the two samples may be of different sizes. As a result, inference procedures for two-sample data differ from those for matched pairs.

We can present two-sample data graphically by a back-to-back stemplot (for small samples) or by side-by-side boxplots (for larger samples). Now we will apply the ideas of formal inference in this setting. When both population distributions are symmetric, and especially when they are at least approximately Normal, a comparison of the mean responses in the two populations is most often the goal of inference.

We have two independent samples, from two distinct populations (such as subjects given the latest Apple iPhone and those given the latest Samsung Galaxy smartphone). The same response variable—say, battery life—is measured for both samples. We will call the variable x_1 in the first population and x_2 in the second because the variable may have different distributions in the two populations. Here is the notation that we will use to describe the two populations:

Population	Variable	Mean	Standard deviation
1	x_1	μ_1	σ_1
2	x_2	μ_2	σ_2

We want to compare the two population means, either by giving a confidence interval for $\mu_1 - \mu_2$ or by testing the hypothesis of no difference, $H_0: \mu_1 = \mu_2$.

Inference is based on two independent SRSs, one from each population. Here is the notation that describes the samples:

Population	Sample size	Sample mean	Sample standard deviation
1	n_1	$\bar{x}_1$	s_1
2	n_2	$\bar{x}_2$	s_2

Throughout this section, the subscripts 1 and 2 show the population to which a parameter or a sample statistic refers.

The two-sample z statistic

The natural estimator of the difference $\mu_1 - \mu_2$ is the difference between the sample means, $\bar{x}_1 - \bar{x}_2$. If we are to base inference on this statistic, we must know its sampling distribution. Here are some facts from our study of probability:

LOOK BACK

addition rule
for means,
p. 254

- The mean of the difference $\bar{x}_1 - \bar{x}_2$ is the difference between the means $\mu_1 - \mu_2$. This follows from the addition rule for means and the fact that the mean of any $\bar{x}$ is the same as the mean μ of the population.

- The variance of the difference $\bar{x}_1 - \bar{x}_2$ is the sum of their variances, which is

$$\frac{\sigma_1^2}{n_1} + \frac{\sigma_2^2}{n_2}$$

LOOK BACK

addition rule
for variances,
p. 258
linear
combination
of Normal
random
variables
p. 304

This follows from the addition rule for variances. Because the samples are independent, their sample means $\bar{x}_1$ and $\bar{x}_2$ are independent random variables.

- If the two population distributions are both Normal, then the distribution of $\bar{x}_1 - \bar{x}_2$ is also Normal. This is true because each sample mean alone is Normally distributed and because a difference between independent Normal random variables is also Normal.

We now know the sampling distribution of $\bar{x}_1 - \bar{x}_2$ when both populations are Normally distributed. The mean and variance of this distribution can be expressed in terms of the parameters of the two populations.

EXAMPLE 7.10

Robert Warren/Getty Images

Heights of 10-year-old girls and boys. A fourth-grade class has 12 girls and 8 boys. The children's heights are recorded on their 10th birthdays. What is the chance that the girls are taller than the boys? Of course, it is very unlikely that all the girls are taller than all the boys. We translate the question into the following: what is the probability that the mean height of the girls is greater than the mean height of the boys?

Based on information from the National Health and Nutrition Examination Survey, we assume that the heights (in inches) of 10-year-old girls are $N(56.9, 2.8)$ and the heights of 10-year-old boys are $N(56.0, 3.5)$.[22] The heights of the students in our class are assumed to be random samples from these populations. The two distributions are shown in Figure 7.11(a).

The difference $\bar{x}_1 - \bar{x}_2$ between the female and male mean heights varies in different random samples. The sampling distribution has mean

$$\mu_1 - \mu_2 = 56.9 - 56.0 = 0.9 \text{ inches}$$

and variance

$$\frac{\sigma_1^2}{n_1} + \frac{\sigma_2^2}{n_2} = \frac{2.8^2}{12} + \frac{3.5^2}{8}$$
$$= 2.18$$

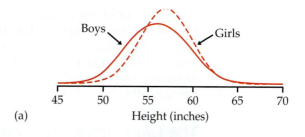

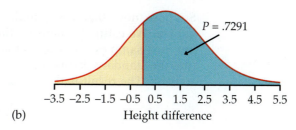

FIGURE 7.11 Distributions, Example 7.10. (a) Distributions of heights of 10-year-old boys and girls. (b) Distribution of the difference between mean heights of 12 girls and 8 boys.

The standard deviation of the difference in sample means is, therefore, $\sqrt{2.18} = 1.48$ inches.

If the heights vary Normally, the difference in sample means is also Normally distributed. The distribution of the difference in heights is shown in Figure 7.11(b). We standardize $\bar{x}_1 - \bar{x}_2$ by subtracting its mean (0.9) and dividing by its standard deviation (1.48). Therefore, the probability that the girls, on average, are taller than the boys is

$$P(\bar{x}_1 - \bar{x}_2 > 0) = P\left(\frac{(\bar{x}_1 - \bar{x}_2) - 0.9}{1.48} > \frac{0 - 0.9}{1.48}\right)$$

$$= P(Z > -0.61) = 0.7291$$

Even though the population mean height of 10-year-old girls is greater than the population mean height of 10-year-old boys, the probability that the sample mean of the girls is greater than the sample mean of the boys in our class is only 73%. *Large samples are needed to see the effects of small differences.*

As Example 7.10 reminds us, any Normal random variable has the $N(0, 1)$ distribution when standardized. We have arrived at a new z statistic.

TWO-SAMPLE z STATISTIC

Suppose that $\bar{x}_1$ is the mean of an SRS of size n_1 drawn from an $N(\mu_1, \sigma_1)$ population and that $\bar{x}_2$ is the mean of an independent SRS of size n_2 drawn from an $N(\mu_2, \sigma_2)$ population. Then the **two-sample z statistic**

$$z = \frac{(\bar{x}_1 - \bar{x}_2) - (\mu_1 - \mu_2)}{\sqrt{\dfrac{\sigma_1^2}{n_1} + \dfrac{\sigma_2^2}{n_2}}}$$

has the standard Normal $N(0, 1)$ sampling distribution.

In the unlikely event that both population standard deviations are known, the two-sample z statistic is the basis for inference about $\mu_1 - \mu_2$. Exact z procedures are seldom used, however, because σ_1 and σ_2 are rarely known. In Chapter 6, we discussed the one-sample z procedures in order to introduce the ideas of inference. Here we move directly to the more useful t procedures.

The two-sample t procedures

Suppose now that the population standard deviations σ_1 and σ_2 are not known. We estimate them by the sample standard deviations s_1 and s_2 from our two samples. Following the pattern of the one-sample case, we substitute the standard errors for the standard deviations used in the two-sample z statistic. The result is the *two-sample t statistic*:

$$t = \frac{(\bar{x}_1 - \bar{x}_2) - (\mu_1 - \mu_2)}{\sqrt{\dfrac{s_1^2}{n_1} + \dfrac{s_2^2}{n_2}}}$$

Unfortunately, this statistic does *not* have a t distribution. A t distribution replaces the $N(0, 1)$ distribution only when a single standard deviation (σ) in a z statistic is replaced by its sample standard deviation (s). In this case, we replace two standard deviations (σ_1 and σ_2) by their estimates (s_1 and s_2), which does not produce a statistic having a t distribution.

df approximation   Nonetheless, we can approximate the distribution of the two-sample t statistic by using the $t(k)$ distribution with an **approximation for the degrees of freedom k**. We use these approximations to find approximate values of t^* for confidence intervals and to find approximate P-values for significance tests. Here are two approximations:

Satterthwaite approximation   1. Use an approximation known as the **Satterthwaite approximation** for the value of k. It is calculated from the data and, in general, will not be a whole number.

2. Use k equal to the smaller of $n_1 - 1$ and $n_2 - 1$.

In practice, the choice of approximation rarely makes a difference in our conclusion. Most statistical software uses the first option to approximate the $t(k)$ distribution for two-sample problems unless the user requests another method. Use of this approximation without software is a bit complicated; we give the details later in this section (see page 447).

If you are not using software, the second approximation is preferred. This approximation is appealing because it is conservative.[23] Margins of error for the level C confidence intervals are a bit larger than they need to be, so the true confidence level is larger than C. For significance testing, the P-values are a bit larger; thus, for tests at a fixed significance level, we are a little less likely to reject H_0 when it is true.

The two-sample t confidence interval

We now apply the basic ideas about t procedures to the problem of comparing two means when the standard deviations are unknown. We start with confidence intervals.

THE TWO-SAMPLE *t* CONFIDENCE INTERVAL

Suppose that an SRS of size n_1 is drawn from a Normal population with unknown mean μ_1 and that an independent SRS of size n_2 is drawn from another Normal population with unknown mean μ_2. The **confidence interval for $\mu_1 - \mu_2$** given by

$$(\bar{x}_1 - \bar{x}_2) \pm t^* \sqrt{\frac{s_1^2}{n_1} + \frac{s_2^2}{n_2}}$$

has confidence level at least C no matter what the population standard deviations may be. The quantity

$$t^* \sqrt{\frac{s_1^2}{n_1} + \frac{s_2^2}{n_2}}$$

is the **margin of error**. Here, t^* is the value for the $t(k)$ density curve with area C between $-t^*$ and t^*. The value of the degrees of freedom k is approximated by software, or we use the smaller of $n_1 - 1$ and $n_2 - 1$. Similarly, we can use either software or the conservative approach with Table D to approximate the value of t^*.

EXAMPLE 7.11

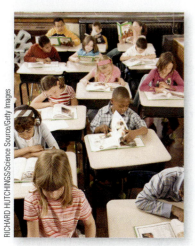

RICHARD HUTCHINGS/Science Source/Getty Images

DRP

Directed reading activities assessment. An educator believes that new directed reading activities in the classroom will help elementary school pupils improve some aspects of their reading ability. She arranges for a third-grade class of 21 students to take part in these activities for an eight-week period. A control classroom of 23 third-graders follows the same curriculum without the activities. At the end of the eight weeks, all students are given a Degree of Reading Power (DRP) test, which measures the aspects of reading ability that the treatment is designed to improve. The data appear in Table 7.3.[24]

The design of the study in Example 7.11 is not ideal. Random assignment of students was not possible in a school environment, so existing third-grade classes were used. The effect of the reading programs is, therefore,

TABLE 7.3	**DRP Scores for Third-Graders**						
Treatment group				Control group			
24	61	59	46	42	33	46	37
43	44	52	43	43	41	10	42
58	67	62	57	55	19	17	55
71	49	54		26	54	60	28
43	53	57		62	20	53	48
49	56	33		37	85	42	

LOOK BACK

confounding,
p. 150

confounded with any other differences between the two classes. The classes were chosen to be as similar as possible—for example, in terms of the social and economic status of the students. Extensive pretesting showed that the two classes were, on the average, quite similar in reading ability at the beginning of the experiment. To avoid the effect of two different teachers, the researcher herself taught reading in both classes during the eight-week period of the experiment. Therefore, we can be somewhat confident that the two-sample test is detecting the effect of the treatment and not some other difference between the classes. This example is typical of many situations in which an experiment is carried out but randomization is not possible.

EXAMPLE 7.12

DRP

Computing an approximate 95% confidence interval for the difference in means. First examine the data:

Control		Treatment
970	1	
860	2	4
773	3	3
8632221	4	3334699
5543	5	23467789
20	6	127
	7	1
5	8	

The back-to-back stemplot suggests that there is a mild outlier in the control group but no deviation from Normality serious enough to forbid use of t procedures. Separate Normal quantile plots for both groups (Figure 7.12) confirm that both distributions are approximately Normal. The scores of

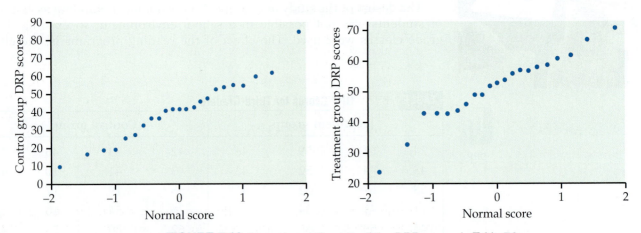

FIGURE 7.12 Normal quantile plots of the DRP scores in Table 7.3.

the treatment group appear to be somewhat higher than those of the control group. The summary statistics are

Group	n	$\bar{x}$	s
Treatment	21	51.48	11.01
Control	23	41.52	17.15

To describe the size of the treatment effect, let's construct a confidence interval for the difference between the treatment group and the control group means. The interval is

$$(\bar{x}_1 - \bar{x}_2) \pm t^* \sqrt{\frac{s_1^2}{n_1} + \frac{s_2^2}{n_2}} = (51.48 - 41.52) \pm t^* \sqrt{\frac{11.01^2}{21} + \frac{17.15^2}{23}}$$

$$= 9.96 \pm 4.31 t^*$$

df = 20

t^*	1.725	2.086	2.197
C	0.90	0.95	0.96

The second degrees of freedom approximation uses the $t(20)$ distribution. Table D gives $t^* = 2.086$. With this approximation, we have

$$9.96 \pm (4.31 \times 2.086) = 9.96 \pm 8.99 = (1.0, 18.9)$$

We estimate the mean improvement to be about 10 points, with a margin of error of almost 9 points. Unfortunately, the data do not allow a very precise estimate of the size of the average improvement.

USE YOUR KNOWLEDGE

7.48 Two-sample t confidence interval. Suppose a study similar to Example 7.11 was performed using two second-grade classes. Assume the summary statistics are $\bar{x}_1 = 46.32$, $\bar{x}_2 = 32.85$, $s_1 = 11.53$, $s_2 = 15.33$, $n_1 = 26$, and $n_2 = 24$. Find a 95% confidence interval for the difference between the treatment (Group 1) and the control (Group 2) means using the second approximation for degrees of freedom. Also write a one-sentence summary of what this confidence interval says about the difference in means.

7.49 Smaller sample sizes. Refer to the previous exercise. Suppose instead that the two classes are smaller, so the summary statistics are $\bar{x}_1 = 46.32$, $\bar{x}_2 = 32.85$, $s_1 = 11.53$, $s_2 = 15.33$, $n_1 = 16$, and $n_2 = 14$. Find a 95% confidence interval for the difference using the second approximation for degrees of freedom. Compare this interval with the one in the previous exercise and discuss the impact smaller sample sizes have on a confidence interval.

The two-sample t significance test

The same ideas that we used for the two-sample t confidence interval also apply to *two-sample t significance tests*. We can use either software or the conservative approach with Table D to approximate the *P*-value.

THE TWO-SAMPLE t SIGNIFICANCE TEST

Suppose that an SRS of size n_1 is drawn from a Normal population with unknown mean μ_1 and that an independent SRS of size n_2 is drawn from another Normal population with unknown mean μ_2. To test the hypothesis $H_0: \mu_1 - \mu_2 = \Delta_0$, compute the **two-sample t statistic**

$$t = \frac{(\bar{x}_1 - \bar{x}_2) - \Delta_0}{\sqrt{\dfrac{s_1^2}{n_1} + \dfrac{s_2^2}{n_2}}}$$

and use P-values or critical values for the $t(k)$ distribution, where the degrees of freedom k either are approximated by software or are the smaller of $n_1 - 1$ and $n_2 - 1$.

EXAMPLE 7.13

DRP

Is there an improvement? For the DRP study described in Example 7.11 (page 437), we hope to show that the treatment (Group 1) performs better than the control (Group 2). For a formal significance test, the hypotheses are

$$H_0: \mu_1 = \mu_2$$
$$H_a: \mu_1 > \mu_2$$

The two-sample t test statistic is

$$t = \frac{(\bar{x}_1 - \bar{x}_2) - 0}{\sqrt{\dfrac{s_1^2}{n_1} + \dfrac{s_2^2}{n_2}}}$$

$$= \frac{51.48 - 41.52}{\sqrt{\dfrac{11.01^2}{21} + \dfrac{17.15^2}{23}}}$$

$$= 2.31$$

The P-value for the one-sided test is $P(T \geq 2.31)$. For the second approximation, the degrees of freedom k are equal to the smaller of

$$n_1 - 1 = 21 - 1 = 20 \quad \text{and} \quad n_2 - 1 = 23 - 1 = 22$$

Comparing 2.31 with the entries in Table D for 20 degrees of freedom, we see that P lies between 0.01 and 0.02.

df = 20

p	0.02	0.01
t^*	2.197	2.528

The data strongly suggest that directed reading activity improves the DRP score ($t = 2.31$, df = 20, $0.01 < P < 0.02$).

USE YOUR KNOWLEDGE

7.50 A two-sample t significance test. Refer to Exercise 7.48. Perform a significance test at the 0.05 level to assess whether the average improvement is five points versus the alternative that it is greater than five points. Write a one-sentence conclusion.

7.51 Interpreting the confidence interval. Refer to the previous exercise and Exercise 7.48. Can the confidence interval in Exercise 7.48 be used to determine whether the significance test of the previous exercise rejects or does not reject the null hypothesis? Explain your answer.

Most statistical software requires the raw data for analysis. A few, like Minitab, will also perform a t test on data in summarized form (such as the summary statistics table in Example 7.12, page 439). It is always preferable to work with the raw data because one can also examine the data through plots such as the back-to-back stemplot and those in Figure 7.12.

EXAMPLE 7.14

DRP

Using software. Figure 7.13 shows JMP and Minitab outputs for the comparison of DRP scores. Both outputs include the 95% confidence interval and the significance test that the means are equal. JMP reports the difference as the mean of treatment minus the mean of control, while Minitab reports the difference in the opposite order.

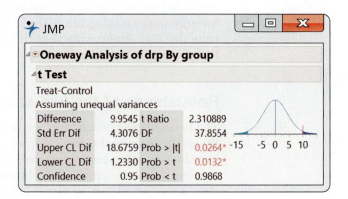

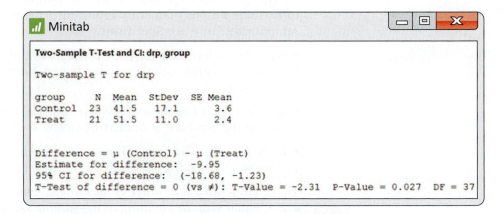

FIGURE 7.13 JMP and Minitab outputs, Example 7.14.

Recall the confidence interval (treatment minus control) is

$$(\bar{x}_1 - \bar{x}_2) \pm t^* \sqrt{\frac{s_1^2}{n_1} + \frac{s_2^2}{n_2}} = (51.48 - 41.52) \pm t^* \sqrt{\frac{11.01^2}{21} + \frac{17.15^2}{23}}$$

$$= 9.96 \pm 4.31 t^*$$

From the JMP output, we see that the degrees of freedom under the first approximation are 37.9. Using these degrees of freedom, the interval is (1.2, 18.7). This interval, as expected, is more narrow than the confidence interval in Example 7.12 (page 439), which uses the conservative approach. The difference, however, is pretty small.

For the significance test, the P-value for the one-sided significance test is $P(T \geq 2.31)$. JMP gives the approximate P-value as 0.0132, again using 37.9 as the degrees of freedom.

Minitab also uses the first degrees of freedom approximation but rounds the degrees of freedom down to the nearest integer (37.9 → 37). As a result, the margin of error is slightly wider than that of JMP and the P-value of the significance test is slightly larger.

In order to get a confidence interval as part of the Minitab output, the two-sided alternative was considered. If your software gives you the P-value for only the two-sided alternative, $2P(T \geq |t|)$, you need to divide the reported value by 2 after checking that the means differ in the direction specified by the alternative hypothesis.

Robustness of the two-sample procedures

The two-sample t procedures are more robust than the one-sample t methods. When the sizes of the two samples are equal and the distributions of the two populations being compared have similar shapes, probability values from the t table are quite accurate for a broad range of distributions when the sample sizes are as small as $n_1 = n_2 = 5$.[25] When the two population distributions have different shapes, larger samples are needed.

The guidelines for the use of one-sample t procedures can be adapted to two-sample procedures by replacing "sample size" with the "sum of the sample sizes" $n_1 + n_2$. Specifically,

- *If $n_1 + n_2$ is less than 15:* Use t procedures if the data are close to Normal. If the data in either sample are clearly non-Normal or if outliers are present, do not use t.

- *If $n_1 + n_2$ is at least 15 and less than 40:* The t procedures can be used except in the presence of outliers or strong skewness.

- *Large samples:* The t procedures can be used even for clearly skewed distributions when the sample is large, roughly $n_1 + n_2 \geq 40$.

These guidelines are rather conservative, especially when the two samples are of equal size. *In planning a two-sample study, choose equal sample sizes if you can.* The two-sample t procedures are most robust against non-Normality in this case, and the conservative probability values are most accurate.

Here is an example with large sample sizes that are almost equal. Even if the distributions are not Normal, we are confident that the sample means will be approximately Normal. The two-sample t test is very robust in this case.

EXAMPLE 7.15

Serge Krouglikoff/Getty Images

Timing of food intake and weight loss. There is emerging evidence of a relationship between timing of feeding and weight regulation. In one study, researchers followed 402 obese or overweight individuals through a 20-week weight-loss treatment.[26] To investigate the timing of food intake, participants were grouped into early eaters and late eaters, based on the timing of their main meal. Here are the summary statistics of their weight loss over the 20 weeks, in kilograms (kg):

Group	n	$\bar{x}$	s
Early eater	202	9.9	5.8
Late eater	200	7.7	6.1

The early eaters lost more weight on average. Can we conclude that these two groups are not the same? Or is this observed difference merely what we could expect to see given the variation among participants?

While other evidence suggests that early eaters should lose more weight, the researchers did not specify a direction for the difference. Thus, the hypotheses are

$$H_0: \mu_1 = \mu_2$$
$$H_a: \mu_1 \neq \mu_2$$

Because the samples are large, we can confidently use the t procedures even though we lack the detailed data and so cannot verify the Normality condition.

The two-sample t statistic is

$$t = \frac{(\bar{x}_1 - \bar{x}_2) - 0}{\sqrt{\dfrac{s_1^2}{n_1} + \dfrac{s_2^2}{n_2}}}$$

$$= \frac{9.9 - 7.7}{\sqrt{\dfrac{5.8^2}{202} + \dfrac{6.1^2}{200}}}$$

$$= 3.71$$

df = 100

p	0.0005
t^*	3.390

The conservative approach finds the P-value by comparing 3.71 to critical values for the $t(199)$ distribution because the smaller sample has 200 observations. Because Table D does not contain a row for 199 degrees of freedom, we will be even more conservative and use the first row in the table with degrees of freedom less than 199. This means we'll use the $t(100)$ distribution to compute the P-value.

Our calculated value of t is larger than the $p = 0.0005$ entry in the table. We must double the table tail area p because the alternative is two-sided, so we conclude that the P-value is less than 0.001. The data give conclusive evidence that early eaters lost more weight, on average, than late eaters ($t = 3.71$, df = 100, $P < 0.001$).

In this example the exact *P*-value is very small because $t = 3.71$ says that the observed difference in means is over 3.5 standard errors above the hypothesized difference of zero ($\mu_1 = \mu_2$). In this study, the researchers also compared energy intake and energy expenditure between late and early eaters. Despite the observed weight loss difference of 2.2 kg, no significant differences in these variables were found.

In this and other examples, we can choose which population to label 1 and which to label 2. After inspecting the data, we chose early eaters as Population 1 because this choice makes the *t* statistic a positive number. This avoids any possible confusion from reporting a negative value for *t*. *Choosing the population labels is* **not** *the same as choosing a one-sided alternative after looking at the data.* Choosing hypotheses after seeing a result in the data is a violation of sound statistical practice.

Inference for small samples

Small samples require special care. We do not have enough observations to examine the distribution shapes, and only extreme outliers stand out. The power of significance tests tends to be low, and the margins of error of confidence intervals tend to be large. Despite these difficulties, we can often draw important conclusions from studies with small sample sizes. If the size of an effect is very large, it should still be evident even if the *n*'s are small.

EXAMPLE 7.16

EATER

Timing of food intake. In the setting of Example 7.15, let's consider a much smaller study that collects weight loss data from only five participants in each eating group. Also, given the results of this past example, we choose the one-sided alternative. The data are

Group	Weight loss (kg)				
Early eater	6.3	15.1	9.4	16.8	10.2
Late eater	7.8	0.2	1.5	11.5	4.6

First, examine the distributions with a back-to-back stemplot (the data are rounded to the nearest integer).

```
        Early      Late
              0 | 02
           96 | 0 | 58
            0 | 1 | 2
           75 | 1 |
```

While there is variation among weight losses within each group, there is also a noticeable separation. The early-eaters group contains four of the five largest losses, and the late-eaters group contains four of the five smallest losses. A significance test can confirm whether this pattern can arise just by chance or if the early-eaters group has a higher mean. We test

$$H_0: \mu_1 = \mu_2$$
$$H_a: \mu_1 > \mu_2$$

The average weight loss is higher in the early-eater group ($t = 2.28$, df $= 7.96$, $P = 0.0262$). The difference in sample means is 6.44 kg.

Figure 7.14 gives outputs for this analysis from several software packages. Although the formats differ, the basic information is the same. All report the sample sizes, the sample means and standard deviations (or variances), the t statistic, and its P-value. All agree that the P-value is small, though some give more detail than others. Software often labels the groups in alphabetical order. Always check the means first and report the statistic (you may need to change the sign) in an appropriate way. Be sure to also mention the size of the effect you observed, such as "The mean weight loss for the early eaters was 6.44 kg higher than for the late eaters."

FIGURE 7.14 SAS, Excel, JMP, and SPSS outputs, Example 7.16.

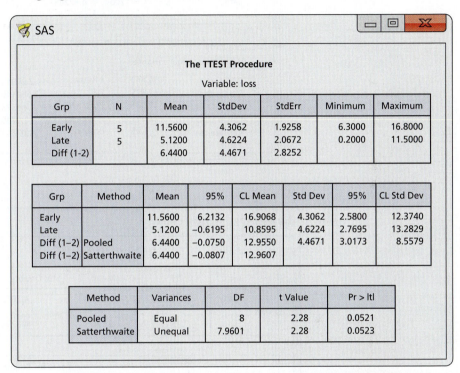

SAS

The TTEST Procedure

Variable: loss

Grp	N	Mean	StdDev	StdErr	Minimum	Maximum
Early	5	11.5600	4.3062	1.9258	6.3000	16.8000
Late	5	5.1200	4.6224	2.0672	0.2000	11.5000
Diff (1-2)		6.4400	4.4671	2.8252		

Grp	Method	Mean	95%	CL Mean	Std Dev	95%	CL Std Dev
Early		11.5600	6.2132	16.9068	4.3062	2.5800	12.3740
Late		5.1200	−0.6195	10.8595	4.6224	2.7695	13.2829
Diff (1–2)	Pooled	6.4400	−0.0750	12.9550	4.4671	3.0173	8.5579
Diff (1–2)	Satterthwaite	6.4400	−0.0807	12.9607			

Method	Variances	DF	t Value	Pr > ItI
Pooled	Equal	8	2.28	0.0521
Satterthwaite	Unequal	7.9601	2.28	0.0523

Excel

	A	B	C
1	t-Test: Two-Sample Assuming Unequal Variances		
2			
3		*Early*	*Late*
4	Mean	11.56	5.12
5	Variance	18.543	21.367
6	Observations	5	5
7	Hypothesized Mean Difference	0	
8	Df	8	
9	t Stat	2.27944966	
10	P(T<=t) one-tail	0.026058036	
11	t Critical one-tail	1.859548033	
12	P(T<=t) two-tail	0.052116073	
13	t Critical two-tail	2.306004133	

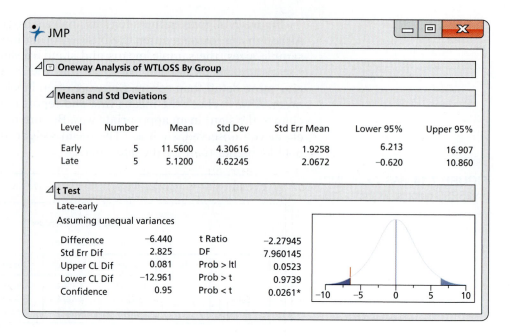

FIGURE 7.14 *Continued*

There are two other things to notice in the outputs. First, SAS and SPSS only give results for the two-sided alternative. To get the *P*-value for the one-sided alternative, we must first check the mean difference to make sure it is in the proper direction. If it is, we divide the given *P*-value by 2. Also, SAS and SPSS report the results of *two t* procedures: a special procedure that assumes that the two population variances are equal and the general two-sample procedure that we have just studied. We don't recommend the "equal-variances" procedures, but we describe them later, in the section on pooled two-sample t procedures.

Software approximation for the degrees of freedom

We noted earlier that the two-sample t statistic does not have a t distribution. Moreover, the distribution changes as the unknown population standard deviations σ_1 and σ_2 change. However, the distribution can be approximated by a t distribution with degrees of freedom given by

$$\text{df} = \frac{\left(\dfrac{s_1^2}{n_1} + \dfrac{s_2^2}{n_2}\right)^2}{\dfrac{1}{n_1 - 1}\left(\dfrac{s_1^2}{n_1}\right)^2 + \dfrac{1}{n_2 - 1}\left(\dfrac{s_2^2}{n_2}\right)^2}$$

This is the approximation used by most statistical software. It is quite accurate when both sample sizes n_1 and n_2 are 5 or larger.

EXAMPLE 7.17

Degrees of freedom for directed reading assessment. For the DRP study of Example 7.11, the following table summarizes the data:

Group	n	$\bar{x}$	s
1	21	51.48	11.01
2	23	41.52	17.15

For greatest accuracy, we will use critical points from the t distribution with degrees of freedom given by the preceding equation:

$$\text{df} = \frac{\left(\dfrac{11.01^2}{21} + \dfrac{17.15^2}{23}\right)^2}{\dfrac{1}{20}\left(\dfrac{11.01^2}{21}\right)^2 + \dfrac{1}{22}\left(\dfrac{17.15^2}{23}\right)^2}$$

$$= \frac{344.486}{9.099} = 37.86$$

This is the value that we reported in Example 7.14 (pages 441–442), where we gave the results produced by software.

The number df given by the preceding approximation is always at least as large as the smaller of $n_1 - 1$ and $n_2 - 1$. On the other hand, the number df is never larger than the sum $n_1 + n_2 - 2$ of the two individual degrees of freedom. The number df is generally not a whole number. There is a t distribution with any positive degrees of freedom, even though Table D contains entries only for whole-number degrees of freedom. When the number df is small and is not a whole number, interpolation between entries in Table D may be needed to obtain an accurate critical value or P-value. Because of this and the need to calculate df, we do not recommend regular use of this approximation if a computer is not doing the arithmetic. With a computer, however, the more accurate procedures are painless.

USE YOUR KNOWLEDGE

7.52 Calculating the degrees of freedom. Assume that $s_1 = 5$, $s_2 = 8$, $n_1 = 25$, and $n_2 = 32$. Find the approximate degrees of freedom.

The pooled two-sample t procedures

There is one situation in which a t statistic for comparing two means has exactly a t distribution. This is when the two Normal population distributions have the *same* standard deviation. As we've done with other t statistics, we will first develop the z statistic and then, from it, the t statistic. In this case, notice that we need to substitute only a single standard error when we go from the z to the t statistic. This is why the resulting t statistic has a t distribution.

Call the common—and still unknown—standard deviation of both populations σ. Both sample variances s_1^2 and s_2^2 estimate σ^2. The best way to combine these two estimates is to average them with weights equal to their degrees of freedom. This gives more weight to the sample variance from the larger sample, which is reasonable. The resulting estimator of σ^2 is

$$s_p^2 = \frac{(n_1 - 1)s_1^2 + (n_2 - 1)s_2^2}{n_1 + n_2 - 2}$$

pooled estimator of σ^2

This is called the **pooled estimator of σ^2** because it combines the information in both samples.

When both populations have variance σ^2, the addition rule for variances says that $\bar{x}_1 - \bar{x}_2$ has variance equal to the *sum* of the individual variances, which is

$$\frac{\sigma^2}{n_1} + \frac{\sigma^2}{n_2} = \sigma^2 \left(\frac{1}{n_1} + \frac{1}{n_2} \right)$$

The standardized difference between means in this equal-variance case is, therefore,

$$z = \frac{(\bar{x}_1 - \bar{x}_2) - (\mu_1 - \mu_2)}{\sigma \sqrt{\dfrac{1}{n_1} + \dfrac{1}{n_2}}}$$

This is a special two-sample z statistic for the case in which the populations have the same σ. Replacing the unknown σ by the estimate s_p gives a t statistic. The degrees of freedom are $n_1 + n_2 - 2$, the sum of the degrees of freedom of the two sample variances. This t statistic is the basis of the pooled two-sample t inference procedures.

THE POOLED TWO-SAMPLE t PROCEDURES

Suppose that an SRS of size n_1 is drawn from a Normal population with unknown mean μ_1 and that an independent SRS of size n_2 is drawn from another Normal population with unknown mean μ_2. Suppose also that the two populations have the same standard deviation. A level C **confidence interval for $\mu_1 - \mu_2$** is

$$(\bar{x}_1 - \bar{x}_2) \pm t^* s_p \sqrt{\frac{1}{n_1} + \frac{1}{n_2}}$$

Here, t^* is the value for the $t(n_1 + n_2 - 2)$ density curve with area C between $-t^*$ and t^*. The quantity

$$t^* s_p \sqrt{\frac{1}{n_1} + \frac{1}{n_2}}$$

is the **margin of error**.

To test the hypothesis $H_0: \mu_1 - \mu_2 = \Delta_0$, compute the **pooled two-sample t statistic**

$$t = \frac{(\bar{x}_1 - \bar{x}_2) - \Delta_0}{s_p \sqrt{\dfrac{1}{n_1} + \dfrac{1}{n_2}}}$$

In terms of a random variable T having the $t(n_1 + n_2 - 2)$ distribution, the P-value for a test of H_0 against

$H_a: \mu_1 - \mu_2 > \Delta_0$ is $P(T \geq t)$

$H_a: \mu_1 - \mu_2 < \Delta_0$ is $P(T \leq t)$

$H_a: \mu_1 - \mu_2 \neq \Delta_0$ is $2P(T \geq |t|)$

EXAMPLE 7.18

Calcium and blood pressure. Does increasing the amount of calcium in our diet reduce blood pressure? Examination of a large sample of people revealed a relationship between calcium intake and blood pressure, but such observational studies do not establish causation. Animal experiments, however, showed that calcium supplements do reduce blood pressure in rats, justifying an experiment with human subjects. A randomized comparative experiment gave one group of 10 black men a calcium supplement for 12 weeks. The control group of 11 black men received a placebo that appeared identical. (In fact, a block design with black and white men as the blocks was used. We will look only at the results for blacks because

TABLE 7.4	Seated Systolic Blood Pressure (mm Hg)				
Calcium Group			Placebo Group		
Begin	**End**	**Decrease**	**Begin**	**End**	**Decrease**
107	100	7	123	124	−1
110	114	−4	109	97	12
123	105	18	112	113	−1
129	112	17	102	105	−3
112	115	−3	98	95	3
111	116	−5	114	119	−5
107	106	1	119	114	5
112	102	10	114	112	2
136	125	11	110	121	−11
102	104	−2	117	118	−1
			130	133	−3

BP_CA

the earlier survey suggested that calcium is more effective for blacks.) The experiment was double-blind. Table 7.4 gives the seated systolic (heart contracted) blood pressure for all subjects at the beginning and end of the 12-week period, in millimeters of mercury (mm Hg). Because the researchers were interested in decreasing blood pressure, Table 7.4 also shows the decrease for each subject. An increase appears as a negative entry.[27]

As usual, we first examine the data. To compare the effects of the two treatments, take the response variable to be the amount of the decrease in blood pressure. Inspection of the data reveals that there are no outliers. Side-by-side boxplots and Normal quantile plots (Figures 7.15 and 7.16) give a more detailed picture. The calcium group has a somewhat short left tail, but there are no severe departures from Normality that will prevent use of *t* procedures. To examine the question of the researchers who collected these data, we perform a significance test.

FIGURE 7.15 Side-by-side boxplots of the decrease in blood pressure from Table 7.4.

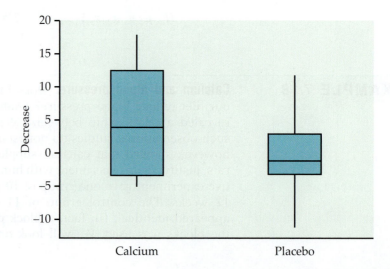

FIGURE 7.16 Normal quantile plots of the change in blood pressure from Table 7.4.

EXAMPLE 7.19

BP_CA

Does increased calcium reduce blood pressure? Take Group 1 to be the calcium group and Group 2 to be the placebo group. The evidence that calcium lowers blood pressure more than a placebo is assessed by testing

$$H_0: \mu_1 = \mu_2$$
$$H_a: \mu_1 > \mu_2$$

Here are the summary statistics for the decrease in blood pressure:

Group	Treatment	n	$\bar{x}$	s
1	Calcium	10	5.000	8.743
2	Placebo	11	−0.273	5.901

The calcium group shows a drop in blood pressure, and the placebo group has a small increase. The sample standard deviations do not rule out equal population standard deviations. A difference this large will often arise by chance in samples this small. We are willing to assume equal population standard deviations. The pooled sample variance is

$$s_p^2 = \frac{(n_1 - 1)s_1^2 + (n_2 - 1)s_2^2}{n_1 + n_2 - 2}$$

$$= \frac{(10 - 1)8.743^2 + (11 - 1)5.901^2}{10 + 11 - 2} = 54.536$$

so that

$$s_p = \sqrt{54.536} = 7.385$$

The pooled two-sample t statistic is

$$t = \frac{(\bar{x}_1 - \bar{x}_2) - 0}{s_p \sqrt{\dfrac{1}{n_1} + \dfrac{1}{n_2}}}$$

$$= \frac{5.000 - (-0.273)}{7.385 \sqrt{\dfrac{1}{10} + \dfrac{1}{11}}}$$

$$= \frac{5.273}{3.227} = 1.634$$

df = 19

p	0.10	0.05
t^*	1.328	1.729

The P-value is $P(T \geq 1.634)$, where T has the $t(19)$ distribution.

From Table D, we can see that P falls between the $\alpha = 0.10$ and $\alpha = 0.05$ levels. Statistical software gives the exact value $P = 0.059$. The experiment found evidence that calcium reduces blood pressure, but the evidence falls a bit short of the traditional 5% and 1% levels.

Sample size strongly influences the P-value of a test. An effect that fails to be significant at a specified level α in a small sample can be significant in a larger sample. In the light of the rather small samples in Example 7.19, the evidence for some effect of calcium on blood pressure is rather good. The published account of the study combined these results for blacks with the results for whites and adjusted for pretest differences among the subjects. Using this more detailed analysis, the researchers were able to report a P-value of 0.008.

Of course, a P-value is almost never the last part of a statistical analysis. To make a judgment regarding the size of the effect of calcium on blood pressure, we need a confidence interval.

EXAMPLE 7.20

BP_CA

How different are the calcium and placebo groups? We estimate that the effect of calcium supplementation is the difference between the sample means of the calcium and the placebo groups, $\bar{x}_1 - \bar{x}_2 = 5.273$ mm Hg. A 90% confidence interval for $\mu_1 - \mu_2$ uses the critical value $t^* = 1.729$ from the $t(19)$ distribution. The interval is

$$(\bar{x}_1 - \bar{x}_2) \pm t^* s_p \sqrt{\frac{1}{n_1} + \frac{1}{n_2}} = [5.000 - (-0.273)] \pm (1.729)(7.385)\sqrt{\frac{1}{10} + \frac{1}{11}}$$

$$= 5.273 \pm 5.579$$

We are 90% confident that the difference in means is in the interval $(-0.306, 10.852)$. The calcium treatment reduced blood pressure by about 5.3 mm Hg more than a placebo on the average, but the margin of error for this estimate is 5.6 mm Hg.

The pooled two-sample t procedures are anchored in statistical theory and so have long been the standard version of the two-sample t in textbooks. *But they require the assumption that the two unknown population*

standard deviations are equal. We discuss methods to assess this condition in Chapter 12.

The pooled *t* procedures are therefore a bit risky. They are reasonably robust against both non-Normality and unequal standard deviations when the sample sizes are nearly the same. When the samples are quite different in size, the pooled *t* procedures become sensitive to unequal standard deviations and should be used with caution unless the samples are large. Unequal standard deviations are quite common. In particular, it is not unusual for the spread of data to increase when the center of the data increases. We recommend regular use of the unpooled *t* procedures because most software automates the Satterthwaite approximation.

USE YOUR KNOWLEDGE

7.53 Timing of food intake revisited. Figure 7.14 (pages 445–446) gives the outputs from four software packages for comparing the weight loss of two groups with different eating schedules. Some of the software reports both pooled and unpooled analyses. Which outputs give the pooled results? What are the pooled *t* and its *P*-value?

7.54 Equal sample sizes. The software outputs in Figure 7.14 give the *same value* for the pooled and unpooled *t* statistics. Do some simple algebra to show that this is always true when the two sample sizes n_1 and n_2 are the same. In other cases, the two *t* statistics usually differ.

SECTION 7.2 SUMMARY

- Significance tests and confidence intervals for the difference between the means μ_1 and μ_2 of two Normal populations are based on the difference $\bar{x}_1 - \bar{x}_2$ between the sample means from two independent SRSs. Because of the central limit theorem, the resulting procedures are approximately correct for other population distributions when the sample sizes are large.

- When independent SRSs of sizes n_1 and n_2 are drawn from two Normal populations with parameters μ_1, σ_1 and μ_2, σ_2 the **two-sample *z* statistic**

$$z = \frac{(\bar{x}_1 - \bar{x}_2) - (\mu_1 - \mu_2)}{\sqrt{\dfrac{\sigma_1^2}{n_1} + \dfrac{\sigma_2^2}{n_2}}}$$

has the $N(0, 1)$ distribution.

- The **two-sample *t* statistic**

$$t = \frac{(\bar{x}_1 - \bar{x}_2) - (\mu_1 - \mu_2)}{\sqrt{\dfrac{s_1^2}{n_1} + \dfrac{s_2^2}{n_2}}}$$

does *not* have a *t* distribution. However, good approximations are available.

- **Conservative inference procedures** for comparing μ_1 and μ_2 are obtained from the two-sample *t* statistic by using the $t(k)$ distribution with degrees of freedom k equal to the smaller of $n_1 - 1$ and $n_2 - 1$.

- **More accurate probability values** can be obtained by estimating the degrees of freedom from the data. This is the usual procedure for statistical software.

- An approximate level C **confidence interval** for $\mu_1 - \mu_2$ is given by

$$(\bar{x}_1 - \bar{x}_2) \pm t^* \sqrt{\frac{s_1^2}{n_1} + \frac{s_2^2}{n_2}}$$

Here, t^* is the value for the $t(k)$ density curve with area C between $-t^*$ and t^*, where k is computed from the data by software or is the smaller of $n_1 - 1$ and $n_2 - 1$. The quantity

$$t^* \sqrt{\frac{s_1^2}{n_1} + \frac{s_2^2}{n_2}}$$

is the **margin of error**.

- Significance tests for $H_0: \mu_1 - \mu_2 = \Delta_0$ use the **two-sample t statistic**

$$t = \frac{(\bar{x}_1 - \bar{x}_2) - \Delta_0}{\sqrt{\dfrac{s_1^2}{n_1} + \dfrac{s_2^2}{n_2}}}$$

The P-value is approximated using the $t(k)$ distribution where k is estimated from the data using software or is the smaller of $n_1 - 1$ and $n_2 - 1$.

- The guidelines for practical use of two-sample t procedures are similar to those for one-sample t procedures. Equal sample sizes are recommended.

- If we can assume that the two populations have equal variances, **pooled two-sample t procedures** can be used. These are based on the **pooled estimator**

$$s_p^2 = \frac{(n_1 - 1)s_1^2 + (n_2 - 1)s_2^2}{n_1 + n_2 - 2}$$

of the unknown common variance and the $t(n_1 + n_2 - 2)$ distribution. We do not recommend this procedure for regular use.

SECTION 7.2 EXERCISES

For Exercises 7.48 and 7.49, see page 439; for Exercises 7.50 and 7.51, see pages 440–441; for Exercise 7.52, see page 448; and for Exercises 7.53 and 7.54, see page 453.

In exercises that call for two-sample t procedures, you may use either of the two approximations for the degrees of freedom that we have discussed: the value given by your software or the smaller of $n_1 - 1$ and $n_2 - 1$. Be sure to state clearly which approximation you have used.

7.55 What is wrong? In each of the following situations, identify what is wrong and then either explain why it is wrong or change the wording of the statement to make it true.

(a) A researcher wants to test $H_0: \bar{x}_1 = \bar{x}_2$ versus the two-sided alternative $H_a: \bar{x}_1 \neq \bar{x}_2$.

(b) A study recorded the IQ scores of 100 college freshmen. The scores of the 56 males in the study were compared with the scores of all 100 freshmen using the two-sample methods of this section.

(c) A two-sample t statistic gave a P-value of 0.94. From this, we can reject the null hypothesis with 90% confidence.

(d) A researcher is interested in testing the one-sided alternative $H_a: \mu_1 < \mu_2$. The significance test gave $t = 2.15$. Because the P-value for the two-sided alternative is 0.036, he concluded that his P-value was 0.018.

7.56 Basic concepts. For each of the following, answer the question and give a short explanation of your reasoning.

(a) A 95% confidence interval for the difference between two means is reported as (0.8, 2.3). What can you conclude about the results of a significance test of the null hypothesis that the population means are equal versus the two-sided alternative?

(b) Will larger samples generally give a larger or smaller margin of error for the difference between two sample means?

7.57 More basic concepts. For each of the following, answer the question and give a short explanation of your reasoning.

(a) A significance test for comparing two means gave $t = -1.97$ with 10 degrees of freedom. Can you reject the null hypothesis that the μ's are equal versus the two-sided alternative at the 5% significance level?

(b) Answer part (a) for the one-sided alternative that the difference between means is negative.

7.58 Physical demands of women's rugby seven matches. Rugby sevens is rapidly growing in popularity and will be included in the 2016 Olympics. Matches are played on a full rugby field and consist of two seven-minute halves. Each team also consists of seven players. To better understand the demands of women's rugby sevens, a group of researchers compared the physical qualities of elite players from the Canadian National team with a university squad. The following table summarizes some of these qualities:[28]

Quality	Elite ($n = 16$)		University ($n = 13$)	
	$\bar{x}$	s	$\bar{x}$	s
Sprint speed (km/hr)	27.3	0.7	26.0	1.5
Peak heart rate (bpm)	192.0	6.0	193.0	6.0
Intermittent recovery test (m)	1160	191	781	129

Carry out the significance tests using $\alpha = 0.05$. Report the test statistic with the degrees of freedom and the P-value. Write a short summary of your conclusion.

7.59 Noise levels in fitness classes. Fitness classes often have very loud music that could affect hearing. One study collected noise levels (decibels) in both high-intensity and low-intensity fitness classes across eight commercial gyms in Sydney, Australia.[29] 🔊 NOISE

(a) Create a histogram or Normal quantile plot for the high-intensity classes. Do the same for the low-intensity

classes. Are the distributions reasonably Normal? Summarize the distributions in words.

(b) Test the equality of means using a two-sided alternative hypothesis and significance level $\alpha = 0.05$.

(c) Are the t procedures appropriate given your observations in part (a)? Explain your answer.

(d) Remove the one low decibel reading for the low-intensity group and redo the significance test. How does this outlier affect the results?

(e) Do you think the results of the significance test from part (b) or (d) should be reported? Explain your answer.

7.60 Noise levels in fitness classes, continued. Refer to the previous exercise. In most countries, the workplace noise standard is 85 db (over eight hours). For every 3 dB increase above that, the amount of exposure time is halved. This means that the exposure time for a dB level of 91 is two hours and for a dB level of 94 it is one hour. 🔊 NOISE

(a) Construct a 95% confidence interval for the mean dB level in high-intensity classes.

(b) Using the interval in part (a), construct a 95% confidence interval for the number of one-hour classes per day an instructor can teach before possibly risking hearing loss. (*Hint*: This is a linear transformation.)

(c) Repeat parts (a) and (b) for low-intensity classes.

(d) Explain how one might use these intervals to determine the staff size of a new gym.

7.61 When is 30/31 days not equal to a month? Time can be expressed on different levels of scale; days, weeks, months, and years. Can the scale provided influence perception of time? For example, if you placed an order over the phone, would it make a difference if you were told the package would arrive in four weeks or one month? To investigate this, two researchers asked a group of 267 college students to imagine their car needed major repairs and would have to stay at the shop. Depending on the group he or she was randomized to, the student was either told it would take one month or 30/31 days. Each student was then asked to give best- and worst-case estimates of when the car would be ready. The interval between these two estimates (in days) was the response. Here are the results:[30]

Group	n	$\bar{x}$	s
30/31 days	177	20.4	14.3
One month	90	24.8	13.9

(a) Given that the interval cannot be less than 0, the distributions are likely skewed. Comment on the appropriateness of using the t procedures.

(b) Test that the average interval is the same for the two groups using the $\alpha = 0.05$ significance level. Report the test statistic, the degrees of freedom, and the P-value. Give a short summary of your conclusion.

7.62 When is 52 weeks not equal to a year? Refer to the previous exercise. The researchers also had 60 marketing students read an announcement about a construction project. The expected duration was either one year or 52 weeks. Each student was then asked to state the earliest and latest completion date.

Group	n	$\bar{x}$	s
52 weeks	30	84.1	55.8
1 year	30	139.6	73.1

Test that the average interval is the same for the two groups using the $\alpha = 0.05$ significance level. Report the test statistic, the degrees of freedom, and the P-value. Give a short summary of your conclusion.

7.63 Trustworthiness and eye color. Why do we naturally tend to trust some strangers more than others? One group of researchers decided to study the relationship between eye color and trustworthiness.[31] In their experiment, the researchers took photographs of 80 students (20 males with brown eyes, 20 males with blue eyes, 20 females with brown eyes, and 20 females with blue eyes), each seated in front of a white background looking directly at the camera with a neutral expression. These photos were cropped so the eyes were horizontal and at the same height in the photo and so the neckline was visible. They then recruited 105 participants to judge the trustworthiness of each student photo. This was done using a 10-point scale, where 1 meant very untrustworthy and 10 very trustworthy. The 80 scores from each participant were then converted to z-scores, and the average z-score of each photo (across all 105 participants) was used for the analysis. Here is a summary of the results:

Eye color	n	$\bar{x}$	s
Brown	40	0.55	1.68
Blue	40	−0.38	1.53

Can we conclude from these data that brown-eyed students appear more trustworthy compared to their blue-eyed counterparts? Test the hypothesis that the average scores for the two groups are the same.

7.64 Facebook use in college. Because of Facebook's rapid rise in popularity among college students, there is a great deal of interest in the relationship between Facebook use and academic performance. One study collected information on $n = 1839$ undergraduate students to look at the relationships among frequency of Facebook use, participation in Facebook activities, time spent preparing for class, and overall GPA.[32]

Students reported preparing for class an average of 706 minutes per week with a standard deviation of 526 minutes. Students also reported spending an average of 106 minutes per day on Facebook with a standard deviation of 93 minutes; 8% of the students reported spending no time on Facebook.

(a) Construct a 95% confidence interval for the average number of minutes per week a student prepares for class.

(b) Construct a 95% confidence interval for the average number of minutes per week a student spends on Facebook. (*Hint:* Be sure to convert from minutes per day to minutes per week.)

(c) Explain why you might expect the population distributions of these two variables to be highly skewed to the right. Do you think this fact makes your confidence intervals invalid? Explain your answer.

7.65 Possible biases? Refer to the previous exercise. The researcher surveyed students at a four-year, public university in the northeastern United States ($N = 3866$). Each student was emailed a link to the survey hosted on SurveyMonkey.com. The researcher also states:

> For the students who did not participate immediately, two additional reminders were sent, one week apart. Participants were offered a chance to enter a drawing to win one of 90 $10 Amazon.com gift cards as incentive. A total of 1839 surveys were completed for an overall response rate of 48%.

Discuss how these factors influence your interpretation of the results of this survey.

7.66 Comparing means. Refer to Exercise 7.64. Suppose that you wanted to compare the average minutes per week spent on Facebook with the average minutes per week spent preparing for class.

(a) Provide an estimate of this difference.

(b) Explain why it is incorrect to use the two-sample t test to see if the means differ.

7.67 Sadness and spending. The "misery is not miserly" phenomenon refers to a person's spending judgment going haywire when the person is sad. In a study, 31 young adults were given $10 and randomly assigned to either a sad or a neutral group. The participants in the sad group watched a video about the death of a boy's mentor (from *The Champ*), and those in the neutral group watched a video on the Great Barrier Reef. After the video, each participant was offered the chance to trade $0.50 increments of the $10 for an insulated water bottle.[33] Here are the data: SADNESS

Group	Purchase price ($)							
Neutral	0.00	2.00	0.00	1.00	0.50	0.00	0.50	
	2.00	1.00	0.00	0.00	0.00	0.00	1.00	
Sad	3.00	4.00	0.50	1.00	2.50	2.00	1.50	0.00	1.00
	1.50	1.50	2.50	4.00	3.00	3.50	1.00	3.50	

(a) Examine each group's prices graphically. Is use of the *t* procedures appropriate for these data? Carefully explain your answer.

(b) Make a table with the sample size, mean, and standard deviation for each of the two groups.

(c) State appropriate null and alternative hypotheses for comparing these two groups.

(d) Perform the significance test at the $\alpha = 0.05$ level, making sure to report the test statistic, degrees of freedom, and *P*-value. What is your conclusion?

(e) Construct a 95% confidence interval for the mean difference in purchase price between the two groups.

7.68 Diet and mood. Researchers were interested in comparing the long-term psychological effects of being on a high-carbohydrate, low-fat (LF) diet versus a high-fat, low-carbohydrate (LC) diet.[34] A total of 106 overweight and obese participants were randomly assigned to one of these two energy-restricted diets. At 52 weeks, 32 LC dieters and 33 LF dieters remained. Mood was assessed using a total mood disturbance score (TMDS), where a lower score is associated with a less negative mood. A summary of these results follows:

Group	n	$\bar{x}$	s
LC	32	47.3	28.3
LF	33	19.3	25.8

(a) Is there a difference in the TMDS at Week 52? Test the null hypothesis that the dieters' average mood in the two groups is the same. Use a significance level of 0.05.

(b) Critics of this study focus on the specific LC diet (that it, the science) and the dropout rate. Explain why the dropout rate is important to consider when drawing conclusions from this study.

7.69 Drive-thru customer service. *QSRMagazine.com* assessed 1855 drive-thru visits at quick-service restaurants.[35] One benchmark assessed was customer service. Responses ranged from "Rude (1)" to "Very Friendly (5)." The following table breaks down the responses according to two of the chains studied. 📊 DRVTHRU

	Rating				
Chain	1	2	3	4	5
Taco Bell	0	5	41	143	119
McDonald's	1	22	55	139	100

(a) A researcher decides to compare the average rating of McDonald's and Taco Bell. Comment on the appropriateness of using the average rating for these data.

(b) Assuming an average of these ratings makes sense, comment on the use of the *t* procedures for these data.

(c) Report the means and standard deviations of the ratings for each chain separately.

(d) Test whether the two chains, on average, have the same customer satisfaction. Use a two-sided alternative hypothesis and a significance level of 5%.

7.70 Comparison of two web page designs. You want to compare the daily number of hits for two different website designs for your indie rock band. You assign the next 30 days to either Design A or Design B, 15 days to each.

(a) Would you use a one-sided or a two-sided significance test for this problem? Explain your choice.

(b) If you use Table D to find the critical value, what are the degrees of freedom using the second approximation?

(c) If you perform the significance test using $\alpha = 0.05$, how large (positive or negative) must the *t* statistic be to reject the null hypothesis that the two designs result in the same average hits?

7.71 Comparison of dietary composition. Refer to Example 7.15 (page 443). That study also broke down the dietary composition of the main meal. The following table summarizes the total fats, protein, and carbohydrates in the main meal (g) for the two groups:

	Early eaters ($n = 202$)		Late eaters ($n = 200$)	
	$\bar{x}$	s	$\bar{x}$	s
Fats	23.1	12.5	21.4	8.2
Protein	27.6	8.6	25.7	6.8
Carbohydrates	64.1	21.0	63.5	20.8

(a) Is it appropriate to use the two-sample *t* procedures that we studied in this section to analyze these data for group differences? Give reasons for your answer.

(b) Describe appropriate null and alternative hypotheses for comparing the two groups in terms of fats consumed.

(c) Carry out the significance test using $\alpha = 0.05$. Report the test statistic with the degrees of freedom and the *P*-value. Write a short summary of your conclusion.

(d) Find a 95% confidence interval for the difference between the two means. Compare the information given by the interval with the information given by the significance test.

7.72 More on dietary composition. Refer to the previous exercise. Repeat parts (b) through (d) for protein and for carbohydrates. Combining these results with the results of Exercise 7.71, write a short summary of your findings.

7.73 Change in portion size. A study of food portion sizes reported that over a 17-year period, the average size of a soft drink consumed by Americans aged two years and older increased from 13.1 ounces (oz) to 19.9 oz. The authors state that the difference is statistically significant with $P < 0.01$.[36] Explain what additional information you would need to compute a confidence interval for the increase, and outline the procedure that you would use for the computations. Do you think that a confidence interval would provide useful additional information? Explain why or why not.

7.74 Beverage consumption. The results in the previous exercise were based on two national surveys with a very large number of individuals. Here is a study that also looked at beverage consumption, but the sample sizes were much smaller. One part of this study compared 20 children who were 7 to 10 years old with 5 children who were 11 to 13.[37] The younger children consumed an average of 8.2 oz of sweetened drinks per day, while the older ones averaged 14.5 oz. The standard deviations were 10.7 oz and 8.2 oz, respectively.

(a) Do you think that it is reasonable to assume that these data are Normally distributed? Explain why or why not. (*Hint:* Think about the 68–95–99.7 rule.)

(b) Using the methods in this section, test the null hypothesis that the two groups of children consume equal amounts of sweetened drinks versus the two-sided alternative. Report all details of the significance-testing procedure with your conclusion.

(c) Give a 95% confidence interval for the difference in means.

(d) Do you think that the analyses performed in parts (b) and (c) are appropriate for these data? Explain why or why not.

(e) The children in this study were all participants in an intervention study at the Cornell Summer Day Camp at Cornell University. To what extent do you think that these results apply to other groups of children?

7.75 Study design is important! Recall Exercise 7.70 (page 457). You are concerned that day of the week may affect the number of hits. So to compare the two web page designs, you choose two successive weeks in the middle of a month. You flip a coin to assign one Monday to the first design and the other Monday to the second. You repeat this for each of the seven days of the week. You now have seven hit amounts for each design. It is *incorrect* to use the two-sample t test to see if the mean hits differ for the two designs. Carefully explain why.

7.76 New hybrid tablet and laptop? The purchasing department has suggested your company switch to a new hybrid tablet and laptop. As CEO, you want data to be assured that employees will like these new hybrids over the old laptops. You designate the next 16 employees needing a new laptop to participate in an experiment in which eight will be randomly assigned to receive the standard laptop and the remainder will receive the new hybrid tablet and laptop. After a month of use, these employees will express their satisfaction with their new computers by responding to the statement "I like my new computer" on a scale from 1 to 5, where 1 represents "strongly disagree," 2 is "disagree," 3 is "neutral," 4 is "agree," and 5 is "strongly agree."

(a) The employees with the hybrid computers have an average satisfaction score of 4.3 with standard deviation 0.7. The employees with the standard laptops have an average of 3.7 with standard deviation 1.5. Give a 95% confidence interval for the difference in the mean satisfaction scores for all employees.

(b) Would you reject the null hypothesis that the mean satisfaction for the two types of computers is the same versus the two-sided alternative at significance level 0.05? Use your confidence interval to answer this question. Explain why you do not need to calculate the test statistic.

7.77 Why randomize? Refer to the previous exercise. A coworker suggested that you give the new hybrid computers to the next eight employees who need new computers and the standard laptop to the following eight. Explain why your randomized design is better.

7.78 Does ad placement matter? Corporate advertising tries to enhance the image of the corporation. A study compared two ads from two sources, the *Wall Street Journal* and the *National Enquirer*. Subjects were asked to pretend that their company was considering a major investment in Performax, the fictitious sportswear firm in the ads. Each subject was asked to respond to the question "How trustworthy was the source in the sportswear company ad for Performax?" on a 7-point scale. Higher values indicated more trustworthiness.[38] Here is a summary of the results:

Ad source	n	$\bar{x}$	s
Wall Street Journal	66	4.77	1.50
National Enquirer	61	2.43	1.64

(a) Compare the two sources of ads using a t test. Be sure to state your null and alternative hypotheses, the test statistic with degrees of freedom, the P-value, and your conclusion.

(b) Give a 95% confidence interval for the difference.

(c) Write a short paragraph summarizing the results of your analyses.

7.79 Size of trees in the northern and southern halves. The study of 584 longleaf pine trees in the Wade Tract in Thomas County, Georgia, had several purposes. Are trees in one part of the tract more or less like trees in any other part of the tract or are there differences? In Example 6.1 (page 342), we examined how the trees were distributed in the tract and found that the pattern was not random. In this exercise, we will examine the sizes of the trees. In Exercise 7.33 (page 429), we analyzed the sizes, measured as diameter at breast height (DBH), for a random sample of 40 trees. Here, we divide the tract into northern and southern halves and take random samples of 30 trees from each half. Here are the diameters in centimeters (cm) of the sampled trees: NSPINES

	27.8	14.5	39.1	3.2	58.8	55.5	25.0	5.4	19.0	30.6
North	15.1	3.6	28.4	15.0	2.2	14.2	44.2	25.7	11.2	46.8
	36.9	54.1	10.2	2.5	13.8	43.5	13.8	39.7	6.4	4.8

	44.4	26.1	50.4	23.3	39.5	51.0	48.1	47.2	40.3	37.4
South	36.8	21.7	35.7	32.0	40.4	12.8	5.6	44.3	52.9	38.0
	2.6	44.6	45.5	29.1	18.7	7.0	43.8	28.3	36.9	51.6

(a) Use a back-to-back stemplot and side-by-side boxplots to examine the data graphically. Describe the patterns in the data.

(b) Is it appropriate to use the methods of this section to compare the mean DBH of the trees in the north half of the tract with the mean DBH of the trees in the south half? Give reasons for your answer.

(c) What are appropriate null and alternative hypotheses for comparing the two samples of tree DBHs? Give reasons for your choices.

(d) Perform the significance test. Report the test statistic, the degrees of freedom, and the P-value. Summarize your conclusion.

(e) Find a 95% confidence interval for the difference in mean DBHs. Explain how this interval provides additional information about this problem.

7.80 Size of trees in the eastern and western halves. Refer to the previous exercise. The Wade Tract can also be divided into eastern and western halves. Here are the DBHs of 30 randomly selected longleaf pine trees from each half: EWPINES

	23.5	43.5	6.6	11.5	17.2	38.7	2.3	31.5	10.5	23.7
East	13.8	5.2	31.5	22.1	6.7	2.6	6.3	51.1	5.4	9.0
	43.0	8.7	22.8	2.9	22.3	43.8	48.1	46.5	39.8	10.9

	17.2	44.6	44.1	35.5	51.0	21.6	44.1	11.2	36.0	42.1
West	3.2	25.5	36.5	39.0	25.9	20.8	3.2	57.7	43.3	58.0
	21.7	35.6	30.9	40.6	30.7	35.6	18.2	2.9	20.4	11.4

Using the questions in the previous exercise, analyze these data.

7.81 Sales of a small appliance across months. A market research firm supplies manufacturers with estimates of the retail sales of their products from samples of retail stores. Marketing managers are prone to look at the estimate and ignore sampling error. Suppose that an SRS of 60 stores this month shows mean sales of 53 units of a small appliance, with standard deviation 12 units. During the same month last year, an SRS of 58 stores gave mean sales of 50 units, with standard deviation 10 units. An increase from 50 to 53 is a rise of 6%. The marketing manager is happy because sales are up 6%.

(a) Use the two-sample t procedure to give a 95% confidence interval for the difference in mean number of units sold at all retail stores.

(b) Explain in language that the manager can understand why he cannot be certain that sales rose by 6%, and that in fact sales may even have dropped.

7.82 An improper significance test. A friend has performed a significance test of the null hypothesis that two means are equal. His report states that the null hypothesis is rejected in favor of the alternative that the first mean is larger than the second. In a presentation on his work, he notes that the first sample mean was larger than the second mean and this is why he chose this particular one-sided alternative.

(a) Explain what is wrong with your friend's procedure and why.

(b) Suppose that he reported $t = 1.93$ with a P-value of 0.06. What is the correct P-value that he should report?

7.83 Breast-feeding versus baby formula. A study of iron deficiency among infants compared samples of infants following different feeding regimens. One group contained breast-fed infants, while the infants in another group were fed a standard baby formula without any iron supplements. Here are summary results on blood hemoglobin levels at 12 months of age:[39]

Group	n	$\bar{x}$	s
Breast-fed	23	13.3	1.7
Formula	19	12.4	1.8

(a) Is there significant evidence that the mean hemoglobin level is higher among breast-fed babies? State H_0 and H_a and carry out a t test. Give the P-value. What is your conclusion?

(b) Give a 95% confidence interval for the mean difference in hemoglobin level between the two populations of infants.

(c) State the assumptions that your procedures in parts (a) and (b) require in order to be valid.

7.84 Revisiting the sadness and spending study. In Exercise 7.67 (page 456), the purchase price of a water bottle was analyzed using the two-sample t procedures that do not assume equal standard deviations. Compare the means using a significance test and find the 95% confidence interval for the difference using the pooled methods. How do the results compare with those you obtained in Exercise 7.67? **SADNESS**

7.85 Revisiting the diet and mood study. In Exercise 7.68 (page 457), the total mood disturbance score means were compared using the two-sample t procedures that do not assume equal standard deviations. Compare the means using a significance test and find the 95% confidence interval for the difference using the pooled methods. How do the results compare with those you obtained in Exercise 7.68?

7.86 Revisiting dietary composition. In Exercise 7.71 (page 457), the total amount of fats was analyzed using the two-sample t procedures that do not assume equal standard deviations. Compare the means using a significance test and find the 95% confidence interval for the difference using the pooled methods. How do the results compare with those you obtained in Exercise 7.71?

7.87 Revisiting the size of trees. Refer to the Wade Tract DBH data in Exercise 7.79, where we compared a sample of trees from the northern half of the tract with a sample from the southern half. Because the standard deviations for the two samples are quite close, it is reasonable to analyze these data using the pooled procedures. Perform the significance test and find the 95% confidence interval for the difference in means using these methods. Summarize your results and compare them with what you found in Exercise 7.79. **NSPINES**

7.88 Revisiting the food-timing study. Example 7.15 (page 443) gives summary statistics for weight loss in early eaters and late eaters. The two sample standard deviations are quite similar, so we may be willing to assume equal population standard deviations. Calculate the pooled t test statistic and its degrees of freedom from the summary statistics. Use Table D to assess significance. How do your results compare with the unpooled analysis in the example?

7.89 Computing the degrees of freedom. Use the Wade Tract data in Exercise 7.79 to calculate the software approximation to the degrees of freedom using the formula on page 447. Verify your calculation with software. **NSPINES**

7.90 Again computing the degrees of freedom. Use the Wade Tract data in Exercise 7.80 to calculate the software approximation to the degrees of freedom using the formula on page 447. Verify your calculation with software. **EWPINES**

7.91 Revisiting the small-sample example. Refer to Example 7.16 (page 444). This is a case where the sample sizes are quite small. With only five observations per group, we have very little information to make a judgment about whether the population standard deviations are equal. The potential gain from pooling is large when the sample sizes are small. Assume that we will perform a two-sided test using the 5% significance level. **EATER**

(a) Find the critical value for the unpooled t test statistic that does not assume equal variances. Use the minimum of $n_1 - 1$ and $n_2 - 1$ for the degrees of freedom.

(b) Find the critical value for the pooled t test statistic.

(c) How does comparing these critical values show an advantage of the pooled test?

7.92 Two-sample test of equivalence. In Section 7.1, we were introduced to the one-sample test of equivalence (page 421). Using those same concepts, describe how to perform a two-sample test of equivalence.

7.3 Additional Topics on Inference

When you complete this section, you will be able to:	• Compute the sample size n needed for a desired margin of error for a mean μ.

• Compute the sample size n needed for a desired margin of error for a mean μ.

• Define the power of a significance test.

• Calculate the power of the one-sample t-test to detect an alternative for a given sample size n.

• Determine the sample size necessary to have adequate power to detect a scaled difference in means of size δ.

• Identify alternative strategies of inference for non-Normal populations.

In this section, we discuss two topics that are related to the procedures we have learned for inference about population means. First, we focus on planning a study—in particular, choosing the sample size. *A wise user of statistics does not plan for inference without at the same time planning data collection.* The second topic introduces us to various inference methods for non-Normal populations. These would be used when our populations are clearly non-Normal and we do not think that the sample size is large enough to rely on the robustness of the t procedures.

Choosing the sample size

We describe sample size procedures for both confidence intervals and significance tests. For anyone planning to design a study, a general understanding of these procedures is necessary. While the actual formulas are a bit technical, statistical software now makes it trivial to get sample size results.

Sample size for confidence intervals We can arrange to have both high confidence and a small margin of error by choosing an appropriate sample size. Let's first focus on the one-sample t confidence interval. Its margin of error is

$$m = t^*\text{SE}_{\bar{x}} = t^*\frac{s}{\sqrt{n}}$$

Besides the confidence level C and sample size n, this margin of error depends on the sample standard deviation s. Because we don't know the value of s until we collect the data, we guess a value to use in the calculations. Because s is our estimate of the population standard deviation σ, this value can also be considered our guess of the population standard deviation.

We will call this guessed value s^*. We typically guess at this value using results from a pilot study or from similar published studies. *It is always better to use a value of the standard deviation that is a little larger than what is expected.* This may result in a sample size that is a little larger than needed, but it helps avoid the situation where the resulting margin of error is larger than desired.

Given an estimate for s and the desired margin of error m, we can find the sample size by plugging everything into the margin of error formula and solving for n. The one complication, however, is that t^* depends not only on the confidence level C, but also on the sample size n. Here are the details.

SAMPLE SIZE FOR DESIRED MARGIN OF ERROR FOR A MEAN μ

The level C confidence interval for a mean μ will have an expected margin of error less than or equal to a specified value m when the sample size is such that

$$m \geq t^*s^*/\sqrt{n}$$

Here t^* is the critical value for confidence level C with $n - 1$ degrees of freedom, and s^* is the guessed value for the population standard deviation.

Finding the smallest sample size n that satisfies this requirement can be done using the following iterative search:

1. Get an initial sample size by replacing t^* with z^*. Compute $n = (z^*s^*/m)^2$ and round up to the nearest integer.

2. Use this sample size to obtain t^*, and check if $m \geq t^*s^*/\sqrt{n}$.

3. If the requirement is satisfied, then this n is the needed sample size. If the requirement is not satisfied, increase n by 1 and return to Step 2.

Notice that this method makes no reference to the size of the *population*. It is the size of the *sample* that determines the margin of error. The size of the population does not influence the sample size we need as long as the population is much larger than the sample. Here is an example.

EXAMPLE 7.21

Planning a survey of college students. In Example 7.1 (page 411), we calculated a 95% confidence interval for the mean hours per week a college student watches traditional television. The margin of error based on an SRS of $n = 8$ students was 12.42 hours. Suppose that a new study is being planned and the goal is to have a margin of error of five hours. How many students need to be sampled?

The sample standard deviation in Example 7.1 is $s = 14.854$ hours. To be conservative, we'll guess that the population standard deviation is 17.5 hours.

1. To compute an initial n, we replace t^* with z^*. This results in

$$n = \left(\frac{z^*s^*}{m}\right)^2 = \left[\frac{1.96(17.5)}{5}\right]^2 = 47.06$$

Round up to get $n = 48$.

2. We now check to see if this sample size satisfies the requirement when we switch back to t^*. For $n = 48$, we have $n - 1 = 47$ degrees of freedom and $t^* = 2.011$. Using this value, the expected margin of error is

$$2.011(17.5)/\sqrt{48} = 5.08$$

This is larger than $m = 5$, so the requirement is not satisfied.

3. The following table summarizes these calculations for some larger values of n.

n	$t^*s^*/\sqrt{n}$
49	5.03
50	4.97
51	4.92

The requirement is first satisfied when $n = 50$. Thus, we need to sample at least $n = 50$ students for the expected margin of error to be no more than five hours.

Figure 7.17 shows the Minitab input window used to do these calculations. Because the default confidence level is 95%, only the desired margin of error m and the estimate for s need to be entered.

FIGURE 7.17 Minitab input window used to compute the sample size for the desired margin of error, Example 7.21.

Note that the $n = 50$ refers to the *expected* margin of error being no more than five hours. This does not guarantee that the margin of error for the collected sample will be less than five hours. That is because the sample standard deviation s varies sample to sample and these calculations are treating it as a fixed quantity. More advanced sample size procedures ask you to also specify the probability of obtaining a margin of error less than the desired value. For our approach, this probability is roughly 50%. For a probability closer to 100%, the sample size will need to be larger. For example, suppose we wanted this probability to be roughly 80%. In SAS, we'd perform these calculations using the command

```
proc power;
    onesamplemeans CI=t stddev=17.5 halfwidth=5 probwidth=0.80 ntotal=.;
run;
```

The needed sample size increases from $n = 50$ to $n = 57$.

LOOK BACK

nonresponse, p. 196

Unfortunately, the actual number of usable observations is often less than that planned at the beginning of a study. This is particularly true of data collected in surveys or studies that involve a time commitment from the participants. Careful study designers often assume a nonresponse rate or dropout rate that specifies what proportion of the originally planned sample will fail to provide data. We use this information to calculate the sample size to be used at the start of the study. For example, if, in the preceding survey, we expect only 40% of those students to respond, we would need to start with a sample size of $2.5 \times 50 = 125$ to obtain usable information from 50 students.

These sample size calculations also do not account for collection costs. In practice, taking observations costs time and money. There are times when the required sample size may be impossibly expensive. In those situations, one might consider a larger margin of error and/or a lower confidence level to be acceptable.

For the two-sample t confidence interval, the margin of error is

$$m = t^* \sqrt{\frac{s_1^2}{n_1} + \frac{s_2^2}{n_2}}$$

A similar type of iterative search can be used to determine the sample sizes n_1 and n_2, but now we need to guess both standard deviations and decide on an estimate for the degrees of freedom. We suggest taking the conservative approach and using the smaller of $n_1 - 1$ and $n_2 - 1$ for the degrees of freedom. Another approach is to consider the standard deviations and sample sizes are equal, so the margin of error is

$$m = t^* \sqrt{\frac{2s^2}{n}}$$

and use degrees of freedom $2(n - 1)$. That is the approach most statistical software take.

EXAMPLE 7.22

Planning a new blood pressure study. In Example 7.20 (page 452), we calculated a 90% confidence interval for the mean difference in blood pressure. The 90% margin of error was roughly 5.6 mm Hg. Suppose that a new study is being planned and the desired margin of error at 90% confidence is 2.8 mm Hg. How many subjects per group do we need?

The pooled sample standard deviation in Example 7.20 is 7.385. To be a bit conservative, we'll guess that the two population standard deviations are both 8.0. To compute an initial n, we replace t^* with z^*. This results in

$$n = \left(\frac{\sqrt{2}z^*s^*}{m}\right)^2 = \left[\frac{\sqrt{2}(1.645)(8)}{2.8}\right]^2 = 44.2$$

We round up to get $n = 45$. The following table summarizes the margin of error for this and some larger values of n.

n	$t^*s^*/\sqrt{2/n}$
45	2.834
46	2.801
47	2.770

The requirement is first satisfied when $n = 47$. In SAS, we'd perform these calculations using the command

```
proc power;
  twosamplemeans CI=diff alpha=0.1 stddev=8 halfwidth=2.8
                 probwidth=0.50 npergroup=.;
run;
```

This sample size is roughly 4.5 times the sample size used in Example 7.20. This researcher may not be able to recruit this large a sample. If so, we should consider a larger margin of error.

USE YOUR KNOWLEDGE

7.93 Starting salaries. In a recent survey by the National Association of Colleges and Employers, the average starting salary for college graduate with a computer and information sciences degree was reported to be $62,194.[40] You are planning to do a survey of starting salaries for recent computer science majors from your university. Using an estimated standard deviation of $11,605, what sample size do you need to have a margin of error equal to $5000 with 95% confidence?

7.94 Changes in sample size. Suppose that, in the setting of the previous exercise, you have the resources to contact 35 recent graduates. If all respond, will your margin of error be larger or smaller than $5000? What if only 50% respond? Verify your answers by performing the calculations.

The power of the one-sample t test The power of a statistical test measures its ability to detect deviations from the null hypothesis. In practice, we carry out the test in the hope of showing that the null hypothesis is false, so high power is important. Power calculations are a way to assess whether or not a sample size is sufficiently large to answer the research question.

The power of the one-sample t test against a specific alternative value of the population mean μ is the probability that the test will reject the null hypothesis when this alternative is true. To calculate the power, we assume a fixed level of significance, usually $\alpha = 0.05$.

Calculation of the exact power of the t test takes into account the estimation of σ by s and requires a new distribution. We will describe that calculation when discussing the power of the two-sample t test. Fortunately, an approximate calculation that is based on assuming that σ is known is almost always adequate for planning a study in the one-sample case. This calculation is very much like that for the z test, presented in Section 6.4. The steps are

LOOK BACK

power
calculation,
p. 392

1. Write the event, in terms of $\bar{x}$, that the test rejects H_0.

2. Find the probability of this event when the population mean has the alternative value.

Here is an example.

EXAMPLE 7.23

Is the sample size large enough? Recall Example 7.2 (page 413) on the average time that U.S. college students spend watching traditional television. The sample mean of $n = 8$ students was four hours lower than the U.S. average of 18- to 24-year-olds but not found significantly different. Suppose a new study is being planned using a sample size of $n = 50$ students. Does this study have adequate power when the population mean is four hours less than the U.S. average?

We wish to compute the power of the t test for

$$H_0\text{: } \mu = 18.5$$
$$H_a\text{: } \mu < 18.5$$

against the alternative that $\mu = 18.5 - 4 = 14.5$ when $n = 50$. This gives us most of the information we need to compute the power. The other important piece is a rough guess of the size of σ. In planning a large study, a pilot study is often run for this and other purposes. In this case, we can use the standard deviation from the earlier survey. Similar to Example 7.21, we will round up and use $\sigma = 17.5$ and $s = 17.5$ in the approximate calculation.

Step 1. The t test with 50 observations rejects H_0 at the 5% significance level if the t statistic

$$t = \frac{\bar{x} - 18.5}{s/\sqrt{50}}$$

is less than the lower 5% point of $t(49)$, which is -1.677. Taking $s = 17.5$, the event that the test rejects H_0 is, therefore,

$$t = \frac{\bar{x} - 18.5}{17.5/\sqrt{50}} \leq -1.677$$

$$\bar{x} \leq 18.5 - 1.677\frac{17.5}{\sqrt{50}}$$

$$\bar{x} \leq 14.35$$

Step 2. The power is the probability that $\bar{x} \leq 14.351$ when $\mu = 14.5$. Taking $\sigma = 17.5$, we find this probability by standardizing $\bar{x}$:

$$P(\bar{x} \leq 14.35 \text{ when } \mu = 14.5) = P\left(\frac{\bar{x} - 14.5}{17.5/\sqrt{50}} \leq \frac{14.35 - 14.5}{17.5/\sqrt{50}}\right)$$

$$= P(Z \leq -0.061)$$

$$= 0.4761$$

A mean value of 14.5 hours per week will produce significance at the 5% level in only 47.6% of all possible samples. Figure 7.18 shows Minitab output for the exact power calculation. It is about 48% and is represented by a dot on the power curve at a difference of -4. This curve is very informative. For many studies, 80% is considered the standard value for desirable power. We see that with a sample size of 50, the power is greater than 80% only for reductions larger than 6.25 hours per week. If we want to detect a reduction of only four hours, we definitely need to increase the sample size.

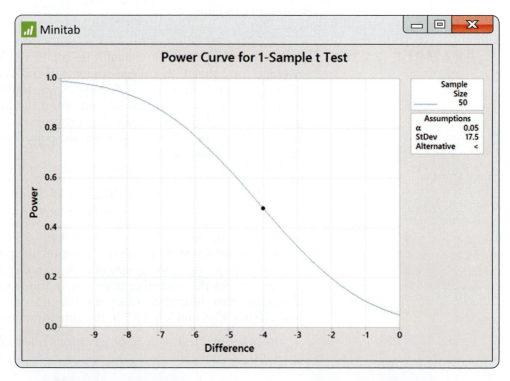

FIGURE 7.18 Minitab output (a power curve) for the one-sample power calculation, Example 7.23.

Power calculations are used in planning studies to ensure that we have a reasonable chance of detecting effects of interest. They give us some guidance in selecting a sample size. In making these calculations, we need assumptions about the standard deviation and the alternative of interest. In our example, we assumed that the standard deviation would be 17.5, but in practice, we are hoping that the value will be somewhere around this value. Similarly, we have used a somewhat arbitrary alternative of 14.5. This is a guess based on the results of the previous study. *Beware of putting too much trust in fine details of the results of these calculations.* They serve as a guide, not a mandate.

USE YOUR KNOWLEDGE

7.95 Power for other values of μ. If you repeat the calculation in Example 7.23 for values of μ that are smaller than 14.5, would you expect the power to be higher or lower than 0.4761? Why?

7.96 Another power calculation. Verify your answer to the previous exercise by doing the calculation for the alternative $\mu = 12$ hours per week.

The power of the two-sample t test The two-sample t test is one of the most used statistical procedures. Unfortunately, because of inadequate planning, users frequently fail to find evidence for the effects that they believe to be present. This is often the result of an inadequate sample size. Power calculations, performed prior to running the experiment, will help avoid this occurrence.

We just learned how to approximate the power of the one-sample t test. The basic idea is the same for the two-sample case, but we will describe the exact method rather than an approximation again. The exact power calculation involves a new distribution, the **noncentral t distribution**. This calculation is not practical by hand but is easy with software that calculates probabilities for this distribution.

noncentral t distribution

We consider only the common case where the null hypothesis is "no difference," $\mu_1 - \mu_2 = 0$. We illustrate the calculation for the pooled two-sample t test. A simple modification is needed when we do not pool. The unknown parameters in the pooled t setting are μ_1, μ_2, and a single common standard deviation σ. To find the power for the pooled two-sample t test, follow these steps.

Step 1. Specify these quantities:

(a) An alternative value for $\mu_1 - \mu_2$ that you consider important to detect.

(b) The sample sizes, n_1 and n_2.

(c) A fixed significance level α, often $\alpha = 0.05$.

(d) An estimate of the standard deviation σ from a pilot study or previous studies under similar conditions.

Step 2. Find the degrees of freedom df $= n_1 + n_2 - 2$ and the value of t^* that will lead to rejecting H_0 at your chosen level α.

noncentrality parameter   **Step 3.** Calculate the **noncentrality parameter**

$$\delta = \frac{|\mu_1 - \mu_2|}{\sigma \sqrt{\dfrac{1}{n_1} + \dfrac{1}{n_2}}}$$

Step 4. The power is the probability that a noncentral t random variable with degrees of freedom df and noncentrality parameter δ will be greater than t^*. Use software to calculate this probability. In SAS, the command is 1 - PROBT(tstar, df, delta). In R the commmand is 1-pt(tstar, df, delta). If you do not have software that can perform this calculation, you can approximate the power as the probability that a standard Normal random variable is greater than $t^* - \delta$, that is, $P(Z > t^* - \delta)$. Use Table A or software for standard Normal probabilities.

Note that the denominator in the noncentrality parameter,

$$\sigma \sqrt{\frac{1}{n_1} + \frac{1}{n_2}}$$

is our guess at the standard error for the difference in the sample means. Therefore, if we wanted to assess a possible study in terms of the margin of error for the estimated difference, we would examine t^* times this quantity.

If we do not assume that the standard deviations are equal, we need to guess both standard deviations and then combine these to get an estimate of the standard error:

$$\sqrt{\frac{\sigma_1^2}{n_1} + \frac{\sigma_2^2}{n_2}}$$

This guess is then used in the denominator of the noncentrality parameter. Use the conservative value, the smaller of $n_1 - 1$ and $n_2 - 1$, for the degrees of freedom.

EXAMPLE 7.24

Planning a new study of calcium versus placebo groups. In Example 7.19 (page 451), we examined the effect of calcium on blood pressure by comparing the means of a treatment group and a placebo group using a pooled two-sample t test. The P-value was 0.059, failing to achieve the usual standard of 0.05 for statistical significance. Suppose that we wanted to plan a new study that would provide convincing evidence—say, at the 0.01 level—with high probability. Let's examine a study design with 45 subjects in each group ($n_1 = n_2 = 45$) to see if this meets our goals.

Step 1. Based on our previous results, we choose $\mu_1 - \mu_2 = 5$ as an alternative that we would like to be able to detect with $\alpha = 0.01$. For σ we use 7.4, our pooled estimate from Example 7.19.

Step 2. The degrees of freedom are $n_1 + n_2 - 2 = 88$, which leads to $t^* = 2.37$ for the significance test.

Step 3. The noncentrality parameter is

$$\delta = \frac{5}{7.4 \sqrt{\dfrac{1}{45} + \dfrac{1}{45}}} = \frac{5}{1.56} = 3.21$$

Step 4. Software gives the power as 0.7965, or 80%. The Normal approximation gives 0.7983, a very accurate result.

With this choice of sample sizes, we are just barely below 80% power. If we judge this to be large enough power, we can proceed to the recruitment of our samples.

With software it is often very easy to examine the effects of variations in a study design. For example, Figure 7.19 shows the JMP power calculator for the two-sample t test. You input values for α, σ, $n_1 + n_2$, and δ (Step 1) and it computes the power (Steps 2–4). Figure 7.19 shows the results of the calculations for Example 7.24. The JMP calculator only considers the two-sided alternative so to get the power for a one-sided alternative, the significance level must be input as 2α. Most other software, such as Minitab, provides the option to choose the alternative.

USE YOUR KNOWLEDGE

7.97 Power and the choice of alternative. If you were to repeat the calculation in Example 7.24 for the two-sided alternative, would the power increase or decrease? Explain your answer.

7.98 Power and the standard deviation. If the true population standard deviation were 8 instead of the 7.4 hypothesized in Example 7.24, would the power increase or decrease? Explain.

7.99 Power and statistical software. Refer to the two previous exercises. Use statistical software to compute the exact power of each scenario.

FIGURE 7.19 JMP input/output window for the two-sample power calculation, Example 7.24.

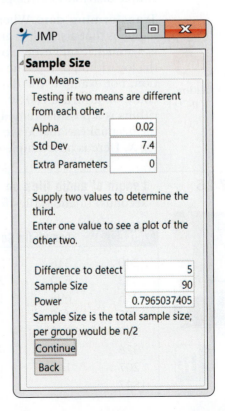

Inference for non-Normal populations

We have not discussed how to do inference about the mean of a clearly non-Normal distribution based on a small sample. If you face this problem, you should consult an expert. Three general strategies are available:

• In some cases, a distribution other than a Normal distribution describes the data well. There are many non-Normal models for data, and inference procedures for these models are available.

• Because skewness is the chief barrier to the use of t procedures on data without outliers, you can attempt to transform skewed data so that the distribution is symmetric and as close to Normal as possible. Confidence levels and P-values from the t procedures applied to the transformed data will be quite accurate for even moderate sample sizes. Methods are generally available for transforming the results back to the original scale.

distribution-free procedures
nonparametric procedures

• Use a **distribution-free** inference procedure. Such procedures do not assume that the population distribution has any specific form, such as Normal. Distribution-free procedures are often called **nonparametric procedures**. Chapter 15 discusses several of these procedures.

Each of these strategies can be effective, but each quickly carries us beyond the basic practice of statistics. We emphasize procedures based on Normal distributions because they are the most common in practice, because their robustness makes them widely useful, and (most important) because we are first of all concerned with understanding the principles of inference. Therefore, we will not discuss procedures for non-Normal continuous distributions. We will be content with illustrating by example the use of a transformation and of a simple distribution-free procedure.

LOOK BACK

log transformation, p. 91

Transforming data When the distribution of a variable is skewed, it often happens that a simple transformation results in a variable whose distribution is symmetric and even close to Normal. The most common transformation is the logarithm, or log. The logarithm tends to pull in the right tail of a distribution. For example, the data 2, 3, 4, 20 show an outlier in the right tail. Their common logarithms 0.30, 0.48, 0.60, 1.30 are much less skewed. Taking logarithms is a possible remedy for right-skewness. Instead of analyzing values of the original variable X, we compute their logarithms and analyze the values of $\log X$. Here is an example of this approach.

Justin Sullivan/Staff/Getty Images

SONGS

EXAMPLE 7.25

Length of audio files on an iPod. Table 7.5 presents data on the length (in seconds) of audio files found on an iPod. There was a total of 10,003 audio

TABLE 7.5	Length (in Seconds) of Audio Files Sampled from an iPod					
240	316	259	46	871	411	1366
233	520	239	259	535	213	492
315	696	181	357	130	373	245
305	188	398	140	252	331	47
309	245	69	293	160	245	184
326	612	474	171	498	484	271
207	169	171	180	269	297	266
1847						

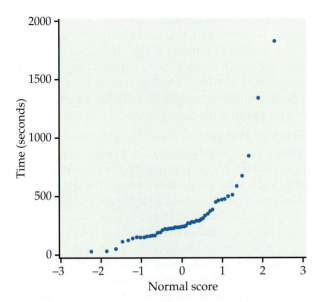

FIGURE 7.20 Normal quantile plot of audio file length, Example 7.25. This sort of pattern occurs when a distribution is skewed to the right.

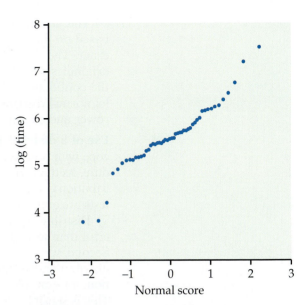

FIGURE 7.21 Normal quantile plot of the logarithms of the audio file lengths, Example 7.25. This distribution appears approximately Normal.

files, and 50 files were randomly selected using the "shuffle songs" command.[41] We would like to give a confidence interval for the average audio file length μ for this iPod.

A Normal quantile plot of the audio data from Table 7.5 (Figure 7.20) shows that the distribution is skewed to the right. Because there are no extreme outliers, the sample mean of the 50 observations will nonetheless have an approximately Normal sampling distribution. The t procedures could be used for approximate inference. For more exact inference, we will transform the data so that the distribution is more nearly Normal. Figure 7.21 is a Normal quantile plot of the natural logarithms of the time measurements. The transformed data are very close to Normal, so t procedures will give quite exact results.

The application of the t procedures to the transformed data is straightforward. Call the original length values from Table 7.5 the variable X. The transformed data are values of $X_{new} = \log X$. In most software packages, it is an easy task to transform data in this way and then analyze the new variable.

EXAMPLE 7.26

SONGS

Software output of audio length data. Analysis of the natural log of the length values in Minitab produces the following output:

N	Mean	StDev	SE Mean	95.0% C.I.
50	5.6315	0.6840	0.0967	(5.4371, 5.8259)

For comparison, the 95% t confidence interval for the original mean μ is found from the original data as follows:

N	Mean	StDev	SE Mean	95.0% C.I.
50	354.1	307.9	43.6	(266.6, 441.6)

The advantage of analyzing transformed data is that use of procedures based on the Normal distributions is better justified and the results are more exact. The disadvantage is that a confidence interval for the mean μ in the original scale (in our example, seconds) cannot be easily recovered from the confidence interval for the mean of the logs. One approach based on the lognormal distribution[42] results in an interval of (285.5, 435.5), which is narrower and slightly asymmetric compared with the t interval.

Use of a distribution-free procedure Perhaps the most straightforward way to cope with non-Normal data is to use a *distribution-free,* or *nonparametric,* procedure. As the name indicates, these procedures do not require the population distribution to have any specific form, such as Normal. Distribution-free significance tests are quite simple and are available in most statistical software packages.

Distribution-free tests have two drawbacks. First, they are generally less powerful than tests designed for use with a specific distribution, such as the t test. Second, we must often modify the statement of the hypotheses in order to use a distribution-free test. A distribution-free test concerning the center of a distribution, for example, is usually stated in terms of the *median* rather than the mean. This is sensible when the distribution may be skewed. But the distribution-free test does not ask the same question (Has the mean changed?) that the t test does.

sign test

The simplest distribution-free test, and one of the most useful, is the **sign test**. The test gets its name from the fact that we look only at the signs of the differences, not their actual values. The following example illustrates this test.

EXAMPLE 7.27

GEPARTS

LOOK BACK

binomial
distribution,
p. 312

The effect of altering a software parameter. Example 7.7 (page 419) describes an experiment to compare the measurements obtained from two software algorithms. In that example, we used the matched pairs t test on these data, despite some skewness, which makes the P-value only roughly correct. The sign test is based on the following simple observation: of the 51 parts measured, 29 had a larger measurement with the option off and 22 had a larger measurement with the option on.

To perform a significance test based on these counts, let p be the probability that a randomly chosen part would have a larger measurement with the option turned on. The null hypothesis of "no effect" says that these two measurements are just repeat measurements, so the measurement with the option off is equally likely to be larger or smaller than the measurement with the option on. Therefore, we want to test

$$H_0: p = 1/2$$
$$H_a: p \neq 1/2$$

The 51 parts are independent trials, so the number that had larger measurements with the option off has the binomial distribution $B(51, 1/2)$ if H_0 is true. The P-value for the observed count 29 is, therefore, $2P(X \geq 29)$, where X has the $B(51, 1/2)$ distribution. You can compute this probability with software or the Normal approximation to the binomial:

$$2P(X \geq 29) = 2P\left(Z \geq \frac{29 - 25.5}{\sqrt{12.75}}\right)$$
$$= 2P\,(Z \geq 0.98)$$
$$= 2(0.1635)$$
$$= 0.3270$$

As in Example 7.7, there is not strong evidence that the two measurements are different.

There are several varieties of sign test, all based on counts and the binomial distribution. The sign test for matched pairs is the most useful. The null hypothesis of "no effect" is then always $H_0: p = 1/2$. The alternative can be one-sided in either direction or two-sided, depending on the type of change we are considering.

SIGN TEST FOR MATCHED PAIRS

Ignore pairs with difference 0; the number of trials n is the count of the remaining pairs. The test statistic is the count X of pairs with a positive difference. P-values for X are based on the binomial $B(n, 1/2)$ distribution.

The matched pairs t test in Example 7.7 tested the hypothesis that the mean of the distribution of differences is 0. The sign test in Example 7.27 is, in fact, testing the hypothesis that the *median* of the differences is 0. If p is the probability that a difference is positive, then $p = 1/2$ when the median is 0. This is true because the median of the distribution is the point with probability 1/2 lying to its right. As Figure 7.22 illustrates, $p > 1/2$ when the median is greater than 0, again because the probability to the right of the median is always 1/2. The sign test of $H_0: p = 1/2$ against $H_a: p > 1/2$ is a test of

$$H_0: \text{population median} = 0$$

$$H_a: \text{population median} > 0$$

The sign test in Example 7.27 makes no use of the actual scores—it just counts how many parts had a larger measurement with the option off. Any parts that did not have different measurements would be ignored altogether. Because the sign test uses so little of the available information, it is much less powerful than the t test when the population is close to Normal. Chapter 15 describes other distribution-free tests that are more powerful than the sign test.

USE YOUR KNOWLEDGE

7.100 Sign test for the oil-free frying comparison. Exercise 7.10 (page 422) gives data on the taste of hash browns made using a hot-oil fryer and an oil-free fryer. Is there evidence that the medians are different? State the hypotheses, carry out the sign test, and report your conclusion.

FIGURE 7.22 Why the sign test tests the median difference: when the median is greater than 0, the probability p of a positive difference is greater than 1/2, and vice versa.

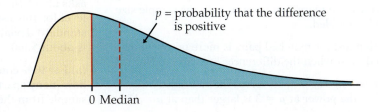

p = probability that the difference is positive

0 Median

SECTION 7.3 SUMMARY

- The **sample size** required to obtain a confidence interval with an expected margin of error no larger than m for a population mean satisfies the constraint

$$m \geq t^* s^* / \sqrt{n}$$

where t^* is the critical value for the desired level of confidence with $n - 1$ degrees of freedom, and s^* is the guessed value for the population standard deviation.

- The sample sizes necessary for a two-sample confidence interval can be obtained using a similar constraint, but guesses of both standard deviations and an estimate for the degrees of freedom are required. We suggest using the smaller of $n_1 - 1$ and $n_2 - 1$ for degrees of freedom.

- The **power** of the one-sample t test can be calculated like that of the z test, using an approximate value for both σ and s.

- The **power** of the two-sample t test is found by first finding the critical value for the significance test, the degrees of freedom, and the **noncentrality parameter** for the alternative of interest. These are used to calculate the power from a **noncentral t distribution**. A Normal approximation works quite well. Calculating margins of error for various study designs and conditions is an alternative procedure for evaluating designs.

- The **sign test** is a **distribution-free test** because it uses probability calculations that are correct for a wide range of population distributions.

- The sign test for "no treatment effect" in matched pairs counts the number of positive differences. The P-value is computed from the $B(n, 1/2)$ distribution, where n is the number of non-0 differences. The sign test is less powerful than the t test in cases where use of the t test is justified.

SECTION 7.3 EXERCISES

For Exercise 7.93 and 7.94, see pages 464–465; for Exercises 7.95 and 7.96, see page 467; for Exercises 7.97 through 7.99, see page 469; and for Exercise 7.100, see page 473.

7.101 What is wrong? In each of the following situations, identify what is wrong, and then either explain why it is wrong or change the wording of the statement to make it true.

(a) To reduce the margin of error in half, the sample size needs to be doubled.

(b) The sign test for matched pairs is more powerful than the paired t test when the differences are close to Normal.

(c) When testing H_0: $\mu = 10$ versus the two-sided alternative, the power at $\mu = 3$ is larger than at $\mu = 17$.

(d) Increasing sample size increases the power for all alternatives and decreases the probability of a Type I error.

7.102 Apartment rental rates. You hope to rent an unfurnished one-bedroom apartment in Dallas next year. You call a friend who lives there and ask him to give you an estimate of the mean monthly rate. Having taken a statistics course recently, the friend asks about the desired margin of error and confidence level for this estimate. He also tells you that the standard deviation of monthly rents for one-bedrooms is about $300.

(a) For 95% confidence and a margin of error of $150, how many apartments should the friend randomly sample from the local newspaper?

(b) Suppose that you want the margin of error to be no more than $50. How many apartments should the friend sample?

(c) Why is the sample size in part (b) not just nine times larger than the sample size in part (a)?

7.103 More on apartment rental rates. Refer to the previous exercise. Will the 95% confidence interval include approximately 95% of the rents of all unfurnished one-bedroom apartments in this area? Explain why or why not.

7.104 Average hours per week on the Internet. The *Student Monitor* surveys 1200 undergraduates from 100 colleges semiannually to understand trends among college students.[43] Recently, the *Student Monitor* reported that the average amount of time spent per week on the Internet was 19.0 hours. You suspect that this amount is far too small for your campus and plan a survey.

(a) You feel that a reasonable estimate of the standard deviation is 10.0 hours. What sample size is needed so that the expected margin of error of your estimate is not larger than one hour for 95% confidence?

(b) The distribution of times is likely to be heavily skewed to the right. Do you think that this skewness will invalidate the use of the t confidence interval in this case? Explain your answer.

7.105 Average hours per week listening to the radio. Refer to the previous exercise. The *Student Monitor* also reported that the average amount of time listening to the radio was 11.5 hours.

(a) Given an estimated standard deviation of 5.2 hours, what sample size is needed so that the expected margin of error of your estimate is not larger than one hour for 95% confidence?

(b) If your survey is going to ask about Internet use and radio use, which of the two calculated sample sizes should you use? Explain your answer.

7.106 Accuracy of a laboratory scale. To assess the accuracy of a laboratory scale, a standard weight known to weigh 10 grams is weighed repeatedly. The scale readings are Normally distributed with unknown mean (this mean is 10 grams if the scale has no bias). The standard deviation of the scale readings in the past has been 0.0013 gram.

(a) The weight is measured five times. The mean result is 10.0009 grams. Give a 98% confidence interval for the mean of repeated measurements of the weight.

(b) How many measurements must be averaged to get an expected margin of error no more than 0.001 with 98% confidence?

7.107 Accuracy of a laboratory scale, continued. Refer to the previous exercise. Suppose that instead of a confidence interval, the researchers want to perform a test (with $\alpha = 0.05$) that the scale is unbiased ($\mu = 10$).

(a) What sample size n is necessary to have at least 90% power when the alternative mean is $\mu = 10.001$?

(b) Suppose they can only perform a maximum of $n = 10$ measurements. Based on your answer in part (a), will the power be more or less than 90%? Explain your answer.

(c) Verify your answer in part (b), by computing the power when $n = 10$.

7.108 Sample size calculations. You are designing a study to test the null hypothesis that $\mu = 0$ versus the alternative that μ is positive. Assume that σ is 20. Suppose that it would be important to be able to detect the alternative $\mu = 4$. What sample size is needed to detect this alternative with power of at least 0.80?

7.109 Power of the comparison of DXA machine operators. Suppose that the bone researchers in Exercise 7.45 (page 431) want to be able to detect an alternative mean difference of 0.002. Find the power for this alternative for a sample size of 20 patients. Make sure to explain the reasoning of your choice of standard deviation in these calculations.

7.110 Determining the sample size. Consider Example 7.23 (page 465). What is the minimum sample size needed for the power to be greater than 80% when $\mu = 14.5$?

7.111 Changing the significance level. In Example 7.24 (page 468), we assessed the power of a new study of calcium on blood pressure assuming $n_1 = n_2 = 45$ subjects. The power was based on $\alpha = 0.01$. Suppose that we wanted to use $\alpha = 0.05$ instead.

(a) Would the power increase or decrease? Explain your answer in terms someone unfamiliar with power calculations can understand.

(b) Verify your answer by computing the power.

7.112 Planning a study to compare tree size. In Exercise 7.79 (page 459), DBH data for longleaf pine trees in two parts of the Wade Tract are compared. Suppose that you are planning a similar study in which you will measure the diameters of longleaf pine trees. Based on Exercise 7.79, you are willing to assume that the standard deviation for both halves is 20 cm. Suppose that a difference in mean DBH of 10 cm or more would be important to detect. You will use a t statistic and a two-sided alternative for the comparison.

(a) Find the power if you randomly sample 20 trees from each area to be compared.

(b) Repeat the calculations for 60 trees in each sample.

(c) If you had to choose between the 20 and 60 trees per sample, which would you choose? Give reasons for your answer.

7.113 More on planning a study to compare tree size. Refer to the previous exercise. Find the two standard deviations from Exercise 7.79. Do the same for the data in Exercise 7.80 (page 459), which is a similar setting. These are somewhat smaller than the assumed value that you used in the previous exercise. Explain why it is generally a better idea to assume a standard deviation that is larger than you expect than one that is smaller. Repeat the power calculations for some other reasonable values of σ and comment on the impact of the size of σ for planning the new study.

7.114 Planning a study to compare ad placement. Refer to Exercise 7.78 (page 458), where we compared trustworthiness ratings for ads from two different publications. Suppose that you are planning a similar study using two different publications that are not expected to show the differences seen when comparing the *Wall Street Journal* with the *National Enquirer*. You would like to detect a difference of 1.5 points using a two-sided significance test with a 5% level of significance. Based on Exercise 7.78, it is reasonable to use 1.6 as the value of the common standard deviation for planning purposes.

(a) What is the power if you use sample sizes similar to those used in the previous study—for example, 65 for each publication?

(b) Repeat the calculations for 100 in each group.

(c) What sample size would you recommend for the new study?

7.115 Sign test for potential insurance fraud. The differences in the repair estimates in Exercise 7.40 (page 430) can also be analyzed using a sign test. Set up the appropriate null and alternative hypotheses, carry out the test, and summarize the results. How do these results compare with those that you obtained in Exercise 7.40? JOCKO

7.116 Sign test for the comparison of operators. The differences in the TBBMC measures in Exercise 7.45 (page 431) can also be analyzed using a sign test. Set up the appropriate null and alternative hypotheses, carry out the test, and summarize the results. How do these results compare with those that you obtained in Exercise 7.45? TBBMC

7.117 Sign test for fuel efficiency comparison. Use the sign test to assess whether the computer calculates a higher mpg than the driver in Exercise 7.41 (page 430). State the hypotheses, give the *P*-value using the binomial table (Table C), and report your conclusion. MPGDIFF

7.118 Insulation study. A manufacturer of electric motors tests insulation at a high temperature (250° C) and records the number of hours until the insulation fails.[44] The data for five specimens are

$$446 \quad 326 \quad 372 \quad 377 \quad 310$$

The small sample size makes judgment from the data difficult, but engineering experience suggests that the logarithm of the failure time will have a Normal distribution. Take the logarithms of the five observations and use *t* procedures to give a 90% confidence interval for the mean of the log failure time for insulation of this type. INSULAT

CHAPTER 7 EXERCISES

7.119 LSAT scores. The scores of four senior roommates on the Law School Admission Test (LSAT) are

$$153 \quad 162 \quad 166 \quad 133$$

Find the mean, the standard deviation, and the standard error of the mean. Is it appropriate to calculate a confidence interval based on these data? Explain why or why not. LSAT

7.120 Converting a two-sided *P*-value. You use statistical software to perform a significance test of the null hypothesis that two means are equal. The software reports a *P*-value for the two-sided alternative. Your alternative is that the first mean is greater than the second mean.

(a) The software reports $t = 1.85$ with a *P*-value of 0.075. Would you reject H_0 at $\alpha = 0.05$? Explain your answer.

(b) The software reports $t = -1.85$ with a *P*-value of 0.075. Would you reject H_0 at $\alpha = 0.05$? Explain your answer.

7.121 Degrees of freedom and *t**. As the degrees of freedom increase, the *t* distributions get closer and closer to the z ($N(0, 1)$) distribution. One way to see this is to look at how the value of *t** for a 95% confidence interval changes with the degrees of freedom.

(a) Make a plot with degrees of freedom from 10 to 100 by 10 on the x axis and *t** on the y axis. Also draw a horizontal line on the plot corresponding to the value of $z^* = 1.96$.

(b) Summarize the main features of the plot.

(c) Describe how this plot would change if you considered a 90% confidence interval.

7.122 Sample size and margin of error. The margin of error for a confidence interval for μ depends on the confidence level, the sample standard deviation s, and the sample size. Fix the confidence level at 95% and the sample standard deviation at $s = 1$ to examine the effect of the sample size. Find the margin of error for sample sizes of 11 to 101 by 10s—that is, let $n = 11, 21, 31, \ldots,$ 101. Plot the margins of error versus the sample size and summarize the relationship.

7.123 Which design? The following situations all require inference about a mean or means. Identify each as (1) a single sample, (2) matched pairs, or (3) two independent samples. Explain your answers.

(a) Your customers are college students. You are interested in comparing the interest in a new product that you are developing between those students who live in the dorms and those who live elsewhere.

(b) Your customers are college students. You are interested in finding out which of two new product labels is more appealing.

(c) Your customers are college students. You are interested in assessing their interest in a new product.

7.124 Which design? The following situations all require inference about a mean or means. Identify each as (1) a single sample, (2) matched pairs, or (3) two independent samples. Explain your answers.

(a) You want to estimate the average age of your store's customers.

(b) You do an SRS survey of your customers every year. One of the questions on the survey asks about customer satisfaction on a seven-point scale with the response 1 indicating "very dissatisfied" and 7 indicating "very satisfied." You want to see if the mean customer satisfaction has improved from last year.

(c) You ask an SRS of customers their opinions on each of two new floor plans for your store.

7.125 Number of critical food violations. The results of a major city's restaurant inspections are available through its online newspaper.[45] Critical food violations are those that put patrons at risk of getting sick and must immediately be corrected by the restaurant. An SRS of $n = 200$ inspections from the more than 16,000 inspections since January 2012 were collected, resulting in $\bar{x} = 0.995$ violations and $s = 1.822$ violations.

(a) Test the hypothesis that the average number of critical violations is less than 1.5 using a significance level of 0.05. State the two hypotheses, the test statistic, and P-value.

(b) Construct a 95% confidence interval for the average number of critical violations and summarize your result.

(c) Which of the two summaries (significance test versus confidence interval) do you find more helpful in this case? Explain your answer.

(d) These data are integers ranging from 0 to 10. The data are also skewed to the right, with 79% of the values either a 0 or a 1. Given this information, do you think use of the t procedures is appropriate? Explain your answer.

7.126 Two-sample t test versus matched pairs t test. Consider the following data set. The data were actually collected in pairs, and each row represents a pair. **PAIRED**

Group 1	Group 2
48.86	48.88
50.60	52.63
51.02	52.55
47.99	50.94
54.20	53.02
50.66	50.66
45.91	47.78
48.79	48.44
47.76	48.92
51.13	51.63

(a) Suppose that we ignore the fact that the data were collected in pairs and mistakenly treat this as a two-sample problem. Compute the sample mean and variance for each group. Then compute the two-sample t statistic, degrees of freedom, and P-value for the two-sided alternative.

(b) Now analyze the data in the proper way. Compute the sample mean and variance of the differences. Then compute the t statistic, degrees of freedom, and P-value.

(c) Describe the differences in the two test results.

7.127 Two-sample t test versus matched pairs t test, continued. Refer to the previous exercise. Perhaps an easier way to see the major difference in the two analysis approaches for these data is by computing 95% confidence intervals for the mean difference.

(a) Compute the 95% confidence interval using the two-sample t confidence interval.

(b) Compute the 95% confidence interval using the matched pairs t confidence interval.

(c) Compare the estimates (that is, the centers of the intervals) and margins of error. What is the major difference between the two approaches for these data?

7.128 Average service time. Recall the drive-thru study in Exercise 7.69 (page 457). Another benchmark that was measured was the service time. A summary of the results (in seconds) for two of the chains is shown below.

Chain	n	$\bar{x}$	s
Taco Bell	308	158.03	33.8
McDonald's	317	189.49	41.3

(a) Is there a difference in the average service time between these two chains? Test the null hypothesis that the chains' average service time is the same. Use a significance level of 0.05.

(b) Construct a 95% confidence interval for the difference in average service time.

(c) Lex plans to go to Taco Bell and Sam to McDonald's. Does the interval in part (b) contain the difference in their service times that they're likely to encounter? Explain your answer.

7.129 Interracial friendships in college. A study utilized the random roommate assignment process of a small college to investigate the interracial mix of friends among students in college.[46] As part of this study, the researchers looked at 238 white students who were randomly assigned a roommate in their first year and recorded the proportion of their friends (not including the first-year roommate) who were black. The following table summarizes the results, broken down by roommate race, for the middle of the first and third years of college.

Middle of First Year			
Randomly assigned	n	$\bar{x}$	s
Black roommate	41	0.085	0.134
White roommate	197	0.063	0.112

Middle of Third Year			
Randomly assigned	n	$\bar{x}$	s
Black roommate	41	0.146	0.243
White roommate	197	0.062	0.154

(a) Proportions are not Normally distributed. Explain why it may still be appropriate to use the t procedures for these data.

(b) For each year, state the null and alternative hypotheses for comparing these two groups.

(c) For each year, perform the significance test at the $\alpha = 0.05$ level, making sure to report the test statistic, degrees of freedom, and P-value.

(d) Write a one-paragraph summary of your conclusions from these two tests.

7.130 Interracial friendships in college, continued. Refer to the previous exercise. For each year, construct a 95% confidence interval for the difference in means $\mu_1 - \mu_2$ and describe how these intervals can be used to test the null hypotheses in part (b) of the previous exercise.

7.131 Alcohol consumption and body composition. Individuals who consume large amounts of alcohol do not use the calories from this source as efficiently as calories from other sources. One study examined the effects of moderate alcohol consumption on body composition and the intake of other foods. Fourteen subjects participated in a crossover design where they either drank wine for the first six weeks and then abstained for the next six weeks or vice versa.[47] During the period when they drank wine, the subjects, on average, lost 0.4 kilogram (kg) of body weight; when they did not drink wine, they lost an average of 1.1 kg. The standard deviation of the difference between the weight lost under these two conditions is 8.6 kg. During the wine period, they consumed an average of 2589 calories; with no wine, the mean consumption was 2575. The standard deviation of the difference was 210.

(a) Compute the differences in means and the standard errors for comparing body weight and caloric intake under the two experimental conditions.

(b) A report of the study indicated that there were no significant differences in these two outcome measures. Verify this result for each measure, giving the test statistic, degrees of freedom, and the P-value.

(c) One concern with studies such as this, with a small number of subjects, is that there may not be sufficient power to detect differences that are potentially important. Address this question by computing 95% confidence intervals for the two measures and discuss the information provided by the intervals.

(d) Here are some other characteristics of the study. The study periods lasted for six weeks. All subjects were males between the ages of 21 and 50 years who weighed between 68 and 91 kg. They were all from the same city. During the wine period, subjects were told to consume two 135-milliliter (ml) servings of red wine per day and no other alcohol. The entire six-week supply was given to each subject at the beginning of the period. During the other period, subjects were instructed to refrain from any use of alcohol. All subjects reported that they complied with these instructions except for three subjects, who said that they drank no more than three to four 12-ounce bottles of beer during the no-alcohol period. Discuss how these factors could influence the interpretation of the results.

7.132 The wine makes the meal? In one study, 39 diners were given a free glass of cabernet sauvignon wine to accompany a French meal.[48] Although the wine was identical, half the bottle labels claimed the wine was from California and the other half claimed it was from North Dakota. The following table summarizes the grams of entrée and wine consumed during the meal.

	Wine label	n	Mean	St. Dev
Entrée	California	24	499.8	87.2
	North Dakota	15	439.0	89.2
Wine	California	24	100.8	23.3
	North Dakota	15	110.4	9.0

Did the patrons who thought that the wine was from California consume more? Analyze the data and write a report summarizing your work. Be sure to include details regarding the statistical methods you used, your assumptions, and your conclusions.

7.133 Can mockingbirds learn to identify specific humans? A central question in urban ecology is why some animals adapt well to the presence of humans and others do not. The following results summarize part of a study of the northern mockingbird (*Mimus polyglottos*) that took place on a campus of a large university.[49] For four consecutive days, the same human approached a nest and stood 1 meter away for 30 seconds, placing his or her hand on the rim of the nest. On the fifth day, a new person did the same thing. Each day, the distance of the human from the nest when the bird flushed was recorded. This was repeated for 24 nests. The human intruder varied his or her appearance (that is, wore different clothes) over the four days. We report results for only Days 1, 4, and 5 here. The response variable is flush distance measured in meters.

Day	Mean	s
1	6.1	4.9
4	15.1	7.3
5	4.9	5.3

(a) Explain why this should be treated as a matched design.

(b) Unfortunately, the research article does not provide the standard error of the difference, only the standard error of the mean flush distance for each day. However, we can use the general addition rule for variances (page 258) to approximate it. If we assume that the correlation between the flush distance at Day 1 and Day 4 for each nest is $\rho = 0.40$, what is the standard deviation for the difference in distance?

(c) Using your result in part (b), test the hypothesis that there is no difference in the flush distance across these two days. Use a significance level of 0.05.

(d) Repeat parts (b) and (c) but now compare Day 1 and Day 5, assuming a correlation between flush distances for each nest of $\rho = 0.30$.

(e) Write a brief summary of your conclusions.

7.134 Sign test for assessment of a foreign-language institute. Use the sign test to assess whether the summer institute of Exercise 7.47 (page 432) improves French listening skills. State the hypotheses, give the *P*-value using the binomial table (Table C), and report your conclusion. SUMLANG

7.135 Study design information. Refer to Exercise 7.132. In this study, diners were seated alone or in groups of two, three, four, and, in one case, nine (for a total of $n = 16$ tables). Also, each table, not each patron, was randomly assigned a particular wine label. Does this information alter how you might do the analysis in the previous problem? Explain your answer.

7.136 Analysis of tree size using the complete data set. The data used in Exercises 7.33 (page 429), 7.79, and 7.80 (page 459) were obtained by taking SRSs from the 584 longleaf pine trees that were measured in the Wade Tract. The entire data set is given in the WADE data set. Find the 95% confidence interval for the mean DBH using the entire data set, and compare this interval with the one that you calculated in Exercise 7.33. Write a report about these data. Include comments on the effect of the sample size on the margin of error, the distribution of the data, the appropriateness of the Normality-based methods for this problem, and the generalizability of the results to other similar stands of longleaf pine or other kinds of trees in this area of the United States and other areas. WADE

7.137 Can snobby salespeople boost retail sales? Researchers asked 180 women to read a hypothetical shopping experience where they entered a luxury store (e.g., Louis Vuitton, Gucci, Burberry) and ask a salesperson for directions to the items they seek. For half the women, the salesperson was condescending while doing this. The other half were directed in a neutral manner. After reading the experience, participants were asked various questions, including what price they were willing to pay (in dollars) for a particular product from the brand.[50] Here is a summary of the results.

Chain	n	$\bar{x}$	s
Condescending	90	4.44	3.98
Neutral	90	3.95	2.88

Were the participants who were treated rudely willing to pay more for the product? Analyze the data, and write a report summarizing your work. Be sure to include details regarding the statistical methods you used, your assumptions, and your conclusions.

7.138 A comparison of female high school students. A study was performed to determine the prevalence of the female athlete triad (low energy availability, menstrual dysfunction, and low bone mineral density) in high school students.[51] A total of 80 high school athletes and 80 sedentary students were assessed. The following table summarizes several measured characteristics:

Characteristic	Athletes $\bar{x}$	s	Sedentary $\bar{x}$	s
Body fat (%)	25.61	5.54	32.51	8.05
Body mass index	21.60	2.46	26.41	2.73
Calcium deficit (mg)	297.13	516.63	580.54	372.77
Glasses of milk/day	2.21	1.46	1.82	1.24

(a) For each of the characteristics, test the hypothesis that the means are the same in the two groups. Use a significance level of 0.05 for each test.

(b) Write a short report summarizing your results.

7.139 More on snobby salespeople. Refer to Exercise 7.137. Researchers also asked a different 180 women to read the same hypothetical shopping experience, but now they entered a mass market (e.g., Gap, American Eagle, H&M). Here are those results (in dollars) for the two conditions:

Chain	n	$\bar{x}$	s
Condescending	90	2.90	3.28
Neutral	90	2.98	3.24

Were the participants who were treated rudely willing to pay more for the product? Analyze the data, and write a report summarizing your work. Be sure to include details regarding the statistical methods you used, your assumptions, and your conclusions. Also compare these results with the ones from Exercise 7.137.

7.140 Transforming the response. Refer to Exercises 7.137 and 7.139. The researchers state that they took the natural log of the willingness to pay variable in order to "normalize the distribution" prior to analysis. Thus, their test results are based on log dollar measurements. For the t procedures used in the previous two exercises, do you feel this transformation is necessary? Explain your answer.

7.141 Competitive prices? A retailer entered into an exclusive agreement with a supplier who guaranteed to provide all products at competitive prices. The retailer eventually began to purchase supplies from other vendors who offered better prices. The original supplier filed a legal action claiming violation of the agreement. In defense, the retailer had an audit performed on a random sample of invoices. For each audited invoice, all purchases made from other suppliers were examined and the prices were compared with those offered by the original supplier. For each invoice, the percent of purchases for which the alternate supplier offered a lower price than the original supplier was recorded.[52] Here are the data:

0	100	0	100	33	34	100	48	78	100	77	100	38
68	100	79	100	100	100	100	100	100	89	100	100	

Report the average of the percents with a 95% margin of error. Do the sample invoices suggest that the original supplier's prices are not competitive on the average? **COMPETE**

7.142 Weight-loss programs. In a study of the effectiveness of weight-loss programs, 47 subjects who were at least 20% overweight took part in a group support program for 10 weeks. Private weighings determined each subject's weight at the beginning of the program and six months after the program's end. The matched pairs t test was used to assess the significance of the average weight loss. The paper reporting the study said, "The subjects lost a significant amount of weight over time, $t(46) = 4.68$, $p < 0.01$." It is common to report the results of statistical tests in this abbreviated style.[53]

(a) Why was the matched pairs statistic appropriate?

(b) Explain to someone who knows no statistics but is interested in weight-loss programs what the practical conclusion is.

(c) The paper follows the tradition of reporting significance only at fixed levels such as $\alpha = 0.01$. In fact, the results are more significant than "$p < 0.01$" suggests. What can you say about the P-value of the t test?

7.143 Do women perform better in school? Some research suggests that women perform better than men in school, but men score higher on standardized tests. Table 1.3 (page 26) presents data on a measure of school performance, grade point average (GPA), and a standardized test, IQ, for 78 seventh-grade students. Do these data lend further support to the previously found gender differences? Give graphical displays of the data and describe the distributions. Use significance tests and confidence intervals to examine this question, and prepare a short report summarizing your findings. **GRADES**

7.144 Self-concept and school performance. Refer to the previous exercise. Although self-concept in this study was measured on a scale with values in the data set ranging from 20 to 80, many prefer to think of this kind of variable as having only two possible values: low self-concept or high self-concept. Find the median of the self-concept scores in Table 1.3, and define those students with scores at or below the median to be low-self-concept students and those with scores above the median to be high-self-concept students. Do high-self-concept students have GPAs that differ from those of low-self-concept students? What about IQ? Prepare a report addressing these questions. Be sure to include graphical and numerical summaries and confidence intervals, and state clearly the details of significance tests. GRADES

7.145 Behavior of pet owners. On the morning of March 5, 1996, a train with 14 tankers of propane derailed near the center of the small Wisconsin town of Weyauwega. Six of the tankers were ruptured and burning when the 1700 residents were ordered to evacuate the town. Researchers study disasters like this so that effective relief efforts can be designed for future disasters. About half the households with pets did not evacuate all their pets. A study conducted after the derailment focused on problems associated with retrieval of the pets after the evacuation and characteristics of the pet owners. One of the scales measured "commitment to adult animals," and the people who evacuated all or some of their pets were compared with those who did not evacuate any of their pets. Higher scores indicate that the pet owner is more likely to take actions that benefit the pet.[54] Here are the data summaries:

Group	n	$\bar{x}$	s
Evacuated all or some pets	116	7.95	3.62
Did not evacuate any pets	125	6.26	3.56

Analyze the data and prepare a short report describing the results.

7.146 Sample size calculation. Example 7.10 (page 434) tells us that the mean height of 10-year-old girls is $N(56.9, 2.8)$ and for boys it is $N(56.0, 3.5)$. The null hypothesis that the mean heights of 10-year-old boys and girls are equal is clearly false. The difference in mean heights is $56.9 - 56.0 = 0.9$ inch. Small differences such as this can require large sample sizes to detect. To simplify our calculations, let's assume that the standard deviations are the same—say, $\sigma = 3.2$—and that we will measure the heights of an equal number of girls and boys. How many would we need to measure to have a 90% chance of detecting the (true) alternative hypothesis?

7.147 Different methods of teaching reading. In the READ data set, the response variable Post3 is to be compared for three methods of teaching reading. The Basal method is the standard, or control, method, and the two new methods are DRTA and Strat. We can use the methods of this chapter to compare Basal with DRTA and Basal with Strat. Note that to make comparisons among three treatments it is more appropriate to use the procedures that we will learn in Chapter 12. READ

(a) Is the mean reading score with the DRTA method higher than that for the Basal method? Perform an analysis to answer this question, and summarize your results.

(b) Answer part (a) for the Strat method in place of DRTA.

7.148 Designing a new stress management survey. Refer to Exercise 6.17 (page 358). Suppose you want to draw a new SRS of millenials such that the expected margin of error with 99% confidence is 0.2 points. What sample size do you need?

7.149 Conditions for inference. Suppose that your state contains 85 school corporations and each corporation reports its expenditures per pupil. Is it proper to apply the one-sample t method to these data to give a 95% confidence interval for the average expenditure per pupil? Explain your answer.

Hans-Peter Merten/Getty Images

Inference for Proportions

Introduction

We frequently collect data on *categorical variables,* such as whether or not a person is employed, the brand name of a cell phone, or the country where a college student studies abroad. When we record categorical variables, our data consist of *counts* or of *percents* obtained from counts.

In these settings, our goal is to say something about the corresponding *population proportions.* Just as in the case of inference about population means, we may be concerned with a single population or with comparing two populations. Inference about one or two proportions is very similar to inference about means, which we discussed in Chapter 7. In particular, inference for both means and proportions is based on sampling distributions that are approximately Normal.

We begin in Section 8.1 with inference about a single population proportion. Section 8.2 concerns methods for comparing two proportions.

8.1 Inference for a Single Proportion

8.2 Comparing Two Proportions

8.1 Inference for a Single Proportion

When you complete this section, you will be able to:	• Identify the sample proportion, the sample size, and the count for a single sample. Use this information to estimate the population proportion.
	• Describe the relationship between the population proportion and the sample proportion.
	• Identify the standard error for a sample proportion and the margin of error for confidence level C.
	• Apply the guidelines for when to use the large-sample confidence interval for a population proportion.
	• Find and interpret the large-sample confidence interval for a single proportion.
	• Apply the guidelines for when to use the large-sample significance test for a population proportion.
	• Use the large-sample significance test to test a null hypothesis about a population proportion.
	• Find the sample size needed for a desired margin of error.
	• Find the sample size needed for a significance test.

LOOK BACK

simple random sample, p. 191

sampling distribution of a count, p. 314

We want to estimate the proportion p of some characteristic in a large population. For example, we may want to know the proportion of likely voters who approve of the president's conduct in office. We select a simple random sample (SRS) of size n from the population and record the count X of "successes" (such as Yes answers to a question about the president). A "success" response represents the characteristic of interest in this example.

In statistical terms, we are concerned with inference about the probability p of a success in the binomial setting. The sample proportion of successes $\hat{p} = X/n$ estimates the unknown population proportion p. If the population is much larger than the sample (at least 20 times as large), the count X has approximately the binomial distribution $B(n, p)$.[1]

EXAMPLE 8.1

ROBOT

Robotics and jobs. A Pew survey asked a panel of experts whether or not they thought that networked, automated, artificial intelligence (AI), and robotic devices will have displaced more jobs than they have created (net jobs) by 2025.[2]

The sample size is the number of experts who responded to the Pew survey question, $n = 1896$. The report on the survey tells us that 48% of the respondents said they "believe net jobs will decrease by 2025 due to networked, automated, artificial intelligence (AI), and robotic devices." Thus, the sample proportion is $\hat{p} = 0.48$. We can calculate the count X from the information given; it is the sample size times the proportion responding Yes, $X = n\hat{p} = 1896(0.48) = 910$.

8.1 Smartphones and purchases. A Google research study asked 5013 smartphone users about how they used their phones. In response to a question about purchases, 2657 reported that they purchased an item after using their smartphone to search for information about the item.[3]

(a) What is the sample size n for this survey?

(b) In this setting, describe the population proportion p in a short sentence.

(c) What is the value of the count X? Describe the count in a short sentence.

(d) Find the sample proportion $\hat{p}$.

8.2 Coca-Cola and demographics. A Pew survey interviewed 162 CEOs from U.S. companies. The report of the survey quotes Muhtar Kent, Coca-Cola Company chairman and CEO, on the importance of demographics in developing customer strategies. Kent notes that the population of the United States is aging and that there is a need to provide products that appeal to this segment of the market. The survey found that 52% of the CEOs in the sample are planning to change their customer growth and retention strategies.

(a) How many CEOs participated in the survey? What is the sample size n for the survey?

(b) What is the count X of those who said that they are planning to change their customer growth and retention strategies?

(c) Find the sample proportion $\hat{p}$.

(d) The quotes from Muhtar Kent in the report could be viewed as anecdotal data. Do you think that these quotes are useful to explain and interpret the results of the survey? Write a short paragraph discussing your answer.

anecdotal
data,
p. 164

Normal
approximations
for counts and
proportions,
p. 322

If the sample size n is very small, we must base tests and confidence intervals for p on the binomial distributions. These are awkward to work with because of the discreteness of the binomial distributions.[4] But we know that when the these counts are large, both the count X and the sample proportion $\hat{p}$ are approximately Normal. We will consider only inference procedures based on the Normal approximation. These procedures are similar to those for inference about the mean of a Normal distribution.

Large-sample confidence interval for a single proportion

The unknown population proportion p is estimated by the sample proportion $\hat{p} = X/n$. If the sample size n is sufficiently large, the sampling distribution of $\hat{p}$ is approximately Normal, with mean $\mu_{\hat{p}} = p$ and standard deviation $\sigma_{\hat{p}} = \sqrt{p(1-p)/n}$. This means that approximately 95% of the time $\hat{p}$ will be within $2\sqrt{p(1-p)/n}$ of the unknown population proportion p.

Note that the standard deviation $\sigma_{\hat{p}}$ depends upon the unknown parameter p. To estimate this standard deviation using the data, we replace p in the formula by the sample proportion $\hat{p}$. As we did in Chapter 7, we use the term *standard error* for the standard deviation of a statistic that is estimated from data. Here is a summary of the procedure.

LOOK BACK

standard
error,
p. 408

LARGE-SAMPLE CONFIDENCE INTERVAL FOR A POPULATION PROPORTION

Choose an SRS of size n from a large population with an unknown proportion p of successes. The **sample proportion** is

$$\hat{p} = \frac{X}{n}$$

where X is the number of successes. The **standard error of $\hat{p}$** is

$$SE_{\hat{p}} = \sqrt{\frac{\hat{p}(1 - \hat{p})}{n}}$$

and the **margin of error** for confidence level C is

$$m = z^*SE_{\hat{p}}$$

where the critical value z^* is the value for the standard Normal density curve with area C between $-z^*$ and z^*.

An **approximate level C confidence interval** for p is

$$\hat{p} \pm m$$

Use this interval for 90% ($z^* = 1.645$), 95% ($z^* = 1.96$), or 99% ($z^* = 2.576$) confidence when the number of successes and the number of failures are both at least 10.

Table D includes a line at the bottom with values of z^* for selected values of C. Use Table A for other values of C.

EXAMPLE 8.2

Inference for robotics and jobs. The sample survey in Example 8.1 found that 910 of a sample of 1896 experts reported that they think net jobs will decrease by 2025 because of robots and related technology developments. Thus, the sample size is $n = 1896$ and the count is $X = 910$. The sample proportion is

$$\hat{p} = \frac{X}{n} = \frac{910}{1896} = 0.47996$$

The standard error is

$$SE_{\hat{p}} = \sqrt{\frac{\hat{p}(1 - \hat{p})}{n}} = \sqrt{\frac{0.47996(1 - 0.47996)}{1896}} = 0.011474$$

The z critical value for 95% confidence is $z^* = 1.96$, so the margin of error is

$$m = 1.96SE_{\hat{p}} = (1.96)(0.011474) = 0.022489$$

The confidence interval is

$$\hat{p} \pm m = 0.480 \pm 0.022$$

We are 95% confident that between 45.8% and 50.2% of CEOs would report that they think net jobs will decrease by 2025 because of robots and related technology developments.

In performing these calculations, we have kept a large number of digits for our intermediate calculations. However, when reporting the results, we prefer to use rounded values. For example, "48.0% with a margin of error of 2.2%." *You should always focus on what is important. Reporting extra digits that are not needed can divert attention from the main point of your summary.* There is no additional information to be gained by reporting $\hat{p} = 0.47996$ with a margin of error of 0.022489. Do you think it would be better to report 48% with a 2% margin of error?

Remember that the margin of error in any confidence interval includes only random sampling error. If people do not respond honestly to the questions asked, for example, your estimate is likely to miss by more than the margin of error. Likewise, if the response rate is low, your estimate and standard error may be biased.

Although the calculations for statistical inference for a single proportion are relatively straightforward and can be done with a calculator or in a spreadsheet, we prefer to use software.

EXAMPLE 8.3

Robotics and jobs confidence interval using software. Figure 8.1 shows a spreadsheet for the robotics and jobs example that could be used as input for statistical software. Note that there are 1896 experts who expressed an opinion in this example. The sheet specifies a value for each of these 1896 cases: there are 910 cases with the value Yes and 986 cases with the value No. An alternative spreadsheet would not summarize the responses but rather would list all 1896 cases with the response for each case.

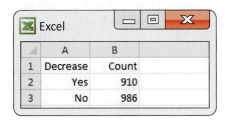

FIGURE 8.1 The robotics and jobs data in an Excel spreadsheet for the confidence interval, Example 8.3.

Figure 8.2 gives output from JMP and Minitab for these data. There are differences in the displays, but it is easy to find the 95% confidence interval. For JMP, the confidence interval is on the line with "Level" equal to Yes under the headings "Lower CI" and "Upper CI." Minitab gives the output

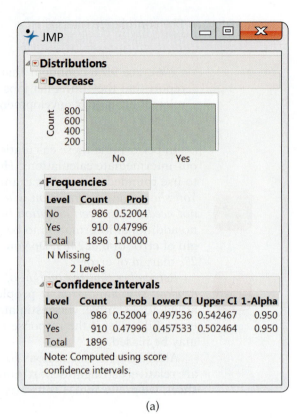

(a)

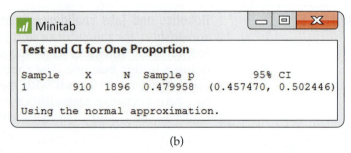

(b)

FIGURE 8.2 (a) JMP and (b) Minitab output for the robotics and jobs survey, Example 8.3.

in the form of an interval under the heading "95% CI." Notice that the confidence intervals are similar but not identical. Minitab notes that the Normal approximation is used. This is the large-sample method that we described. JMP notes that an alternative method, using score functions, is used.

As usual, the output reports more digits than are useful. *When you use software, be sure to think about how many digits are meaningful for your purposes. Do not clutter your report with information that is not meaningful.*

We recommend the large-sample confidence interval for 90%, 95%, and 99% confidence whenever the number of successes and the number of failures are both at least 10. For smaller sample sizes, we recommend exact methods that use the binomial distribution. These, as well as other alternative procedures, such as the score function, are available as the default or as options in many statistical software packages. We do not cover them here. There is also an intermediate case between large samples and very small samples where a slight modification of the large-sample approach works quite well. This method is called the "plus four" procedure and is described next.

USE YOUR KNOWLEDGE

8.3 Smartphones and purchases. Refer to Exercise 8.1 (page 485).

(a) Find $SE_{\hat{p}}$, the standard error of $\hat{p}$.

(b) Give the 95% confidence interval for p in the form of estimate plus or minus the margin of error.

(c) Give the confidence interval as an interval of percents.

(d) State your conclusion and interpret the meaning of the confidence interval in part (c).

8.4 Coca-Cola and demographics. Refer to Exercise 8.2 (page 485).

(a) Find $SE_{\hat{p}}$, the standard error of $\hat{p}$.

(b) Give the 95% confidence interval for p in the form of estimate plus or minus the margin of error.

(c) Give the confidence interval as an interval of percents.

(d) State your conclusion and interpret the meaning of the confidence interval in part (c).

BEYOND THE BASICS

The Plus Four Confidence Interval for a Single Proportion

Computer studies reveal that confidence intervals based on the large-sample approach can be quite inaccurate when the number of successes and the number of failures are not at least 10. When this occurs, a simple adjustment to the confidence interval works very well in practice. The adjustment is based on assuming that the sample contains four additional observations, two of which are successes and two of which are failures. The estimator of the population proportion based on this *plus four* rule is

$$\tilde{p} = \frac{X + 2}{n + 4}$$

plus four estimate

This estimate was first suggested by Edwin Bidwell Wilson in 1927, and it is sometimes called the Agresti-Coull interval.[5] We call it the **plus four estimate**. The confidence interval is based on the z statistic obtained by standardizing the plus four estimate $\tilde{p}$. Because $\tilde{p}$ is the sample proportion for our modified sample of size $n + 4$, it isn't surprising that the distribution of $\tilde{p}$ is close to the Normal distribution with mean p and standard deviation $\sqrt{p(1 - p)/(n + 4)}$. To get a confidence interval, we estimate p by $\tilde{p}$ in this standard deviation to get the standard error of $\tilde{p}$. Here is an example.

EXAMPLE 8.4

Percent of equol producers. Research has shown that there are many health benefits associated with a diet that contains soy foods. Substances in soy called isoflavones are known to be responsible for these benefits. When soy foods are consumed, some subjects produce a chemical called equol, and it is thought that production of equol is a key factor in the health benefits of a soy diet. Unfortunately, not all people are equol producers; there appear to be two distinct subpopulations: equol producers and equol nonproducers.

A nutrition researcher planning some bone health experiments would like to include some equol producers and some nonproducers among her subjects. A preliminary sample of 12 female subjects were measured, and four were found to be equol producers. We would like to estimate the proportion of equol producers in the population from which this researcher will draw her subjects.

The plus four estimate of the proportion of equol producers is

$$\tilde{p} = \frac{4 + 2}{12 + 4} = \frac{6}{16} = 0.375$$

For a 95% confidence interval, we use Table D to find $z^* = 1.96$. We first compute the standard error

$$SE_{\tilde{p}} = \sqrt{\frac{\tilde{p}(1 - \tilde{p})}{n + 4}}$$

$$= \sqrt{\frac{(0.375)(1 - 0.375)}{16}}$$

$$= 0.12103$$

and then the margin of error

$$m = z^* SE_{\tilde{p}}$$

$$= (1.96)(0.12103)$$

$$= 0.237$$

So the confidence interval is

$$\tilde{p} \pm m = 0.375 \pm 0.237$$

$$= (0.138, 0.612)$$

We estimate with 95% confidence that between 14% and 61% of women from this population are equol producers. Note that the interval is very wide because the sample size is very small. Compare this result with the large-sample confidence interval.

If the true proportion of equol users is near 14%, the lower limit of this interval, there may not be a sufficient number of equol producers in the study if subjects are tested only after they are enrolled in the experiment. It may be necessary to determine whether or not a potential subject is an equol producer. The study could then be designed to have the same number of equol producers and nonproducers.

Significance test for a single proportion

LOOK BACK

Normal
approximation
for proportions,
p. 322

Recall that the sample proportion $\hat{p} = X/n$ is approximately Normal, with mean $\mu_{\hat{p}} = p$ and standard deviation $\sigma_{\hat{p}} = \sqrt{p(1-p)/n}$. For confidence intervals, we substitute $\hat{p}$ for p in the last expression to obtain the standard error. When performing a significance test, however, the null hypothesis specifies a value for p, and we assume that this is the true value when calculating the P-value. Therefore, when we test $H_0: p = p_0$, we substitute p_0 into the expression for $\sigma_{\hat{p}}$ and then standardize $\hat{p}$. Here are the details.

LARGE-SAMPLE SIGNIFICANCE TEST FOR A POPULATION PROPORTION

Draw an SRS of size n from a large population with an unknown proportion p of successes. To test the hypothesis $H_0: p = p_0$, compute the **z statistic**

$$z = \frac{\hat{p} - p_0}{\sqrt{\dfrac{p_0(1 - p_0)}{n}}}$$

In terms of a standard Normal random variable Z, the approximate P-value for a test of H_0 against

$H_a: p > p_0$ is $P(Z \geq z)$

$H_a: p < p_0$ is $P(Z \leq z)$

$H_a: p \neq p_0$ is $2P(Z \geq |z|)$

We recommend the large-sample z significance test as long as the expected number of successes, np_0, and the expected number of failures, $n(1 - p_0)$, are both at least 10.

LOOK BACK

sign test for
matched pairs,
p. 473

If the numbers of successes and failures are not both at least 10, or if the population is less than 20 times as large as the sample, other procedures should be used. One such approach is to use the binomial distribution as we did with the sign test. Here is a large-sample example.

EXAMPLE 8.5

SUNBL

Comparing two sunblock lotions. Your company produces a sunblock lotion designed to protect the skin from both UVA and UVB exposure to the sun. You hire a company to compare your product with the product sold by your major competitor. The testing company exposes skin on the backs of a sample of 20 people to UVA and UVB rays and measures the protection provided by each product. For 13 of the subjects, your

Fancy/Alamy

product provided better protection, while for the other 7 subjects, your competitor's product provided better protection. Do you have evidence to support a commercial claiming that your product provides superior UVA and UVB protection? For the data we have $n = 20$ subjects and $X = 13$ successes. The parameter p is the proportion of people who would receive superior UVA and UVB protection from your product. To answer the claim question, we test

$$H_0: p = 0.5$$
$$H_a: p \neq 0.5$$

The expected numbers of successes (your product provides better protection) and failures (your competitor's product provides better protection) are $20 \times 0.5 = 10$ and $20 \times 0.5 = 10$. Both are at least 10, so we can use the z test. The sample proportion is

$$\hat{p} = \frac{X}{n} = \frac{13}{20} = 0.65$$

The test statistic is

$$z = \frac{\hat{p} - p_0}{\sqrt{\dfrac{p_0(1 - p_0)}{n}}} = \frac{0.65 - 0.5}{\sqrt{\dfrac{(0.5)(0.5)}{20}}} = 1.34$$

From Table A, we find $P(Z < 1.34) = 0.9099$, so the probability in the upper tail is $1 - 0.9099 = 0.0901$. The P-value is the area in both tails, $P = 2 \times 0.0901 = 0.1802$.

We conclude that the sunblock testing data are compatible with the hypothesis of no difference between your product and your competitor's product ($\hat{p} = 0.65$, $z = 1.34$, $P = 0.18$). The data do not support your proposed advertising claim.

Note that we have used the two-sided alternative for this example. In settings like this, we must start with the view that either product could be better if we want to prove a claim of superiority. Thinking or hoping that your product is superior cannot be used to justify a one-sided test.

Although these calculations are not particularly difficult to do using a calculator, we prefer to use software. Here are some details.

EXAMPLE 8.6

SUNBL

Sunblock significance tests using software. JMP and Minitab outputs for the analysis in Example 8.5 appear in Figure 8.3. First, JMP uses a slightly different way of reporting the results. Two ways of performing the significance test are labeled in the column "Test." The one that corresponds to the procedure that we have described is on the second line, labeled "Pearson." The P-value under the heading "Prob>Chisq" is 0.1797, which is very close to the 0.1802 that we calculated using Table A. Minitab reports the value of the test statistic z, and the P-value is rounded to 0.180.

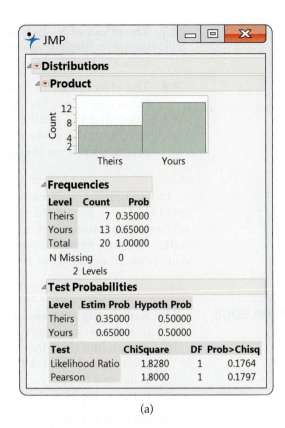

(a)

(b)

FIGURE 8.3 (a) JMP and (b) Minitab output for comparing sunblock lotions, Example 8.5.

USE YOUR KNOWLEDGE

8.5 Draw a picture. Draw a picture of a standard Normal curve and shade the tail areas to illustrate the calculation of the *P*-value for Example 8.5.

8.6 What does the confidence interval tell us? Inspect the outputs in Figure 8.3. Report the confidence interval for the percent of people who would get better sun protection from your product than from your competitor's. Be sure to convert from proportions to percents and to round appropriately. Interpret the confidence interval and compare this way of analyzing data with the significance test.

8.7 The effect of X. In Example 8.5 (page 491), suppose that your product provided better UVA and UVB protection for 15 of the 20 subjects. Perform the significance test and summarize the results.

8.8 The effect of n. In Example 8.5 (page 491), consider what would have happened if you had paid for twice as many subjects to be tested. Assume that the results would be similar to those in Example 8.5, that is, 65% of the subjects had better UVA and UVB protection with your product. Perform the significance test and summarize the results.

In Example 8.5, we treated an outcome as a success whenever your product provided better sun protection. Would we get the same results if we defined success as an outcome where your competitor's product was superior? In this setting, the null hypothesis is still H_0: $p = 0.5$. You will find that the z test statistic is unchanged except for its sign and that the P-value remains the same.

USE YOUR KNOWLEDGE

8.9 Redefining success. In Example 8.5 (page 491), we performed a significance test to compare your product with your competitor's. Success was defined as the outcome where your product provided better protection. Now, take the viewpoint of your competitor where success is defined to be the outcome where your competitor's product provides better protection. In other words, n remains the same, but X is now 7.

(a) Perform the two-sided significance test and report the results. How do these compare with what we found in Example 8.5?

(b) Find the 95% confidence interval for this setting, and compare it with the interval calculated when success is defined as the outcome where your product provides better protection.

We do not often use significance tests for a single proportion because it is uncommon to have a situation where there is a precise p_0 that we want to test. For physical experiments such as coin tossing or drawing cards from a well-shuffled deck, probability arguments lead to an ideal p_0. Even here, however, it can be argued, for example, that no real coin has a probability of heads *exactly* equal to 0.5. Data from past large samples can sometimes provide a p_0 for the null hypothesis of a significance test. In some types of epidemiology research, for example, "historical controls" from past studies serve as the benchmark for evaluating new treatments. Medical researchers argue about the validity of these approaches, because the past never quite resembles the present. In general, we prefer comparative studies whenever possible. The matched pairs study of Example 8.5 is an example of a comparative study that involved a single proportion.

Choosing a sample size for a confidence interval

choosing
sample size,
p. 353

In Chapter 6, we showed how to choose the sample size n to obtain a confidence interval with specified margin of error m for a Normal mean. We also discussed the effect of the sample size on the power of a significance test for a Normal mean. Because we are using a Normal approximation for inference about a population proportion, sample size selection proceeds in much the same way.

Recall that the margin of error for the large-sample confidence interval for a population proportion is

$$m = z^* \text{SE}_{\hat{p}} = z^* \sqrt{\frac{\hat{p}(1 - \hat{p})}{n}}$$

Choosing a confidence level C fixes the critical value z^*. The margin of error also depends on the value of $\hat{p}$ and the sample size n. Because we don't know the value of $\hat{p}$ until we gather the data, we must guess a value to use in the calculations. We will call the guessed value p^*. There are two common ways to get p^*:

1. Use the sample estimate from a pilot study or from similar studies done earlier.

2. Use $p^* = 0.5$. Because the margin of error is largest when $\hat{p} = 0.5$, this choice gives a sample size that is somewhat larger than we really need for the confidence level we choose. It is a safe choice no matter what the data later show.

Once we have chosen p^* and the margin of error m that we want, we can find the n we need to achieve this margin of error. Here is the result.

SAMPLE SIZE FOR DESIRED MARGIN OF ERROR

The level C confidence interval for a proportion p will have a margin of error approximately equal to a specified value m when the sample size satisfies

$$n = \left(\frac{z^*}{m}\right)^2 p^*(1 - p^*)$$

Here, z^* is the critical value for confidence level C, and p^* is a guessed value for the proportion of successes in the future sample.

The margin of error will be less than or equal to m if p^* is chosen to be 0.5. Substituting $p^* = 0.5$ into the formula above gives

$$n = \frac{1}{4}\left(\frac{z^*}{m}\right)^2$$

The value of n obtained by this method is not particularly sensitive to the choice of p^* when p^* is fairly close to 0.5. However, if the value of p is likely to be smaller than about 0.3 or larger than about 0.7, use of $p^* = 0.5$ may result in a sample size that is much larger than needed.

EXAMPLE 8.7

Planning a survey of students. A large university is interested in assessing student satisfaction with the overall campus environment. The plan is to distribute a questionnaire to an SRS of students, but before proceeding, the university wants to determine how many students to sample. The questionnaire asks about a student's degree of satisfaction with various student services, each measured on a five-point scale. The university is interested in the proportion p of students who are satisfied (that is, who choose either "satisfied" or "very satisfied," the two highest levels on the five-point scale).

The university wants to estimate p with 95% confidence and a margin of error less than or equal to 3%, or 0.03. For planning purposes, it is willing to use $p^* = 0.5$. To find the sample size required,

$$n = \frac{1}{4}\left(\frac{z^*}{m}\right)^2 = \frac{1}{4}\left(\frac{1.96}{0.03}\right)^2 = 1067.1$$

Round up to get $n = 1068$. (Always round up. Rounding down would give a margin of error slightly greater than 0.03.)

Similarly, for a 2.5% margin of error, we have (after rounding up)

$$n = \frac{1}{4}\left(\frac{1.96}{0.025}\right)^2 = 1537$$

and for a 2% margin of error,

$$n = \frac{1}{4}\left(\frac{1.96}{0.02}\right)^2 = 2401$$

News reports frequently describe the results of surveys with sample sizes between 1000 and 1500 and a margin of error of about 3%. These surveys generally use sampling procedures more complicated than simple random sampling, so the calculation of confidence intervals is more involved than what we have studied in this section. The calculations in Example 8.7 show in principle how such surveys are planned.

In practice, many factors influence the choice of a sample size. The following example illustrates one set of factors.

EXAMPLE 8.8

iStockphoto

Assessing interest in Pilates classes. The Division of Recreational Sports (Rec Sports) at a major university is responsible for offering comprehensive recreational programs, services, and facilities to the students. Rec Sports is continually examining its programs to determine how well it is meeting the needs of the students. Rec Sports is considering adding some new programs and would like to know how much interest there is in a new exercise program based on the Pilates method.[6] It will take a survey of undergraduate students. In the past, Rec Sports emailed short surveys to all undergraduate students. The response rate obtained in this way was about 5%. This time, it will send emails to a simple random sample of the students and will follow up with additional emails and eventually a phone call to get a higher response rate. Because of limited staff and the work involved with the follow-up, it would like to use a sample size of about 200 responses. It assumes that the new procedures will improve the response rate to 90%, so it will contact 225 students in the hope that these will provide at least 200 valid responses. One of the questions it will ask is, "Have you ever heard about the Pilates method of exercise?"

The primary purpose of the survey is to estimate various sample proportions for undergraduate students. Will the proposed sample size of $n = 200$ be adequate to provide Rec Sports with the needed information? To address this question, we calculate the margins of error of 95% confidence intervals for various values of $\hat{p}$.

EXAMPLE 8.9

Margins of error. In the Rec Sports survey, the margin of error of a 95% confidence interval for any value of $\hat{p}$ and $n = 200$ is

$$m = z^*\text{SE}_{\hat{p}}$$

$$= 1.96\sqrt{\frac{\hat{p}(1 - \hat{p})}{200}}$$

$$= 0.139\sqrt{\hat{p}(1 - \hat{p})}$$

The results for various values of $\hat{p}$ are

$\hat{p}$	m	$\hat{p}$	m
0.05	0.030	0.60	0.068
0.10	0.042	0.70	0.064
0.20	0.056	0.80	0.056
0.30	0.064	0.90	0.042
0.40	0.068	0.95	0.030
0.50	0.070		

Rec Sports judged these margins of error to be acceptable, and it contacted 225 students, hoping to achieve a sample size of 200 for its survey.

The table in Example 8.9 illustrates two points. First, the margins of error for *phat and 1-phat* are the same. Second, the margin of error varies between only 0.064 and 0.070 as $\hat{p}$ varies from 0.3 to 0.7, and the margin of error is greatest when $\hat{p} = 0.5$, as we claimed earlier (page 495). It is true in general that the margin of error will vary relatively little for values of $\hat{p}$ between 0.3 and 0.7. Therefore, when planning a study, it is not necessary to have a very precise guess for p. If $p^* = 0.5$ is used and the observed $\hat{p}$ is between 0.3 and 0.7, the actual interval will be a little shorter than needed, but the difference will be small.

Again it is important to emphasize that these calculations consider only the effects of sampling variability that are quantified in the margin of error. Other sources of error, such as nonresponse and possible misinterpretation of questions, are not included in the table of margins of error for Example 8.9. Rec Sports is trying to minimize these kinds of errors. It performed a pilot study using a small group of current users of its facilities to check the wording of the questions, and for the final survey it devised a careful plan to follow up with the students who did not respond to the initial email.

8.10 Confidence level and sample size. Refer to Example 8.7 (page 495). Suppose that the university was interested in a 95% confidence interval with margin of error 0.04. Would the required sample size be smaller or larger than 1068 students? Verify this by performing the calculation.

8.11 Make a plot. Use the values for $\hat{p}$ and m given in Example 8.9 to draw a plot of the sample proportion versus the margin of error. Summarize the major features of your plot.

Choosing a sample size for a significance test

LOOK BACK

power, p. 391

In Chapter 6, we also introduced the idea of power for a significance test. These ideas apply to the significance test for a proportion that we studied in this section. There are some more complicated details, but the basic ideas are the same. Fortunately, software can take care of the details, and we can concentrate on the input and output.

To find the required sample size, we need to specify

- The value of p_0 in the null hypothesis $H_0: p = p_0$.

- The alternative hypothesis, two-sided ($H_a: p \neq p_0$), one-sided ($H_a: p > p_0$ or $H_a: p < p_0$).

- A value of p for the alternative hypothesis.

- The Type I error (α, the probability of rejecting the null hypothesis when it is true); usually we choose 5% ($\alpha = 0.05$) for the Type I error.

- Power (probability of rejecting the null hypothesis when it is false); usually we choose 80% (0.80) for power.

EXAMPLE 8.10

Sample size for comparing two sunblock lotions. In Example 8.5 (page 491), we performed the significance test for comparing two sunblock lotions in a setting where each subject used the two lotions and the product that provided better protection was recorded. Although your product performed better 13 times in 20 trials, the value of $\hat{p} = 13/20 = 0.65$ was not sufficiently far from the null hypothesized value of $p_0 = 0.5$ for us to reject the H_0, ($p = 0.18$). Let's suppose that the true percent of the time that your lotion would perform better is $p_0 = 0.65$, and we plan to test the null hypothesis $H_0: p = 0.5$ versus the two-sided alternative $H_a: p \neq 0.5$ using a Type I error probability of 0.05.

What sample size n should we choose if we want to have an 80% chance of rejecting H_0? Outputs from JMP and Minitab are given in Figure 8.4. JMP indicates that $n = 89$ should be used, while Minitab suggests $n = 85$. The difference is due to the different methods that can be used for these calculations.

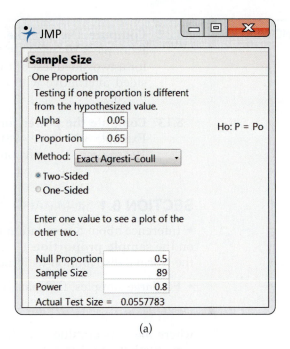

(a)

(b)

FIGURE 8.4 (a) JMP and (b) Minitab output for sample size needed to compare sunblock lotions, Example 8.10.

Note that Minitab provides a graph as a function of the value of the proportion for the alternative hypothesis. Similar plots can be produced by JMP. In some situations, you might want to specify the sample size n and have software compute the power. This option is available in JMP, Minitab, and other software.

8.12 Compute the sample size for a different alternative. Refer to Example 8.10 (page 498). Use software to find the sample size needed for a two-sided test of the null hypothesis that $p = 0.5$ versus the two-sided alternative with $\alpha = 0.05$ and 80% power if the alternative is $p = 0.7$.

8.13 Compute the power for a given sample size. Consider the setting in Example 8.10 (page 498). You have a budget that will allow you to test 100 subjects. Use software to find the power of the test for this value of n.

SECTION 8.1 SUMMARY

- Inference about a population proportion p from an SRS of size n is based on the **sample proportion** $\hat{p} = X/n$. When n is large, $\hat{p}$ has approximately the Normal distribution with mean p and standard deviation $\sqrt{p(1-p)/n}$.

- For large samples, the **margin of error for confidence level C** is

$$m = z^*\mathrm{SE}_{\hat{p}}$$

where the critical value z^* is the value for the standard Normal density curve with area C between $-z^*$ and z^*, and the **standard error of $\hat{p}$** is

$$\mathrm{SE}_{\hat{p}} = \sqrt{\frac{\hat{p}(1-\hat{p})}{n}}$$

- The **level C large-sample confidence interval** is

$$\hat{p} \pm m$$

We recommend using this interval for 90%, 95%, and 99% confidence whenever the number of successes and the number of failures are both at least 10. When sample sizes are smaller, alternative procedures such as the **plus four estimate of the population proportion** are recommended.

- The **sample size** required to obtain a confidence interval of approximate margin of error m for a proportion is found from

$$n = \left(\frac{z^*}{m}\right)^2 p^*(1-p^*)$$

where p^* is a guessed value for the proportion and z^* is the standard Normal critical value for the desired level of confidence. To ensure that the margin of error of the interval is less than or equal to m no matter what $\hat{p}$ may be, use

$$n = \frac{1}{4}\left(\frac{z^*}{m}\right)^2$$

- Tests of H_0: $p = p_0$ are based on the **z statistic**

$$z = \frac{\hat{p} - p_0}{\sqrt{\dfrac{p_0(1-p_0)}{n}}}$$

with P-values calculated from the $N(0, 1)$ distribution. Use this procedure when the expected number of successes, np_0, and the expected number of failures, $n(1 - p_0)$, are both greater than 10.

- Software can be used to determine the sample sizes for significance tests.

SECTION 8.1 EXERCISES

For Exercises 8.1 and 8.2, see page 485; for Exercises 8.3 and 8.4, see page 489; for Exercises 8.5 through 8.8, see pages 493–494; for Exercise 8.9, see page 494; for Exercises 8.10 and 8.11, see page 498; and for Exercises 8.12 and 8.13, see page 500.

8.14 How did you use your cell phone? A Pew Internet poll asked cell phone owners about how they used their cell phones. One question asked whether or not during the past 30 days they had used their phone while in a store to call a friend or family member for advice about a purchase they were considering. The poll surveyed 1003 adults living in the United States by telephone. Of these, 462 responded that they had used their cell phone while in a store within the last 30 days to call a friend or family member for advice about a purchase they were considering.[7]

(a) Identify the sample size and the count.

(b) Calculate the sample proportion.

(c) Explain the relationship between the population proportion and the sample proportion.

8.15 Do you eat breakfast? A random sample of 300 students from your college is asked if they regularly eat breakfast. One hundred and nine students responded that they did eat breakfast regularly.

(a) Identify the sample size and the count.

(b) Calculate the sample proportion.

(c) Explain the relationship between the population proportion and the sample proportion.

8.16 Would you recommend the service to a friend? An automobile dealership asks all its customers who used its service department in a given two-week period if they would recommend the service to a friend. A total of 250 customers used the service during the two-week period, and 210 said that they would recommend the service to a friend.

(a) Identify the sample size and the count.

(b) Calculate the sample proportion.

(c) Explain the relationship between the population proportion and the sample proportion.

8.17 How did you use your cell phone? Refer to Exercise 8.14.

(a) Report the sample proportion, the standard error of the sample proportion, and the margin of error for 95% confidence.

(b) Are the guidelines for when to use the large-sample confidence interval for a population proportion satisfied in this setting? Explain your answer.

(c) Find the 95% large-sample confidence interval for the population proportion.

(d) Write a short statement explaining the meaning of your confidence interval.

8.18 Do you eat breakfast? Refer to Exercise 8.15.

(a) Report the sample proportion, the standard error of the sample proportion, and the margin of error for 95% confidence.

(b) Are the guidelines for when to use the large-sample confidence interval for a population proportion satisfied in this setting? Explain your answer.

(c) Find the 95% large-sample confidence interval for the population proportion.

(d) Write a short statement explaining the meaning of your confidence interval.

8.19 Would you recommend the service to a friend? Refer to Exercise 8.16.

(a) Report the sample proportion, the standard error of the sample proportion, and the margin of error for 95% confidence.

(b) Are the guidelines for when to use the large-sample confidence interval for a population proportion satisfied in this setting? Explain your answer.

(c) Find the 95% large-sample confidence interval for the population proportion.

(d) Write a short statement explaining the meaning of you confidence interval.

8.20 Whole grain versus regular grain? A study of young children was designed to increase their intake of whole-grain, rather than regular-grain, snacks. At the end of the study, the 86 children who participated in the study were presented with a choice between a regular-grain snack and a whole-grain alternative. The whole-grain alternative was chosen by 48 children. You want to examine the possibility that the children are equally likely to choose each type of snack.

(a) Formulate the null and alternative hypotheses for this setting.

(b) Are the guidelines for using the large-sample significance test satisfied for testing this null hypothesis? Explain your answer.

(c) Perform the significance test and summarize your results in a short paragraph.

8.21 Find the sample size. You are planning a survey similar to the one about cell phone use described in Exercise 8.14. You will report your results with a large-sample confidence interval. How large a sample do you need to be sure that the margin of error will not be greater than 0.05? Show your work.

8.22 What's wrong? Explain what is wrong with each of the following:

(a) An approximate 90% confidence interval for an unknown proportion p is $\hat{p}$ plus or minus its standard error.

(b) You can use a significance test to evaluate the hypothesis H_0: $\hat{p} = 0.3$ versus the one-sided alternative.

(c) The large-sample significance test for a population proportion is based on a t statistic.

8.23 What's wrong? Explain what is wrong with each of the following:

(a) A student project used a confidence interval to describe the results in a final report. The confidence level was 115%.

(b) The margin of error for a confidence interval used for an opinion poll takes into account the fact that people who did not answer the poll questions may have had different responses from those who did answer the questions.

(c) If the P-value for a significance test is 0.50, we can conclude that the null hypothesis has a 50% chance of being true.

8.24 Draw some pictures. Consider the binomial setting with $n = 800$ and $p = 0.3$.

(a) The sample proportion $\hat{p}$ will have a distribution that is approximately Normal. Give the mean and the standard deviation of this Normal distribution.

(b) Draw a sketch of this Normal distribution. Mark the location of the mean.

(c) Find a value p^* for which the probability is 95% that $\hat{p}$ will be between $\pm p^*$. Mark these two values on your sketch.

8.25 Country food and Inuits. Country food includes seals, caribou, whales, ducks, fish, and berries and is an important part of the diet of the aboriginal people called Inuits who inhabit Inuit Nunangat, the northern region of what is now called Canada. A survey of Inuits in Inuit Nunangat reported that 3274 out of 5000 respondents said that at least half of the meat and fish that they eat is country food.[8] Find the sample proportion and a 95% confidence interval for the population proportion of Inuits whose meat and fish consumption consists of at least half country food.

8.26 Soft drink consumption in New Zealand. A survey commissioned by the Southern Cross Healthcare Group reported that 16% of New Zealanders consume five or more servings of soft drinks per week. The data were obtained by an online survey of 2006 randomly selected New Zealanders over 15 years of age.[9]

(a) What number of survey respondents reported that they consume five or more servings of soft drinks per week? You will need to round your answer. Why?

(b) Find a 95% confidence interval for the proportion of New Zealanders who report that they consume five or more servings of soft drinks per week.

(c) Convert the estimate and your confidence interval to percents.

(d) Discuss reasons why the estimate might be biased.

8.27 Violent video games. A survey of 1050 parents who have a child under the age of 18 living at home asked about their opinions regarding violent video games. A report describing the results of the survey stated that 89% of parents say that violence in today's video games is a problem.[10]

(a) What number of survey respondents reported that they thought that violence in today's video games is a problem? You will need to round your answer. Why?

(b) Find a 95% confidence interval for the proportion of parents who think that violence in today's video games is a problem.

(c) Convert the estimate and your confidence interval to percents.

(d) Discuss reasons why the estimate might be biased.

8.28 Bullying. Refer to the previous exercise. The survey also reported that 93% of the parents surveyed said that bullying contributes to violence in the United States. Answer the questions in the previous exercise for this item on the survey.

8.29 $\hat{p}$ and the Normal distribution. Consider the binomial setting with $n = 40$. You are testing the null hypothesis that $p = 0.4$ versus the two-sided alternative with a 5% chance of rejecting the null hypothesis when it is true.

(a) Find the values of the sample proportion $\hat{p}$ that will lead to rejection of the null hypothesis.

(b) Repeat part (a) assuming a sample size of $n = 80$.

(c) Make a sketch illustrating what you have found in parts (a) and (b). What does your sketch show about the effect of the sample size in this setting?

8.30 Students doing community service. In a sample of 159,949 first-year college students, the National Survey of Student Engagement reported that 39% participated in community service or volunteer work.[11]

(a) Find the margin of error for 99% confidence.

(b) Here are some facts from the report that summarizes the survey. The students were from 617 four-year colleges and universities. The response rate was 36%. Institutions paid a participation fee of between $1800 and $7800 based on the size of their undergraduate enrollment. Discuss these facts as possible sources of error in this study. How do you think these errors would compare with the margin of error that you calculated in part (a)?

8.31 Plans to study abroad. The survey described in the previous exercise also asked about items related to academics. In response to one of these questions, 42% of first-year students reported that they plan to study abroad.

(a) Based on the information available, how many students plan to study abroad?

(b) Give a 99% confidence interval for the population proportion of first-year college students who plan to study abroad.

8.32 Student credit cards. In a survey of 1430 undergraduate students, 1087 reported that they had one or more credit cards.[12] Give a 95% confidence interval for the proportion of all college students who have at least one credit card.

8.33 How many credit cards? The summary of the survey described in the previous exercise reported

that 43% of undergraduates had four or more credit cards. Give a 95% confidence interval for the proportion of all college students who have four or more credit cards.

8.34 How would the confidence interval change? Refer to Exercise 8.33.

(a) Would a 80% confidence interval be wider or narrower than the one that you found in Exercise 8.33? Verify your answer by computing the interval.

(b) Would a 98% confidence interval be wider or narrower than the one that you found in that exercise? Verify your results by computing the interval.

8.35 Do students report Internet sources? The National Survey of Student Engagement found that 87% of students report that their peers at least "sometimes" copy information from the Internet in their papers without reporting the source.[13] Assume that the sample size is 430,000.

(a) Find the margin of error for 99% confidence.

(b) Here are some items from the report that summarizes the survey. More than 430,000 students from 730 four-year colleges and universities participated. The average response rate was 43% and ranged from 15% to 89%. Institutions pay a participation fee of between $3000 and $7500 based on the size of their undergraduate enrollment. Discuss these facts as possible sources of error in this study. How do you think these errors would compare with the error that you calculated in part (a)?

8.36 Can we use the z test? In each of the following cases, state whether or not the Normal approximation to the binomial should be used for a significance test on the population proportion p. Explain your answers.

(a) $n = 30$ and H_0: $p = 0.3$.

(b) $n = 60$ and H_0: $p = 0.2$.

(c) $n = 100$ and H_0: $p = 0.12$.

(d) $n = 150$ and H_0: $p = 0.04$.

8.37 Long sermons. The National Congregations Study collected data in a one-hour interview with a key informant—that is, a minister, priest, rabbi, or other staff person or leader.[14] One question concerned the length of the typical sermon. For this question, 390 out of 1191 congregations reported that the typical sermon lasted more than 30 minutes.

(a) Use the large-sample inference procedures to estimate the true proportion for this question with a 95% confidence interval.

(b) The respondents to this question were not asked to use a stopwatch to record the lengths of a random sample of sermons at their congregations. They responded based on their impressions of the sermons. Do you think that ministers, priests, rabbis, or other staff persons or leaders might perceive sermon lengths differently from the people listening to the sermons? Discuss how your ideas would influence your interpretation of the results of this study.

8.38 Instant versus fresh-brewed coffee. A matched pairs experiment compares the taste of instant with fresh-brewed coffee. Each subject tastes two unmarked cups of coffee, one of each type, in random order and states which he or she prefers. Of the 60 subjects who participate in the study, 21 prefer the instant coffee. Let p be the probability that a randomly chosen subject prefers fresh-brewed coffee to instant coffee. (In practical terms, p is the proportion of the population who prefer fresh-brewed coffee.)

(a) Test the claim that a majority of people prefer the taste of fresh-brewed coffee. Report the large-sample z statistic and its P-value.

(b) Draw a sketch of a standard Normal curve and mark the location of your z statistic. Shade the appropriate area that corresponds to the P-value.

(c) Is your result significant at the 5% level? What is your practical conclusion?

8.39 Tossing a coin 10,000 times! The South African mathematician John Kerrich, while a prisoner of war during World War II, tossed a coin 10,000 times and obtained 5067 heads.

(a) Is this significant evidence at the 5% level that the probability that Kerrich's coin comes up heads is not 0.5? Use a sketch of the standard Normal distribution to illustrate the P-value.

(b) Use a 95% confidence interval to find the range of probabilities of heads that would not be rejected at the 5% level.

8.40 Is there interest in a new product? One of your employees has suggested that your company develop a new product. You decide to take a random sample of your customers and ask whether or not there is interest in the new product. The response is on a 1 to 5 scale with 1 indicating "definitely would not purchase"; 2, "probably would not purchase"; 3, "not sure"; 4, "probably would purchase"; and 5, "definitely would purchase." For an initial analysis, you will record the responses 1, 2, and 3 as No and 4 and 5 as Yes. What sample size would you use if you wanted the 95% margin of error to be 0.25 or less?

8.41 More information is needed. Refer to the previous exercise. Suppose that after reviewing the results of

the previous survey, you proceeded with preliminary development of the product. Now you are at the stage where you need to decide whether or not to make a major investment to produce and market it. You will use another random sample of your customers, but now you want the margin of error to be smaller. What sample size would you use if you wanted the 95% margin of error to be 0.015 or less?

8.42 Sample size needed for an evaluation. You are planning an evaluation of a semester-long alcohol awareness campaign at your college. Previous evaluations indicate that about 20% of the students surveyed will respond Yes to the question "Did the campaign alter your behavior toward alcohol consumption?" How large a sample of students should you take if you want the margin of error for 95% confidence to be about 0.07?

8.43 Sample size needed for an evaluation, continued. The evaluation in the previous exercise will also have questions that have not been asked before, so you do not have previous information about the possible value of p. Repeat the preceding calculation for the following values of p^*: 0.1, 0.2, 0.3, 0.4, 0.5, 0.6, 0.7, 0.8, and 0.9. Summarize the results in a table and graphically. What sample size will you use?

8.44 Are the customers dissatisfied? An automobile manufacturer would like to know what proportion of its customers are dissatisfied with the service received from their local dealer. The customer relations department will survey a random sample of customers and compute a 95% confidence interval for the proportion who are dissatisfied. From past studies, it believes that this proportion will be about 0.25. Find the sample size needed if the margin of error of the confidence interval is to be no more than 0.035.

8.45 Sample size for coffee. Refer to Exercise 8.38 where we analyzed data from a matched pairs study that compared preferences for instant versus fresh-brewed coffee. Suppose that you want to design a similar study. The null hypothesis is that instant and fresh-brewed are equally likely to be preferred and the alternative is two-sided. You will use $\alpha = 0.05$. What is the sample size needed to detect a preference of 60% for fresh-brewed with 0.80 probability?

8.46 Sample size for tossing a coin. Refer to Exercise 8.39 where we analyzed the 10,000 coin tosses made by John Kerrich. Suppose that you want to design a study that would test the hypothesis that a coin is fair versus the alternative that the probability of a head is 0.51. Using a two-sided test with $\alpha = 0.05$. what sample size would be needed to have 0.80 power to detect this alternative?

8.2 Comparing Two Proportions

When you complete this section, you will be able to:

- Identify the counts and sample sizes for a comparison between two proportions, compute the sample proportions, and find their difference.
- Apply the guidelines for when to use the large-sample confidence interval for a difference between two proportions.
- Apply the large-sample method to find the confidence interval for a difference between two proportions and interpret the confidence interval.
- Apply the guidelines for when to use the large-sample significance test for a difference between two proportions.
- Apply the large-sample method to perform a significance test for comparing two proportions and interpret the results of the significance test.
- Find the sample size needed for a desired margin of error for the difference in proportions.
- Find the sample size needed for a significance test for comparing two proportions.
- Calculate and interpret the relative risk.

Because comparative studies are so common, we often want to compare the proportions of two groups (such as men and women) that have some characteristic. In the previous section, we learned how to estimate a single proportion. Our problem now concerns the comparison of two proportions.

We call the two groups being compared Population 1 and Population 2 and the two population proportions of "successes" p_1 and p_2. The data consist of two independent SRSs, of size n_1 from Population 1 and size n_2 from Population 2. The proportion of successes in each sample estimates the corresponding population proportion. Here is the notation we will use in this section:

Population	Population proportion	Sample size	Count of successes	Sample proportion
1	p_1	n_1	X_1	$\hat{p}_1 = X_1/n_1$
2	p_2	n_2	X_2	$\hat{p}_2 = X_2/n_2$

To compare the two populations, we use the difference between the two sample proportions:

$$D = \hat{p}_1 - \hat{p}_2$$

When both sample sizes are sufficiently large, the sampling distribution of the difference D is approximately Normal.

Inference procedures for comparing proportions are z procedures based on the Normal approximation and on standardizing the difference D. The first

LOOK BACK

addition rule
for means,
p. 254

addition rule
for variances,
p. 258

step is to obtain the mean and standard deviation of D. By the addition rule for means, the mean of D is the difference of the means:

$$\mu_D = \mu_{\hat{p}_1} - \mu_{\hat{p}_2} = p_1 - p_2$$

That is, the difference $D = \hat{p}_1 - \hat{p}_2$ between the sample proportions is an unbiased estimator of the population difference $p_1 - p_2$. Similarly, the addition rule for variances tells us that the variance of D is the *sum* of the variances:

$$\sigma_D^2 = \sigma_{\hat{p}_1}^2 + \sigma_{\hat{p}_2}^2$$
$$= \frac{p_1(1 - p_1)}{n_1} + \frac{p_2(1 - p_2)}{n_2}$$

Therefore, when n_1 and n_2 are large, D is approximately Normal with mean $\mu_D = p_1 - p_2$ and standard deviation

$$\sigma_D = \sqrt{\frac{p_1(1 - p_1)}{n_1} + \frac{p_2(1 - p_2)}{n_2}}$$

USE YOUR KNOWLEDGE

8.47 Rules for means and variances. Suppose that $p_1 = 0.4$, $n_1 = 25$, $p_2 = 0.5$, $n_2 = 30$. Find the mean and the standard deviation of the sampling distribution of $p_1 - p_2$.

8.48 Effect of the sample sizes. Suppose that $p_1 = 0.4$, $n_1 = 100$, $p_2 = 0.5$, $n_2 = 120$.

(a) Find the mean and the standard deviation of the sampling distribution of $p_1 - p_2$.

(b) The sample sizes here are four times as large as those in the previous exercise while the population proportions are the same. Compare the results for this exercise with those that you found in the previous exercise. What is the effect of multiplying the sample sizes by 4?

8.49 Rules for means and variances. It is quite easy to verify the formulas for the mean and standard deviation of the difference D.

(a) What are the means and standard deviations of the two sample proportions $\hat{p}_1$ and $\hat{p}_2$?

(b) Use the addition rule for means of random variables: what is the mean of $D = \hat{p}_1 - \hat{p}_2$?

(c) The two samples are independent. Use the addition rule for variances of random variables: what is the variance of D?

Large-sample confidence interval for a difference in proportions

To obtain a confidence interval for $p_1 - p_2$, we once again replace the unknown parameters in the standard deviation with estimates to obtain an estimated standard deviation, or standard error. Here is the confidence interval we want.

LARGE-SAMPLE CONFIDENCE INTERVAL FOR COMPARING TWO PROPORTIONS

Choose an SRS of size n_1 from a large population having proportion p_1 of successes and an independent SRS of size n_2 from another population having proportion p_2 of successes. The estimate of the difference in the population proportions is

$$D = \hat{p}_1 - \hat{p}_2$$

The **standard error of D** is

$$SE_D = \sqrt{\frac{\hat{p}_1(1 - \hat{p}_1)}{n_1} + \frac{\hat{p}_2(1 - \hat{p}_2)}{n_2}}$$

and the **margin of error** for confidence level C is

$$m = z^* SE_D$$

where the critical value z^* is the value for the standard Normal density curve with area C between $-z^*$ and z^*. An **approximate level C confidence interval** for $p_1 - p_2$ is

$$D \pm m$$

Use this method for 90%, 95%, or 99% confidence when the number of successes and the number of failures in each sample are both 10 or more.

EXAMPLE 8.11

INSTAG

Who uses Instagram? A recent study compared the proportions of young women and men who use Instagram.[15] A total of 1069 young women and men were surveyed. These are the cases for the study. The response variable is User with values Yes and No. The explanatory variable is Sex with values "Men" and "Women." Here are the data:

Sex	n	X	$\hat{p} = X/n$
Women	537	328	0.6108
Men	532	234	0.4398
Total	1069	562	0.5257

In this table, the $\hat{p}$ column gives the sample proportions of women and men who use Instagram. The proportion for the total sample is given in the last entry in this column.

Let's find a 95% confidence interval for the difference between the proportions of women and of men who use Instagram. We first find the difference in the proportions:

$$
\begin{aligned}
D &= \hat{p}_1 - \hat{p}_2 \\
&= 0.6108 - 0.4398 \\
&= 0.1710
\end{aligned}
$$

Then we calculate the standard error of D:

$$\text{SE}_D = \sqrt{\frac{\hat{p}_1(1-\hat{p}_1)}{n_1} + \frac{\hat{p}_2(1-\hat{p}_2)}{n_2}}$$

$$= \sqrt{\frac{(0.6108)(1-0.6108)}{537} + \frac{(0.4398)(1-0.4398)}{532}}$$

$$= 0.0301$$

For 95% confidence, we have $z^* = 1.96$, so the margin of error is

$$\begin{aligned} m &= z^*\text{SE}_D \\ &= (1.96)(0.0301) \\ &= 0.0590 \end{aligned}$$

The 95% confidence interval is

$$\begin{aligned} D \pm m &= 0.1710 \pm 0.0590 \\ &= (0.112, 0.230) \end{aligned}$$

With 95% confidence, we can say that the difference in the proportions is between 0.112 and 0.230. Alternatively, we can report that the difference between the percent of women who are Instagram users and the percent of men who are Instagram users is 17.1%, with a 95% margin of error of 5.9%.

In this example, men and women were not sampled separately. The sample sizes are, in fact, random and reflect the gender distributions of the subjects who responded to the survey. Two-sample significance tests and confidence intervals are still approximately correct in this situation.

In the preceding example, we chose women to be the first population. Had we chosen men to be the first population, the estimate of the difference would be negative (-0.1710). Because it is easier to discuss positive numbers, we generally choose the first population to be the one with the higher proportion.

EXAMPLE 8.12

INSTAG

Instagram confidence interval from software. Figure 8.5 shows a spreadsheet that can be used as input to software. Output from JMP and Minitab is given in Figure 8.6. Compare these outputs with the calculations that we performed in Example 8.11.

Excel

	A	B	C
1	Sex	User	Count
2	1Women	Yes	328
3	1Women	No	209
4	2Men	Yes	234
5	2Men	No	298

FIGURE 8.5 Spreadsheet that can be used as input to software that computes the confidence interval for the Instagram data, Example 8.11.

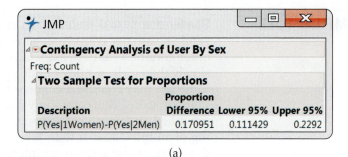

(a)

```
 ·|| Minitab                              —  ▫  X

   Sample    X     N    Sample p
   1        328   537   0.610801
   2        234   532   0.439850


   Difference = p (1) - p (2)
   Estimate for difference:   0.170951
   95% CI for difference:    (0.111963, 0.229940)
```

(b)

FIGURE 8.6 (a) JMP and (b) Minitab output for the Instagram confidence interval, Example 8.11.

USE YOUR KNOWLEDGE

8.50 Gender and commercial preference. A study was designed to compare two energy drink commercials. Each participant was shown the commercials in random order and asked to select the better one. Commercial A was selected by 54 out of 115 women and 83 out of 145 men. Give an estimate of the difference in gender proportions that favored Commercial A. Also construct a large-sample 95% confidence interval for this difference.

8.51 Gender and commercial preference, revisited. Refer to Exercise 8.50. Construct a 95% confidence interval for the difference in proportions that favor Commercial B. Explain how you could have obtained these results from the calculations you did in Exercise 8.50.

BEYOND THE BASICS

The Plus Four Confidence Interval for a Difference in Proportions

Just as in the case of estimating a single proportion, a small modification of the sample proportions can greatly improve the accuracy of confidence intervals.[16] As before, we add two successes and two failures to the actual data, but now we divide them equally between the two samples. That is, we *add one success and one failure to each sample*. This method can be used for 90%, 95%, or 99% confidence when both sample sizes are at least five. Here is an example.

EXAMPLE 8.13

Gender and sexual maturity. In studies that look for a difference between genders, a major concern is whether or not apparent differences are due to other variables that are associated with gender. Because boys mature more slowly than girls, a study of adolescents that compares boys and girls of the same age may confuse a gender effect with an effect of sexual maturity. The "Tanner score" is a commonly used measure of sexual maturity.[17] Subjects are asked to determine their score by placing a mark next to a rough drawing of an individual at their level of sexual maturity. There are five different drawings, so the score is an integer between 1 and 5.

A pilot study included 12 girls and 12 boys from a population that will be used for a large experiment. Four of the boys and three of the girls had Tanner scores of 4 or 5, a high level of sexual maturity. Let's find a 95% confidence interval for the difference between the proportions of boys and girls who have high (4 or 5) Tanner scores in this population. The numbers of successes and failures in both groups are not all at least 10, so the large-sample approach is not recommended. On the other hand, the sample sizes are both at least 5, so the plus four method is appropriate.

The plus four estimate of the population proportion for boys is

$$\tilde{p}_1 = \frac{X_1 + 1}{n_1 + 2} = \frac{4 + 1}{12 + 2} = 0.3571$$

For girls, the estimate is

$$\tilde{p}_2 = \frac{X_2 + 1}{n_2 + 2} = \frac{3 + 1}{12 + 2} = 0.2857$$

Therefore, the estimate of the difference is

$$\tilde{D} = \tilde{p}_1 - \tilde{p}_2 = 0.3571 - 0.2857 = 0.071$$

The standard error of $\tilde{D}$ is

$$\mathrm{SE}_{\tilde{D}} = \sqrt{\frac{\tilde{p}_1(1 - \tilde{p}_1)}{n_1 + 2} + \frac{\tilde{p}_2(1 - \tilde{p}_2)}{n_2 + 2}}$$

$$= \sqrt{\frac{(0.3571)(1 - 0.3571)}{12 + 2} + \frac{(0.2857)(1 - 0.2857)}{12 + 2}}$$

$$= 0.1760$$

For 95% confidence, $z^* = 1.96$ and the margin of error is

$$m = z^* \mathrm{SE}_{\tilde{D}} = (1.96)(0.1760) = 0.345$$

The confidence interval is

$$\tilde{D} \pm m = 0.071 \pm 0.345$$
$$= (-0.274, 0.416)$$

With 95% confidence, we can say that the difference in the proportions is between -0.274 and 0.416. Alternatively, we can report that the difference in the proportions of boys and girls with high Tanner scores in this population is 7.1% with a 95% margin of error of 34.5%.

The very large margin of error in this example indicates that either boys or girls could be more sexually mature in this population and that the difference could be quite large. *Although the interval includes the possibility that there is no difference, corresponding to $p_1 = p_2$ or $p_1 - p_2 = 0$, we should not conclude that there is* **no** *difference in the proportions.* With small sample sizes such as these, the data do not provide us with a lot of information for our inference. This fact is expressed quantitatively through the very large margin of error.

Significance test for a difference in proportions

Although we prefer to compare two proportions by giving a confidence interval for the difference between the two population proportions, it is sometimes useful to test the null hypothesis that the two population proportions are the same.

We standardize $D = \hat{p}_1 - \hat{p}_2$ by subtracting its mean $p_1 - p_2$ and then dividing by its standard deviation

$$\sigma_D = \sqrt{\frac{p_1(1 - p_1)}{n_1} + \frac{p_2(1 - p_2)}{n_2}}$$

If n_1 and n_2 are large, the standardized difference is approximately $N(0, 1)$. For the large-sample confidence interval we used sample estimates in place of the unknown population values in the expression for σ_D. Although this approach would lead to a valid significance test, we instead adopt the more common practice of replacing the unknown σ_D with an estimate that takes into account our null hypothesis $H_0: p_1 = p_2$. If these two proportions are equal, then we can view all the data as coming from a single population. Let p denote the common value of p_1 and p_2; then the standard deviation of $D = \hat{p}_1 - \hat{p}_2$ is

$$\sigma_D = \sqrt{\frac{p(1 - p)}{n_1} + \frac{p(1 - p)}{n_2}}$$

$$= \sqrt{p(1 - p)\left(\frac{1}{n_1} + \frac{1}{n_2}\right)}$$

We estimate the common value of p by the overall proportion of successes in the two samples:

$$\hat{p} = \frac{\text{number of successes in both samples}}{\text{number of observations in both samples}} = \frac{X_1 + X_2}{n_1 + n_2}$$

pooled estimate of p   This estimate of p is called the **pooled estimate** because it combines, or pools, the information from both samples.

To estimate σ_D under the null hypothesis, we substitute $\hat{p}$ for p in the expression for σ_D. The result is a standard error for D that assumes $H_0: p_1 = p_2$:

$$SE_{Dp} = \sqrt{\hat{p}(1 - \hat{p})\left(\frac{1}{n_1} + \frac{1}{n_2}\right)}$$

The subscript on SE_{Dp} reminds us that we pooled data from the two samples to construct the estimate.

SIGNIFICANCE TEST FOR COMPARING TWO PROPORTIONS

To test the hypothesis

$$H_0: p_1 = p_2$$

compute the **z statistic**

$$z = \frac{\hat{p}_1 - \hat{p}_2}{SE_{Dp}}$$

where the **pooled standard error** is

$$SE_{Dp} = \sqrt{\hat{p}(1 - \hat{p})\left(\frac{1}{n_1} + \frac{1}{n_2}\right)}$$

and where the **pooled estimate** of the common value of p_1 and p_2 is

$$\hat{p} = \frac{X_1 + X_2}{n_1 + n_2}$$

In terms of a standard Normal random variable Z, the approximate P-value for a test of H_0 against

$H_a: p_1 > p_2$ is $P(Z \geq z)$

$H_a: p_1 < p_2$ is $P(Z \leq z)$

$H_a: p_1 \neq p_2$ is $2P(Z \geq |z|)$

This z test is based on the Normal approximation to the binomial distribution. As a general rule, we will use it when the number of successes and the number of failures in each of the samples are at least 5.

EXAMPLE 8.14

INSTAG

Sex and Instagram use: The z test. Are young women and men equally likely to say they use Instagram? We examine the data in Example 8.11 (page 507) to answer this question. Here is the data summary:

Sex	n	X	$\hat{p} = X/n$
Women	537	328	0.6108
Men	532	234	0.4398
Total	1069	562	0.5257

The sample proportions are certainly quite different, but we will perform a significance test to see if the difference is large enough to lead us to believe that the population proportions are not equal. Formally, we test the hypotheses

$$H_0: p_1 = p_2$$
$$H_a: p_1 \neq p_2$$

The pooled estimate of the common value of p is

$$\hat{p} = \frac{328 + 234}{537 + 532} = \frac{562}{1069} = 0.5257$$

Note that this is the estimate on the bottom line of the preceding data summary. The test statistic is calculated as follows:

$$SE_{Dp} = \sqrt{(0.5257)(1 - 0.5257)\left(\frac{1}{537} + \frac{1}{532}\right)} = 0.03055$$

$$z = \frac{\hat{p}_1 - \hat{p}_2}{SE_{Dp}} = \frac{0.6108 - 0.4398}{0.03055}$$

$$= 5.60$$

The P-value is $2P(Z \geq 5.60)$. Note that the largest value for z in Table A is 3.49. Therefore, from Table A, we can conclude that $P < 2(1 - 0.9998) = 0.0004$, although we know that the true P value is smaller.

Here is our summary: 61% of the women and 44% of the men are Instagram users; the difference is statistically significant ($z = 5.60$, $P < 0.0004$).

EXAMPLE 8.15

Output for the Instagram significance test. We prefer to use software to obtain the significance test results for comparing the Instagram use of young women and men. Output from JMP and Minitab is given in Figure 8.7. JMP reports the significance tests for the two-sided alternative and for the two one-sided alternatives. We are interested in the two-sided alternative.

FIGURE 8.7 (a) JMP and (b) Minitab output for the Instagram significance test, Example 8.15.

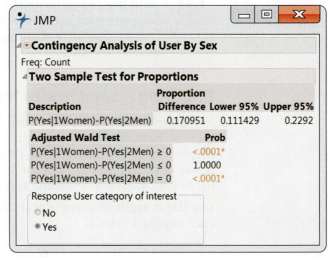

(a)

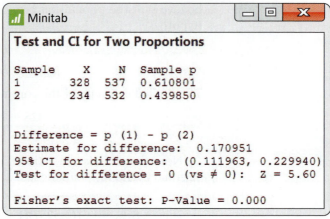

FIGURE 8.7 *Continued*   (b)

Therefore, we report the *P*-value as < 0.0001. Minitab reports the test statistic, $z = 5.60$, and gives the *P*-value as 0.000 (this means $P < 0.0005$) for the Fisher exact test. This test is an alternative to the large-sample significance test that we have discussed. It is preferred by many, particularly for small sample sizes.

Do you think that we could have argued that the proportion would be higher for women than for men before looking at the data in this example? This would allow us to use the one-sided alternative $H_a: p_1 > p_2$. The *P*-value would be half of the value obtained for the two-sided test. Do you think that this approach is justified?

USE YOUR KNOWLEDGE

8.52 Gender and commercial preference: the z test. Refer to Exercise 8.50 (page 509). Test whether the proportions of women and men who liked Commercial A are the same versus the two-sided alternative at the 5% level.

8.53 Changing the alternative hypothesis. Refer to the previous exercise. Does your conclusion change if you test whether the proportion of men who favor Commercial A is larger than the proportion of females? Explain.

Choosing a sample size for two sample proportions

In Section 8.1, we studied methods for determining the sample size using two settings. First, we used the margin of error for a confidence interval for a single proportion as the criterion for choosing n (page 495). Second, we used the power of the significance test for a single proportion as the determining factor (page 498). We follow the same approach here for comparing two proportions.

Use the margin of error Recall that the large-sample estimate of the difference in proportions is

$$D = \hat{p}_1 - \hat{p}_2 = \frac{X_1}{n_1} - \frac{X_2}{n_2}$$

the standard error of the difference is

$$SE_D = \sqrt{\frac{\hat{p}_1(1 - \hat{p}_1)}{n_1} + \frac{\hat{p}_2(1 - \hat{p}_2)}{n_2}}$$

and the margin of error for confidence level C is

$$m = z^* SE_D$$

where z^* is the value for the standard Normal density curve with area C between $-z^*$ and z^*.

For a single proportion, we guessed a value for the true proportion and computed the margins of error for various choices of n. Here, we use the same idea but we need to guess values for the two proportions. We can display the results in a table, as in Example 8.9 (page 497), or in a graph, as in Exercise 8.43 (page 504).

SAMPLE SIZE FOR DESIRED MARGIN OF ERROR

The level C confidence interval for a difference in two proportions will have a margin of error approximately equal to a specified value m when the sample size for each of the two proportions is

$$n = \left(\frac{z^*}{m}\right)^2 (p_1^*(1 - p_1^*) + p_2^*(1 - p_2^*))$$

Here, z^* is the critical value for confidence C, and p_1^* and p_2^* are guessed values for p_1 and p_2, the proportions of successes in the future sample.

The margin of error will be less than or equal to m if p_1^* and p_2^* are chosen to be 0.5. The common sample size required is then given by

$$n = \left(\frac{1}{2}\right)\left(\frac{z^*}{m}\right)^2$$

Note that to use the confidence interval, which is based on the Normal approximation, we still require that the number of successes and the number of failures in each of the samples are at least 10.

EXAMPLE 8.16

Confidence interval–based sample sizes for preferences of women and men. Consider the setting in Exercise 8.50, where we compared the preferences of women and men for two commercials. Suppose we want to do a study in which we perform a similar comparison using a 95% confidence interval that will have a margin of error of 0.1 or less. What should we choose for our sample size? Using $m = 0.1$ and z^* in our formula, we have

$$n = \left(\frac{1}{2}\right)\left(\frac{z^*}{m}\right)^2 = \left(\frac{1}{2}\right)\left(\frac{1.96}{0.1}\right)^2 = 192.08$$

We would include 192 women and 192 men in our study.

Note that we have rounded the calculated value, 192.08, down because it is very close to 192. The normal procedure would be to round the calculated value up to the next larger integer.

8.54 What would the margin of error be? Consider the setting in Exercise 8.50.

(a) Compute the margins of error for $n_1 = 24$ and $n_2 = 24$ for each of the following scenarios: $p_1 = 0.6$, $p_2 = 0.5$; $p_1 = 0.7$, $p_2 = 0.5$; and $p_1 = 0.8$, $p_2 = 0.5$.

(b) If you think that any of these scenarios is likely to fit your study, should you reconsider your choice of $n_1 = 24$ and $n_2 = 24$? Explain your answer.

Use the power of the significance test When we studied using power to compute the sample size needed for a significance test for a single proportion, we used software. We will do the same for the significance test for comparing two proportions.

Some software allows us to consider significance tests that are a little more general than the version we studied in this section. Specifically, we used the null hypothesis H_0: $p_1 = p_2$, which we can rewrite as H_0: $p_1 - p_2 = 0$. The generalization allows us to use values different from zero in the alternative way of writing H_0. Therefore, we write H_0: $p_1 - p_2 = \Delta_0$ for the null hypothesis, and we will need to specify $\Delta_0 = 0$ for the significance test that we studied.

Here is a summary of the inputs needed for software to perform the calculations:

- The value of Δ_0 in the null hypothesis H_0: $p_1 - p_2 = \Delta_0$.

- The alternative hypothesis, two-sided (H_a: $p_1 \neq p_2$) or one-sided (H_a: $p_1 > p_2$ or H_a: $p_1 < p_2$).

- Values for p_1 and p_2 in the alternative hypothesis.

- The Type I error (α, the probability of rejecting the null hypothesis when it is true); usually we choose 5% ($\alpha = 0.05$) for the Type I error.

- Power (probability of rejecting the null hypothesis when it is false); usually we choose 80% (0.80) for power.

EXAMPLE 8.17

Sample sizes for preferences of women and men. Refer to Example 8.16 where we used the margin of error to find the sample sizes for comparing the preferences of women and men for two commercials. Let's find the sample sizes required for a significance test that the two proportions who prefer Commercial A are equal ($\Delta_0 = 0$) using a two-sided alternative with $p_1 = 0.6$ and $p_2 = 0.4$, $\alpha = 0.05$, and 80% (0.80) power. Outputs from JMP and Minitab are given in Figure 8.8. We need $n_1 = 97$ women and $n_2 = 97$ men for our study.

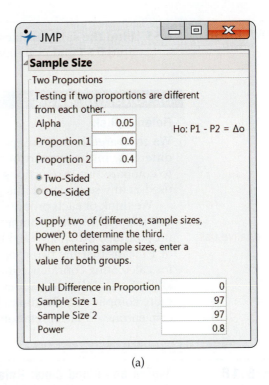

(a)

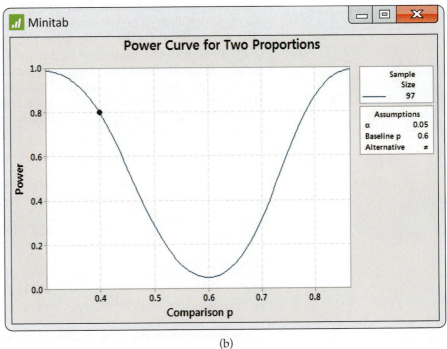

FIGURE 8.8 (a) JMP and (b) Minitab output for finding the sample size, Example 8.17.

(b)

Note that the Minitab output [Figure 8.8(b)] gives the power curve for different alternatives. All of these have $p_1 = 0.6$, which Minitab calls the "Baseline p," while p_2, the Comparison p, varies from 0.3 to 0.9. We see that the power is essentially 100% (1) at these extremes. It is 0.05, the type I error, at $p_2 = 0.6$, which corresponds to the null hypothesis.

USE YOUR KNOWLEDGE

8.55 Find the sample sizes. Consider the setting in Example 8.17. Change p_1 to 0.85 and p_2 to 0.90. Find the required sample sizes.

BEYOND THE BASICS

Relative Risk

We compared Instagram use for women and men by reporting the difference in the proportions with a confidence interval. Another way to compare two proportions is to take the ratio. This approach can be used in any setting and it is particularly common in medical settings.

risk

We think of each proportion as a **risk** that something (usually bad) will happen. We then compare these two risks with the ratio of the two

relative risk

proportions, which is called the **relative risk** (RR). Note that a relative risk of 1 means that the two proportions, $\hat{p}_1$ and $\hat{p}_2$, are equal. The procedure for calculating confidence intervals for relative risk is based on the same kind of principles that we have studied, but the details are somewhat more complicated. Fortunately, we can leave the details to software and concentrate on interpretation and communication of the results.

EXAMPLE 8.18

Aspirin and blood clots: Relative risk. A study of patients who had blood clots (venous thromboembolism) and had completed the standard treatment were randomly assigned to receive a low-dose aspirin or a placebo treatment. The 822 patients in the study were randomized to the treatments, 411 to each. Patients were monitored for several years for the occurrence of several related medical conditions. Counts of patients who experienced one or more of these conditions were reported for each year after the study began.[18] The following table gives the data for a composite of events, termed "major vascular events." Here, X is the number of patients who had a major event.

Population	n	X	$\hat{p} = X/n$
1 (aspirin)	411	45	0.1095
2 (placebo)	411	73	0.1776
Total	822	118	0.1436

The relative risk is

$$\text{RR} = \frac{\hat{p}_1}{\hat{p}_2} = \frac{45/411}{73/411} = 0.6164$$

Software gives the 95% confidence interval as 0.4364 to 0.8707. Taking aspirin has reduced the occurrence of major events to 62% of what it is for patients taking the placebo. The 95% confidence interval is 44% to 87%.

Note that the confidence interval is not symmetric about the estimate. Relative risk is one of many situations where this occurs.

SECTION 8.2 SUMMARY

- The **large-sample estimate of the difference in two population proportions** is

$$D = \hat{p}_1 - \hat{p}_2$$

where $\hat{p}_1$ and $\hat{p}_2$ are the sample proportions:

$$\hat{p}_1 = \frac{X_1}{n_1} \quad \text{and} \quad \hat{p}_2 = \frac{X_2}{n_2}$$

- The **standard error of the difference D** is

$$\text{SE}_D = \sqrt{\frac{\hat{p}_1(1 - \hat{p}_1)}{n_1} + \frac{\hat{p}_2(1 - \hat{p}_2)}{n_2}}$$

- The **margin of error for confidence level C** is

$$m = z^*\text{SE}_D$$

where z^* is the value for the standard Normal density curve with area C between $-z^*$ and z^*. The **large-sample level C confidence interval** is

$$D \pm m$$

We recommend using this interval for 90%, 95%, or 99% confidence when the number of successes and the number of failures in both samples are all at least 10. When sample sizes are smaller, alternative procedures such as the **plus four estimate of the difference in two population proportions** are recommended.

- Significance tests of H_0: $p_1 = p_2$ use the **z statistic**

$$z = \frac{\hat{p}_1 - \hat{p}_2}{\text{SE}_{Dp}}$$

with P-values from the $N(0, 1)$ distribution. In this statistic,

$$\text{SE}_{Dp} = \sqrt{\hat{p}(1 - \hat{p})\left(\frac{1}{n_1} + \frac{1}{n_2}\right)}$$

and $\hat{p}$ is the **pooled estimate** of the common value of p_1 and p_2:

$$\hat{p} = \frac{X_1 + X_2}{n_1 + n_2}$$

Use this test when the number of successes and the number of failures in each of the samples are at least 5.

- **Relative risk** is the ratio of two sample proportions:

$$\text{RR} = \frac{\hat{p}_1}{\hat{p}_2}$$

Confidence intervals for relative risk are often used to summarize the comparison of two proportions.

SECTION 8.2 EXERCISES

For Exercises 8.47, 8.48, and 8.49, see page 506; for Exercises 8.50 and 8.51, see page 509; for Exercises 8.52 and 8.53, see page 514; for Exercise 8.54, see page 516; and for Exercise 8.55, see page 518.

8.56 Identify the key elements. For each of the following scenarios, identify the populations, the counts, and the sample sizes; compute the two proportions and find their difference.

(a) A study of tipping behaviors examined the relationship between the color of the shirt worn by the server and whether or not the customer left a tip.[19] There were 418 male customers in the study; 40 of the 69 who were served by a server wearing a red shirt left a tip. Of the 349 who were served by a server wearing a different colored shirt, 130 left a tip.

(b) A sample of 40 runners will be used to compare two new routines for stretching. The runners will be randomly assigned to one of the routines which they will follow for two weeks. Satisfaction with the routines will be measured using a questionnaire at the end of the two-week period. For the first routine, nine runners said that they were satisfied or very satisfied. For the second routine, six runners said that they were satisfied or very satisfied.

8.57 Apply the confidence interval guidelines. Refer to the previous exercise. For each of the scenarios, determine whether or not the guidelines for using the large-sample method for a 95% confidence interval are satisfied. Explain your answers.

8.58 Find the 95% confidence interval. Refer to Exercise 8.56. For each scenario, find the large-sample 95% confidence interval for the difference in proportions and use the scenario to explain the meaning of the confidence interval.

8.59 Apply the significance test guidelines. Refer to Exercise 8.56. For each of the scenarios, determine whether or not the guidelines for using the large-sample significance test are satisfied. Explain your answers.

8.60 Perform the significance test. Refer to Exercise 8.56. For each scenario, perform the large-sample significance test and use the scenario to explain the meaning of the significance test.

8.61 Find the relative risk. Refer to Exercise 8.56. For each scenario, find the relative risk. Be sure to give a justification for your choice of proportions to use in the numerator and the denominator of the ratio. Use the scenarios to explain the meaning of the relative risk.

8.62 Teeth and military service. In 1898, the United States and Spain fought a war over the U.S. intervention in the Cuban War of Independence. At that time, the U.S. military was concerned about the nutrition of its recruits. Many did not have a sufficient number of teeth to chew the food provided to soldiers. As a result, it was likely that they would be undernourished and unable to fulfill their duties as soldiers. The requirements at that time specified that a recruit must have "at least four sound double teeth, one above and one below on each side of the mouth, and so opposed" so that they could chew food. Of the 58,952 recruits who were under the age of 20, 68 were rejected for this reason. For the 43,786 recruits who were 40 or over, 3801 were rejected.[20]

(a) Find the proportion of rejects for each age group.

(b) Find a 99% confidence interval for the difference in the proportions.

(c) Use a significance test to compare the proportions. Write a short paragraph describing your results and conclusions.

(d) Are the guidelines for the use of the large-sample approach satisfied for your work in parts (b) and (c)? Explain your answers.

8.63 Physical education requirements. In the 1920s, about 97% of U.S. colleges and universities required a physical education course for graduation. Today, about 40% require such a course. A recent study of physical education requirements included 354 institutions: 225 private and 129 public. Among the private institutions, 60 required a physical education course, while among the public institutions, 101 required a course.[21]

(a) What are the explanatory and response variables for this exercise? Justify your answers.

(b) What are the populations?

(c) What are the statistics?

(d) Use a 95% confidence interval to compare the private and the public institutions with regard to the physical education requirement.

(e) Use a significance test to compare the private and the public institutions with regard to the physical education requirement.

(f) For parts (d) and (e), verify that the guidelines for using the large-sample methods are satisfied.

(g) Summarize your analysis of these data in a short paragraph.

8.64 Exergaming in Canada. Exergames are active video games such as rhythmic dancing games, virtual bicycles, balance board simulators, and virtual sports simulators that require a screen and a console. A study of exergaming practiced by students from grades 10 and 11 in Montreal, Canada, examined many factors related to participation in exergaming.[22] Of the 358 students who reported that they stressed about their health, 29.9% said that they were exergamers. Of the 851 students who reported that they did not stress about their health, 20.8% said that they were exergamers.

(a) Define the two populations to be compared for this exercise.

(b) What are the counts, the sample sizes, and the proportions?

(c) Are the guidelines for the use of the large-sample confidence interval satisfied?

(d) Are the guidelines for the use of the large-sample significance test satisfied?

8.65 Confidence interval for exergaming in Canada. Refer to the previous exercise. Find the 95% confidence interval for the difference in proportions. Write a short statement interpreting this result.

8.66 Significance test for exergaming in Canada. Refer to Exercise 8.64. Use a significance test to compare the proportions. Write a short statement interpreting this result.

8.67 Adult gamers versus teen gamers. A Pew Internet Project Data Memo presented data comparing adult gamers with teen gamers with respect to the devices on which they play. The data are from two surveys. The adult survey had 1063 gamers, while the teen survey had 1064 gamers. The memo reports that 54% of adult gamers played on game consoles (Xbox, PlayStation, Wii, etc.), while 89% of teen gamers played on game consoles.[23]

(a) Refer to the table that appears on page 505. Fill in the numerical values of all quantities that are known.

(b) Find the estimate of the difference between the proportion of teen gamers who played on game consoles and the proportion of adults who played on these devices.

(c) Is the large-sample confidence interval for the difference between two proportions appropriate to use in this setting? Explain your answer.

(d) Find the 95% confidence interval for the difference.

(e) Convert your estimated difference and confidence interval to percents.

(f) The adult survey was conducted between October and December 2008, whereas the teen survey was conducted between November 2007 and February 2008. Do you think that this difference should have any effect on the interpretation of the results? Be sure to explain your answer.

8.68 Significance test for gaming on computers. Refer to the previous exercise. Test the null hypothesis that the two proportions are equal. Report the test statistic with the P-value and summarize your conclusion.

8.69 Gamers on computers. The report described in Exercise 8.67 also presented data from the same surveys for gaming on computers (desktops or laptops). These devices were used by 73% of adult gamers and by 76% of teen gamers. Answer the questions given in Exercise 8.67 for gaming on computers.

8.70 Significance test for gaming on consoles. Refer to the previous exercise. Test the null hypothesis that the two proportions are equal. Report the test statistic with the P-value and summarize your conclusion.

8.71 Can we compare gaming on consoles with gaming on computers? Refer to the previous four exercises. Do you think that you can use the large-sample confidence intervals for a difference in proportions to compare teens' use of computers with teens' use of consoles? Write a short paragraph giving the reason for your answer. (*Hint:* Look carefully at the assumptions needed for this procedure on page 512.)

8.72 What's wrong? For each of the following, explain what is wrong and why.

(a) A z statistic is used to test the null hypothesis that $\hat{p}_1 = \hat{p}_2$.

(b) If two sample proportions are equal, then the sample counts are equal.

(c) A 95% confidence interval for the difference in two proportions includes errors due to nonresponse.

8.73 Find the power. Consider testing the null hypothesis that two proportions are equal versus the two-sided alternative with $\alpha = 0.05$, 80% power, and equal sample sizes in the two groups.

(a) For each of the following situations, find the required sample size: (i) $p_1 = 0.1$ and $p_2 = 0.2$ (ii) $p_1 = 0.2$ and $p_2 = 0.3$, (iii) $p_1 = 0.3$ and $p_2 = 0.4$, (iv) $p_1 = 0.4$ and $p_2 = 0.5$, (v) $p_1 = 0.5$ and $p_2 = 0.6$, (vi) $p_1 = 0.6$ and $p_2 = 0.7$, (vii) $p_1 = 0.7$ and $p_2 = 0.8$, and (viii) $p_1 = 0.8$ and $p_2 = 0.9$.

(b) Write a short summary describing your results.

CHAPTER 8 EXERCISES

8.74 The future of gamification. Gamification is an interactive design that includes rewards such as points, payments, and gifts. A Pew survey of 1021 technology stakeholders and critics was conducted to predict the future of gamification. A report on the survey said that 42% of those surveyed thought that there would be no major increases in gamification by 2020. On the other hand, 53% said that they believed that there would be significant advances in the adoption and use of gamification by 2020.[24] Analyze these data using the methods that you learned in this chapter and write a short report summarizing your work.

8.75 Where do you get your news? A report produced by the Pew Research Center's Project for Excellence in Journalism summarized the results of a survey on how people get their news. Of the 2342 people in the survey who own a desktop or laptop, 1639 reported that they get their news from the desktop or laptop.[25]

(a) Identify the sample size and the count.

(b) Find the sample proportion and its standard error.

(c) Find and interpret the 95% confidence interval for the population proportion.

(d) Are the guidelines for use of the large-sample confidence interval satisfied? Explain your answer.

8.76 Is the calcium intake adequate? Young children need calcium in their diet to support the growth of their bones. The Institute of Medicine provides guidelines for how much calcium should be consumed by people of different ages.[26] One study examined whether or not a sample of children consumed an adequate amount of calcium based on these guidelines. Because there are different guidelines for children aged 5 to 10 years and those aged 11 to 13 years, the children were classified into these two age groups. Each student's calcium intake was classified as meeting or not meeting the guideline. There were 2029 children in the study. Here are the data:[27]

Met requirement	Age (years)	
	5 to 10	11 to 13
No	194	557
Yes	861	417

Identify the populations, the counts, and the sample sizes for comparing the extent to which the two age groups of children met the calcium intake requirement.

8.77 Use a confidence interval for the comparison. Refer to the previous exercise. Use a 95% confidence interval for the comparison and explain what the confidence interval tells us. Be sure to include a justification for the use of the large-sample procedure for this comparison.

8.78 Use a significance test for the comparison. Refer to Exercise 8.76. Use a significance test to make the comparison. Interpret the result of your test. Be sure to include a justification for the use of the large-sample procedure for this comparison.

8.79 Confidence interval or significance test? Refer to Exercises 8.76, 8.77, and 8.78. Do you prefer to use the confidence interval or the significance test for this comparison? Give reasons for your answer.

8.80 Changing majors during college. In a random sample of 975 students from a large public university, it was found that 463 of the students changed majors during their college years.

(a) Give a 95% confidence interval for the proportion of students at this university who change majors.

(b) Express your results from part (a) in terms of the *percent* of students who change majors.

(c) University officials concerned with counseling students are interested in the number of students who change majors rather than the proportion. The university has 37,500 undergraduate students. Convert the confidence interval you found in part (a) to a confidence interval for the *number* of students who change majors during their college years.

8.81 Facebook users. A Pew survey of 1802 Internet users found that 67% use Facebook.[28]

(a) How many of those surveyed used Facebook?

(b) Give a 95% confidence interval for the proportion of Internet users who use Facebook.

(c) Convert the confidence interval that you found in part (b) to a confidence interval for the percent of Internet users who use Facebook.

8.82 Twitter users. Refer to the previous exercise. The same survey reported that 16% of Internet users use Twitter. Answer the questions in the previous exercise for Twitter use.

8.83 Facebook versus Twitter. Refer to Exercises 8.81 and 8.82. Can you use the data provided in these two exercises to compare the proportion of Facebook users with the proportion of Twitter users? If your answer is Yes, do the comparison. If your answer is No, explain why you cannot make the comparison.

8.84 Video game genres. U.S. computer and video game software sales were $13.26 billion in 2012.[29]

A survey of 1102 teens collected data about video game use by teens. According to the survey, the following are the most popular game genres:[30]

Genre	Examples	Percent who play
Racing	NASCAR, Mario Kart, Burnout	74
Puzzle	Bejeweled, Tetris, Solitaire	72
Sports	Madden, FIFA, Tony Hawk	68
Action	Grand Theft Auto, Devil May Cry, Ratchet and Clank	67
Adventure	Legend of Zelda, Tomb Raider	66
Rhythm	Guitar Hero, Dance Dance Revolution, Lumines	61

Give a 95% confidence interval for the proportion who play games in each of these six genres.

8.85 Too many errors. Refer to the previous exercise. The chance that each of the six intervals that you calculated includes the true proportion for that genre is approximately 95%. In other words, the chance that your interval misses the true value is approximately 5%.

(a) Explain why the chance that at least one of your intervals does not contain the true value of the parameter is greater than 5%.

(b) One way to deal with this problem is to adjust the confidence level for each interval so that the overall probability of at least one miss is 5%. One simple way to do this is to use a **Bonferroni procedure**. Here is the basic idea: You have an error budget of 5% and you choose to spend it equally on six intervals. Each interval has a budget of $0.05/6 = 0.008$. So, each confidence interval should have a 0.8% chance of missing the true value. In other words, the confidence level for each interval should be $1 - 0.008 = 0.992$. Use Table A to find the value of z^* for a large-sample confidence interval for a single proportion corresponding to 99.2% confidence.

(c) Calculate the six confidence intervals using the Bonferroni procedure.

8.86 Changes in credit card usage by undergraduates. In Exercise 8.32 (page 503), we looked at data from a survey of 1430 undergraduate students and their credit card use. In the sample, 43% said that they had four or more credit cards. A similar study performed four years earlier by the same organization reported that 32% of the sample said that they had four or more credit cards.[31] Assume that the sample sizes for the two studies are the same. Find a 95% confidence interval for the change in the percent of undergraduates who report having four or more credit cards.

8.87 Do the significance test for the change. Refer to the previous exercise. Perform the significance test for

comparing the two proportions. Report your test statistic, the *P*-value, and summarize your conclusion.

8.88 We did not know the sample size. Refer to the previous two exercises. We did not report the sample size for the earlier study, but it is reasonable to assume that it is close to the sample size for the later study.

(a) Suppose that the sample size for the earlier study was only 800. Redo the confidence interval and significance test calculations for this scenario.

(b) Suppose that the sample size for the earlier study was 2500. Redo the confidence interval and significance test calculations for this scenario.

(c) Compare your results for parts (a) and (b) of this exercise with the results that you found in the previous two exercises. Write a short paragraph about the effects of assuming a value for the sample size on your conclusions.

8.89 Student employment during the school year. A study of 1530 undergraduate students reported that 1006 work 10 or more hours a week during the school year. Give a 95% confidence interval for the proportion of all undergraduate students who work 10 or more hours a week during the school year.

8.90 Examine the effect of the sample size. Refer to the previous exercise. Assume a variety of different scenarios where the sample size changes, but the proportion in the sample who work 10 or more hours a week during the school year remains the same. Write a short report summarizing your results and conclusions. Be sure to include numerical and graphical summaries of what you have found.

8.91 Gender and soft drink consumption. Refer to Exercise 8.26 (page 502). This survey found that 16% of the 2006 New Zealanders surveyed reported that they consumed five or more servings of soft drinks per week. The corresponding percents for men and women were 17% and 15%, respectively. Assuming that the numbers of men and women in the survey are approximately equal, do the data suggest that the proportions vary by gender? Explain your methods, assumptions, results, and conclusions.

8.92 Examine the effect of the sample size. Refer to the previous exercise. Assume the following values for the total sample size: 1000, 4000, 10,000. Also assume that the sample proportions do not change. For each of these scenarios, redo the calculations that you performed in the previous exercise. Write a short paragraph summarizing the effect of the sample size on the results.

8.93 Sample size and the *P*-value. In this exercise, we examine the effect of the sample size on the significance test for comparing two proportions. In each case, suppose that $\hat{p}_1 = 0.65$ and $\hat{p}_2 = 0.45$, and take n to be the common value of n_1 and n_2. Use the z statistic to test $H_0: p_1 = p_2$ versus the alternative $H_a: p_1 \neq p_2$. Compute the statistic and the associated P-value for the following values of n: 60, 70, 80, 100, 400, 500, and 1000. Summarize the results in a table. Explain what you observe about the effect of the sample size on statistical significance when the sample proportions $\hat{p}_1$ and $\hat{p}_2$ are unchanged.

8.94 Sample size and the margin of error. In Section 8.1, we studied the effect of the sample size on the margin of error of the confidence interval for a single proportion. In this exercise, we perform some calculations to observe this effect for the two-sample problem. Suppose that $\hat{p}_1 = 0.7$ and $\hat{p}_2 = 0.5$ and n represents the common value of n_1 and n_2. Compute the 95% margins of error for the difference between the two proportions for $n = 60$, 70, 80, 100, 400, 500, and 1000. Present the results in a table and with a graph. Write a short summary of your findings.

8.95 Calculating sample sizes for the two-sample problem. For a single proportion, the margin of error of a confidence interval is largest for any given sample size n and confidence level C when $\hat{p} = 0.5$. This led us to use $p^* = 0.5$ for planning purposes. The same kind of result is true for the two-sample problem. The margin of error of the confidence interval for the difference between two proportions is largest when $\hat{p}_1 = \hat{p}_2 = 0.5$. You are planning a survey and will calculate a 95% confidence interval for the difference between two proportions when the data are collected. You would like the margin of error of the interval to be less than or equal to 0.055. You will use the same sample size n for both populations.

(a) How large a value of n is needed?

(b) Give a general formula for n in terms of the desired margin of error m and the critical value z^*.

8.96 A corporate liability trial. A major court case on the health effects of drinking contaminated water took place in the town of Woburn, Massachusetts. A town well in Woburn was contaminated by industrial chemicals. During the period that residents drank water from this well, there were 16 birth defects among 414 births. In years when the contaminated well was shut off and water

was supplied from other wells, there were three birth defects among 228 births. The plaintiffs suing the firm responsible for the contamination claimed that these data show that the rate of birth defects was higher when the contaminated well was in use.[32] How statistically significant is the evidence? What assumptions does your analysis require? Do these assumptions seem reasonable in this case?

8.97 Statistics and the law. *Castaneda v. Partida* is an important court case in which statistical methods were used as part of a legal argument.[33] When reviewing this case, the Supreme Court used the phrase "two or three standard deviations" as a criterion for statistical significance. This Supreme Court review has served as the basis for many subsequent applications of statistical methods in legal settings. (The two or three standard deviations referred to by the Court are values of the z statistic and correspond to P-values of approximately 0.05 and 0.0026.) In *Castaneda,* the plaintiffs alleged that the method for selecting juries in a county in Texas was biased against Mexican Americans. For the period of time at issue, there were 181,535 persons eligible for jury duty, of whom 143,611 were Mexican Americans. Of the 870 people selected for jury duty, 339 were Mexican Americans.

(a) What proportion of eligible jurors were Mexican Americans? Let this value be p_0.

(b) Let p be the probability that a randomly selected juror is a Mexican American. The null hypothesis to be tested is $H_0: p = p_0$. Find the value of $\hat{p}$ for this problem, compute the z statistic, and find the P-value. What do you conclude? (A finding of statistical significance in this circumstance does not constitute proof of discrimination. It can be used, however, to establish a prima facie case. The burden of proof then shifts to the defense.)

(c) We can reformulate this exercise as a two-sample problem. Here we wish to compare the proportion of Mexican Americans among those selected as jurors with the proportion of Mexican Americans among those not selected as jurors. Let p_1 be the probability that a randomly selected juror is a Mexican American and let p_2 be the probability that a randomly selected nonjuror is a Mexican American. Find the z statistic and its P-value. How do your answers compare with your results in part (b)?

© Image Source/Alamy

Inference for Categorical Data

Introduction

We continue our study of methods for analyzing categorical data in this chapter. Inference about proportions in one-sample and two-sample settings was the focus of Chapter 8. We now study how to compare two or more populations when the response variable has two or more categories and how to test whether two categorical variables are independent. A single statistical test handles both of these cases.

The first section of this chapter gives the basics of statistical inference that are appropriate in this setting. A goodness-of-fit test is presented in the second section. The methods in this chapter answer questions such as:

• Are men and women equally likely to suffer lingering fear symptoms after watching scary movies like *10 Cloverfield Lane* and *The Boy* at a young age?

• Is there an association between texting while driving and automobile accidents?

• Does political preference predict whether a person makes contributions online?

9.1 Inference for Two-Way Tables

9.2 Goodness of Fit

9.1 Inference for Two-Way Tables

When you complete this section, you will be able to:

- Translate a problem from a comparison of two proportions to an analysis of a 2×2 table.
- Find the joint distribution, the marginal distributions, and the conditional distributions for a two-way table of counts.
- Identify the joint distribution, the marginal distributions, and the conditional distributions for a two-way table from software output.
- Choose appropriate conditional distributions to describe relationships in a two-way table.
- Compute expected counts from the counts in a two-way table.
- Compute the chi-square statistic, and the *P*-value from the expected counts in a two-way table. Find the degrees of freedom and use the *P*-value to draw your conclusion.
- Identify the chi-square statistic, the degrees of freedom, and the *P*-value for a two-way table from software output. Use the *P*-value to draw your conclusion.
- For a 2×2 table, explain the relationship between the chi-square test and the *z* test for comparing two proportions.

When we studied inference for two proportions in Chapter 8, we started summarizing the raw data by giving the number of observations in each population (*n*) and how many of these were classified as "successes" (*X*).

EXAMPLE 9.1

Franckreporter/Getty Images

INSTAG

Who uses Instagram? In Example 8.11 (page 507), we compared the proportions of young women and men who use Instagram. The following table summarizes the data used in this comparison:

Population	n	X	$\hat{p} = X/n$
1 (women)	537	328	0.6108
2 (men)	532	234	0.4398
Total	1069	562	0.5257

These data suggest that the percent of women who use Instagram is 17.1% larger than the percent for men, with a 95% margin of error of 5.9%.

In this chapter, we consider a different summary of the data. Rather than recording just the count of those who use Instagram, we record counts of all the outcomes in a two-way table.

LOOK BACK

two-way table, p. 136

EXAMPLE 9.2

Two-way table for Instagram users. Here is the two-way table classifying women and men by their Instagram usage:

INSTAG

Two-way table for Instagram users

User	Male	Female	Total
	Sex		
No	298	209	507
Yes	234	328	562
Total	532	537	1069

r × c table

We use the term **$r \times c$ table** to describe a two-way table of counts with r rows and c columns. The two categorical variables in the 2×2 table of Example 9.2 are "User" and "Sex." "User" is the row variable, with values "No" and "Yes," and "Sex" is the column variable, with values "Male" and "Female." Because the objective in this example is to compare the sexes, we view "Sex" as an explanatory variable. Just as in Chapter 2 where we used the x-axis for the explanatory variable (page 87), here we use Sex as the column variable. The next example presents another two-way table.

EXAMPLE 9.3

VACCINE

Vaccinations and political party preference. Should parents be able to decide whether or not to vaccinate their children or should all vaccinations be required for all children? A Pew Internet survey asked this question of U.S. adults aged 18 and over.[1] The following table breaks down these results by political party preference:

Observed numbers of adults

Required	Democratic	Republican	Total
	Party		
No	230	258	488
Yes	729	479	1208
Total	959	737	1696

The two categorical variables in Example 9.3 are "Required," with values "No" and "Yes," and "Party," with values "Democrat" and "Republican." We view "Party" as an explanatory variable and "Required" as a categorical response variable.

In Chapter 2, we discussed two-way tables and the basics about joint, marginal, and conditional distributions. We now view those sample distributions as estimates of the corresponding population distributions. Let's look at some software output that gives these distributions.

EXAMPLE 9.4

VACCINE

Software output for vaccinations and political party. Figure 9.1 shows the output from JMP, Minitab, and SPSS for the vaccination data of Example 9.3. For now, we will just concentrate on the different distributions. Later, we will explore other parts of the output.

The three packages use similar displays for the distributions. In the cells of the 2×2 table, we find the counts, the conditional distributions

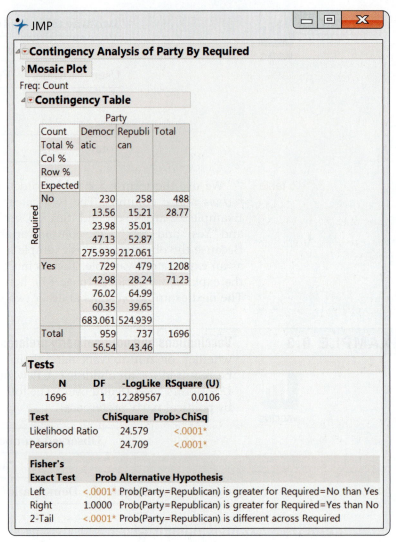

FIGURE 9.1 Computer output from (a) JMP, (b) Minitab, and (c) SPSS, Examples 9.3 and 9.4.

(a) JMP

of the column variable for each value of the row variable, the conditional distributions of the row variable for each value of the column variable, and the joint distribution. All of these are expressed as percents rather than proportions.

Let's look at the entries in the upper-left cell of the JMP output. We see that there are 230 Democrats who think vaccinations should not be required. These 230 represent 13.56% of the study participants. They represent 23.98% of the Democrats in the study. And they represent 47.13% of the respondents who think vaccinations should not be required. The marginal distributions are in the rightmost column and the bottom row. Minitab and SPSS give the same information but not necessarily in the same order.

LOOK BACK

conditional distributions, p. 140

In Chapter 2, we learned that the key to examining the relationship between two categorical variables is to look at conditional distributions. Let's do that for the vaccination data.

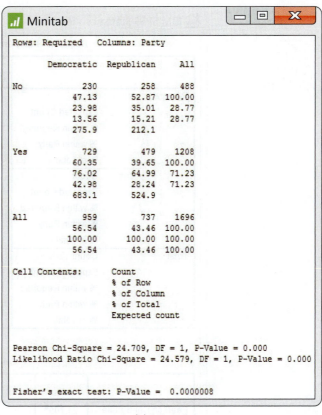

FIGURE 9.1 (*Continued*)   (b) Minitab

EXAMPLE 9.5

VACCINE

Two-way table of vaccination opinions and political party preference. To compare the frequency of vaccination opinions across political party preference, we examine column percents. Here they are, rounded from the output in Figure 9.1 for clarity:

Column percents for political party		
	Party	
Required	Democratic	Republican
No	24%	35%
Yes	76%	65%
Total	100%	100%

The "Total" row reminds us that 100% of the Democrats and Republicans have been classified as either thinking that vaccinations should be required or not. (The sums sometimes differ slightly from 100% because of roundoff error.) The bar graphs in Figure 9.2 compare the percents. The difference between the percents of adults who think vaccinations should not be required is reasonably large (24% for Democrats versus 35% percent for Republicans).

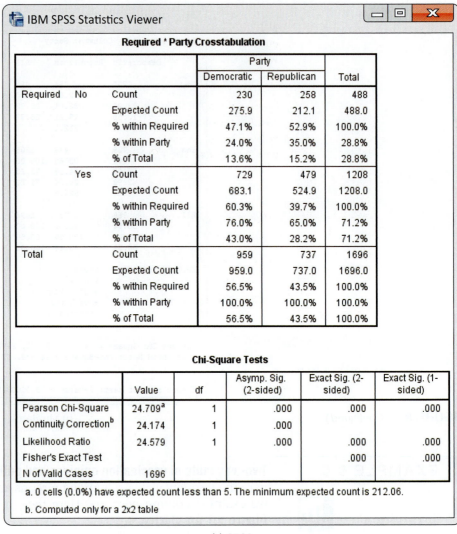

FIGURE 9.1 (*Continued*)   (c) SPSS

FIGURE 9.2 Bar graph of the percents of adults who believe vaccinations should not be required (no) and who believe that vaccinations should be required (yes), by political party preference, Example 9.5.

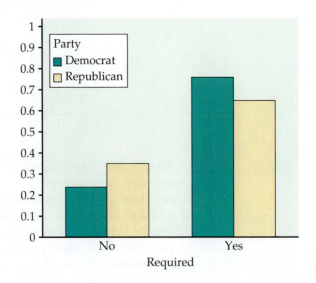

A statistical test will tell us whether or not this difference can be plausibly attributed to chance. Specifically, if there is no association between party preference and opinions about requiring vaccinations, how likely is it that a sample would show a difference as large or larger than that displayed in Figure 9.2? In the last part of this section, we discuss the significance test to examine this question.

Note that Figure 9.2 shows the percents favoring required vaccinations (yes) as well as percents opposed (no). In a description of the results, we would choose one of these for our main story. For tables with more than two columns, we would normally plot the percents for all columns. Here is another way to display the data in a two-way table.

EXAMPLE 9.6

VACCINE

LOOK BACK

mosaic plot,
p. 143

Mosaic plot for vaccination opinions and political party preference. Figure 9.3 displays the joint distribution and the two marginal distributions in a single plot, called a mosaic plot. The sizes of the four rectangles are proportional to the four probabilities of the joint distribution. The bar at the right side gives the marginal distribution of the required variable while the widths of the vertical bars give the marginal distribution of the variable party.

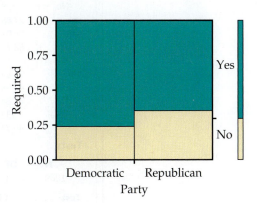

FIGURE 9.3 Mosaic plot for the vaccinations and political party data, Example 9.6.

USE YOUR KNOWLEDGE

INSTAG

9.1 **Find two conditional distributions for the Instagram data.** Figure 9.4 shows JMP output for the Instagram data of Example 9.2 (page 526). Use this output to answer the following questions.

(a) Find the conditional distribution of Instagram use for females.

(b) Do the same for males.

(c) Graphically display the two conditional distributions.

(d) Write a short summary interpreting the two conditional distributions.

INSTAG

9.2 **Condition on Instagram user.** Refer to the previous exercise. Use the output in Figure 9.4 to answer the following questions.

(a) Find the conditional distribution of sex for Instagram users.

(b) Do the same for those who do not use Instagram.

(c) Graphically display the two conditional distributions.

(d) Write a short summary interpreting the two conditional distributions.

9.3 **Which conditional distributions should you use?** Refer to your answers to the two previous exercises. Which of these distributions do you prefer for interpreting these data? Give reasons for your answer.

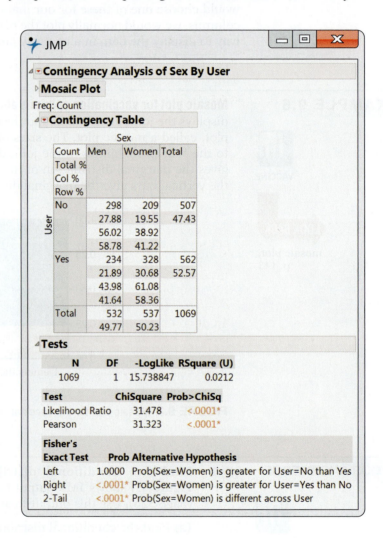

FIGURE 9.4 Computer output for Instagram users, Exercises 9.1, 9.2, and 9.3.

The hypothesis: No association

The null hypothesis H_0 of interest in a two-way table is "There is *no association* between the row variable and the column variable." In Example 9.3, this null hypothesis says that there is no association between political party preference and belief that vaccinations should be required. The alternative hypothesis H_a is that there is an association between these two variables. The alternative H_a does not specify any particular direction for the association. For two-way tables in general, the alternative includes many different possibilities. Because it includes all sorts of possible associations, we cannot describe H_a as either one-sided or two-sided.

In our example, the hypothesis H_0 that there is no association between political party preference and opinions about requiring vaccinations is equivalent to the statement that the variables "required" and "party" are independent. For other two-way tables, where the columns correspond to independent samples from c distinct populations, there are c distributions for the row variable, one for each population. The null hypothesis then says that the c distributions of the row variable are identical. The alternative hypothesis is that the distributions are not all the same.

Expected cell counts

expected cell counts

To test the null hypothesis in $r \times c$ tables, we compare the observed cell counts with **expected cell counts** calculated under the assumption that the null hypothesis is true. A numerical summary of the comparison will be our test statistic.

EXAMPLE 9.7

VACCINE

Expected counts from software. The observed and expected counts for the vaccine example appear in the JMP, Minitab, and SPSS computer outputs shown in Figure 9.1 (pages 528–530). The expected counts are given as the last entry in each cell for JMP and Minitab and as the second entry in each cell for SPSS. For example, in the cell for Democrats who do not think that vaccinations should be required, the observed count is 230 and the expected count is 275.939 (JMP) or 275.9 (Minitab and SPSS).

How is this expected count obtained? Look at the percents in the right margin of the tables in Figure 9.1. We see that 28.77% of all adults thought that vaccinations should not be required. If the null hypothesis of no relation between party and required is true, we expect this overall percent to apply to both Democrats and Republicans. In particular, we expect 28.77% of the Democrats to be opposed to making vaccinations required. Because there are 959 Democrats, the expected count is 28.77% of 959, or 275.9. The other expected counts are calculated in the same way.

The reasoning of Example 9.7 leads to a simple formula for calculating expected cell counts. To compute the expected count of Democrats opposed to requiring vaccinations, we multiplied the proportion of adults opposed to requiring vaccinations (488/1696) by the number of Democrats (959). From Figure 9.1, we see that the numbers 488 and 959 are the row and column totals for the cell of interest and that 1696 is n, the total number of observations for the table. The expected cell count is, therefore, the product of the row and column totals divided by the table total.

EXPECTED CELL COUNTS

$$\text{expected count} = \frac{\text{row total} \times \text{column total}}{n}$$

In Figure 9.3 (page 531), we used a mosaic plot to display the data for the vaccination and political party preference data. Looking at the two columns, we can see that the proportion in the lower region, corresponding to being opposed to required vaccinations, is smaller for the Democrats than for the Republicans.

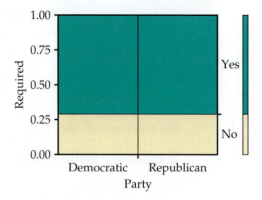

FIGURE 9.5 Mosaic plot for the vaccinations and political party scenario with expected counts in place of observed counts.

This illustrates graphically the difference in the conditional distributions of required for the two parties. What would the mosaic plot look like if there was no difference? If there was no difference in the conditional distributions, then the two variables would be independent, and the observed counts would be equal to the expected counts. If we rerun the analysis with the expected counts in place of the observed counts, we obtain the mosaic plot in Figure 9.5. Notice that the proportions of each party responding yes are now equal.

The chi-square test

To test the H_0 that there is no association between the row and column classifications, we use a statistic that compares the entire set of observed counts with the set of expected counts. To compute this statistic,

- First, take the difference between each observed count and its corresponding expected count, and square these values so that they are all 0 or positive.

standardizing, p. 59

- Because a large difference means less if it comes from a cell that is expected to have a large count, divide each squared difference by the expected count. This is a type of standardization.

- Finally, sum over all cells.

The result is called the *chi-square statistic* X^2. The chi-square statistic was proposed by the English statistician Karl Pearson (1857–1936) in 1900. It is the oldest inference procedure still used in its original form.

CHI-SQUARE STATISTIC

The **chi-square statistic** is a measure of how much the observed cell counts in a two-way table diverge from the expected cell counts. The formula for the statistic is

$$X^2 = \sum \frac{(\text{observed count} - \text{expected count})^2}{\text{expected count}}$$

where "observed" represents an observed cell count, "expected" represents the expected count for the same cell, and the sum is over all $r \times c$ cells in the table.

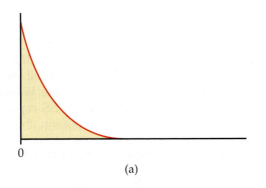
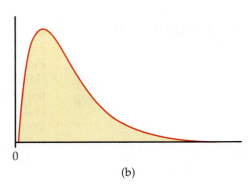

FIGURE 9.6 (a) The $\chi^2(2)$ density curve. (b) The $\chi^2(4)$ density curve.

(a)

(b)

If the expected counts and the observed counts are very different, a large value of X^2 will result. Large values of X^2 provide evidence against the null hypothesis. To obtain a P-value for the test, we need the sampling distribution of X^2 under the assumption that H_0 (no association between the row and column variables) is true. The distribution is called the **chi-square distribution**, which we denote by χ^2 (χ is the lowercase Greek letter chi).

chi-square distribution χ^2

LOOK BACK

degrees of freedom, p. 40

Like the t distributions, the χ^2 distributions form a family described by a single parameter, the degrees of freedom. We use $\chi^2(\text{df})$ to indicate a particular member of this family. Figure 9.6 displays the density curves of the $\chi^2(2)$ and $\chi^2(4)$ distributions. As you can see in the figure, χ^2 distributions take only positive values and are skewed to the right. Table F in the back of the book gives upper critical values for the χ^2 distributions.

CHI-SQUARE TEST FOR TWO-WAY TABLES

The null hypothesis H_0 is that there is no association between the row and column variables in a two-way table. The alternative hypothesis is that these variables are related.

If H_0 is true, the chi-square statistic X^2 has approximately a χ^2 distribution with $(r - 1)(c - 1)$ degrees of freedom.

The P-value for the chi-square test is

$$P(\chi^2 \geq X^2)$$

where χ^2 is a random variable having the $\chi^2(\text{df})$ distribution with df $= (r - 1)(c - 1)$. For tables larger than 2×2, we will use this approximation whenever the average of the expected counts is 5 or more and the smallest expected count is 1 or more. For 2×2 tables, we require all four cell counts to be 5 or more.[2]

The chi-square test always uses the upper tail of the χ^2 distribution because any deviation from the null hypothesis makes the statistic larger. The approximation of the distribution of X^2 by χ^2 becomes more accurate as the cell counts increase. Moreover, it is more accurate for tables larger than 2×2 tables.

EXAMPLE 9.8

VACCINE

Chi-square significance test from software. The results of the chi-square significance test for the vaccination example appear in the computer outputs in Figure 9.1 (pages 928–930), labeled Pearson (JMP) or Pearson Chi-square (Minitab and SPSS). Because all the expected cell counts are large (5 or more), the χ^2 distribution provides an accurate P-value. We see that $X^2 = 12.29$, df = 1, and $P < 0.0001$. Note that Minitab and SPSS report the P-value as 0.000 or .000. These are rounded numbers and potentially misleading. The P-value is small, but it is not zero. For this reason, we prefer to report $P < 0.0001$. As a check, we verify that the degrees of freedom are correct for a 2×2 table:

$$\text{df} = (r - 1)(c - 1) = (2 - 1)(2 - 1) = 1$$

The chi-square test confirms that the data provide evidence against the null hypothesis that there is no relationship between political party preference and vaccination opinion. Under H_0, the chance of obtaining a value of X^2 greater than or equal to the calculated value of 12.29 is small, less than 0.0001—fewer than 1 time in 10,000.

The outputs in Figure 9.1 also report results for testing the hypothesis of no association using alternatives to the chi-square significance test. Fisher's exact test is preferred by many, particularly when the counts are small and the chi-square approximation is not very accurate.

The test does not provide insight into the nature of the relationship between the variables. It is up to us to see that the data show that Republicans are more likely to believe that vaccinations should not be required. You should always accompany a chi-square test by percents such as those in Example 9.5 and Figure 9.3 and by a description of the nature of the relationship.

Observational studies such as the one in Example 9.3 cannot tell us whether or not an explanatory variable is a *cause* of a pattern in a response variable. For the party and vaccine scenario, a causal association does not seem plausible. Often, association can be explained by confounding with other variables.

LOOK BACK

confounding, p. 150

Computations

The calculations required to analyze a two-way table are straightforward but tedious. In practice, we recommend using software, but it is possible to do the work with a calculator, and some insight can be gained by examining the details. Here is an outline of the steps required.

COMPUTATIONS FOR TWO-WAY TABLES

1. Calculate descriptive statistics that convey the important information in the table. Usually, these will be column or row percents.

2. Find the expected counts and use these to compute the X^2 statistic.

3. Use chi-square critical values from Table F to find the approximate P-value.

4. Draw a conclusion about the association between the row and column variables.

The next few examples illustrate these steps.

EXAMPLE 9.9

HEALTH

Health habits of college students. Physical activity generally declines when students leave high school and enroll in college. This suggests that college is an ideal setting to promote physical activity. One study examined the level of physical activity and other health-related behaviors in a sample of 1184 college students.[3] Let's look at the data for physical activity and consumption of fruits. We categorize physical activity as low, moderate, or vigorous and fruit consumption as low, medium, or high. Here is the two-way table that summarizes the data:

| | Physical activity | | | |
Fruit consumption	Low	Moderate	Vigorous	Total
Low	69	206	294	569
Medium	25	126	170	321
High	14	111	169	294
Total	108	443	633	1184

The table in Example 9.9 is a 3 × 3 table, to which we have added the marginal totals obtained by summing across rows and columns. For example, the first-row total is 69 + 206 + 294 = 569. The grand total, the number of students in the study, can be computed by summing the row totals (569 + 321 + 294 = 1184), or the column totals (108 + 443 + 633 = 1184). *It is easy to make an error in these calculations, so it is a good idea to do both as a check on your arithmetic.*

Computing conditional distributions

First, we summarize the observed relation between physical activity and fruit consumption. We expect a positive association, but there is no clear distinction between an explanatory variable and a response variable in this setting. If we have such a distinction, then the clearest way to describe the relationship is to compare the conditional distributions of the response variable for each value of the explanatory variable. Otherwise, we can compute the conditional distribution each way and then decide which gives a better description of the data.

EXAMPLE 9.10

Health habits of college students: Conditional distributions. Let's look at the data in the first column of the table in Example 9.9. There were 108 students with low physical activity. Of these, there were 69 with low fruit consumption. Therefore, the column proportion for this cell is

$$\frac{69}{108} = 0.639$$

That is, 63.9% of the low physical activity students had low fruit consumption. Similarly, 25 of the low physical activity students has moderate fruit consumption. This percent is 23.1%.

$$\frac{25}{108} = 0.231$$

In all, we calculate nine percents. Here are the results:

Column percents for fruit consumption and physical activity

	Physical activity			
Fruit consumption	Low	Moderate	Vigorous	Total
Low	63.9	46.5	46.4	48.1
Medium	23.1	28.4	26.9	27.1
High	13.0	25.1	26.7	24.8
Total	100.0	100.0	100.0	100.0

In addition to the conditional distributions of fruit consumption for each level of physical activity, the table also gives the marginal distribution of fruit consumption. These percents appear in the rightmost column, labeled "Total."

The sum of the percents in each column should be 100, except for possible small roundoff errors. *It is good practice to calculate each percent separately and then sum each column as a check.* In this way, we can find arithmetic errors that would not be uncovered if, for example, we calculated the column percent for the "High" row by subtracting the sum of the percents for "Low" and "Medium" from 100.

Figure 9.7 compares the distributions of fruit consumption for each of the three physical activity levels. For each activity level, the highest percent is for students who consume low amounts of fruit. For low physical activity, there is a clear decrease in the percent when moving from low to medium to high fruit consumption. The patterns for moderate physical activity and vigorous physical activity are similar. Low fruit consumption is still dominant, but the

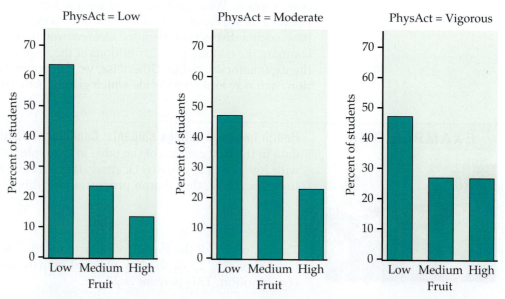

FIGURE 9.7 Comparison of the distribution of fruit consumption for different levels of physical activity, Example 9.10.

percents for medium and high fruit consumption are about the same for the moderate and vigorous activity levels. The percent of low fruit consumption is highest for the low physical activity students compared with those who have moderate or vigorous physical activity. These plots suggest that there is an association between these two variables.

USE YOUR KNOWLEDGE

HEALTH

HEALTH

9.4   **Examine the row percents.** Refer to the health habits data that we examined in Example 9.9 (page 537). For the row percents, make a table similar to the one in Example 9.10 (page 537).

9.5   **Make some plots.** Refer to the previous exercise. Make plots of the row percents similar to those in Figure 9.7.

9.6   **Compare the conditional distributions.** Compare the plots you made in the previous exercise with those given in Figure 9.7. Which set of plots do you think gives a better graphical summary of the relationship between these two categorical variables? Give reasons for your answer. Note that there is not a clear right or wrong answer for this exercise. You need to make a choice and to explain your reasons for making it.

We observe a clear relationship between physical activity and fruit consumption in this study. The chi-square test assesses whether this observed association is statistically significant, that is, too strong to occur often just by chance. The test confirms only that there is some relationship. The percents we have compared describe the nature of the relationship.

The chi-square test does not in itself tell us what population our conclusion describes. The subjects in this study were college students from four midwestern universities. The researchers could argue that these findings apply to college students in general. This type of inference is important, but it is based on expert judgment and is beyond the scope of the statistical inference that we have been studying.

EXAMPLE 9.11

HEALTH

The chi-square significance test for health habits of college students. The first step in performing the significance test is to calculate the expected cell counts. Let's start with the cell for students with low fruit consumption and low physical activity. Using the formula on page 533, we need three quantities: (1) the corresponding row total, 569, the number of students who have low fruit consumption; (2) the column total, 108, the number of students who have low physical activity; and (3) the total number of students, 1184. The expected cell count is, therefore,

$$\frac{(108)(569)}{1184} = 51.90$$

Note that although any observed count of the number of students must be a whole number, an expected count need not be.

Calculations for the other eight cells in the 3×3 table are performed in the same way. With these nine expected counts, we are now ready to use the formula for the X^2 statistic on page 534. The first term in the sum comes from the cell for students with low fruit consumption and low physical activity. The observed count is 69 and the expected count is 51.90. Therefore, the contribution to the X^2 statistic for this cell is

$$\frac{(69 - 51.90)^2}{51.90} = 5.63$$

When we add the terms for each of the nine cells, the result is

$$X^2 = 14.15$$

Because there are $r = 3$ levels of fruit consumption and $c = 3$ levels of physical activity, the degrees of freedom for this statistic are

$$df = (r - 1)(c - 1) = (3 - 1)(3 - 1) = 4$$

Under the null hypothesis that fruit consumption and physical activity are independent, the test statistic X^2 has a $\chi^2(4)$ distribution. To obtain the P-value, look at the df = 4 row in Table F.

The calculated value $X^2 = 14.15$ lies between the critical points for probabilities 0.01 and 0.005. The P-value is, therefore, between 0.01 and 0.005. (Software gives the value as 0.0068.) There is strong evidence ($X^2 = 14.15$, $df = 4$, $P < 0.01$) that there is a relationship between fruit consumption and physical activity.

df = 4

p	0.01	0.005
χ^2	13.28	14.86

We can check our work by adding the expected counts to obtain the row and column totals, as in the table. These totals are the same as those in the table of observed counts except for small roundoff errors.

USE YOUR KNOWLEDGE

HEALTH

9.7   Find the expected counts. Refer to Example 9.11. Compute the expected counts and display them in a 3 × 3 table. Check your work by adding the expected counts to obtain row and column totals. These should be the same as those in the table of observed counts except for small roundoff errors.

HEALTH

9.8   Find the X^2 statistic. Refer to the previous exercise. Use the formula on page 534 to compute the contributions to the chi-square statistic for each cell in the table. Verify that their sum is 14.15.

9.9   Find the P-value. For each of the following give the degrees of freedom and an appropriate bound on the P-value for the X^2 statistic.

(a) $X^2 = 13.00$ for a 4 × 3 table.

(b) $X^2 = 13.00$ for a 3 × 4 table.

(c) $X^2 = 4.00$ for a 2 × 3 table.

(d) $X^2 = 8.00$ for a 2 × 3 table.

INSTAG

9.10   Instagram users: The chi-square test. Refer to Example 9.2 (page 526). Use the chi-square test to assess the relationship between sex and Instagram use. State your conclusion.

The chi-square test and the z test

A comparison of the proportions of "successes" in two populations leads to a 2 × 2 table. We can compare two population proportions either by the chi-square test or by the two-sample z test from Section 8.2. In fact, *these tests always give exactly the same result* because the X^2 statistic is equal to the square of the z statistic and $\chi^2(1)$ critical values are equal to the squares of the corresponding $N(0, 1)$ critical values. The advantage of the z test is that we can test either one-sided or two-sided alternatives. The chi-square test always tests the two-sided alternative. Of course, the chi-square test can compare more than two populations, whereas the z test compares only two.

USE YOUR KNOWLEDGE

COMP

9.11 Comparison of conditional distributions. Consider the following 2×2 table.

Observed counts			
	Explanatory variable		
Response variable	1	2	Total
Yes	75	95	170
No	135	115	250
Total	210	210	420

(a) Compute the conditional distribution of the response variable for each of the two explanatory-variable categories.

(b) Display the distributions graphically.

(c) Write a short paragraph describing the two distributions and how they differ.

COMP

9.12 Expected cell counts and the chi-square test. Refer to Exercise 9.11. You consider using the chi-square test to compare these two conditional distributions.

(a) Find the expected counts for all cells. Are they large enough to justify use of the chi-square test for these data?

(b) Computer software gives you $X^2 = 3.95$. What are the degrees of freedom for this statistic?

(c) Using Table F, give an appropriate bound on the P-value.

COMP

9.13 Compare the chi-square test with the z test. Refer to the previous two exercises and the significance test for comparing two proportions (page 512).

(a) Set up the problem as a comparison between two proportions. Describe the population proportions, state the null and alternative hypotheses, and give the sample proportions.

(b) Carry out the significance test to compare the two proportions. Report the z statistic, the P-value, and your conclusion.

(c) Compare the P-value for this significance test with the one that you reported in the previous exercise.

(d) Verify that the square of the z statistic is the X^2 statistic given in the previous exercise.

BEYOND THE BASICS

Meta-Analysis

meta-analysis

Policymakers wanting to make decisions based on research are sometimes faced with the problem of summarizing the results of many studies. These studies may show effects of different magnitudes, some highly significant and some not significant. What *overall conclusion* can we draw? **Meta-analysis** is a collection of statistical techniques designed to combine information from different but similar studies. Each individual study must be examined with care to ensure that its design and data quality are adequate. The basic idea is to compute a measure of the effect of interest for each study. These are then combined, usually by taking some sort of weighted average, to produce a summary measure for all of the studies. Of course, a confidence interval for the summary is included in the results. Here is an example.

EXAMPLE 9.12

© Juice Images/Alamy Stock Photo

LOOK BACK

relative risk, p. 518

Do we eat too much salt? Evidence from a variety of sources suggests that diets high in salt are associated with risks to human health. To investigate the relationship between salt intake and stroke, information from 14 studies was combined in a meta-analysis.[4] Subjects were classified based on the amount of salt in their normal diet. They were followed for several years and then classified according to whether or not they had developed cardiovascular disease (CVD). A total of 104,933 subjects were studied, and 5161 of them developed CVD. Here are the data from one of the studies:[5]

	Low salt	High salt
CVD	88	112
No CVD	1081	1134
Total	1169	1246

Let's look at the relative risk for this study. We first find the proportion of subjects who developed CVD in each group. For the subjects with a low salt intake, the proportion who developed CVD is

$$\frac{88}{1169} = 0.0753$$

or 75 per thousand; for the high-salt group, the proportion is

$$\frac{112}{1246} = 0.0899$$

or 90 per thousand. We can now compute the relative risk as the ratio of these two proportions. We choose to put the high-salt group in the numerator. The relative risk is

$$\frac{0.0899}{0.0753} = 1.19$$

Relative risk greater than 1 means that the high-salt group developed more CVD than the low-salt group. For this study, the association is not statistically significant.

When the data from all 14 studies were combined, the relative risk was reported as 1.17 with a 95% confidence interval of (1.02, 1.32). Because this interval does not include the value 1, corresponding to equal proportions in the two groups, we conclude that the higher CVD rates are not the same for the two diets ($P < 0.05$). The high-salt diet is associated with a 17% higher rate of CVD than the low-salt diet. Note that the relative risk for the individual study in this example was not statistically significant, even though it was higher than the overall estimate (1.19 versus 1.17). This illustrates the value of the meta analysis where the conclusion is based on combining results from several studies.

USE YOUR KNOWLEDGE

9.14 A different view of the relative risk. In the previous example, we computed the relative risk for the high-salt group relative to the low-salt group. Now, compute the relative risk for the low-salt group relative to the high-salt group by inverting the relative risk reported in the meta-analysis in Example 9.15—that is, compute 1/1.17. Then restate the last paragraph of the exercise with this change. (*Hint:* For the lower confidence limit, use 1 divided by the upper limit for the original ratio and do a similar calculation for the upper limit.)

SECTION 9.1 SUMMARY

• The **null hypothesis** for $r \times c$ tables of count data is that there is no relationship between the row variable and the column variable.

• **Expected cell counts** under the null hypothesis are computed using the formula

$$\text{expected count} = \frac{\text{row total} \times \text{column total}}{n}$$

• The null hypothesis is tested by the **chi-square statistic**, which compares the observed counts with the expected counts:

$$X^2 = \sum \frac{(\text{observed} - \text{expected})^2}{\text{expected}}$$

Under the null hypothesis, X^2 has approximately the χ^2 distribution with $(r - 1)(c - 1)$ degrees of freedom. The P-value for the test is

$$P\,(\chi^2 \geq X^2)$$

where χ^2 is a random variable having the $\chi^2(\text{df})$ distribution with df = $(r - 1)(c - 1)$.

• The chi-square approximation is adequate for practical use when the average expected cell count is 5 or greater and all individual expected counts are 1 or greater, except in the case of 2×2 tables. All four expected counts in a 2×2 table should be 5 or greater.

• For two-way tables, we first compute percents or proportions that describe the relationship of interest. Then, we compute expected counts, the X^2 statistic, and the P-value.

• Two different models for generating $r \times c$ tables lead to the chi-square test. In the first model, independent simple random samples (SRSs) are drawn from each of c populations, and each observation is classified according to a categorical variable with r possible values. The null hypothesis is that the distributions of the row categorical variable are the same for all c populations. In the second model, a single SRS is drawn from a population, and observations are classified according to two categorical variables having r and c possible values. In this model, H_0 states that the row and column variables are independent.

SECTION 9.1 EXERCISES

For Exercises 9.1, 9.2, and 9.3, see pages 531–532; for Exercises 9.4, 9.5, and 9.6, see page 539; for Exercises 9.7 through 9.10, see page 540; for Exercises 9.11, 9.12, and 9.13, see page 541; and for Exercise 9.14, see page 543.

9.15 Adult gamers versus teen gamers. In Exercise 8.67 (page 521), you analyzed data from a study that compared adult gamers with teen gamers with respect to the devices on which they play. The study surveyed 1063 adult gamers and 1064 teen gamers. For the adults, 54% played on game consoles (Xbox, PlayStation Wii, etc.), while 89% of the teen gamers played on game consoles. Your analysis in Exercise 8.67 focused on the comparison of two proportions. Use these data to construct a two-way table for analysis.

9.16 Physical education requirements. In Exercise 8.63 (page 520), you analyzed data from a study that included 354 higher education institutions: 225 private and 129 public. Among the private institutions, 60 required a physical education course, while among the public institutions, 101 required a course. Your analysis in that exercise focused on the comparison of two proportions. Use these data to construct a two-way table for analysis.

9.17 Adult gamers versus teen gamers. Refer to Exercise 9.15. Find the joint distribution, the marginal distributions, and the conditional distributions. Which conditional distribution do you prefer to explain the results of your analysis? Give a reason for you answer.

9.18 Physical education requirements. Refer to Exercise 9.16. Find the joint distribution, the marginal distributions, and the conditional distributions. Which conditional distribution do you prefer to explain the results of your analysis? Give a reason for you answer.

9.19 Adult gamers versus teen gamers. Refer to Exercise 9.15. Find the expected counts.

9.20 Physical education requirements. Refer to Exercise 9.16. Find the expected counts.

9.21 Adult gamers versus teen gamers. Refer to Exercise 9.15. Find the chi-square statistic and the P-value. What do you conclude?

9.22 Physical education requirements. Refer to Exercise 9.16. Find the chi-square statistic and the P-value. What do you conclude?

9.23 Adult gamers versus teen gamers. Refer to Exercise 9.15. Show that the chi-square statistic that you found in Exercise 9.21 is the square of the z statistic that you found in Exercise 8.68.

9.24 Physical education requirements. Refer to Exercise 9.16. Show that the chi-square statistic that you found in Exercise 9.22 is the square of the z statistic that you found in Exercise 8.63.

9.25 Sexual harassment in middle and high schools. A nationally representative survey of students in grades 7 to 12 asked about the experience of these students with respect to sexual harassment.[6] One question asked how many times the student had witnessed sexual harassment in school. The two-way table for this exercise is given in Figure 9.8. Use the figure to find the joint distribution, the two marginal distributions, and the conditional distributions. Which conditional distribution do you prefer to explain the results of your analysis? Give a reason for your answer. HARAS1

9.26 Remote deposit capture. The Federal Reserve has called remote deposit capture (RDC) "the most important development the [U.S.] banking industry has seen in years." This service allows users to scan checks and to transmit the scanned images to a bank for posting.[7] In its annual survey of community banks, the American Bankers Association asked banks whether or not they offered this service.[8] The two-way table for this exercise is given in Figure 9.9. Use the figure to find the joint distribution, the two marginal distributions, and the conditional distributions. Which conditional distribution do you prefer to explain the results of your analysis? Give a reason for your answer. RDC

9.27 Sexual harassment in middle and high schools. Refer to Exercise 9.25. Use the output in Figure 9.8 to find the chi-square statistic, the degrees of freedom, and the P value. What do you conclude from this analysis? HARAS1

9.28 Remote deposit capture. Refer to Exercise 9.26. Use the output in Figure 9.9 to find the chi-square statistic, the degrees of freedom, and the P value. What do you conclude from this analysis? RDCA

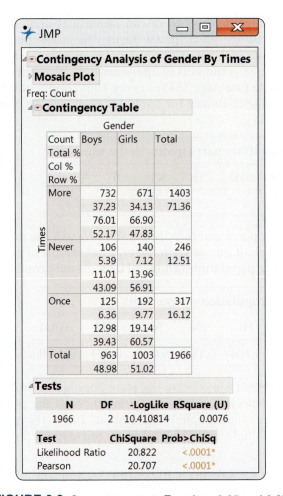

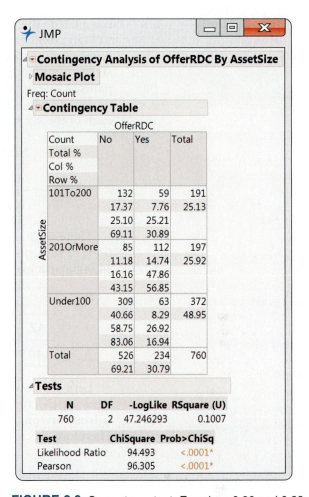

FIGURE 9.8 Computer output, Exercises 9.25 and 9.27.   **FIGURE 9.9** Computer output, Exercises 9.26 and 9.28.

9.2 Goodness of Fit

When you complete this section, you will be able to:	• Compute expected counts given a sample size and the probabilities specified by a null hypothesis for a chi-square goodness-of-fit test. • Find the chi-square test statistic and its *P*-value. • Interpret the results of a chi-square goodness-of-fit significance test.

In the last section, we discussed the use of the chi-square test to compare categorical-variable distributions of *c* populations. We now consider a slight variation on this scenario where we compare a sample from one population with a hypothesized distribution. Here is an example that illustrates the basic ideas.

EXAMPLE 9.13

ACT

Sampling in the Adequate Calcium Today (ACT) study. The ACT study was designed to examine relationships among bone growth patterns, bone development, and calcium intake. Participants were more than 14,000 adolescents from six states: Arizona (AZ), California (CA), Hawaii (HI), Indiana (IN), Nevada (NV), and Ohio (OH). After the major goals of the study were completed, the investigators decided to do an additional analysis of the written comments made by the participants during the study. Because the number of participants was so large, a sampling plan was devised to select sheets containing the written comments of approximately 10% of the participants. A systematic sample (see page 364) of every 10th comment sheet was retrieved from each storage container for analysis.[9] Here are the counts for each of the six states:

Number of study participants in the sample						
AZ	CA	HI	IN	NV	OH	Total
167	257	257	297	107	482	1567

There were 1567 study participants in the sample. We will use the proportions of students from each of the states in the original sample of more than 14,000 participants as the population values.[10] Here are the proportions:

Population proportions						
AZ	CA	HI	IN	NV	OH	Total
0.105	0.172	0.164	0.188	0.070	0.301	100.000

Let's see how well our sample reflects the state population proportions. We start by computing expected counts. Because 10.5% of the population is from Arizona, we expect the sample to have about 10.5% from Arizona. Therefore, because the sample has 1567 subjects, our expected count for Arizona is

$$\text{expected count for Arizona} = 0.105(1567) = 164.535$$

Here are the expected counts for all six states:

Expected counts						
AZ	CA	HI	IN	NV	OH	Total
164.54	269.52	256.99	294.60	109.69	471.67	1567.01

USE YOUR KNOWLEDGE

ACT

9.29 Why is the sum 1567.01? Refer to the table of expected counts in Example 9.13. Explain why the sum of the expected counts is 1567.01 and not 1567.

9.30 Calculate the expected counts. Refer to Example 9.13. Find the expected counts for the other five states. Report your results with three places after the decimal as we did for Arizona.

As we saw with the expected counts in the analysis of two-way tables in Section 9.1, we do not really expect the observed counts to be *exactly* equal to the expected counts. Different samples under the same conditions would give different counts. We expect the average of these counts to be equal to the expected counts when the null hypothesis is true. How close do we think the counts and the expected counts should be?

We can think of our table of observed counts in Example 9.13 as a one-way table with six cells, each with a count of the number of subjects sampled from a particular state. Our question of interest is translated into a null hypothesis that says that the observed proportions of students in the six states can be viewed as random samples from the subjects in the ACT study. The alternative hypothesis is that the process generating the observed counts, a form of systematic sampling in this case, does not provide samples that are compatible with this hypothesis. In other words, the alternative hypothesis says that there is some bias in the way we selected the subjects whose comments we will examine.

Our analysis of these data is very similar to the analyses of two-way tables that we studied in Section 9.1. We have already computed the expected counts. We now construct a chi-square statistic that measures how far the observed counts are from the expected counts. Here is a summary of the procedure.

THE CHI-SQUARE GOODNESS-OF-FIT TEST

Data for n observations of a categorical variable with k possible outcomes are summarized as observed counts, $n_1, n_2, \ldots, n_k$, in k cells. The null hypothesis specifies probabilities $p_1, p_2, \ldots, p_k$ for the possible outcomes. The alternative hypothesis says that the true probabilities of the possible outcomes are not the probabilities specified in the null hypothesis.

For each cell, multiply the total number of observations n by the specified probability to determine the expected counts:

$$\text{expected count} = np_i$$

The **chi-square statistic** measures how much the observed cell counts differ from the expected cell counts. The formula for the statistic is

$$X^2 = \sum \frac{(\text{observed count} - \text{expected count})^2}{\text{expected count}}$$

The degrees of freedom are $k - 1$, and P-values are computed from the chi-square distribution.

Use this procedure when the expected counts are all 5 or more.

EXAMPLE 9.14

ACT

The goodness-of-fit test for the ACT study. For Arizona, the observed count is 167. In Example 9.13, we calculated the expected count, 164.535. The contribution to the chi-square statistic for Arizona is

$$\frac{(\text{observed count} - \text{expected count})^2}{\text{expected count}} = \frac{(167 - 164.535)^2}{164.535} = 0.0369$$

We use the same approach to find the contributions to the chi-square statistic for the other five states. The expected counts are all at least 5, so we can proceed with the significance test.

The sum of these six values is the chi-square statistic,

$$X^2 = 0.93$$

The degrees of freedom are the number of cells minus 1, df = 6 − 1 = 5. We calculate the *P*-value using Table F or software. From Table F, we can determine $P > 0.25$. We conclude that the observed counts are compatible with the hypothesized proportions. The data do not provide any evidence that our systematic sample was biased with respect to selection of subjects from different states.

USE YOUR KNOWLEDGE

ACT

9.31 Compute the chi-square statistic. For each of the other five states, compute the contribution to the chi-square statistic using the method illustrated for Arizona in Example 9.14. (You can use the expected counts that you found in Exercise 9.30 for these calculations.) Show that the sum of these values is the chi-square statistic.

EXAMPLE 9.15

ACT

The goodness-of-fit test from software. Software output from Minitab, SPSS, and JMP for this problem is given in Figure 9.10. Minitab and SPSS report the *P*-value as 0.968. JMP gives an additional place after the decimal, 0.9679. Note that the SPSS output includes a column titled "Residual." For tables of counts, a residual for a cell is defined as

$$\text{residual} = \frac{\text{observed count} - \text{expected count}}{\sqrt{\text{expected count}}}$$

that the residual reported by SPSS is the numerator of this ratio. The chi-square statistic is the sum of the squares of these residuals.

Some software packages do not provide routines for computing the chi-square goodness-of-fit test. However, there is a very simple trick that can be used to produce the results from software that can analyze two-way tables. Make a two-way table in which the first column contains *k* cells with the observed counts. Add a second column with counts that correspond *exactly* to the probabilities specified by the null hypothesis, with a very large number of observations. Then perform the chi-square significance test for two-way tables.

FIGURE 9.10 (a) Minitab, (b) SPSS, and (c) JMP output, Example 9.15.

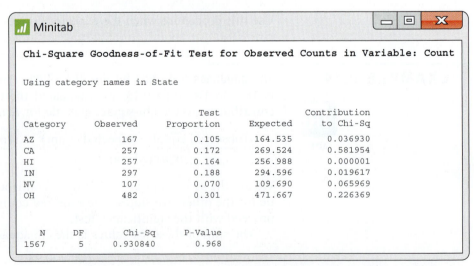

Chi-Square Goodness-of-Fit Test for Observed Counts in Variable: Count

Using category names in State

Category	Observed	Test Proportion	Expected	Contribution to Chi-Sq
AZ	167	0.105	164.535	0.036930
CA	257	0.172	269.524	0.581954
HI	257	0.164	256.988	0.000001
IN	297	0.188	294.596	0.019617
NV	107	0.070	109.690	0.065969
OH	482	0.301	471.667	0.226369

N	DF	Chi-Sq	P-Value
1567	5	0.930840	0.968

(a) Minitab

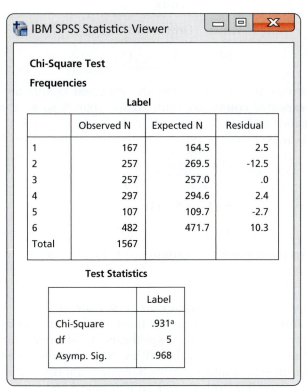

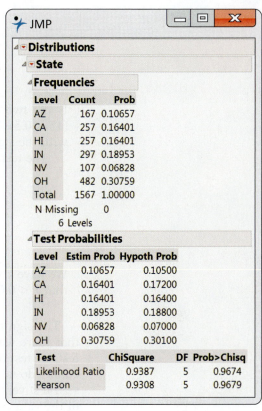

(b) SPSS   (c) JMP

FIGURE 9.10 (*Continued*)

USE YOUR KNOWLEDGE

MM

9.32 **Distribution of M&M colors.** M&M Mars Company has varied the mix of colors for M&M'S Plain Chocolate Candies over the years. These changes in color blends are the result of consumer preference tests. Most recently, the color distribution is reported to be 13% brown, 14% yellow, 13% red, 20% orange, 24% blue, and 16% green.[11] You open up a 14-ounce bag of M&M's and find 61 brown, 59 yellow, 49 red, 77 orange, 141 blue, and 88 green. Use a goodness of fit test to examine how well this bag fits the percents stated by the M&M Mars Company.

EXAMPLE 9.16

The sign test as a goodness-of-fit test. A study of the effect of the full moon on aggressive behaviors of dementia patients included 15 patients, 14 of whom exhibited a greater number of aggressive behaviors on moon days than on other days. The sign test (page 472) tests the null hypothesis that patients are equally likely to exhibit more aggressive behaviors on moon days than on other days. Because $n = 15$, the sample proportion is $\hat{p} = 14/15$ and the null hypothesis is H_0: $p = 0.5$.

To look at these data from the viewpoint of goodness of fit, we think of the data as two counts: patients who had a greater number of aggressive behaviors on moon days and patients who had a greater number of aggressive behaviors on other days.

Counts		
Moon	**Other**	**Total**
14	1	15

If the two outcomes are equally likely, the expected counts are both 7.5 (15 × 0.5). The expected counts are both greater than 5, so we can proceed with the significance test.

The test statistic is

$$X^2 = \frac{(14 - 7.5)^2}{7.5} + \frac{(1 - 7.5)^2}{7.5}$$

$$= 5.633 + 5.633$$

$$= 11.27$$

We have $k = 2$, so the degrees of freedom are 1. From Table F, we conclude that $P < 0.001$.

The sign test can test the null hypothesis versus the one-sided alternative that there was a "moon effect." Within the framework of the goodness of fit test, we test only the general alternative hypothesis that the distribution of the counts do not follow the specified probabilities. Note that the P-value for the sign test versus the one-sided alternative is 0.000488, approximately one-half of the value that we reported from Table F in Example 9.16.

SECTION 9.2 SUMMARY

• The **chi-square goodness-of-fit test** is used to compare the sample distribution of a categorical variable from a population with a hypothesized distribution. The data for n observations with k possible outcomes are summarized as observed counts, $n_1, n_2, \ldots, n_k$, in k cells. The **null hypothesis** specifies probabilities $p_1, p_2, \ldots, p_k$ for the possible outcomes.

• The analysis of these data is similar to the analyses of two-way tables discussed in Section 9.1. For each cell, the **expected count** is determined by multiplying the total number of observations n by the specified probability p_i. The null hypothesis is tested by the usual **chi-square statistic**, which compares the observed counts, n_i, with the expected counts. Under the null hypothesis, X^2 has approximately the χ^2 distribution with df $= k - 1$.

SECTION 9.2 EXERCISES

For Exercises 9.29 and 9.30, see page 546; for Exercise 9.31, see page 548; and for Exercise 9.32, see page 549.

9.33 Is the coin fair? In Example 4.3 (page 218), we learned that the South African statistician John Kerrich tossed a coin 10,000 times while imprisoned by the Germans during World War II. The coin came up heads 5067 times.

(a) Formulate the question about whether or not the coin was fair as a goodness-of-fit hypothesis.

(b) Perform the chi-square significance test and write a short summary of the results.

9.34 Goodness of fit to a standard Normal distribution. Computer software generated 500 random numbers that should look as if they are from the standard Normal distribution. They are categorized into five groups: (1) less than or equal to −0.6, (2) greater than −0.6 and less than or equal to −0.1, (3) greater than −0.1 and less than or equal to 0.1, (4) greater than 0.1

and less than or equal to 0.6, and (5) greater than 0.6. The counts in the five groups are 140, 101, 43, 76, and 140, respectively. Find the probabilities for these five intervals using Table A. Then compute the expected number for each interval for a sample of 500. Finally, perform the goodness-of-fit test and summarize your results.

9.35 More on the goodness of fit to a standard Normal distribution. Refer to the previous exercise. Use software to generate your own sample of 500 standard Normal random variables and perform the goodness-of-fit test. Choose a different set of intervals than the ones used in the previous exercise.

9.36 Goodness of fit to a Poisson distribution. Refer to Example 5.30 (page 329) where a Poisson distribution

is described as a model for the number of dropped calls on your cellphone per day. The mean number of calls is 2.1. In this setting, the probabilities for 0, 1, 2, and 3 or more dropped calls are 0.1225, 0.2572, 0.2700, and 0.3503, respectively. Suppose that you record the number of dropped calls per day for the next 100 days. Your observed counts of dropped calls are 11, 22, 28, and 39, respectively. Use a chi-square goodness of fit test to test the hypothesis that your calls are distributed according to this Poisson distribution.

9.37 More on the goodness of fit to a Poisson distribution. Refer to the previous exercise. Repeat the analysis using 10, 55, 22, and 53 as the observed counts. What do you conclude?

CHAPTER 9 EXERCISES

9.38 Translate each problem into a $r \times c$ table. In each of the following scenarios, translate the problem into one that can be analyzed using a $r \times c$ table. Give the values of r and c, the table, and its entries.

(a) Three website designs are being compared. Sixty students have agreed to be subjects for the study, and they are randomly assigned to watch one of the designs for as long as they like. For each student, the study directors record whether or not the website is watched for more than a minute. For the first design, 16 students watched for more than a minute; for the second, 5 watched for more than a minute; and for the third, 10 students watched for more than a minute.

(b) A sample of undergraduate students were asked whether or not they were in favor of a new proposed core curriculum. For the first-year students, 95 said Yes and 286 said No. For the fourth-year students 127 said Yes and 114 said No.

9.39 Sexual harassment online or in person. In the study described in Exercise 9.25, the students were also asked whether or not they were harassed in person and whether or not they were harassed online. Here are the data for the girls: **HARASG**

Harassed in person	Harassed online Yes	No
Yes	321	200
No	40	441

(a) Analyze these data using the method presented in Chapter 8 for comparing two proportions (page 512).

(b) Analyze these data using the method presented in this chapter for examining a relationship between two categorical variables in a 2 × 2 table.

(c) Use this example to explain the relationship between the chi-square test and the z test for comparing two proportions.

(d) The number of girls reported in this exercise is not the same as the number reported for Exercise 9.25. Suggest a possible reason for this difference.

9.40 Data for the boys. Refer to the previous exercise. Here are the corresponding data for boys: **HARASB**

Harassed in person	Harassed online Yes	No
Yes	183	154
No	48	578

Using these data, repeat the analyses that you performed for the girls in Exercise 9.39. How do the results for the boys differ from those that you found for girls?

9.41 Repeat your analysis. In part (a) of Exercise 9.39, you had to decide which variable was explanatory and which variable was response when you computed the proportions to be compared. **HARASG**

(a) Did you use harassed online or harassed in person as the explanatory variable? Explain the reasons for your choice.

(b) Repeat the analysis that you performed in Exercise 9.39 with the other choice for the explanatory variable.

(c) Summarize what you have learned from comparing the results of using the different choices for analyzing these data.

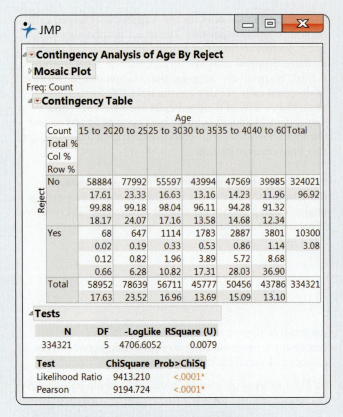

FIGURE 9.11 Computer output, Exercise 9.42.

9.42 Read the output for teeth. Exercise 8.62 (page 520) gives data on individuals rejected for military service in the Cuban War of Independence in 1898 because they did not have enough teeth. In that exercise, you compared the rejection rate for those under the age of 20 with the rejection rate for those over 40. Figure 9.11 gives software output for the table that classifies the recruits into six age categories. Use the output to find the joint distribution, the marginal distributions, and the conditional distributions for these data. [⊞] TEETH

9.43 Is the die fair? You suspect that a die has been altered so that the outcomes of a roll, the numbers 1 to 6, are not equally likely. You toss the die 600 times and obtain the following results: [⊞] DIE

Outcome	1	2	3	4	5	6
Count	87	80	125	117	100	91

Compute the expected counts that you would need to use in a goodness-of-fit test for these data.

9.44 Perform the significance test Refer to the previous exercise. Find the chi-square test statistic and its P-value and write a short summary of your conclusions.

9.45 Health care fraud. Most errors in billing insurance providers for health care services involve honest mistakes by patients, physicians, or others involved in the health care system. However, fraud is a serious problem. The National Health Care Anti-fraud Association estimates that approximately $68 billion is lost to health care fraud each year.[12] When fraud is suspected, an audit of randomly selected billings is often conducted. The selected claims are then reviewed by experts, and each claim is classified as allowed or not allowed. The distributions of the amounts of claims are frequently highly skewed, with a large number of small claims and a small number of large claims. Because simple random sampling would likely be overwhelmed by small claims and would tend to miss the large claims, stratification is often used. See the section on stratified sampling in Chapter 3 (page 195). Here are data from an audit that used three strata based on the sizes of the claims (small, medium, and large):[13] [⊞] BILLER

Stratum	Sampled claims	Number not allowed
Small	57	6
Medium	17	5
Large	5	1

(a) Construct the 3 × 2 table of counts for these data that includes the marginal totals.

(b) Find the percent of claims that were not allowed in each of the three strata.

(c) To perform a significance test, combine the medium and large strata. Explain why we do this.

(d) State an appropriate null hypothesis to be tested for these data.

(e) Perform the significance test and report your test statistic with degrees of freedom and the P-value. State your conclusion.

9.46 Population estimates. Refer to the previous exercise. One reason to do an audit such as this is to estimate the number of claims that would not be allowed if all claims in a population were examined by experts. We have an estimate of the proportion of unallowed claims from each stratum based on our sample. We know the corresponding population proportion for each stratum. Therefore, if we take the sample proportions of unallowed claims and multiply by the population sizes, we would have the estimates that we need. Here are the population sizes for the three strata:

Stratum	Claims in strata
Small	3342
Medium	246
Large	58

(a) For each stratum, estimate the total number of claims that would not be allowed if all claims in the strata had been audited.

(b) Give margins of error for your estimates. ([*Hint*: you first need to find standard errors for your sample estimates using material presented in Chapter 8 (page 486). Then you need to use the rules for variances from Chapter 4 (page 258) to find the standard errors for the population estimates. Finally, you need to multiply by z^* to determine the margins of error.])

9.47 DFW rates. One measure of student success for colleges and universities is the percent of admitted students who graduate. Studies indicate that a key issue in retaining students is their performance in so-called gateway courses. These are courses that serve as prerequisites for other key courses that are essential for student success. One measure of student performance in these courses is the DFW rate, the percent of students who receive grades of D, F, or W (withdraw). A major project was undertaken to improve the DFW rate in a gateway course at a large midwestern university. The course curriculum was revised to make it more relevant to the majors of the students taking the course, a small group of excellent teachers taught the course, technology (including clickers and online homework) was introduced, and student support outside of the classroom was increased. The following table gives data on the DFW rates for the course over three years.[14] In Year 1, the traditional course was given; in Year 2, a few changes were introduced; and in Year 3, the course was substantially revised.

Year	DFW rate	Number of students taking course
Year 1	42.3%	2408
Year 2	24.9%	2325
Year 3	19.9%	2126

Do you think that the changes in this gateway course had an impact on the DFW rate? Write a report giving your answer to this question. Support your answer by an analysis of the data.

9.48 Lying to a teacher. One of the questions in a survey of high school students asked about lying to teachers.[15] The following table gives the numbers of students who said that they lied to a teacher at least once during the past year, classified by sex: LIE

	Sex	
Lied at least once	Male	Female
Yes	3,228	10,295
No	9,659	4,620

(a) Add the marginal totals to the table.

(b) Calculate appropriate percents to describe the results of this question.

(c) Summarize your findings in a short paragraph.

(d) Test the null hypothesis that there is no association between sex and lying to teachers. Give the test statistic and the *P*-value (with a sketch similar to the one on page 535) and summarize your conclusion. Be sure to include numerical and graphical summaries.

(e) The survey asked students if they lied, but we do not know if they answered the question truthfully. How does this fact affect the conclusions that you can draw from these data?

9.49 When do Canadian students enter private career colleges? A survey of 13,364 Canadian students who enrolled in private career colleges was conducted to understand student participation in the private, postsecondary educational system.[16] In one part of the survey, students were asked about their field of study and about when they entered college. Here are the results: CANF

Field of study	Number of students	Time of entry	
		Right after high school	Later
Trades	942	34%	66%
Design	584	47%	53%
Health	5085	40%	60%
Media/IT	3148	31%	69%
Service	1350	36%	64%
Other	2255	52%	48%

In this table, the second column gives the number of students in each field of study. The next two columns give the marginal distribution of time of entry for each field of study.

(a) Use the data provided to make the 6 × 2 table of counts for this problem.

(b) Analyze the data.

(c) Write a summary of your conclusions. Be sure to include the results of your significance testing as well as a graphical summary.

9.50 Government loans for Canadian students in private career colleges. Refer to the previous exercise. The survey also asked about how these college students paid for their education. A major source of funding was government loans. Here are the survey percents of Canadian private students who use government loans to finance their education by field of study: CANGOV

Field of study	Number of students	Percent using government loans
Trades	942	45%
Design	599	53%
Health	5234	55%
Media/IT	3238	55%
Service	1378	60%
Other	2300	47%

(a) Construct the 6×2 table of counts for this exercise.

(b) Test the null hypothesis that the percent of students using government loans to finance their education does not vary with field of study. Be sure to provide all the details of your significance test.

(c) Summarize your analysis and conclusions. Be sure to include a graphical summary.

(d) The number of students reported in this exercise is not the same as the number reported in Exercise 9.49. Suggest a possible reason for this difference.

9.51 Are Mexican Americans less likely to be selected as jurors? Refer to Exercise 8.97 (page 524) concerning *Castaneda v. Partida,* the case where the Supreme Court review used the phrase "two or three standard deviations" as a criterion for statistical significance. Recall that there were 181,535 persons eligible for jury duty, of whom 143,611 were Mexican Americans. Of the 870 people selected for jury duty, 339 were Mexican Americans. We are interested in finding out if there is an association between being a Mexican American and being selected as a juror. Formulate this problem using a two-way table of counts. Construct the 2×2 table using the variables Mexican American or not and juror or not. Find the X^2 statistic and its *P*-value. Square the *z* statistic that you obtained in Exercise 8.97 and verify that the result is equal to the X^2 statistic.

9.52 Goodness of fit to the uniform distribution. Computer software generated 500 random numbers that should look as if they are from the uniform distribution on the interval 0 to 1 (see page 240). They are categorized into five groups: (1) less than or equal to 0.2, (2) greater than 0.2 and less than or equal to 0.4, (3) greater than 0.4 and less than or equal to 0.6, (4) greater than 0.6 and less than or equal to 0.8, and (5) greater than 0.8. The counts in the five groups are 114, 92, 108, 101, and 85, respectively. The probabilities for these five intervals are all the same. What is this probability? Compute the expected number for each interval for a sample of 500. Finally, perform the goodness of fit test and summarize your results.

9.53 More on goodness of fit to the uniform distribution. Refer to the previous exercise. Use software to generate your own sample of 800 uniform random variables on the interval from 0 to 1, and

perform the goodness of fit test. Choose a different set of intervals than the ones used in the previous exercise.

 9.54 Suspicious results? An instructor who assigned an exercise similar to the one described in the previous exercise received homework from a student who reported a *P*-value of 0.999. The instructor suspected that the student did not use the computer for the assignment but just made up some numbers for the homework. Why was the instructor suspicious? How would this scenario change if there were 2000 students in the class?

9.55 Is there a random distribution of trees? In Example 6.1 (page 342), we examined data concerning the longleaf pine trees in the Wade Tract and concluded that the distribution of trees in the tract was not random. Here is another way to examine the same question. First, we divide the tract into four equal parts, or quadrants, in the east–west direction. Call the four parts Q_1 through Q_4. Then we take a random sample of 100 trees and count the number of trees in each quadrant. Here are the data: TREEQ

Quadrant	Q_1	Q_2	Q_3	Q_4
Count	18	22	39	21

(a) If the trees are randomly distributed, we expect to find 25 trees in each quadrant. Why? Explain your answer.

(b) We do not really expect to get *exactly* 25 trees in each quadrant. Why? Explain your answer.

(c) Perform the goodness-of-fit test for these data to determine if these trees are randomly scattered. Write a short report giving the details of your analysis and your conclusion.

9.56 McNemar's test. In Exercise 9.39 (page 551), you examined the relationship between being harassed online and being harassed in person for a sample of 1002 girls. An additional question can be asked about these data. Suppose we wanted to compare the proportions of girls who were harassed online and the proportion who were harassed in person. This is very much like the type of question that we studied in Section 8.2 (page 505). There, however, we used the assumption that the two samples used to calculate the proportions were independent. This assumption is not valid for our harassment data because the proportions are calculated from data provided by the same girls. **McNemar's test** is the recommended procedure. The null hypothesis is that the two population proportions are equal and the alternative is two-sided. The test examines the counts in the cells where the two responses do not agree. In our case, these are 200 and 40. Note that if these two counts are equal, then the proportions will be equal for any possible values of counts in the other two cells. McNemar's test is equivalent to the goodness-of-fit test that we examined in Example 9.16. Find the sample proportions, report the results of the significance test, and write a short summary of your conclusions.

Peathegee Inc/Getty Images

Inference for Regression

Introduction

In this chapter, we continue our study of relationships between variables and describe methods for inference when there is a quantitative response variable and a single quantitative explanatory variable. The descriptive tools we learned in Chapter 2—scatterplots, least-squares regression, and correlation—are essential preliminaries to these methods and also provide a foundation for confidence intervals and significance tests.

We first met the sample mean $\bar{x}$ in Chapter 1 as a measure of the center of a collection of observations. Later, we learned that when the data are a random sample from a population, the sample mean is an unbiased estimate of the population mean μ. In Chapters 6 and 7, we used $\bar{x}$ as the basis for confidence intervals and significance tests for inference about μ.

Now we take this same approach for the problem of fitting straight lines to data. In Chapter 2, we met the least-squares regression line $\hat{y} = b_0 + b_1 x$ as a description of a straight-line relationship between a response variable y and an explanatory variable x. At that point, however, we did not distinguish between sample and population. In this chapter, we will now think of the least-squares line computed from the sample as an estimate of the *true* regression line for the population.

Following the common practice of using Greek letters for population parameters, we write the population line as $\beta_0 + \beta_1 x$. This notation reminds us

10.1 Simple Linear Regression

10.2 More Detail about Simple Linear Regression

that the intercept of the fitted line b_0 estimates the intercept of the population line β_0, and the fitted slope b_1 estimates the slope of the population line β_1.

The methods detailed in this chapter will help us answer questions such as

• For female college students, is a higher level of physical activity (average number of steps per day) associated with a lower body mass index? How strong is the predictive relationship?

• Is the trend in the annual number of tornadoes reported in the United States approximately linear? If so, what is the average yearly increase in the number of tornadoes? How many are predicted for next year?

• Is there a strong positive correlation between a state's adult binge-drinking rate and the prevalence of underage drinking?

10.1 Simple Linear Regression

When you complete this section, you will be able to:

• Describe the simple linear regression model in terms of a population regression line and the distribution of deviations of the response variable y from this line.

• Use linear regression output from statistical software to find the least-squares regression line and estimated model standard deviation.

• Distinguish the model deviations ϵ_i from the residuals e_i that are obtained from a least-squares fit to a data set.

• Use plots to visually check the assumptions of the simple linear regression model.

• Construct and interpret a level C confidence interval for the population intercept and for the population slope.

• Perform a level α significance test for the population intercept and for the population slope.

• Construct and interpret a level C confidence interval for a mean response and a level C prediction interval for a future observation when $x = x^*$.

Statistical model for linear regression

Simple linear regression studies the relationship between a response variable y and a single explanatory variable x. We expect that different values of x will produce different mean responses for y. We encountered a similar but simpler situation in Chapter 7 when we discussed methods for comparing two population means. Figure 10.1 illustrates the statistical model for a comparison of blood pressure change in two groups of experimental subjects. Group 2 subjects were provided extra servings of fruits and vegetables in a calorie-controlled diet, while Group 1 subjects were not. We can think of the extra servings of fruits and vegetables (yes or no) as the explanatory variable in this example. This model has two important parts:

• The mean change in blood pressure may be different in the two populations. These means are labeled μ_1 and μ_2 in Figure 10.1.

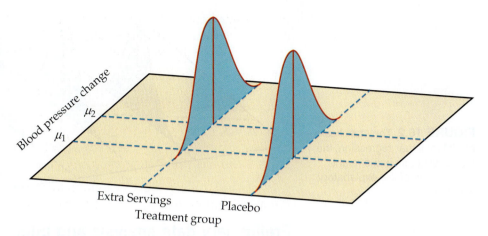

FIGURE 10.1 The statistical model for comparing responses to two treatments; the mean response varies with the treatment.

- Individual changes vary within each population according to a Normal distribution. The two Normal curves in Figure 10.1 describe these responses. These Normal distributions have the same spread, indicating that the population standard deviations are equal.

In linear regression, the explanatory variable x is quantitative and can have many different values. Imagine, for example, giving a different number of servings of fruits and vegetables x to different groups of college students. We can think of the values of x as defining different **subpopulations**, one for each possible value of x. Each subpopulation consists of all individuals in the college student population having the same value of x. If we gave $x = 5$ servings to some students, $x = 7$ servings to some others, $x = 9$ servings to some others, and $x = 11$ servings to the rest, these four groups of students would be considered samples from the corresponding four subpopulations.

> subpopulations

The statistical model for simple linear regression assumes that for each value of x, the observed values of the response variable y are Normally distributed with a mean that depends on x. We use μ_y to represent these means. In general, the means μ_y can change as x changes according to any sort of pattern. In **simple linear regression**, the means all lie on a line when plotted against x. To summarize, this model also has two important parts:

> simple linear regression

- The mean blood pressure change is different for the different subpopulations of x. The means all lie on a straight line. That is, $\mu_y = \beta_0 + \beta_1 x$.

- Individual blood pressure responses y with the same servings x vary according to a Normal distribution. This variation, measured by the standard deviation σ, is the same for all values of x.

The simple linear regression model is pictured in Figure 10.2. The line describes how the mean response μ_y changes with x. This is the **population regression line**. The four Normal curves show how the response y will vary for four different values of the explanatory variable x. Each curve is centered at its mean response μ_y. All four curves have the same spread, measured by their common standard deviation σ.

> population regression line

FIGURE 10.2 The statistical model for linear regression; the mean response is a straight-line function of the explanatory variable.

In the figure: y, x, $\mu_y = \beta_0 + \beta_1 x$

Preliminary data analysis and inference considerations

The data for a linear regression are observed values of y and x. The model takes each x to be a known quantity, like the number of servings of fruits and vegetables. The response y for a given x is assumed to be a Normal random variable. The linear regression model describes the mean and standard deviation of this random variable.

We use the following example to explain the fundamentals of simple linear regression. Because regression calculations in practice are done by statistical software, we rely on computer output for the arithmetic. In Section 10.2, we give an example that illustrates how to do the work with a calculator or spreadsheet.

EXAMPLE 10.1

PABMI

LOOK BACK

simple random sample, p. 191

Relationship between BMI and physical activity. Decrease in physical activity is considered to be a major contributor to the increase in prevalence of overweight and obesity in the general adult population. Because the prevalence of physical inactivity among college students is similar to that of the adult population, researchers have tried to understand college students' physical activity perceptions and behaviors.

In several studies, researchers have looked at the relationship between physical activity (PA) and body mass index (BMI).[1] For this study, each participant wore a FitBit Flex™ for a week, and the average number of steps taken per day (in thousands) was recorded. Various body composition variables, including BMI in kilograms per square meter, kg/m², were also measured. We consider a sample of 100 female undergraduates.

Before starting our analysis, it is appropriate to consider the extent to which the results can reasonably be generalized. In the original study, undergraduate volunteers were obtained at a large southeastern public university through classroom announcements and campus flyers. *The potential for bias should always be considered when obtaining volunteers.* In this case, the participants were screened, and those with severe health issues, as well as varsity athletes, were excluded. As a result, the researchers considered these volunteers as a simple random sample (SRS) from the population of undergraduates at this university. However, they acknowledged the risks of generalizing further, stating that similar investigations at universities of different sizes and in other climates of the United States are needed.

Another issue to consider in this example is the fact that the explanatory variable is not exactly known but, instead, estimated using a FitBit Flex worn over a one-week period. *If there is error in measuring x and it is large relative to the spread of the x's, more advanced inference methods are needed.* In this case, the measurement error is expected to be relatively small. Subjects were instructed to continue normal activities over the one-week period and to notify the researchers of any deviations. Also, this estimate of physical activity represents an average over seven days.

EXAMPLE 10.2

PABMI

LOOK BACK

scatterplot,
p. 86

Graphical display of BMI and physical activity. We start our analysis with a scatterplot of the data. Figure 10.3 is a plot of BMI versus physical activity for our sample of 100 participants. We use the variable names BMI and PA. The least-squares regression line is also shown in the plot. There is a negative association between BMI and PA that appears approximately linear. There is also a considerable amount of scatter about this least-squares line.

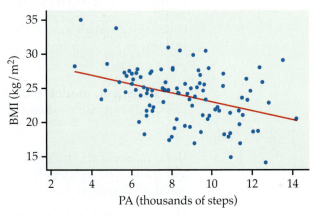

FIGURE 10.3 Scatterplot of body mass index (BMI) versus physical activity (PA) with the least-squares line, Example 10.2.

LOOK BACK

outliers and
influential
observations,
p. 128

Always start with a graphical display of the data. *There is no point in fitting a linear model if the relationship is not approximately linear.* A graphical display can also be used to assess the direction and strength of the relationship and to identify outliers and influential observations.

In this example, subpopulations are defined by the explanatory variable, physical activity. The considerable amount of scatter about the least-squares regression line suggests a large amount of variation of BMI in each subpopulation. Why might this occur? Consider sampling women from your university, each averaging the same number of steps per day—say, 9000. Even though the average number of steps per day is the same, you would not expect all these women to have the same BMI. Variation in other factors such as genetic makeup, lifestyle, and diet should all contribute to the variation of BMI.

The statistical model for linear regression assumes that these BMI values are Normally distributed with a mean μ_y that depends upon x in a linear way. Specifically,

$$\mu_y = \beta_0 + \beta_1 x$$

This was displayed in Figure 10.2 with the line and the four Normal curves. The line is the population regression line, which gives the average BMI for all values of x. The Normal curves provide a description of the variation of BMI about these means. The following equation expresses this idea:

$$\text{DATA} = \text{FIT} + \text{RESIDUAL}$$

The FIT part of the model consists of the subpopulation means, given by the expression $\beta_0 + \beta_1 x$. The RESIDUAL part represents deviations of the data from the line of population means. We assume that these deviations are Normally distributed with mean 0 and standard deviation σ.

We use ϵ (the lowercase Greek letter epsilon) to stand for the RESIDUAL part of the statistical model. A response y is the sum of its mean and a chance deviation ϵ from the mean. These model deviations ϵ represent "noise," that is, variation in y due to other causes that prevent the observed (x, y)-values from forming a perfectly straight line on the scatterplot.

SIMPLE LINEAR REGRESSION MODEL

Given n observations of the explanatory variable x and the response variable y,

$$(x_1, y_1), (x_2, y_2), \ldots, (x_n, y_n)$$

the **statistical model for simple linear regression** states that the observed response y_i when the explanatory variable takes the value x_i is

$$y_i = \beta_0 + \beta_1 x_i + \epsilon_i$$

Here, $\beta_0 + \beta_1 x_i$ is the mean response when $x = x_i$. The deviations ϵ_i are assumed to be independent and Normally distributed with mean 0 and standard deviation σ.

The **parameters of the model** are β_0, β_1, and σ.

USE YOUR KNOWLEDGE

10.1 Understanding a linear regression model. Consider a linear regression model for the decrease in blood pressure (mmHg) over a four-week period with $\mu_y = 2.8 + 0.8x$ and standard deviation $\sigma = 3.2$. The explanatory variable x is the number of servings of fruits and vegetables in a calorie-controlled diet.

(a) What is the slope of the population regression line?

(b) Explain clearly what this slope says about the change in the mean of y for a change in x.

(c) What is the subpopulation mean when $x = 7$ servings per day?

(d) The decrease in blood pressure y will vary about this subpopulation mean. What is the distribution of y for this subpopulation?

(e) Using the 68–95–99.7 rule (page 57), between what two values would approximately 95% of the observed responses, y, fall when $x = 7$?

Estimating the regression parameters

LOOK BACK

least-squares
regression,
p. 111

The method of least squares presented in Chapter 2 fits a line to summarize a relationship between the observed values of an explanatory variable and a response variable. Now we want to use the least-squares line as a basis for inference about a population from which our observations are a sample. In that setting, the slope b_1 and intercept b_0 of the least-squares line

$$\hat{y} = b_0 + b_1 x$$

estimate the slope β_1 and the intercept β_0 of the population regression line.

This inference should only be done when the statistical model assumptions just presented are reasonably met. Model checks are needed and some judgment is required. Because additional methods to check model assumptions rely on first fitting the model to the data, let's briefly review the methods of Chapter 2 concerning least-squares regression.

Using the formulas from Chapter 2 (page 112), the slope of the least-squares line is

$$b_1 = r \frac{s_y}{s_x}$$

and the intercept is

$$b_0 = \bar{y} - b_1 \bar{x}$$

LOOK BACK

correlation,
p. 101

Here, r is the correlation between y and x, s_y is the standard deviation of y, and s_x is the standard deviation of x. Notice that if the slope is 0, so is the correlation, and vice versa. We discuss this relationship more later in the chapter.

The predicted value of y for a given value x^* of x is the point on the least-squares line $\hat{y} = b_0 + b_1 x^*$. This is an unbiased estimator of the mean response μ_y when $x = x^*$. The **residual** is

residual

$$e_i = \text{observed response} - \text{predicted response}$$
$$= y_i - \hat{y}_i$$
$$= y_i - b_0 - b_1 x_i$$

The residuals e_i correspond to the linear regression model deviations ϵ_i. The e_i sum to 0, and the ϵ_i come from a population with mean 0. Because we do not observe the ϵ_i, we use the residuals to check the model assumptions of the ϵ_i.

Recall that the least-squares line is the line that minimizes the sum of the squares of the residuals. The least-squares regression line also always passes through the point $(\bar{x}, \bar{y})$. These are helpful facts to remember when considering the fit of this line to a data set.

The remaining parameter to be estimated is σ, which measures the variation of y about the population regression line. Because this parameter is the standard deviation of the model deviations, it should come as no surprise that we use the residuals to estimate it. As usual, we work first with the variance and take the square root to obtain the standard deviation.

For simple linear regression, the estimate of σ^2 is the average squared residual

LOOK BACK

sample variance,
p. 38

$$s^2 = \frac{\sum e_i^2}{n - 2}$$
$$= \frac{\sum (y_i - \hat{y}_i)^2}{n - 2}$$

We average by dividing the sum by $n - 2$ in order to make s^2 an unbiased estimate of σ^2 (the sample variance of n observations uses the divisor $n - 1$ for this same reason). The quantity $n - 2$ is called the degrees of freedom for s^2. The estimate of the **model standard deviation** σ is given by

model standard deviation σ

$$s = \sqrt{s^2}$$

We now use statistical software to calculate the regression for predicting BMI from physical activity for Example 10.1. In entering the data, we chose the names PA for the explanatory variable and BMI for the response. *It is good practice to use names, rather than just x and y, to remind yourself which variables the output describes.*

EXAMPLE 10.3

PABMI

Statistical software output for BMI and physical activity. Figure 10.4 gives the outputs from three commonly used statistical software packages and Excel. Other software will give similar information. The SPSS output reports estimates of our three parameters as $b_0 = 29.578$, $b_1 = -0.655$, and $s = 3.6549$. Be sure that you can find these values in this output and the corresponding values in the other outputs.

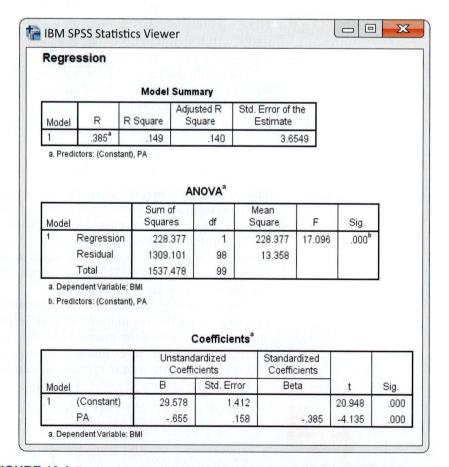

FIGURE 10.4 Regression output from SPSS, Minitab, Excel, and JMP, Example 10.3.

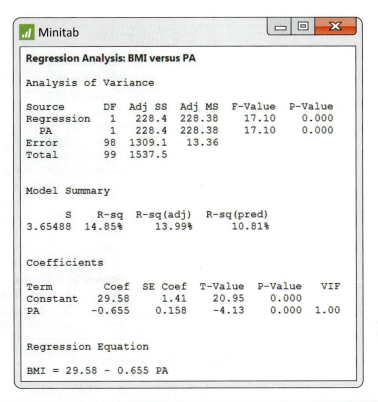

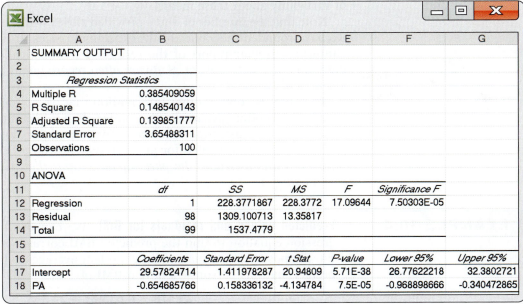

FIGURE 10.4 (*Continued*)

The least-squares regression line is the straight line that is plotted in Figure 10.3. We would report it as

$$\widehat{BMI} = 29.578 - 0.655PA$$

with an estimated model standard deviation of $s = 3.655$. It says that for each increase of 1000 steps per day, we expect the average BMI to be 0.655 smaller.

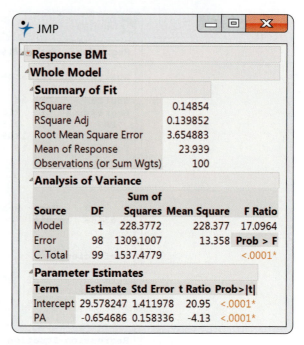

FIGURE 10.4 (*Continued*)

The large estimated model standard deviation, however, suggests there is a lot of variability about these means.

Note that the number of digits provided varies with the software used, and we have rounded the values to three decimal places. It is important to avoid cluttering up your report of the results of a statistical analysis with many digits that are not relevant. *Software often reports many more digits than are meaningful or useful.*

The outputs contain other information that we will ignore for now. Computer outputs often give more information than we want or need. This is done to reduce user frustration when a software package does not print out the particular statistics wanted for an analysis. *The experienced user of statistical software learns to ignore the parts of the output that are not needed for the current analysis.*

EXAMPLE 10.4

Predicted values and residuals for BMI. We can use the least-squares regression equation to find the predicted BMI corresponding to any value of PA. Suppose that a female college student averages 8000 steps per day. We predict that this person will have a BMI of

$$29.578 - 0.655(8) = 24.338$$

If her actual BMI is 25.655, then the residual would be

$$y - \hat{y} = 25.655 - 24.338 = 1.317$$

Because the means μ_y lie on the line $\mu_y = \beta_0 + \beta_1 x$, they are all determined by β_0 and β_1. Thus, once we have estimates of β_0 and β_1, the linear relationship determines the estimates of μ_y for all values of x. Linear regression allows us to do inference not only for subpopulations for which we have data but also

extrapolation,
p. 110

for those corresponding to *x*'s not present in the data. These *x*-values can be both within and outside the range of observed *x*'s. *However, extreme caution must be taken when performing inference for an x-value outside the range of the observed x's because there is no assurance that the same linear relationship between* μ_y *and x holds.*

USE YOUR KNOWLEDGE

10.2 **More on BMI and physical activity.** Refer to Examples 10.3 (page 562) and 10.4 (page 564).

(a) What is the predicted BMI for a woman who averages 9500 steps per day?

(b) If an observed BMI at $x = 9.5$ were 24.3, what would be the residual?

(c) Suppose that you wanted to use the estimated population regression line to examine the predicted BMI for a woman who averages 4000, 10,000, or 16,000 steps per day. Discuss the appropriateness of using the least-squares equation to predict BMI for each of these activity levels.

Checking model assumptions

Now that we have fitted a line, we can further check the conditions that the simple linear regression model imposes on this fit. These checks are *very important* to do and often overlooked. *There is no point in trying to do statistical inference if the data do not, at least approximately, meet the conditions that are the foundation for the inference.* Misleading or incorrect conclusions can result.

These conditions concern the population, but we can observe only our sample. Thus, in doing inference, we act as if the **sample is an SRS from the population**. When the data are collected through some sort of random sampling, this assumption is often easy to justify. In other settings, this assumption requires more thought, and the justification is often debatable. For example, in Example 10.1 (page 558), the sample was a collection of volunteers and the researchers argued they could be considered an SRS from the population of college students at that university.

The remaining model conditions can be checked through a visual examination of the residuals. The first condition is that there is a **linear relationship** in the population, and the second is that **the standard deviation of the responses about the population line is the same** for all values of the explanatory variable. It is common to plot the residuals both against the case number (especially if this reflects the order in which the observations were collected) and against the explanatory variable. These *residual plots* are preferred to scatterplots of *y* versus *x* because they better magnify patterns.

residual plots,
p. 125

scatterplot
smoothers,
p. 94

Figure 10.5 is a plot of the residuals versus physical activity with a smooth-function fit. The scatterplot smoothers are another helpful tool to detect patterns. This smooth function suggests that the average residual is greater than 0 at both low and high physical activity levels. This could mean that a curved relationship between BMI and physical activity would better fit the data. It also could just be the result of chance variation. Notice that there is a large positive residual near each end of the physical activity range that is

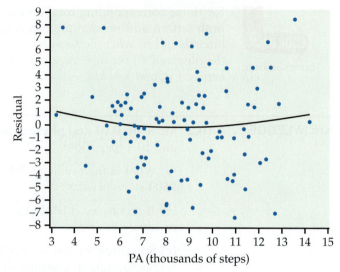

FIGURE 10.5 Plot of residuals versus physical activity (PA) with a smooth function for the physical activity example.

LOOK BACK

Normal quantile plot, p. 66

likely pulling up each end of the smoothed curve. Because the effect does not appear large, we attribute this pattern to chance variation. We do, however, investigate this further in Exercise 11.25 (page 636).

To check the assumption of a common standard deviation, we look at the spread of the residuals across the range of x. In Figure 10.5, the spread is roughly uniform across the range of PA, suggesting that this assumption is reasonable. There also do not appear to be any outliers or influential observations.

The final condition is that **the response varies Normally about the population regression line**. That is, the model deviations vary Normally about 0. Figure 10.6 is a Normal quantile plot of the residuals. Because the

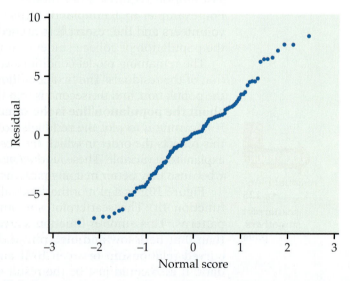

FIGURE 10.6 Normal quantile plot of the residuals for the physical activity example.

plot looks fairly straight, we are confident that we do not have a serious violation of the assumption that the model deviations are Normally distributed.

LINEAR REGRESSION MODEL CONDITIONS

To use the least-squares line as a basis for inference about a population, each of the following conditions should be approximately met:

- The sample is an SRS from the population.

- There a linear relationship between x and y.

- The standard deviation of the responses y about the population regression line is the same for all x.

- The model deviations are Normally distributed.

In summary, the data of Example 10.1 give no reason to doubt the simple linear regression model. We can comfortably proceed to inference about parameters, or functions of parameters, such as

- The slope β_1 and the intercept β_0 of the population regression line.

- The mean response μ_y for a given value of x.

- An individual future response y for a given value of x.

If these assumption checks were to raise doubts, it is best to consult an expert, as a more sophisticated regression model is likely needed. There is, however, one relatively simple remedy that may be worth investigation. This is described at the end of this section.

Confidence intervals and significance tests

Chapter 7 presented confidence intervals and significance tests for means and differences in means. In each case, inference rested on the standard errors of estimates and on t distributions. Inference in simple linear regression is similar in principle. For example, the confidence intervals have the form

$$\text{estimate} \pm t^* \text{SE}_{\text{estimate}}$$

where t^* is a critical point of a t distribution. The formulas for the estimate and standard error, however, are more complicated.

As a consequence of the model assumptions about the deviations ϵ_i, the sampling distributions of b_0 and b_1 are Normally distributed with means β_0 and β_1 and standard deviations that are multiples of σ, the model parameter that describes the variability about the true regression line. In fact, even if the ϵ_i are not Normally distributed, a general form of the central limit theorem tells us that the distributions of b_0 and b_1 will be approximately Normal.

central limit theorem,
p. 298

Because we do not know σ, we use the estimated model standard deviation s, which measures the variability of the data about the least-squares line. When we do this, we move from the Normal distribution to t distributions with degrees of freedom $n-2$, the degrees of freedom of s. We give formulas for the standard errors SE_{b_1} and SE_{b_0} in Section 10.2. For now, we concentrate on the basic ideas and let the computer do the computations.

CONFIDENCE INTERVAL AND SIGNIFICANCE TEST FOR THE REGRESSION SLOPE

A **level C confidence interval for the slope** β_1 is

$$b_1 \pm t^* \text{SE}_{b_1}$$

In this expression, t^* is the value for the $t(n-2)$ density curve with area C between $-t^*$ and t^*.

To test the hypothesis H_0: $\beta_1 = 0$, compute the **test statistic**

$$t = \frac{b_1}{\text{SE}_{b_1}}$$

The **degrees of freedom** are $n - 2$. In terms of a random variable T having the $t(n-2)$ distribution, the P-value for a test of H_0 against

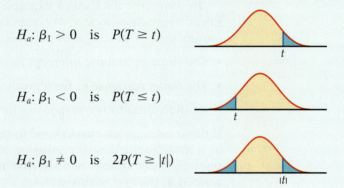

$$H_a: \beta_1 > 0 \quad \text{is} \quad P(T \geq t)$$

$$H_a: \beta_1 < 0 \quad \text{is} \quad P(T \leq t)$$

$$H_a: \beta_1 \neq 0 \quad \text{is} \quad 2P(T \geq |t|)$$

Formulas for confidence intervals and significance tests for the intercept β_0 are exactly the same, replacing b_1 and SE_{b_1} by b_0 and its standard error SE_{b_0}. Although computer outputs often include a test of H_0: $\beta_0 = 0$, this information usually has little practical value. From the equation for the population regression line, $\mu_y = \beta_0 + \beta_1 x$, we see that β_0 is the mean response corresponding to $x = 0$. In many practical situations, this subpopulation does not exist or is not interesting.

On the other hand, the test of H_0: $\beta_1 = 0$ is quite useful. When we substitute $\beta_1 = 0$ in the model, the x term drops out and we are left with

$$\mu_y = \beta_0$$

This equation says that the mean of y does *not* vary with x. In other words, all the y's come from a single population with mean β_0, which we would estimate by $\bar{y}$. The hypothesis H_0: $\beta_1 = 0$ therefore says that there is *no straight-line relationship between y and x* and that linear regression of y on x is of no value for predicting y.

EXAMPLE 10.5

Statistical software output, continued. The computer outputs in Figure 10.4 (pages 562–564) for the physical activity study contain the information needed for inference about the regression slope and intercept. Let's look at the JMP output. The column labeled "Std Error" gives the standard errors of the estimates. The value of SE_{b_1} appears on the line labeled with the variable name for the explanatory variable, PA. Rounding to three decimal places, it is given as 0.158. In a summary, we would report that the regression coefficient for the average number of steps per day is -0.655 with a standard error of 0.158.

The t statistic and P-value for the test of $H_0: \beta_1 = 0$ against the two-sided alternative $H_a: \beta_1 \neq 0$ appear in the columns labeled "t Ratio" and "Prob>|t|." We can verify the t calculation from the formula for the standardized estimate:

$$t = \frac{b_1}{SE_{b_1}} = \frac{-0.654696}{0.158336} = -4.13$$

The P-value is given as <0.0001. The other outputs in Figure 10.4 also indicate that the P-value is very small. Less than one chance in 10,000 is sufficiently small for us to decisively reject the null hypothesis.

We have found a statistically significant linear relationship between physical activity and BMI. The estimated slope is more than 4 standard deviations away from zero. Because this is highly unlikely to happen if the true slope is zero, we have strong evidence for our claim.

Note, however, that this is not the same as concluding that we have found a strong linear relationship between the response and explanatory variables in this example. We saw in Figure 10.3 that there is a lot of scatter about the regression line. *A very small P-value for the significance test for a zero slope does not necessarily imply that we have found a strong relationship.*

A confidence interval provides additional information about the linear relationship. For most statistical software, these intervals are optional output and must be requested. We can also construct them by hand from the default output.

EXAMPLE 10.6

Confidence interval for the slope. A confidence interval for β_1 requires a critical value t^* from the $t(n-2) = t(98)$ distribution. In Table D, there are entries for 80 and 100 degrees of freedom. The values for these rows are very similar. To be conservative, we will use the larger critical value, for 80 degrees of freedom. Find the confidence level values at the bottom of the table. In the 95% confidence column, the entry for 80 degrees of freedom is $t^* = 1.990$.

To compute the 95% confidence interval for β_1, we combine the estimate of the slope with the margin of error:

$$b_1 \pm t^* SE_{b_1} = -0.655 \pm (1.990)(0.158)$$
$$= -0.655 \pm 0.314$$

The interval is $(-0.969, -0.341)$. As expected, this is slightly wider than the interval given by software (see Excel output in Figure 10.4). We estimate that, on average, an increase of 1000 steps per day is associated with a decrease in BMI of between 0.341 and 0.969 kg/m^2.

Note that the intercept in this example is not of practical interest. It estimates average BMI when the activity level is 0, a value that isn't realistic.

For this reason, we do not compute a confidence interval for β_0 or discuss the significance test available in the software.

USE YOUR KNOWLEDGE

10.3 Significance test for the slope. Test the null hypothesis that the slope is zero versus the two-sided alternative in each of the following settings using the $\alpha = 0.05$ significance level:

(a) $n = 20$, $\hat{y} = 28.5 + 1.4x$, and $SE_{b_1} = 0.65$.

(b) $n = 30$, $\hat{y} = 30.8 + 2.1x$, and $SE_{b_1} = 1.05$.

(c) $n = 100$, $\hat{y} = 29.3 + 2.1x$, and $SE_{b_1} = 1.05$.

10.4 95% confidence interval for the slope. For each of the settings in the previous exercise, find the 95% confidence interval for the slope and explain what the interval means.

Confidence intervals for mean response

Besides performing inference about the slope (and sometimes the intercept) in a linear regression, we may want to use the estimated regression line to make predictions about the response y at certain values of x. We may be interested in the mean response for different subpopulations or in the response of future observations at different values of x. In either case, we would want an estimate and associated margin of error.

For any specific value of x, say x^*, the mean of the response y in this subpopulation is given by

$$\mu_y = \beta_0 + \beta_1 x^*$$

To estimate this mean from the sample, we substitute the estimates b_0 and b_1 for β_0 and β_1:

$$\hat{\mu}_y = b_0 + b_1 x^*$$

A confidence interval for μ_y adds to this estimate a margin of error based on the standard error $SE_{\hat{\mu}}$. The formula for the standard error is given in Section 10.2.

CONFIDENCE INTERVAL FOR A MEAN RESPONSE

A **level C confidence interval for the mean response μ_y** when x takes the value x^* is

$$\hat{\mu}_y \pm t^* SE_{\hat{\mu}}$$

where t^* is the value for the $t(n - 2)$ density curve with area C between $-t^*$ and t^*.

Many computer programs calculate confidence intervals for the mean response corresponding to each of the x-values in the data. Some can calculate an interval for any value x^* of the explanatory variable. We will use a plot to illustrate these intervals.

EXAMPLE 10.7

PABMI

Confidence intervals for the mean response. Figure 10.7 shows the upper and lower confidence limits on a graph with the data and the least-squares line. The 95% confidence limits appear as dashed curves. For any x^*, the confidence interval for the mean response extends from the lower dashed curve to the upper dashed curve. The intervals are narrowest for values of x^* near the mean of the observed x's and widen as x^* moves away from $\bar{x}$.

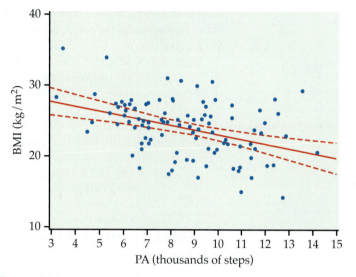

FIGURE 10.7 The 95% confidence limits (dashed curves) for the mean response for the physical activity study, Example 10.7.

Some software will do these calculations directly if you input a value for the explanatory variable. Other software will calculate the intervals for each value of x in the data set. Creating a new data set with an additional observation with x equal to the value of interest and y missing will often work.

EXAMPLE 10.8

Confidence interval for an average of 9000 steps per day. Let's find the confidence interval for the average BMI at $x = 9.0$. Our predicted BMI is

$$\widehat{BMI} = 29.578 - 0.655PA$$
$$= 29.578 - 0.655(9.0)$$
$$= 23.7$$

Software tells us that the 95% confidence interval for the mean response is 23.0 to 24.4 kg/m^2.

If we sampled many women who averaged 9000 steps per day, we would expect their average BMI to be between 23.0 and 24.4 kg/m^2. Note that many of the observations in Figure 10.7 lie outside the confidence bands. *These confidence intervals do not tell us what BMI to expect for a single observation at a particular average steps per day.* We need a different kind of interval, a prediction interval, for this purpose.

Prediction intervals

In the last example, we predicted the *average* BMI for $x^* = 9000$ steps per day. Suppose that we now want to predict an *observation* of BMI for a woman averaging 9000 steps per day. The predicted response y for an individual case with a specific value x^* of the explanatory variable x is

$$\hat{y} = b_0 + b_1 x^*$$

This is the same as the expression for $\hat{\mu}_y$. That is, the fitted line is used both to estimate the mean response when $x = x^*$ and to predict a single future response. We use the two notations $\hat{\mu}_y$ and $\hat{y}$ to remind ourselves of these two distinct uses.

This means our best guess for the BMI of this woman averaging 9000 steps per day is what we obtained using the regression equation, 23.7 kg/m^2. A useful prediction, however, also needs a margin of error (or interval) to indicate its precision. The interval used to predict a future observation is called a **prediction interval** · prediction interval. Although the response y that is being predicted is a random variable, the interpretation of a prediction interval is similar to that for a confidence interval.

Consider doing the following many times:

- Draw a sample of n observations (x_i, y_i) and then one additional observation (x^*, y).

- Calculate the 95% prediction interval for y when $x = x^*$ using the sample of size n.

Then, 95% of the prediction intervals will contain the value of y for the additional observation. In other words, the probability that this method produces an interval that contains the value of a future observation is 0.95.

The form of the prediction interval is very similar to that of the confidence interval for the mean response. The difference is that the standard error SE$_{\hat{y}}$ used in the prediction interval includes both the variability due to the fact that the least-squares line is not exactly equal to the true regression line *and* the variability of the future response variable y around the subpopulation mean. The formula for SE$_{\hat{y}}$ appears in Section 10.2.

PREDICTION INTERVAL FOR A FUTURE OBSERVATION

A **level C prediction interval for a future observation** on the response variable y from the subpopulation corresponding to x^* is

$$\hat{y} \pm t^* \text{SE}_{\hat{y}}$$

where t^* is the value for the $t(n-2)$ density curve with area C between $-t^*$ and t^*.

Again, we use a graph to illustrate the results.

EXAMPLE 10.9	**Prediction intervals for BMI.** Figure 10.8 shows the upper and lower prediction limits, along with the data and the least-squares line. The 95% prediction limits are indicated by the dashed curves. Compare this figure with Figure 10.7, which shows the 95% confidence limits drawn to the same

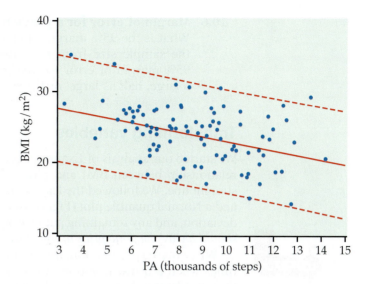

FIGURE 10.8 The 95% prediction limits (dashed curves) for individual responses for the physical activity study, Example 10.9. Compare with Figure 10.7. The limits are wider because the margins of error incorporate the variability about the subpopulation means.

PABMI

scale. The upper and lower limits of the prediction intervals are much farther away from the least-squares line than are the confidence limits. This results in most, but not all, of the observations in Figure 10.8 lying within the prediction bands.

The comparison of Figures 10.7 and 10.8 reminds us that the interval for a single future observation must be larger than an interval for the mean of its subpopulation.

EXAMPLE 10.10

Prediction interval for 9000 steps per day. Let's find the prediction interval for a future observation of BMI for a college-aged woman who averages 9000 steps per day. The predicted value is the same as the estimate of the average BMI that we calculated in Example 10.8, that is, 23.7 kg/m^2. Software tells us that the 95% prediction interval is 16.4 to 31.0 kg/m^2. This interval is extremely wide, covering BMI values that are classified as underweight and obese. Because of the large amount of scatter about the regression line, prediction intervals here are relatively useless.

Although a larger sample would better estimate the population regression line, it would not reduce the degree of scatter about the line. This means that prediction intervals for BMI, given activity level, will *always* be wide. This example clearly demonstrates that a very small P-value for the significance test for a zero slope does not necessarily imply that we have found a strong predictive relationship.

USE YOUR KNOWLEDGE

10.5 Margin of error for the predicted mean. Refer to Figure 10.7 and Example 10.8 (page 571). What is the 95% margin of error of $\hat{\mu}_y$ when $x = 9.0$? Would you expect the margin of error to be larger, smaller, or the same for $x = 11.0$? Explain your answer.

10.6 Margin of error for a predicted response. Refer to Example 10.10. What is the 95% margin of error of $\hat{y}$ when $x = 9.0$? If you increased the sample size from $n = 100$ to $n = 400$, would you expect the 95% margin of error for the predicted response to be roughly twice a large, half as large, or the same for $x = 9.0$? Explain your answer.

Transforming variables

We started our analysis of Example 10.1 with a scatterplot to check whether the relationship between BMI and physical activity could be summarized with a straight line. We followed the least-squares fit with a residual plot (Figure 10.5) and a Normal quantile plot (Figure 10.6) to check Normality, constant standard deviation, and any remaining patterns in the data. *A check of model assumptions should always be done prior to inference.*

LOOK BACK

log transformation, p. 91

When there is a violation, it is best to consult an expert. However, there are times when a transformation of one or both variables will remedy the situation. In Chapter 2, we discussed the use of the log transformation to describe a curved relationship between x and y. Here is an example where the log transformation has more of an impact on other model assumptions.

EXAMPLE 10.11

ENTRE

The relationship between income and education for entrepreneurs. Numerous studies have shown that better-educated employees have higher incomes. Is this also true for entrepreneurs? Do more years of formal education translate into higher incomes? One study explored this question using the National Longitudinal Survey of Youth (NLSY), which followed a large group of individuals aged 14 to 22 for roughly 10 years.[2] The researchers looked at both employees and entrepreneurs, but we just focus on entrepreneurs here. We consider a random sample of 100 entrepreneurs.

The researchers defined *entrepreneurs* to be those who were self-employed or who were the owner/director of an incorporated business. For each of these individuals, they recorded the education level and income. The education level was defined as the years of completed schooling prior to starting the business. The income level was the average annual total earnings since starting the business.

EXAMPLE 10.12

ENTRE

Graphical display of the income and education relationship. Figure 10.9 is a plot of income versus eduction for our sample of 100 entrepreneurs. We use the variable names INC and EDUC. The least-squares regression line and a smoothed curve are also included.

The most striking feature of the plot is not the lack of linearity (there is some suggested curvature between y and x), but rather the distribution of income about the least-squares line. Instead of incomes being Normally distributed, the observations are skewed to the right. That is, for each subpopulation, defined by years of education, there are many small incomes and just a few large incomes. In fact, there are several very large incomes that one might consider to be outliers.

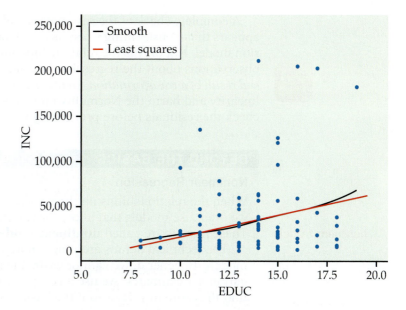

FIGURE 10.9 Scatterplot of income versus education with a smooth function and the least-squares line, Example 10.12. Instead of incomes being Normally distributed, the observations are skewed to the right.

A common remedy for a strongly skewed variable is to consider transforming the variable prior to fitting the model. In this example, the researchers considered the natural logarithm of income (LOGINC).

EXAMPLE 10.13

ENTRE

Is this linear regression model reasonable? Figure 10.10 is a scatterplot of these transformed data with the new least-squares line and smoothed curve. Notice that the smoothed curve is now practically the same as the least-squares line. More importantly, the observations are now more equally dispersed above and below the line and those very large incomes don't look unusual anymore.

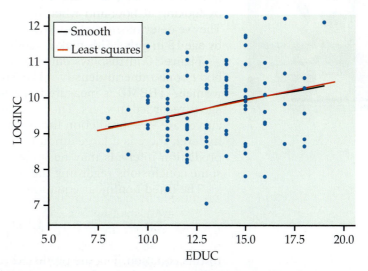

FIGURE 10.10 Scatterplot of income versus education with the new least squares line and smoothed curve, Example 10.13. The smoothed curve is now practically the same as the least-squares line.

A complete check of the residuals is still needed (see Exercise 10.11), but it appears that transforming *y* results in a data set that satisfies the linear regression model. Not only is the relationship more linear, but the distribution of the observations about the regression line is more Normal. *This is not always the end result of a transformation.* In other cases, transforming a variable may help linearity and harm the Normality and constant variance assumptions. Always check the residuals before proceeding with inference.

BEYOND THE BASICS

Nonlinear Regression

nonlinear models

When the relationship is not linear and a transformation does not work, we often use models that allow for various types of curved relationships. These models are called **nonlinear models**.

The technical details are much more complicated for nonlinear models. In general, we cannot write down simple formulas for the parameter estimates; we use a computer to solve systems of equations to find the estimates. However, the basic principles are those that we have already learned. For example,

$$\text{DATA} = \text{FIT} + \text{RESIDUAL}$$

still applies. The FIT is a nonlinear (curved) function, and the residuals are assumed to be an SRS from the $N(0, \sigma)$ distribution. The nonlinear function contains parameters that must be estimated from the data. Approximate standard errors for these estimates are part of the standard output provided by software. Here is an example.

EXAMPLE 10.14

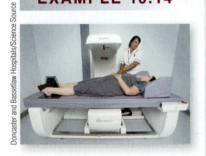

Doncaster and Bassetlaw Hospitals/Science Source

Investing in one's bone health. As we age, our bones become weaker and are more likely to break. Osteoporosis (or weak bones) is the major cause of bone fractures in older women. Various researchers have studied this problem by looking at how and when bone mass is accumulated by young women. They've determined that up to 90% of a person's peak bone mass is acquired by age 18 in girls.[3] This makes youth the best time to invest in stronger bones.

Figure 10.11 displays data for a measure of bone strength, called "total body bone mineral density" (TBBMD), and age for a sample of 256 young women.[4] TBBMD is measured in grams per square centimeter (g/cm^2), and age is recorded in years. The solid curve is the nonlinear fit, and the dashed curves are 95% prediction limits. Similar to our example of BMI and activity level, there is a large amount of scatter about the fitted curve. Although prediction intervals may be useless in this case, the researchers can draw some conclusions regarding the relationship.

The fitted nonlinear equation is

$$\hat{y} = 1.162 \frac{e^{-1.162 + 0.28x}}{1 + e^{-1.162 + 0.28x}}$$

In this equation, $\hat{y}$ is the predicted value of TBBMD, the response variable, and x is age, the explanatory variable. A straight line would not do a very good job of summarizing the relationship between TBBMD and age. At first, TBBMD increases with age, but then it levels off as age increases. The value

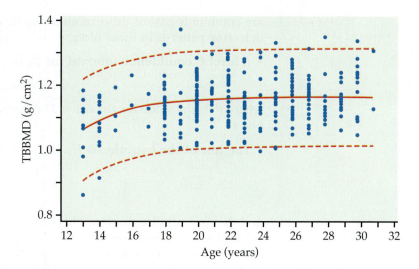

FIGURE 10.11 Plot of total body bone mineral density versus age, Example 10.14.

of the function where it is level is called "peak bone mass"; it is a parameter in the nonlinear model. The estimate is 1.162 and the standard error is 0.008. Software gives the 95% confidence interval as (1.146, 1.178). Other calculations could be done to determine the age by which 90% of this peak bone mass is acquired.

The long-range goals of the researchers who conducted this study include developing intervention programs (exercise and increasing calcium intake have been shown to be effective) for young women that will increase their TBBMD.

SECTION 10.1 SUMMARY

• The statistical model for **simple linear regression** assumes that the means of the response variable y fall on a line when plotted against x, with the observed y's varying Normally about these means. For n observations, this model can be written

$$y_i = \beta_0 + \beta_1 x_i + \epsilon_i$$

where $i = 1, 2, \ldots, n$, and the ϵ_i are assumed to be independent and Normally distributed with mean 0 and standard deviation σ. Here $\beta_0 + \beta_1 x_i$ is the mean response when $x = x_i$. The **parameters** of the model are β_0, β_1, and σ.

• The **population regression line** intercept and slope, β_0 and β_1, are estimated by the intercept and slope of the **least-squares regression line**, b_0 and b_1. The **model standard deviation σ** is estimated by

$$s = \sqrt{\frac{\sum e_i^2}{n - 2}}$$

where the e_i are the **residuals**

$$e_i = y_i - \hat{y}_i$$

• Prior to inference, always examine the residuals for Normality, constant variance, and any other remaining patterns in the data. **Plots of the residuals** both against the case number and against the explanatory variable are commonly part of this examination. Scatterplot smoothers are helpful in detecting patterns in these plots.

• A **level C confidence interval for β_1** is

$$b_1 \pm t^* SE_{b_1}$$

where t^* is the value for the $t(n - 2)$ density curve with area C between $-t^*$ and t^*.

• The **test of the hypothesis H_0: $\beta_1 = 0$** is based on the **t statistic**

$$t = \frac{b_1}{SE_{b_1}}$$

and the $t(n - 2)$ distribution. This tests whether there is a straight-line relationship between y and x. There are similar formulas for confidence intervals and tests for β_0, but these are meaningful only in special cases.

• The **estimated mean response** for the subpopulation corresponding to the value x^* of the explanatory variable is

$$\hat{\mu}_y = b_0 + b_1 x^*$$

• A **level C confidence interval for the mean response** is

$$\hat{\mu}_y \pm t^* SE_{\hat{\mu}}$$

where t^* is the value for the $t(n - 2)$ density curve with area C between $-t^*$ and t^*.

• The **estimated value of the response variable** y for a future observation from the subpopulation corresponding to the value x^* of the explanatory variable is

$$\hat{y} = b_0 + b_1 x^*$$

• A **level C prediction interval** for the estimated response is

$$\hat{y} \pm t^* SE_{\hat{y}}$$

where t^* is the value for the $t(n - 2)$ density curve with area C between $-t^*$ and t^*. The standard error for the prediction interval is larger than the confidence interval because it also includes the variability of the future observation around its subpopulation mean.

• Sometimes, a **transformation** of one or both of the variables can make their relationship linear. However, these transformations can harm the assumptions of Normality and constant variance, so it is important to examine the residuals.

SECTION 10.1 EXERCISES

For Exercise 10.1, see page 560; for Exercise 10.2, see page 565; for Exercises 10.3 and 10.4, see page 570; for Exercises 10.5 and 10.6, see pages 573–574.

10.7 What's wrong? For each of the following, explain what is wrong and why.

(a) The parameters of the simple linear regression model are b_0, b_1, and s.

(b) To test H_0: $b_1 = 0$, use a t test.

(c) For a particular value of the explanatory variable x, the confidence interval for the mean response will be wider than the prediction interval for a future observation.

10.8 What's wrong? For each of the following, explain what is wrong and why.

(a) The slope describes the change in x for a change in y.

(b) The population regression line is $y = b_0 + b_1 x$.

(c) A 95% confidence interval for the mean response is the same width regardless of x.

10.9 Importance of Normal model deviations. A general form of the central limit theorem tells us that the sampling distributions of b_0 and b_1 will be approximately Normal even if the model deviations are not Normally distributed. Using this fact, explain why the Normal distribution assumption is much more important for a prediction interval than for the confidence interval of the mean response at $x = x^*$.

10.10 Complete check of the residuals. In Example 10.12 (page 574), we checked model assumptions using a scatterplot (Figure 10.9). Let's consider assessing the model assumptions using the residuals. ENTRE

(a) Fit the (EDUC, INC) data using least-squares regression and obtain the residuals. Write down the least-squares regression line.

(b) Generate a plot of the residuals versus EDUC and comment on the pattern. Does a linear fit appear reasonable? Does there appear to be constant variance? Are there any unusual observations? Explain your answers.

(c) Construct a histogram and a Normal quantile plot of the residuals. Do the residuals appear Normal? Explain your answer.

(d) Analysis of the residuals is typically done because patterns in the residuals are easier to see. Do you think the plots in parts (b) and (c) magnify the violations of

assumptions better than the scatterplot in Figure 10.9? Write a short paragraph comparing the scatterplot with the residual plots.

10.11 Complete check of the residuals, continued. Refer to the previous exercise. In Example 10.13 (page 575), we checked model assumptions using a scatterplot (Figure 10.10) after log transforming the response variable. ENTRE

(a) Repeat parts (a) through (c) of the previous exercise using LOGINC and EDUC.

(b) Do you think we can comfortably perform inference using the log transformed y? Explain your answer.

10.12 College debt versus adjusted in-state costs. Kiplinger's "Best Values in Public Colleges" provides a ranking of U.S. public colleges based on a combination of various measures of academics and affordability.[5] Let's focus on the relationship between the average debt in dollars at graduation (AveDebt) and the in-state cost per year after need-based aid (InCostAid). A scatterplot with least-squares regression line is shown in Figure 10.12 for a sample of 25 colleges from Kiplinger's 2015 report. BESTVAL

(a) Does a linear relationship between InCostAid and AveDebt seem reasonable? Explain your answer.

(b) Are there any unusual cases in this sample? If yes, state which ones they are and how they may be affecting the least-squares model fit.

10.13 Can we consider this an SRS? Refer to the previous exercise. The report states that Kiplinger's rankings focus on traditional four-year public colleges with broad-based curricula. Each year, they start with

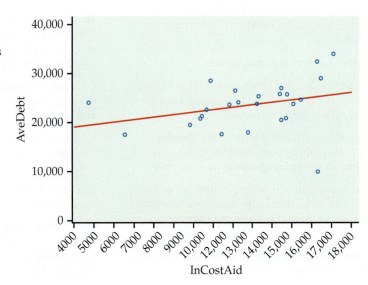

FIGURE 10.12 Scatterplot with least-squares regression line for a sample of 25 colleges from Kiplinger's 2015 report, Exercise 10.12.

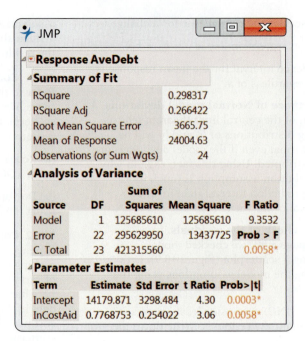

FIGURE 10.13 JMP output for the simple linear regression, Exercise 10.14.

more than 500 schools and then narrow the list down to roughly 120 based on academic quality before ranking them. The data set in the previous exercise is an SRS from Kiplinger's published list of 100 schools. As far as investigating the relationship between the average debt and the in-state cost after adjusting for need-based aid, is it reasonable to consider this to be an SRS from the population of more than 500 schools? Write a short paragraph explaining your answer.

10.14 Predicting college debt. Refer to Exercise 10.12. Baruch College has substantially less average debt compared to the other schools with similar in-state costs. Figure 10.13 contains JMP output for the simple linear regression of AveDebt on InCostAid with this case removed. BESTVAL

(a) State the least-squares regression line.

(b) The University of North Florida is one school in this sample. It has an in-state cost of $11,421 and average debt of $17,617. What is the residual?

(c) Construct a 95% confidence interval for the slope. What does this interval tell you about the change in average debt for a $1000 change in the in-state cost?

(d) Penn State University is reported to have an adjusted in-state cost of $23,053. Discuss the appropriateness of using this data set to predict the average debt for this university.

10.15 More on predicting college debt. Refer to the previous exercise. Appalachian State University has an

in-state cost of $7372, and Texas A&M University has an in-state cost of $10,566.

(a) Using your answer to part (a) of the previous exercise, what is the predicted average debt for a student at Appalachian State University?

(b) What is the predicted average debt for a student at Texas A&M University?

(c) Without doing any calculations, would the 95% margin of error for the predicted average debt be larger for Appalachian State University or Texas A&M University? Explain your answer.

10.16 Impact of an unusual observation. Refer to Exercise 10.14. Baruch College was removed from this analysis because it was deemed an outlier. Let's investigate its impact on the fit. BESTVAL

(a) Refit the model using the entire sample of 25 schools. Create a table that summarizes the model estimates with and without this case.

(b) Describe the impact this observation has on the fit of the linear regression model.

(c) If you were writing a report for publication, would you include the fit with or without this case? Explain your answer.

10.17 Predicting college debt: Other measures. Refer to Exercise 10.12. Let's look at AveDebt and its relationship with the other explanatory variables in the data set. In addition to the in-state cost after aid

(InCostAid), there is the admittance rate (Admit), the four-year graduation rate (GradRate), and out-of-state cost after aid (OutCostAid). **BESTVAL**

(a) Generate scatterplots of each explanatory variable and AveDebt. Do all these relationships look linear? Describe what you see. Does Baruch College still look unusual?

(b) Fit each of the explanatory variables separately and create a table that lists the explanatory variable, estimated model standard deviation s, and the P-value for the test of a linear association. For each analysis, make sure to specify whether you removed Baruch College or not.

(c) Which variable do you think is the best single explanatory variable of average debt? Explain your answer.

10.18 Are the two fuel efficiency measurements similar? Refer to Exercise 7.32 (page 429). In addition to the computer calculating miles per gallon (mpg), the driver also measured mpg by dividing the miles driven by the number of gallons at fill-up. The driver wants to determine if these calculations are similar. **MPGDIFF**

Fill-up	1	2	3	4	5	6	7	8	9	10
Computer	41.5	50.7	36.6	37.3	34.2	45.0	48.0	43.2	47.7	42.2
Driver	36.5	44.2	37.2	35.6	30.5	40.5	40.0	41.0	42.8	39.2

Fill-up	11	12	13	14	15	16	17	18	19	20
Computer	43.2	44.6	48.4	46.4	46.8	39.2	37.3	43.5	44.3	43.3
Driver	38.8	44.5	45.4	45.3	45.7	34.2	35.2	39.8	44.9	47.5

(a) Consider the driver's mpg calculations as the explanatory variable. Plot the data and describe the relationship. Are there any outliers or unusual values? Does a linear relationship seem reasonable?

(b) Run the simple linear regression and state the least-squares regression line.

(c) Summarize the results. Does it appear that the computer and driver calculations are the same? Explain your answer.

10.19 Is the number of tornadoes increasing? The Storm Prediction Center of the National Oceanic and Atmospheric Administration maintains a database of tornadoes, floods, and other weather phenomena. Table 10.1 summarizes the annual number of tornadoes in the United States between 1953 and 2014.[6] **TWISTER**

(a) Make a plot of the total number of tornadoes by year. Does a linear trend over years appear reasonable? Are there any outliers or unusual patterns? Explain your answer.

(b) Run the simple linear regression and report the least-squares regression line.

(c) A friend of yours thinks you made a mistake fitting the model because b_0 is a large negative value. Explain to him why this is not a mistake.

(d) Obtain the residuals and plot them versus year. Are there any unusual patterns or cases that you did not discuss in part (a)? If so, comment on them.

TABLE 10.1 Annual Number of Tornadoes in The United States Between 1953 and 2014

Year	Number of tornadoes	Year	Number of tornadoes	Year	Number of tornadoes	Year	Number of tornadoes
1953	421	1969	608	1985	684	2001	1215
1954	550	1970	653	1986	764	2002	934
1955	593	1971	888	1987	656	2003	1374
1956	504	1972	741	1988	702	2004	1817
1957	856	1973	1102	1989	856	2005	1265
1958	564	1974	947	1990	1133	2006	1103
1959	604	1975	920	1991	1132	2007	1096
1960	616	1976	835	1992	1298	2008	1692
1961	697	1977	852	1993	1176	2009	1156
1962	657	1978	788	1994	1082	2010	1282
1963	464	1979	852	1995	1235	2011	1691
1964	704	1980	866	1996	1173	2012	938
1965	906	1981	783	1997	1148	2013	907
1966	585	1982	1046	1998	1449	2014	888
1967	926	1983	931	1999	1340		
1968	660	1984	907	2000	1075		

(e) Are the residuals approximately Normal? Justify your answer.

(f) Based on the these residual checks, are you confident proceeding with inference? Explain your answer.

10.20 Annual increase? Refer to the previous exercise. Let's proceed with inference. [📊] TWISTER

(a) Do these data support a linear trend in the number of tornadoes? Justify your answer.

(b) Construct a 95% confidence interval for the average annual increase in the number of tornadoes. Explain how this interval can be used to justify your response in part (a).

(c) What is the predicted number of tornadoes in 2015?

(d) Provide an interval that should contain the actual count 95% of the time.

10.21 Computer memory. The capacity of memory commonly available at retail has increased rapidly over time.[7] [📊] MEM

(a) Make a scatterplot of the data. The growth is much faster than linear.

(b) Compute the logarithm of capacity and plot it against year. Are these points closer to a straight line?

(c) Fit the simple linear regression model with logarithm of capacity as the response and year as the explanatory variable. Give a 90% confidence interval for the slope of the population regression line.

(d) Write a brief summary describing the change in memory capacity over time using the confidence interval from part (c).

10.22 Alternative model. Refer to Exercise 10.19. The number of tornadoes in 2004 is much larger than expected and the number in 2014 is much smaller than expected. In fact, most of the large positive and negative deviations occur later in time. This suggests there may not be constant variance. Because the response variable is a count, one can argue the variance is not constant (for example, see the Poisson distribution, page 329). [📊] TWISTER

(a) Take the natural logarithm of the count and refit the model. What is the least-squares regression line?

(b) Check the residuals of this model. Does the linear regression model fit these data? Explain your answer.

(c) When the response y is on the log scale, the slope approximates the percent change in y for a unit increase in x. Construct an approximate 95% confidence interval for the annual percent change.

(d) Does this model also support the hypothesis that tornadoes have increased over time? Explain your answer.

(e) Construct a prediction interval for the predicted number of tornadoes in 2015 and compare it with the interval from part (d) of Exercise 10.19. (*Note:* An approximate interval can be constructed by first obtaining a prediction interval for log y and then taking the antilog (inverse function of log) of each interval endpoint.)

(f) Which of the two models (and prediction) do you prefer? Explain why.

10.2 More Detail about Simple Linear Regression

When you complete this section, you will be able to:

- Construct a linear regression analysis of variance (ANOVA) table.
- Use an ANOVA table to perform the ANOVA F test and draw appropriate conclusions regarding H_0: $\beta_1 = 0$.
- Use an ANOVA table to compute the square of the sample correlation and provide an interpretation of it in terms of explained variation.
- Perform, using a calculator, inference in simple linear regression when a computer is not available.
- Distinguish the formulas for the standard error that we use for a confidence interval for the mean response and the standard error that we use for a prediction interval when $x = x^*$.
- Test the hypothesis that there is no linear association in the population and summarize the results.
- Explain the close connection between the tests H_0: $\beta_1 = 0$ and H_0: $\rho = 0$.

In this section, we study three topics. The first is analysis of variance for regression. If you plan to read Chapter 11 on multiple regression, you should study this material. The second topic concerns computations for regression inference. The section we just completed assumes that you have access to software or a statistical calculator. Here we present and illustrate the use of formulas for the inference procedures. Finally, we discuss inference for correlation.

Analysis of variance for regression

The usual computer output for regression includes additional calculations called **analysis of variance**. Analysis of variance, often abbreviated ANOVA, is essential for multiple regression (Chapter 11) and for comparing several means (Chapters 12 and 13). Analysis of variance summarizes information about the sources of variation in the data. It is based on the

analysis of variance

$$\text{DATA} = \text{FIT} + \text{RESIDUAL}$$

framework (page 560).

The total variation in the response y is expressed by the deviations $y_i - \bar{y}$. If these deviations were all 0, all observations would be equal and there would be no variation in the response. There are two reasons the individual observations y_i are not all equal to their mean $\bar{y}$.

1. The responses y_i correspond to different values of the explanatory variable x and will differ because of that. The fitted value $\hat{y}_i$ estimates the mean response for x_i. The differences $\hat{y}_i - \bar{y}$ reflect the variation in mean response due to differences in the x_i. This variation is accounted for by the regression line because the $\hat{y}$'s lie exactly on the line.

2. Individual observations will vary about their mean because of variation within the subpopulation of responses for a fixed x_i. This variation is represented by the residuals $y_i - \hat{y}_i$ that record the scatter of the actual observations about the fitted line.

The overall deviation of any y observation from the mean of the y's is the sum of these two deviations:

$$(y_i - \bar{y}) = (\hat{y}_i - \bar{y}) + (y_i - \hat{y}_i)$$

In terms of deviations, this equation expresses the idea that DATA = FIT + RESIDUAL.

Several times, we have measured variation by an average of squared deviations. If we square each of the preceding three deviations and then sum over all n observations, it can be shown that the sums of squares add:

$$\sum(y_i - \bar{y})^2 = \sum(\hat{y}_i - \bar{y})^2 + \sum(y_i - \hat{y}_i)^2$$

We rewrite this equation as

$$\text{SST} = \text{SSM} + \text{SSE}$$

where

$$\text{SST} = \sum(y_i - \bar{y})^2$$
$$\text{SSM} = \sum(\hat{y}_i - \bar{y})^2$$
$$\text{SSE} = \sum(y_i - \hat{y}_i)^2$$

sum of squares

The SS in each abbreviation stands for **sum of squares,** and the T, M, and E stand for total, model, and error, respectively. ("Error" here stands for deviations from the line, which might better be called "residual" or "unexplained variation.") The total variation, as expressed by SST, is the sum of the variation due to the straight-line model (SSM) and the variation due to deviations from this model (SSE). This partition of the variation in the data between two sources is the heart of analysis of variance.

If H_0: $\beta_1 = 0$ were true, there would be no subpopulations, and all of the y's should be viewed as coming from a single population with mean μ_y. The variation of the y's would then be described by the sample variance

$$s_y^2 = \frac{\sum(y_i - \bar{y})^2}{n - 1}$$

The numerator in this expression is SST. The denominator is the total degrees of freedom, or simply DFT.

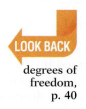

LOOK BACK

degrees of
freedom,
p. 40

Just as the total sum of squares SST is the sum of SSM and SSE, the total degrees of freedom DFT is the sum of DFM and DFE, the degrees of freedom for the model and for the error:

$$DFT = DFM + DFE$$

The model has one explanatory variable x, so the degrees of freedom for this source are DFM = 1. Because DFT = $n - 1$, this leaves DFE = $n - 2$ as the degrees of freedom for error.

mean square

For each source, the ratio of the sum of squares to the degrees of freedom is called the **mean square**, or simply MS. The general formula for a mean square is

$$MS = \frac{\text{sum of squares}}{\text{degrees of freedom}}$$

Each mean square is an average squared deviation. MST is just s_y^2, the sample variance that we would calculate if all of the data came from a single population. MSE is also familiar to us:

$$MSE = s^2 = \frac{\sum(y_i - \hat{y}_i)^2}{n - 2}$$

It is our estimate of σ^2, the variance about the population regression line.

SUMS OF SQUARES, DEGREES OF FREEDOM, AND MEAN SQUARES

Sums of squares represent variation present in the responses. They are calculated by summing squared deviations. **Analysis of variance** partitions the total variation between two sources.

The sums of squares are related by the formula

$$SST = SSM + SSE$$

That is, the total variation is partitioned into two parts, one due to the model and one due to deviations from the model.

Degrees of freedom are associated with each sum of squares. They are related in the same way:

$$DFT = DFM + DFE$$

To calculate **mean squares**, use the formula

$$MS = \frac{\text{sum of squares}}{\text{degrees of freedom}}$$

interpretation of r^2   In Section 2.4 (page 116), we noted that r^2 is the fraction of variation in the values of y that is explained by the least-squares regression of y on x. The sums of squares make this interpretation precise. Recall that SST = SSM + SSE. It is an algebraic fact that

$$r^2 = \frac{\text{SSM}}{\text{SST}} = \frac{\sum(\hat{y}_i - \bar{y})^2}{\sum(y_i - \bar{y})^2}$$

Because SST is the total variation in y and SSM is the variation due to the regression of y on x, this equation is the precise statement of the fact that r^2 is the fraction of variation in y explained by x in the linear regression.

The ANOVA *F* test

The null hypothesis H_0: $\beta_1 = 0$ that y is not linearly related to x can be tested by comparing MSM with MSE. The ANOVA test statistic is an **F statistic**,

F statistic

$$F = \frac{\text{MSM}}{\text{MSE}}$$

When H_0 is true, this statistic has an F distribution with 1 degree of freedom in the numerator and $n - 2$ degrees of freedom in the denominator. These degrees of freedom are those of MSM and MSE. When $\beta_1 \neq 0$, MSM tends to be large relative to MSE. So, large values of F are evidence against H_0 in favor of the two-sided alternative.

F distributions   The **F distributions** are a family of distributions with two parameters: the degrees of freedom of the mean square in the numerator and denominator of the F statistic. The F distributions are another of R. A. Fisher's contributions to statistics and are called F in his honor. Fisher introduced F statistics for comparing several means. We meet these useful statistics in Chapters 14 and 15.

The numerator degrees of freedom are always mentioned first. Interchanging the degrees of freedom changes the distribution, so the order is important. Our brief notation will be $F(j, k)$ for the F distribution with j degrees of freedom in the numerator and k in the denominator. The F distributions are not symmetric but are right-skewed. The density curve in Figure 10.14 illustrates the shape. Because mean squares cannot be negative, the F statistic takes only positive values, and the F distribution has no probability below 0. The peak of the F density curve is near 1; values much greater than 1 provide evidence against the null hypothesis.

We recommend using statistical software for calculations involving F distributions. We do, however, supply a table of critical values similar to Table D.

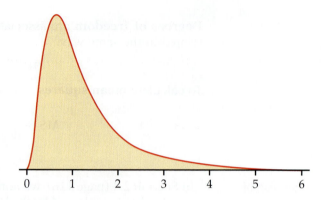

FIGURE 10.14 The density curve for the $F(9,10)$ distribution. The F distributions are skewed to the right.

Tables of F critical values are awkward because a separate table is needed for every pair of degrees of freedom j and k. Table E in the back of the book gives upper p critical values of the F distributions for $p = 0.10, 0.05, 0.025, 0.01$, and 0.001.

ANALYSIS OF VARIANCE F TEST

In the simple linear regression model, the hypotheses

$$H_0: \beta_1 = 0$$
$$H_a: \beta_1 \neq 0$$

are tested by the **F statistic**

$$F = \frac{\text{MSM}}{\text{MSE}}$$

The P-value is the probability that a random variable having the $F(1, n-2)$ distribution is greater than or equal to the calculated value of the F statistic.

The F statistic tests the same null hypothesis as one of the t statistics that we encountered earlier in this chapter, so it is not surprising that the two are related. It is an algebraic fact that $t^2 = F$ in this case. For linear regression with one explanatory variable, we prefer the t form of the test because it more easily allows us to test one-sided alternatives and is closely related to the confidence interval for β_1.

The ANOVA calculations are displayed in an *analysis of variance table*, often abbreviated **ANOVA table**. Here is the format of the table for simple linear regression:

ANOVA table

Source	Degrees of freedom	Sum of squares	Mean square	F
Model	1	$\sum(\hat{y}_i - \bar{y})^2$	SSM/DFM	MSM/MSE
Error	$n - 2$	$\sum(y_i - \hat{y}_i)^2$	SSE/DFE	
Total	$n - 1$	$\sum(y_i - \bar{y})^2$	SST/DFT	

EXAMPLE 10.15

PABMI

Interpreting SAS output for BMI and physical activity. The output generated by SAS for the physical activity study in Example 10.3 is given in Figure 10.15. Note that SAS uses the labels Model and Error but replaces Total with Corrected Total. Other statistical software packages may use slightly different labels. The F statistic is 17.10; the P-value is given as < 0.0001. There is strong evidence against the null hypothesis that there is no relationship between BMI and average number of steps per day (PA).

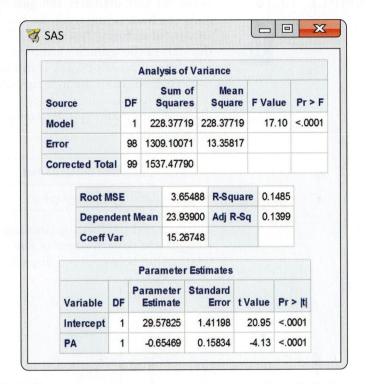

Analysis of Variance

Source	DF	Sum of Squares	Mean Square	F Value	Pr > F
Model	1	228.37719	228.37719	17.10	<.0001
Error	98	1309.10071	13.35817		
Corrected Total	99	1537.47790			

Root MSE	3.65488	R-Square	0.1485
Dependent Mean	23.93900	Adj R-Sq	0.1399
Coeff Var	15.26748		

Parameter Estimates

Variable	DF	Parameter Estimate	Standard Error	t Value	Pr > |t|
Intercept	1	29.57825	1.41198	20.95	<.0001
PA	1	-0.65469	0.15834	-4.13	<.0001

FIGURE 10.15 SAS output for the physical activity study, Example 10.15.

Now look at the output for the regression coefficients. The t statistic for PA is given as -4.13. If we square this number, we obtain the F statistic (accurate up to roundoff error). The value of r^2 is also given in the output. Average number of steps per day explains only 14.9% of the variability in BMI. *Strong evidence against the null hypothesis that there is no relationship does not imply that a large percentage of the total variability is explained by the model.*

USE YOUR KNOWLEDGE

10.23 Reading linear regression outputs. Figure 10.4 (pages 562–564) shows the regression output from three other software packages and Excel. Create a table that lists the labels each output uses in its ANOVA table, the F statistic, its P-value, and r^2. Which of the outputs do you prefer? Explain your answer.

Calculations for regression inference

We recommend using statistical software for regression calculations. With time and care, however, the work is feasible with a calculator. We will use the following example to illustrate how to perform inference for regression analysis using a calculator.

EXAMPLE 10.16

Umbilical cord diameter and gestational age. Knowing the gestational age (GA) of a fetus is important for biochemical screening tests and planning for successful delivery. Typically GA is calculated as the number of days since the start of the woman's last menstrual period (LMP). However, for women with irregular periods, GA is difficult to compute, and ultrasound imaging is often used. In the search for helpful ultrasound measurements, a group of Nigerian researchers looked at the relationship between umbilical cord diameter (mm) and gestational age based on LMP (weeks).[8] Here is a small subset of the data:

Umbilical cord diameter (x)	2	6	9	14	21	23
Gestational age (y)	16	18	26	33	28	39

The data and the least-squares line are plotted in Figure 10.16. The strong straight-line pattern suggests that we can use linear regression to model the relationship between cord diameter and gestational age.

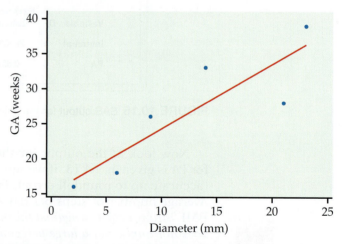

FIGURE 10.16 Scatterplot and regression line, Example 10.16.

We begin our regression calculations by fitting the least-squares line. Fitting the line gives estimates b_1 and b_0 of the model parameters β_1 and β_0. Next we examine the residuals from the fitted line and obtain an estimate s of the remaining parameter σ. These calculations are preliminary to inference. Finally, we use s to obtain the standard errors needed for the various interval estimates and significance tests. *Roundoff errors that accumulate during these calculations can ruin the final results. Be sure to carry many significant digits and check your work carefully.*

Preliminary calculations After examining the scatterplot (Figure 10.16) to verify that the data show a straight-line pattern, we begin our calculations.

EXAMPLE 10.17

GADIA

Summary statistics for gestational age study. We start by making a table with the mean and standard deviation for each of the variables, the correlation, and the sample size. These calculations should be familiar from Chapters 1 and 2. Here is the summary:

Variable	Mean	Standard deviation	Correlation	Sample size
Diameter	$\bar{x} = 12.5$	$s_x = 8.36062$	$r = 0.87699$	$n = 6$
Gestational age	$\bar{y} = 26.66667$	$s_y = 8.75595$		

These quantities are the building blocks for our calculations.

We will need one additional quantity for the calculations to follow. It is the expression $\sum(x_i - \bar{x})^2$. We obtain this quantity as an intermediate step when we calculate s_x. You could also find it using the fact that $\sum(x_i - \bar{x})^2 = (n - 1)s_x^2$. You should verify that the value for our example is

$$\sum(x_i - \bar{x})^2 = (2 - 12.5)^2 + (6 - 12.5)^2 + \cdots + (23 - 12.5)^2 = 349.5$$

Our first task is to find the least-squares line. This is easy with the building blocks that we have assembled.

EXAMPLE 10.18

Computing the least-squares regression line. The slope of the least-squares line is

$$b_1 = r\frac{s_y}{s_x}$$

$$= 0.87699\frac{8.75595}{8.36062}$$

$$= 0.91846$$

The intercept is

$$b_0 = \bar{y} - b_1\bar{x}$$

$$= 26.66667 - (0.91846)(12.5)$$

$$= 15.18592$$

The equation of the least-squares regression line is therefore

$$\hat{y} = 15.1859 + 0.9185x$$

This is the line shown in Figure 10.16.

We now have estimates of the first two parameters, β_0 and β_1, of our linear regression model. Next, we find the estimate of the third parameter, σ: the standard deviation s about the fitted line. To do this we need to find the predicted values and then the residuals.

EXAMPLE 10.19

GADIA

Computing the predicted values and residuals. The first observation is a diameter of $x = 2$. The corresponding predicted value of gestational age is

$$\hat{y}_1 = b_0 + b_1 x_1$$
$$= 15.1859 + (0.9185)(2)$$
$$= 17.023$$

and the residual is

$$e_1 = y_1 - \hat{y}_1$$
$$= 16 - (17.023)$$
$$= -1.023$$

The residuals for the other diameters are calculated in the same way. They are $-2.697, 2.548, 4.955, -6.474$ and 2.689, respectively. Notice that the sum of these six residuals is zero (except for some roundoff error). When doing these calculations by hand, it is always helpful to check that the sum of the residuals is zero.

EXAMPLE 10.20

Computing s^2. The estimate of σ^2 is s^2, the sum of the squares of the residuals divided by $n - 2$. The estimated standard deviation about the line is the square root of this quantity.

$$s^2 = \frac{\sum e_i^2}{n - 2}$$
$$= \frac{(-1.023)^2 + (-2.697)^2 + \cdots + (2.689)^2}{4}$$
$$= 22.127$$

So the estimate of the standard deviation about the line is

$$s = \sqrt{22.12702} = 4.704$$

USE YOUR KNOWLEDGE

GADIA

10.24 Computing residuals. Refer to Examples 10.17, 10.18, and 10.19.

(a) Show that the least squares line goes through $(\bar{x}, \bar{y})$.

(b) Verify that the other five residuals are as stated in Example 10.19.

Inference for slope and intercept Confidence intervals and significance tests for the slope β_1 and intercept β_0 of the population regression line make use of the estimates b_1 and b_0 and their standard errors. Some algebra and the rules for variances establishes that the standard deviation of b_1 is

LOOK BACK

rules for
variances,
p. 258

$$\sigma_{b_1} = \frac{\sigma}{\sqrt{\sum(x_i - \bar{x})^2}}$$

Similarly, the standard deviation of b_0 is

$$\sigma_{b_0} = \sigma \sqrt{\frac{1}{n} + \frac{\bar{x}^2}{\sum(x_i - \bar{x})^2}}$$

To estimate these standard deviations, we need only replace σ by its estimate s.

STANDARD ERRORS FOR ESTIMATED REGRESSION COEFFICIENTS

The **standard error of the slope** b_1 of the least-squares regression line is

$$SE_{b_1} = \frac{s}{\sqrt{\sum(x_i - \bar{x})^2}}$$

The **standard error of the intercept** b_0 is

$$SE_{b_0} = s \sqrt{\frac{1}{n} + \frac{\bar{x}^2}{\sum(x_i - \bar{x})^2}}$$

The plot of the regression line with the data in Figure 10.16 shows a very strong relationship, but our sample size is small. We assess the situation with a significance test for the slope.

EXAMPLE 10.21

Testing the slope. First we need the standard error of the estimated slope:

$$SE_{b_1} = \frac{s}{\sqrt{\sum(x_i - \bar{x})^2}}$$

$$= \frac{4.704}{\sqrt{349.5}}$$

$$= 0.2516$$

To test

$$H_0: \beta_1 = 0$$
$$H_a: \beta_1 \neq 0$$

calculate the t statistic:

$$t = \frac{b_1}{SE_{b_1}}$$

$$= \frac{0.9185}{0.2516} = 3.65$$

Using Table D with $n - 2 = 4$ degree of freedom, we conclude that $0.02 < P < 0.04$. The exact P-value obtained from software is 0.022. The data provide evidence in favor of a linear relationship between gestational age and umbilical cord diameter ($t = 3.65$, df $= 4$, $0.02 < P < 0.04$).

Two things are important to note about this example. First, *it is important to remember that we need to have a very large effect if we expect to detect a slope different from zero with a small sample size.* The estimated slope is more than

3.5 standard deviations away from zero, but we are not much below the 0.05 standard for statistical significance. Second, because we expect gestational age to increase with increasing diameter, a one-sided significance test would be justified in this setting.

The significance test tells us that the data provide sufficient information to conclude that gestational age and umbilical cord diameter are linearly related. We use the estimate b_1 and its confidence interval to further describe the relationship.

EXAMPLE 10.22

Computing a 95% confidence interval for the slope. Let's find a 95% confidence interval for the slope β_1. The degrees of freedom are $n - 2 = 4$, so t^* from Table D is 2.776. We compute

$$b_1 \pm t^* \mathrm{SE}_{b_1} = 0.9185 \pm (2.776)(0.2516)$$
$$= 0.9185 \pm 0.6984$$

The interval is (0.220, 1.617). For each additional millimeter in diameter, the gestational age of the fetus is expected to be 0.220 to 1.617 weeks older.

In this example, the intercept β_0 does not have a meaningful interpretation. An umbilical cord diameter of zero millimeters is not realistic. For problems where inference for β_0 is appropriate, the calculations are performed in the same way as those for β_1. Note that there is a different formula for the standard error, however.

Confidence intervals for the mean response and prediction intervals for a future observation When we substitute a particular value x^* of the explanatory variable into the regression equation and obtain a value of $\hat{y}$, we can view the result in two ways:

1. We have estimated the mean response μ_y.

2. We have predicted a future value of the response y.

The margins of error for these two uses are often quite different. Prediction intervals for an individual response are wider than confidence intervals for estimating a mean response. We now proceed with the details of these calculations. Once again, standard errors are the essential quantities. And once again, these standard errors are multiples of s, our basic measure of the variability of the responses about the fitted line.

STANDARD ERRORS FOR $\hat{\mu}$ AND $\hat{Y}$

The standard error of $\hat{\mu}$ is

$$\mathrm{SE}_{\hat{\mu}} = s \sqrt{\frac{1}{n} + \frac{(x^* - \bar{x})^2}{\sum(x_i - \bar{x})^2}}$$

The standard error for predicting an individual response $\hat{y}$ is

$$\mathrm{SE}_{\hat{y}} = s \sqrt{1 + \frac{1}{n} + \frac{(x^* - \bar{x})^2}{\sum(x_i - \bar{x})^2}}$$

Note that the only difference between the formulas for these two standard errors is the extra 1 under the square root sign in the standard error for prediction. This standard error is larger due to the additional variation of individual responses about the mean response. This additional variation remains regardless of the sample size n and is the reason that prediction intervals are wider than the confidence intervals for the mean response.

For the gestational age example, we can think about the average gestational age for a particular subpopulation, defined by the umbilical cord diameter. The confidence interval would provide an interval estimate of this subpopulation mean. On the other hand, we might want to predict the gestational age for a new fetus. A prediction interval attempts to capture this new observation.

EXAMPLE 10.23

Computing a confidence interval for μ. Let's find a 95% confidence interval for the average gestational age when the umbilical cord diameter is 10 millimeters. The estimated mean age is

$$\hat{\mu} = b_0 + b_1 x$$
$$= 15.1859 + (0.9185)(10)$$
$$= 24.371$$

The standard error is

$$SE_{\hat{\mu}} = s\sqrt{\frac{1}{n} + \frac{(x^* - \bar{x})^2}{\sum(x_i - \bar{x})^2}}$$

$$= 4.704\sqrt{\frac{1}{6} + \frac{(10.0 - 12.5)^2}{349.5}}$$

$$= 2.021$$

To find the 95% confidence interval, we compute

$$\hat{\mu} \pm t^*SE_{\hat{\mu}} = 24.371 \pm (2.776)(2.021)$$
$$= 24.371 \pm 5.610$$

The interval is 18.761 to 29.981 weeks of age. This is a pretty wide interval given gestation lasts for about 40 weeks.

Calculations for the prediction intervals are similar. The only difference is the use of the formula for $SE_{\hat{y}}$ in place of $SE_{\hat{\mu}}$. This results in a much wider interval. In fact, the interval is slightly more than 28 weeks in width. Even though a linear relationship was found statistically significant, it does not appear umbilical cord diameter is a precise predictor of gestational age.

LOOK BACK

correlation, p. 101

Inference for correlation

The correlation coefficient is a measure of the strength and direction of the *linear* association between two variables. Correlation does not require an explanatory-response relationship between the variables. We can consider the sample correlation r as an estimate of the correlation in the population and base inference about the population correlation on r.

population correlation

The correlation between the variables x and y when they are measured for every member of a population is the **population correlation**. As usual, we use Greek letters to represent population parameters. In this case ρ (the Greek letter rho) is the population correlation.

When $\rho = 0$, there is no linear association in the population. In the important case where the two variables x and y are both Normally distributed, the condition $\rho = 0$ is equivalent to the statement that x and y are independent. That is, there is no association of any kind between x and y. (Technically, the condition required is that x and y be **jointly Normal**. This means that the distribution of x is Normal and also that the conditional distribution of y, given any fixed value of x, is Normal.) We, therefore, may wish to test the null hypothesis that a population correlation is 0.

jointly Normal variables

TEST FOR A ZERO POPULATION CORRELATION

To test the hypothesis H_0: $\rho = 0$, compute the *t* **statistic**

$$t = \frac{r\sqrt{n - 2}}{\sqrt{1 - r^2}}$$

where n is the sample size and r is the sample correlation.

In terms of a random variable T having the $t(n - 2)$ distribution, the P-value for a test of H_0 against

H_a: $\rho > 0$	is	$P(T \geq t)$

H_a: $\rho < 0$	is	$P(T \leq t)$

H_a: $\rho \neq 0$	is	$2P(T \geq |t|)$

Most computer packages have routines for calculating correlations, and some will provide the significance test for the null hypothesis that ρ is zero.

EXAMPLE 10.24

Correlation in the physical activity study. The Minitab output for the physical activity example appears in Figure 10.17. The sample correlation between BMI and the average number of steps per day (PA) is $r = -0.385$. Minitab calls this a Pearson correlation to distinguish it from other kinds of correlations that it can calculate. The P-value for a two-sided test of H_0: $\rho = 0$ is given as 0.000. This means that the actual P-value is less than 0.0005. We conclude that there is a nonzero correlation between BMI and PA.

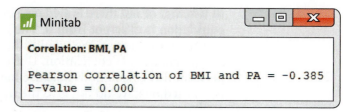

FIGURE 10.17 Minitab output for the physical activity study, Example 10.24.

If we wanted to test the one-sided alternative that the population correlation is negative, we divide the *P*-value in the output by 2, after checking that the sample coefficient is in fact negative.

If your software does not give the significance test, you can do the computations easily with a calculator.

EXAMPLE 10.25

Correlation test using a calculator. The correlation between BMI and PA is $r = -0.385$. Recall that $n = 100$. The *t* statistic for testing the null hypothesis that the population correlation is zero is

$$t = \frac{r\sqrt{n-2}}{\sqrt{1-r^2}}$$

$$= \frac{-0.385\sqrt{100-2}}{\sqrt{1-(-0.385)^2}}$$

$$= -4.13$$

The degrees of freedom are $n - 2 = 98$. From Table D, we conclude that $P < 0.0001$. This agrees with the Minitab output in Figure 10.17, where the *P*-value is given as 0.000. The data provide clear evidence that BMI and PA are related.

There is a close connection between the significance test for a correlation and the test for the slope in a linear regression. Recall that

$$b_1 = r\frac{s_y}{s_x}$$

From this fact, we see that if the slope is 0, so is the correlation, and vice versa. It should come as no surprise to learn that the procedures for testing $H_0: \beta_1 = 0$ and $H_0: \rho = 0$ are also closely related. In fact, the *t* statistics for testing these hypotheses are numerically equal. That is,

$$\frac{b_1}{SE_{b_1}} = \frac{r\sqrt{n-2}}{\sqrt{1-r^2}}$$

Check that this holds in both of our examples.

In our examples, the conclusion that there is a statistically significant correlation between the two variables would not come as a surprise to anyone familiar with the meaning of these variables. The significance test simply tells us whether or not there is evidence in the data to conclude that the population correlation is different from 0. The actual size of the correlation is of considerably more interest. We would, therefore, like to give a confidence interval for the population correlation. Unfortunately, most software packages do not perform this calculation. Because hand calculation of the confidence interval is very tedious, we do not give the method here.[9]

SECTION 10.2 SUMMARY

• The **ANOVA table** for a linear regression gives the degrees of freedom, sum of squares, and mean squares for the model, error, and total sources of variation. The **ANOVA F statistic** is the ratio MSM/MSE. Under H_0: $\beta_1 = 0$, this statistic has an $F(1, n - 2)$ distribution and is used to test H_0 versus the two-sided alternative.

• The **square of the sample correlation** can be expressed as

$$r^2 = \frac{\text{SSM}}{\text{SST}}$$

and is interpreted as the proportion of the variability in the response variable y that is explained by the explanatory variable x in the linear regression.

• The **standard errors for b_0 and b_1** are

$$\text{SE}_{b_0} = s\sqrt{\frac{1}{n} + \frac{\bar{x}^2}{\sum(x_i - \bar{x})^2}}$$

$$\text{SE}_{b_1} = \frac{s}{\sqrt{\sum(x_i - \bar{x})^2}}$$

• The **standard error that we use for a confidence interval** for the estimated mean response for the subpopulation corresponding to the value x^* of the explanatory variable is

$$\text{SE}_{\hat{\mu}} = s\sqrt{\frac{1}{n} + \frac{(x^* - \bar{x})^2}{\sum(x_i - \bar{x})^2}}$$

• The **standard error that we use for a prediction interval** for a future observation from the subpopulation corresponding to the value x^* of the explanatory variable is

$$\text{SE}_{\hat{y}} = s\sqrt{1 + \frac{1}{n} + \frac{(x^* - \bar{x})^2}{\sum(x_i - \bar{x})^2}}$$

• When the variables y and x are jointly Normal, the sample correlation is an estimate of the **population correlation ρ**. The test of H_0: $\rho = 0$ is based on the **t statistic**

$$t = \frac{r\sqrt{n - 2}}{\sqrt{1 - r^2}}$$

which has a $t(n - 2)$ distribution under H_0. This test statistic is numerically identical to the t statistic used to test H_0: $\beta_1 = 0$.

SECTION 10.2 EXERCISES

For Exercise 10.23, see page 587; and for Exercise 10.24, see page 590.

10.25 What's wrong? For each of the following, explain what is wrong and why.

(a) In simple linear regression, the null hypothesis of the ANOVA F test is $H_0: \beta_0 = 0$.

(b) In an ANOVA table, the mean squares add. In other words, MST = MSM + MSE.

(c) The smaller the P-value for the ANOVA F test, the greater the explanatory power of the model.

(d) The total degrees of freedom in an ANOVA table are equal to the number of observations n.

10.26 What's wrong? For each of the following, explain what is wrong and why.

(a) In simple linear regression, the standard error for a future observation is s, the measure of spread about the regression line.

(b) In an ANOVA table, SSE is the sum of the deviations.

(c) There is a close connection between the correlation r and the intercept of the regression line.

(d) The squared correlation r^2 is equal to MSM/MST.

10.27 Research and development obligations. The National Science Foundation collects data on the research and development obligations for science and engineering to universities and colleges in the United States.[10] Here are the data for the years 2006, 2008, 2010, and 2012:

Year	2006	2008	2010	2012
Spending (billions of dollars)	31.0	35.2	39.4	41.7

Do the following by hand or with a calculator and verify your results with a software package of Excel. [NSF]

(a) Make a scatterplot that shows the increase in research and development obligations over time. Does the pattern suggest that the obligations are increasing linearly over time?

(b) Find the equation of the least-squares regression line for predicting obligations from year. Add this line to your scatterplot.

(c) For each of the four years, find the residual. Use these residuals to calculate the estimated model standard error s.

(d) Write the regression model for this setting. What are your estimates of the unknown parameters in this model?

(e) Compute a 95% confidence interval for the slope and summarize what this interval tells you about the increase in obligations over time.

10.28 Food neophobia. Food neophobia is a personality trait associated with avoiding unfamiliar foods. In one study of 564 children who were two to six years of age, food neophobia and the frequency of consumption of different types of food were measured.[11] Here is a summary of the correlations:

Type of food	Correlation
Vegetables	−0.27
Fruit	−0.16
Meat	−0.15
Eggs	−0.08
Sweet/fatty snacks	0.04
Starchy staples	−0.02

Perform the significance test for each correlation and write a summary about food neophobia and the consumption of different types of food.

10.29 Correlation between the prevalences of adult binge drinking and underage drinking. A group of researchers compiled data on the prevalence of adult binge drinking and the prevalence of underage drinking in 42 states.[12] A correlation of 0.32 was reported.

(a) Test the null hypothesis that the population correlation $\rho = 0$ against the alternative $\rho > 0$. Are the results significant at the 5% level?

(b) Explain this correlation in terms of the direction of the association and the percent of variability in the prevalence of underage drinking that is explained by the prevalence of adult binge drinking.

(c) The researchers collected information from 42 of 50 states, so almost all the data available was used in the analysis. Provide an argument for the use of statistical inference in this setting.

10.30 Grade inflation. The average undergraduate GPA for American colleges and universities was estimated based on a sample of institutions that published this information.[13] Here are the data for public schools in that report:

Year	1992	1996	2002	2007
GPA	2.85	2.90	2.97	3.01

Do the following by hand or with a calculator and verify your results with a software package. [GRADEUP]

(a) Make a scatterplot that shows the increase in GPA over time. Does a linear increase appear reasonable?

(b) Find the equation of the least-squares regression line for predicting GPA from year. Add this line to your scatterplot.

(c) Compute a 95% confidence interval for the slope and summarize what this interval tells you about the increase in GPA over time.

10.31 Completing an ANOVA table. How are returns on common stocks in overseas markets related to returns in U.S. markets? Consider measuring U.S. returns by the annual rate of return on the Standard & Poor's 500 stock index and overseas returns by the annual rate of return on the Morgan Stanley Europe, Australasia, Far East (EAFE) index.[14] Both are recorded in percents. We will regress the EAFE returns on the S&P 500 returns for the years 1989 to 2014. Here is part of the Minitab output for this regression:

The regression equation is

EAFE = −3.19 + 0.813 S&P

Analysis of Variance

Source	DF	SS	MS	F
Regression	1	5552.9		
Residual Error				
Total		10077.9		

Using the ANOVA table format on page 586 as a guide, complete the analysis of variance table.

10.32 Interpreting statistical software output. Refer to the previous exercise. What are the values of the estimated model standard error s and the squared correlation r^2?

10.33 Confidence intervals for the slope and intercept. Refer to the previous two exercises. The mean and standard deviation of the S&P 500 returns for these years is 12.04% and 18.33%, respecitvely. From this and your work in the previous exercise:

(a) Find the standard error for the least-squares slope b_1.

(b) Give a 95% confidence interval for the slope β_1 of the population regression line.

(c) Explain why the intercept β_0 is meaningful in this example.

(d) Find the standard error for the least-squares intercept b_0 and use it to construct a 95% confidence interval.

CHAPTER 10 EXERCISES

10.34 School budget and number of students. Suppose that there is a linear relationship between the number of students x in a school system and the annual budget y. Write a population regression model to describe this relationship.

(a) Which parameter in your model is the fixed cost in the budget (for example, the salary of the principals and some administrative costs) that does not change as x increases?

(b) Which parameter in your model shows how total cost changes when there are more students in the system? Do you expect this number to be greater than 0 or less than 0?

(c) Actual data from various school systems will not fit a straight line exactly. What term in your model allows variation among schools of the same size x?

10.35 Interpreting a residual plot. Figure 10.18 shows four plots of residuals versus x. For each plot, comment on the regression model conditions necessary for inference. Which plots suggest a reasonable fit to the linear regression model?

10.36 The relationship between cell phone use and academic performance. College students are the most rapid adopters of cell phone technology. They use the phone to surf the Internet, watch videos, listen to music, email, and play video games. Because a cell phone is almost always nearby, researchers have begun studying the relationship between cell phone use and various attitudes and behaviors. In one study, researchers assessed the relationship between cell phone use (CPU), cumulative GPA, anxiety, and general life satisfaction (GLS) using 496 students.[15]

(a) Participants were undergraduates from a large midwestern university. They were recruited during class time from courses in Sociology, General Biology, American Politics, Human Nutrition, and World History. The researchers argued these courses attracted students from a diversity of majors. To participate, students had to consent to have their GPA retrieved. What do you think about this recruitment process? Can we feel comfortable assuming this is an SRS from the population of undergraduates? Write a short summary of your opinions.

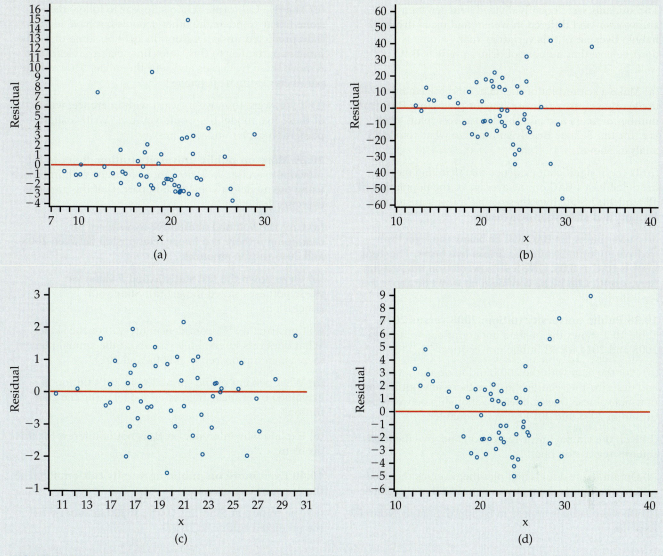

FIGURE 10.18 Four plots of residuals versus *x*, Exercise 10.35.

(b) The following table summarizes the pairwise correlations among the four variables. For each pair of variables, test the null hypothesis that the correlation is zero. Make sure to state the test statistic, degrees of freedom, and *P*-value.

	GPA	Anxiety	GLS
CPU	−0.203	0.096	0.012
GPA		0.004	0.207
Anxiety			−0.221

(c) Write a short paragraph that summarizes your findings.

10.37 Beer and blood alcohol. How well does the number of beers a student drinks predict his or her blood alcohol content (BAC)? Sixteen student volunteers at Ohio State University drank a randomly assigned number of 12-ounce cans of beer. Thirty minutes after consuming their last beer, a police officer measured their BAC. Here are the data:[16]

Student	1	2	3	4	5	6	7	8
Beers	5	2	9	8	3	7	3	5
BAC	0.10	0.03	0.19	0.12	0.04	0.095	0.07	0.06

Student	9	10	11	12	13	14	15	16
Beers	3	5	4	6	5	7	1	4
BAC	0.02	0.05	0.07	0.10	0.085	0.09	0.01	0.05

The students were equally divided between men and women and differed in weight and usual drinking habits. Because of this variation, many students don't believe that number of drinks predicts BAC well. [iii] BAC

(a) Make a scatterplot of the data. Find the equation of the least-squares regression line for predicting BAC from number of beers and add this line to your plot. What is r^2 for these data? Briefly summarize what your data analysis shows.

(b) Is there significant evidence that drinking more beers increases BAC on the average in the population of all students? State hypotheses, give a test statistic and P-value, and state your conclusion.

(c) Steve thinks his BAC will be below the legal limit to drive 30 minutes after he drinks five beers. The legal limit is BAC = 0.08. Give a 90% prediction interval for Steve's BAC. Can he be confident he won't be above the legal limit?

10.38 Public university tuition: 2008 versus 2014. Table 10.2 shows the in-state undergraduate tuition in 2008 and 2014 for 33 public universities.[17] [iii] TUIT

(a) Plot the data with the 2008 tuition on the x axis and describe the relationship. Are there any outliers or unusual values? Does a linear relationship between the tuition in 2008 and 2014 seem reasonable?

(b) Run the simple linear regression and give the least-squares regression line.

(c) Obtain the residuals and plot them versus the 2008 tuition amount. Is there anything unusual in the plot?

(d) Do the residuals appear to be approximately Normal? Explain.

(e) The five California schools appear to follow the same linear trend as the other schools but have higher-than-predicted in-state tuition in 2014. Assume that this jump is particular to this state (financial troubles?) and remove these five cases and refit the model. How do the parameter estimates change?

(f) If you were to move forward with inference, which of these two model fits would you use? Write a short paragraph explaining your answer.

10.39 More on public university tuition. Refer to the previous exercise. We'll now move forward with inference using the model fit you chose in part (f) of the previous exercise. [iii] TUIT

(a) Give the null and alternative hypotheses for examining if there is a linear relationship between 2008 and 2014 tuition amounts.

(b) Write down the test statistic and P-value for the hypotheses stated in part (a). State your conclusions.

(c) Construct a 95% confidence interval for the slope. What does this interval tell you about the annual percent increase in tuition between 2008 and 2014?

(d) What percent of the variability in 2014 tuition is explained by a linear regression model using the 2008 tuition?

(e) Explain why inference on β_0 is not of interest for this problem.

10.40 Even more on public university tuition. Refer to the previous two exercises. [iii] TUIT

(a) The tuition at Skinflint U was $8800 in 2008. What is the predicted tuition in 2014?

TABLE 10.2	In-state Tuition and Fees (in Dollars) for 33 Public Universities							
School	2008	2014	School	2008	2014	School	2008	2014
Penn State	13706	17955	Ohio State	8679	10995	Texas	8532	11094
Pittsburgh	13642	18075	Virginia	9300	13373	Nebraska	6584	8724
Michigan	11037	14126	Cal-Davis	8635	15589	Iowa	6544	8807
Rutgers	11540	14297	Cal-Berkeley	7656	14421	Colorado	7278	10388
Michigan State	10214	13771	Cal-Irvine	8046	14686	Iowa State	6360	8430
Maryland	8005	9734	Purdue	7750	10868	North Carolina	5397	8616
Illinois	12106	15938	Cal-San Diego	8062	14785	Kansas	7042	10760
Minnesota	10756	14889	Oregon	6435	10254	Arizona	5542	11205
Missouri	8467	10186	Wisconsin	7564	11429	Florida	3778	6748
Buffalo	6285	8784	Washington	6802	13757	Georgia Tech	6040	11094
Indiana	8231	10991	UCLA	7551	14224	Texas A&M	7844	9461

(b) The tuition at I.O.U. was $15,700 in 2008. What is the predicted tuition in 2014?

(c) Discuss the appropriateness of using the fitted equation to predict tuition for each of these universities.

10.41 Predicting public university tuition: 2000 versus 2014. Refer to Exercise 10.39. The data file also includes the in-state undergraduate tuition for the year 2000. TUIT

(a) Run the simple linear regression using year 2000 in place of year 2008. What is the least-squares line?

(b) Obtain the residuals and check model assumptions.

(c) If you had to choose between the model using 2008 tuition and the model using 2000 tuition, which would you choose? Give reasons for your answers.

10.42 U.S. versus overseas stock returns. Returns on common stocks in the United States and overseas appear to be growing more closely correlated as economies become more interdependent. Suppose that the following population regression line connects the total annual returns (in percent) on two indexes of stock prices:

MEAN OVERSEAS RETURN = −0.08 + 0.20 × U.S.RETURN

(a) What is β_0 in this line? What does this number say about overseas returns when the U.S. market is flat (0% return)?

(b) What is β_1 in this line? What does this number say about the relationship between U.S. and overseas returns?

(c) We know that overseas returns will vary in years when U.S. returns do not vary. Write the regression model based on the population regression line given above. What part of this model allows overseas returns to vary when U.S. returns remain the same?

10.43 Performance bonuses. In the National Football League (NFL), performance bonuses now account for roughly 25% of player compensation.[18] Does tying a player's salary into performance bonuses result in better individual or team success on the field? Focusing on linebackers, let's look at the relationship between a player's end-of-year production rating and the percent of his salary devoted to incentive payments in that same year. PERFPAY

(a) Use numerical and graphical methods to describe the two variables and summarize your results.

(b) Both variable distributions are non-Normal. Does this necessarily pose a problem for performing linear regression? Explain.

(c) Construct a scatterplot of the data and describe the relationship. Are there any outliers or unusual values? Does a linear relationship between the percent of salary and the player rating seem reasonable? Is it a very strong relationship? Explain.

(d) Run the simple linear regression and state the least-squares regression line.

(e) Obtain the residuals and assess whether the assumptions for the linear regression analysis are reasonable. Include all plots and numerical summaries used in doing this assessment.

10.44 Performance bonuses, continued. Refer to the previous exercise. PERFPAY

(a) Now run the simple linear regression for the variables sqrt(rating) and percent of salary devoted to incentive payments.

(b) Obtain the residuals and assess whether the assumptions for the linear regression analysis are reasonable. Include all plots and numerical summaries used in doing this assessment.

(c) Construct a 95% confidence interval for the square root increase in rating given a 1% increase in the percent of salary devoted to incentive payments.

(d) Consider the values 0%, 20%, 40%, 60%, and 80% salary devoted to incentives. Compute the predicted rating for this model and for the one in the previous exercise. For the model in this problem, you will need to square the predicted value to get back to the original units.

(e) Plot the predicted values versus the percent and connect those values from the same model. For which regions of percent do the predicted values from the two models differ the most?

(f) Based on the comparison of regression models (both predicted values and residuals), which model do you prefer? Explain.

10.45 Sales price versus assessed value. Real estate is typically reassessed annually for property tax purposes. This assessed value, however, is not necessarily the same as the fair market value of the property. Table 10.3 summarizes an SRS of 35 homes recently sold in a midwestern city.[19] Both variables are measured in thousands of dollars. SALES

(a) Inspect the data. How many homes have a sales price greater than the assessed value? Do you think this trend would be true for the larger population of all homes recently sold? Explain your answer.

(b) Make a scatterplot with assessed value on the horizontal axis. Briefly describe the relationship between assessed value and sales price.

(c) Based on the scatterplot, there is one distinctly unusual observation. State which property it is, and describe the impact you expect this observation has on the least-squares line.

TABLE 10.3	Sales Price and Assessed Value (in Thousands of $) of 35 Homes in a Midwestern City							
Property	Sales price	Assessed value	Property	Sales price	Assessed value	Property	Sales price	Assessed value
1	83.0	87.0	13	249.9	192.0	25	146.0	121.1
2	129.9	103.8	14	112.0	117.4	26	230.5	212.1
3	125.0	111.0	15	133.0	117.2	27	360.0	167.9
4	245.0	157.4	16	177.5	116.6	28	127.9	110.2
5	100.0	127.5	17	162.5	143.7	29	205.0	183.2
6	134.7	127.7	18	238.0	198.2	30	163.5	93.6
7	106.0	110.9	19	120.9	93.4	31	225.0	156.2
8	91.5	90.8	20	142.5	92.3	32	335.0	278.1
9	170.0	160.7	21	299.0	279.0	33	192.0	151.0
10	295.0	250.5	22	82.5	90.4	34	232.0	178.8
11	179.0	160.9	23	152.5	103.2	35	197.9	172.4
12	230.0	213.2	24	139.9	114.9			

(d) Report the least-squares regression line for predicting selling price from assessed value using all 35 properties. What is the estimated model standard error?

(e) Now remove the unusual observation and fit the data again. Report the least-squares regression line and estimated model standard error.

(f) Compare the two sets of results. Describe the impact this unusual observation has on the results.

(g) Do you think it is more appropriate to consider all 35 properties for linear regression analysis or just consider the 34 properties? Explain your decision.

10.46 Sales price versus assessed value, continued. Refer to the previous exercise. Let's consider linear regression analysis using just 34 properties. SALES

(a) Obtain the residuals and plot them versus assessed value. Is there anything unusual to report? If so, explain.

(b) Do the residuals appear to be approximately Normal? Describe how you assessed this.

(c) Based on your answers to parts (a) and (b), do you think the assumptions for statistical inference are reasonably satisfied? Explain your answer.

(d) Construct a 95% confidence interval for the slope and summarize the results.

(e) Using the result from part (d), compare the estimated regression line with $y = x$, which says that, on average, the selling price is equal to the assessed value. Is there evidence that this model is not reasonable? In other words, is the selling price typically larger or smaller than the assessed value? Explain your answer.

10.47 Gambling and alcohol use by first-year college students. Gambling and alcohol use are problematic behaviors for many college students. One study looked at 908 first-year students from a large northeastern university.[20] Each participant was asked to fill out the 10-item Alcohol Use Disorders Identification Test (AUDIT) and a seven-item inventory used in prior gambling research among college students. AUDIT assesses alcohol consumption and other alcohol-related risks and problems (a higher score means more risks). A correlation of 0.29 was reported between the frequency of gambling and the AUDIT score.

(a) What percent of the variability in AUDIT score is explained by frequency of gambling?

(b) Test the null hypothesis that the correlation between the gambling frequency and the AUDIT score is zero.

(c) The sample in this study represents 45% of the students contacted for the online study. To what extent do you think these results apply to all first-year students at this university? To what extent do you think these results apply to all first-year students? Give reasons for your answers.

10.48 Predicting water quality. The index of biotic integrity (IBI) is a measure of the water quality in streams. IBI and land use measures for a collection of streams in the Ozark Highland ecoregion of Arkansas were collected as part of a study.[21] Table 10.4 gives the data for IBI, the percent of the watershed that was forest, and the area of the watershed in square kilometers for streams in the original sample with watershed area less than or equal to 70 km². IBI

(a) Use numerical and graphical methods to describe the variable IBI. Do the same for area. Summarize your results.

TABLE 10.4   **Watershed Area (km², Percent Forest, and Index of Biotic Integrity**

Area	Forest	IBI	Area	Forest	IBI	Area	Forest	IBI	Area	Forest	IBI	Area	Forest	IBI
21	0	47	29	0	61	31	0	39	32	0	59	34	0	72
34	0	76	49	3	85	52	3	89	2	7	74	70	8	89
6	9	33	28	10	46	21	10	32	59	11	80	69	14	80
47	17	78	8	17	53	8	18	43	58	21	88	54	22	84
10	25	62	57	31	55	18	32	29	19	33	29	39	33	54
49	33	78	9	39	71	5	41	55	14	43	58	9	43	71
23	47	33	31	49	59	18	49	81	16	52	71	21	52	75
32	59	64	10	63	41	26	68	82	9	75	60	54	79	84
12	79	83	21	80	82	27	86	82	23	89	86	26	90	79
16	95	67	26	95	56	26	100	85	28	100	91			

(b) Plot the data and describe the relationship between IBI and area. Are there any outliers or unusual patterns?

(c) Give the statistical model for simple linear regression for this problem.

(d) State the null and alternative hypotheses for examining the relationship between IBI and area.

(e) Run the simple linear regression and summarize the results.

(f) Obtain the residuals and plot them versus area. Is there anything unusual in the plot?

(g) Do the residuals appear to be approximately Normal? Give reasons for your answer.

(h) Do the assumptions for the analysis of these data using the model you gave in part (c) appear to be reasonable? Explain your answer.

10.49 More on predicting water quality. The researchers who conducted the study described in the previous exercise also recorded the percent of the watershed area that was forest for each of the streams. These data are also given in Table 10.4. Analyze these data using the questions in the previous exercise as a guide. IBI

10.50 Comparing the analyses. In Exercises 10.48 and 10.49, you used two different explanatory variables to predict IBI. Summarize the two analyses and compare the results. If you had to choose between the two explanatory variables for predicting IBI, which one would you prefer? Give reasons for your answer. IBI

10.51 How an outlier can affect statistical significance. Consider the data in Table 10.4 and the relationship between IBI and the percent of watershed

area that was forest. The relationship between these two variables is almost significant at the 0.05 level. In this exercise, you will demonstrate the potential effect of an outlier on statistical significance. Investigate what happens when you decrease the IBI to 0.0 for (1) an observation with 0% forest and (2) an observation with 100% forest. Write a short summary of what you learn from this exercise. IBI

10.52 Predicting water quality for an area of 40 km². Refer to Exercise 10.48. IBI

(a) Find a 95% confidence interval for the mean response corresponding to an area of 40 km².

(b) Find a 95% prediction interval for a future response corresponding to an area of 40 km².

(c) Write a short paragraph interpreting the meaning of the intervals in terms of Ozark Highland streams.

(d) Do you think that these results can be applied to other streams in Arkansas or in other states? Explain why or why not.

10.53 Compare the predictions. Refer to Exercise 10.50. Another way to compare analyses is to compare predictions. Consider Case 37 in Table 10.4 (8th row, 2nd column). For this case, the area is 10 km² and the percent forest is 63%. Calculate the predicted index of biotic integrity based on area and the predicted index of biotic integrity based on percent forest. Compare these two predictions and explain why they differ. Use the idea of a prediction interval to interpret these results. IBI

10.54 Reading test scores and IQ. For a study of reading ability in schoolchildren, researchers collected reading test scores and IQ scores for a sample of 60 fifth-grade children.[22] READIQ

(a) Run the regression and summarize the results of the significance tests.

(b) Rerun the analysis with the four possible outliers removed. Summarize your findings, paying particular attention to the effects of removing the outliers.

10.55 Leaning Tower of Pisa. The Leaning Tower of Pisa is an architectural wonder. Engineers concerned about the tower's stability have done extensive studies of its increasing tilt. Measurements of the lean of the tower over time provide much useful information. The following table gives measurements for the years 1975 to 1987. The variable "lean" represents the difference between where a point on the tower would be if the tower were straight and where it actually is. The data are coded as tenths of a millimeter in excess of 2.9 meters, so that the 1975 lean, which was 2.9642 meters, appears in the table as 642. Only the last two digits of the year were entered into the computer.[23] PISA

Year	75	76	77	78	79	80	81	82	83	84	85	86	87
Lean	642	644	656	667	673	688	696	698	713	717	725	742	757

(a) Plot the data. Does the trend in lean over time appear to be linear?

(b) What is the equation of the least-squares line? What percent of the variation in lean is explained by this line?

(c) Give a 99% confidence interval for the average rate of change (tenths of a millimeter per year) of the lean.

10.56 More on the Leaning Tower of Pisa. Refer to the previous exercise. PISA

(a) In 1918 the lean was 2.9071 meters. (The coded value is 71.) Using the least-squares equation for the years 1975 to 1987, calculate a predicted value for the lean in 1918. (Note that you must use the coded value 18 for year.)

(b) Although the least-squares line gives an excellent fit to the data for 1975 to 1987, this pattern did not extend back to 1918. Write a short statement explaining why this conclusion follows from the information available. Use numerical and graphical summaries to support your explanation.

10.57 Predicting the lean in 2016. Refer to the previous two exercises. PISA

(a) How would you code the explanatory variable for the year 2016?

(b) The engineers working on the Leaning Tower of Pisa were most interested in how much the tower would lean if no corrective action was taken. Use the least-squares equation to predict the tower's lean in the year 2016. (*Note*: The tower was renovated in 2001 to make sure it does not fall down.)

(c) To give a margin of error for the lean in 2016, would you use a confidence interval for a mean response or a prediction interval? Explain your choice.

10.58 Does a math pretest predict success? Can a pretest on mathematics skills predict success in a statistics course? The 62 students in an introductory statistics class took a pretest at the beginning of the semester. The least-squares regression line for predicting the score y on the final exam from the pretest score x was $\hat{y} = 13.8 + 0.81x$. The standard error of b_1 was 0.43.

(a) Test the null hypothesis that there is no linear relationship between the pretest score and the score on the final exam against the two-sided alternative.

(b) Would you reject this null hypothesis versus the one-sided alternative that the slope is positive? Explain your answer.

10.59 Significance test of the correlation. A study reported a correlation $r = 0.5$ based on a sample size of $n = 15$; another reported the same correlation based on a sample size of $n = 25$. For each, perform the test of the null hypothesis that $\rho = 0$. Describe the results and explain why the conclusions are different.

10.60 State and college binge drinking. Excessive consumption of alcohol is associated with numerous adverse consequences. In one study, researchers analyzed binge-drinking rates from two national surveys, the Harvard School of Public Health College Alcohol Study (CAS) and the Centers for Disease Control and Prevention's Behavioral Risk Factor Surveillance System (BRFSS).[24] The CAS survey was used to provide an estimate of the college binge-drinking rate in each state, and the BRFSS was used to determine the adult binge-drinking rate in each state. A correlation of 0.43 was reported between these two rates for their sample of $n = 40$ states. The college binge-drinking rate had a mean of 46.5% and standard deviation 13.5%. The adult binge-drinking rate had a mean of 14.88% and standard deviation 3.8%.

(a) Find the equation of the least-squares line for predicting the college binge-drinking rate from the adult binge-drinking rate.

(b) Give the results of the significance test for the null hypothesis that the slope is 0. (*Hint:* What is the relation between this test and the test for a zero correlation?)

10.61 SAT versus ACT. The SAT and the ACT are the two major standardized tests that colleges use to evaluate candidates. Most students take just one of these tests. However, some students take both. Consider the scores of 60 students who did this. How can we relate the two tests? **SATACT**

(a) Plot the data with SAT on the x axis and ACT on the y axis. Describe the overall pattern and any unusual observations.

(b) Find the least-squares regression line and draw it on your plot. Give the results of the significance test for the slope.

(c) What is the correlation between the two tests?

10.62 SAT versus ACT, continued. Refer to the previous exercise. Find the predicted value of ACT for each observation in the data set. **SATACT**

(a) What is the mean of these predicted values? Compare it with the mean of the ACT scores.

(b) Compare the standard deviation of the predicted values with the standard deviation of the actual ACT scores. If least-squares regression is used to predict ACT scores for a large number of students such as these, the average predicted value will be accurate, but the variability of the predicted scores will be too small.

(c) Find the SAT score for a student who is 1 standard deviation above the mean ($z = (x - \bar{x})/s_x = 1$). Find the predicted ACT score and standardize this score. (Use the means and standard deviations from this set of data for these calculations.)

(d) Repeat part (c) for a student whose SAT score is 1 standard deviation below the mean ($z = -1$).

(e) What do you conclude from parts (c) and (d)? Perform additional calculations for different z's if needed.

10.63 Matching standardized scores. Refer to the previous two exercises. An alternative to the least-squares method is based on matching standardized scores. Specifically, we set

$$\frac{(y - \bar{y})}{s_y} = \frac{(x - \bar{x})}{s_x}$$

and solve for y. Let's use the notation $y = a_0 + a_1x$ for this line. The slope is $a_1 = s_y/s_x$ and the intercept is $a_0 = \bar{y} - a_1\bar{x}$. Compare these expressions with the formulas for the least-squares slope and intercept (page 561). **SATACT**

(a) Using the data in the previous exercise, find the values of a_0 and a_1.

(b) Plot the data with the least-squares line and the new prediction line.

(c) Use the new line to find predicted ACT scores. Find the mean and the standard deviation of these scores. How do they compare with the mean and standard deviation of the ACT scores?

10.64 Index of biotic integrity. Refer to the data on the index of biotic integrity and area in Exercise 10.48 (page 602) and the additional data on percent watershed area that was forest in Exercise 10.49. Find the correlations among these three variables, perform the test of statistical significance, and summarize the results. Which of these test results could have been obtained from the analyses that you performed in Exercises 10.48 and 10.49? **IBI**

10.65 A mechanistic explanation of popularity. Previous experimental work has suggested that the serotonin system plays an important and causal role in social status. In other words, genes may predispose individuals to be popular/likable. As part of a recent study on adolescents, an experimenter looked at the relationship between the expression of a particular serotonin receptor gene, a person's "popularity," and the person's rule-breaking (RB) behaviors.[25] RB was measured by both a questionnaire and video observation. The composite score is an equal combination of these two assessments. Here is a table of the correlations:

Rule-breaking measure	Popularity	Gene expression
Sample 1 ($n = 123$)		
RB.composite	0.28	0.26
RB.questionnaire	0.22	0.23
RB.video	0.24	0.20
Sample 1 Caucasians only ($n = 96$)		
RB.composite	0.22	0.23
RB.questionnaire	0.16	0.24
RB.video	0.19	0.16

For each correlation, test the null hypothesis that the corresponding true correlation is zero. Reproduce the table and mark the correlations that have $P < 0.001$ with ***, those that have $P < 0.01$ with **, and those that have $P < 0.05$ with *. Write a summary of the results of your significance tests.

10.66 Resting metabolic rate and exercise. Metabolic rate, the rate at which the body consumes energy, is important in studies of weight gain, dieting, and exercise. The following table gives data on the lean body mass and resting metabolic rate for 12 women and seven men who are subjects in a study of dieting. Lean body mass, given in kilograms, is a person's weight leaving out all fat. Metabolic rate is measured in calories burned per 24 hours, the same calories used to describe the energy content of foods. The researchers believe that lean body mass is an important influence on metabolic rate. METRATE

Subject	Sex	Mass	Rate	Subject	Sex	Mass	Rate
1	M	62.0	1792	11	F	40.3	1189
2	M	62.9	1666	12	F	33.1	913
3	F	36.1	995	13	M	51.9	1460
4	F	54.6	1425	14	F	42.4	1124
5	F	48.5	1396	15	F	34.5	1052
6	F	42.0	1418	16	F	51.1	1347
7	M	47.4	1362	17	F	41.2	1204
8	F	50.6	1502	18	M	51.9	1867
9	F	42.0	1256	19	M	46.9	1439
10	M	48.7	1614				

(a) Make a scatterplot of the data, using different symbols or colors for men and women. Summarize what you see in the plot.

(b) Run the regression to predict metabolic rate from lean body mass for the women in the sample and summarize the results. Do the same for the men.

10.67 Resting metabolic rate and exercise, continued. Refer to the previous exercise. It is tempting to conclude that there is a strong linear relationship for the women but no relationship for the men. Let's look at this issue a little more carefully. METRATE

(a) Find the confidence interval for the slope in the regression equation that you ran for the females. Do the same for the males. What do these suggest about the possibility that these two slopes are the same? (The formal method for making this comparison is a bit complicated and is beyond the scope of this chapter.)

(b) Examine the formula for the standard error of the regression slope given on page 591. The term in the denominator is $\sqrt{\Sigma(x_i - \bar{x})^2}$. Find this quantity for the females; do the same for the males. How do these calculations help to explain the results of the significance tests?

(c) Suppose that you were able to collect additional data for males. How would you use lean body mass in deciding which subjects to choose?

Barry Austin Photography/Getty Images

Multiple Regression

Introduction

In Chapter 10, we presented methods for inference in the setting of a linear relationship between a response variable y and a *single* explanatory variable x. In this chapter, we use *more than one* explanatory variable to explain or predict a single response variable.

Many of the ideas that we encountered in our study of simple linear regression carry over to the multiple linear regression setting. For example, the descriptive tools we learned in Chapter 2—scatterplots, least-squares regression, and correlation—are still essential preliminaries to inference and also provide a foundation for confidence intervals and significance tests.

The introduction of several explanatory variables leads to many additional considerations. In this short chapter, we cannot explore all these issues. Rather, we will outline some basic facts about inference in the multiple regression setting and then illustrate the analysis with a case study whose purpose was to predict success in college based on standardized tests and several high school achievement scores.

11.1 Inference for Multiple Regression

11.2 A Case Study

11.1 Inference for Multiple Regression

When you complete these two sections, you will be able to:

- Describe the multiple linear regression model in terms of a population regression line and the distribution of deviations of the response variable y from this line.
- Interpret statistical software regression output to obtain the least-squares regression equation and estimated model standard deviation, multiple correlation coefficient, ANOVA F test, and individual regression coefficient t tests.
- Explain the difference between the ANOVA F test and the t tests for individual coefficients.
- Interpret a level C confidence interval and a significance test for a regression coefficient.
- Use residual plots to check the assumptions of the multiple linear regression model.

Population multiple regression equation

The simple linear regression model assumes that the mean of the response variable y depends on the explanatory variable x according to a linear equation

$$\mu_y = \beta_0 + \beta_1 x$$

For any fixed value of x, the response y varies Normally about this mean and has a standard deviation σ that is the same for all values of x.

In the multiple regression setting, the response variable y depends on p explanatory variables, which we will denote by $x_1, x_2, \ldots, x_p$. The mean response depends on these explanatory variables according to a linear function

$$\mu_y = \beta_0 + \beta_1 x_1 + \beta_2 x_2 + \cdots + \beta_p x_p$$

population regression equation   Similar to simple linear regression, this expression is the **population regression equation**, and the observed values y vary about their means given by this equation.

Just as we did in simple linear regression, we can also think of this model in terms of subpopulations of responses. Here, each subpopulation corresponds to a particular set of values for *all* the explanatory variables $x_1, x_2, \ldots, x_p$. In each subpopulation, y varies Normally with a mean given by the population regression equation. The regression model assumes that the standard deviation σ of the responses is the same in all subpopulations.

EXAMPLE 11.1

GPA

Predicting early success in college. Our case study is based on data collected on science majors at a large university.[1] The purpose of the study was to attempt to predict success in the early university years. Success was measured using the cumulative grade point average (GPA) after three semesters. The explanatory variables were achievement scores available at the time of enrollment in the university. These included their average high school grades in mathematics (HSM), science (HSS), and English (HSE).

We will use high school grades to predict the response variable GPA. There are $p = 3$ explanatory variables: $x_1 =$ HSM, $x_2 =$ HSS, and $x_3 =$ HSE. The high school grades are coded on a scale from 1 to 10, with 10 corresponding to A, 9 to A−, 8 to B+, and so on. These grades define the subpopulations. For example, the straight-C students are the subpopulation defined by HSM = 4, HSS = 4, and HSE = 4.

One possible multiple regression model for the subpopulation mean GPAs is

$$\mu_{\text{GPA}} = \beta_0 + \beta_1 \text{HSM} + \beta_2 \text{HSS} + \beta_3 \text{HSE}$$

For the straight-C subpopulation of students, the model gives the subpopulation mean as

$$\mu_{\text{GPA}} = \beta_0 + \beta_1 4 + \beta_2 4 + \beta_3 4$$

Data for multiple regression

The data for a simple linear regression problem consist of observations (x_i, y_i) of the two variables. Because there are several explanatory variables in multiple regression, the notation needed to describe the data is more elaborate. Each observation or case consists of a value for the response variable and for each of the explanatory variables. Call x_{ij} the value of the jth explanatory variable for the ith case. The data are then

$$\text{Case 1: } (y_1, x_{11}, x_{12}, \ldots, x_{1p})$$
$$\text{Case 2: } (y_2, x_{21}, x_{22}, \ldots, x_{2p})$$
$$\vdots$$
$$\text{Case } n: (y_n, x_{n1}, x_{n2}, \ldots, x_{np})$$

Here, n is the number of cases and p is the number of explanatory variables. Data are often entered into computer regression programs in this format. Each row is a case and each column corresponds to a different variable.

The data for Example 11.1, with several additional explanatory variables, appear in this format in the GPA data file. Figure 11.1 shows the first six rows entered into an Excel spreadsheet. Grade point average (GPA) is the response variable, followed by $p = 7$ explanatory variables, six achievement scores and sex. There are a total of $n = 150$ students in this data set.

	A	B	C	D	E	F	G	H	I
1	OBS	GPA	HSM	HSS	HSE	SATM	SATCR	SATW	SEX
2	1	3.84	10	10	10	630	570	590	1
3	2	3.97	10	10	10	750	700	630	0
4	3	3.49	8	10	9	570	510	490	1
5	4	1.95	6	4	8	640	600	610	0
6	5	2.59	8	10	9	510	490	490	1
7	6	3	7	10	10	660	680	630	0

FIGURE 11.1 Format of data set for Example 11.1 in an Excel spreadsheet.

The six achievement scores are all quantitative explanatory variables. SEX is an **indicator variable** using the numeric values 0 and 1 to represent male and female, respectively. Indicator variables are used frequently in multiple regression to represent the levels or groups of a categorical explanatory variable. See Exercise 11.22 (page 635) for more discussion of their use in a multiple regression model.

indicator variable

USE YOUR KNOWLEDGE

11.1 Describing a multiple regression. To minimize the negative impact of math anxiety on achievement in a research design course, a group of researchers implemented a series of feedback sessions, in which the teacher went over the small-group assignments and discussed the most frequently committed errors.[2] At the end of the course, data from 166 students were obtained. The researchers investigated how students' final exam scores were explained by math course anxiety, math test anxiety, numerical task anxiety, enjoyment, self-confidence, motivation, and perceived usefulness of the feedback sessions.

(a) What is the response variable?

(b) What are the cases and what is n, the number of cases?

(c) What is p, the number of explanatory variables?

(d) What are the explanatory variables?

Multiple linear regression model

Similar to simple linear regression, we combine the population regression equation and the assumptions about how the observed y vary about their means to construct the multiple linear regression model. The subpopulation means describe the FIT part of our conceptual model

$$\text{DATA} = \text{FIT} + \text{RESIDUAL}$$

The RESIDUAL part represents the variation of observed y about their means.

We will use the same notation for the residual that we used in the simple linear regression model. The symbol ϵ represents the deviation of an individual observation from its subpopulation mean. We assume that these deviations are Normally distributed with mean 0 and an unknown model standard deviation σ that does not depend on the values of the x variables. *These are assumptions that we can check by examining the residuals in the same way that we did for simple linear regression.*

checking model assumptions, p. 565

MULTIPLE LINEAR REGRESSION MODEL

The **statistical model for multiple linear regression** is

$$y_i = \beta_0 + \beta_1 x_{i1} + \beta_2 x_{i2} + \cdots + \beta_p x_{ip} + \epsilon_i$$

for $i = 1, 2, \ldots, n$.

The **mean response** μ_y is a linear function of the explanatory variables:

$$\mu_y = \beta_0 + \beta_1 x_1 + \beta_2 x_2 + \cdots + \beta_p x_p$$

The **deviations** ϵ_i are assumed to be independent and Normally distributed with mean 0 and standard deviation σ. In other words, they are a simple random sample (SRS) from the $N(0, \sigma)$ distribution.

The **parameters of the model** are $\beta_0, \beta_1, \beta_2, \ldots, \beta_p$, and σ.

The assumption that the subpopulation means are related to the regression coefficients β by the equation

$$\mu_y = \beta_0 + \beta_1 x_1 + \beta_2 x_2 + \cdots + \beta_p x_p$$

implies that we can estimate all subpopulation means from estimates of the β's. To the extent that this equation is accurate, we have a useful tool for describing how the mean of y varies with the collection of x's.

We do, however, need to be cautious when interpreting each of the regression coefficients in a multiple regression. First, the β_0 coefficient represents the mean of y when *all* the x variables equal zero. Even more so than in simple linear regression, this subpopulation is rarely of interest. Second, the description provided by the regression coefficient of each x variable is similar to that provided by the slope in simple linear regression but only in a specific situation—namely, *when all other x variables are held constant.* We need this extra condition because with multiple x variables, it is quite possible that a unit change in one x variable may be associated with changes in other x variables. If that occurs, then the overall change in the mean of y is not described by just a single regression coefficient.

USE YOUR KNOWLEDGE

11.2 Understanding the fitted regression line. The fitted regression equation for a multiple regression is

$$\hat{y} = -10.8 + 3.2x_1 + 2.8x_2$$

(a) If $x_1 = 4$ and $x_2 = 2$, what is the predicted value of y?

(b) For the answer to part (a) to be valid, is it necessary that the values $x_1 = 4$ and $x_2 = 2$ correspond to a case in the data set? Explain why or why not.

(c) If you hold x_1 at a fixed value, what is the effect of an increase of three units of x_2 on the predicted value of y?

Estimation of the multiple regression parameters

LOOK BACK

least squares,
p. 112

Similar to simple linear regression, we use the method of least squares to obtain estimators of the regression coefficients β. Let

$$b_0, b_1, b_2, \ldots, b_p$$

denote the estimators of the parameters

$$\beta_0, \beta_1, \beta_2, \ldots, \beta_p$$

For the ith observation, the predicted response is

$$\hat{y}_i = b_0 + b_1 x_{i1} + b_2 x_{i2} + \cdots + b_p x_{ip}$$

The ith residual, the difference between the observed and the predicted response, is, therefore,

$$e_i = \text{observed response} - \text{predicted response}$$
$$= y_i - \hat{y}_i$$
$$= y_i - b_0 - b_1 x_{i1} - b_2 x_{i2} - \cdots - b_p x_{ip}$$

The method of least squares chooses the values of the b's that make the sum of the squared residuals as small as possible. In other words, the parameter estimates $b_0, b_1, b_2, \ldots, b_p$ minimize the quantity

$$\sum (y_i - b_0 - b_1 x_{i1} - b_2 x_{i2} - \cdots - b_p x_{ip})^2$$

The formula for the least-squares estimates in multiple regression is complicated. We will be content to understand the principle on which it is based and to let software do the computations.

The parameter σ^2 measures the variability of the responses about the population regression equation. As in the case of simple linear regression, we estimate σ^2 by an average of the squared residuals. The estimator is

$$s^2 = \frac{\sum e_i^2}{n - p - 1}$$

$$= \frac{\sum (y_i - \hat{y}_i)^2}{n - p - 1}$$

LOOK BACK

degrees of
freedom,
p. 40

The quantity $n - p - 1$ is the degrees of freedom associated with s^2. The degrees of freedom equal the sample size, n, minus $(p + 1)$, the number of β's we must estimate to fit the model. In the simple linear regression case, there is just one explanatory variable, so $p = 1$ and the degrees of freedom are $n - 2$. To estimate the model standard deviation σ, we use

$$s = \sqrt{s^2}$$

Confidence intervals and significance tests for regression coefficients

We can obtain confidence intervals and perform significance tests for each of the regression coefficients β_j as we did in simple linear regression. The standard errors of the b's have more complicated formulas, but all are multiples of the estimated model standard deviation s. We again rely on statistical software to do the calculations.

CONFIDENCE INTERVALS AND SIGNIFICANCE TESTS FOR β_J

A **level C confidence interval** for β_j is

$$b_j \pm t^* SE_{b_j}$$

where SE_{b_j} is the standard error of b_j and t^* is the value for the $t(n - p - 1)$ density curve with area C between $-t^*$ and t^*.

To test the hypothesis $H_0: \beta_j = 0$, compute the **t statistic**

$$t = \frac{b_j}{SE_{b_j}}$$

In terms of a random variable T having the $t(n - p - 1)$ distribution, the P-value for a test of H_0 against

H_a: $\beta_j > 0$ is $P(T \geq t)$

H_a: $\beta_j < 0$ is $P(T \leq t)$

H_a: $\beta_j \neq 0$ is $2P(T \geq |t|)$

Be very careful in your interpretation of the t *tests and confidence intervals for individual regression coefficients.* In *simple linear regression*, the model says that $\mu_y = \beta_0 + \beta_1 x$. The null hypothesis H_0: $\beta_1 = 0$ says that regression on x is of no value for predicting the response y or, alternatively, that there is no straight-line relationship between x and y. The corresponding null hypothesis for the *multiple regression* model $\mu_y = \beta_0 + \beta_1 x_1 + \beta_2 x_2 + \cdots + \beta_p x_p$ says that x_1 is of no value for predicting y, ***given that*** $\mathbf{x_2}$, $\mathbf{x_3}$, $\mathbf{\ldots}$, $\mathbf{x_p}$ ***are in the model***. In other words, the predictor x_1 contains no additional information over or above that contained in $x_2, x_3, \ldots, x_p$ that is useful for predicting y. This a very important difference and one we will return to when analyzing the data of Example 11.1.

Because regression is often used for prediction, we may wish to use multiple regression models to construct confidence intervals for a mean response and prediction intervals for a future observation. The basic ideas are the same as in the simple linear regression case.

In most software systems, the same commands that give confidence and prediction intervals for simple linear regression work for multiple regression. The only difference is that we specify a list of explanatory variables rather than a single variable. Modern software allows us to perform these rather complex calculations without an intimate knowledge of all the computational details. This frees us to concentrate on the meaning and appropriate use of the results.

LOOK BACK

confidence intervals for mean response, p. 570

ANOVA table for multiple regression

LOOK BACK

ANOVA F test, p. 586

In simple linear regression, the F test from the ANOVA table is equivalent to the two-sided t test of the hypothesis that the slope of the regression line is 0. For multiple regression, there is a corresponding ANOVA F test, but it tests the hypothesis that *all* the regression coefficients (with the exception of the intercept) are 0. Here is the general form of the ANOVA table for multiple regression:

Source	Degrees of freedom	Sum of squares	Mean square	F
Model	p	$\sum(\hat{y}_i - \bar{y})^2$	SSM/DFM	MSM/MSE
Error	$n - p - 1$	$\sum(y_i - \hat{y}_i)^2$	SSE/DFE	
Total	$n - 1$	$\sum(y_i - \bar{y})^2$	SST/DFT	

The ANOVA table is similar to that for simple linear regression. The degrees of freedom for the model increase from 1 to p to reflect the fact that we now have p explanatory variables rather than just one. As a consequence, the degrees of freedom for error decrease by the same amount. *It is always a good idea to calculate the degrees of freedom by hand and then check that your software agrees with your calculations.* This ensures that you have not made some serious error in specifying the model or in entering the data.

The sums of squares represent sources of variation. Once again, both the sums of squares and their degrees of freedom add:

$$SST = SSM + SSE$$

$$DFT = DFM + DFE$$

LOOK BACK

F statistic, p. 585

The estimate of the variance σ^2 for our model is again given by the MSE in the ANOVA table. That is, $s^2 = MSE$.

The ratio MSM/MSE is an F statistic for testing the null hypothesis

$$H_0: \beta_1 = \beta_2 = \cdots = \beta_p = 0$$

against the alternative hypothesis

$$H_a: \text{at least one of the } \beta_j \text{ is not } 0$$

The null hypothesis says that none of the explanatory variables are predictors of the response variable when used in the form expressed by the multiple regression equation. The alternative states that *at least one* of them is a predictor of the response variable.

As in simple linear regression, large values of F give evidence against H_0. When H_0 is true, F has the $F(p, n - p - 1)$ distribution. The degrees of freedom for the F distribution are those associated with the model and error in the ANOVA table.

ANALYSIS OF VARIANCE *F* TEST

In the multiple regression model, the hypothesis

$$H_0: \beta_1 = \beta_2 = \cdots = \beta_p = 0$$

is tested against the alternative hypothesis

$$H_a: \text{at least one of the } \beta_j \text{ is not } 0$$

by the analysis of variance ***F* statistic**

$$F = \frac{MSM}{MSE}$$

The *P*-value is the probability that a random variable having the $F(p, n - p - 1)$ distribution is greater than or equal to the calculated value of the F statistic.

A common error in the use of multiple regression is to assume that all the regression coefficients are statistically different from zero whenever the F *statistic has a small P-value. Be sure that you understand the difference between the* F *test and the* t *tests for individual coefficients in the multiple regression setting.*

Squared multiple correlation R^2

LOOK BACK

r^2 in regression
p. 116

For simple linear regression, we noted that the square of the sample correlation could be written as the ratio of SSM to SST and could be interpreted as the proportion of variation in y explained by x. The ratio of SSM to SST is routinely calculated for multiple regression and still can be interpreted as the proportion of explained variation. The difference is that it relates to the collection of explanatory variables in the model.

THE SQUARED MULTIPLE CORRELATION

The statistic

$$R^2 = \frac{\text{SSM}}{\text{SST}} = \frac{\sum (\hat{y}_i - \bar{y})^2}{\sum (y_i - \bar{y})^2}$$

is the proportion of the variation of the response variable y that is explained by the explanatory variables $x_1, x_2, \ldots, x_p$ in a multiple linear regression.

---

multiple correlation coefficient

Often, R^2 is multiplied by 100 and expressed as a percent. The square root of R^2, called the **multiple correlation coefficient**, is the correlation between the observations y_i and the predicted values $\hat{y}_i$.

SECTION 11.1 SUMMARY

• **Data for multiple linear regression** consist of the values of a response variable y and p explanatory variables $x_1, x_2, \ldots, x_p$ for n cases. We write the data and enter them into software in the form

		Variables			
Individual	y	x_1	x_2	$\ldots$	x_p
1	y_1	x_{11}	x_{12}	$\ldots$	x_{1p}
2	y_2	x_{21}	x_{22}	$\ldots$	x_{2p}
$\vdots$					
n	y_n	x_{n1}	x_{n2}	$\ldots$	x_{np}

• The statistical model for **multiple linear regression** with response variable y and p explanatory variables $x_1, x_2, \ldots, x_p$ is

$$y_i = \beta_0 + \beta_1 x_{i1} + \beta_2 x_{i2} + \cdots + \beta_p x_{ip} + \epsilon_i$$

where $i = 1, 2, \ldots, n$. The ϵ_i are assumed to be independent and Normally distributed with mean 0 and standard deviation σ. The **parameters** of the model are $\beta_0, \beta_1, \beta_2, \ldots, \beta_p$, and σ.

• The **multiple regression equation** predicts the response variable by a linear relationship with all the explanatory variables:

$$\hat{y} = b_0 + b_1 x_1 + b_2 x_2 + \cdots + b_p x_p$$

- The β's are estimated by $b_0, b_1, b_2, \ldots, b_p$, which are obtained by the **method of least squares**. The model standard deviation σ is estimated by

$$s = \sqrt{\text{MSE}} = \sqrt{\frac{\sum e_i^2}{n - p - 1}}$$

where the e_i are the **residuals**,

$$e_i = y_i - \hat{y}_i$$

- A **level C confidence interval** for β_j is

$$b_j \pm t^* \text{SE}_{b_j}$$

where t^* is the value for the $t(n - p - 1)$ density curve with area C between $-t^*$ and t^*.

- The test of the hypothesis $H_0: \beta_j = 0$ is based on the **t statistic**

$$t = \frac{b_j}{\text{SE}_{b_j}}$$

and the $t(n - p - 1)$ distribution.

- The **ANOVA table** for a multiple linear regression gives the degrees of freedom; sum of squares; and mean squares for the model, error, and total sources of variation. The **ANOVA F statistic** is the ratio MSM/MSE and is used to test the null hypothesis

$$H_0: \beta_1 = \beta_2 = \cdots = \beta_p = 0$$

If H_0 is true, this statistic has an $F(p, n - p - 1)$ distribution.

- The **squared multiple correlation** is given by the expression

$$R^2 = \frac{\text{SSM}}{\text{SST}}$$

and is interpreted as the proportion of the variability in the response variable y that is explained by the explanatory variables $x_1, x_2, \ldots, x_p$ in the multiple linear regression.

SECTION 11.1 EXERCISES

For Exercise 11.1, see page 610 and for Exercise 11.2, see page 611.

11.3 What's wrong? In each of the following situations, explain what is wrong and why.

(a) A small *P*-value for the ANOVA *F* test implies that all explanatory variables are significantly different from zero.

(b) R^2 is the proportion of variation explained by the collection of explanatory variables. It can obtained by squaring the correlations between y and each x_i and summing them up.

(c) In a multiple regression with a sample size of 45 and six explanatory variables, the test statistic for the null hypothesis $H_0: b_2 = 0$ is a t statistic that follows the $t(38)$ distribution when the null hypothesis is true.

11.4 What's wrong? In each of the following situations, explain what is wrong and why.

(a) One of the assumptions for multiple regression is that the distribution of each explanatory variable is Normal.

(b) The null hypothesis $H_0: \beta_3 = 0$ in a multiple regression involving three explanatory variables implies there is no linear association between x_3 and y.

(c) The multiple correlation coefficient gives the average correlation between the response variable and each explanatory variable in the model.

11.5 95% confidence intervals for regression coefficients. In each of the following settings, give a 95% confidence interval for the coefficient of x_1.

(a) $n = 25$, $\hat{y} = 1.6 + 6.4x_1 + 5.7x_2$, $SE_{b_1} = 3.3$

(b) $n = 43$, $\hat{y} = 1.6 + 6.4x_1 + 5.7x_2$, $SE_{b_1} = 2.9$

(c) $n = 25$, $\hat{y} = 1.6 + 4.8x_1 + 3.2x_2 + 5.2x_3$, $SE_{b_1} = 2.7$

(d) $n = 104$, $\hat{y} = 1.6 + 4.8x_1 + 3.2x_2 + 5.2x_3$, $SE_{b_1} = 1.8$

11.6 Significance tests for regression coefficients. For each of the settings in the previous exercise, test the null hypothesis that the coefficient of x_1 is zero versus the two-sided alternative.

11.7 Constructing the ANOVA table. Six explanatory variables are used to predict a response variable using a multiple regression. There are 183 observations.

(a) Write the statistical model that is the foundation for this analysis. Also include a description of all assumptions.

(b) Outline the analysis of variance table giving the sources of variation and numerical values for the degrees of freedom.

11.8 More on constructing the ANOVA table. A multiple regression analysis of 57 cases was performed with four explanatory variables. Suppose that SSM = 16.5 and SSE = 100.8.

(a) Find the value of the F statistic for testing the null hypothesis that the coefficients of all the explanatory variables are zero.

(b) What are the degrees of freedom for this statistic?

(c) Find bounds on the P-value using Table E. Show your work.

(d) What proportion of the variation in the response variable is explained by the explanatory variables?

11.9 Significance tests for regression coefficients. Refer to Exercise 11.1 (page 610). The following table contains the estimated coefficients and standard errors of their multiple regression fit. Each explanatory variable is an average of several five-point Likert scale questions.

Variable	Estimate	SE
Intercept	1.316	0.651
Math course anxiety	−0.212	0.114
Math test anxiety	−0.155	0.119
Numerical task anxiety	−0.094	0.116
Enjoyment	0.176	0.114
Self-confidence	0.118	0.114
Motivation	0.097	0.115
Feedback usefulness	0.644	0.194

(a) Look at the signs of the coefficients (positive and negative). Is this what you would expect in this setting? Explain your answer.

(b) What are the degrees of freedom for the model and error?

(c) Test the significance of each coefficient and state your conclusions.

11.10 ANOVA table for multiple regression. Use the following information and the general form of the ANOVA table for multiple regression on page 613 to perform the ANOVA F test and compute R^2.

Source	Degrees of freedom	Sum of squares	Mean square	F
Model	4	70		
Error				
Total	33	524		

11.11 Game-day spending. Game-day spending (ticket sales and food and beverage purchases) is critical for the sustainability of many professional sports teams. In the National Hockey League (NHL), nearly half the franchises generate more than two-thirds of their annual income from game-day spending. Understanding and possibly predicting this spending would allow teams to respond with appropriate marketing and pricing strategies. To investigate this possibility, a group of researchers looked at data from one NHL team over a three-season period ($n = 123$ home games).[3] The following table summarizes the multiple regression used to predict ticket sales. Each explanatory variable is an *indicator variable* taking the value 1 for the condition specified and 0 otherwise.

Explanatory variables	b	t
Constant	12,493.47	12.13
Division	−788.74	−2.01
Nonconference	−474.83	−1.04
November	−1800.81	−2.65
December	−559.24	−0.82
January	−925.56	−1.54
February	−35.59	−0.05
March	−131.62	−0.21
Weekend	2992.75	8.48
Night	1460.31	2.13
Promotion	2162.45	5.65
Season 2	−754.56	−1.85
Season 3	−779.81	−1.84

(a) Which of the explanatory variables significantly aid prediction in the presence of all the explanatory variables? Show your work.

(b) The overall F statistic was 11.59. What are the degrees of freedom and P-value of this statistic?

(c) The value of R^2 is 0.52. What percent of the variance in ticket sales is explained by these explanatory variables?

(d) The constant predicts the number of tickets sold for a nondivisional, conference game with no promotions played during the day on a weekday in October of Season 1. What is the predicted number of tickets sold for a divisional conference game with no promotions played on a weekend evening in March during Season 3?

(e) Would a 95% confidence interval for the mean response or a 95% prediction interval be more appropriate to include with your answer to part (d)? Explain your reasoning.

11.12 Discrimination at work? A survey of 457 engineers in Canada was performed to identify the relationship of race, language proficiency, and location of training in finding work in the engineering field. In addition, each participant completed the Workplace Prejudice and Discrimination Inventory (WPDI), which is designed to measure perceptions of prejudice on the job, primarily due to race or ethnicity. The score of the WPDI ranged from 16 to 112, with higher scores indicating more perceived discrimination. The following table summarizes two multiple regression models used to predict an engineer's WPDI score. The first explanatory variable indicates whether the engineer was foreign trained ($x = 1$) or locally trained ($x = 0$). The next set of seven variables indicate race and the last six are demographic variables.

	Model 1		Model 2	
Explanatory variables	b	$s(b)$	b	$s(b)$
Foreign trained	0.55	0.21	0.58	0.22
Chinese			0.06	0.24
South Asian			−0.06	0.19
Black			−0.03	0.52
Other Asian			−0.38	0.34
Latin American			0.20	0.46
Arab			0.56	0.44
Other (not white)			0.05	0.38
Mechanical	−0.19	0.25	−0.16	0.25
Other (not electrical)	−0.14	0.20	−0.13	0.21
Masters/PhD	0.32	0.18	0.37	0.18
30–39 years old	−0.03	0.22	−0.06	0.22
40 or older	0.32	0.25	0.25	0.26
Female	−0.02	0.19	−0.05	0.19
R^2		0.10		0.11

(a) The F statistics for these two models are 7.12 and 3.90, respectively. What are the degrees of freedom and P-value of each statistic?

(b) The F statistics for the multiple regressions are highly significant, but the R^2 are relatively low. Explain to a statistical novice how this can occur.

(c) Do foreign-trained engineers perceive more discrimination than do locally trained engineers? To address this, test if the first coefficient in each model is equal to zero versus the greater than alternative. Summarize your results.

11.2 A Case Study

In this section, we illustrate multiple regression by analyzing the data from the study described in Example 11.1. There are data for $n = 150$ students. The response variable is the cumulative GPA, on a four-point scale, after three semesters. The explanatory variables previously mentioned are average high school grades, represented by HSM, HSS, and HSE. We also consider SAT Mathematics (SATM), SAT Critical Reading (SATCR), and SAT Writing (SATW) scores as explanatory variables. We leave the inclusion of gender (SEX) as an explanatory variable to the exercises.

Before starting the analysis, we first consider the extent to which our results can be generalized. For this study, all the available data are being analyzed. There is no random sampling from the population of science majors. In this setting, we often justify the use of inference by viewing the data as coming from some sort of process. Here, we consider this collection of students as a sample of all the science majors who will attend this university. Still, opinions may vary as to the extent to which these data can be considered an SRS sample of current and future students. For example, schools seem to consistently brag that their new batch of first-year students is the smartest and most accomplished group they've ever had.

Preliminary analysis

As with any statistical analysis, we begin our multiple regression with a careful examination of the data. We first look at each variable separately, then at relationships among the variables. In both cases, we continue our practice of combining plots and numerical descriptions. We use JMP, Excel, SAS, Minitab, and SPSS to illustrate the outputs that are given by most software.

Means, standard deviations, and minimum and maximum values appear in Figure 11.2. The minimum value for high school mathematics (HSM) appears to be rather extreme; it is $|2.00 - 8.59|/1.46 = 4.51$ standard deviations below the mean. Similarly, the minimum value for GPA is 3.43 standard deviations below the mean. We do not discard either of these cases at this time but will take care in our subsequent analyses to see if they have an excessive influence on our results.

The mean for the SATM score is higher than the means for the Critical Reading (SATCR) and Writing (SATW) scores, as we might expect for a group of science majors. The three SAT standard deviations are all about the same.

Although mathematics scores were higher on the SAT, the means and standard deviations of the three high school grade variables are very similar. Because the level and difficulty of high school courses vary within and across schools, this may not be that surprising. The mean GPA is 2.842 on a four-point scale, with standard deviation 0.818.

Because the variables GPA, SATM, SATCR, and SATW have many possible values, we could use stemplots or histograms to examine the shapes of their distributions. Normal quantile plots indicate whether or not the distributions look Normal. *It is important to note that the multiple regression model does not require any of these distributions to be Normal.* Only the deviations of the responses *y* from their means are assumed to be Normal.

The purpose of examining these plots is to understand something about each variable alone before attempting to use it in a complicated model. *Extreme values of any variable should be noted and checked for accuracy.* If found to be correct, the cases with these values should be carefully examined

SAS					
The MEANS Procedure					
Variable	**N**	**Mean**	**Std Dev**	**Minimum**	**Maximum**
GPA	150	2.8421333	0.8178992	0.0300000	4.0000000
HSM	150	8.5866667	1.4617571	2.0000000	10.0000000
HSS	150	8.8000000	1.3951017	4.0000000	10.0000000
HSE	150	8.8333333	1.2660601	4.0000000	10.0000000
SATM	150	623.6000000	74.8356589	460.0000000	800.0000000
SATCR	150	573.8000000	87.6208274	330.0000000	800.0000000
SATW	150	562.6000000	80.0874522	350.0000000	770.0000000

FIGURE 11.2 Descriptive statistics for the College of Science student case study.

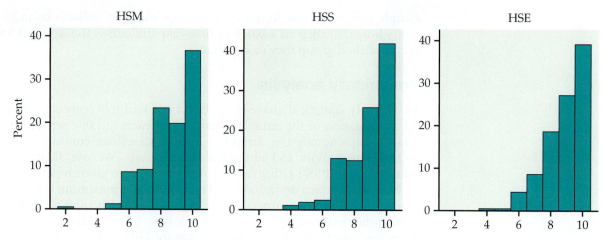

FIGURE 11.3 Bar plots using relative frequencies.

to see if they are truly exceptional and perhaps do not belong in the same analysis with the other cases. When our data on science majors are examined in this way, no obvious problems are evident.

The high school grade variables HSM, HSS, and HSE take only integer values. The bar plots using relative frequencies are shown in Figure 11.3. The distributions are all skewed, with a large proportion of high grades (10 = A and 9 = A−). Again we emphasize that these distributions need not be Normal.

Relationships between pairs of variables

LOOK BACK

correlation, p. 101

The second step in our analysis is to examine the relationships between all pairs of variables. Scatterplots and correlations are our tools for studying two-variable relationships. The correlations appear in Figure 11.4. The output includes the P-value for the test of the null hypothesis that the population correlation is 0 versus the two-sided alternative for each pair. Thus, we see that the correlation between GPA and HSM is 0.420, with a P-value of 0.000 (that is, $P < 0.0005$), whereas the correlation between GPA and SATW is 0.223, with a P-value of 0.006. Because of the large sample size, even somewhat weak associations are found to be statistically significant.

As we might expect, math and science grades have the highest correlation with GPA ($r = 0.420$ and $r = 0.443$), followed by English grades (0.359) and then SAT Mathematics (0.330). SAT Critical Reading (SATCR) and SAT Writing (SATW) have comparable, somewhat weak, correlations with GPA. On the other hand, SATCR and SATW have a high correlation with each other (0.734). The high school grades also correlate well with each other (0.485 to 0.695). SATM correlates well with the other SAT scores (0.579 and 0.551), somewhat with HSM (0.325), less with HSS (0.215), and poorly with HSE (0.134). SATCR and SATW do not correlate well with any of the high school grades (0.072 to 0.259).

It is important to keep in mind that, by examining pairs of variables, we are seeking a better understanding of the data. *The fact that the correlation of a particular explanatory variable with the response variable does not achieve statistical significance does not necessarily imply that it will not be a useful (and statistically significant) predictor in a multiple regression model.*

IBM SPSS Statistics Viewer

Correlations

		GPA	HSM	HSS	HSE	SATM	SATCR	SATW
GPA	Pearson Correlation	1	.420**	.443**	.359**	.330**	.251**	.223**
	Sig. (2-tailed)		.000	.000	.000	.000	.002	.006
	N	150	150	150	150	150	150	150
HSM	Pearson Correlation	.420**	1	.670**	.485**	.325**	.150	.072
	Sig. (2-tailed)	.000		.000	.000	.000	.067	.383
	N	150	150	150	150	150	150	150
HSS	Pearson Correlation	.443**	.670**	1	.695**	.215**	.215**	.161*
	Sig. (2-tailed)	.000	.000		.000	.008	.008	.048
	N	150	150	150	150	150	150	150
HSE	Pearson Correlation	.359**	.485**	.695**	1	.134	.259**	.185*
	Sig. (2-tailed)	.000	.000	.000		.102	.001	.023
	N	150	150	150	150	150	150	150
SATM	Pearson Correlation	.330**	.325**	.215**	.134	1	.579**	.551**
	Sig. (2-tailed)	.000	.000	.008	.102		.000	.000
	N	150	150	150	150	150	150	150
SATCR	Pearson Correlation	.251**	.150	.215**	.259**	.579**	1	.734**
	Sig. (2-tailed)	.002	.067	.008	.001	.000		.000
	N	150	150	150	150	150	150	150
SATW	Pearson Correlation	.223**	.072	.161*	.185*	.551**	.734**	1
	Sig. (2-tailed)	.006	.383	.048	.023	.000	.000	
	N	150	150	150	150	150	150	150

**. Correlation is significant at the 0.01 level (2-tailed).
*. Correlation is significant at the 0.05 level (2-tailed).

FIGURE 11.4 Correlations among the case study variables.

Numerical summaries such as correlations are useful, but plots are generally more informative when seeking to understand data. Plots tell us whether the numerical summary gives a fair representation of the data.

For a multiple regression, each pair of variables should be plotted. For the seven variables in our case study, this means that we should examine 21 plots. In general, there are $p + 1$ variables in a multiple regression analysis with p explanatory variables, so that $p(p + 1)/2$ plots are required. *Multiple regression is a complicated procedure. If we do not do the necessary preliminary work, we are in serious danger of producing useless or misleading results.*

USE YOUR KNOWLEDGE

GPA

11.13 Pairwise relationships among variables in the GPA data set. Most statistical software packages have the option to create a "scatterplot matrix" of all $p(p + 1)/2$ scatterplots. For example, in JMP, there is the option "Scatterplot Matrix" under the Graph menu. Figure 11.5 is the scatterplot matrix for the GPA data, including the least-squares lines. Comment on any unusual patterns or observations. Do you think the pairwise correlations or scatterplot matrix better describe the pairwise relationships? Explain your answer.

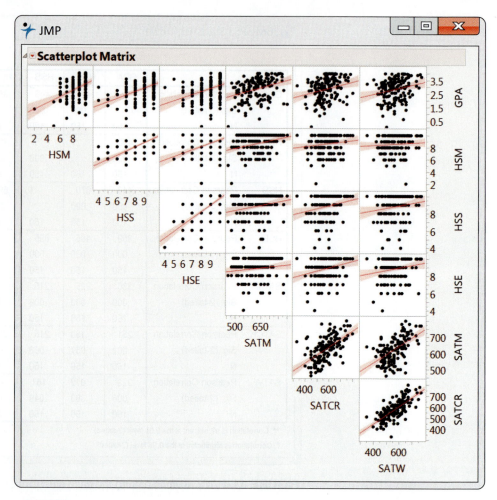

FIGURE 11.5 Scatterplot matrix for the GPA data, including the least-squares lines, Exercise 11.13.

Regression on high school grades

To explore the relationship between the explanatory variables and our response variable GPA, we run several multiple regressions. The explanatory variables fall into three classes. High school grades are represented by HSM, HSS, and HSE; standardized tests are represented by the three SAT scores; and sex of the student is represented by SEX. We begin our analysis by using the high school grades to predict GPA. Figure 11.6 gives the multiple regression output.

The output contains an ANOVA table, some additional fit statistics, and information about the parameter estimates. Because there are $n = 150$ cases, we have DFT $= n - 1 = 149$. The three explanatory variables give DFM $= p = 3$ and DFE $= n - p - 1 = 150 - 3 - 1 = 146$.

The ANOVA F statistic is 14.35, with a P-value of <0.0001. Under the null hypothesis

$$H_0: \beta_1 = \beta_2 = \beta_3 = 0$$

the F statistic has an $F(3, 146)$ distribution. According to this distribution, the chance of obtaining an F statistic of 14.35 or larger is less than 0.0001.

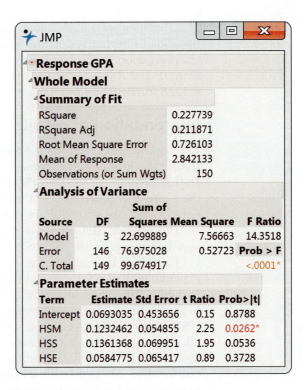

FIGURE 11.6 Multiple regression output for regression using high school grades to predict GPA.

Therefore, we conclude that at least one of the three regression coefficients for the high school grades is different from 0 in the population regression equation.

In the fit statistics that precede the ANOVA table, we find that Root MSE is 0.726. This value is the square root of the MSE given in the ANOVA table and is s, the estimate of the parameter σ of our model. The value of R^2 is 0.23. That is, 23% of the observed variation in the GPA scores is explained by linear regression on high school grades.

Although the P-value of the F test is very small, the model does not explain very much of the variation in GPA. Remember, a small P-value does not necessarily tell us that we have a strong predictive relationship, particularly when the sample size is large.

From the Parameter Estimates section of the computer output, we obtain the fitted regression equation

$$\widehat{\text{GPA}} = 0.069 + 0.123\text{HSM} + 0.136\text{HSS} + 0.058\text{HSE}$$

Let's find the predicted GPA for a student with an A− average in HSM, B+ in HSS, and B in HSE. The explanatory variables are HSM = 9, HSS = 8, and HSE = 7. The predicted GPA is

$$\widehat{\text{GPA}} = 0.069 + 0.123(9) + 0.136(8) + 0.058(7)$$
$$= 2.67$$

Recall that the t statistics for testing the regression coefficients are obtained by dividing the estimates by their standard errors. Thus, for the coefficient of HSM, we obtain the t-value given in the output by calculating

$$t = \frac{b}{\text{SE}_b} = \frac{0.12325}{0.05486} = 2.25$$

The *P*-values appear in the last column. Note that these *P*-values are for the two-sided alternatives. HSM has a *P*-value of 0.0262, and we conclude that the regression coefficient for this explanatory variable is significantly different from 0. The *P*-values for the other explanatory variables (0.0536 for HSS and 0.3728 for HSE) do not achieve statistical significance.

Interpretation of results

The significance tests for the individual regression coefficients seem to contradict the impression obtained by examining the correlations in Figure 11.4. In that display, we see that the correlation between GPA and HSS is 0.44 and the correlation between GPA and HSE is 0.36. The *P*-values for both of these correlations are < 0.0005. In other words, if we used HSS alone in a regression to predict GPA, or if we used HSE alone, we would obtain statistically significant regression coefficients.

This phenomenon is not unusual in multiple regression analysis. Part of the explanation lies in the correlations between HSM and the other two explanatory variables. These are rather high (at least compared with most other correlations in Figure 11.4). The correlation between HSM and HSS is 0.67 and that between HSM and HSE is 0.49. Thus, when we have a regression model that contains all three high school grades as explanatory variables, there is considerable *overlap of the predictive information* contained in these variables. This is called **collinearity** or **multicollinearity**. In extreme cases, collinearity can cause numerical instabilities that result in very imprecise parameter estimates.

collinearity
multicollinearity

As mentioned earlier, *the significance tests for individual regression coefficients assess the significance of each predictor variable assuming that all other predictors are included in the regression equation.* Given that we use a model with HSM and HSS as predictors, the coefficient of HSE is not statistically significant. Similarly, given that we have HSM and HSE in the model, HSS does not have a significant regression coefficient. HSM, however, adds significantly to our ability to predict GPA even after HSS and HSE are already in the model.

Unfortunately, we cannot conclude from this analysis that the *pair* of explanatory variables HSS and HSE contribute nothing significant to our model for predicting GPA once HSM is in the model. Questions like these require fitting additional models.

The impact of relations among the several explanatory variables on fitting models for the response is the most important new phenomenon encountered in moving from simple linear regression to multiple regression. In this chapter, we can only illustrate some of the many complicated problems that can arise.

Examining the residuals

As in simple linear regression, we should always examine the residuals as an aid to determining whether the multiple regression model is appropriate for the data. Because there are several explanatory variables, we must examine several residual plots. It is usual to plot the residuals versus the predicted values $\hat{y}$ and also versus each of the explanatory variables. Look for outliers, influential observations, evidence of a curved (rather than linear) relation, and anything else unusual. We leave the task of making these plots for the case study as an exercise. We find the plots all show more or less random noise above and below the center value of 0.

If the deviations ϵ in the model are Normally distributed, the residuals should be Normally distributed. Figure 11.7 presents a Normal quantile plot

residual plots, p. 125

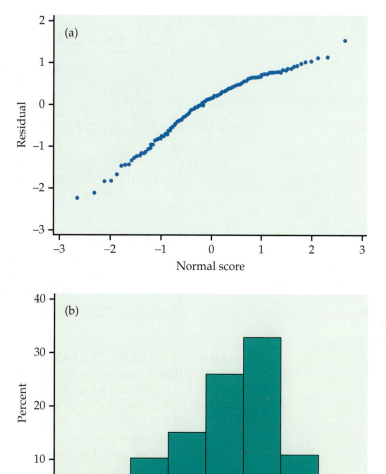

FIGURE 11.7 (a) Normal quantile plot and (b) histogram of the residuals from the high school grades model. There are no important deviations from Normality.

and histogram of the residuals. Both suggest some skewness (shorter right tail) in the distribution. However, given our large sample size, we do not think this skewness is strong enough to invalidate this analysis.

USE YOUR KNOWLEDGE

GPA

11.14 Residual plots for the GPA analysis. Using a statistical package, fit the linear model with HSM, HSS, and HSE as predictors and obtain the residuals and predicted values. Plot the residuals versus the predicted values, HSM, HSS, and HSE. Are the residuals more or less randomly dispersed around zero? Comment on any unusual patterns.

Refining the model

Because the variable HSE has the largest P-value of the three explanatory variables (see Figure 11.6) and, therefore, appears to contribute the least to our explanation of GPA, we rerun the regression using only HSM and HSS as

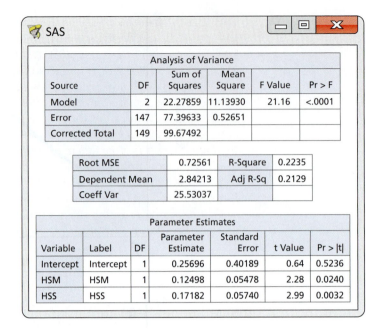

Analysis of Variance					
Source	DF	Sum of Squares	Mean Square	F Value	Pr > F
Model	2	22.27859	11.13930	21.16	<.0001
Error	147	77.39633	0.52651		
Corrected Total	149	99.67492			

Root MSE	0.72561	R-Square	0.2235
Dependent Mean	2.84213	Adj R-Sq	0.2129
Coeff Var	25.53037		

Parameter Estimates						
Variable	Label	DF	Parameter Estimate	Standard Error	t Value	Pr > |t|
Intercept	Intercept	1	0.25696	0.40189	0.64	0.5236
HSM	HSM	1	0.12498	0.05478	2.28	0.0240
HSS	HSS	1	0.17182	0.05740	2.99	0.0032

FIGURE 11.8 Multiple regression output for regression using HSM and HSS to predict GPA.

explanatory variables. The SAS output appears in Figure 11.8. The F statistic indicates that we can reject the null hypothesis that the regression coefficients for the two explanatory variables are both 0. The P-value is still < 0.0001. The value of R^2 has dropped very slightly compared with our previous run, from 0.2277 to 0.2235. Thus, dropping HSE from the model resulted in the loss of very little explanatory power.

The estimated model standard deviation s (Root MSE in the printout) is nearly identical for the two regressions, another indication that we lose very little when we drop HSE. The t statistics for the individual regression coefficients indicate that HSM is still significant ($P = 0.0240$), while the statistic for HSS is larger than before (2.99 versus 1.95) and is now statistically significant ($P = 0.0032$).

Comparison of the fitted equations for the two multiple regression analyses tells us something more about the intricacies of this procedure. For the first run, we have

$$\widehat{\text{GPA}} = 0.069 + 0.123\text{HSM} + 0.136\text{HSS} + 0.058\text{HSE}$$

whereas the second gives us

$$\widehat{\text{GPA}} = 0.257 + 0.125\text{HSM} + 0.172\text{HSS}$$

Eliminating HSE from the model changes the regression coefficients for all the remaining variables and the intercept. This phenomenon occurs quite generally in multiple regression. *Individual regression coefficients, their standard errors, and significance tests are meaningful only when interpreted in the context of the other explanatory variables in the model.*

Regression on SAT scores

We now turn to the problem of predicting GPA using the three SAT scores. Figure 11.9 gives the Minitab output. The fitted model is

$$\widehat{\text{GPA}} = 0.458 + 0.00301\text{SATM} + 0.00080\text{SATCR} + 0.00008\text{SATW}$$

The degrees of freedom are as expected: 3, 146, and 149. The F statistic is 6.28, with a P-value of < 0.0005. We conclude that the regression coefficients for SATM, SATCR, and SATW are not all 0. Recall that we obtained the P-value < 0.0001 when we used high school grades to predict GPA. Both multiple regression equations are highly significant, but this obscures the fact that the two models have quite different explanatory power. For the SAT regression, $R^2 = 0.1143$, whereas for the high school grades model even with only HSM and HSS (Figure 11.8), we have $R^2 = 0.2235$, a value almost twice as large. *Stating that we have a statistically significant result is quite different from saying that an effect is large or important.*

Further examination of the output in Figure 11.9 reveals that the coefficient of SATM is significant ($t = 2.81$, $P = 0.006$) and that those of SATCR ($t = 0.71$, $P = 0.477$) and SATW ($t = 0.07$, $P = 0.948$) are not. For a complete analysis, we should carefully examine the residuals. Also, we might want to run the analysis without SATW and the analysis with SATM as the only explanatory variable.

Regression using all variables

We have seen that fitting a model using either the high school grades or the SAT scores results in a highly significant regression equation. The mathematics component of each of these groups of explanatory variables appears to be a key predictor. Comparing the values of R^2 for the two models indicates

FIGURE 11.9 Multiple regression output for regression using SAT scores to predict GPA.

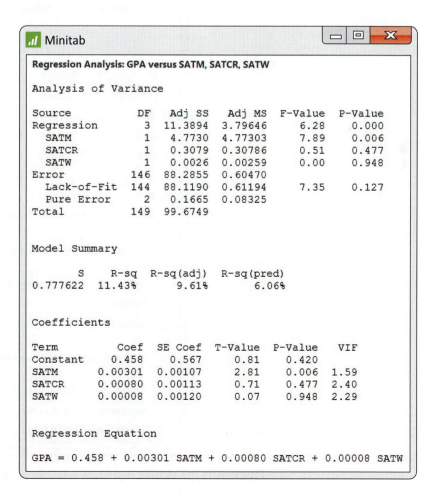

Regression Analysis: GPA versus SATM, SATCR, SATW

Analysis of Variance

Source	DF	Adj SS	Adj MS	F-Value	P-Value
Regression	3	11.3894	3.79646	6.28	0.000
SATM	1	4.7730	4.77303	7.89	0.006
SATCR	1	0.3079	0.30786	0.51	0.477
SATW	1	0.0026	0.00259	0.00	0.948
Error	146	88.2855	0.60470		
Lack-of-Fit	144	88.1190	0.61194	7.35	0.127
Pure Error	2	0.1665	0.08325		
Total	149	99.6749			

Model Summary

S	R-sq	R-sq(adj)	R-sq(pred)
0.777622	11.43%	9.61%	6.06%

Coefficients

Term	Coef	SE Coef	T-Value	P-Value	VIF
Constant	0.458	0.567	0.81	0.420	
SATM	0.00301	0.00107	2.81	0.006	1.59
SATCR	0.00080	0.00113	0.71	0.477	2.40
SATW	0.00008	0.00120	0.07	0.948	2.29

Regression Equation

GPA = 0.458 + 0.00301 SATM + 0.00080 SATCR + 0.00008 SATW

that high school grades are better predictors than SAT scores. Can we get a better prediction equation using all the explanatory variables together in one multiple regression?

To address this question, we run the regression with all six explanatory variables. The output from SAS, Excel, and Minitab appears in Figure 11.10. Although the format and organization of outputs differ among software packages, the basic results that we need are easy to find.

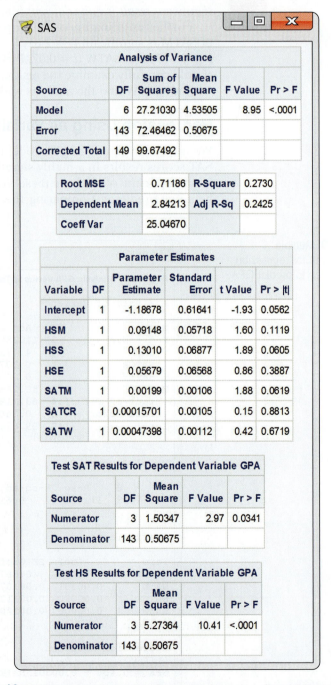

FIGURE 11.10 Multiple regression output for regression using all variables to predict GPA.

Excel						
	A	B	C	D	E	F	G

	A	B	C	D	E	F	G
1	SUMMARY OUTPUT						
2							
3	*Regression Statistics*						
4	Multiple R	0.522484872					
5	R Square	0.272990441					
6	Adjusted R Square	0.242486543					
7	Standard Error	0.711860645					
8	Observations	150					
9							
10	ANOVA						
11		*df*	*SS*	*MS*	*F*	*Significance F*	
12	Regression	6	27.21029964	4.53505	8.949363	2.69075E-08	
13	Residual	143	72.46461769	0.506746			
14	Total	149	99.67491733				
15							
16		*Coefficients*	*Standard Error*	*t Stat*	*P-value*	*Lower 95%*	*Upper 95%*
17	Intercept	-1.18678259	0.616408136	-1.92532	0.056174	-2.40523174	0.031666565
18	HSM	0.091476942	0.057181116	1.599775	0.111855	-0.021552525	0.204506408
19	HSS	0.130096576	0.068767787	1.891824	0.060536	-0.005836172	0.266029324
20	HSE	0.056790708	0.065681417	0.864639	0.388685	-0.073041235	0.186622652
21	SATM	0.001988735	0.001056716	1.881996	0.061869	-0.000100067	0.004077537
22	SATCR	0.000157014	0.001049462	0.149614	0.88128	-0.001917449	0.002231478
23	SATW	0.000473983	0.001116795	0.424414	0.671902	-0.001733577	0.002681542

FIGURE 11.10 (*Continued*)

The degrees of freedom are as expected: 6, 143, and 149. The F statistic is 8.95, with a P-value <0.0001, so at least one of our explanatory variables has a nonzero regression coefficient. This result is not surprising, given that we have already seen that HSM and SATM are strong predictors of GPA. The value of R^2 is 0.2730, which is about 0.05 higher than the value of 0.2235 that we found for the high school grades regression.

Examination of the t statistics and the associated P-values for the individual regression coefficients reveals a surprising result. None of the variables are significant! At first, this result may appear to contradict the ANOVA results. How can the model explain more than 27% of the variation and have t tests that suggest none of the variables make a significant contribution?

Once again, it is important to understand that these t tests assess the contribution of each variable when it is added to a model that already has the other five explanatory variables. This result does not necessarily mean that the regression coefficients for the six explanatory variables are *all* 0. It simply means that the contribution of each variable overlaps considerably with the contribution of the other five variables already in the model.

model selection

When a model has a large number of insignificant variables, it is common to refine the model. This is often termed **model selection**. We prefer smaller models to larger models because they are easier to work with and understand.

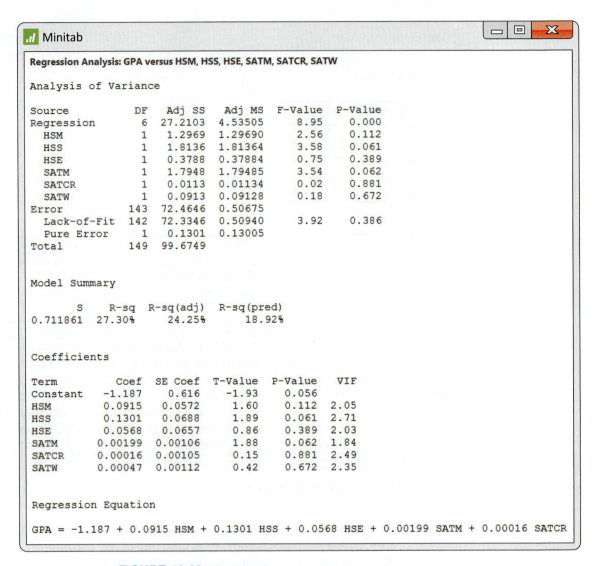

```
Minitab                                                    ▭ ▢ ✕

Regression Analysis: GPA versus HSM, HSS, HSE, SATM, SATCR, SATW

Analysis of Variance

Source             DF    Adj SS    Adj MS   F-Value   P-Value
Regression          6   27.2103   4.53505      8.95     0.000
  HSM               1    1.2969   1.29690      2.56     0.112
  HSS               1    1.8136   1.81364      3.58     0.061
  HSE               1    0.3788   0.37884      0.75     0.389
  SATM              1    1.7948   1.79485      3.54     0.062
  SATCR             1    0.0113   0.01134      0.02     0.881
  SATW              1    0.0913   0.09128      0.18     0.672
Error             143   72.4646   0.50675
  Lack-of-Fit     142   72.3346   0.50940      3.92     0.386
  Pure Error        1    0.1301   0.13005
Total             149   99.6749


Model Summary

        S    R-sq  R-sq(adj)  R-sq(pred)
 0.711861   27.30%     24.25%      18.92%


Coefficients

Term          Coef  SE Coef  T-Value  P-Value   VIF
Constant    -1.187    0.616    -1.93    0.056
HSM         0.0915   0.0572     1.60    0.112  2.05
HSS         0.1301   0.0688     1.89    0.061  2.71
HSE         0.0568   0.0657     0.86    0.389  2.03
SATM       0.00199  0.00106     1.88    0.062  1.84
SATCR      0.00016  0.00105     0.15    0.881  2.49
SATW       0.00047  0.00112     0.42    0.672  2.35


Regression Equation

GPA = -1.187 + 0.0915 HSM + 0.1301 HSS + 0.0568 HSE + 0.00199 SATM + 0.00016 SATCR
```

FIGURE 11.10 (*Continued*)

However, given the many complications that can arise in multiple regression, there is no universal "best" approach to refine a model. There is also no guarantee that there is just one acceptable refined model.

Many statistical software packages now provide the capability of summarizing all possible models from a set of *p* variables. We suggest using this capability when possible to reduce the number of candidate models (for example, there are a total of 63 models when $p = 6$) and then carefully studying the remaining models before making a decision as to a best model or set of best models. If in doubt, consult an expert.

For example, Figure 11.11 contains the output from the SAS commands

```
proc reg;
    model GPA = HSM HSS HSE SATM SATCR SATW/selection=rsquare
    mse best=2;
run;
```

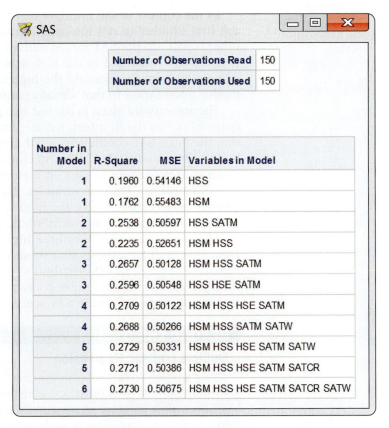

| Number of Observations Read | 150 |
| Number of Observations Used | 150 |

Number in Model	R-Square	MSE	Variables in Model
1	0.1960	0.54146	HSS
1	0.1762	0.55483	HSM
2	0.2538	0.50597	HSS SATM
2	0.2235	0.52651	HSM HSS
3	0.2657	0.50128	HSM HSS SATM
3	0.2596	0.50548	HSS HSE SATM
4	0.2709	0.50122	HSM HSS HSE SATM
4	0.2688	0.50266	HSM HSS SATM SATW
5	0.2729	0.50331	HSM HSS HSE SATM SATW
5	0.2721	0.50386	HSM HSS HSE SATM SATCR
6	0.2730	0.50675	HSM HSS HSE SATM SATCR SATW

FIGURE 11.11 Output from SAS summarizing the fit of different regression models in terms of R^2 and MSE.

It shows the two best models in terms of highest R^2, when the number of explanatory variables in the model are $p = 1$ through $p = 6$. A list like this can be very helpful in reducing the number of models to consider. In this case, there's very little difference in R^2 once the model has at least three explanatory variables, so we might consider further studying just the four models listed with $p = 3$ or $p = 4$.

adjusted R^2

The list also contains the estimated model variance (MSE). Finding a model (or models) that minimizes this quantity is another model selection approach. It is equivalent to choosing a model based on **adjusted R^2**. Unlike R^2, the adjusted R^2 and MSE take into account the number of parameters in the model and thus penalize larger models. In Figure 11.11, we see a model with $p = 3$ and a model with $p = 4$ have the two smallest MSEs. Pending an examination of the residuals, these are the two best refined models based on this selection method.

Test for a collection of regression coefficients

Many statistical software packages also provide the capability for testing whether a collection of regression coefficients in a multiple regression model are *all* 0. We use this approach to address two interesting questions about our data set. We did not discuss such tests in the outline that opened this section, but the basic idea is quite simple and is discussed in Exercise 11.20 (page 635).

In the context of the multiple regression model with all six predictors, we ask first whether or not the coefficients for the three SAT scores are all 0. In other words, do the SAT scores add any significant predictive information to that already contained in the high school grades? To be fair, we also ask the complementary question: do the high school grades add any significant predictive information to that already contained in the SAT scores?

The answers are given in the last two parts of the SAS output in Figure 11.10 (page 628). For the first test, we see that $F = 2.97$. Under the null hypothesis that the three SAT coefficients are 0, this statistic has an $F(3,143)$ distribution and the P-value is 0.0341. We conclude that the SAT scores (as a group) are significant predictors of GPA in a regression that already contains the high school scores as predictor variables. This means that we cannot just focus on refined models that involve the high school grades. Both high school grades and SAT scores appear to contribute to our explanation of GPA.

The test statistic for the three high school grade variables is $F = 10.41$. Under the null hypothesis that these three regression coefficients are 0, the statistic has an $F(3,143)$ distribution and the P-value is <0.0001. Again, this means that high school grades contain useful information for predicting GPA that is not contained in the SAT scores.

BEYOND THE BASICS

Multiple Logistic Regression

Many studies have yes/no or success/failure response variables. A surgery patient lives or dies; a consumer does or does not purchase a product after viewing an advertisement. Because the response variable in a multiple regression is assumed to have a Normal distribution, this methodology is not suitable for predicting such responses. However, there are models that apply the ideas of regression to response variables with only two possible outcomes.

logistic regression

One type of model that can be used is called **logistic regression**. We think in terms of a binomial model for the two possible values of the response variable and use one or more explanatory variables to explain the probability of success. Details are more complicated than those for multiple regression and are given in Chapter 14. However, the fundamental ideas are very much the same. Here is an example.

EXAMPLE 11.2

Tipping behavior in Canada. The Consumer Report on Eating Share Trends (CREST) contains data spanning all provinces of Canada and details away-from-home food purchases by roughly 4000 households per quarter. Some researchers accessed these data but restricted their attention to restaurants at which tips would normally be given.[4] From a total of 73,822 observations, "high" and "low" tipping variables were created based on whether the observed tip rate was above 20% or below 10%, respectively. They then used logistic regression to identify explanatory variables associated with either "high" or "low" tips.

The model consisted of more than 25 explanatory variables, grouped as "control" variables and "stereotype-related" variables. The stereotype-related explanatory variables were x_1, a variable having the value 1 if the age of the diner was greater than 65 years, and 0 otherwise; x_2, coded as 1 if the meal was on Sunday, and 0 otherwise; x_3, coded as 1 to indicate English was a second language; x_4, a variable coded 1 if the diner was a French-speaking Canadian; x_5, a variable coded 1 if alcoholic drinks were served with the meal; and x_6, a variable coded 1 if the meal involved a lone male.

LOOK BACK

chi-square distribution, p. 535

Similar to the F test in multiple regression, there is a chi-square test for multiple logistic regression that tests the null hypothesis that *all* coefficients of the explanatory variables are zero. These results were not presented in the article because the focus was more on comparing the high- and low-tip models. In place of the t tests for individual coefficients in multiple regression, chi-square tests, each with 1 degree of freedom, are used to test whether individual coefficients are zero. The article does report these tests. A majority of the variables considered in the models have P-values less than 0.01.

Interpretation of the coefficients is a little more difficult in multiple logistic regression because of the form of the model. For example, the high-tip model (using only the stereotype-related variables) is

$$\log\left(\frac{p}{1-p}\right) = \beta_0 + \beta_1 x_1 + \beta_2 x_2 + \cdots + \beta_6 x_6$$

odds

odds ratio

The expression $p/(1-p)$ is the **odds** that the tip was above 20%. Logistic regression models the "log odds" as a linear combination of the explanatory variables. Positive coefficients are associated with a higher probability that the tip is high. These coefficients are often transformed back (e^{β_j}) to the odds scale, giving us an **odds ratio**. An odds ratio greater than 1 is associated with a higher probability that the tip is high. Here is the table of odds ratios reported in the article for the high-tip model:

Explanatory variable	Odds ratio
Senior adult	0.7420*
Sunday	0.9970
English as second language	0.7360*
French-speaking Canadian	0.7840*
Alcoholic drinks	1.1250*
Lone male	1.0220*

The starred values were significant at the 0.01 level. We see that the probability of a high tip is reduced (odds ratio less than 1) when the diner is over 65 years old, speaks English as a second language, and is a French-speaking Canadian. The probability of a high tip is increased (odds ratio greater than 1) if alcohol is served with the meal.

SECTION 11.2 SUMMARY

• **Multiple linear regression** should always begin with a careful examination of the data. This involves looking at each variable separately and then at pairs of variables. Cases with extreme values should be noted and examined carefully throughout the analysis.

• Multiple linear regression has the same model conditions as simple linear regression. Prior to inference, always examine the **distribution of the residuals** and plot them against each of the explanatory variables to make sure there are no remaining patterns or nonconstant variance.

• The estimate b_j of β_j and the test and confidence interval for β_j are all based on a specific multiple linear regression model. The results of all these procedures change if other explanatory variables are added to or deleted from the model.

SECTION 11.2 EXERCISES

For Exercise 11.13, see page 621; and for Exercise 11.14, see page 625.

11.15 Refining the GPA model using all six explanatory variables: Residual checks. Figure 11.11 (page 631) provides a list of the top models based on R^2. Let's look more closely at the four models listed with $p = 3$ and $p = 4$. Fit each of these models to the data and obtain the residuals. Do the data, at least approximately, meet the conditions of the multiple regression model? Provide some plots to support your opinion. [GPA]

11.16 Refining the GPA model using all six explanatory variables: Inference. Refer to the previous exercise. For each of the four models considered in the previous exercise, report the least-squares equation, estimated model standard deviation s, and the P-values for each of the individual coefficients. Based on these results and the residuals checks of the previous exercise, which model do you think provides the "best" fit? Explain your answer. [GPA]

11.17 A mechanistic explanation of popularity. In Exercise 10.65 (page 605), correlations between an adolescent's "popularity," expression of a serotonin receptor gene, and rule-breaking behaviors were assessed. An additional portion of the analysis looked at the relationship between the gene expression level and popularity, after adjusting for rule-breaking (RB) behaviors. This adjustment was necessary because RB is positively associated both with this gene expression and with popularity in adolescents. The following summarizes these regression analyses using the composite (questionnaire and video) RB score. A total of 202 individuals were included in this analysis.

	b	$s(b)$
Model 1		
Gene expression	0.204	0.066
Model 2		
Gene expression	0.161	0.066
RB.composite	0.100	0.030

For all analysis use the 0.05 significance level.

(a) What are the error degrees of freedom for Model 1 and Model 2?

(b) Test the null hypothesis that the serotonin gene receptor coefficient is equal to 0 in Model 1. State the test statistic and P-value.

(c) Perform both individual-variable t tests for Model 2. Again state the test statistics and P-values.

(d) Is there still a positive relationship between the serotonin gene receptor expression level and popularity after adjusting for RB? If yes, compare the increase in popularity for a unit increase in gene expression (while RB remains unchanged) in the two models.

Results such as these suggest not only that adolescents with high serotonin receptor gene expression are predisposed to increased RB behaviors, but also that such behaviors are socially advantageous.

11.18 Predicting college debt: Multiple regression. Refer to Exercises 10.12 (page 579) and 10.17 (page 580) for a description of the problem. Let's now consider fitting a model using Admit, GradRate, InCostAid, and OutCostAid as the explanatory variables. [BESTVAL]

(a) Write out the statistical model for this analysis, making sure to specify all assumptions.

(b) Run the multiple regression model and specify the fitted regression equation.

(c) Obtain the residuals from part (b) and check assumptions. Is Baruch College still an unusual case? Provide a brief summary.

(d) Run the same multiple regression model but this time without Baruch College. Again comment on the residuals.

(e) Should we proceed with inference using the entire data set? Or the data set without Baruch College? Explain your reasoning.

11.19 Predicting college debt: Inference. Refer to the previous exercise. Let's proceed using the data set without Baruch College. [BESTVAL]

(a) Report the least-squares equation using all four variables.

(b) What percent of the variability in average debt is explained by this model?

(c) Report the F statistic, its degrees of freedom, and the P-value. What do you conclude based on this test result?

(d) Using this F test and the individual parameter t tests, write a one-paragraph summary of this model's fit to the data.

11.20 Testing a collection of variables. Refer to the previous exercise. For the model that included all $p = 4$ explanatory variables, only InCostAid is found significant using the individual parameter t tests. This raises the question whether these other three variables further contribute to the prediction of average debt given in-state cost is in the model.

In this chapter, we discussed the F test for a collection of regression coefficients. In most cases, this capability is provided by the software. When it is not, the test can be performed using the R^2-values from the larger (full) and smaller (reduced) models. The test statistic is

$$F = \left(\frac{n - p - 1}{q}\right)\left(\frac{R_1^2 - R_2^2}{1 - R_1^2}\right)$$

with q and $n - p - 1$ degrees of freedom. R_1^2 is the value for the full model, and R_2^2 is the value for the reduced model. Here $n = 24$ schools, $p = 4$ variables in the full model, and $q = 3$ variables were removed to form the reduced model. Plug in the values of R^2 from part (b) of the previous exercise and the R^2 value from Figure 10.13 (page 580). Compute the test statistic and P-value. Do Admit, GradRate, and OutCostAid combined add any significant predictive information beyond what is already contained in InCostAid?

11.21 Comparison of prediction intervals. Refer to the previous exercise. Another way to compare these two models is in terms of prediction. The Ohio State University has Admit = 56, GradRate = 59, InCostAid = 12,103, and OutCostAid = 28,603. Use statistical software to construct.

(a) a 95% prediction interval based on the model with all $p = 4$ predictors.

(b) a 95% prediction interval based on the model using just InCostAid.

(c) Compare the two intervals. Do the two models give similar predictions? Which provides a more narrow prediction interval?

11.22 Consider the sex of the students. Refer to Exercises 11.15 and 11.16. The seventh explanatory variable provided in the GPA data set is a sex indicator

variable. This variable (SEX) takes the value 0 for males and 1 for females. If we include it in our model involving the other six variables, it allows the intercept to differ for the two genders. Using b_7 to represent the fitted coefficient for the SEX variable, the estimated male intercept is $b_0 + b_7(0) = b_0$ and the estimated female intercept is $b_0 + b_7(1) = b_0 + b_7$. The difference between these two intercept estimates is $(b_0 + b_7) - b_0 = b_7$, so the coefficient is also an estimate of the difference in intercepts. GPA

(a) Include the variable SEX with the other six explanatory variables and refit the model. Compare the fit of this model, using R^2 and s, with the model in Figure 11.10 (pages 627–629).

(b) Does this indicator variable appear to contribute to our explanation of GPA? Report the test results.

(c) Does the coefficient suggest males or females have higher GPA scores? Explain your answer.

11.23 Predicting energy-drink consumption. Energy-drink advertising consistently emphasizes a physically active lifestyle and often features extreme sports and risk taking. Are these typical characteristics of an energy-drink consumer? A researcher decided to examine the links between energy-drink consumption, sport-related (jock) identity, and risk taking.[5] She invited more than 1500 undergraduate students enrolled in large introductory-level courses at a public university to participate. Each participant had to complete a 45-minute anonymous questionnaire. From this questionnaire, jock identity and risk-taking scores were obtained, where the higher the score, the stronger the trait. She ended up with 795 respondents. The following table summarizes the results of a multiple regression analysis using the frequency of energy-drink consumption in the past 30 days as the response variable:

Explanatory variable	b
Age	−0.02
Sex (1 = female, 0 = male)	−0.11**
Race (1 = nonwhite, 0 = white)	−0.02
Ethnicity (1 = Hispanic, 0 = non-Hispanic)	0.10**
Parental education	0.02
College GPA	−0.01
Jock identity	0.05
Risk taking	0.19***

A superscript of ** means that the individual coefficient t test had a P-value less than 0.01, and a superscript of *** means that the test had a P-value less than 0.001. All other P-values were greater than 0.05.

(a) The overall F statistic is reported to be 8.11. What are the degrees of freedom associated with this statistic?

(b) R is reported to be 0.28. What percent of the variation in energy-drink consumption is explained by the model? Is this a highly predictive model? Explain.

(c) Interpret each of the regression coefficients that are significant.

(d) The researcher states, "Controlling for gender, age, race, ethnicity, parental educational achievement, and college GPA, each of the predictors (risk taking and jock identity) was positively associated with energy-drink consumption frequency." Explain what is meant by "controlling for" these variables and how this helps strengthen her assertion that jock identity and risk taking are positively associated with energy-drink consumption.

11.24 Is the number of tornadoes increasing? In Exercise 10.19, data on the number of tornadoes in the United States between 1953 and 2014 were analyzed to see if there was a linear trend over time. Some argue that it's not the number of tornadoes increasing over time, but rather the probability of sighting them because there are more people living in the United States. Let's investigate this by including the U.S. census count as an additional explanatory variable. **TWISTER**

(a) Using numerical and graphical summaries, describe the relationship between each pair of variables.

(b) Perform a multiple regression using both year and census count as explanatory variables. Write down the fitted model.

(c) Obtain the residuals from part (b). Plot them versus the two explanatory variables and generate a Normal quantile plot. What do you conclude?

(d) Test the hypothesis that there is a linear increase over time. State the null and alternative hypotheses, test statistic, and P-value. What is your conclusion?

CHAPTER 11 EXERCISES

11.25 Checking for a polynomial relationship. When looking at the residuals from the simple linear model of BMI versus physical activity (PA), Figure 10.5 (page 566) suggested a possible curvilinear relationship. Let's investigate this further. Multiple regression can be used to fit the polynomial curve of degree q, $y = \beta_0 + \beta_1 x + \beta_2 x^2 + \cdots + \beta_q x^q$, through the creation of additional explanatory variables x^2, x^3, etc. Let's investigate a quadratic fit ($q = 2$) for the physical activity problem. **PABMI**

(a) It is often best to subtract the sample mean $\bar{x}$ before creating the necessary explanatory variables. In this case, the average number of steps per day is 8.614. Create new explanatory variables $x_1 = (PA - 8.614)$ and $x_2 = (PA - 8.614)^2$ and run a multiple regression for BMI using the explanatory variables x_1 and x_2. Write down the fitted regression line.

(b) The regression model that included only PA had a $R^2 = 14.9\%$. What is R^2 with the inclusion of this quadratic term?

(c) Obtain the residuals from part (a) and check the multiple regression assumptions. Are there any remaining patterns in the data? Are the residuals approximately Normal? Explain.

(d) Test the hypothesis that the coefficient of the variable $(PA - 8.614)^2$ is equal to 0. Report the t statistic, degrees of freedom, and P-value. Does the quadratic term contribute significantly to the fit? Explain your answer.

11.26 Architectural firm billings. A summary of firms engaged in commercial architecture in the Indianapolis, Indiana, area provides firm characteristics, including total annual billing in the current year, total annual billing in the previous year, the number of architects, the number of engineers, and the number of staff employed in the firm.[6] Consider developing a model to predict current total billing using the other four variables. **ARCH**

(a) Using numerical and graphical summaries, describe the distribution of current and past year total billing and the number of architects, engineers, and staff.

(b) For each of the 10 pairs of variables, use graphical and numerical summaries to describe the relationship.

(c) Carry out a multiple regression. Report the fitted regression equation and the value of the regression standard error s.

(d) Analyze the residuals from the multiple regression. Are there any concerns?

(e) A firm did not report its current total billing but had $1 million in billing last year and employs three architects, one engineer, and 17 staff members. What is the predicted total billing for this firm?

(f) This analysis utilized the data from all commercial firms in the Indianapolis area that responded to the survey. Provide justification for the use of inference under this setting.

The following six exercises use the MOVIES data file. This data set contains an SRS of 43 movies released four to five years ago to guarantee they are no longer in the theaters. This sample was collected from the Internet Movie Database (IMDb) to see if information available soon after a movie's theatrical release can successfully predict total U.S. revenue.[7] All dollar amounts are measured in millions of U.S. dollars. MOVIES

11.27 Predicting movie revenue—preliminary analysis.
The response variable is a movie's total U.S. revenue (USRevenue). Let's consider as explanatory variables the movie's budget (Budget); opening-weekend revenue (Opening); the number of theaters (Theaters) the movie was in for the opening weekend; and the movie's IMDb rating (Ratings), which is on a 1 to 10 scale (10 being best). While this rating is updated continuously, we'll assume that the current rating is the rating at the end of the first week.

(a) Using numerical and graphical summaries, describe the distribution of each explanatory variable. Are there any unusual observations that should be monitored?

(b) Using numerical and/or graphical summaries, describe the relationship between each pair of explanatory variables.

11.28 Predicting movie revenue—simple linear regressions.
Now let's look at the response variable and its relationship with each explanatory variable.

(a) Using numerical and graphical summaries, describe the distribution of the response variable, USRevenue.

(b) This variable is not Normally distributed. Does this violate one of the key model assumptions? Explain.

(c) Generate scatterplots of each explanatory variable and USRevenue. Do all these relationships look linear? Explain what you see

11.29 Predicting movie revenue—multiple linear regression.
Now consider fitting a model using all the explanatory variables.

(a) Write out the statistical model for this analysis, making sure to specify all assumptions.

(b) Run the multiple regression model and specify the fitted regression equation.

(c) Obtain the residuals from part (b) and check assumptions. Comment on any unusual residuals or patterns in the residuals.

(d) What percent of the variability in USRevenue is explained by this model?

11.30 A simpler model.
In the multiple regression analysis using all four explanatory variables, Theaters and Budget appear to be the least helpful (given that the other two explanatory variables are in the model).

(a) Perform a new analysis using only the movie's opening-weekend revenue and IMDb rating. Give the estimated regression equation for this analysis.

(b) What percent of the variability in USRevenue is explained by this model?

(c) Test the null hypothesis that Theaters and Budget combined add no additional predictive information beyond what is already contained in Opening and Opinion?

11.31 Predicting U.S. movie revenue.
The movie *Kick-Ass* was released during this same time period. It had a budget of $30.0 million and was shown in 3065 theaters, grossing $19.83 million during the first weekend. Use software to construct the following. MOVIES

(a) A 95% prediction interval based on the model with all three explanatory variables.

(b) A 95% prediction interval based on the model using only opening-weekend revenue and budget.

(c) Compare the two intervals. Do the models give similar predictions and standard errors?

11.32 Considering the log transformation.
Refer to Exercise 11.29. Variables like income often have very skewed distributions. This can result in certain cases strongly influencing the fit of the model. A common remedy is to take the log before analysis. Create a new response variable by taking the log of U.S. Revenue and fit the model using all four predictors. Obtain the residuals and assess the model conditions. Do these data fit the linear regression model better than the untransformed data? Explain your answer.

The following three exercises use the RANKINGS data file. Since 2004, The Times Higher Education Supplement has provided an annual ranking of the world universities. A total score for each university is calculated based on the scores for the following explanatory variables: Teaching (30%), Research (30%), Citations (30%), Industry Income (2.5%), and International Outlook (7.5%) The percents represent the contributions of each score to the total. For our purposes, we will assume that these weights are unknown and will focus on the development of a model for the total score based on the first three explanatory variables. The report includes a table for the top 200 universities.[8] The RANKINGS data file contains a random sample of 55 of these universities. This is not a random sample of all universities, but for our purposes here, we will consider it to be. 📊 RANKINGS

11.33 Annual ranking of world universities. Let's consider developing a model to predict total score (Overall) based on the teaching, research, and citations scores.

(a) Using numerical and graphical summaries, describe the distribution of each explanatory variable.

(b) Using numerical and graphical summaries, describe the relationship between each pair of explanatory variables.

11.34 Looking at the simple linear regressions. Now let's look at the relationship between each explanatory variable and the total score.

(a) Generate scatterplots for each explanatory variable and the total score. Do these relationships all look linear?

(b) Compute the correlation between each explanatory variable and the total score. Are certain explanatory variables more strongly associated with the total score?

11.35 Multiple linear regression model. Now consider a regression model using all three explanatory variables.

(a) Write out the statistical model for this analysis, making sure to specify all assumptions.

(b) Run the multiple regression model and specify the fitted regression equation.

(c) Generate a 95% confidence interval for each coefficient. Should any of these intervals contain 0? Explain.

(d) What percent of the variation in total score is explained by this model? What is the estimate for σ?

11.36 Predicting GPA of seventh-graders. Refer to the educational data for 78 seventh-grade students given in Table 1.3 (page 26). We view GPA as the response variable. IQ, gender, and self-concept are the explanatory variables. 📊 SEVENGR

(a) Find the correlation between GPA and each of the explanatory variables. What percent of the total variation in student GPAs can be explained by the straight-line relationship with each of the explanatory variables?

(b) The importance of IQ in explaining GPA is not surprising. The purpose of the study is to assess the influence of self-concept on GPA. So we will include IQ in the regression model and ask, "How much does self-concept contribute to explaining GPA after the effect of IQ on GPA is taken into account?" Give a model that can be used to answer this question.

(c) Run the model and report the fitted regression equation. What percent of the variation in GPA is explained by the explanatory variables in your model?

(d) Translate the question of interest into appropriate null and alternative hypotheses about the model parameters. Give the value of the test statistic and its *P*-value. Write a short summary of your analysis with an emphasis on your conclusion.

*The following three exercises use the HAPPY data file. The World Database of Happiness is an online registry of scientific research on the subjective appreciation of life. It is available at **worlddatabaseofhappiness.eur.nl**, and the project is directed by Dr. Ruut Veenhoven, Erasmus University, Rotterdam. One inventory presents the "average happiness" score for various nations. This average is based on individual responses from numerous general population surveys to a general life satisfaction (well-being) question. Scores range from 0 (dissatisfied) to 10 (satisfied). The Nation-Master website, **www .nationmaster.com**, contains a collection of statistics associated with various nations. For our analysis, we will consider the GINI index, which measures the degree of inequality in the distribution of income (higher score = greater inequality), the degree of corruption in government (higher score = less corruption), average life expectancy, and the degree of democracy (higher score = more civil and political liberties).* 📊 HAPPY

11.37 Predicting a nation's "average happiness" score. Consider the five statistics for each nation: LSI, the average life-satisfaction score; GINI, the GINI index; CORRUPT, the degree of government corruption; LIFE, the average life expectancy; and DEMOCRACY, a measure of civil and political liberties.

(a) Using numerical and graphical summaries, describe the distribution of each variable.

(b) Using numerical and graphical summaries, describe the relationship between each pair of variables.

11.38 Building a multiple linear regression model. Let's now build a model to predict the life-satisfaction score, LSI.

(a) Consider a simple linear regression using GINI as the explanatory variable. Run the regression and summarize the results. Be sure to check assumptions.

(b) Now consider a model using GINI and LIFE. Run the multiple regression and summarize the results. Again be sure to check assumptions.

(c) Now consider a model using GINI, LIFE, and DEMOCRACY. Run the multiple regression and summarize the results. Again be sure to check assumptions.

(d) Now consider a model using all four explanatory variables. Again summarize the results and check assumptions.

11.39 Selecting from among several models. Refer to the results from the previous exercise.

(a) Make a table giving the estimated regression coefficients, standard errors, t statistics, and P-values.

(b) Describe how the coefficients and P-values change for the four models.

(c) Based on the table of coefficients, suggest another model. Run that model, summarize the results, and compare it with the other ones. Which model would you choose to explain LSI? Explain.

The following six exercises use the BIOMARK data file. Healthy bones are continually being renewed by two processes. Through bone formation, new bone is built; through bone resorption, old bone is removed. If one or both of these processes are disturbed—by disease, aging, or space travel, for example—bone loss can be the result. The variables VO+ and VO− measure bone formation and bone resorption, respectively. Osteocalcin (OC) is a biochemical marker for bone formation: higher levels of bone formation are associated with higher levels of OC. A blood sample is used to measure OC, and it is much less expensive to obtain than direct measures of bone formation. The units are milligrams of OC per milliliter of blood (mg/ml). Similarly, tartrate-resistant acid phosphatase (TRAP) is a biochemical marker for bone resorption that is also measured in blood. It is measured in units per liter (U/l). These variables were measured in a study of 31 healthy women aged 11 to 32 years.[9] Variables with the first letter "L" are the logarithms of the measured variables. **BIOMARK**

11.40 Bone formation and resorption. Consider the following four variables: VO+, a measure of bone formation; VO−, a measure of bone resorption; OC, a biomarker of bone formation; and TRAP, a biomarker of bone resorption.

(a) Using numerical and graphical summaries, describe the distribution of each of these variables.

(b) Using numerical and graphical summaries, describe the relationship between each pair of variables.

11.41 Predicting bone formation. Let's use regression methods to predict VO+, the measure of bone formation.

(a) Because OC is a biomarker of bone formation, we start with a simple linear regression using OC as the explanatory variable. Run the regression and summarize the results. Be sure to include an analysis of the residuals.

(b) Because the processes of bone formation and bone resorption are highly related, it is possible that there is some information in the bone resorption variables that can tell us something about bone formation. Use a model with both OC and TRAP, the biomarker of bone resorption, to predict VO+. Summarize the results. In the context of this model, it appears that TRAP is a better predictor of bone formation, VO+, than the biomarker of bone formation, OC. Is this view consistent with the pattern of relationships that you described in the previous exercise? One possible explanation is that, although all these variables are highly related, TRAP is measured with more precision than OC.

11.42 More on predicting bone formation. Now consider a regression model for predicting VO+ using OC, TRAP, and VO−.

(a) Write out the statistical model for this analysis including all assumptions.

(b) Run the multiple regression to predict VO+ using OC, TRAP, and VO−. Summarize the results.

(c) Make a table giving the estimated regression coefficients, standard errors, and t statistics with P-values for this analysis and for the two that you ran in the previous exercise. Describe how the coefficients and the P-values differ for the three analyses.

(d) Give the percent of variation in VO+ explained by each of the three models and the estimate of σ. Give a short summary.

(e) The results you found in part (b) suggest another model. Run that model, summarize the results, and compare them with the results in part (b).

11.43 Predicting bone formation using transformed variables. Because the distributions of VO+, VO−, OC, and TRAP tend to be skewed, it is common to work with logarithms rather than the measured values. Using the questions in the previous three exercises as a guide, analyze the log data.

11.44 Predicting bone resorption. Refer to Exercises 11.40, 11.41, and 11.42. Answer these questions with the roles of VO+ and VO− reversed; that is, run models to predict VO−, with VO+ as an explanatory variable.

11.45 Predicting bone resorption using transformed variables. Refer to the previous exercise. Rerun using logs.

The following 11 exercises use the PCB data file. Polychlorinated biphenyls (PCBs) are a collection of synthetic compounds, called congeners, that are particularly toxic to fetuses and young children. Although

PCBs are no longer produced in the United States, they are still found in the environment. Because human exposure to these PCBs is primarily through the consumption of fish, the Environmental Protection Agency (EPA) monitors the PCB levels in fish. Unfortunately, there are 209 different congeners and measuring all of them in a fish specimen is an expensive and time-consuming process. You've been asked to see if the total amount of PCBs in a specimen can be estimated with only a few, easily quantifiable congeners.[10] If this can be done, costs can be greatly reduced. 🏬 PCB

11.46 Relationships among PCB congeners. Consider the following variables: PCB (the total amount of PCB) and four congeners: PCB52, PCB118, PCB138, and PCB180.

(a) Using numerical and graphical summaries, describe the distribution of each of these variables.

(b) Using numerical and graphical summaries, describe the relationship between each pair of variables.

11.47 Predicting the total amount of PCB. Use the four congeners PCB52, PCB118, PCB138, and PCB180 in a multiple regression to predict PCB.

(a) Write the statistical model for this analysis. Include all assumptions.

(b) Run the regression and summarize the results.

(c) Examine the residuals. Do they appear to be approximately Normal? When you plot them versus each of the explanatory variables, are any patterns evident?

11.48 Adjusting the analysis for potential outliers. The examination of the residuals in part (c) of the previous exercise suggests that there may be two outliers, one with a high residual and one with a low residual.

(a) Because of safety issues, we are more concerned about underestimating PCB in a specimen than about overestimating. Give the specimen number for each of the two suspected outliers. Which one corresponds to an overestimate of PCB?

(b) Rerun the analysis with the two suspected outliers deleted, summarize these results, and compare them with those you obtained in the previous exercise.

11.49 More on predicting the total amount of PCB. Run a regression to predict PCB using the variables PCB52, PCB118, and PCB138. Note that this is similar to the analysis that you did in Exercise 11.47, with the change that PCB180 is not included as an explanatory variable.

(a) Summarize the results.

(b) In this analysis, the regression coefficient for PCB118 is not statistically significant. Give the estimate of the coefficient and the associated *P*-value.

(c) Find the estimate of the coefficient for PCB118 and the associated *P*-value for the model analyzed in Exercise 11.47.

(d) Using the results in parts (b) and (c), write a short paragraph explaining how the inclusion of other variables in a multiple regression can have an effect on the estimate of a particular coefficient and the results of the associated significance test.

11.50 Multiple regression model for total TEQ. Dioxins and furans are other classes of chemicals that can cause undesirable health effects similar to those caused by PCB. The three types of chemicals are combined using toxic equivalent scores (TEQs), which attempt to measure the health effects on a common scale. The PCB data file contains TEQs for PCB, dioxins, and furans. The variables are called TEQPCB, TEQDIOXIN, and TEQFURAN. The data file also includes the total TEQ, defined to be the sum of these three variables.

(a) Consider using a multiple regression to predict TEQ using the three components TEQPCB, TEQDIOXIN, and TEQFURAN as explanatory variables. Write the multiple regression model in the form

$$TEQ = \beta_0 + \beta_1 TEQPCB + \beta_2 TEQDIOXIN + \beta_3 TEQFURAN + \epsilon$$

Give numerical values for the parameters β_0, β_1, β_2, and β_3.

(b) The multiple regression model assumes that the ϵ's are Normal with mean zero and standard deviation σ. What is the numerical value of σ?

(c) Use software to run this regression and summarize the results.

🔧 **11.51 Multiple regression model for total TEQ, continued.** The information summarized in TEQ is used to assess and manage risks from these chemicals. For example, the World Health Organization (WHO) has established the tolerable daily intake (TDI) as one to four TEQs per kilogram of body weight per day. Therefore, it would be very useful to have a procedure for estimating TEQ using just a few variables that can be measured cheaply. Use the four PCB congeners PCB52, PCB118, PCB138, and PCB180 in a multiple regression to predict TEQ. Give a description of the model and assumptions, summarize the results, examine the residuals, and write a summary of what you have found.

🔧 **11.52 Predicting total amount of PCB using transformed variables.** Because distributions of variables such as PCB, the PCB congeners, and TEQ tend to be skewed, researchers frequently analyze the logarithms of the measured variables. Create a data set that has the logs of each of the variables in the PCB data file.

Note that zero is a possible value for PCB126; most software packages will eliminate these cases when you request a log transformation.

(a) If you do not do anything about the 16 zero values of PCB126, what does your software do with these cases? Is there an error message of some kind?

(b) If you attempt to run a regression to predict the log of PCB using the log of PCB126 and the log of PCB52, are the cases with the zero values of PCB126 eliminated? Do you think that this is a good way to handle this situation?

(c) The smallest nonzero value of PCB126 is 0.0052. One common practice when taking logarithms of measured values is to replace the zeros by one-half of the smallest observed value. Create a logarithm data set using this procedure; that is, replace the 16 zero values of PCB126 by 0.0026 before taking logarithms. Use numerical and graphical summaries to describe the distributions of the log variables.

11.53 Predicting total amount of PCB using transformed variables, continued. Refer to the previous exercise.

(a) Use numerical and graphical summaries to describe the relationships between each pair of log variables.

(b) Compare these summaries with the summaries that you produced in Exercise 11.46 for the measured variables.

11.54 Even more on predicting total amount of PCB using transformed variables. Use the log data set that you created in Exercise 11.52 to find a good multiple regression model for predicting the log of PCB. Use only log PCB variables for this analysis. Write a report summarizing your results.

11.55 Predicting total TEQ using transformed variables. Use the log data set that you created in Exercise 11.52 to find a good multiple regression model for predicting the log of TEQ. Use only log PCB variables for this analysis. Write a report summarizing your results and comparing them with the results that you obtained in the previous exercise.

11.56 Interpretation of coefficients in log PCB regressions. Use the results of your analysis of the log PCB data in Exercise 11.54 to write an explanation of how regression coefficients, standard errors of regression coefficients, and tests of significance for explanatory variables can change depending on what other explanatory variables are included in the multiple regression analysis.

The following nine exercises use the CHEESE data file. As cheddar cheese matures, a variety of chemical processes take place. The taste of matured cheese is related to the concentration of several chemicals in the final product. In a study of cheddar cheese from the LaTrobe Valley of Victoria, Australia, samples of cheese were analyzed for their chemical composition and were subjected to taste tests. The variable "Case" is used to number the observations from 1 to 30. "Taste" is the response variable of interest. The taste scores were obtained by combining the scores from several tasters. Three of the chemicals whose concentrations were measured were acetic acid, hydrogen sulfide, and lactic acid. For acetic acid and hydrogen sulfide, (natural) log transformations were taken. Thus, the explanatory variables are the transformed concentrations of acetic acid ("Acetic") and hydrogen sulfide ("H2S") and the untransformed concentration of lactic acid ("Lactic").[11] 📊 CHEESE

11.57 Describing the explanatory variables. For each of the four variables in the CHEESE data file, find the mean, median, standard deviation, and interquartile range. Display each distribution by means of a stemplot and use a Normal quantile plot to assess Normality of the data. Summarize your findings. Note that when doing regressions with these data, we do not assume that these distributions are Normal. Only the residuals from our model need to be (approximately) Normal. The careful study of each variable to be analyzed is, nonetheless, an important first step in any statistical analysis.

11.58 Pairwise scatterplots of the explanatory variables. Make a scatterplot for each pair of variables in the CHEESE data file (you will have six plots). Describe the relationships. Calculate the correlation for each pair of variables and report the *P*-value for the test of zero population correlation in each case.

11.59 Simple linear regression model of Taste. Perform a simple linear regression analysis using Taste as the response variable and Acetic as the explanatory variable. Be sure to examine the residuals carefully. Summarize your results. Include a plot of the data with the least-squares regression line. Plot the residuals versus each of the other two chemicals. Are any patterns evident? (The concentrations of the other chemicals are lurking variables for the simple linear regression.)

11.60 Another simple linear regression model of Taste. Repeat the analysis of Exercise 11.59 using Taste as the response variable and H2S as the explanatory variable.

11.61 The final simple linear regression model of Taste. Repeat the analysis of Exercise 11.59 using Taste as the response variable and Lactic as the explanatory variable.

11.62 Comparing the simple linear regression models. Compare the results of the regressions performed in the three previous exercises. Construct a table with values of the F statistic, its P-value, R^2, and the estimate s of the standard deviation for each model. Report the three regression equations. Why are the intercepts in these three equations different?

11.63 Multiple regression model of Taste. Carry out a multiple regression using Acetic and H2S to predict Taste. Summarize the results of your analysis. Compare the statistical significance of Acetic in this model with its significance in the model with Acetic alone as a predictor (Exercise 11.59). Which model do you prefer? Give a simple explanation for the fact that Acetic alone appears to be a good predictor of Taste, but with H2S in the model, it is not.

11.64 Another multiple regression model of Taste. Carry out a multiple regression using H2S and Lactic to predict Taste. When we compare the results of this analysis with the simple linear regressions using each of these explanatory variables alone, it is evident that a better result is obtained by using both predictors in a model. Support this statement with explicit information obtained from your analysis.

11.65 The final multiple regression model of Taste. Use the three explanatory variables Acetic, H2S, and Lactic in a multiple regression to predict Taste. Write a short summary of your results, including an examination of the residuals. Based on all the regression analyses you have carried out on these data, which model do you prefer and why?

11.66 Finding a multiple regression model on the Internet. Search the Internet to find an example of the use of multiple regression. Give the setting of the example, describe the data, give the model, and summarize the results. Explain why the use of multiple regression in this setting was appropriate or inappropriate.

© Monkey Business Images Ltd/Dreamstime.com

One-Way Analysis of Variance

Introduction

Many of the most effective statistical studies are comparative. For example, we may wish to compare customer satisfaction of men and women who use an online fantasy football site or compare the responses to various treatments in a clinical trial. With a quantitative response, we display these comparisons with back-to-back stemplots or side-by-side boxplots, and we measure them with five-number summaries or with means and standard deviations.

When only two groups are compared, Chapter 7 provides the tools we need to answer the question, "Is the difference between groups statistically significant?" Two-sample t procedures compare the means of two Normal populations, and we saw that these procedures are sufficiently robust to be widely useful.

In this chapter, we will compare any number of means by techniques that generalize the two-sample t test and share its robustness and usefulness. These methods will allow us to address comparisons such as

• Does a user's number of Facebook friends affect his or her social attractiveness?

• On average, which of five brands of automobile tires wears longest?

• How do five coffeehouses around campus differ in the demographics of their customers?

12.1 Inference for One-Way Analysis of Variance

12.2 Comparing the Means

12.1 Inference for One-Way Analysis of Variance

When you complete this section, you will be able to:

- Describe the one-way ANOVA model and when it is used for inference.
- Describe the underlying idea of the ANOVA F test in terms of the variation between population means and the variation within populations.
- Summarize what the ANOVA F test can tell you about the population means and what it cannot.
- Construct an ANOVA table with sources of variation and degrees of freedom. Compute mean squares and the F statistic when provided various sums of squares.
- Use ANOVA output to obtain the ANOVA F test results and the coefficient of determination.
- Use residual plots and sample statistics to check the assumptions of the one-way ANOVA model.

LOOK BACK

comparing
two means,
p. 432

ANOVA

one-way ANOVA
factor

two-way ANOVA

When comparing different populations or treatments, the data are subject to sampling variability. For example, we would not expect to observe exactly the same sales data if we mailed an advertising offer to different random samples of households. We also wouldn't expect a new group of cancer patients to provide the same set of progression-free survival times. We therefore pose the question for inference in terms of the *mean* response.

In Chapter 7, we met procedures for comparing the means of two populations. We now extend those methods to problems involving more than two populations. The statistical methodology for comparing several means is called **analysis of variance**, or simply **ANOVA**. In this and the following section, we will examine the basic ideas and assumptions that are needed for ANOVA. Although the details differ, many of the concepts are similar to those discussed in the two-sample case.

We will consider two ANOVA techniques. When there is only one way to classify the populations of interest, we use **one-way ANOVA** to analyze the data. We call this categorical explanatory variable a **factor**. For example, to compare the average tread lifetimes of five specific brands of tires, we use one-way ANOVA with tire brand as our factor. This chapter presents the details for one-way ANOVA.

In many other comparative studies, there is more than one way to classify the populations. For the tire study, the researcher may also want to consider temperature. Are there brands that do relatively better in a cooler environment? Analyzing the effect of two factors, brand and temperature, requires **two-way ANOVA**. This technique will be discussed in Chapter 13.

Data for one-way ANOVA

One-way analysis of variance is a statistical method for comparing several population means. We draw a simple random sample (SRS) from each population and use the data to test the null hypothesis that the population means are all equal. Consider the following two examples.

EXAMPLE 12.1

Does haptic feedback improve performance? A group of technology students is interested in whether haptic feedback (forces and vibrations applied through a joystick) is helpful in navigating a simulated game environment they created. To investigate this, they randomly assign 20 students to each of three joystick controller types and record the time it takes to complete a navigation mission. The joystick types are (1) a standard video game joystick, (2) a game joystick with force feedback, and (3) a game joystick with vibration feedback.

EXAMPLE 12.2

Average age of coffeehouse customers. How do five coffeehouses around campus differ in the demographics of their customers? Are certain coffeehouses more popular among graduate students? Do professors tend to favor one coffeehouse? A market researcher asks 50 customers of each coffeehouse to respond to a questionnaire. One variable of interest is the customer's age.

These two examples are similar in that

• There is a single quantitative response variable measured on many units; the units are students in the first example and customers in the second.

• The goal is to compare several populations: students using different joystick types in the first example and customers of five coffeehouses in the second.

observation
versus experiment,
p. 168
bias and
variability,
p. 287

There is, however, an important difference. Example 12.1 describes an experiment in which each student is randomly assigned to a type of joystick. Example 12.2 is an observational study in which customers are selected during a particular time period and not all agree to provide data. These samples of customers are not random samples, but we will treat them as such because we believe that the selective sampling and nonresponse are ignorable sources of bias and variability. This will not always be the case. *Always consider the sampling design and various sources of bias in an observational study.*

In both examples, we will use ANOVA to compare the mean responses. The same ANOVA methods apply to data from randomized experiments and to data from random samples. *However, it is important to keep the data-production method in mind when interpreting the results.* A strong case for causation is best made by a randomized experiment.

Comparing means

The question we ask in ANOVA is "Do all groups have the same population mean?" We will often use the term *groups* for the populations to be compared in a one-way ANOVA. To answer this question we compare the sample means. Figure 12.1 displays the sample means for Example 12.1. It appears that a joystick with force feedback has the shortest average completion time. But is the observed difference in sample means just the result of chance variation? We should not expect sample means to be equal even if the population means are all identical.

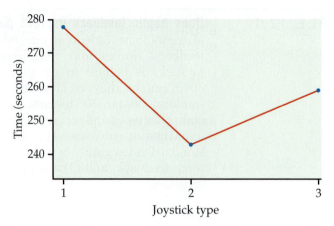

FIGURE 12.1 Average completion time for three different joystick types, Example 12.1.

LOOK BACK

standard
deviation of $\bar{x}$,
p. 297

The purpose of ANOVA is to assess whether the observed differences among sample means are *statistically significant*. Could a variation among the three sample means this large be plausibly due to chance, or is it good evidence for a difference among the population means? This question can't be answered from the sample means alone. Because the standard deviation of a sample mean $\bar{x}$ is the population standard deviation σ divided by $\sqrt{n}$, the answer also depends upon both the variation within the groups of observations and the sizes of the samples.

Side-by-side boxplots help us see the within-group variation. Compare Figures 12.2(a) and 12.2(b). The sample medians are the same in both figures, but the large variation within the groups in Figure 12.2(a) suggests that the differences among the sample medians could be due simply to chance variation. The data in Figure 12.2(b) are much more convincing evidence that the populations differ.

Even the boxplots omit essential information, however. To assess the observed differences, we must also know how large the samples are. Nonetheless, boxplots are a good preliminary display of the data.

Although ANOVA compares means and boxplots display medians, these two measures of center will be close together for distributions that are nearly

FIGURE 12.2 (a) Side-by-side boxplots for three groups with large within-group variation. The differences among centers may be just chance variation. (b) Side-by-side boxplots for three groups with the same centers as in panel (a) but with small within-group variation. The differences among centers are more likely to be significant.

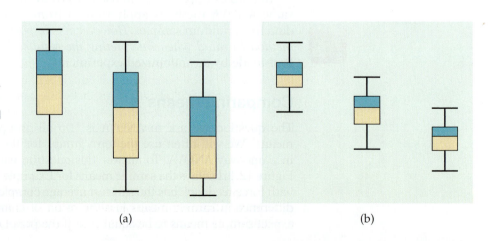

(a)   (b)

symmetric. This is something we will need to check prior to inference. If the distributions are strongly skewed, we may consider a transformation prior to displaying and analyzing the data.

transforming data, p. 470

pooled two-sample t statistic, p. 449

The two-sample t statistic

Two-sample t statistics compare the means of two populations. If the two populations are assumed to have equal but unknown standard deviations and the sample sizes are both equal to n, the t statistic is

$$t = \frac{\bar{x}_1 - \bar{x}_2}{s_p \sqrt{\dfrac{1}{n} + \dfrac{1}{n}}} = \frac{\sqrt{\dfrac{n}{2}}\,(\bar{x}_1 - \bar{x}_2)}{s_p}$$

The square of this t statistic is

$$t^2 = \frac{\dfrac{n}{2}(\bar{x}_1 - \bar{x}_2)^2}{s_p^2}$$

If we use ANOVA to compare two populations, the ANOVA F statistic is exactly equal to this t^2. Thus, we learn something about how ANOVA works by looking carefully at the statistic in this form.

between-group variation

The numerator in the t^2 statistic measures the variation **between** (or among) the groups in terms of the difference between their sample means $\bar{x}_1$ and $\bar{x}_2$ and the common sample size n. The numerator can be large because of a large difference between the sample means or because the common sample size is large.

within-group variation

The denominator measures the variation **within** groups by s_p^2, the pooled estimator of the common variance. If the within-group variation is small, the same variation between the groups produces a larger statistic and thus a more significant result.

Although the general form of the F statistic is more complicated, the idea is the same. To assess whether several populations all have the same mean, we compare the variation *among* the means of several groups with the variation *within* groups. Because we are comparing variation, the method is called *analysis of variance.*

An overview of ANOVA

ANOVA tests the null hypothesis that the population means are *all* equal. The alternative is that they are not all equal. This alternative could be true because all the means are different or simply because one of them differs from the rest. This is a more complex situation than comparing just two populations. If we reject the null hypothesis, we need to perform some further analysis to draw conclusions about which population means differ from which others and by how much.

The computations needed for ANOVA are more lengthy than those for the t test. For this reason, we generally use computer programs to perform the calculations. Automating the calculations frees us from the burden of arithmetic and allows us to concentrate on interpretation.

We should always start our ANOVA with a careful examination of the data using graphical and numerical summaries. *Complicated computations do not guarantee a valid statistical analysis.* Just as in linear regression, outliers and extreme deviations from Normality can invalidate the computed results.

EXAMPLE 12.3

FRIENDS

Number of Facebook friends. A feature of each Facebook user's profile is the number of Facebook "friends," an indicator of the user's social network connectedness. Among college students on Facebook, the average number of Facebook friends has been estimated to be around 650.[1]

Offline, having more friends is associated with higher ratings of positive attributes such as likability and trustworthiness. Is this also the case with Facebook friends?

An experiment was run to examine the relationship between the number of Facebook friends and the user's perceived social attractiveness.[2] A total of 134 undergraduate participants were randomly assigned to observe one of five Facebook profiles. Everything about the profile was the same except the number of friends, which appeared on the profile as 102, 302, 502, 702, or 902.

After viewing the profile, each participant was asked to fill out a questionnaire on the physical and social attractiveness of the profile user. Each attractiveness score is an average of several seven-point questionnaire items, ranging from 1 (strongly disagree) to 7 (strongly agree). Here is a summary of the data for the social attractiveness score:

Number of friends	n	$\bar{x}$	s
102	24	3.82	1.00
302	33	4.88	0.85
502	26	4.56	1.07
702	30	4.41	1.43
902	21	3.99	1.02

LOOK BACK

guidelines for two-sample t procedures, p. 442

Histograms for the five groups are given in Figure 12.3. Note that the heights of the bars in the histograms are percents rather than counts. This is commonly done when the group sample sizes vary. Figure 12.4 gives side-by-side boxplots for these data. We see that the scores covered the entire range of possible values, from 1.0 to 7.0. We also see a lot of overlap in scores across groups. The histograms are relatively symmetric, and with the group sample sizes all more than 15, we can feel confident that the sample means are approximately Normal.

The five sample means are plotted in Figure 12.5 (page 650). They rise and then fall as the number of friends increases. This suggests that having too many Facebook friends can harm a user's social attractiveness. However, given the variability in the data, this pattern could also just be the result of chance variation. We will use ANOVA to make this determination.

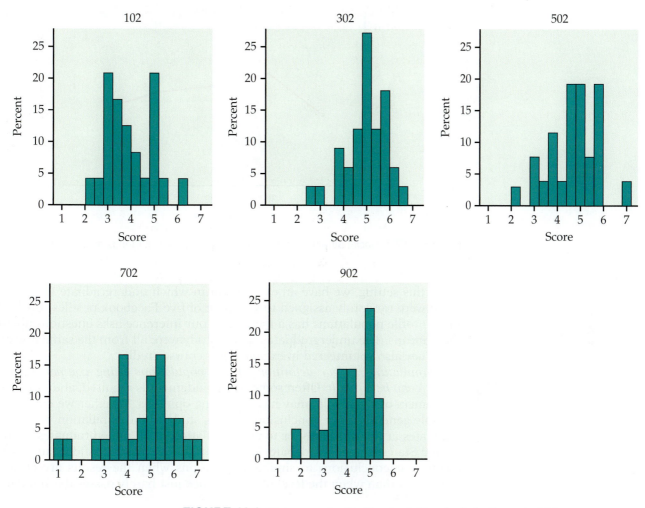

FIGURE 12.3 Histograms for the Facebook friends study, Example 12.3.

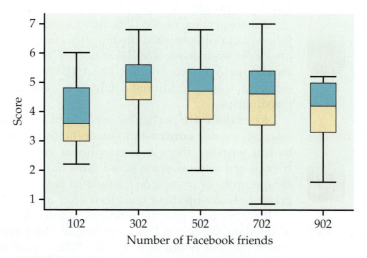

FIGURE 12.4 Side-by-side boxplots for the Facebook friends study, Example 12.3.

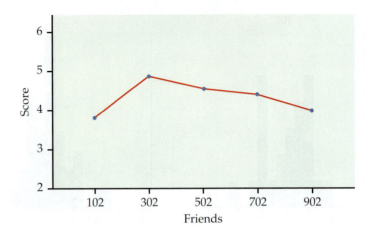

FIGURE 12.5 Social attractiveness means for the Facebook friends study, Example 12.3.

 In this setting, we have an experiment in which undergraduate Facebook users were randomly assigned to view one of five Facebook profiles. Each of these profile populations has a mean, and our inference asks questions about these means. The undergraduates in this study were all from the same university. They also volunteered in exchange for course credit.

Formulating a clear definition of the populations being compared with ANOVA can be difficult. Often some expert judgment is required, and different consumers of the results may have differing opinions. Whether we can comfortably generalize our conclusions of this study to the population of undergraduates at the university or to the population of all undergraduates in the United States is open for debate. Regardless, we are more confident in generalizing our conclusions to similar populations when the results are clearly significant than when the level of significance just barely passes the standard of $P = 0.05$.

We first ask whether or not there is sufficient evidence in the data to conclude that the corresponding population means are not all equal. Our null hypothesis here states that the population mean score is the same for all five Facebook profiles. The alternative is that they are not all the same.

 Our inspection of the data for our example suggests that the means may follow a curvilinear relationship. *Rejecting the null hypothesis that the means are all the same using ANOVA is not the same as concluding that all the means are different from one another.* The ANOVA null hypothesis can be false in many different ways. Additional analysis is required to distinguish among these possibilities.

When there are particular versions of the alternative hypothesis that are of interest, we use **contrasts** to examine them. In our example, we might want to test whether there is a curvilinear relationship between the number of friends and attractiveness score. *Note that, to use contrasts, it is necessary that the questions of interest be formulated before examining the data.* It is inappropriate to make up these questions after analyzing the data.

contrasts

multiple-comparisons

If we have no specific relations among the means in mind before looking at the data, we instead use a **multiple-comparisons** procedure to determine which pairs of population means differ significantly. In the next section, we explore both contrasts and multiple comparisons in detail.

USE YOUR KNOWLEDGE

12.1 What's wrong? In each of the following, identify what is wrong and then either explain why it is wrong or change the wording of the statement to make it true.

(a) ANOVA tests the null hypothesis that the sample means are all equal.

(b) A strong case for causation is best made in an observational study.

(c) You use ANOVA to compare the variances of the populations.

(d) A multiple-comparisons procedure is used to compare a relation among means that was specified prior to looking at the data.

12.2 What's wrong? For each of the following, explain what is wrong and why.

(a) In rejecting the null hypothesis, one can conclude that all the means are different from one another.

(b) A one-way ANOVA can be used only when there are two means to be compared.

(c) The ANOVA F statistic will be large when the within-group variation is much larger than the between-group variation.

(d) ANOVA is insensitive to outliers and extreme departures from Normality.

The ANOVA model

When analyzing data, the following equation reminds us that we look for an overall pattern and deviations from it:

$$\text{DATA} = \text{FIT} + \text{RESIDUAL}$$

LOOK BACK
DATA = FIT + RESIDUAL, p. 560

In the regression model of Chapter 10, the FIT was the population regression line, and the RESIDUAL represented the deviations of the data from this line. We now apply this framework to describe the statistical models used in ANOVA. These models provide a convenient way to summarize the assumptions that are the foundation for our analysis. They also give us the necessary notation to describe the calculations needed.

First, recall the statistical model for a random sample of observations from a single Normal population with mean μ and standard deviation σ. If the observations are

LOOK BACK
Normal distributions, p. 56

$$x_1, x_2, \ldots, x_n$$

we can describe this model by saying that the x_j are an SRS from the $N(\mu, \sigma)$ distribution. Another way to describe the same model is to think of the x's varying about their population mean. To do this, write each observation x_j as

$$x_j = \mu + \epsilon_j$$

The ϵ_j are then an SRS from the $N(0, \sigma)$ distribution. Because μ is unknown, the ϵ's cannot actually be observed. This form more closely corresponds to our

$$\text{DATA} = \text{FIT} + \text{RESIDUAL}$$

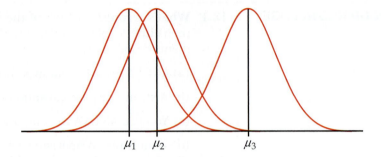

FIGURE 12.6 Model for one-way ANOVA with three groups. The three populations have Normal distributions with the same standard deviation.

way of thinking. The FIT part of the model is represented by μ. It is the systematic part of the model, like the line in a regression. The RESIDUAL part is represented by ϵ_j. It represents the deviations of the data from the fit and is due to random, or chance, variation.

There are two unknown parameters in this statistical model: μ and σ. We estimate μ by $\bar{x}$, the sample mean, and σ by s, the sample standard deviation. The differences $e_j = x_j - \bar{x}$ are the residuals and correspond to the ϵ_j in the statistical model.

The model for one-way ANOVA is very similar. We take random samples from each of I different populations. The sample size is n_i for the ith population. Let x_{ij} represent the jth observation from the ith population. The I population means are the FIT part of the model and are represented by μ_i. The random variation, or RESIDUAL, part of the model is represented by the deviations ϵ_{ij} of the observations from the means.

THE ONE-WAY ANOVA MODEL

Consider SRSs from each of I populations, with the sample from the ith population having n_i observations denoted $x_{i1}, x_{i2}, \ldots, x_{in_i}$. The **one-way ANOVA model** is

$$x_{ij} = \mu_i + \epsilon_{ij}$$

for $i = 1, \ldots, I$ and $j = 1, \ldots, n_i$. The ϵ_{ij} are assumed to be from an $N(0, \sigma)$ distribution. The **parameters of the model** are the population means $\mu_1, \mu_2, \ldots, \mu_I$ and the common standard deviation σ.

Note that the sample sizes n_i may differ, but the standard deviation σ is assumed to be the same in all the populations. Figure 12.6 pictures this model for $I = 3$. The three population means μ_i are different, but the shapes of the three Normal distributions are the same, reflecting the assumption that all three populations have the same standard deviation.

EXAMPLE 12.4

ANOVA model for the Facebook friends study. In the Facebook friends study, there are five profiles that we want to compare, so $I = 5$. The population means $\mu_1, \mu_2, \ldots, \mu_5$ are the mean social attractiveness scores for the profiles with 102, 302, 502, 702, and 902 friends, respectively. The sample sizes n_i are 24, 33, 26, 30, and 21. It is common to use numerical subscripts to distinguish the different means, and some software requires that levels of factors in ANOVA be specified as numerical values. In this situation, it is very important to keep track of what each numerical value represents when drawing conclusions. In our example, we could use numerical values to suggest the actual groups by replacing μ_1 with μ_{102}, μ_2 with μ_{302}, and so on.

The observation $x_{1,1}$, for example, is the social attractiveness score for the first participant who observed the profile with 102 friends (Profile 1). The data for the other participants assigned to this profile are denoted by $x_{1,2}, x_{1,3}, \ldots, x_{1,24}$. Similarly, the data for the other four profile groups have a first subscript indicating the profile and a second subscript indicating the participant assigned to that profile.

According to our model, the score for the first participant in Profile 1 is $x_{1,1} = \mu_1 + \epsilon_{1,1}$, where μ_1 is the average score for *all* undergraduates after viewing Profile 1 and $\epsilon_{1,1}$ is the chance variation due to this particular participant. Similarly, the score for the last participant in Profile 5 is $x_{5,21} = \mu_5 + \epsilon_{5,21}$, where μ_5 is the average score for all undergraduates after viewing Profile 5 and $\epsilon_{5,21}$ is the chance variation due to this participant.

The ANOVA model assumes that these chance variations ϵ_{ij} are independent and Normally distributed with mean 0 and standard deviation σ. For our example, we have clear evidence that the data are non-Normal. The observed scores are numbers ranging from 1.0 to 7.0 by increments of 0.2. However, because our inference is based on the sample means, which will be approximately Normally distributed, we are not overly concerned about this violation of model assumptions.

LOOK BACK

central limit theorem, p. 298

USE YOUR KNOWLEDGE

12.3 **Time to complete a navigation mission.** Example 12.1 (page 645) describes a study designed to compare different joystick types on the time it takes to complete a navigation mission. Write out the ANOVA model for this study. Be sure to give specific values for I and the n_i. List all the parameters of the model.

12.4 **Ages of customers at different coffeehouses.** In Example 12.2 (page 645), the ages of customers at different coffeehouses are compared. Write out the ANOVA model for this study. Be sure to give specific values for I and the n_i. List all the parameters of the model.

Estimates of population parameters

The unknown parameters in the statistical model for ANOVA are the I population means μ_i and the common population standard deviation σ. To estimate μ_i, we use the sample mean for the ith group:

$$\overline{x}_i = \frac{1}{n_i}\sum_{j=1}^{n_i} x_{ij}$$

LOOK BACK

residuals, p. 561

The residuals $e_{ij} = x_{ij} - \overline{x}_i$ reflect the variation about the sample means that we see in the data and are used in the calculations of the sample standard deviations

$$s_i = \sqrt{\frac{\sum_{j=1}^{n_i}(x_{ij} - \overline{x}_i)^2}{n_i - 1}}$$

The ANOVA model assumes that the population standard deviations are all equal. Before estimating σ, it is important to check this equality assumption using the sample standard deviations. Most statistical software provide at least one test for the equality of standard deviations. Unfortunately, many of these tests lack robustness against non-Normality.

ANOVA procedures, however, are not extremely sensitive to unequal standard deviations provided the group sample sizes are the same or similar. Thus, we do not recommend a formal test of equality of standard deviations as a preliminary to the ANOVA. Instead, we will use the following rule as a guideline.

RULE FOR EXAMINING STANDARD DEVIATIONS IN ANOVA

If the largest standard deviation is less than twice the smallest standard deviation, we can use methods based on the assumption of equal standard deviations, and our results will still be approximately correct.[3]

If it appears that we have unequal standard deviations, we generally try to transform the data so that they are approximately equal. We might, for example, work with $\sqrt{x_{ij}}$ or $\log x_{ij}$. Fortunately, we can often find a transformation that *both* makes the group standard deviations more nearly equal and also makes the distributions of observations in each group more nearly Normal. If the standard deviations are markedly different and cannot be made similar by a transformation, inference requires different methods such as nonparametric methods described in Chapter 15 and the bootstrap described in Chapter 16.

EXAMPLE 12.5

Are the standard deviations equal? In the Facebook friends study, there are $I = 5$ groups and the sample standard deviations are $s_1 = 1.00$, $s_2 = 0.85$, $s_3 = 1.07$, $s_4 = 1.43$, and $s_5 = 1.02$. Because the largest standard deviation (1.43) is less than twice the smallest ($2 \times 0.85 = 1.70$), our rule indicates that we can use the assumption of equal population standard deviations.

When we assume that the population standard deviations are equal, each sample standard deviation is an estimate of σ. To combine these into a single estimate, we use a generalization of the pooling method introduced in Chapter 7 (page 448).

POOLED ESTIMATOR OF σ

Suppose that we have sample variances $s_1^2, s_2^2, \ldots, s_I^2$ from I independent SRSs of sizes $n_1, n_2, \ldots, n_I$ from populations with common variance σ^2. The **pooled sample variance**

$$s_p^2 = \frac{(n_1 - 1)s_1^2 + (n_2 - 1)s_2^2 + \cdots + (n_I - 1)s_I^2}{(n_1 - 1) + (n_2 - 1) + \cdots + (n_I - 1)}$$

is an unbiased estimator of σ^2. The **pooled standard deviation**

$$s_p = \sqrt{s_p^2}$$

is the estimate of σ.

Pooling gives more weight to groups with larger sample sizes. If the sample sizes are equal, s_p^2 is just the average of the I sample variances. *Note that* s_p *is not the average of the* I *sample standard deviations.*

EXAMPLE 12.6

The common standard deviation estimate. In the Facebook friends study, there are $I = 5$ groups and the sample sizes are $n_1 = 24$, $n_2 = 33$, $n_3 = 26$, $n_4 = 30$, and $n_5 = 21$. The sample standard deviations are $s_1 = 1.00$, $s_2 = 0.85$, $s_3 = 1.07$, $s_4 = 1.43$, and $s_5 = 1.02$.

The pooled variance estimate is

$$s_p^2 = \frac{(n_1 - 1)s_1^2 + (n_2 - 1)s_2^2 + (n_3 - 1)s_3^2 + (n_4 - 1)s_4^2 + (n_5 - 1)s_5^2}{(n_1 - 1) + (n_2 - 1) + (n_3 - 1) + (n_4 - 1) + (n_5 - 1)}$$

$$= \frac{(23)(1.00)^2 + (32)(0.85)^2 + (25)(1.07)^2 + (29)(1.43)^2 + (20)(1.02)^2}{23 + 32 + 25 + 29 + 20}$$

$$= \frac{154.85}{129} = 1.20$$

The pooled standard deviation is

$$s_p = \sqrt{1.20} = 1.10$$

This is our estimate of the common standard deviation σ of the social attractiveness scores for the five profiles.

USE YOUR KNOWLEDGE

12.5 Joystick types. Example 12.1 (page 645) describes a study designed to compare different joystick types on the time it takes to complete a navigation mission. In Exercise 12.3 (page 653), you described the ANOVA model for this study. The three joystick types are designated 1, 2, and 3. The following table summarizes the time (seconds) data.

Joystick	$\bar{x}$	s	n
1	279	78	20
2	245	68	20
3	258	80	20

(a) Is it reasonable to pool the standard deviations for these data? Explain your answer.

(b) For each parameter in your model from Exercise 12.3, give the estimate.

12.6 Ages of customers at different coffeehouses. In Example 12.2 (page 645) the ages of customers at different coffeehouses are compared, and you described the ANOVA model for this study in

Exercise 12.4 (page 653). Here is a summary of the ages of the customers:

Store	$\bar{x}$	s	n
A	38	8	50
B	48	13	50
C	42	11	50
D	28	7	50
E	35	10	50

(a) Is it reasonable to pool the standard deviations for these data? Explain your answer.

(b) For each parameter in your model from Exercise 12.4, give the estimate.

FRIENDS

12.7 An alternative Normality check. Figure 12.3 (page 649) displays separate histograms for the five profile groups. An alternative procedure is to make one histogram (or single Normal quantile plot) using the *residuals* $e_{ij} = x_{ij} - \bar{x}_i$ for all five groups together. Construct this histogram and summarize what it shows.

12.8 An alternative Normality check, continued. Refer to the previous exercise. Describe the benefits and drawbacks of checking Normality using all the residuals together versus looking at each population sample separately. Which approach do you prefer and why?

Testing hypotheses in one-way ANOVA

Comparison of several means is accomplished by using an F statistic to compare the variation among groups with the variation within groups. We now show how the F statistic expresses this comparison. Calculations are organized in an ANOVA table, which contains numerical measures of the variation among groups and within groups.

LOOK BACK

ANOVA table, p. 586

First, we must specify our hypotheses for one-way ANOVA. As before, I represents the number of populations to be compared.

HYPOTHESES FOR ONE-WAY ANOVA

The **null and alternative hypotheses** for one-way ANOVA are

$$H_0: \mu_1 = \mu_2 = \cdots = \mu_I$$
$$H_a: \text{not all of the } \mu_i \text{ are equal}$$

We now use the Facebook friends study to illustrate how to do a one-way ANOVA. Because the calculations are generally performed using statistical software, we focus on interpretation of the output.

EXAMPLE 12.7

FRIENDS

Reading software output. Figure 12.7 gives descriptive statistics generated by SPSS for the ANOVA of the Facebook friends example. Summaries for each profile are given on the first five lines. In addition to the sample size, the mean, and the standard deviation, this output also gives the minimum and maximum observed value, standard error of the mean, and the 95% confidence interval for the mean of each profile. The five sample means $\bar{x}_i$ given in the output are estimates of the five unknown population means μ_i.

IBM SPSS Statistics Viewer

Descriptives

Score

	N	Mean	Std. Deviation	Std. Error	95% Confidence Interval for Mean Lower Bound	Upper Bound	Minimum	Maximum
102	24	3.817	.9990	.2039	3.395	4.239	2.2	6.0
302	33	4.879	.8514	.1482	4.577	5.181	2.6	6.4
502	26	4.562	1.0704	.2099	4.129	4.994	2.0	6.8
702	30	4.407	1.4283	.2608	3.873	4.940	1.0	7.0
902	21	3.990	1.0227	.2232	3.525	4.456	1.6	5.2
Total	134	4.382	1.1463	.0990	4.186	4.578	1.0	7.0

FIGURE 12.7 SPSS output with descriptive statistics for the Facebook friends study, Example 12.7.

The output gives the estimates of the standard deviations, s_i, for each group but does not provide s_p, the pooled estimate of the model standard deviation, σ. We could perform the calculation using a calculator, as we did in Example 12.6. We will see an easier way to obtain this quantity from the ANOVA table.

Some software packages report s_p as part of the standard ANOVA output. *Sometimes you are not sure whether or not a quantity given by software is what you think it is.* A good way to resolve this dilemma is to do a sample calculation with a simple example to check the numerical results.

Note that s_p *is not the standard deviation given in the "Total" row of Figure 12.7.* This quantity is the standard deviation that we would obtain if we viewed the data as a single sample of 134 participants and ignored the possibility that the profile means could be different. As we have mentioned many times before, it is important to use care when reading and interpreting software output.

EXAMPLE 12.8

FRIENDS

Reading software output, continued. Additional output generated by SPSS for the ANOVA of the Facebook friends example is given in Figure 12.8. We will discuss the construction of this output later. For now, we observe that the results of our significance test are given in the last two columns of the output. The null hypothesis that the five population means are the same is tested by the statistic $F = 4.142$, and the associated P-value is reported as $P = 0.003$. The data provide clear evidence to support the claim that there are some differences among the five profile population means.

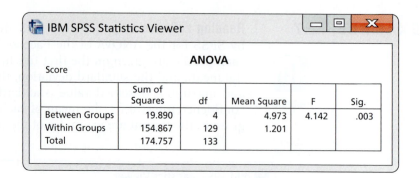

FIGURE 12.8 SPSS output giving the ANOVA table for the Facebook friends study, Example 12.3.

The ANOVA table

The information in an analysis of variance is organized in an ANOVA table. To understand the table, it is helpful to think in terms of our

$$DATA = FIT + RESIDUAL$$

view of statistical models. For one-way ANOVA, this corresponds to

$$x_{ij} = \mu_i + \epsilon_{ij}$$

We can think of these three terms as sources of variation. The ANOVA table separates the variation in the data into two parts: the part due to the fit and the remainder, which we call residual.

EXAMPLE 12.9

ANOVA table for the Facebook friends study. The SPSS output in Figure 12.8 gives the sources of variation in the first column. Here, FIT is called Between Groups, RESIDUAL is called Within Groups, and DATA is the last entry, Total. Different software packages use different terms for these sources of variation but the basic concept is common to all. In place of FIT, some software packages use Between Groups, Model, or the name of the factor. Similarly, terms like Within Groups or Error are frequently used in place of RESIDUAL.

variation among groups
The Between Groups row in the table gives information related to the **variation among group means**. In writing ANOVA tables, for this row we will use the generic label "groups" or some other term that describes the factor being studied.

variation within groups
The Within Groups row in the table gives information related to the **variation within groups**. We noted that the term "error" is frequently used for this source of variation, particularly for more general statistical models. This label is most appropriate for experiments in the physical sciences where the observations within a group differ because of measurement error. In business and the biological and social sciences, on the other hand, the within-group variation is often due to the fact that not all firms or plants or people are the same. This sort of variation is not due to errors and is better described as "residual" or "within-group" variation. Nevertheless, we will use the generic label "error" for this source of variation in writing ANOVA tables.

Finally, the Total row in the ANOVA table corresponds to the DATA term in our DATA = FIT + RESIDUAL framework. So, for analysis of variance,

$$DATA = FIT + RESIDUAL$$

translates into

Total variation = Variation between groups + Variation within groups

LOOK BACK

sum of
squares,
p. 584

The second column in the software output given in Figure 12.8 is labeled Sum of Squares. As you might expect, each sum of squares is a sum of squared deviations. We use SSG, SSE, and SST for the entries in this column, corresponding to groups, error, and total. Each sum of squares measures a different type of variation. SST measures variation of the data around the overall mean, $x_{ij} - \bar{x}$. Variation of the group means around the overall mean, $\bar{x}_i - \bar{x}$, is measured by SSG. Finally, SSE measures variation of each observation around its group mean, $x_{ij} - \bar{x}_i$.

EXAMPLE 12.10

ANOVA table for the Facebook friends study, continued. The Sum of Squares column in Figure 12.8 gives the values for the three sums of squares. They are

$$SST = 174.757$$
$$SSG = 19.890$$
$$SSE = 154.867$$

Verify that SST = SSG + SSE for this example.

This fact is true in general. The total variation is always equal to the among-group variation plus the within-group variation. Note that software output frequently gives many more digits than we need, as in this case.

In this example, it appears that most of the variation is coming from within groups. However, to assess whether the observed differences in sample means are statistically significant, some additional calculations are needed.

Associated with each sum of squares is a quantity called the degrees of freedom. Because SST measures the variation of all N observations around the overall mean, its degrees of freedom are DFT = $N - 1$. This is the same as the degrees of freedom for the ordinary sample variance with sample size N. Similarly, because SSG measures the variation of the I sample means around the overall mean, its degrees of freedom are DFT = $I - 1$. Finally, SSE is the sum of squares of the deviations $x_{ij} - \bar{x}_i$. Here we have N observations being compared with I sample means, and DFE = $N - I$.

LOOK BACK

degrees of
freedom,
p. 40

EXAMPLE 12.11

Degrees of freedom for the Facebook friends study. In the Facebook friends example, we have $I = 5$ and $N = 134$. Therefore,

$$DFT = N - 1 = 134 - 1 = 133$$
$$DFG = I - 1 = 5 - 1 = 4$$
$$DFE = N - I = 134 - 5 = 129$$

These are the entries in the df column of Figure 12.8.

Note that the degrees of freedom add in the same way that the sums of squares add. That is, DFT = DFG + DFE.

For each source of variation, the mean square is the sum of squares divided by the degrees of freedom. You can verify this by doing the divisions for the values given on the output in Figure 12.8. We compare these mean squares to test whether the population means are all the same.

mean square, p. 584

SUMS OF SQUARES, DEGREES OF FREEDOM, AND MEAN SQUARES

Sums of squares represent variation present in the data. They are calculated by summing squared deviations. In one-way ANOVA, there are three **sources of variation:** groups, error, and total. The sums of squares are related by the formula

$$\text{SST} = \text{SSG} + \text{SSE}$$

Thus, the total variation is composed of two parts, one due to groups and one due to error.

Degrees of freedom are related to the deviations that are used in the sums of squares. The degrees of freedom are related in the same way as the sums of squares are:

$$\text{DFT} = \text{DFG} + \text{DFE}$$

To calculate each **mean square**, divide the corresponding sum of squares by its degrees of freedom.

We can use the mean square for error to find s_p, the pooled estimate of the parameter σ of our model. It is true in general that

$$s_p^2 = \text{MSE} = \frac{\text{SSE}}{\text{DFE}}$$

In other words, the mean square for error is an estimate of the within-group variance, σ^2. The estimate of σ is, therefore, the square root of this quantity. So,

$$s_p = \sqrt{\text{MSE}}$$

EXAMPLE 12.12

The pooled estimate for σ. From the SPSS output in Figure 12.8 we see that the MSE is reported as 1.201. The pooled estimate of σ is therefore

$$s_p = \sqrt{\text{MSE}}$$
$$= \sqrt{1.201} = 1.10$$

This estimate is equal to our calculations of s_p in Example 12.6.

The *F* test

If H_0 is true, there are no differences among the group means. The ratio MSG/MSE is a statistic that is approximately 1 if H_0 is true and tends to be larger if H_a is true. This is the ANOVA F statistic. In our example, MSG = 4.973 and MSE = 1.201, so the ANOVA F statistic is

$$F = \frac{\text{MSG}}{\text{MSE}} = \frac{4.973}{1.201} = 4.142$$

When H_0 is true, the F statistic has an F distribution that depends upon two numbers: the *degrees of freedom for the numerator* and the *degrees of freedom for the denominator.* These degrees of freedom are those associated with the mean squares in the numerator and denominator of the F statistic. For one-way ANOVA, the degrees of freedom for the numerator are DFG = $I - 1$, and the degrees of freedom for the denominator are DFE = $N - I$. We use the notation $F(I - 1, N - I)$ for this distribution.

The *One-Way ANOVA* applet is an excellent way to see how the value of the F statistic and the P-value depend upon the variability of the data within the groups, the sample sizes, and the differences between the means. See Exercises 12.42, 12.43, and 12.44 (page 687) for use of this applet.

The ANOVA F test shares the robustness of the two-sample t test. It is relatively insensitive to moderate non-Normality and unequal variances, especially when the sample sizes are similar. The constant variance assumption is more important when we compare means using contrasts and multiple comparisons, additional analyses that are generally performed after the ANOVA F test. We discuss these analyses in the next section.

THE ANOVA F TEST

To test the null hypothesis in a one-way ANOVA, calculate the **F statistic**

$$F = \frac{\text{MSG}}{\text{MSE}}$$

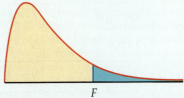

When H_0 is true, the F statistic has the $F(I - 1, N - I)$ distribution. When H_a is true, the F statistic tends to be large. We reject H_0 in favor of H_a if the F statistic is sufficiently large.

The **P-value** of the F test is the probability that a random variable having the $F(I - 1, N - I)$ distribution is greater than or equal to the calculated value of the F statistic.

Tables of F critical values are available for use when software does not give the P-value. Table E in the back of the book contains the F critical values for probabilities $p = 0.100, 0.050, 0.025, 0.010$, and 0.001. For one-way ANOVA we use critical values from the table corresponding to $I - I$ degrees of freedom in the numerator and $N - I$ degrees of freedom in the denominator. *When determining the P-value, remember that the* F *test is always one-sided because any differences among the group means tend to make* F *large.*

EXAMPLE 12.13

The ANOVA F test for the Facebook friends study. In the Facebook friends study, we found $F = 4.14$. (Note that it is standard practice to round F statistics to two places after the decimal point.) There were five populations, so the degrees of freedom in the numerator are DFG = $I - 1 = 4$. For this example, the degrees of freedom in the denominator are DFE = $N - I = 134 - 5 = 129$.

Software provided a *P*-value of 0.003, so at the 0.05 significance level, we reject H_0 and conclude that the population means are not all the same.

Suppose that $P = 0.003$ was not provided. We'll now run through the process of using the table of F critical values to approximate the *P*-value. Although you will rarely need to do this in practice, the process will help you to understand the *P*-value calculation.

In Table E, we first find the column corresponding to 4 degrees of freedom in the numerator. For the degrees of freedom in the denominator, we see that there are entries for 100 and 200. The values for these entries are very close. To be conservative, we use critical values corresponding to 100 degrees of freedom in the denominator because these are slightly larger.

We have $F = 4.14$. This is in between the critical value for $P = 0.010$ and $P = 0.001$. Using the table, we can conclude only that $0.001 < P < 0.010$.

df = (4, 100)

p	Critical value
0.100	2.00
0.050	2.46
0.025	2.92
0.010	3.51
0.001	5.02

The following display shows the general form of a one-way ANOVA table with the F statistic. The formulas in the sum of squares column can be used for calculations in small problems. There are other formulas that are more efficient for hand or calculator use, but ANOVA calculations are usually done by computer software.

Source	Degrees of freedom	Sum of squares	Mean square	F
Groups	$I - 1$	$\sum_{\text{groups}} n_i(\overline{x}_i - \overline{x})^2$	SSG/DFG	MSG/MSE
Error	$N - I$	$\sum_{\text{groups}} (n_i - 1)s_i^2$	SSE/DFE	
Total	$N - 1$	$\sum_{\text{obs}} (x_{ij} - \overline{x})^2$		

coefficient of determination

One other item given by some software for ANOVA is worth noting. For an analysis of variance, we define the **coefficient of determination** as

$$R^2 = \frac{\text{SSG}}{\text{SST}}$$

LOOK BACK

squared multiple correlation, p. 615

The coefficient of determination plays the same role as the squared multiple correlation R^2 in a multiple regression. We can easily calculate the value from the ANOVA table entries.

EXAMPLE 12.14

Coefficient of determination for the Facebook friends study. The software-generated ANOVA table for the Facebook friends study is given in Figure 12.8 (page 658). From that display, we see that SSG = 19.890 and SST = 174.757. The coefficient of determination is

$$R^2 = \frac{\text{SSG}}{\text{SST}} = \frac{19.890}{174.757} = 0.11$$

About 11% of the variation in social attractiveness scores is explained by the different profiles. The other 89% of the variation is due to participant-to-participant variation within each of the profile groups. We can see this in the histograms of Figure 12.3 (page 649). Each of the groups has a large amount of variation, and there is a substantial amount of overlap in the distributions. *The fact that we have strong evidence (P < 0.003) against the null hypothesis that the five population means are all the same does not tell us that the distributions of values are far apart.*

USE YOUR KNOWLEDGE

12.9 What's wrong? In each of the following, identify what is wrong and then either explain why it is wrong or change the wording of the statement to make it true.

(a) Within-group variation is the variation in the data due to the differences in the sample means.

(b) The mean squares in an ANOVA table will add; that is, MST = MSG + MSE.

(c) The pooled estimate s_p is a parameter of the ANOVA model.

(d) A very small P-value implies that the group distributions of responses are far apart.

12.10 Determining the critical value of F. For each of the following situations, state how large the F statistic needs to be for rejection of the null hypothesis at the 0.05 level.

(a) Compare three groups with four observations per group.

(b) Compare three groups with five observations per group.

(c) Compare four groups with five observations per group.

(d) Summarize what you have learned about F distributions from this exercise.

Software

We have used SPSS to illustrate the analysis of the Facebook friends study. Other statistical software gives similar output, and you should be able to read it without any difficulty. Here's an example with output from three software packages.

EXAMPLE 12.15

Thinkstock

EYES

Do eyes affect ad response? Research from a variety of fields has found significant effects of eye gaze and eye color on emotions and perceptions such as arousal, attractiveness, and honesty. These findings suggest that a model's eyes may play a role in a viewer's response to an ad.

In one study, students in marketing and management classes of a southern, predominantly Hispanic, university were each presented with one of four portfolios.[4] Each portfolio contained a target ad for a fictional product, Sparkle Toothpaste. Students were asked to view the ad and then respond to questions concerning their attitudes and emotions about the ad and product. All questions were from advertising-effects questionnaires previously used in the literature. Each response was on a seven-point scale.

Although the researchers investigated nine attitudes and emotions, we will focus on the viewer's "attitudes toward the brand." This response was obtained by averaging 10 survey questions.

The target ads were created using two digital photographs of a model. In one picture, the model is looking directly at the camera so the eyes can be seen. This picture was used in three target ads. The only difference was the model's eyes, which were made to be either brown, blue, or green. In the second picture, the model is in virtually the same pose but looking downward so the eyes are not visible. A total of 222 surveys were used for analysis. Outputs from Excel, SAS, and Minitab are given in Figure 12.9.

FIGURE 12.9 Excel, SAS, and Minitab outputs for the advertising study, Example 12.15.

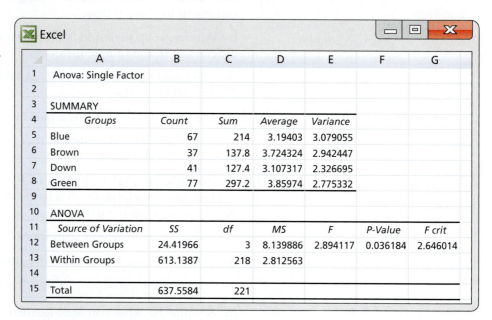

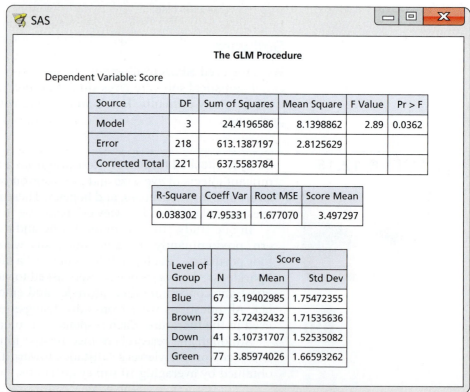

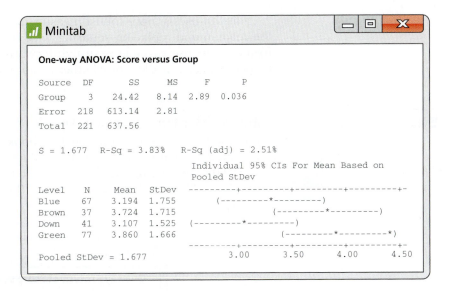

```
Minitab

One-way ANOVA: Score versus Group

Source   DF       SS      MS      F       P
Group     3     24.42    8.14    2.89   0.036
Error   218    613.14    2.81
Total   221    637.56

S = 1.677   R-Sq = 3.83%    R-Sq (adj) = 2.51%

                                 Individual 95% CIs For Mean Based on
                                 Pooled StDev
Level    N    Mean    StDev  ---------+---------+---------+---------+-
Blue    67    3.194   1.755          (---------*---------)
Brown   37    3.724   1.715                      (---------*---------)
Down    41    3.107   1.525  (---------*---------)
Green   77    3.860   1.666                  (---------*---------*)
                            ---------+---------+---------+---------+-
Pooled StDev = 1.677           3.00      3.50      4.00      4.50
```

FIGURE 12.9 *(Continued)*

There is evidence at the 5% significance level to reject the null hypothesis that the four groups have equal means ($P = 0.036$). In Exercises 12.35 and 12.36 (page 686), you are asked to perform further inference using contrasts.

BEYOND THE BASICS

Testing the Equality of Spread

While the standard deviation is a natural measure of spread for Normal distributions, it is not for distributions in general. In fact, because skewed distributions have unequally spread tails, no single numerical measure is adequate to describe the spread of a skewed distribution. Because of this, we recommend caution when testing equal standard deviations and interpreting the results.

Most formal tests for equal standard deviations are extremely sensitive to non-Normal populations. Of the tests commonly available in software packages, we suggest using the *modified Levene's* (or Brown-Forsythe) test due to its simplicity and robustness against non-Normal data.[5] The test involves performing a one-way ANOVA on a transformation of the response variable, constructed to measure the spread in each group. If the populations have the same standard deviation, then the average deviation from the population center should also be the same.

MODIFIED LEVENE'S TEST FOR EQUALITY OF STANDARD DEVIATIONS

To test for the equality of the I population standard deviations, perform a one-way ANOVA using the transformed response

$$y_{ij} = |x_{ij} - M_i|$$

where M_i is the sample median for population i. We reject the assumption of equal spread if the P-value of this test is less than the significance level α.

This test uses a more robust measure of deviation replacing the mean with the median and replacing squaring with the absolute value. Also, the transformed response variable is straightforward to create, so this test can easily be performed regardless of whether or not your software specifically has it.

EXAMPLE 12.16

FRIENDS

Are the Standard Deviations Equal? Figure 12.10 shows output of the modified Levene's test for the Facebook friends study. In JMP, the test is called Brown-Forsythe. The *P*-value is 0.0554, which is larger than $\alpha = 0.05$, suggesting we cannot reject the null hypothesis that they are the same. This result further supports the use of ANOVA for the Facebook friends study.

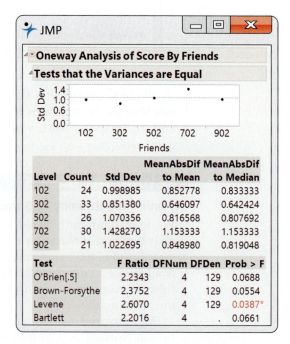

FIGURE 12.10 JMP output for the test that the variances are equal, Example 12.16.

Remember that our rule of thumb (page 654) is used to assess whether different standard deviations will impact the ANOVA results. *It's not a formal test that the standard deviations are equal.* There will be times when the rule and formal test do not agree.

SECTION 12.1 SUMMARY

• **One-way analysis of variance (ANOVA)** is used to compare several population means based on independent SRSs from each population. The populations are assumed to be Normal with possibly different means and the same standard deviation.

- To do an analysis of variance, first examine the data. Side-by-side boxplots give an overview of the data. Examine Normal quantile plots (either for each group separately or for the residuals) to detect outliers or extreme deviations from Normality.

- In addition to Normality, the ANOVA model assumes equal population standard deviations. Compute the ratio of the largest to the smallest sample standard deviation. If this ratio is less than 2 and the Normality assumption is reasonable, ANOVA can be performed.

- If the data do not support equal standard deviations, consider transforming the response variable. This often makes the group standard deviations more nearly equal and makes the group distributions more Normal.

- The **null hypothesis** is that the population means are *all* equal. The **alternative hypothesis** is true if there are *any* differences among the population means.

- ANOVA is based on separating the total variation observed in the data into two parts: variation **among group means** and variation **within groups**. If the variation among groups is large relative to the variation within groups, we have evidence against the null hypothesis.

- An **analysis of variance table** organizes the ANOVA calculations. **Degrees of freedom, sums of squares, and mean squares** appear in the table. The **F statistic** and its **P-value** are used to test the null hypothesis.

- The ANOVA *F* test shares the **robustness** of the two-sample *t* test. It is relatively insensitive to moderate non-Normality and unequal variances, especially when the sample sizes are similar.

SECTION 12.1 EXERCISES

For Exercises 12.1 and 12.2, see page 651 for Exercises 12.3 and 12.4, see page 653 for Exercises 12.5 through 12.8, see pages 655–656 and for Exercises 12.9 and 12.10, see page 663.

12.11 A one-way ANOVA example. A study compared five groups with six observations per group. An *F* statistic of 4.81 was reported.

(a) Give the degrees of freedom for this statistic and the entries from Table E that correspond to this distribution.

(b) Sketch a picture of this *F* distribution with the information from the table included.

(c) Based on the table information, how would you report the *P*-value?

(d) Can you conclude that all pairs of group means are different? Explain your answer.

12.12 Visualizing the ANOVA model. For each of the following settings, draw a picture of the ANOVA model similar to Figure 12.6 (page 652). To sketch the Normal curves, you may want to review the 68–95–99.7 rule on page 57.

(a) $\mu_1 = 17$, $\mu_2 = 13$, $\mu_3 = 12$, and $\sigma = 2$.

(b) $\mu_1 = 17$, $\mu_2 = 13$, $\mu_3 = 12$, and $\sigma = 4$.

(c) $\mu_1 = 20$, $\mu_2 = 12$, $\mu_3 = 10$, and $\sigma = 3$.

12.13 Visualizing the ANOVA model, continued. Refer to the previous exercise. If SRSs of size $n = 5$ were obtained from each of the three populations, under which setting would you most likely obtain a significant ANOVA *F* test? Explain your answer.

12.14 Calculating the ANOVA F test P-value. For each of the following situations, find the degrees of freedom for the F statistic and then use Table E to approximate the P-value.

(a) Six groups are being compared with five observations per group. The value of the F statistic is 2.47.

(b) Four groups are being compared with 11 observations per group. The value of the F statistic is 5.03.

(c) Five groups are being compared using 65 total observations. The value of the F statistic is 3.11.

12.15 Calculating the ANOVA F test P-value, continued. For each of the following situations, find the F statistic and the degrees of freedom. Then draw a sketch of the distribution under the null hypothesis and shade in the portion corresponding to the P-value. State how you would report the P-value.

(a) Compare three groups with 21 observations per group, MSE = 50, and MSG = 340.

(b) Compare eight groups with six observations per group, SSG = 77, and SSE = 190.

12.16 Calculating the pooled standard deviation. An experiment was run to compare three groups. The sample sizes were 28, 33, and 102, and the corresponding estimated standard deviations were 2.7, 2.6, and 4.8.

(a) Is it reasonable to use the assumption of equal standard deviations when we analyze these data? Give a reason for your answer.

(b) Give the values of the variances for the three groups.

(c) Find the pooled variance.

(d) What is the value of the pooled standard deviation?

(e) Explain why your answer in part (d) is much closer to the standard deviation for the third group than to either of the other two standard deviations.

12.17 Describing the ANOVA model. For each of the following situations, identify the response variable and the populations to be compared, and give I, n_i, and N.

(a) A poultry farmer is interested in reducing the cholesterol level in his marketable eggs. He wants to compare two different cholesterol-lowering drugs added to the hens' standard diet as well as an all-vegetarian diet. He assigns 25 of his hens to each of the three treatments.

(b) A researcher is interested in students' opinions regarding an additional annual fee to support non-income-producing varsity sports. Students were asked to rate their acceptance of this fee on a seven-point scale. She received 94 responses, of which 31 were from students who attend varsity football or basketball games only, 18 were from students who also attend other varsity competitions, and 45 were from students who did not attend any varsity games.

(c) A professor wants to evaluate the effectiveness of his teaching assistants. In one class period, the 42 students were randomly divided into three equal-sized groups, and each group was taught power calculations from one of the assistants. At the beginning of the next class, each student took a quiz on power calculations, and these scores were compared.

12.18 Describing the ANOVA model, continued. For each of the following situations, identify the response variable and the populations to be compared, and give I, n_i, and N.

(a) A developer of a virtual-reality (VR) teaching tool for the deaf wants to compare the effectiveness of different navigation methods. A total of 40 children were available for the experiment, of which equal numbers were randomly assigned to use a joystick, wand, dancemat, or gesture-based pinch gloves. The time (in seconds) to complete a designed VR path is recorded for each child.

(b) To study the effects of pesticides on birds, an experimenter randomly (and equally) allocated 65 chicks to five diets (a control and four with a different pesticide included). After a month, the calcium content (milligrams) in a 1-centimeter length of bone from each chick was measured.

(c) A university sandwich shop wants to compare the effects of providing free food with a sandwich order on sales. The experiment will be conducted from 11:00 A.M. to 2:00 P.M. for the next 20 weekdays. On each day, customers will be offered one of the following: a free drink, free chips, a free cookie, or nothing. Each option will be offered five times.

12.19 Determining the degrees of freedom. Refer to Exercise 12.17. For each situation, give the following:

(a) Degrees of freedom for group, for error, and for the total.

(b) Null and alternative hypotheses.

(c) Numerator and denominator degrees of freedom for the F statistic.

12.20 Determining the degrees of freedom, continued. Refer to Exercise 12.18. For each situation, give the following:

(a) Degrees of freedom for group, for error, and for the total.

(b) Null and alternative hypotheses.

(c) Numerator and denominator degrees of freedom for the *F* statistic.

12.21 Data collection and the interpretation of results. Refer to Exercise 12.17. For each situation, discuss the method of obtaining the data and how this will affect the extent to which the results can be generalized.

12.22 Data collection, continued. Refer to Exercise 12.18. For each situation, discuss the method of obtaining the data and how this will affect the extent to which the results can be generalized.

12.23 The effects of two stimulant drugs. An experimenter was interested in investigating the effects of two stimulant drugs (labeled A and B). She divided 25 rats equally into five groups (placebo, Drug A low, Drug A high, Drug B low, and Drug B high) and, 20 minutes after injection of the drug, recorded each rat's activity level (higher score is more active). The following table summarizes the results:

Treatment	$\bar{x}$	s^2
Placebo	11.80	17.20
Low A	15.25	13.10
High A	18.55	10.25
Low B	16.15	7.75
High B	17.10	12.50

(a) Plot the means versus the type of treatment. Does there appear to be a difference in the activity level? Explain.

(b) Is it reasonable to assume that the variances are equal? Explain your answer, and if reasonable, compute s_p.

(c) Give the degrees of freedom for the *F* statistic.

(d) The *F* statistic is 2.64. Find the associated *P*-value and state your conclusions.

12.24 Perceptions of social media. It is estimated that more than 90% of North American students use social media. This has prompted much research on the mental health impacts of these technologies. In one study, researchers investigated how mental health workers perceive the association between social media and mental disorders. A sample of psychiatrists from Canada completed a questionnaire, from which a perception score was obtained (a higher score indicating a stronger perceived association). The following ANOVA table summarizes a comparison of these scores across three age groups (generations).

Source	DF	SS	MS	F
Age	2	137.78	68.89	0.45
Error	45	6899.54	153.32	
Total	47	7037.32		

(a) How many psychiatrists completed the questionnaire?

(b) What is the estimated common standard deviation?

(c) What is the *P*-value? Make sure to specify the degrees of freedom of the *F* statistic.

(d) State your conclusion using the *P*-value from part (c) and a 5% significance level.

12.25 Pain tolerance among sports teams. Many have argued that sports such as football require the ability to withstand pain from injury for extended periods of time. To see if there is greater pain tolerance among certain sports teams, a group of researchers assessed 183 male Division II athletes from five sports.[6] Each athlete was asked to put his dominant hand and forearm in a 3°C water bath and keep it in there until the pain became intolerable. The total amount of time (in seconds) that each athlete maintained his hand and forearm in the bath was recorded. Following this procedure, each athlete completed a series of surveys on aggression and competitiveness. In their report, the researchers state:

> *A univariate between subjects (sports team) ANOVA was performed on the total amount of time athletes were able to keep their hand and forearm in the water bath, and found it to be statistically significant, F(4,146) = 4.96, p < .001.*

Further analysis revealed that the lacrosse and soccer players tolerated the pain for a significantly longer period of time and swimmers tolerated the pain for a significantly shorter period of time than athletes from the other teams.

(a) Based on the description of the experiment, what should the degrees of freedom be for this analysis?

(b) Assuming that the degrees of freedom reported are correct, data from how many athletes were used in this analysis?

(c) The researchers do not comment on the missing data in their report. List two reasons these data may not have been used, and for each, explain how the omission could impact or bias the results.

12.26 Constructing an ANOVA table Refer to Exercise 12.5 (page 655). Using the table of group means and standard deviations, construct an ANOVA table similar to that on page 662. Based on the *F* statistic and degrees of freedom, compute the *P*-value. What do you conclude?

12.2 Comparing the Means

When you complete this section, you will be able to:

- Distinguish between the use of contrasts to examine particular versions of the alternative hypothesis and the use of a multiple-comparisons method to compare pairs of means.

- Construct a level C confidence interval for a comparison of means expressed as a contrast.

- Perform a t significance test for a contrast and summarize the results.

- Summarize the trade-off of a multiple-comparisons method in terms of controlling false rejections and not detecting true differences in means.

- Describe and use the Bonferroni method to control the probability of a false rejection.

- Interpret statistical software ANOVA output and draw conclusions regarding differences in population means.

- Determine the power of the ANOVA F test for a given set of population means and sample sizes.

The ANOVA F test gives a general answer to a general question: are the differences among observed group means statistically significant? Unfortunately, a small P-value simply tells us that the group means are not all the same. It does not tell us specifically which means differ from each other. Plotting and inspecting the means give us some indication of where the differences lie, but we would like to supplement inspection with formal inference. This section presents two approaches to the task of comparing group means.

Contrasts

In the ideal situation, specific questions regarding comparisons among the means are posed before the data are collected. We can answer specific questions of this kind and attach a level of confidence to the answers we give. We now explore these ideas through a different Facebook study.

EXAMPLE 12.17

FACETYM

How do users spend their time on Facebook? An online study was designed to compare the amount of time a Facebook user devotes to reading positive, negative, and neutral Facebook profiles. Each participant was randomly assigned to one of five Facebook profile groups:

1. Positive female

2. Positive male

3. Negative female

4. Negative male

5. Gender neutral with neutral content

and provided an email link to a survey on Survey Monkey. As part of the survey, the participant was directed to view the assigned Facebook profile page and then answer some additional questions. The amount of time (in minutes) the participant spent viewing the profile was recorded as the response.[7]

We begin our analysis with a check of the data. Time-to-event data (here, the time until the participant begins to answer the additional survey questions) is often skewed to the right. Preliminary analysis of the residuals (Figure 12.11) confirms this for these data.

As a result, we consider the square root of time for analysis. These results are summarized in Figures 12.12 and 12.13. The residuals appear Normal

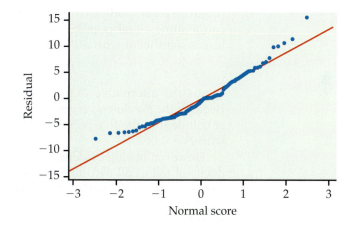

FIGURE 12.11 Normal quantile plot of residuals suggests a skewed distribution, Example 12.17.

FIGURE 12.12 SPSS output giving the ANOVA table for the Facebook profile study after the square root transformation, Example 12.17.

IBM SPSS Statistics Viewer

Report

SqrtTime

Grp	Mean	N	Std. Deviation	Minimum	Maximum
1	2.51753556	21	.850361950	.848528137	3.75765885
2	2.58477147	21	.892414862	1.03923048	4.35315977
3	2.40478314	21	.920798006	1.06301458	4.16533312
4	2.61489644	21	1.04136085	.479583152	4.84974226
5	1.60039238	21	.834075199	.360555128	3.65376518
Total	2.34447580	105	.970876337	.360555128	4.84974226

ANOVA

SqrtTime

	Sum of Squares	df	Mean Square	F	Sig.
Between Groups	15.080	4	3.770	4.545	.002
Within Groups	82.950	100	.830		
Total	98.030	104			

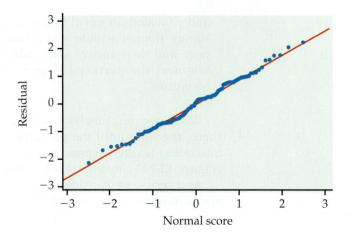

FIGURE 12.13 Normal quantile plot of residuals for the transformed response, Example 12.17.

(Figure 12.13), and our rule for examining standard deviations indicates we can assume equal population standard deviations ($1.041 < 2(0.834)$). The F test is significant with a P-value of 0.002. It's testing the null hypothesis

$$H_0: \mu_1 = \mu_2 = \mu_3 = \mu_4 = \mu_5$$

versus the alternative that the five population means are not all the same. Because the P-value is very small, there is strong evidence against H_0 and we can conclude that the five population means are not all the same ($F(4,100) = 4.55$ with $P = 0.002$).

However, having evidence that the five population means are not the same does not tell us all we'd like to know. We would really like our analysis to provide us with more specific information. For example, the alternative hypothesis is true if

$$\mu_1 < \mu_2 = \mu_3 = \mu_4 = \mu_5$$

or if

$$\mu_1 = \mu_2 > \mu_3 = \mu_4 > \mu_5$$

or if

$$\mu_1 < \mu_3 < \mu_4 < \mu_2 < \mu_5$$

When you reject the ANOVA null hypothesis, additional analyses are required to clarify the nature of the differences between the means.

For this study, the researcher predicted that participants would spend more time viewing the negative Facebook pages compared to the positive or neutral pages because the negative pages would stand out more and thus garner more attention (this is called cognitive salience). How do we take these predictions and translate them into testable hypotheses?

EXAMPLE 12.18

A comparison of interest. The researcher hypothesizes that participants exposed to a negative Facebook profile would spend more time viewing the page than would participants who are exposed to a positive Facebook profile. Because two groups are exposed to negative profiles and two are exposed to positive profiles, we can consider the following null hypothesis:

$$H_{01}: \frac{1}{2}(\mu_3 + \mu_4) = \frac{1}{2}(\mu_1 + \mu_2)$$

versus the two-sided alternative

$$H_{a1}: \frac{1}{2}(\mu_3 + \mu_4) \neq \frac{1}{2}(\mu_1 + \mu_2)$$

We could argue that the one-sided alternative

$$H_{a1}: \frac{1}{2}(\mu_3 + \mu_4) > \frac{1}{2}(\mu_1 + \mu_2)$$

is appropriate for this problem provided other evidence suggests this direction and is not just what the researcher wants to see.

In the preceding example, we used H_{01} and H_{a1} to designate the null and alternative hypotheses. The reason for this is that there is an additional set of hypotheses to assess. We use H_{02} and H_{a2} for this set.

EXAMPLE 12.19

Another comparison of interest. This comparison tests if there is a difference in time between groups exposed to a negative page and the group exposed to the neutral page. Here are the null and alternative hypotheses:

$$H_{02}: \frac{1}{2}(\mu_3 + \mu_4) = \mu_5$$

$$H_{a2}: \frac{1}{2}(\mu_3 + \mu_4) \neq \mu_5$$

Each of H_{01} and H_{02} says that a combination of population means is 0. These combinations of means are called *contrasts* because the coefficients sum to zero. We use ψ, the Greek letter psi, for contrasts among population means. For our first comparison, we have

$$\psi_1 = -\frac{1}{2}(\mu_1 + \mu_2) + \frac{1}{2}(\mu_3 + \mu_4)$$
$$= (-0.5)\mu_1 + (-0.5)\mu_2 + (0.5)\mu_3 + (0.5)\mu_4$$

and for the second comparison

$$\psi_2 = \frac{1}{2}(\mu_3 + \mu_4) - \mu_5$$
$$= (0.5)\mu_3 + (0.5)\mu_4 + (-1)\mu_5$$

sample contrast

LOOK BACK

rules for
variances,
p. 258

In each case, the value of the contrast is 0 when H_0 is true. *Note that we have chosen to define the contrasts so that they will be positive when the alternative of interest (what we expect) is true. Whenever possible, this is a good idea because it makes some computations easier.*

A contrast expresses an effect in the population as a combination of population means. To estimate the contrast, form the corresponding **sample contrast** by using sample means in place of population means. Under the ANOVA assumptions, a sample contrast is a linear combination of independent Normal variables and, therefore, has a Normal distribution (page 304). We can obtain the standard error of a contrast by using the rules for variances. Inference is based on t statistics. Here are the details.

CONTRASTS

A **contrast** is a combination of population means of the form

$$\psi = \sum a_i \mu_i$$

where the coefficients a_i sum to 0. The corresponding **sample contrast** is

$$c = \sum a_i \bar{x}_i$$

The **standard error of c** is

$$SE_c = s_p \sqrt{\sum \frac{a_i^2}{n_i}}$$

To test the null hypothesis

$$H_0: \psi = 0$$

use the t **statistic**

$$t = \frac{c}{SE_c}$$

with degrees of freedom DFE that are associated with s_p. The alternative hypothesis can be one-sided or two-sided.

A **level C confidence interval for ψ** is

$$c \pm t^* SE_c$$

where t^* is the value for the $t(DFE)$ density curve with area C between $-t^*$ and t^*.

LOOK BACK

addition rule for means, p. 254

Because each $\bar{x}_i$ estimates the corresponding μ_i, the addition rule for means tells us that the mean μ_c of the sample contrast c is ψ. In other words, c is an unbiased estimator of ψ. Testing the hypothesis that a contrast is 0 assesses the significance of the effect measured by the contrast. It is often more informative to estimate the size of the effect using a confidence interval for the population contrast.

EXAMPLE 12.20

The contrast coefficients. In our example the coefficients in the contrasts are

$$a_1 = -0.5, a_2 = -0.5, a_3 = 0.5, a_4 = 0.5, a_5 = 0, \text{ for } \psi_1$$

and

$$a_1 = 0, a_2 = 0, a_3 = 0.5, a_4 = 0.5, a_5 = -1, \text{ for } \psi_2$$

where the subscripts 1, 2, 3, 4, and 5 correspond to the profiles listed in Example 12.17, respectively. In each case the sum of the a_i is 0. We look at inference for each of these contrasts in turn.

EXAMPLE 12.21

Testing the first contrast of interest. The sample contrast that estimates ψ_1 is

$$c_1 = (-0.5)\bar{x}_1 + (-0.5)\bar{x}_2 + (0.5)\bar{x}_3 + (0.5)\bar{x}_4$$
$$= -(0.5)2.518 + (-0.5)2.585 + (0.5)2.405 + (0.5)(2.615) = -0.0415$$

with standard error

$$\text{SE}_{c_1} = 0.911\sqrt{\frac{(-0.5)^2}{21} + \frac{(-0.5)^2}{21} + \frac{(0.5)^2}{21} + \frac{(0.5)^2}{21}}$$
$$= 0.1988$$

The t statistic for testing $H_{01}: \psi_1 = 0$ versus $H_{a1}: \psi_1 > 0$ is

$$t = \frac{c_1}{\text{SE}_{c_1}}$$
$$= \frac{-0.0415}{0.1988} = -0.21$$

Because s_p has 100 degrees of freedom, software using the $t(100)$ distribution gives the two-sided P-value as $P = 0.8341$. If we used Table D, we would conclude that $P > 2(0.25) = 0.50$. The P-value is very large, so there is little evidence against H_{01}.

We use the same method for the second contrast.

EXAMPLE 12.22

Testing the second contrast of interest. The sample contrast that estimates $\psi 2$ is

$$c_2 = (0.5)\bar{x}_3 + (0.5)\bar{x}_4 + (-1)\bar{x}_5$$
$$= (0.5)2.405 + (0.5)2.615 + (-1)1.600$$
$$= 1.2025 + 1.3075 - 1.600$$
$$= 0.91$$

with standard error

$$\text{SE}_{c_2} = 0.911\sqrt{\frac{(0.5)^2}{21} + \frac{(0.5)^2}{21} + \frac{(-1)^2}{21}}$$
$$= 0.2435$$

The t statistic for assessing the significance of this contrast is

$$t = \frac{0.91}{0.2435} = 3.74$$

The P-value for the two-sided alternative is 0.0003. If we used Table D, we would conclude that $P < 2(0.0005) = 0.001$. The P-value is very small, so there is strong evidence against H_{02}.

We have strong evidence to conclude that time viewing a negative content page is different from the time viewing a neutral content page. The size of the difference can be described with a confidence interval.

EXAMPLE 12.23

Confidence interval for the second contrast. To find the 95% confidence interval for ψ_2, we combine the estimate with its margin of error:

$$c_2 \pm t^* \text{SE}_{c_2} = 0.91 \pm (1.962)(0.24)$$
$$= 0.91 \pm 0.47$$

The interval is (0.44, 1.38). Unfortunately, this interval is difficult to interpret because the units are $\sqrt{\text{minutes}}$. We can obtain an approximate 95% interval on the orginial units scale by back-transforming (squaring the interval end points). This results in an approximate 95% confidence interval of the difference to be between 0.19 minutes and 1.90 minutes.

SPSS output for the contrasts is given in Figure 12.14. The results agree with the calculations that we performed in Examples 12.21 and 12.22 except for minor differences due to roundoff error in our calculations. Note that the output does not give the confidence interval that we calculated in Example 12.23. This is easily computed, however, from the contrast estimate and standard error provided in the output.

Some statistical software packages report the test statistics associated with contrasts as F statistics rather than t statistics. These F statistics are the squares of the t statistics described previously. As with much statistical software output, P-values for significance tests are reported for the two-sided alternative.

If the software you are using gives P-*values for the two-sided alternative, and you are using the appropriate one-sided alternative, divide the reported* P-*value by 2.* In our example, we argued that a one-sided alternative may be appropriate for the first contrast. The software reported the P-value as 0.836, so we can conclude $P = 0.418$. Dividing this value by 2 has no effect on the conclusion.

Questions about population means are expressed as hypotheses about contrasts. A contrast should express a specific question that we have in mind when designing the study. Because the ANOVA F test answers a very general question, it is less powerful than tests for contrasts designed to answer specific questions.

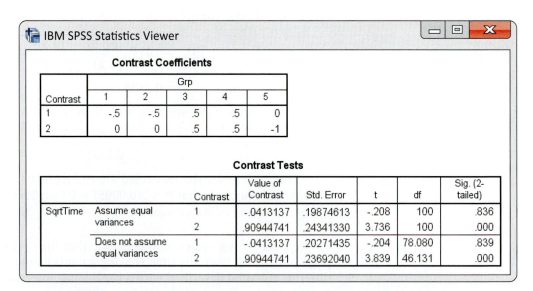

FIGURE 12.14 SPSS output giving the contrast analysis for the Facebook profile study, Example 12.17.

When contrasts are formulated before seeing the data, inference about contrasts is valid whether or not the ANOVA H_0 of equality of means is rejected. Specifying the important questions before the analysis is undertaken enables us to use this powerful statistical technique.

USE YOUR KNOWLEDGE

12.27 Defining a contrast. Refer to Example 12.17 (page 670). Suppose the researcher was also interested in comparing the viewing time between male and female profile pages. Specify the coefficients for this contrast.

12.28 Defining different coefficients. Refer to Example 12.22 (page 675). Suppose we had selected the coefficients $a_1 = 0$, $a_2 = 0$, $a_3 = -1$, $a_4 = -1$, and $a_5 = 2$. Would this choice of coefficients alter our inference in this example? Explain your answer.

Multiple comparisons

In many studies, specific questions cannot be formulated in advance of the analysis. If H_0 is not rejected, we conclude that the population means are indistinguishable on the basis of the data given. On the other hand, if H_0 is rejected, we would like to know which pairs of means differ. **Multiple-comparisons methods** address this issue. It is important to keep in mind that multiple-comparisons methods are used only *after rejecting* the ANOVA H_0.

multiple-comparisons methods

EXAMPLE 12.24

Comparing each pair of groups. Let's return once more to the Facebook friends data with five groups (page 648). We can make 10 comparisons between pairs of means. We can write a t statistic for each of these pairs. For example, the statistic

$$t_{12} = \frac{\bar{x}_1 - \bar{x}_2}{s_p \sqrt{\frac{1}{n_1} + \frac{1}{n_2}}}$$

$$= \frac{3.82 - 4.88}{1.10 \sqrt{\frac{1}{24} + \frac{1}{33}}}$$

$$= -3.59$$

compares profiles with 102 and 302 friends. The subscripts on t specify which groups are compared.

The t statistics for two other pairs are

$$t_{23} = \frac{\bar{x}_2 - \bar{x}_3}{s_p \sqrt{\frac{1}{n_2} + \frac{1}{n_3}}} \qquad\qquad t_{25} = \frac{\bar{x}_2 - \bar{x}_5}{s_p \sqrt{\frac{1}{n_2} + \frac{1}{n_5}}}$$

$$= \frac{4.88 - 4.56}{1.10 \sqrt{\frac{1}{33} + \frac{1}{26}}} \qquad\qquad = \frac{4.88 - 3.99}{1.10 \sqrt{\frac{1}{33} + \frac{1}{21}}}$$

$$= 1.11 \qquad\qquad\qquad\qquad = 2.90$$

two-sample *t*
procedures,
p. 449

These 10 *t* statistics are very similar to the pooled two-sample *t* statistic for comparing two population means. The difference is that we now have more than two populations, so each statistic uses the pooled estimator s_p from all groups rather than the pooled estimator from just the two groups being compared. This additional information about the common σ increases the power of the tests. The degrees of freedom for all these statistics are DFE = 129, those associated with s_p.

Because we do not have any specific ordering of the means in mind as an alternative to equality, we must use a two-sided approach to the problem of deciding which pairs of means are significantly different.

MULTIPLE COMPARISONS

To perform a **multiple-comparisons procedure**, compute ***t* statistics** for all pairs of means using the formula

$$t_{ij} = \frac{\bar{x}_i - \bar{x}_j}{s_p \sqrt{\dfrac{1}{n_i} + \dfrac{1}{n_j}}}$$

If

$$|t_{ij}| \geq t^{**}$$

we declare that the population means μ_i and μ_j are different. Otherwise, we conclude that the data do not distinguish between them. The value of t^{**} depends upon which multiple-comparisons procedure we choose.

One obvious choice for t^{**} is the upper $\alpha/2$ critical value for the $t(\text{DFE})$ distribution. This choice simply carries out as many separate significance tests of fixed level α as there are pairs of means to be compared. The procedure based on this choice is called the **least-significant differences method**, or simply **LSD**.

LSD method

LSD has some undesirable properties, particularly if the number of means being compared is large. Suppose, for example, that there are $I = 20$ groups and we use LSD with $\alpha = 0.05$. There are 190 different pairs of means. If we perform 190 *t* tests, each with an error rate of 5%, our overall error rate will be unacceptably large. We expect about 5% of the 190 to be significant even if the corresponding population means are the same. Because 5% of 190 is 9.5, we expect 9 or 10 false rejections.

The LSD procedure fixes the probability of a false rejection for each single pair of means being compared. It does not control the overall probability of *some* false rejection among all pairs. Other choices of t^{**} control possible errors in other ways. The choice of t^{**} is, therefore, a complex problem, and a detailed discussion of it is beyond the scope of this text. Many choices for t^{**} are used in practice. Most statistical packages provide several to choose from.

Bonferroni
procedure,
p. 391

We will discuss only one of these, called the Bonferroni method. Use of this procedure with $\alpha = 0.05$, for example, guarantees that the probability of any false rejection among all comparisons made is no greater than 0.05. This is much stronger protection than controlling the probability of a false rejection at 0.05 for *each separate* comparison.

EXAMPLE 12.25

Applying the Bonferroni method. We apply the Bonferroni multiple-comparisons procedure with $\alpha = 0.05$ to the data from the Facebook friends study. Given 10 comparisons of interest, the value of t^{**} for this procedure uses $\alpha = 0.05/10 = 0.005$ for each test. From Table D, this value is 2.63. Of the statistics $t_{12} = -3.59$, $t_{23} = 1.11$, and $t_{25} = 2.90$ calculated in Example 12.24, only t_{12} and t_{25} are significant. These two statistics compare the profile of 302 friends with the two extreme levels.

Of course, we prefer to use software for the calculations.

EXAMPLE 12.26

Interpreting software output. The output generated by SPSS for Bonferroni comparisons appears in Figure 12.15. The software uses an asterisk to indicate that the difference in a pair of means is statistically significant. Here, all 10 comparisons are reported. These results agree with the calculations that we performed in Examples 12.24 and 12.25. There are no significant differences except those already mentioned. Note that each comparison is given twice in the output.

The data in the Facebook friends study provide a clear result: the social attractiveness score increases as the number of friends increases to a point and then decreases. Unfortunately with these data, we cannot accurately describe this relationship in more detail. This lack of clarity is not unusual when performing a multiple-comparisons analysis.

Here, the mean associated with 302 friends is significantly different from the means for the 102- and 902-friend profiles, but it is not found significantly different from the means for the profiles with 502 and 702 friends. To complicate things, the means for profiles with 502 and 702 friends were not found significantly different from the means for the 102- and 902-friend profiles.

This kind of apparent contradiction points out dramatically the nature of the conclusions of statistical tests of significance. The conclusion appears to be illogical. If μ_1 is the same as μ_3 and if μ_3 is the same as μ_2, doesn't it follow that μ_1 is the same as μ_2? Logically, the answer must be Yes.

Some of the difficulty can be resolved by noting the choice of words used. In describing the inferences, we talk about failing to detect a difference or concluding that two groups are different. In making logical statements, we say things such as "is the same as." There is a big difference between the two modes of thought. Statistical tests ask, "Do we have adequate evidence to distinguish two means?" It is not illogical to conclude that we have sufficient evidence to distinguish μ_1 from μ_2, but not μ_1 from μ_3 or μ_2 from μ_3.

One way to deal with these difficulties of interpretation is to give confidence intervals for the differences. The intervals remind us that the differences are not known exactly. We want to give *simultaneous confidence intervals*, that

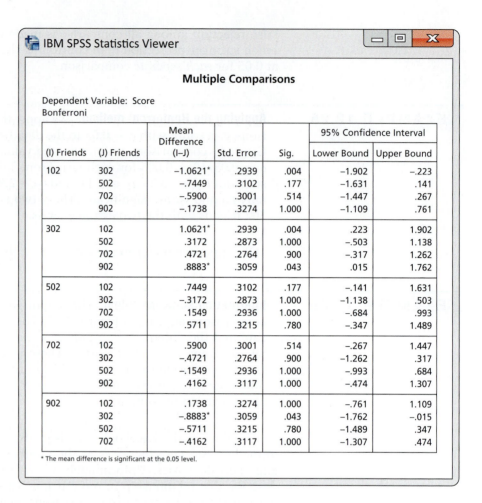

FIGURE 12.15 SPSS output giving the multiple-comparisons analysis for the Facebook friends study, Example 12.26.

is, intervals for *all* differences among the population means at once. Again, we must face the problem that there are many competing procedures—in this case, many methods of obtaining simultaneous intervals.

SIMULTANEOUS CONFIDENCE INTERVALS FOR DIFFERENCES BETWEEN MEANS

Simultaneous confidence intervals for all differences $\mu_i - \mu_j$ between population means have the form

$$(\bar{x}_i - \bar{x}_j) \pm t^{**}s_p \sqrt{\frac{1}{n_i} + \frac{1}{n_j}}$$

The critical values t^{**} are the same as those used for the multiple-comparisons procedure chosen.

The confidence intervals generated by a particular choice of t^{**} are closely related to the multiple-comparisons results for that same method. If one of the confidence intervals includes the value 0, then that pair of means will not be declared significantly different, and vice versa.

EXAMPLE 12.27

Interpreting software output, continued. The SPSS output for the Bonferroni multiple-comparisons procedure given in Figure 12.15 includes the simultaneous 95% confidence intervals. We can see, for example, that the interval for $\mu_1 - \mu_3$ is -1.63 to 0.14. The fact that the interval includes 0 is consistent with the fact that we failed to detect a difference between these two means using this procedure. Note that the interval for $\mu_3 - \mu_1$ is also provided. This is not really a new piece of information because it can be obtained from the other interval by reversing the signs and reversing the order, that is, -0.14 to 1.63. So, in fact, we really have only 10 intervals. Use of the Bonferroni procedure provides us with 95% confidence that *all 10* intervals simultaneously contain the true values of the population mean differences.

USE YOUR KNOWLEDGE

12.29 Why no multiple comparisons? Any pooled two-sample t problem can be run as a one-way ANOVA with $I = 2$. Explain why it is inappropriate to analyze the data using multiple-comparisons procedures in this setting.

12.30 Growth of Douglas fir seedlings. An experiment was conducted to compare the growth of Douglas fir seedlings under three different levels of vegetation control (0%, 50%, and 100%). Sixteen seedlings were randomized to each level of control. The resulting sample means for stem volume were 58, 73, and 105 cubic centimeters (cm^3), respectively, with $s_p = 17$ cm^3. The researcher hypothesized that the average growth at 50% control would be less than the average of the 0% and 100% levels.

(a) What are the coefficients for testing this contrast?

(b) Perform the test and report the test statistic, degrees of freedom, and P-value. Do the data provide evidence to support this hypothesis?

Power

Recall that the power of a test is the probability of rejecting H_0 when H_a is, in fact, true. Power measures how likely a test is to detect a specific alternative. When planning a study in which ANOVA will be used for the analysis, it is important to perform power calculations to check that the sample sizes are adequate to detect differences among means that are judged to be important.

Power calculations also help evaluate and interpret the results of studies in which H_0 was not rejected. We sometimes find that the power of the test was so low against reasonable alternatives that there was little chance of obtaining a significant F.

LOOK BACK

power,
p. 392

In Chapter 7, we found the power for the two-sample t test. One-way ANOVA is a generalization of the two-sample t test, so it is not surprising that the procedure for calculating power is quite similar.

Here are the steps that are needed:

1. Specify

  (a) An alternative (H_a) that you consider important; that is, values for the true population means $\mu_1, \mu_2, \ldots, \mu_I$.

(b) Sample sizes $n_1, n_2, \ldots, n_I$; usually these will all be equal to the common value n.

(c) A level of significance α, usually equal to 0.05.

(d) A guess at the standard deviation σ.

2. Use the degrees of freedom DFG $= I - 1$ and DFE $= N - I$ to find the critical value that will lead to the rejection of H_0. This value, which we denote by F^*, is the upper α critical value for the F(DFG,DFE) distribution.

noncentrality parameter   3. Calculate the **noncentrality parameter**[8]

$$\lambda = \frac{\sum n_i(\mu_i - \overline{\mu})^2}{\sigma^2}$$

where $\overline{\mu}$ is a weighted average of the group means

$$\overline{\mu} = \sum \frac{n_i}{N}\mu_i$$

If the means are all equal (the ANOVA H_0), then $\lambda = 0$. The noncentrality parameter measures how unequal the given set of means is. Large λ points to an alternative far from H_0, and we expect the ANOVA F test to have high power.

4. Find the power, which is the probability of rejecting H_0 when the alternative hypothesis is true; that is, the probability that the observed F is greater than F^*. Under H_a, the F statistic has a distribution known as the **noncentral F distribution**
noncentral *F* distribution   **tral F distribution**. SAS, for example, has a function for this distribution. Using this function, the power is

$$\text{Power} = 1 - \text{PROBF}(F^*, \text{DFG}, \text{DFE}, \lambda)$$

Software makes calculation of the power quite easy. The software does Steps 2, 3, and 4, so our task simplifies to just Step 1. Some software doesn't request the alternative means, but rather a difference in means that is judged important. Most software will also assume a constant sample size. Let's run through an example doing the calculations ourselves and then compare the results with output from two software programs.

EXAMPLE 12.28

Power of a reading comprehension study. Suppose that a study on reading comprehension for three different teaching methods has 10 students in each group. How likely is this study to detect differences in the mean responses? A previous study performed in a different setting found sample means of 41, 47, and 44, and the pooled standard deviation was 7. Based on these results, we will use $\mu_1 = 41$, $\mu_2 = 47$, $\mu_3 = 44$, and $\sigma = 7$ in a calculation of power. The n_i are equal, so $\overline{\mu}$ is simply the average of the μ_i:

$$\overline{\mu} = \frac{41 + 47 + 44}{3} = 44$$

The noncentrality parameter is, therefore,

$$\lambda = \frac{n \sum (\mu_i - \overline{\mu})^2}{\sigma^2}$$

$$= \frac{(10)[(41 - 44)^2 + (47 - 44)^2 + (44 - 44)^2]}{49}$$

$$= \frac{(10)(18)}{49} = 3.67$$

Because there are three groups with 10 observations per group, DFG = 2 and DFE = 27. The critical value for α = 0.05 is F^* = 3.35. The power is, therefore,

$$1 - \text{PROBF}(3.35, 2, 27, 3.67) = 0.3486$$

The chance that we reject the ANOVA H_0 at the 5% significance level given these population means and standard deviation is slightly less than 35%.

Figure 12.16 shows the power calculation output from JMP and Minitab. For JMP, you specify the alternative means, standard deviation, and the total sample size N. The power is calculated once the "Continue" button is clicked.

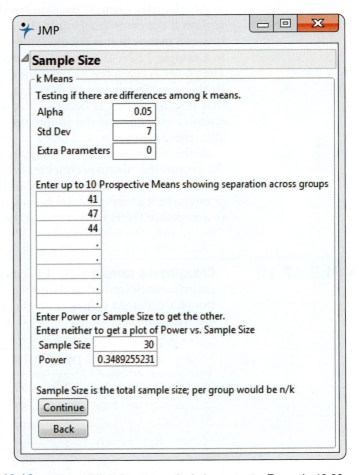

FIGURE 12.16 JMP and Minitab power calculation outputs, Example 12.28.

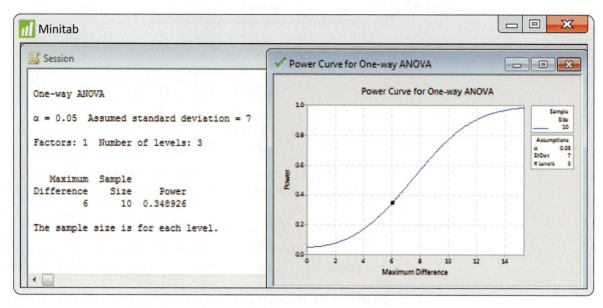

FIGURE 12.16 (*Continued*)

Notice that this result is the same as the result in Example 12.28. For Minitab, you enter the common sample size n, standard deviation σ, and the difference between means that is deemed important. For the alternative means specified in Example 12.28, the largest difference is $6 = 47 - 41$, so that was entered. The power is again the same as the result in Example 12.28. This won't always be the case. Specifying an important difference will often give a power value that is smaller. This is because it computes a noncentrality parameter that is always less than or equal to the noncentrality value based on knowing all the alternative means.

If the assumed values of the μ_i in this example describe differences among the groups that the experimenter wants to detect, then we would want to use more than 10 subjects per group. Although H_0 is false for these μ_i, the chance of rejecting it at the 5% level is only about 35%. This chance can be increased to acceptable levels by increasing the sample sizes.

EXAMPLE 12.29

Changing the sample size. To decide on an appropriate sample size for the experiment described in the previous example, we repeat the power calculation for different values of n, the number of subjects in each group. Here are the results:

n	DFG	DFE	F^*	λ	Power
20	2	57	3.16	7.35	0.65
30	2	87	3.10	11.02	0.84
40	2	117	3.07	14.69	0.93
50	2	147	3.06	18.37	0.97
100	2	297	3.03	36.73	≈ 1

Try using JMP to verify these calculations. With $n = 40$, the experimenters have a 93% chance of rejecting H_0 with $\alpha = 0.05$ and thereby demonstrating that the groups have different means. In the long run, 93 out of every 100 such experiments would reject H_0 at the $\alpha = 0.05$ level of significance. Using 50 subjects per group increases the chance of finding significance to 97%. With 100 subjects per group, the experimenters are virtually certain to reject H_0. The exact power for $n = 100$ is 0.99990. In most real-life situations, the additional cost of increasing the sample size from 50 to 100 subjects per group would not be justified by the relatively small increase in the chance of obtaining statistically significant results.

USE YOUR KNOWLEDGE

12.31 Understanding power calculations. Refer to Example 12.28. Suppose that the researcher decided to use $\mu_1 = 39$, $\mu_2 = 44$, and $\mu_3 = 49$ in the power calculations. With $n = 10$ and $\sigma = 7$, would the power be larger or smaller than 35%? Explain your answer.

12.32 Understanding power calculations, continued. If all the group means are equal (H_0 is true), what is the power of the F test? Explain your answer.

SECTION 12.2 SUMMARY

• The ANOVA F test does not say which of the group means differ. It is, therefore, usual to add comparisons among the means to basic ANOVA.

• Specific questions formulated before examination of the data can be expressed as **contrasts**. Tests and confidence intervals for contrasts provide answers to these questions.

• If no specific questions are formulated before examination of the data and the null hypothesis of equality of population means is rejected, **multiple-comparisons** methods are used to assess the statistical significance of the differences between pairs of means.

• The **power** of the F test depends upon the sample sizes, the variation among population means, and the within-group standard deviation.

SECTION 12.2 EXERCISES

For Exercises 12.27 and 12.28, see page 677 for Exercises 12.29 and 12.30, see page 681 and for Exercises 12.31 and 12.32, see page 685.

12.33 College dining facilities. University and college food service operations have been trying to keep up with the growing expectations of consumers with regard to the overall campus dining experience. Because customer satisfaction has been shown to be associated with repeat patronage and new customers through word-of-mouth, a public university in the Midwest took a sample of patrons from their eating establishments and asked them about

their overall dining satisfaction.[9] The following table summarizes the results for three groups of patrons:

Category	$\bar{x}$	n	s
Student—meal plan	3.44	489	0.804
Faculty—meal plan	4.04	69	0.824
Student—no meal plan	3.47	212	0.657

(a) Is it reasonable to use a pooled standard deviation for these data? Why or why not? If yes, compute it.

(b) The ANOVA F statistic was reported as 17.66. Give the degrees of freedom and either an approximate (from a table) or an exact (from software) P-value. Sketch a picture of the F distribution that illustrates the P-value. What do you conclude?

(c) Prior to performing this survey, food service operations thought that satisfaction among faculty would be higher than satisfaction among students. Use the results in the table to test this contrast. Make sure to specify the null and alternative hypotheses, test statistic, and P-value.

12.34 Writing contrasts. You've been asked to help some administrators analyze survey data on textbook expenditures collected at a large public university. Let μ_1, μ_2, μ_3, and μ_4 represent the population mean expenditures on textbooks for the freshmen, sophomores, juniors, and seniors, respectively.

(a) Because freshman and sophomores take lower-level courses, which might use more expensive introductory textbooks, the administrators want to compare the average of the freshmen and sophomores with the average of the juniors and seniors. Write a contrast that expresses this comparison.

(b) Write a contrast for comparing the freshmen with the sophomores.

(c) Write a contrast for comparing the juniors with the seniors.

12.35 Writing contrasts, continued. Return to the eye study described in Example 12.15 (page 663). Let μ_1, μ_2, μ_3, and μ_4 represent the mean scores for blue, brown, gaze down, and green eyes, respectively.

(a) Because a majority of the population in this study are Hispanic (eye color predominantly brown), we want to compare the average score of the brown eyes with the average score of the other two eye colors. Write a contrast that expresses this comparison.

(b) Write a contrast to compare the average score when the model is looking at you versus the score when looking down.

12.36 Analyzing contrasts. Answer the following questions for the two contrasts that you defined in the previous exercise. [EYES icon]

(a) For each contrast, give H_0 and an appropriate H_a. In choosing the alternatives, you should use information given in the description of the problem, but you may not consider any impressions obtained by inspection of the sample means.

(b) Find the values of the corresponding sample contrasts c_1 and c_2.

(c) Calculate the standard errors SE_{c_1} and SE_{c_2}.

(d) Give the test statistics and approximate P-values for the two significance tests. What do you conclude?

(e) Compute 95% confidence intervals for the two contrasts.

12.37 Two contrasts of interest for the stimulant study. Refer to Exercise 12.23 (page 669). There are two comparisons of interest to the experimenter. They are (1) placebo versus the average of the two low-dose treatments and (2) the difference between High A and Low A versus the difference between High B and Low B.

(a) Express each contrast in terms of the means (μ's) of the treatments.

(b) Give estimates with standard errors for each of the contrasts.

(c) Perform the significance tests for the contrasts. Summarize the results of your tests and your conclusions.

12.38 Multitasking with technology in the classroom. Laptops and other digital technologies with wireless access to the Internet are becoming more and more common in the classroom. While numerous studies have shown that these technologies can be used effectively as part of teaching, there is concern that these technologies can also distract learners if used for off-task behaviors.

In one study that looked at the effects of off-task multitasking with digital technologies in the classroom, a total of 145 undergraduates were randomly assigned to one of seven conditions.[10] Each condition involved a task that was conducted simultaneously with a class lecture. The study consisted of three 20-minute lectures, each followed by a 15-item quiz. The following table summarizes the conditions and quiz results (mean proportion correct).

Condition	n	Lecture 1	Lecture 2	Lecture 3
Texting	21	0.57	0.75	0.56
Email	20	0.52	0.69	0.50
Facebook	20	0.50	0.68	0.43
MSN messaging	21	0.48	0.71	0.42
Natural use control	21	0.50	0.78	0.58
Word-processing control	21	0.55	0.75	0.57
Paper-and-pencil control	21	0.60	0.74	0.53

(a) For this analysis, let's consider the average of the three quizzes as the response. Compute this mean for each condition.

(b) The analysis of these average scores results in SSG = 0.22178 and SSE = 2.00238. Test the null hypothesis that the mean scores across all conditions are equal.

(c) Using the marginal means from part (a) and the Bonferroni multiple-comparisons method, determine which pairs of means differ significantly at the 0.05 significance level. (*Hint:* There are 21 pairwise comparisons, so the critical *t*-value is 3.095. Also, it is best to order the means from smallest to largest to help with pairwise comparisons)

(d) Summarize your results from parts (b) and (c) in a short report.

12.39 Contrasts for multitasking. Refer to the previous exercise. Let $\mu_1, \mu_2, \ldots, \mu_7$ represent the mean scores for the seven conditions. The first four conditions refer to off-task behaviors, while the last three conditions represent different sorts of controls.

(a) The researchers hypothesized that the average score for the off-task behaviors would be lower than that for the paper-and-pencil control condition. Write a contrast that expresses this comparison.

(b) For this contrast, give H_0 and an appropriate H_a.

(c) Calculate the test statistic and approximate *P*-value for the significance test. What do you conclude?

12.40 Power calculations for planning a study. You are planning a new eye gaze study for a different university than that studied in Example 12.15 (page 663). From

Example 12.15, we know that the standard deviations for the four groups considered in that study were 1.75, 1.72, 1.53, and 1.67. In Figure 12.9, we found the pooled standard error to be 1.68. Because the power of the *F* test decreases as the standard deviation increases, use $\sigma = 2.0$ for the calculations in this exercise. This choice leads to sample sizes that are perhaps a little larger than we need but prevents us from choosing sample sizes that are too small to detect the effects of interest. You would like to conclude that the population means are different when $\mu_1 = 3.2$, $\mu_2 = 3.7$, $\mu_3 = 3.0$ and $\mu_4 = 4.0$.

(a) Pick several values for *n* (the number of students that you will select from each group) and calculate the power of the ANOVA *F* test for each of your choices.

(b) Plot the power versus the sample size. Describe the general shape of the plot.

(c) What choice of *n* would you choose for your study? Give reasons for your answer.

12.41 Power for a different alternative. Refer to the previous exercise. Suppose we increase μ_4 to 4.2. For each of the choices of *n* in the previous example, would the power be larger or smaller under this new set of alternative means? Explain your answer.

CHAPTER 12 EXERCISES

12.42 The effect of increased variation within groups. The *One-Way ANOVA* applet lets you see how the *F* statistic and the *P*-value depend on the variability of the data within groups, the sample size, and the differences among the means.

(a) The black dots are at the means of the three groups. Move these up and down until you get a configuration that gives a *P*-value of about 0.01. What is the value of the *F* statistic?

(b) Now increase the variation within the groups by sliding the standard deviation bar to the right. Describe what happens to the *F* statistic and the *P*-value.

(c) Using between- and within-group variation, explain why the *F* statistic and *P*-value change in this way.

12.43 The effect of increased variation between groups. Set the pooled standard error for the *One-Way ANOVA* applet at a middle value. Drag the black dots so that they are approximately equal.

(a) What is the *F* statistic? Give its *P*-value.

(b) Drag the mean of the second group up and the mean of the third group down. Describe the effect on the *F* statistic and its *P*-value. Explain why they change in this way.

12.44 The effect of increased sample size. Set the pooled standard error for the *One-Way ANOVA* applet at a middle value and drag the black dots so that the means are roughly 5.00, 4.50, and 5.25, respectively.

(a) What are the *F* statistic, its degrees of freedom, and the *P*-value?

(b) Slide the sample size bar to the right so $n = 80$. Also drag the black dots back to the values of 5.00, 4.50, and 5.25, respectively. What are the *F* statistic, its degrees of freedom, and the *P*-value?

(c) Explain why the *F* statistic and *P*-value change in this way as *n* increases.

12.45 The multiple-play strategy. Multiple play is a bundling strategy through which multiple services are provided over a single network. A common triple-play service these days is Internet, television, and telephone. The market for this service has become a key battleground among telecommunication, cable, and broadband service providers. A study compared the pricing (average monthly cost in U.S. dollars) among triple-play providers using DSL, cable, or fiber platforms.[11] The following table summarizes the results for 47 providers.

Group	n	$\bar{x}$	s
DSL	19	104.49	26.09
Cable	20	119.98	40.39
Fiber	8	83.87	31.78

(a) Plot the means versus the platform type. Does there appear to be a difference in pricing?

(b) Is it reasonable to assume that the variances are equal? Explain.

(c) The F statistic is 3.39. Give the degrees of freedom and either an approximate (from a table) or an exact (from software) P-value. What do you conclude?

12.46 The two-sample t test and one-way ANOVA. Refer to the diet and mood data in Exercise 7.68 (page 457). Find the two-sample pooled t statistic for comparing the two energy-restricted diets. Then formulate the problem as an ANOVA and report the results of this analysis. Verify that $F = t^2$.

12.47 Winery websites. As part of a study of British Columbia wineries, each of the 193 wineries were classified into one of three categories based on their website features. The Presence stage just had information about the winery. The Portals stage included order placement and online feedback. The Transactions Integration stage included direct payment or payment through a third party online. The researchers then compared the number of market integration features of each winery (for example, in-house touring, a wine shop, a restaurant, in-house wine tasting, gift shop, etc.). Here are the results:[12]

Stage	n	$\bar{x}$	s
Presence	55	3.15	2.264
Portals	77	4.75	2.097
Transactions	61	4.62	2.346

(a) Plot the means versus the stage of website. Does there appear to be a difference in the average number of market integration features?

(b) Is this an observational study or an experiment? Explain your answer.

(c) Is it reasonable to assume the variances are equal? Justify your reasoning.

(d) The data are counts (integer values). Also, based on the means and standard deviations, the distributions are skewed (can't have a negative count). Do you think this lack of Normality poses a problem for ANOVA? Explain your answer.

(e) The F statistic for these data is 9.594. Give the degrees of freedom and P-value. What do you conclude?

12.48 Time levels of scale. Recall Exercise 7.62 (page 456). This experiment actually involved three groups. The last group was told the construction project would last 12 months. Here is a summary of the interval lengths (in days) between the earliest and latest completion dates. TIMESCL

Group	n	$\bar{x}$	s
1: 52 weeks	30	84.1	55.8
2: 12 months	30	104.6	70.1
3: 1 year	30	139.6	73.1

(a) Is this an observational study or an experiment? Explain your answer.

(b) Use graphical methods to describe the three populations.

(c) Examine the conditions necessary for ANOVA. Summarize your findings.

12.49 Time levels of scale, continued. Refer to the previous exercise. TIMESCL

(a) Run the ANOVA and report the results.

(b) Use a multiple-comparisons method to compare the three groups. State your conclusions.

(c) The researchers hypothesized that the more fine-grained the time unit presented to a participant, the smaller the reported interval would be. To test this, they performed a simple linear regression using the group labels 1, 2, and 3 as the predictor variable. They found the slope ($b = 27.8$) significantly different from 0 ($P < 0.005$) and thus concluded the data supported their hypothesis. Do you think this is an appropriate way to test their hypothesis? Explain your answer.

12.50 Facebook recruitment of young adult smokers. Studies about tobacco use have had difficulties recruiting young adults. Because Facebook is visited daily by a very large percent of young adults, researchers decided to investigate the effectiveness of using Facebook to

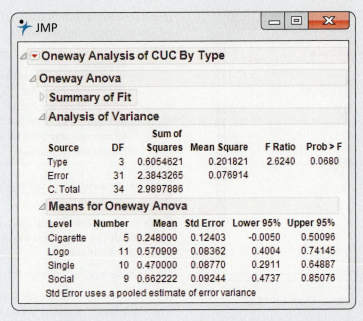

FIGURE 12.17 JMP output giving the ANOVA table for the Facebook recruitment study, Exercise 12.50.

recruit young adults for a smoking cessation trial. Thirty-six ads were run over a seven-week period, with each ad classified based on its image type. The effectiveness of each ad was assessed using the ad cost per unique click (CUC). RECRUIT

(a) Figure 12.17 contains the JMP output for the analysis of the CUCs by image type. State your conclusions based on this output.

(b) The researchers concluded there were significant differences between the means for Cigarette and Logo and between the means for Cigarette and Social. Perform these comparisons using the least-significant differences (LSD) method. Do these comparisons have P-values below $\alpha = 0.05$?

(c) Given your results in part (a), should the researchers report the results in part (b)? Explain your answer.

12.51 Organic foods and morals? Organic foods are often marketed using moral terms such as "honesty" and "purity." Is this just a marketing strategy, or is there a conceptual link between organic food and morality? In one experiment, 62 undergraduates were randomly assigned to one of three food conditions (organic, comfort, and control).[13] First, each participant was given a packet of four food types from the assigned condition and told to rate the desirability of each food on a seven-point scale. Then, each was presented with a list of six moral transgressions and asked to rate each on a seven-point scale ranging from 1 = not at all morally wrong to 7 = very morally wrong. The average of these six scores was used as the response. ORGANIC

(a) Make a table giving the sample size, mean, and standard deviation for each group. Is it reasonable to pool the variances?

(b) Generate a histogram for each of the groups. Can we feel confident that the sample means are approximately Normal? Explain your answer.

12.52 Organic foods and morals, continued. Refer to the previous exercise. ORGANIC

(a) Analyze the scores using analysis of variance. Report the test statistic, degrees of freedom, and P-value.

(b) Assess the assumptions necessary for inference by examining the residuals. Summarize your findings.

(c) Compare the groups using the least-significant differences method.

(d) A higher score is associated with a harsher moral judgment. Using the results from parts (a) and (b), write a short summary of your conclusions.

12.53 Organic foods and friendly behavior? Refer to Exercise 12.51 for the design of the experiment. After rating the moral transgressions, the participants were told "that another professor from another department is also conducting research and really needs volunteers." They were told that they would not receive compensation or course credit for their help and then were asked to write down the number of minutes (out of 30) that they would be willing to volunteer. This sort of question is often used to measure a person's prosocial behavior. ORGANIC

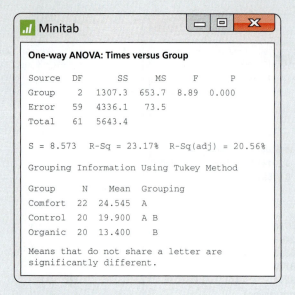

FIGURE 12.18 Minitab output, Exercise 12.53.

(a) Figure 12.18 contains the Minitab output for the analysis of this response variable. Write a one-paragraph summary of your conclusions.

(b) Figure 12.19 contains a residual plot and a Normal quantile plot of the residuals. Are there any concerns regarding the assumptions necessary for inference? Explain your answer.

12.54 Massage therapy for osteoarthritis of the knee. Various studies have shown the benefits of massage to manage pain. In one study, 125 adults suffering from osteoarthritis of the knees were randomly assigned to one of five eight-week regimens.[14] The primary outcome was the change in the Western Ontario and McMaster Universities Arthritis Index (WOMAC-Global). This index is used extensively to assess pain and functioning in those suffering from arthritis. Negative values indicate

improvement. The following table summarizes the results of those completing the study.

Regimen	n	$\bar{x}$	s
30 min massage 1 × /wk	22	−17.4	17.9
30 min massage 2 × /wk	24	−18.4	20.7
60 min massage 1 × /wk	24	−24.0	18.4
60 min massage 2 × /wk	25	−24.0	19.8
Usual care, no massage	24	−6.3	14.6

(a) What proportion of adults dropped out of the study before completion?

(b) Is it reasonable to use the assumption of equal standard deviations when we analyze these data? Give a reason for your answer.

(c) Find the pooled standard deviation.

(d) The SS(Regimen) = 5060.346. Test the null hypothesis that the mean change in WOMAC-Global score is the same for all regimens.

(e) There are 10 pairs of means to compare. For the Bonferroni multiple-comparisons method, the critical t-value is 2.863. Which pairs of means are found to be significantly different? Write a short summary of your analysis.

12.55 Financial incentives for weight loss. The use of financial incentives has shown promise in promoting weight loss and healthy behaviors. In one study, 104 employees of the Children's Hospital of Philadelphia, with BMIs of 30 to 40 kilograms per square meter (kg/m²), were each randomly assigned to one of three weight-loss programs.[15] Participants in the control program were provided a link to weight-control information. Participants in the individual-incentive program received this link but were also told that $100 would be given to them each time they met or exceeded their target monthly weight loss.

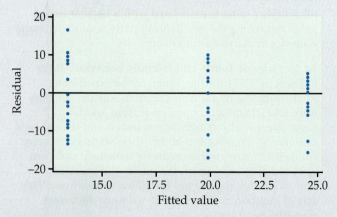

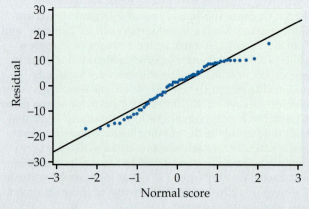

FIGURE 12.19 Residual plot and Normal quantile plot, Exercise 12.53.

Finally, participants in the group-incentive program received similar information and financial incentives as the individual-incentive program but were also told that they were placed in secret groups of five and at the end of each four-week period, those in their group that met their goals throughout the period would equally split an additional $500. The study ran for 24 weeks and the total change in weight (in pounds) was recorded. [📊 LOSS]

(a) Make a table giving the sample size, mean, and standard deviation for each group.

(b) Is it reasonable to pool the variances? Explain your answer.

(c) Generate a histogram for each of the programs. Can we feel confident that the sample means are approximately Normal? Defend your answer.

12.56 Financial incentives for weight loss, continued. Refer to the previous exercise. [📊 LOSS]

(a) Analyze the change in weight using analysis of variance. Report the test statistic, degrees of freedom, P-value, and your conclusions.

(b) Even though you assessed the model assumptions in the previous exercise, let's check the assumptions again by examining the residuals. Summarize your findings.

(c) Compare the groups using the least-significant difference method.

(d) Using the results from parts (a), (b), and (c), write a short summary of your conclusions.

12.57 Changing the response variable. Refer to the previous two exercises, where we compared three weight-loss programs using change in weight measured in pounds. Suppose that you decide to instead make the comparison using change in weight measured in kilograms. [📊 LOSS]

(a) Convert the weight loss from pounds to kilograms by dividing each response by 2.2.

(b) Analyze these new weight changes using analysis of variance. Compare the test statistic, degrees of freedom, and P-value you obtain here with those reported in part (a) of the previous exercise. Summarize what you find.

12.58 Do labels matter? A study was performed to examine the self-identification of college students of Asian descent with various identity categories and assess whether there are attitudinal differences across these categories. Undergraduates at a large midwestern university who had identified themselves as being of Asian descent on their admission application were asked to participate in the study.[16] A total of 620 undergraduates filled out the survey. One question classified the participants into groups by asking them to indicate the option with which they primarily identify: (a) Asian American, (b) specific ethnicity

(for example, Chinese), (c) ethnicity American (for example, Chinese American), and (d) other. The responses to the remaining survey questions were then compared across these four groups. One item was "The campus is supportive of Asian American students." Responses were on a four-point scale (1 = strongly disagree, 4 = strongly agree). A summary of the results follows:

Label	n	$\bar{x}$
Asian American	130	2.93
Specific ethnicity	248	3.00
Ethnicity American	174	3.01
Other	68	3.39

(a) What are the numerator and denominator degrees of freedom for the F test?

(b) Using the formula on page 662 and the preceding results, calculate SSG.

(c) Given SSE = 797.25, use your result from part (b) to compute the F statistic.

(d) Compute the P-value and state your conclusions.

(e) Without doing any additional analysis, describe the pattern in the means that is likely responsible for your conclusions in part (d).

12.59 Do we experience emotions differently? Do people from different cultures experience emotions differently? One study designed to examine this question collected data from 410 college students from five different cultures.[17] The participants were asked to record, on a 1 (never) to 7 (always) scale, how much of the time they typically felt eight specific emotions. These were averaged to produce the global emotion score for each participant. Here is a summary of this measure:

Culture	n	Mean (s)
European American	46	4.39 (1.03)
Asian American	33	4.35 (1.18)
Japanese	91	4.72 (1.13)
Indian	160	4.34 (1.26)
Hispanic American	80	5.04 (1.16)

Note that the convention of giving the standard deviations in parentheses after the means saves a great deal of space in a table such as this.

(a) From the information given, do you think that we need to be concerned that a possible lack of Normality in the data will invalidate the conclusions that we might draw using ANOVA to analyze the data? Give reasons for your answer.

(b) Is it reasonable to use a pooled standard deviation for these data? Why or why not?

(c) The ANOVA F statistic was reported as 5.69. Give the degrees of freedom and either an approximate (from a table) or an exact (from software) P-value. Sketch a picture of the F distribution that illustrates the P-value. What do you conclude?

(d) Without doing any additional formal analysis, describe the pattern in the means that appears to be responsible for your conclusion in part (c). Are there pairs of means that are quite similar?

12.60 The emotions study, continued. Refer to the previous exercise. The experimenters also measured emotions in some different ways. For a period of a week, each participant carried a device that sounded an alarm at random times during a three-hour interval five times a day. When the alarm sounded, participants recorded several mood ratings indicating their emotions for the time immediately preceding the alarm. These responses were combined to form two variables: frequency, the number of emotions recorded, expressed as a percent; and intensity, an average of the intensity scores measured on a scale of 0 to 6. At the end of the one-week experimental period, the subjects were asked to recall the percent of time that they experienced different emotions. This variable was called "recall." Here is a summary of the results:

Culture	n	Frequency mean (s)	Intensity mean (s)	Recall mean (s)
European American	46	82.87 (18.26)	2.79 (0.72)	49.12 (22.33)
Asian American	33	72.68 (25.15)	2.37 (0.60)	39.77 (23.24)
Japanese	91	73.36 (22.78)	2.53 (0.64)	43.98 (22.02)
Indian	160	82.71 (17.97)	2.87 (0.74)	49.86 (21.60)
Hispanic American	80	92.25 (8.85)	3.21 (0.64)	59.99 (24.64)
F statistic		11.89	13.10	7.06

(a) For each response variable, state whether or not it is reasonable to use a pooled standard deviation to analyze these data. Give reasons for your answer.

(b) Give the degrees of freedom for the F statistics and find the associated P-values. Summarize what you can conclude from these ANOVA analyses.

(c) Summarize the means, paying particular attention to similarities and differences across cultures and across variables. Include the means from the previous exercise in your summary.

(d) The European American and Asian American subjects were from the University of Illinois, the Japanese subjects were from two universities in Tokyo, the Indian subjects were from eight universities in or near Kolkata, and the Hispanic American subjects were from California State

University at Fresno. Participants were paid $25 or an equivalent monetary incentive for the Japanese and Indians. Ads were posted on or near the campuses to recruit volunteers for the study. Discuss how these facts influence your conclusions and the extent to which you would generalize the results.

(e) The percents of female students in the samples were as follows: European American, 83%; Asian American, 67%; Japanese, 63%; Indian, 64%; and Hispanic American, 79%. Use a chi-square test to compare these proportions (see Section 9.1, page 536) and discuss how this information influences your interpretation of the results that you have found in this exercise.

12.61 Shopping and bargaining in Mexico. Price haggling and other bargaining behaviors among consumers have been observed for a long time. However, research addressing these behaviors, especially in a real-life setting, remains relatively sparse. A group of researchers performed a small study to determine whether gender or nationality of the bargainer has an effect in the final price obtained.[18] The study took place in Mexico because of the prevalence of price haggling in informal markets. Salespersons working at various informal shops were approached by one of three bargainers looking for a specific product. After an initial price was stated by the vendor, bargaining took place. The response was the difference between the initial and the final price of the product. The bargainers were a Spanish-speaking Hispanic male, a Spanish-speaking Hispanic female, and an Anglo non-Spanish-speaking male. The following table summarizes the results:

Bargainer	n	Average reduction
Hispanic male	40	1.055
Anglo male	40	1.050
Hispanic female	40	2.310

(a) To compare the mean reductions in price, what are the degrees of freedom for the ANOVA F statistic?

(b) The reported test statistic is $F = 8.708$. Give an approximate (from a table) or exact (from software) P-value. What do you conclude?

(c) To what extent do you think the results of this study can be generalized? Give reasons for your answer.

12.62 Restaurant ambiance and consumer behavior. There have been numerous studies investigating the effects of restaurant ambiance on consumer behavior. One study investigated the effects of musical genre on consumer spending.[19] At a single high-end restaurant in England over a three-week period, there were a total of 141 participants; 49 of them were subjected to background pop music (for example, Britney Spears, Culture Club, and Ricky Martin) while dining, 44 to background classical music (for example, Vivaldi, Handel, and Strauss), and 48 to no background music. For each participant, the total food bill, adjusted for time spent dining, was recorded. The following table summarizes the means and standard deviations (in British pounds):

Background music	Mean bill	n	s
Classical	24.130	44	2.243
Pop	21.912	49	2.627
None	21.697	48	3.332
Total	22.531	141	2.969

(a) Plot the means versus the type of background music. Does there appear to be a difference in spending?

(b) Is it reasonable to assume that the variances are equal? Explain.

(c) The F statistic is 10.62. Give the degrees of freedom and either an approximate (from a table) or an exact (from software) P-value. What do you conclude?

(d) Refer back to part (a). Without doing any formal analysis, describe the pattern in the means that is likely responsible for your conclusion in part (c).

(e) To what extent do you think the results of this study can be generalized to other settings? Give reasons for your answer.

12.63 Do isoflavones increase bone mineral density? Kudzu is a plant that was imported to the United States from Japan and now covers over seven million acres in the South. The plant contains chemicals called isoflavones that have been shown to have beneficial effects on bones. One study used three groups of rats to compare a control group with rats that were fed either a low dose or a high dose of isoflavones from kudzu.[20] One of the outcomes examined was the bone mineral density in the femur (in grams per square centimeter). Here are the data: **BMD**

Treatment	Bone mineral density (g/cm²)
Control	0.228 0.207 0.234 0.220 0.217 0.228 0.209 0.221 0.204 0.220 0.203 0.219 0.218 0.245 0.210
Low dose	0.211 0.220 0.211 0.233 0.219 0.233 0.226 0.228 0.216 0.225 0.200 0.208 0.198 0.208 0.203
High dose	0.250 0.237 0.217 0.206 0.247 0.228 0.245 0.232 0.267 0.261 0.221 0.219 0.232 0.209 0.255

(a) Use graphical and numerical methods to describe the data.

(b) Examine the assumptions necessary for ANOVA. Summarize your findings.

(c) Run the ANOVA and report the results.

(d) Use a multiple-comparisons method to compare the three groups.

(e) Write a short report explaining the effect of kudzu isoflavones on the femur of the rat.

12.64 Do poets die young? According to William Butler Yeats, "She is the Gaelic muse, for she gives inspiration to those she persecutes. The Gaelic poets die young, for she is restless, and will not let them remain long on earth." One study designed to investigate this issue examined the age at death for writers from different cultures and genders.[21] Three categories of writers examined were novelists, poets, and nonfiction writers. Most of the writers are from the United States, but Canadian and Mexican writers are also included. **POETS**

(a) Use graphical and numerical methods to describe the data.

(b) Examine the assumptions necessary for ANOVA. Summarize your findings.

(c) Run the ANOVA and report the results.

(d) Use a contrast to compare the poets with the two other types of writers. Do you think that the quotation from Yeats justifies the use of a one-sided alternative for examining this contrast? Explain your answer.

(e) Use another contrast to compare the novelists with the nonfiction writers. Explain your choice for an alternative hypothesis for this contrast.

(f) Use a multiple-comparisons procedure to compare the three means. How do the conclusions from this approach compare with those using the contrasts?

12.65 Exercise and healthy bones. Many studies have suggested that there is a link between exercise and healthy bones. Exercise stresses the bones and this causes them to get stronger. One study examined the effect of jumping on the bone density of growing rats.[22] There were three treatments: a control with no jumping, a low-jump condition (the jump height was 30 centimeters), and a high-jump condition (60 centimeters). After eight weeks of 10 jumps per day, five days per week, the bone density of the rats (expressed in milligrams per cubic centimeter) was measured. Here are the data: **JUMP**

Group	Bone density (mg/cm³)									
Control	611	621	614	593	593	653	600	554	603	569
Low jump	635	605	638	594	599	632	631	588	607	596
High jump	650	622	626	626	631	622	643	674	643	650

(a) Make a table giving the sample size, mean, and standard deviation for each group of rats. Is it reasonable to pool the variances?

(b) Run the analysis of variance. Report the F statistic with its degrees of freedom and P-value. What do you conclude?

12.66 Exercise and healthy bones, continued. Refer to the previous exercise. JUMP

(a) Examine the residuals. Is the Normality assumption reasonable for these data?

(b) Use the Bonferroni or another multiple-comparisons procedure to determine which pairs of means differ significantly. Summarize your results in a short report. Be sure to include a graph.

12.67 Contrasts of interest. Refer to Exercise 12.61 (page 692). Given the group means and F statistic, we can determine that $MSE = 2.421$. Use this value and the other information in Exercise 12.61 to do the following.

(a) Test if there is a difference between the sexes.

(b) Test if there is a difference between nationalities.

(c) Explain why this study would have benefited from also including an Anglo female bargainer.

12.68 Orthogonal polynomial contrasts. Recall the Facebook friends study (page 648). Previous research has shown that the bigger one's social network, the higher one's social attractiveness. In fact, the relationship between the number of friends and social attractiveness is approximately linear. A reasonable question to ask is whether this is same sort of pattern exists within an online social network. With orthogonal polynomial contrasts, we can assess the contributions of different polynomial trends to the overall pattern. Given the five equally spaced levels of the factor in this study, we can investigate up to a quartic (x^4) trend. The derivation of the coefficients is beyond the scope of this book, so we will just investigate the trends here. The coefficients for the linear, quadratic, and cubic trends follow: FRIENDS

Trend	a_1	a_2	a_3	a_4	a_5
Linear	−2	−1	0	1	2
Quadratic	2	−1	−2	−1	2
Cubic	−1	2	0	−2	1

(a) Plot the a_i versus i for the linear trend. Describe the pattern. Suppose that all the μ_i were constant. What would the value of ψ equal?

(b) Plot the a_i versus i for the quadratic trend. Describe the pattern. Suppose that all the μ_i were constant. What would the value of ψ equal? Suppose that $\mu_i = 5_i$ (that is, a linear trend). What would the value of ψ equal?

(c) Test the hypotheses that there is a linear, quadratic, and cubic trend. What do you conclude?

12.69 A comparison of different types of scaffold material. One way to repair serious wounds is to insert some material as a scaffold for the body's repair cells to use as a template for new tissue. Scaffolds made from extracellular material (ECM) are particularly promising for this purpose. Because they are made from biological material, they serve as an effective scaffold and are then resorbed. Unlike biological material that includes cells, however, they do not trigger tissue rejection reactions in the body. One study compared six types of scaffold material.[23] Three of these were ECMs and the other three were made of inert materials (MAT). There were three mice used per scaffold type. The response measure was the percent of glucose phosphated isomerase (Gpi) cells in the region of the wound. A large value is good, indicating that there are many bone marrow cells sent by the body to repair the tissue. ECM

Material	Gpi (%)		
ECM1	55	70	70
ECM2	60	65	65
ECM3	75	70	75
MAT1	20	25	25
MAT2	5	10	5
MAT3	10	15	10

(a) Make a table giving the sample size, mean, and standard deviation for each of the six types of material. Is it reasonable to pool the variances? Note that the sample sizes are small and the data are rounded.

(b) Run the analysis of variance. Report the F statistic with its degrees of freedom and P-value. What do you conclude?

12.70 A comparison of different types of scaffold material, continued. Refer to the previous exercise. ECM

(a) Examine the residuals. Is the Normality assumption reasonable for these data?

(b) Use the Bonferroni or another multiple-comparisons procedure to determine which pairs of means differ significantly. Summarize your results in a short report. Be sure to include a graph.

(c) Use a contrast to compare the three ECM materials with the three other materials. Summarize your conclusions. How do these results compare with those that you obtained from the multiple-comparisons procedure in part (b)?

12.71 Contrasts for the massage study. Refer to Exercise 12.54 (page 690). There are several comparisons of interest in this study. They are (1) usual care versus the average of the massage groups; (2) the average of the two 30-minute massage groups versus the average of the two 60-minute massage groups; and (3) the difference between a 30-minute massage once a week and twice a week versus the difference between a 60-minute massage once a week and twice a week.

(a) Express each contrast in terms of the means (μ's) of the treatments.

(b) Give estimates with standard errors for each of the contrasts.

(c) Perform the significance tests for the contrasts. Summarize the results of your tests and your conclusions.

12.72 A dandruff study. Analysis of variance (ANOVA) methods are often used in clinical trials where the goal is to assess the effectiveness of one or more treatments for a particular medical condition. One such study compared three treatments for dandruff and a placebo. The treatments were 1% pyrithione zinc shampoo (PyrI), the same shampoo but with instructions to shampoo two times (PyrII), 2% ketoconazole shampoo (Keto), and a placebo shampoo (Placebo). After six weeks of treatment, eight sections of the scalp were examined and given a score that measured the amount of scalp flaking on a 0 to 10 scale. The response variable was the sum of these eight scores. An analysis of the baseline flaking measure indicated that randomization of patients to treatments was successful in that no differences were found between the groups. At baseline, there were 112 subjects in each of the three treatment groups and 28 subjects in the Placebo group. During the clinical trial, three dropped out from the PyrII group and six from the Keto group. No patients dropped out of the other two groups. 📊 DANDRUFF

(a) Find the mean, standard deviation, and standard error for the subjects in each group. Summarize these, along with the sample sizes, in a table and make a graph of the means.

(b) Run the analysis of variance on these data. Write a short summary of the results and your conclusion. Be sure to include the hypotheses tested, the test statistic with degrees of freedom, and the *P*-value.

12.73 The dandruff study, continued. Refer to the previous exercise. 📊 DANDRUFF

(a) Plot the residuals versus case number (the first variable in the data set). Describe the plot. Is there any pattern that would cause you to question the assumption that the data are independent?

(b) Examine the standard deviations for the four treatment groups. Is there a problem with the assumption of equal standard deviations for ANOVA in this data set? Explain your answer.

(c) Create Normal quantile plots for each treatment group. What do you conclude from these plots?

(d) Obtain the residuals from the analysis of variance and create a Normal quantile plot of these. What do you conclude?

12.74 Comparing each pair of dandruff treatments. Refer to Exercise 12.72. Use the Bonferroni or another multiple-comparisons procedure that your software provides to compare the individual group means in the dandruff study. Write a short summary of your conclusions. 📊 DANDRUFF

12.75 Testing several contrasts from the dandruff study. Refer to Exercise 12.72. There are several natural contrasts in this experiment that describe comparisons of interest to the experimenters. They are (1) Placebo versus the average of the other three treatments, (2) Keto versus the average of the two Pyr treatments, and (3) PyrI versus PyrII. 📊 DANDRUFF

(a) Express each of these three contrasts in terms of the means (μ's) of the treatments.

(b) Give estimates with standard errors for each of the contrasts.

(c) Perform the significance tests for the contrasts. Summarize the results of your tests and your conclusions.

12.76 Changing the response variable. Refer to Exercise 12.69 (page 694), where we compared six types of scaffold material to repair wounds. The data are given as percents ranging from 5 to 75. 📊 ECM

(a) Convert these percents into their decimal form by dividing by 100. Calculate the transformed means, standard deviations, and standard errors and summarize them, along with the sample sizes, in a table.

(b) Explain how you could have calculated the table entries directly from the table you gave in part (a) of Exercise 12.51.

(c) Analyze the decimal forms of the percents using analysis of variance. Compare the test statistic, degrees of freedom, *P*-value, and conclusion you obtain here with the corresponding values that you found in Exercise 12.51.

12.77 More on changing the response variable. Refer to the previous exercise and Exercise 12.69 (page 694). A calibration error was found with the device that measured Gpi, which resulted in a shifted response. Add 5% to each response and redo the calculations. Summarize the effects of transforming the data by adding a constant to all responses. ECM

12.78 Linear transformation of the response variable. Refer to the previous two exercises. Can you suggest a general conclusion regarding what happens to the test statistic, degrees of freedom, P-value, and conclusion when you perform analysis of variance on data that have been transformed by multiplying the raw data by a constant and then adding another constant? (That is, if y is the original data, we analyze y^*, where $y^* = a + by$ and a and $b \neq 0$ are constants.)

12.79 More on the Facebook friends study. Refer to the Facebook friends study that we began to examine in Example 12.3 (page 648). The explanatory variable in this study is the number of Facebook friends, with possible values of 102, 302, 502, 702, and 902. When using ANOVA, we treat the explanatory variable as categorical. An alternative analysis is to use simple linear regression. Perform this analysis and summarize the results. Plot the residuals from the regression model versus the number of Facebook friends. What do you conclude? FRIENDS

12.80 Using the table of group means and standard deviations. Refer to Exercise 12.6 (page 655). Using the table of group means and standard deviations, construct an ANOVA table similar to that on page 662.

(a) Based on the F statistic and degrees of freedom, compute the P-value. What do you conclude?

(b) Perform pairwise comparisons using the LSD method to determine which coffeehouses have different average ages.

12.81 Planning another emotions study. Scores on an emotional scale were compared for five different cultures in Exercise 12.59 (page 691). Suppose that you are planning a new study using the same outcome variable. Your study will use European American, Asian American, and Hispanic American students from a large university.

(a) Explain how you would select the students to participate in your study.

(b) Use the data from Exercise 12.59 to perform power calculations to determine sample sizes for your study.

(c) Write a report that could be understood by someone with limited background in statistics and that describes your proposed study and why you think it is likely that you will obtain interesting results.

12.82 Planning another isoflavone study. Exercise 12.63 (page 693) gave data for a bone health study that examined the effect of isoflavones on rat bone mineral density. In this study, there were three groups. Controls received a placebo, and the other two groups received either a low or a high dose of isoflavones from kudzu. You are planning a similar study of a new kind of isoflavone. Use the results of the study described in Exercise 12.63 to plan your study. Write a proposal explaining why your study should be funded.

12.83 Planning another restaurant ambiance study. Exercise 12.62 (page 692) gave data for a study that examined the effect of background music on total food spending at a high-end restaurant. You are planning a similar study but intend to look at total food spending at a more casual restaurant. Use the results of the study described in Exercise 12.62 to plan your study.

©Jupiter Images/ Getty Images

Two-Way Analysis of Variance

13.1 The Two-Way
ANOVA Model

13.2 Inference for
Two-Way ANOVA

CHAPTER

13

Introduction

In this chapter, we move from one-way ANOVA, which compares the means of several populations, to two-way ANOVA. Two-way ANOVA compares the means of populations that can be classified in two ways or the mean responses in two-factor experiments.

Many of the key concepts are similar to those of one-way ANOVA, but the presence of more than one classification factor also introduces some new ideas. We once more assume that the data are approximately Normal and that although groups may have different means, they have the same standard deviation; we again pool to estimate the variance; and we again use *F* statistics for significance tests.

The major difference between one-way and two-way ANOVA is in the FIT part of the model. We will carefully study this term, and we will find much that is both new and useful.

13.1 The Two-Way ANOVA Model

When you complete this section, you will be able to:

- Discuss the advantages of a two-way ANOVA design.
- Describe the two-way ANOVA model and when it is used for inference.
- Interpret the relationship between two factors in terms of main effects and interaction.
- Construct an interaction plot and determine whether it shows that there is interaction among the factors.

We begin with a discussion of the advantages of the two-way ANOVA design, illustrated through some examples. Then we discuss the model.

Advantages of two-way ANOVA

LOOK BACK

factor,
p. 644

In one-way ANOVA, we classify populations according to one categorical variable, or factor. In the two-way ANOVA model, there are two factors, each with its own number of levels. When we are interested in the effects of two factors, a two-way design offers great advantages over several single-factor studies.

EXAMPLE 13.1

Design 1: Does haptic feedback improve performance? In Example 12.1 (page 645), a group of technology students wanted to see if haptic feedback is helpful in navigating a simulated game environment. To do this, they plan to randomly assign 20 students to each of three joystick controller types and record the time it takes to complete a navigation mission.

It turns out that their simulated game has four difficulty levels. Suppose that a second experiment is planned to compare these levels when using the standard joystick. A similar experimental design will be used, with the four difficulty levels randomly assigned equally among the 60 students.

Here is a picture of the designs of the first and second experiments with the sample sizes:

Joystick	n
1	20
2	20
3	20
Total	60

Difficulty	n
1	15
2	15
3	15
4	15
Total	60

In the first experiment, 20 students were assigned to each level of the factor for a total of 60 students. In the second experiment, 15 students were assigned to each level of the factor for a total of 60 students. If each experiment takes one week, the total amount of time for the two experiments is two weeks.

Each experiment will be analyzed using one-way ANOVA. The factor in the first experiment is joystick type with three levels, and the factor in the second experiment is game difficulty with four levels. Let's now consider combining the two experiments into one.

EXAMPLE 13.2

Design 2: Does haptic feedback improve performance regardless of difficulty level? Suppose that we use a two-way approach for the simulated game problem. There are two factors, joystick type and difficulty. Because joystick type has three levels and difficulty has four levels, this is a 3×4 design. This gives a total of 12 possible combinations of type and difficulty. With a total of 60 students, we could assign each combination of type and difficulty to five students. The time it takes to complete a navigation mission is the outcome variable.

Here is a picture of the two-way design with the sample sizes:

		Difficulty			
Joystick	1	2	3	4	Total
1	5	5	5	5	20
2	5	5	5	5	20
3	5	5	5	5	20
Total	15	15	15	15	60

cell

Each combination of the factors in a two-way design corresponds to a **cell**. The 3×4 ANOVA for the haptic feedback experiment has 12 cells, each corresponding to a particular combination of joystick type and difficulty level.

With the two-way design, notice that we have 20 students assigned to each joystick type, the same as we had for the one-way experiment for type alone. Similarly, there are still 15 students assigned to each level of difficulty. Thus, the two-way design gives us the *same amount* of information for estimating the completion time for each level of each factor as we had with the two one-way designs. The difference is that we can collect all the information in only one experiment. This experiment lasts one week (instead of a combined two weeks) and involves a single observation from each of the 60 students. By combining the two factors into one experiment, we have increased our efficiency by reducing the amount of data to be collected by half.

EXAMPLE 13.3

The effect of a limited time offer on purchase intent. Starbucks' Pumpkin Spice Latte (PSL) is the company's most popular seasonal item. Why is this? Is it the unique flavor? Or could it be because it is only available for a limited time each year? To investigate this, some students surveyed 100 Starbucks consumers about their intent to purchase a PSL when it is offered in the fall.[1] Half of the surveys included the upcoming PSL advertisement. The other half included the same advertisement with the additional words "Limited Time Offer" above the image of the drink. Because purchase intent may depend on how frequently a consumer visits Starbucks, the students included a survey question about this. The question was used to classify each customer as either a "light" or "heavy" user of Starbucks.

The factors for the two-way ANOVA are advertisement type with two levels and user status with two levels. There are $2 \times 2 = 4$ cells in their study. The outcome variable purchase intent is measured on a 1 to 7 scale.

Here is a table of sample sizes that summarizes their design:

	User status		
Advertisement	Light	Heavy	Total
Regular	27	23	50
Added wording	19	31	50
Total	46	54	100

The students were not able to control the number of subjects in each cell of the study because they did not know user status until the survey was administered.

This example illustrates another advantage of two-way designs. Although the students are primarily interested in the effect of adding the words "Limited Time Offer" on purchase intent, they also included user status because they suspected that the wording effect might be different in light and heavy users.

Consider an alternative one-way design where we ignore user status. With this design, we will have the same number of customers at each of the ad type levels, so in this way, it is similar to our two-way design.

However, suppose that there are, in fact, differences due to user status. In this case, the one-way ANOVA would assign this variation to the RESIDUAL (within groups) part of the model. In the two-way ANOVA, user status is included as a factor; therefore, this variation is included in the FIT part of the model. *Whenever we can move variation from RESIDUAL to FIT, we reduce the σ of our model and increase the power of our tests.*

LOOK BACK

DATA = FIT + RESIDUAL, p. 560

Professor Pietro M. Motta / Science Source

EXAMPLE 13.4

Vitamin D and osteoporosis. Osteoporosis is a disease primarily of the elderly. People with osteoporosis have low bone mass and an increased risk of bone fractures. More than 10 million people in the United States, 1.4 million Canadians, and many millions throughout the world have this disease. Adequate calcium in the diet is necessary for strong bones, but vitamin D is also needed for the body to efficiently use calcium. High doses of calcium in the diet will not prevent osteoporosis unless there is adequate vitamin D. Exposure of the skin to the ultraviolet rays in sunlight enables our bodies to make vitamin D. However, elderly people often don't go outside as much as younger people do, and in northern areas such as Canada, there is not sufficient ultraviolet light for the body to make vitamin D, particularly in the winter months.

Suppose that we wanted to see if calcium supplements will increase bone mass (or prevent a decrease in bone mass) in an elderly Canadian population. Because of the vitamin D complication, we will make this a factor in our design. We will use a 2 × 2 design for our osteoporosis study. The two factors are calcium and vitamin D. The levels of each factor will be zero (placebo) and an amount that is expected to be adequate, 800 milligrams per day (mg/d) for calcium and 300 international units per day (IU/d) for vitamin D.

Women between the ages of 70 and 80 will be recruited as subjects. Bone mineral density (BMD) will be measured at the beginning of the study, and supplements will be taken for one year. The change in BMD over the one-year period is the outcome variable. We expect a dropout rate of 20%, and we would like to have about 20 subjects providing data in each group at the end of the study. We will, therefore, recruit 100 subjects and randomly assign 25 to each treatment combination.

Here is a table that summarizes the design with the sample sizes at the start of the study:

	Vitamin D		
Calcium	Placebo	300 IU/d	Total
Placebo	25	25	50
800 mg/d	25	25	50
Total	50	50	100

This example illustrates a third reason for using two-way designs. The effectiveness of the calcium supplement on BMD may differ across the two levels of vitamin D. We call this an **interaction**. In contrast, the average values for the calcium effect and the vitamin D effect are represented as **main effects**. The two-way model represents FIT as the sum of a main effect for each of the two factors *and* an interaction. One-way designs that vary a single factor and hold other factors fixed cannot discover interactions. We will discuss interactions more fully later.

These examples illustrate several reasons two-way designs are preferable to one-way designs.

ADVANTAGES OF TWO-WAY ANOVA

1. It is more efficient to study two factors simultaneously rather than separately.

2. We can reduce the residual variation in a model by including a second factor thought to influence the response.

3. We can investigate interactions between factors.

These considerations also apply to study designs with more than two factors. We will be content, however, to explore only the two-way case in this chapter. Remember that the choice of sampling or experimental design is fundamental to any statistical study. *Factors and levels must be carefully selected by an individual or team who understands both the statistical models and the issues that the study will address.*

The two-way ANOVA model

When discussing two-way models in general, we will use the labels A and B for the two factors. *For particular examples and when using statistical software, it is better to use meaningful names for these categorical variables.* Thus, in Example 13.2 (page 699), we would say that the factors are joystick type and difficulty level, and in Example 13.4, we would say that the factors are the calcium and vitamin D.

The numbers of levels of the factors are often used to describe the model. Again using our earlier examples, we would say that Example 13.2 represents a 3×4 ANOVA, and Example 13.4 illustrates a 2×2 ANOVA. In general, Factor A will have I levels, and Factor B will have J levels. Therefore, we call the general two-way problem an $I \times J$ ANOVA.

In a two-way design, every level of A appears in combination with every level of B, so that $I \times J$ groups are compared. The sample size for level i of Factor A and level j of Factor B is n_{ij}. In Examples 13.2 and 13.4 the n_{ij} have been equal but this is not required.[2] The total number of observations is

$$N = \sum n_{ij}$$

ASSUMPTIONS FOR TWO-WAY ANOVA

We have independent simple random samples (SRSs) of size n_{ij} from each of $I \times J$ Normal populations. The population means μ_{ij} may differ, but all populations have the same standard deviation σ. The μ_{ij} and σ are unknown parameters.

Let x_{ijk} represent the kth observation from the population having Factor A at level i and Factor B at level j. The statistical model is

$$x_{ijk} = \mu_{ij} + \epsilon_{ijk}$$

for $i = 1, \ldots, I$ and $j = 1, \ldots, J$ and $k = 1, \ldots, n_{ij}$, where the deviations ϵ_{ijk} are from an $N(0, \sigma)$ distribution.

LOOK BACK

estimates of
population
parameters
p. 653

Similar to the one-way model, the FIT part is the group means μ_{ij}, and the RESIDUAL part is the deviations ϵ_{ijk} of the individual observations from their group means. To estimate a group mean μ_{ij}, we use the sample mean of the observations in the samples from this group:

$$\bar{x}_{ij} = \frac{1}{n_{ij}} \sum_k x_{ijk}$$

The k below the $\sum$ means that we sum the n_{ij} observations that belong to the (i, j)th sample.

The RESIDUAL part of the model contains the unknown σ. We first calculate the sample variances for each SRS. Provided it is reasonable to consider a common standard deviation (page 654), we pool these to estimate σ^2:

$$s_p^2 = \frac{\sum (n_{ij} - 1)s_{ij}^2}{\sum (n_{ij} - 1)}$$

Just as in one-way ANOVA, the numerator in this fraction is SSE and the denominator is DFE. Also, DFE is the total number of observations minus the number of groups. That is, DFE $= N - IJ$. The estimator of σ is s_p.

USE YOUR KNOWLEDGE

13.1 Limited-time offer effect on purchase intent. Example 13.3 (page 699) describes a study designed to compare different advertisement types and user status on purchase intent. Write out the ANOVA model for this study. Be sure to give specific values for I, J, and the n_{ij}. List all the parameters of the model.

13.2 Limited-time offer effect on purchase intent, continued. Refer to the previous exercise. The following table summarizes the group means and standard deviations.

Advertisement	Light user $\bar{x}$	Light user s	Heavy user $\bar{x}$	Heavy user s
Regular	4.56	1.75	5.00	1.79
Added wording	5.74	1.19	5.19	1.91

(a) Is it reasonable to pool the standard deviations for these data? Explain your answer.

(b) For each parameter in your model from Exercise 13.1, give the estimate.

Main effects and interactions

In this section, we will further explore the FIT part of the two-way ANOVA, which is represented in the model by the population means μ_{ij}. The two-way design gives some structure to the set of means μ_{ij}.

So far, because we have independent samples from each of $I \times J$ groups, we have presented the problem as a one-way ANOVA with IJ groups. Each population mean μ_{ij} is estimated by the corresponding sample mean $\bar{x}_{ij}$, and we can calculate sums of squares and degrees of freedom as in one-way ANOVA. In accordance with the conventions used by many computer software packages, we use the term *model* when discussing the sums of squares and degrees of freedom calculated as in one-way ANOVA with IJ groups. Thus, SSM is a model sum of squares constructed from deviations of the form $\bar{x}_{ij} - \bar{x}$, where

$\bar{x}$ is the average of all the observations and $\bar{x}_{ij}$ is the mean of the (i, j)th group. Similarly, DFM is simply $IJ - 1$.

In two-way ANOVA, the terms SSM and DFM can be further broken down into terms corresponding to a main effect for A, a main effect for B, and an AB interaction. Each of SSM and DFM is then a sum of terms:

$$SSM = SSA + SSB + SSAB$$

and

$$DFM = DFA + DFB + DFAB$$

The term SSA represents variation among the means for the different levels of Factor A. Because there are I such means, DFA $= I - 1$ degrees of freedom. Similarly, SSB represents variation among the means for the different levels of Factor B, with DFB $= J - 1$.

Interactions are a bit more involved. We can see that SSAB, which is SSM − SSA − SSB, represents the variation in the model that is not accounted for by the main effects. By subtraction we see that its degrees of freedom are

$$DFAB = (IJ - 1) - (I - 1) - (J - 1) = (I - 1)(J - 1)$$

There are many kinds of interactions. The easiest way to study them is through examples.

EXAMPLE 13.5

Investigating differences in sugar-sweetened beverage consumption. Consumption of sugar-sweetened beverages has been linked to Type 2 diabetes and obesity. One study used data from the National Health and Nutrition Examination Survey (NHANES) to estimate consumption of these beverages among children. More than 14,000 individuals provided data for this study. Individuals were divided into three age categories: preschoolers (two to five years old), preadolescents (6 to 11 years old), and adolescents (12 to 19 years old).[3] Here are the means for the number of calories in sugar-sweetened beverages consumed per day during 2003 to 2006 and 2007 to 2010:

	Year		
Group	2006	2010	Mean
Preschoolers	170	130	150
Preadolescents	214	192	203
Adolescents	341	295	318
Mean	242	206	224

The table in Example 13.5 includes averages of the means in the rows and columns. For example, in 2006 the mean of calories consumed per day is

$$\frac{170 + 214 + 341}{3} = 241.67$$

which is rounded to 242 in the table. Similarly, the corresponding value for 2010 is

$$\frac{130 + 192 + 295}{3} = 205.67$$

marginal means which is rounded to 206 in the table. These averages are called **marginal means** (because of their location at the *margins* of such tabulations). The grand mean (224 in this case) can be obtained by averaging either set of marginal means.

Figure 13.1 is a plot of the group means. From the plot, we see that fewer calories from sugar-sweetened beverages were consumed by each group in 2010 than in 2006. In statistical language, there is a main effect for year. We also see that the means are different across age categories. This means there is a main effect for age. These main effects can be described by differences between the marginal means. For example, the mean for 2006 is 242 calories and decreases 36 calories to 206 calories in 2010. Similarly, the mean for preschoolers is 150, it increases 53 calories to 203 for preadolescents, and then increases 115 calories to 318 for adolescents.

To examine two-way ANOVA data for a possible interaction, always construct a plot similar to Figure 13.1. When no interaction is present, the marginal means provide a reasonable description of the two-way table of means. This will be reflected in the plot by profiles that are roughly parallel. In this case, it is debatable whether the two profiles (the collections of marginal means for a given year) should be considered parallel.

When there is an interaction, the marginal means do not tell the whole story. For example, with these data, the marginal mean difference between years is 36 calories. This is smaller than the difference in calories for the preschoolers (170 − 130 = 40) and adolescents (341 − 295 = 46) and larger than the change in the preadolescents (214 − 192 = 22). If differences of roughly 20 calories per day are scientifically meaningful, then we would say that it appears there is an interaction. Inference is still needed to confirm that these differences are not likely the result of chance variation.

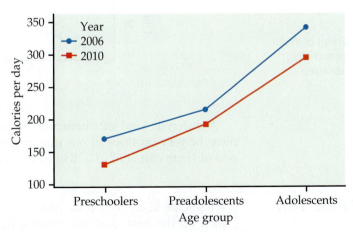

FIGURE 13.1 Plot of the mean calories in sugar-sweetened beverages consumed per day in 2003 to 2006 and 2007 to 2010 for different age groups, Example 13.5.

Interactions come in many shapes and forms. *When we find an interaction, a careful examination of the means is needed to properly interpret the data.* Simply stating that interactions are significant tells us very little. Plots of the group means, called **interaction plots**, are essential. Here is another example.

EXAMPLE 13.6

Eating in groups. Some research has shown that people eat more when they eat in groups. One possible mechanism for this phenomenon is that they may spend more time eating when in a larger group. A study designed to examine this idea measured the length of time spent (in minutes) eating lunch in different settings.[4] Here are some data from this study:

| | Number of people eating | | | | | |
Lunch setting	1	2	3	4	5 or more	Mean
Workplace	12.6	23.0	33.0	41.1	44.0	30.7
Fast-food restaurant	10.7	18.2	18.4	19.7	21.9	17.8
Mean	11.6	20.6	25.7	30.4	32.9	24.2

Figure 13.2 gives the plot of the means for this example. The patterns are not parallel, so it appears that we have an interaction. Meals take longer when there are more people present, but this phenomenon is much greater for the meals consumed at work. For fast-food eating, the meal durations are fairly similar when there is more than one person present.

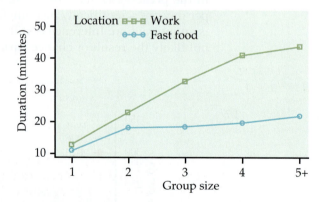

FIGURE 13.2 Plot of mean meal duration versus lunch setting and group size, Example 13.6.

A different kind of interaction is present in the next example. Here, we must be very cautious in our interpretation of the main effects because either one of them can lead to a distorted conclusion.

EXAMPLE 13.7

We got the beat? When we hear music that is familiar to us, we can quickly pick up the beat, and our mind synchronizes with the music. However, if the music is unfamiliar, it takes us longer to synchronize. In a study that investigated the theoretical framework for this phenomenon, French and

Tunisian nationals listened to French and Tunisian music.[5] Each subject was asked to tap in time with the music being played. A synchronization score, recorded in milliseconds, measured how well the subjects synchronized with the music. A higher score indicates better synchronization. Six songs of each music type were used. Here are the means:

Nationality	Music French	Tunisian	Mean
French	950	750	850
Tunisian	760	1090	925
Mean	855	920	887

The means are plotted in Figure 13.3. In the study, the researchers were not interested in main effects. Their theory predicted the interaction that we see in the figure. Subjects synchronize better with music from their own culture. The main effects, on the other hand, suggest that Tunisians sychronize better than the French (regardless of music type) and that it is easier to synchronize to Tunisian music (regardless of nationality).

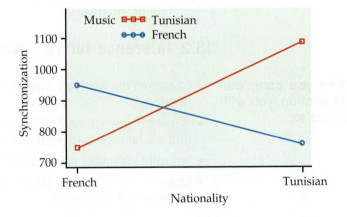

FIGURE 13.3 Plot of mean synchronization score versus type of music for French and Tunisian nationals, Example 13.7.

The interaction in Figure 13.3 is very different from those that we saw in Figures 13.1 and 13.2. These examples illustrate the point that it is necessary to plot the means and carefully describe the patterns when interpreting an interaction.

The design of the study in Example 13.7 allows us to examine two main effects and an interaction. However, this setting does not meet all the assumptions needed for statistical inference using the two-way ANOVA framework of this chapter. *As with one-way ANOVA, we require that observations be independent.*

In this study, we have a design that has each subject contributing data for two types of music, so these two scores will be dependent. The framework is similar to the matched pairs setting (page 182). The design is called a **repeated-measures design**. More advanced texts on statistical methods cover this important design.

repeated-measures design

13.3 What's wrong? In each of the following, identify what is wrong and then either explain why it is wrong or change the wording of the statement to make it true.

(a) A two-way ANOVA is used when the outcome variable can take only two possible values.

(b) In a 2×3 ANOVA, each level of Factor A appears with two levels of Factor B.

(c) The FIT part of the model in a two-way ANOVA represents the variation that is sometimes called error or residual.

(d) In an $I \times J$ ANOVA, $DFAB = IJ - 1$.

13.4 What's wrong? In each of the following, identify what is wrong and then either explain why it is wrong or change the wording of the statement to make it true.

(a) Parallel profiles of cell means imply that a strong interaction is present.

(b) You can perform a two-way ANOVA only when the sample sizes are the same in all cells.

(c) The estimate s_p^2 is obtained by pooling the marginal sample variances.

(d) When interaction is present, the marginal means are always uninformative.

13.2 Inference for Two-Way ANOVA

When you complete this section, you will be able to:

- Construct the two-way ANOVA table in terms of sources and degrees of freedom.
- Summarize what the ANOVA table F tests can tell you about main effects and interactions and what they cannot without further analysis.
- Interpret statistical software ANOVA output for a two-way ANOVA.
- Use residual plots and sample statistics to check the assumptions of the two-way ANOVA model.

Inference for two-way ANOVA involves F statistics for each of the two main effects and an additional F statistic for the interaction. As with one-way ANOVA, the calculations are organized in an ANOVA table.

The ANOVA table for two-way ANOVA

Two-way ANOVA is the statistical analysis for a two-way design with a quantitative response variable. The results of a two-way ANOVA are summarized in an ANOVA table based on splitting the total variation SST and the total degrees of freedom DFT among the two main effects and the interaction. Both the sums of squares (which measure variation) and the degrees of freedom add:

$$SST = SSA + SSB + SSAB + SSE$$
$$DFT = DFA + DFB + DFAB + DFE$$

The sums of squares are always calculated in practice by statistical software. *When the n_{ij} are not all equal, there are different ways to calculate the sums of squares, and some can give sums of squares that do not add.* The degrees of freedom, on the other hand, will always add.

From each sum of squares and its degrees of freedom we find the mean square in the usual way:

$$\text{mean square} = \frac{\text{sum of squares}}{\text{degrees of freedom}}$$

The significance of each of the main effects and the interaction is assessed by an F statistic that compares the variation due to the effect of interest with the within-group variation. Each F statistic is the mean square for the source of interest divided by MSE. Here is the general form of the two-way ANOVA table:

Source	Degrees of freedom	Sum of squares	Mean square	F
A	$I-1$	SSA	SSA/DFA	MSA/MSE
B	$J-1$	SSB	SSB/DFB	MSB/MSE
AB	$(I-1)(J-1)$	SSAB	SSAB/DFAB	MSAB/MSE
Error	$N-IJ$	SSE	SSE/DFE	
Total	$N-1$	SST		

There are three null hypotheses in two-way ANOVA, with an F test for each. We can test for significance of the main effect of A, the main effect of B, and the AB interaction. *It is generally good practice to examine the test for interaction first because the presence of a strong interaction may influence the interpretation of the main effects.* Be sure to plot the means as an aid to interpreting the results of the significance tests.

SIGNIFICANCE TESTS IN TWO-WAY ANOVA

To test the main effect of A, use the F statistic

$$F_A = \frac{\text{MSA}}{\text{MSE}}$$

To test the main effect of B, use the F statistic

$$F_B = \frac{\text{MSB}}{\text{MSE}}$$

To test the interaction of A and B, use the F statistic

$$F_{AB} = \frac{\text{MSAB}}{\text{MSE}}$$

The P-value is the probability that a random variable having an F distribution with numerator degrees of freedom corresponding to the effect and denominator degrees of freedom equal to DFE is greater than or equal to the calculated F statistic.

The following example illustrates how to do a two-way ANOVA. As with the one-way ANOVA, we focus our attention on interpretation of the computer output.

EXAMPLE 13.8

HRTRATE

A study of cardiovascular risk factors. A study of cardiovascular risk factors compared runners who averaged at least 15 miles per week with a control group described as "generally sedentary." Both men and women were included in the study.[6] The design is a 2 × 2 ANOVA with the factors group and sex. There were 200 subjects in each of the four combinations. One of the variables measured was the heart rate after six minutes of exercise on a treadmill. SAS computer analysis produced the outputs in Figure 13.4 and Figure 13.5.

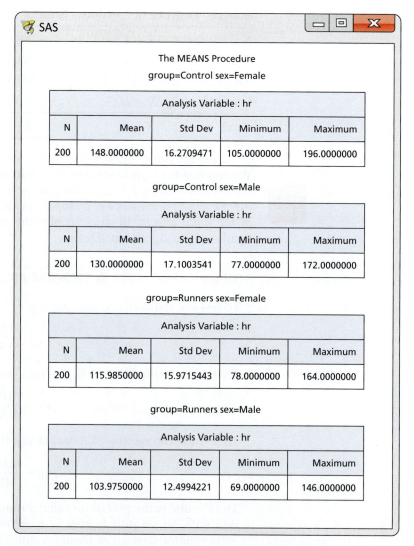

FIGURE 13.4 Summary statistics for heart rates in the four groups of a 2 × 2 ANOVA, Example 13.8.

SAS

The GLM Procedure

Dependent Variable: hr

Source	DF	Sum of Squares	Mean Square	F Value	Pr > F
Model	3	215256.0900	71752.0300	296.35	<.0001
Error	796	192729.8300	242.1229		
Corrected Total	799	407985.9200			

R-Square	Coeff Var	Root MSE	hr Mean
0.527607	12.49924	15.56030	124.4900

Source	DF	Type I SS	Mean Square	F Value	Pr > F
group	1	168432.0800	168432.0800	696.65	<.0001
sex	1	45030.0050	45030.0050	185.98	<.0001
group*sex	1	1794.0050	1794.0050	7.41	0.0066

FIGURE 13.5 Two-way ANOVA output for heart rates, Example 13.8.

We begin with the usual preliminary examination. From Figure 13.4, we see that the ratio of the largest to the smallest standard deviation is less than 2. Therefore, we are not concerned about violating the assumption of equal population standard deviations. Normal quantile plots (not shown) do not reveal any outliers, and the data appear to be reasonably Normal.

The ANOVA table at the top of the output in Figure 13.5 is, in effect, a one-way ANOVA with four groups: female control, female runner, male control, and male runner. In this analysis, Model has 3 degrees of freedom and Error has 796 degrees of freedom. *Because we will be relying on software to do all these calculations, it is always a good idea to do some quick arithmetic checks like degrees of freedom to make sure things make sense.* The *F* test and its associated *P*-value for this analysis refer to the hypothesis that all four groups have the same population mean. We are interested in the main effects and interaction, so we ignore this test.

The sums of squares for the group and sex main effects and the group-by-sex interaction appear at the bottom of Figure 13.5 under the heading "Type I SS." These sum to the sum of squares for Model. Similarly, the degrees of freedom for these sums of squares sum to the degrees of freedom for Model. Two-way ANOVA splits the variation among the means (expressed by the Model sum of squares) into three parts that reflect the two-way layout.

Because the degrees of freedom are all 1 for the main effects and the interaction, the mean squares are the same as the sums of squares. The *F* statistics

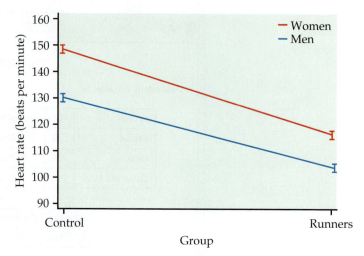

FIGURE 13.6 Plot of the group means, with standard errors indicated, for heart rates in the 2 × 2 ANOVA, Example 13.8.

for the three effects appear in the column labeled "F Value," and the *P*-values are under the heading "Pr > F." For the group main effect, we verify the calculation of *F* as follows:

$$F = \frac{\text{MSG}}{\text{MSE}} = \frac{168{,}432}{242.12} = 695.65$$

All three effects are statistically significant. The group effect has the largest *F*, followed by the sex effect and then the group-by-sex interaction. To interpret these results, we examine the plot of means, with bars indicating one standard error, in Figure 13.6. Note that the standard errors are quite small due to the large sample sizes. The significance of the main effect for group is due to the fact that the controls have higher average heart rates than the runners for both sexes. This is the largest effect evident in the plot.

The significance of the main effect for sex is due to the fact that the females have higher heart rates than the men in both groups. The differences are not as large as those for the group effect, and this is reflected in the smaller value of the *F* statistic.

The analysis indicates that a complete description of the average heart rates requires consideration of the interaction in addition to the main effects. The two lines in the plot are not parallel. This interaction can be described in two ways. The female-male difference in average heart rates is greater for the controls than for the runners. Alternatively, the difference in average heart rates between controls and runners is greater for women than for men. As the plot suggests, the interaction is not large. It is statistically significant because there were 800 subjects in the study.

Two-way ANOVA output for other software is similar to that given by SAS. Figure 13.7 gives the analysis of the heart rate data using Excel and Minitab.

FIGURE 13.7 Excel and Minitab two-way ANOVA outputs for the heart rate study, Example 13.8.

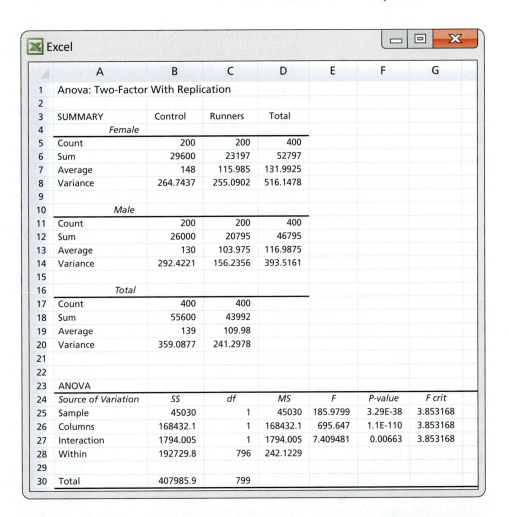

Excel

Anova: Two-Factor With Replication

SUMMARY	Control	Runners	Total
Female			
Count	200	200	400
Sum	29600	23197	52797
Average	148	115.985	131.9925
Variance	264.7437	255.0902	516.1478
Male			
Count	200	200	400
Sum	26000	20795	46795
Average	130	103.975	116.9875
Variance	292.4221	156.2356	393.5161
Total			
Count	400	400	
Sum	55600	43992	
Average	139	109.98	
Variance	359.0877	241.2978	

ANOVA

Source of Variation	SS	df	MS	F	P-value	F crit
Sample	45030	1	45030	185.9799	3.29E-38	3.853168
Columns	168432.1	1	168432.1	695.647	1.1E-110	3.853168
Interaction	1794.005	1	1794.005	7.409481	0.00663	3.853168
Within	192729.8	796	242.1229			
Total	407985.9	799				

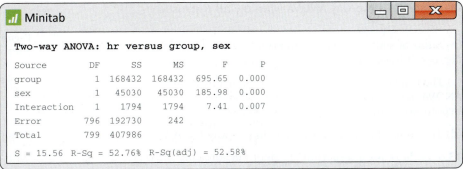

Minitab

```
Two-way ANOVA: hr versus group, sex

Source        DF      SS      MS       F      P
group          1  168432  168432  695.65  0.000
sex            1   45030   45030  185.98  0.000
Interaction    1    1794    1794    7.41  0.007
Error        796  192730     242
Total        799  407986

S = 15.56  R-Sq = 52.76%  R-Sq(adj) = 52.58%
```

CHAPTER 13 SUMMARY

• **Two-way analysis of variance** (ANOVA) is used to compare population means when populations are classified according to two factors.

• We assume that independent SRSs are drawn from each population and that the responses from each population are Normal with possibly different means but the same standard deviation.

- As with one-way ANOVA, these assumptions should be assessed. Preliminary analysis includes examination of means, standard deviations, residual plots, and Normal quantile plots.

- **Marginal means** are calculated by taking averages of the cell means, either across rows or down columns. These means can be used in an **interaction plot** to aid in the interpretation of results.

- Similar to one-way ANOVA, the total variation is separated into parts for the **model** and **error.** Pooling is also used to estimate the error, or within-group variance. However, given that there are now two factors, the model variation is separated into parts for each of the **main effects** and the **interaction**.

- The calculations are organized into an **ANOVA table**. F statistics and P-values are used to test hypotheses about the main effects and the interaction.

- Careful inspection of the means is necessary to interpret significant main effects and interactions. Plots are a useful aid.

CHAPTER 13 EXERCISES

For Exercises 13.1 and 13.2, see page 703 and for Exercises 13.3 and 13.4, see page 708.

13.5 What's wrong? In each of the following, identify what is wrong and then either explain why it is wrong or change the wording of the statement to make it true.

(a) You should reject the null hypothesis that there is no interaction in a two-way ANOVA when the AB F statistic is small.

(b) Sums of squares are equal to mean squares divided by degrees of freedom.

(c) The test statistics for the main effects in a two-way ANOVA have a chi-square distribution when the null hypothesis is true.

(d) The sums of squares always add in two-way ANOVA.

13.6 Is there an interaction? Each of the following tables gives means for a two-way ANOVA. Make a plot of the means with the levels of Factor A on the x axis. State whether or not there is an interaction, and if there is, describe it.

(a)

	Factor A		
Factor B	1	2	3
1	11	21	31
2	6	11	16

(b)

	Factor A		
Factor B	1	2	3
1	10	5	15
2	20	15	25

(c)

	Factor A		
Factor B	1	2	3
1	10	15	20
2	15	20	25

(d)

	Factor A		
Factor B	1	2	3
1	10	15	20
2	20	15	10

13.7 Describing a two-way ANOVA model. A 3 × 3 ANOVA was run with four observations per cell.

(a) Give the degrees of freedom for the F statistic that is used to test for interaction in this analysis and the entries from Table E that correspond to this distribution.

(b) Sketch a picture of this distribution with the information from the table included.

(c) The calculated value of this F statistic is 3.03. Report the P-value and state your conclusion.

(d) Based on your answer to part (c), would you expect an interaction plot to have mean profiles that look parallel? Explain your answer.

13.8 Determining the critical value of F. For each of the following situations, state how large the F statistic needs to be for rejection of the null hypothesis at the 5% level. Sketch each distribution and indicate the region where you would reject.

(a) The main effect for the first factor in a 2×5 ANOVA with three observations per cell.

(b) The interaction in a 3×2 ANOVA with six observations per cell.

(c) The interaction in a 2×3 ANOVA with six observations per cell.

13.9 Identifying the factors of a two-way ANOVA model. For each of the following situations, identify both factors and the response variable. Also, state the number of levels for each factor (I and J) and the total number of observations (N).

(a) A child psychologist is interested in studying how a child's percent of pretend play differs with sex and age (4, 8, and 12 months). There are 11 infants assigned to each cell of the experiment.

(b) Brewers malt is produced from germinating barley. A home brewer wants to determine the best conditions for germinating the barley. Thirty lots of barley seed were equally and randomly assigned to 10 germination conditions. The conditions are combinations of the week after harvest (1, 3, 6, 9, or 12 weeks) and the amount of water used in the process (4 or 8 milliliters). The percent of seeds germinating is the outcome variable.

(c) The strength of concrete depends upon the formula used to prepare it. An experiment compares six different mixtures. Nine specimens of concrete are poured from each mixture. Three of these specimens are subjected to 0 cycles of freezing and thawing, three are subjected to 100 cycles, and three are subjected to 500 cycles. The strength of each specimen is then measured.

(d) A marketing experiment compares four different colors of for-sale tags at an outlet mall. Each color tag is used for one week. Shoppers are classified as impulse buyers or not through a survey instrument. The total dollar amount each of the 138 shoppers spent on sale items is recorded.

13.10 Determining the degrees of freedom. For each part in Exercise 13.9, outline the ANOVA table, giving the sources of variation and the degrees of freedom.

13.11 Smart shopping carts. Smart shopping carts are shopping carts equipped with scanners that track the total price of the items in the cart (real-time feedback). To help understand the smart shopping cart's influence on spending behavior, a group of researchers designed a two-factor study. Each participant was randomly assigned to either be on or not on a budget of $35. Also, each participant's cart was equipped with or without real-time feedback. The total amount spent on a common grocery list was the response.[7] SMART2

(a) Construct a plot of the means and describe the main features of the plot.

(b) Analyze the data using a two-way ANOVA. Report the F statistics, degrees of freedom, and P-values. Because the n_{ij} are not equal, different software may give slightly different F statistics and P-values.

(c) Write a short summary of your findings.

13.12 Using makeup. A study was performed in which 44 women participated as models. Each model was photographed after applying makeup as if she was going on a "night out." Software was then used to create a sequence of 21 images ranging from 50% makeup to 150% makeup. Another set of observers (consisting of both sexes) was asked to look at each sequence of images and select the image that they felt was most attractive, what they felt was most attractive to other women, and what they felt was most attractive to other men. The average percent makeup over the 44 models was the response. Figure 13.8 replicates one of the plots used in their summary.[8]

(a) Does there appear to be interaction between the sex of the observer and attractiveness category? Explain your answer.

(b) Describe what you see in terms of the main effects, making sure to relate these means to 100% (the value that represents what the models applied themselves).

13.13 Writing about testing worries and exam performance. For many students, self-induced worries and pressure to perform well on exams cause them to perform below their ability. This is because these worries compete with the working memory available for performance. Expressive writing has been shown to be an effective technique to overcome traumatic or emotional experiences. Thus, a group of researchers decided to investigate whether expressive writing prior to test-taking may help performance.[9]

The small study involved 20 subjects. Half the subjects were assigned to the expressive-writing group and the others to a control group. Each subject took two short mathematics exams. Prior to the first exam, students were told just to perform their best. Prior to the second exam, students were told that they each had been paired with another student and if the members

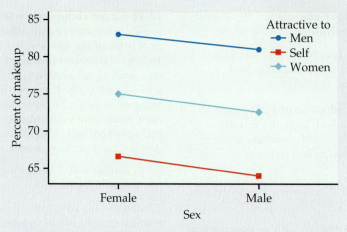

FIGURE 13.8 Interaction plot, Exercise 13.12.

of a pair both performed well on the exam, the pair would receive a monetary reward. Each student was then told privately that his or her partner had already scored well. This was done to create a high-stakes testing environment for the second exam. Those in the control group sat quietly for 10 minutes prior to taking the second exam. Those in the expressive-writing group had 10 minutes to write about their thoughts and feelings regarding the exam. The following table summarizes the test results (% correct):

Group	First exam		Second exam	
	$\bar{x}$	s	$\bar{x}$	s
Control	83.4	11.5	70.1	14.3
Expressive-writing	86.2	6.3	90.1	5.8

(a) Explain why this is a repeated-measures design and not a standard two-way ANOVA design.

(b) Generate a plot to look at changes in score across time and across group. Describe what you see in terms of the main effects and interaction.

(c) Because exam scores can run only between 0% and 100%, variances for populations with means near 0% or 100% may be smaller and the distribution of scores may be skewed. Does it appear reasonable here to pool variances? Explain your answer.

13.14 Influence of age and sex on motor performance. The slowing of motor performance as humans age is well established. Differences across the sexes, however, are less so. A recent study assessed the motor performance of 246 healthy adults.[10] One task was to tap the thumb and forefinger of the right hand together 20 times as quickly as possible. The following

table summarizes the results (in seconds) for seven age classes and two sexes:

Age class (years)	Males			Females		
	n	$\bar{x}$	s	n	$\bar{x}$	s
41–50	19	4.72	1.31	19	5.88	0.82
51–55	12	4.10	1.62	12	5.93	1.13
56–60	12	4.80	1.04	12	5.85	0.87
61–65	24	5.08	0.98	24	5.81	0.94
66–70	17	5.47	0.85	17	6.50	1.23
71–75	23	5.84	1.44	23	6.12	1.04
> 75	16	5.86	1.00	16	6.19	0.91

Generate a plot to look at changes in the time across age class for each sex. Describe what you see in terms of the main effects for age and sex as well as their interaction.

13.15 Influence of age and sex on motor performance, continued. Refer to the previous exercise.

(a) In their article, the researchers state that each of their response variables was assessed for Normality prior to performing a two-way ANOVA. Is it necessary for the 246 time measurements to be Normally distributed? Explain your answer.

(b) Is it reasonable to pool the variances?

(c) Suppose for these data that SS(sex) = 44.66, SS(age) = 31.97, SS(interaction) = 13.22, and SSE = 280.95. Construct an ANOVA table and state your conclusions.

13.16 Ecological effects of pharmaceuticals on fish. Drugs used to treat anxiety persist in wastewater effluent, resulting in relatively high concentrations of these drugs in our rivers and streams. To understand the impacts of these anxiety drugs on fish, researchers commonly expose fish to various levels of a drug in a laboratory setting and observe their behavior.[11] In one 2×2 experiment, researchers considered exposure to oxazepan through the water (Y/N) and through the diet (Y/N). Ten one-year-old perch were assigned to each of the four treatment combinations. After seven days of exposure, each fish was observed for activity. This was recorded as the number of swimming bouts (defined as movement exceeding 3.5 cm) over a 10-minute period. **FISH**

(a) The response is the number of movements in 10 minutes, which can only be whole number. Should we be concerned about violating the assumption of Normality? Explain your answer.

(b) Construct an interaction plot and comment on the main effects of exposure through diet and water and their interaction.

(c) Analyze the count of swimming bouts using analysis of variance. Report the test statistics, degrees of freedom, and *P*-values.

(d) Use the residuals to check the model assumptions. Are there any concerns? Explain your answer.

(e) Based on parts (c) and (d), write a short paragraph summarizing your findings.

13.17 Where are your eyes? The objectifying gaze, often referred to as "ogling" or "checking out," can have have many adverse consequences. A group of researchers used eye-tracking technology to better understand the nature and causes for this gaze. They asked 29 women and 36 men to look at images of college-aged women. Each woman had the same clothes and neutral expression but varied in body shape (ideal, average, and below average). Prior to looking at the images, each participant was told to focus on either the appearances or personalities of the women. Here is a summary of the amount of time (in milliseconds) the eyes focused on the chest of the women:[12]

	Sex			
	Male		Female	
Focus	$\bar{x}$	SE	$\bar{x}$	SE
Appearance	448.25	35.98	463.22	48.09
Personality	338.78	54.25	276.48	46.06

(a) Plot the means. Do you think there is an interaction? Explain your answer.

(b) Do you think the marginal means would be useful for understanding the results of this study? Explain why or why not.

(c) The researchers broke these results down further using body shape as a third factor. Describe why the inclusion of this factor complicates the analysis. In other words, why is this not a standard $2 \times 2 \times 3$ experiment?

13.18 Ecological effects of pharmaceuticals on fish, continued. Refer to Exercise 13.16.

(a) Often with a count as the response, one considers taking the square root of the count and performing ANOVA on this transformed response. Explain why a transformation might be useful here.

(b) Using the response SqrtCnt, repeat parts (b) through (e) of Exercise 13.16.

(c) Which analysis do you prefer here? Explain your answer.

13.19 The influences of transaction history and a thank-you statement. A service failure is defined as any service-related problem (real or perceived) that transpires during a customer's experience with a firm. In the hotel industry, there is a high human component, so these sorts of failures commonly occur regardless of extensive training and established policies. As a result, hotel firms must learn to effectively react to these failures. A recent study investigated the relationship between a consumer's transaction history (levels: long and short) and an employee thank-you statement (levels: yes and no) on a consumer's repurchase intent.[13] Each subject was randomly assigned to one of the four treatment groups and asked to read some service failure/resolution scenarios and respond accordingly. Repurchase intent was measured using a nine-point scale. Here is a summary of the means:

	Thank-you	
History	No	Yes
Short	5.69	6.80
Long	7.53	7.37

(a) Plot the means. Do you think there is an interaction? If yes, describe the interaction in terms of the two factors.

(b) Find the marginal means. Are they useful for understanding the results of this study? Explain your answer.

13.20 Transaction history and a thank-you statement, continued. Refer to the previous exercise. The numbers of subjects in the cells were not equal, so the researchers used linear regression to analyze the data. This was done by creating an indicator variable for each factor and the interaction. Following is a partial ANOVA table. Complete it and state your conclusions regarding the main effects and interaction described in the previous exercise.

Source	DF	SS	MS	F	P-value
Transaction history		61.445			
Thank-you statement		21.810			
Interaction		15.404			
Error	160	759.904			

13.21 The effects of proximity and visibility on food intake. A study investigated the influence that proximity and visibility of food have on food intake.[14] A total of 40 secretaries from the University of Illinois participated in the study. A candy dish full of individually wrapped chocolates was placed either at the desk of the participant or at a location 2 meters from the participant. The candy dish was either a clear (candy visible) or opaque (candy not visible) covered bowl. After a week, the researchers noted not only the number of candies consumed per day, but also the self-reported number of candies consumed by each participant. The following table summarizes the mean difference between these two values (reported minus actual):

Proximity	Visibility Clear	Opaque
Proximate	−1.2	−0.8
Less proximate	0.5	0.4

(a) Make a plot of the means and describe the patterns that you see. Does the plot suggest an interaction between visibility and proximity?

(b) This study actually took four weeks, with each participant being observed at each treatment combination in a random order. Explain why a repeated-measures design like this may be beneficial.

13.22 Bilingualism. Not only does speaking two languages have many practical benefits in this globalized world, but there is also growing evidence that it appears to help with brain functioning as we age. In one study, 80 participants were divided equally among 4 groups: younger adult bilinguals, older adult bilinguals, younger adult monolinguals, and older adult monolinguals.[15] Each participant was asked to complete a series of color-shape task-switching tests. For our analysis, we'll focus on the total reaction time (in microseconds) for these experiments. The shorter the reaction time, the better. BILING

(a) Make a table giving the sample size, mean, and standard deviation for each group. Is it reasonable to pool the variances?

(b) Generate a histogram for each of the groups. Can we feel confident that the sample means are approximately Normal? Explain your answer.

13.23 Bilingualism, continued. Refer to the previous exercise. BILING

(a) If bilingualism helps with brain functioning as we age, explain why we'd expect to find an interaction between age and lingualism. Also create an interaction plot of what sort of pattern we'd expect.

(b) Analyze the reaction times using analysis of variance. Report the test statistics, degrees of freedom, and P-values.

(c) Based on part (b), write a short paragraph summarizing your findings.

13.24 Hypotension and endurance exercise. In sedentary individuals, low blood pressure (hypotension) often occurs after a single bout of aerobic exercise and lasts nearly two hours. This can cause dizziness, light-headedness, and possibly fainting upon standing. It is thought that endurance exercise training can reduce the degree of postexercise hypotension. To test this, researchers studied 16 endurance-trained and 16 sedentary men and women.[16] The following table summarizes the postexercise systolic arterial pressure (mm Hg) after 60 minutes of upright cycling:

Group	n	$\bar{x}$	SE
Women, sedentary	8	100.7	3.4
Women, endurance	8	105.3	3.6
Men, sedentary	8	114.2	3.8
Men, endurance	8	110.2	2.3

(a) Make a plot similar to Figure 13.3 (page 707) with the systolic blood pressure on the y axis and training level on the x axis. Describe the pattern you see.

(b) From the table, one can show that SSA = 677.12, SSB = 0.72, SSAB = 147.92, and SSE = 2478, where A is the sex effect and B is the training level. Construct the ANOVA table with F statistics and degrees of freedom, and state your conclusions regarding main effects and interaction.

(c) The researchers also measured the before-exercise systolic blood pressure of the participants and looked at a model that incorporated both the pre- and postexercise values. Explain why it is likely to be beneficial to incorporate both measurements in the study.

13.25 The effect of humor. In advertising, humor is often used to overcome sales resistance and stimulate customer purchase behavior. One experiment looked at the use of humor to offset the negative feelings often associated with website encounters.[17] The setting of the experiment was an online travel agency, and the researchers used a three-factor design, each factor with two levels. The factors were humor (used, not used), process (favorable, unfavorable), and outcome (favorable, unfavorable). For the humor condition, cartoons and jokes of the day about skiing were presented on the site. For the no humor condition, standard pictures of ski sites were used. Two hundred and forty-one business students from a large Dutch university participated in the experiment. Each was randomly assigned to one of the eight treatment conditions. The students were asked to book a skiing holiday and then rate their perceived enjoyment and satisfaction with the process. All responses were measured on a seven-point Likert scale. A summary of the results for satisfaction follows:

Treatment	n	$\bar{x}$	s
No humor—favorable process— unfavorable outcome	27	3.04	0.79
No humor—favorable process— favorable outcome	29	5.36	0.47
No humor—unfavorable process— unfavorable outcome	26	2.84	0.59
No humor—unfavorable process— favorable outcome	31	3.08	0.59
Humor—favorable process— unfavorable outcome	32	5.06	0.59
Humor—favorable process— favorable outcome	30	5.55	0.65
Humor—unfavorable process— unfavorable outcome	36	1.95	0.52
Humor—unfavorable process— favorable outcome	30	3.27	0.71

(a) Plot the means of the four treatments without humor. Do you think there is an interaction? If yes, describe the interaction in terms of the process and outcome factors.

(b) Plot the means of the four treatments that used humor. Do you think there is an interaction? If yes, describe the interaction in terms of the process and outcome factors.

(c) The three-factor interaction can be assessed by looking at the two interaction plots created in parts (a) and (b). If the relationship between process and outcome is different across the two humor conditions, there is evidence of an interaction among all three factors. Do you think there is a three-factor interaction? Explain your answer.

13.26 Pooling the standard deviations. Refer to the previous exercise. Find the pooled estimate of the

standard deviation for these data. What are its degrees of freedom? Using the rule from Chapter 12 (page 654), is it reasonable to use a pooled standard deviation for the analysis? Explain your answer.

13.27 Describing the effects. Refer to Exercise 13.25. The P-values for all main effects and two-factor interactions are significant at the 0.05 level. Using the table, find the marginal means (that is, the mean for the no humor treatment, the mean for the no humor and unfavorable process treatment combination, etc.) and use them to describe these effects.

13.28 Acceptance of functional foods. Functional foods are foods that are fortified with health-promoting supplements, like calcium-enriched orange juice or vitamin-enriched cereal. Although the number of functional foods is growing in the marketplace, very little is known about how the next generation of consumers views these foods. Because of this, a questionnaire was given to college students from the United States, Canada, and France.[18] This questionnaire measured the students' attitudes and beliefs about general food and functional food. One of the response variables collected concerned cooking enjoyment. This variable was the average of numerous items, each measured on a 10-point scale, where 1 = most negative value and 10 = most positive value. Here are the means:

	Culture		
Sex	Canada	United States	France
Female	7.70	7.36	6.38
Male	6.39	6.43	5.69

(a) Make a plot of the means and describe the patterns that you see.

(b) Does the plot suggest that there is an interaction between culture and sex? If your answer is Yes, describe the interaction.

13.29 Estimating the within-group variance. Refer to the previous exercise. Here are the cell standard deviations and sample sizes for cooking enjoyment:

	Culture					
	Canada		United States		France	
Sex	s	n	s	n	s	n
Female	1.668	238	1.736	178	2.024	82
Male	1.909	125	1.601	101	1.875	87

Find the pooled estimate of the standard deviation for these data. Use the rule for examining standard deviations in ANOVA from Chapter 12 (page 654) to determine if it is reasonable to use a pooled standard deviation for the analysis of these data.

13.30 Comparing the groups. Refer to Exercises 13.28 and 13.29. The researchers presented a table of means with different superscripts indicating pairs of means that differed at the 0.05 significance level, using the Bonferroni method.

(a) What denominator degrees of freedom would be used here?

(b) How many pairwise comparisons are there for this problem?

(c) Perform these comparisons using $t^{**} = 2.94$ and summarize your results.

13.31 More on acceptance of functional foods. Refer to Exercise 13.28. The means for four of the response variables associated with functional foods are as follows:

	General attitude			Product benefits		
	Culture			Culture		
Sex	Canada	United States	France	Canada	United States	France
Female	4.93	4.69	4.10	4.59	4.37	3.91
Male	4.50	4.43	4.02	4.20	4.09	3.87

	Credibility of information			Purchase intention		
	Culture			Culture		
Sex	Canada	United States	France	Canada	United States	France
Female	4.54	4.50	3.76	4.29	4.39	3.30
Male	4.23	3.99	3.83	4.11	3.86	3.41

For each of the four response variables, give a graphical summary of the means. Use this summary to discuss any interactions that are evident. Write a short report summarizing any differences in culture and sex with respect to the response variables measured.

13.32 Interpreting the results. The goal of the study in the previous exercise was to understand cultural and sex differences in functional food attitudes and behaviors among young adults, the next generation of food consumers. The researchers used a sample of undergraduate students and had each participant fill out the survey during class time. How reasonable is it to generalize these results to the young adult population in these countries? Explain your answer.

13.33 What can you conclude? Analysis of data for a 3×3 ANOVA with three observations per cell gave the F statistics in the following table:

Effect	F
A	3.25
B	4.49
AB	2.14

What can you conclude from the information given?

13.34 What can you conclude? A study reported the following results for data analyzed using the methods that we studied in this chapter:

Effect	F	P-value
A	8.20	0.001
B	2.06	0.123
AB	3.68	0.006

(a) What can you conclude from the information given?

(b) What additional information would you need to write a summary of the results for this study?

13.35 Conspicuous consumption and men's testosterone levels. It is argued that conspicuous consumption is a means by which men communicate their social status to prospective mates. One study looked at changes in a male's testosterone level in response to fluctuations in his status created by the consumption of a product.[19] The products considered were a new and luxurious sports car and an old family sedan. Participants were asked to drive on either an isolated highway or a busy city street. A table of cell means and standard deviations for the change (post minus pre) in testosterone level follows:

	Location			
	Highway		City	
Car	$\bar{x}$	s	$\bar{x}$	s
Old sedan	0.03	0.12	−0.03	0.12
New sports car	0.15	0.14	0.13	0.13

(a) Make a plot of the means and describe the patterns that you see. Does the plot suggest an interaction between location and type of car?

(b) Compute the pooled standard error s_p, assuming equal sample sizes.

(c) The researchers wanted to test the following hypotheses:

(1) Testosterone levels will increase more in men who drive the new car.

(2) For men driving the new car, testosterone levels will increase more in men who drive in the city.

(3) For men driving the old car, testosterone levels will decrease less in men who drive the old car on the highway.

Write out the contrasts for each of these hypotheses.

(d) This study actually involved each male participating in all four combinations. Half of them drove the sedan first and the other half drove the sports car first. Explain why a repeated-measures design like this may be beneficial.

13.36 The effects of peer pressure on mathematics achievement. Researchers were interested in comparing the relationship between high achievement in mathematics and peer pressure across several countries.[20] They hypothesized that in countries where high achievement is not valued highly, considerable peer pressure may exist. A questionnaire was distributed to 14-year-olds from three countries (Germany, Canada, and Israel). One of the questions asked students to rate how often they fear being called a nerd or teacher's pet on a four-point scale (1 = never, 4 = frequently). The following table summarizes the response:

Country	Sex	n	$\bar{x}$
Germany	Female	336	1.62
Germany	Male	305	1.39
Israel	Female	205	1.87
Israel	Male	214	1.63
Canada	Female	301	1.91
Canada	Male	304	1.88

(a) The P-values for the interaction and the main effects for country and for sex are 0.016, 0.068, and 0.108, respectively. Using the table and P-values, summarize the results both graphically and numerically.

(b) The researchers contend that Germany does not value achievement as highly as Canada and Israel. Do the results from part (a) allow you to address their primary hypothesis? Explain.

(c) The students were also asked to indicate their current grade in mathematics on a six-point scale (1 = excellent, 6 = insufficient). How might both responses be used to address the researchers' primary hypothesis?

13.37 The effect of chromium on insulin metabolism. The amount of chromium in the diet has an effect on

the way the body processes insulin. In an experiment designed to study this phenomenon, four diets were fed to male rats. There were two factors. Chromium had two levels: low (L) and normal (N). The rats were allowed to eat as much as they wanted (M), or the total amount that they could eat was restricted (R). We call the second factor Eat. One of the variables measured was the amount of an enzyme called GITH.[21] The means for this response variable appear in the following table:

	Eat	
Chromium	M	R
L	4.545	5.175
N	4.425	5.317

(a) Make a plot of the mean GITH for these diets, with the factor Chromium on the x axis and GITH on the y axis. For each Eat group, connect the points for the two Chromium means.

(b) Describe the patterns you see. Does the amount of chromium in the diet appear to affect the GITH mean? Does restricting the diet rather than letting the rats eat as much as they want appear to have an effect? Is there an interaction?

(c) Compute the marginal means. Compute the differences between the M and R diets for each level of Chromium. Use this information to summarize numerically the patterns in the plot.

13.38 Use of animated agents in a multimedia environment. Multimedia learning environments are designed to enhance learning by providing a more hands-on and exploratory investigation of a topic. Often, animated agents (human-like characters) are used with the hope of enhancing social interaction with the software and thus improving learning. One group of researchers decided to investigate whether the presence of an agent and the type of verbal feedback provided were actually helpful.[22] To do this, they recruited 135 college students and randomly divided them among four groups: agent/simple feedback, agent/elaborate feedback, no agent/simple feedback, and no agent/elaborate feedback. The topic of the software was thermodynamics. The change in score on a 20-question test taken before and after using the software was the response. 📊 AGENT

(a) Make a table giving the sample size, mean, and standard deviation for each group.

(b) Use these means to construct an interaction plot. Describe the main effects for agent presence and for feedback type as well as their interaction.

(c) Analyze the change in score using ANOVA. Report the test statistics, degrees of freedom, and P-values.

(d) Use the residuals to check model assumptions. Are there any concerns? Explain your answer.

(e) Based on parts (b) and (c), write a short paragraph summarizing your findings.

13.39 Trust of individuals and groups. Trust is an essential element in any exchange of goods or services. The following trust game is often used to study trust experimentally:

> A sender starts with $X and can transfer any amount $x \leq X$ to a responder. The responder then gets $3x$ and can transfer any amount $y \leq 3x$ back to the sender. The game ends with final amounts $X - x + y$ and $3x - y$ for the sender and responder, respectively.

The value x is taken as a measure of the sender's trust, and the value $y/3x$ indicates the responder's trustworthiness. A recent study used this game to study the dynamics between individuals and groups of three.[23] The following table summarizes the average amount x sent by senders starting with $100:

Sender	Responder	n	$\bar{x}$	s
Individual	Individual	32	65.5	36.4
Individual	Group	25	76.3	31.2
Group	Individual	25	54.0	41.6
Group	Group	27	43.7	42.4

(a) Find the pooled estimate of the standard deviation for this study and its degrees of freedom.

(b) Is it reasonable to use a pooled standard deviation for the analysis? Explain your answer.

(c) Compute the marginal means.

(d) Plot the means. Do you think there is an interaction? If yes, describe it.

(e) The F statistics for sender, responder, and interaction are 9.05, 0.001, and 2.08, respectively. Compute the P-values and state your conclusions.

13.40 Does the type of cooking pot affect iron content? Iron-deficiency anemia is the most common form of malnutrition in developing countries, affecting about 50% of children and women and 25% of men. Iron pots for cooking foods had traditionally been used in many of these countries, but they have been largely replaced by aluminum pots, which are cheaper and lighter. Some research has suggested that food cooked in iron pots will contain more iron than food cooked in other types of pots. One study designed to investigate this issue compared the iron content of some Ethiopian foods cooked in aluminum, clay, and iron pots.[24] Foods considered were *yesiga wet'*, beef cut into small pieces and prepared with several Ethiopian spices; *shiro wet'*, a legume-based mixture of chickpea flour and Ethiopian spiced pepper; and *ye-atkilt allych'a*, a lightly spiced vegetable casserole. Four samples of each food were cooked in each type of pot. The iron in the food is measured in milligrams of iron per 100 grams of cooked food. The data are shown in Table 13.1. COOK

(a) Make a table giving the sample size, mean, and standard deviation for each type of pot. Is it reasonable to pool the variances? Although the standard deviations vary more than we would like, this is partially due to the small sample sizes, and we will proceed with the analysis of variance.

(b) Plot the means. Give a short summary of how the iron content of foods depends upon the cooking pot.

(c) Run the ANOVA. Give the ANOVA table, the F statistics with degrees of freedom and P-values, and your conclusions regarding the hypotheses about main effects and interactions.

13.41 Interpreting the results. Refer to the previous exercise. Although there is a statistically significant interaction, do you think that these data support the conclusion that foods cooked in iron pots contain more iron than foods cooked in aluminum or clay pots? Discuss.

13.42 Analysis using a one-way ANOVA. Refer to Exercise 13.40. Rerun the analysis as a one-way ANOVA with nine groups and four observations per group. Report the results of the F test. Examine differences in means using a multiple-comparisons procedure. Summarize

TABLE 13.1   Iron Content (mg/100g) of Food Cooked in Different Pots

Type of pot	Meat				Legumes				Vegetables			
Aluminum	1.77	2.36	1.96	2.14	2.40	2.17	2.41	2.34	1.03	1.53	1.07	1.30
Clay	2.27	1.28	2.48	2.68	2.41	2.43	2.57	2.48	1.55	0.79	1.68	1.82
Iron	5.27	5.17	4.06	4.22	3.69	3.43	3.84	3.72	2.45	2.99	2.80	2.92

your results and compare them with those you obtained in Exercise 13.40.

13.43 Examination of a drilling process. One step in the manufacture of large engines requires that holes of very precise dimensions be drilled. The tools that do the drilling are regularly examined and are adjusted to ensure that the holes meet the required specifications. Part of the examination involves measurement of the diameter of the drilling tool. A team studying the variation in the sizes of the drilled holes selected this measurement procedure as a possible cause of variation in the drilled holes. They decided to use a designed experiment as one part of this examination. Some of the data are given in Table 13.2. The diameters in millimeters (mm) of five tools were measured by the same operator at three times (8:00 A.M., 11:00 A.M., and 3:00 P.M.). Three measurements were taken on each tool at each time. The person taking the measurements could not tell which tool was being measured, and the measurements were taken in random order.[25] **DRILL**

(a) Make a table of means and standard deviations for each of the 5 × 3 combinations of the two factors.

(b) Plot the means and describe how the means vary with tool and time. Note that we expect the tools to have slightly different diameters. These will be adjusted as needed. It is the process of measuring the diameters that is important.

(c) Use a two-way ANOVA to analyze these data. Report the test statistics, degrees of freedom, and P-values for the significance tests.

(d) Write a short report summarizing your results.

TABLE 13.2   Tool Diameter Data

Tool	Time	Diameter (mm)		
1	1	25.030	25.030	25.032
1	2	25.028	25.028	25.028
1	3	25.026	25.026	25.026
2	1	25.016	25.018	25.016
2	2	25.022	25.020	25.018
2	3	25.016	25.016	25.016
3	1	25.005	25.008	25.006
3	2	25.012	25.012	25.014
3	3	25.010	25.010	25.008
4	1	25.012	25.012	25.012
4	2	25.018	25.020	25.020
4	3	25.010	25.014	25.018
5	1	24.996	24.998	24.998
5	2	25.006	25.006	25.006
5	3	25.000	25.002	24.999

13.44 Examination of a drilling process, continued. Refer to the previous exercise. Multiply each measurement by 0.04 to convert from millimeters to inches. Redo the plots and rerun the ANOVA using the transformed measurements. Summarize what parts of the analysis have changed and what parts have remained the same. **DRILL**

13.45 Do left-handed people live shorter lives than right-handed people? A study of this question examined a sample of 949 death records and contacted next of kin to determine handedness.[26] Note that there are many possible definitions of "left-handed." The researchers examined the effects of different definitions on the results of their analysis and found that their conclusions were not sensitive to the exact definition used. For the results presented here, people were defined to be right-handed if they wrote, drew, and threw a ball with the right hand. All others were defined to be left-handed. People were classified by sex (female or male) and handedness (left or right), and a 2 × 2 ANOVA was run with the age at death as the response variable. The F statistics were 22.36 (handedness), 37.44 (sex), and 2.10 (interaction). The following marginal mean ages at death (in years) were reported: 77.39 (females), 71.32 (males), 75.00 (right-handed), and 66.03 (left-handed).

(a) For each of the F statistics given, find the degrees of freedom and an approximate P-value. Summarize the results of these tests.

(b) Using the information given, write a short summary of the results of the study.

13.46 A radon exposure study. Scientists believe that exposure to the radioactive gas radon is associated with some types of cancers in the respiratory system. Radon from natural sources is present in many homes in the United States. A group of researchers decided to study the problem in dogs because dogs get similar types of cancers and are exposed to environments similar to those of their owners. Radon detectors are available for home monitoring, but the researchers wanted to obtain actual measures of the exposure of a sample of dogs. To do this, they placed the detectors in holders and attached them to the collars of the dogs. One problem was that the holders might, in some way, affect the radon readings. The researchers, therefore, devised a laboratory experiment to study the effects of the holders. Detectors from four series of production were available, so they used a two-way ANOVA design (series with four levels and holder with two, representing the presence or absence of a holder). All detectors were exposed to the same radon source and the radon measure in picocuries per liter was recorded.[27] The F statistic for the effect of series is 7.02, for holder it is 1.96, for the interaction it is 1.24, and N = 69.

(a) Using Table E or statistical software, find approximate P-values for the three test statistics. Summarize the results of these three significance tests.

(b) The mean radon readings for the four series were 330, 303, 302, and 295. The results of the significance test for series were of great concern to the researchers. Explain why.

13.47 A comparison of plant species under low water conditions. The PLANTS1 data file gives the percent of nitrogen in four different species of plants grown in a laboratory. The species are *Leucaena leucocephala*, *Acacia saligna*, *Prosopis juliflora*, and *Eucalyptus citriodora*. The researchers who collected these data were interested in commercially growing these plants in parts of the country of Jordan where there is very little rainfall. To examine the effect of water, they varied the amount per day from 50 millimeters (mm) to 650 mm in 100 mm increments. There were nine plants per species-by-water combination. Because the plants are to be used primarily for animal food, with some parts that can be consumed by people, a high nitrogen content is very desirable. PLANTS1

(a) Find the means for each species-by-water combination. Plot these means versus water for the four species, connecting the means for each species by lines. Describe the overall pattern.

(b) Find the standard deviations for each species-by-water combination. Is it reasonable to pool the standard deviations for this problem? Note that with sample sizes of size 9, we expect these standard deviations to be quite variable.

(c) Run the two-way analysis of variance. Give the results of the hypothesis tests for the main effects and the interaction.

13.48 Examination of the residuals. Refer to the previous exercise. Examine the residuals. Are there any unusual patterns or outliers? If you think that there are one or more points that are somewhat extreme, rerun the two-way analysis without these observations. Does this change the results in any substantial way? PLANTS1

13.49 Analysis using multiple one-way ANOVAs. Refer to Exercise 13.47. Run a separate one-way analysis of variance for each water level. If there is evidence that the species are not all the same, use a multiple-comparisons procedure to determine which pairs of species are significantly different. In what way, if any, do the differences appear to vary by water level? Write a short summary of your conclusions. PLANTS1

13.50 More on the analysis using multiple one-way ANOVAs. Refer to Exercise 13.47. Run a separate one-way analysis of variance for each species and summarize the results. Because the amount of water is a quantitative factor,

we can also analyze these data using regression. Run simple linear regressions separately for each species to predict nitrogen percent from water. Use plots to determine whether or not a line is a good way to approximate this relationship. Summarize the regression results and compare them with the one-way ANOVA results. PLANTS1

13.51 Another comparison of plant species under low water conditions. Refer to Exercise 13.47. Additional data collected by the same researchers according to a similar design are given in the PLANTS2 data file. Here, there are two response variables. They are fresh biomass and dry biomass. High values for both of these variables are desirable. The same four species and seven levels of water are used for this experiment. Here, however, there are four plants per species-by-water combination. Analyze each of the response variables in the PLANTS2 data file using the outline from Exercise 13.47. PLANTS2

13.52 Examination of the residuals. Perform the tasks described in Exercise 13.48 for the two response variables in the PLANTS2 data set. PLANTS2

13.53 Analysis using multiple one-way ANOVAs. Perform the tasks described in Exercise 13.49 for the two response variables in the PLANTS2 data set. PLANTS2

13.54 More on the analysis using multiple one-way ANOVAs. Perform the tasks described in Exercise 13.50 for the two response variables in the PLANTS2 data set. PLANTS2

13.55 Are insects more attracted to male plants? Some scientists wanted to determine if there are sex-related differences in the level of herbivory for the jack-in-the-pulpit, a spring-blooming perennial plant common in deciduous forests. A study was conducted in southern Maryland in forests associated with the Smithsonian Environmental Research Center (SERC).[28] To determine the effects of flowering and floral characteristics on herbivory, the researchers altered the floral morphology of male and female plants. The three levels of floral characteristics were (1) the spathes were completely removed; (2) in females, a gap was created in the base of the spathe, and in males, the gap was closed; (3) plants were not altered (control). The percent of leaf area damaged by thrips (an order of insects) between early May and mid-June was recorded for each of 30 plants per combination of sex and floral characteristic. Here is a table of means and standard deviations (in parentheses):

	Floral characteristic level		
Sex	1	2	3
Males	0.11 (0.081)	1.28 (0.088)	1.63 (0.382)
Females	0.02 (0.002)	0.58 (0.321)	0.20 (0.035)

(a) Give the degrees of freedom for the F statistics that are used to test for sex, floral characteristic, and the interaction.

(b) Describe the main effects and interaction using appropriate graphs.

(c) The researchers used the natural logarithm of percent area as the response in their analysis. Using the relationship between the means and standard deviations, explain why this was done.

13.56 Change-of-majors study: HSS. Refer to the data given for the change-of-majors study in the data file MAJORS. Consider sex and whether the students changed majors as the two factors. Analyze the data for HSS, the high school science grades. Your analysis should include a table of sample sizes, means, and standard deviations; Normal quantile plots; a plot of the means; and a two-way ANOVA using sex and major as the factors. Write a short summary of your conclusions. [ll] MAJORS

13.57 Change-of-majors study: HSE. Refer to the previous exercise. Analyze the data for HSE, the high school English grades. Your analysis should include a table of sample sizes, means, and standard deviations; Normal quantile plots; a plot of the means; and a two-way ANOVA using sex and major as the factors. Write a short summary of your conclusions. [ll] MAJORS

13.58 Change-of-majors study: GPA. Refer to Exercise 13.56. Analyze the data for GPA, the college grade point average. Your analysis should include a table of sample sizes, means, and standard deviations; Normal quantile plots; a plot of the means; and a two-way ANOVA using sex and major as the factors. Write a short summary of your conclusions. [ll] MAJORS

13.59 Change-of-majors study: SATV. Refer to Exercise 13.56. Analyze the data for SATV, the SAT Verbal score. Your analysis should include a table of sample sizes, means, and standard deviations; Normal quantile plots; a plot of the means; and a two-way ANOVA using sex and major as the factors. Write a short summary of your conclusions. [ll] MAJORS

13.60 Search the Internet. Search the Internet or your library to find a study that is interesting to you and uses a two-way ANOVA to analyze the data. First describe the question or questions of interest, and then give the details of how ANOVA was used to provide answers. Be sure to include how the study authors examined the assumptions for the analysis. Evaluate how well the authors used ANOVA in this study. If your evaluation finds the analysis deficient, make suggestions for how it could be improved.

Tables

Table A Standard Normal Probabilities

Table B Random Digits

Table C Binomial Probabilities

Table D t Distribution Critical Values

Table E F Distribution Critical Values

Table F χ^2 Distribution Critical Values

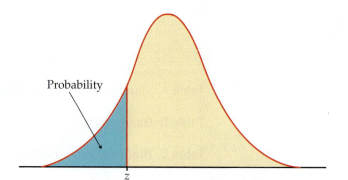

Table entry for z is the area
under the standard Normal
curve to the left of z.

TABLE A	Standard Normal Probabilities									
z	.00	.01	.02	.03	.04	.05	.06	.07	.08	.09
−3.4	.0003	.0003	.0003	.0003	.0003	.0003	.0003	.0003	.0003	.0002
−3.3	.0005	.0005	.0005	.0004	.0004	.0004	.0004	.0004	.0004	.0003
−3.2	.0007	.0007	.0006	.0006	.0006	.0006	.0006	.0005	.0005	.0005
−3.1	.0010	.0009	.0009	.0009	.0008	.0008	.0008	.0008	.0007	.0007
−3.0	.0013	.0013	.0013	.0012	.0012	.0011	.0011	.0011	.0010	.0010
−2.9	.0019	.0018	.0018	.0017	.0016	.0016	.0015	.0015	.0014	.0014
−2.8	.0026	.0025	.0024	.0023	.0023	.0022	.0021	.0021	.0020	.0019
−2.7	.0035	.0034	.0033	.0032	.0031	.0030	.0029	.0028	.0027	.0026
−2.6	.0047	.0045	.0044	.0043	.0041	.0040	.0039	.0038	.0037	.0036
−2.5	.0062	.0060	.0059	.0057	.0055	.0054	.0052	.0051	.0049	.0048
−2.4	.0082	.0080	.0078	.0075	.0073	.0071	.0069	.0068	.0066	.0064
−2.3	.0107	.0104	.0102	.0099	.0096	.0094	.0091	.0089	.0087	.0084
−2.2	.0139	.0136	.0132	.0129	.0125	.0122	.0119	.0116	.0113	.0110
−2.1	.0179	.0174	.0170	.0166	.0162	.0158	.0154	.0150	.0146	.0143
−2.0	.0228	.0222	.0217	.0212	.0207	.0202	.0197	.0192	.0188	.0183
−1.9	.0287	.0281	.0274	.0268	.0262	.0256	.0250	.0244	.0239	.0233
−1.8	.0359	.0351	.0344	.0336	.0329	.0322	.0314	.0307	.0301	.0294
−1.7	.0446	.0436	.0427	.0418	.0409	.0401	.0392	.0384	.0375	.0367
−1.6	.0548	.0537	.0526	.0516	.0505	.0495	.0485	.0475	.0465	.0455
−1.5	.0668	.0655	.0643	.0630	.0618	.0606	.0594	.0582	.0571	.0559
−1.4	.0808	.0793	.0778	.0764	.0749	.0735	.0721	.0708	.0694	.0681
−1.3	.0968	.0951	.0934	.0918	.0901	.0885	.0869	.0853	.0838	.0823
−1.2	.1151	.1131	.1112	.1093	.1075	.1056	.1038	.1020	.1003	.0985
−1.1	.1357	.1335	.1314	.1292	.1271	.1251	.1230	.1210	.1190	.1170
−1.0	.1587	.1562	.1539	.1515	.1492	.1469	.1446	.1423	.1401	.1379
−0.9	.1841	.1814	.1788	.1762	.1736	.1711	.1685	.1660	.1635	.1611
−0.8	.2119	.2090	.2061	.2033	.2005	.1977	.1949	.1922	.1894	.1867
−0.7	.2420	.2389	.2358	.2327	.2296	.2266	.2236	.2206	.2177	.2148
−0.6	.2743	.2709	.2676	.2643	.2611	.2578	.2546	.2514	.2483	.2451
−0.5	.3085	.3050	.3015	.2981	.2946	.2912	.2877	.2843	.2810	.2776
−0.4	.3446	.3409	.3372	.3336	.3300	.3264	.3228	.3192	.3156	.3121
−0.3	.3821	.3783	.3745	.3707	.3669	.3632	.3594	.3557	.3520	.3483
−0.2	.4207	.4168	.4129	.4090	.4052	.4013	.3974	.3936	.3897	.3859
−0.1	.4602	.4562	.4522	.4483	.4443	.4404	.4364	.4325	.4286	.4247
−0.0	.5000	.4960	.4920	.4880	.4840	.4801	.4761	.4721	.4681	.4641

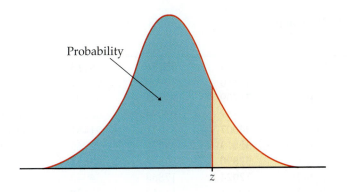

Table entry for z is the area
under the standard Normal
curve to the left of z.

Probability

z

TABLE A	Standard Normal Probabilities (continued)								
z	.00	.01	.02	.03	.04	.05	.06	.07	.08	.09
0.0	.5000	.5040	.5080	.5120	.5160	.5199	.5239	.5279	.5319	.5359
0.1	.5398	.5438	.5478	.5517	.5557	.5596	.5636	.5675	.5714	.5753
0.2	.5793	.5832	.5871	.5910	.5948	.5987	.6026	.6064	.6103	.6141
0.3	.6179	.6217	.6255	.6293	.6331	.6368	.6406	.6443	.6480	.6517
0.4	.6554	.6591	.6628	.6664	.6700	.6736	.6772	.6808	.6844	.6879
0.5	.6915	.6950	.6985	.7019	.7054	.7088	.7123	.7157	.7190	.7224
0.6	.7257	.7291	.7324	.7357	.7389	.7422	.7454	.7486	.7517	.7549
0.7	.7580	.7611	.7642	.7673	.7704	.7734	.7764	.7794	.7823	.7852
0.8	.7881	.7910	.7939	.7967	.7995	.8023	.8051	.8078	.8106	.8133
0.9	.8159	.8186	.8212	.8238	.8264	.8289	.8315	.8340	.8365	.8389
1.0	.8413	.8438	.8461	.8485	.8508	.8531	.8554	.8577	.8599	.8621
1.1	.8643	.8665	.8686	.8708	.8729	.8749	.8770	.8790	.8810	.8830
1.2	.8849	.8869	.8888	.8907	.8925	.8944	.8962	.8980	.8997	.9015
1.3	.9032	.9049	.9066	.9082	.9099	.9115	.9131	.9147	.9162	.9177
1.4	.9192	.9207	.9222	.9236	.9251	.9265	.9279	.9292	.9306	.9319
1.5	.9332	.9345	.9357	.9370	.9382	.9394	.9406	.9418	.9429	.9441
1.6	.9452	.9463	.9474	.9484	.9495	.9505	.9515	.9525	.9535	.9545
1.7	.9554	.9564	.9573	.9582	.9591	.9599	.9608	.9616	.9625	.9633
1.8	.9641	.9649	.9656	.9664	.9671	.9678	.9686	.9693	.9699	.9706
1.9	.9713	.9719	.9726	.9732	.9738	.9744	.9750	.9756	.9761	.9767
2.0	.9772	.9778	.9783	.9788	.9793	.9798	.9803	.9808	.9812	.9817
2.1	.9821	.9826	.9830	.9834	.9838	.9842	.9846	.9850	.9854	.9857
2.2	.9861	.9864	.9868	.9871	.9875	.9878	.9881	.9884	.9887	.9890
2.3	.9893	.9896	.9898	.9901	.9904	.9906	.9909	.9911	.9913	.9916
2.4	.9918	.9920	.9922	.9925	.9927	.9929	.9931	.9932	.9934	.9936
2.5	.9938	.9940	.9941	.9943	.9945	.9946	.9948	.9949	.9951	.9952
2.6	.9953	.9955	.9956	.9957	.9959	.9960	.9961	.9962	.9963	.9964
2.7	.9965	.9966	.9967	.9968	.9969	.9970	.9971	.9972	.9973	.9974
2.8	.9974	.9975	.9976	.9977	.9977	.9978	.9979	.9979	.9980	.9981
2.9	.9981	.9982	.9982	.9983	.9984	.9984	.9985	.9985	.9986	.9986
3.0	.9987	.9987	.9987	.9988	.9988	.9989	.9989	.9989	.9990	.9990
3.1	.9990	.9991	.9991	.9991	.9992	.9992	.9992	.9992	.9993	.9993
3.2	.9993	.9993	.9994	.9994	.9994	.9994	.9994	.9995	.9995	.9995
3.3	.9995	.9995	.9995	.9996	.9996	.9996	.9996	.9996	.9996	.9997
3.4	.9997	.9997	.9997	.9997	.9997	.9997	.9997	.9997	.9997	.9998

TABLE B	Random Digits						
Line							
101	19223	95034	05756	28713	96409	12531	42544	82853
102	73676	47150	99400	01927	27754	42648	82425	36290
103	45467	71709	77558	00095	32863	29485	82226	90056
104	52711	38889	93074	60227	40011	85848	48767	52573
105	95592	94007	69971	91481	60779	53791	17297	59335
106	68417	35013	15529	72765	85089	57067	50211	47487
107	82739	57890	20807	47511	81676	55300	94383	14893
108	60940	72024	17868	24943	61790	90656	87964	18883
109	36009	19365	15412	39638	85453	46816	83485	41979
110	38448	48789	18338	24697	39364	42006	76688	08708
111	81486	69487	60513	09297	00412	71238	27649	39950
112	59636	88804	04634	71197	19352	73089	84898	45785
113	62568	70206	40325	03699	71080	22553	11486	11776
114	45149	32992	75730	66280	03819	56202	02938	70915
115	61041	77684	94322	24709	73698	14526	31893	32592
116	14459	26056	31424	80371	65103	62253	50490	61181
117	38167	98532	62183	70632	23417	26185	41448	75532
118	73190	32533	04470	29669	84407	90785	65956	86382
119	95857	07118	87664	92099	58806	66979	98624	84826
120	35476	55972	39421	65850	04266	35435	43742	11937
121	71487	09984	29077	14863	61683	47052	62224	51025
122	13873	81598	95052	90908	73592	75186	87136	95761
123	54580	81507	27102	56027	55892	33063	41842	81868
124	71035	09001	43367	49497	72719	96758	27611	91596
125	96746	12149	37823	71868	18442	35119	62103	39244
126	96927	19931	36089	74192	77567	88741	48409	41903
127	43909	99477	25330	64359	40085	16925	85117	36071
128	15689	14227	06565	14374	13352	49367	81982	87209
129	36759	58984	68288	22913	18638	54303	00795	08727
130	69051	64817	87174	09517	84534	06489	87201	97245
131	05007	16632	81194	14873	04197	85576	45195	96565
132	68732	55259	84292	08796	43165	93739	31685	97150
133	45740	41807	65561	33302	07051	93623	18132	09547
134	27816	78416	18329	21337	35213	37741	04312	68508
135	66925	55658	39100	78458	11206	19876	87151	31260
136	08421	44753	77377	28744	75592	08563	79140	92454
137	53645	66812	61421	47836	12609	15373	98481	14592
138	66831	68908	40772	21558	47781	33586	79177	06928
139	55588	99404	70708	41098	43563	56934	48394	51719
140	12975	13258	13048	45144	72321	81940	00360	02428
141	96767	35964	23822	96012	94591	65194	50842	53372
142	72829	50232	97892	63408	77919	44575	24870	04178
143	88565	42628	17797	49376	61762	16953	88604	12724
144	62964	88145	83083	69453	46109	59505	69680	00900
145	19687	12633	57857	95806	09931	02150	43163	58636
146	37609	59057	66967	83401	60705	02384	90597	93600
147	54973	86278	88737	74351	47500	84552	19909	67181
148	00694	05977	19664	65441	20903	62371	22725	53340
149	71546	05233	53946	68743	72460	27601	45403	88692
150	07511	88915	41267	16853	84569	79367	32337	03316

TABLE B   Random Digits (continued)

Line								
151	03802	29341	29264	80198	12371	13121	54969	43912
152	77320	35030	77519	41109	98296	18984	60869	12349
153	07886	56866	39648	69290	03600	05376	58958	22720
154	87065	74133	21117	70595	22791	67306	28420	52067
155	42090	09628	54035	93879	98441	04606	27381	82637
156	55494	67690	88131	81800	11188	28552	25752	21953
157	16698	30406	96587	65985	07165	50148	16201	86792
158	16297	07626	68683	45335	34377	72941	41764	77038
159	22897	17467	17638	70043	36243	13008	83993	22869
160	98163	45944	34210	64158	76971	27689	82926	75957
161	43400	25831	06283	22138	16043	15706	73345	26238
162	97341	46254	88153	62336	21112	35574	99271	45297
163	64578	67197	28310	90341	37531	63890	52630	76315
164	11022	79124	49525	63078	17229	32165	01343	21394
165	81232	43939	23840	05995	84589	06788	76358	26622
166	36843	84798	51167	44728	20554	55538	27647	32708
167	84329	80081	69516	78934	14293	92478	16479	26974
168	27788	85789	41592	74472	96773	27090	24954	41474
169	99224	00850	43737	75202	44753	63236	14260	73686
170	38075	73239	52555	46342	13365	02182	30443	53229
171	87368	49451	55771	48343	51236	18522	73670	23212
172	40512	00681	44282	47178	08139	78693	34715	75606
173	81636	57578	54286	27216	58758	80358	84115	84568
174	26411	94292	06340	97762	37033	85968	94165	46514
175	80011	09937	57195	33906	94831	10056	42211	65491
176	92813	87503	63494	71379	76550	45984	05481	50830
177	70348	72871	63419	57363	29685	43090	18763	31714
178	24005	52114	26224	39078	80798	15220	43186	00976
179	85063	55810	10470	08029	30025	29734	61181	72090
180	11532	73186	92541	06915	72954	10167	12142	26492
181	59618	03914	05208	84088	20426	39004	84582	87317
182	92965	50837	39921	84661	82514	81899	24565	60874
183	85116	27684	14597	85747	01596	25889	41998	15635
184	15106	10411	90221	49377	44369	28185	80959	76355
185	03638	31589	07871	25792	85823	55400	56026	12193
186	97971	48932	45792	63993	95635	28753	46069	84635
187	49345	18305	76213	82390	77412	97401	50650	71755
188	87370	88099	89695	87633	76987	85503	26257	51736
189	88296	95670	74932	65317	93848	43988	47597	83044
190	79485	92200	99401	54473	34336	82786	05457	60343
191	40830	24979	23333	37619	56227	95941	59494	86539
192	32006	76302	81221	00693	95197	75044	46596	11628
193	37569	85187	44692	50706	53161	69027	88389	60313
194	56680	79003	23361	67094	15019	63261	24543	52884
195	05172	08100	22316	54495	60005	29532	18433	18057
196	74782	27005	03894	98038	20627	40307	47317	92759
197	85288	93264	61409	03404	09649	55937	60843	66167
198	68309	12060	14762	58002	03716	81968	57934	32624
199	26461	88346	52430	60906	74216	96263	69296	90107
200	42672	67680	42376	95023	82744	03971	96560	55148

TABLE C   Binomial Probabilities

Entry is $P(X = k) = \binom{n}{k} p^k (1-p)^{n-k}$

						p				
n	k	.01	.02	.03	.04	.05	.06	.07	.08	.09
2	0	.9801	.9604	.9409	.9216	.9025	.8836	.8649	.8464	.8281
	1	.0198	.0392	.0582	.0768	.0950	.1128	.1302	.1472	.1638
	2	.0001	.0004	.0009	.0016	.0025	.0036	.0049	.0064	.0081
3	0	.9703	.9412	.9127	.8847	.8574	.8306	.8044	.7787	.7536
	1	.0294	.0576	.0847	.1106	.1354	.1590	.1816	.2031	.2236
	2	.0003	.0012	.0026	.0046	.0071	.0102	.0137	.0177	.0221
	3				.0001	.0001	.0002	.0003	.0005	.0007
4	0	.9606	.9224	.8853	.8493	.8145	.7807	.7481	.7164	.6857
	1	.0388	.0753	.1095	.1416	.1715	.1993	.2252	.2492	.2713
	2	.0006	.0023	.0051	.0088	.0135	.0191	.0254	.0325	.0402
	3			.0001	.0002	.0005	.0008	.0013	.0019	.0027
	4									.0001
5	0	.9510	.9039	.8587	.8154	.7738	.7339	.6957	.6591	.6240
	1	.0480	.0922	.1328	.1699	.2036	.2342	.2618	.2866	.3086
	2	.0010	.0038	.0082	.0142	.0214	.0299	.0394	.0498	.0610
	3		.0001	.0003	.0006	.0011	.0019	.0030	.0043	.0060
	4						.0001	.0001	.0002	.0003
	5									
6	0	.9415	.8858	.8330	.7828	.7351	.6899	.6470	.6064	.5679
	1	.0571	.1085	.1546	.1957	.2321	.2642	.2922	.3164	.3370
	2	.0014	.0055	.0120	.0204	.0305	.0422	.0550	.0688	.0833
	3		.0002	.0005	.0011	.0021	.0036	.0055	.0080	.0110
	4					.0001	.0002	.0003	.0005	.0008
	5									
	6									
7	0	.9321	.8681	.8080	.7514	.6983	.6485	.6017	.5578	.5168
	1	.0659	.1240	.1749	.2192	.2573	.2897	.3170	.3396	.3578
	2	.0020	.0076	.0162	.0274	.0406	.0555	.0716	.0886	.1061
	3		.0003	.0008	.0019	.0036	.0059	.0090	.0128	.0175
	4				.0001	.0002	.0004	.0007	.0011	.0017
	5								.0001	.0001
	6									
	7									
8	0	.9227	.8508	.7837	.7214	.6634	.6096	.5596	.5132	.4703
	1	.0746	.1389	.1939	.2405	.2793	.3113	.3370	.3570	.3721
	2	.0026	.0099	.0210	.0351	.0515	.0695	.0888	.1087	.1288
	3	.0001	.0004	.0013	.0029	.0054	.0089	.0134	.0189	.0255
	4			.0001	.0002	.0004	.0007	.0013	.0021	.0031
	5							.0001	.0001	.0002
	6									
	7									
	8									

TABLE C	Binomial Probabilities (continued)

Entry is $P(X = k) = \binom{n}{k}p^k(1-p)^{n-k}$

						p				
n	k	.10	.15	.20	.25	.30	.35	.40	.45	.50
2	0	.8100	.7225	.6400	.5625	.4900	.4225	.3600	.3025	.2500
	1	.1800	.2550	.3200	.3750	.4200	.4550	.4800	.4950	.5000
	2	.0100	.0225	.0400	.0625	.0900	.1225	.1600	.2025	.2500
3	0	.7290	.6141	.5120	.4219	.3430	.2746	.2160	.1664	.1250
	1	.2430	.3251	.3840	.4219	.4410	.4436	.4320	.4084	.3750
	2	.0270	.0574	.0960	.1406	.1890	.2389	.2880	.3341	.3750
	3	.0010	.0034	.0080	.0156	.0270	.0429	.0640	.0911	.1250
4	0	.6561	.5220	.4096	.3164	.2401	.1785	.1296	.0915	.0625
	1	.2916	.3685	.4096	.4219	.4116	.3845	.3456	.2995	.2500
	2	.0486	.0975	.1536	.2109	.2646	.3105	.3456	.3675	.3750
	3	.0036	.0115	.0256	.0469	.0756	.1115	.1536	.2005	.2500
	4	.0001	.0005	.0016	.0039	.0081	.0150	.0256	.0410	.0625
5	0	.5905	.4437	.3277	.2373	.1681	.1160	.0778	.0503	.0313
	1	.3280	.3915	.4096	.3955	.3602	.3124	.2592	.2059	.1563
	2	.0729	.1382	.2048	.2637	.3087	.3364	.3456	.3369	.3125
	3	.0081	.0244	.0512	.0879	.1323	.1811	.2304	.2757	.3125
	4	.0004	.0022	.0064	.0146	.0284	.0488	.0768	.1128	.1562
	5		.0001	.0003	.0010	.0024	.0053	.0102	.0185	.0312
6	0	.5314	.3771	.2621	.1780	.1176	.0754	.0467	.0277	.0156
	1	.3543	.3993	.3932	.3560	.3025	.2437	.1866	.1359	.0938
	2	.0984	.1762	.2458	.2966	.3241	.3280	.3110	.2780	.2344
	3	.0146	.0415	.0819	.1318	.1852	.2355	.2765	.3032	.3125
	4	.0012	.0055	.0154	.0330	.0595	.0951	.1382	.1861	.2344
	5	.0001	.0004	.0015	.0044	.0102	.0205	.0369	.0609	.0937
	6			.0001	.0002	.0007	.0018	.0041	.0083	.0156
7	0	.4783	.3206	.2097	.1335	.0824	.0490	.0280	.0152	.0078
	1	.3720	.3960	.3670	.3115	.2471	.1848	.1306	.0872	.0547
	2	.1240	.2097	.2753	.3115	.3177	.2985	.2613	.2140	.1641
	3	.0230	.0617	.1147	.1730	.2269	.2679	.2903	.2918	.2734
	4	.0026	.0109	.0287	.0577	.0972	.1442	.1935	.2388	.2734
	5	.0002	.0012	.0043	.0115	.0250	.0466	.0774	.1172	.1641
	6		.0001	.0004	.0013	.0036	.0084	.0172	.0320	.0547
	7				.0001	.0002	.0006	.0016	.0037	.0078
8	0	.4305	.2725	.1678	.1001	.0576	.0319	.0168	.0084	.0039
	1	.3826	.3847	.3355	.2670	.1977	.1373	.0896	.0548	.0313
	2	.1488	.2376	.2936	.3115	.2965	.2587	.2090	.1569	.1094
	3	.0331	.0839	.1468	.2076	.2541	.2786	.2787	.2568	.2188
	4	.0046	.0185	.0459	.0865	.1361	.1875	.2322	.2627	.2734
	5	.0004	.0026	.0092	.0231	.0467	.0808	.1239	.1719	.2188
	6		.0002	.0011	.0038	.0100	.0217	.0413	.0703	.1094
	7			.0001	.0004	.0012	.0033	.0079	.0164	.0312
	8					.0001	.0002	.0007	.0017	.0039

(*Continued*)

TABLE C   Binomial Probabilities (continued)

Entry is $P(X = k) = \binom{n}{k} p^k (1 - p)^{n - k}$

n	k	.01	.02	.03	.04	.05	.06	.07	.08	.09
						p				
9	0	.9135	.8337	.7602	.6925	.6302	.5730	.5204	.4722	.4279
	1	.0830	.1531	.2116	.2597	.2985	.3292	.3525	.3695	.3809
	2	.0034	.0125	.0262	.0433	.0629	.0840	.1061	.1285	.1507
	3	.0001	.0006	.0019	.0042	.0077	.0125	.0186	.0261	.0348
	4			.0001	.0003	.0006	.0012	.0021	.0034	.0052
	5						.0001	.0002	.0003	.0005
	6									
	7									
	8									
	9									
10	0	.9044	.8171	.7374	.6648	.5987	.5386	.4840	.4344	.3894
	1	.0914	.1667	.2281	.2770	.3151	.3438	.3643	.3777	.3851
	2	.0042	.0153	.0317	.0519	.0746	.0988	.1234	.1478	.1714
	3	.0001	.0008	.0026	.0058	.0105	.0168	.0248	.0343	.0452
	4			.0001	.0004	.0010	.0019	.0033	.0052	.0078
	5					.0001	.0001	.0003	.0005	.0009
	6									.0001
	7									
	8									
	9									
	10									
12	0	.8864	.7847	.6938	.6127	.5404	.4759	.4186	.3677	.3225
	1	.1074	.1922	.2575	.3064	.3413	.3645	.3781	.3837	.3827
	2	.0060	.0216	.0438	.0702	.0988	.1280	.1565	.1835	.2082
	3	.0002	.0015	.0045	.0098	.0173	.0272	.0393	.0532	.0686
	4		.0001	.0003	.0009	.0021	.0039	.0067	.0104	.0153
	5				.0001	.0002	.0004	.0008	.0014	.0024
	6							.0001	.0001	.0003
	7									
	8									
	9									
	10									
	11									
	12									
15	0	.8601	.7386	.6333	.5421	.4633	.3953	.3367	.2863	.2430
	1	.1303	.2261	.2938	.3388	.3658	.3785	.3801	.3734	.3605
	2	.0092	.0323	.0636	.0988	.1348	.1691	.2003	.2273	.2496
	3	.0004	.0029	.0085	.0178	.0307	.0468	.0653	.0857	.1070
	4		.0002	.0008	.0022	.0049	.0090	.0148	.0223	.0317
	5			.0001	.0002	.0006	.0013	.0024	.0043	.0069
	6						.0001	.0003	.0006	.0011
	7								.0001	.0001
	8									
	9									
	10									
	11									
	12									
	13									
	14									
	15									

TABLE C   Binomial Probabilities (continued)

Entry is $P(X = k) = \binom{n}{k}p^k(1 - p)^{n - k}$

n	k	.10	.15	.20	.25	.30	.35	.40	.45	.50
9	0	.3874	.2316	.1342	.0751	.0404	.0207	.0101	.0046	.0020
	1	.3874	.3679	.3020	.2253	.1556	.1004	.0605	.0339	.0176
	2	.1722	.2597	.3020	.3003	.2668	.2162	.1612	.1110	.0703
	3	.0446	.1069	.1762	.2336	.2668	.2716	.2508	.2119	.1641
	4	.0074	.0283	.0661	.1168	.1715	.2194	.2508	.2600	.2461
	5	.0008	.0050	.0165	.0389	.0735	.1181	.1672	.2128	.2461
	6	.0001	.0006	.0028	.0087	.0210	.0424	.0743	.1160	.1641
	7			.0003	.0012	.0039	.0098	.0212	.0407	.0703
	8				.0001	.0004	.0013	.0035	.0083	.0176
	9						.0001	.0003	.0008	.0020
10	0	.3487	.1969	.1074	.0563	.0282	.0135	.0060	.0025	.0010
	1	.3874	.3474	.2684	.1877	.1211	.0725	.0403	.0207	.0098
	2	.1937	.2759	.3020	.2816	.2335	.1757	.1209	.0763	.0439
	3	.0574	.1298	.2013	.2503	.2668	.2522	.2150	.1665	.1172
	4	.0112	.0401	.0881	.1460	.2001	.2377	.2508	.2384	.2051
	5	.0015	.0085	.0264	.0584	.1029	.1536	.2007	.2340	.2461
	6	.0001	.0012	.0055	.0162	.0368	.0689	.1115	.1596	.2051
	7		.0001	.0008	.0031	.0090	.0212	.0425	.0746	.1172
	8			.0001	.0004	.0014	.0043	.0106	.0229	.0439
	9					.0001	.0005	.0016	.0042	.0098
	10							.0001	.0003	.0010
12	0	.2824	.1422	.0687	.0317	.0138	.0057	.0022	.0008	.0002
	1	.3766	.3012	.2062	.1267	.0712	.0368	.0174	.0075	.0029
	2	.2301	.2924	.2835	.2323	.1678	.1088	.0639	.0339	.0161
	3	.0852	.1720	.2362	.2581	.2397	.1954	.1419	.0923	.0537
	4	.0213	.0683	.1329	.1936	.2311	.2367	.2128	.1700	.1208
	5	.0038	.0193	.0532	.1032	.1585	.2039	.2270	.2225	.1934
	6	.0005	.0040	.0155	.0401	.0792	.1281	.1766	.2124	.2256
	7		.0006	.0033	.0115	.0291	.0591	.1009	.1489	.1934
	8		.0001	.0005	.0024	.0078	.0199	.0420	.0762	.1208
	9			.0001	.0004	.0015	.0048	.0125	.0277	.0537
	10					.0002	.0008	.0025	.0068	.0161
	11						.0001	.0003	.0010	.0029
	12								.0001	.0002
15	0	.2059	.0874	.0352	.0134	.0047	.0016	.0005	.0001	.0000
	1	.3432	.2312	.1319	.0668	.0305	.0126	.0047	.0016	.0005
	2	.2669	.2856	.2309	.1559	.0916	.0476	.0219	.0090	.0032
	3	.1285	.2184	.2501	.2252	.1700	.1110	.0634	.0318	.0139
	4	.0428	.1156	.1876	.2252	.2186	.1792	.1268	.0780	.0417
	5	.0105	.0449	.1032	.1651	.2061	.2123	.1859	.1404	.0916
	6	.0019	.0132	.0430	.0917	.1472	.1906	.2066	.1914	.1527
	7	.0003	.0030	.0138	.0393	.0811	.1319	.1771	.2013	.1964
	8		.0005	.0035	.0131	.0348	.0710	.1181	.1647	.1964
	9		.0001	.0007	.0034	.0116	.0298	.0612	.1048	.1527
	10			.0001	.0007	.0030	.0096	.0245	.0515	.0916
	11				.0001	.0006	.0024	.0074	.0191	.0417
	12					.0001	.0004	.0016	.0052	.0139
	13						.0001	.0003	.0010	.0032
	14								.0001	.0005
	15									

(Continued)

TABLE C	Binomial Probabilities (continued)								

						p				
n	k	.01	.02	.03	.04	.05	.06	.07	.08	.09
20	0	.8179	.6676	.5438	.4420	.3585	.2901	.2342	.1887	.1516
	1	.1652	.2725	.3364	.3683	.3774	.3703	.3526	.3282	.3000
	2	.0159	.0528	.0988	.1458	.1887	.2246	.2521	.2711	.2818
	3	.0010	.0065	.0183	.0364	.0596	.0860	.1139	.1414	.1672
	4		.0006	.0024	.0065	.0133	.0233	.0364	.0523	.0703
	5			.0002	.0009	.0022	.0048	.0088	.0145	.0222
	6				.0001	.0003	.0008	.0017	.0032	.0055
	7						.0001	.0002	.0005	.0011
	8								.0001	.0002
	9									
	10									
	11									
	12									
	13									
	14									
	15									
	16									
	17									
	18									
	19									
	20									

						p				
n	k	.10	.15	.20	.25	.30	.35	.40	.45	.50
20	0	.1216	.0388	.0115	.0032	.0008	.0002	.0000	.0000	.0000
	1	.2702	.1368	.0576	.0211	.0068	.0020	.0005	.0001	.0000
	2	.2852	.2293	.1369	.0669	.0278	.0100	.0031	.0008	.0002
	3	.1901	.2428	.2054	.1339	.0716	.0323	.0123	.0040	.0011
	4	.0898	.1821	.2182	.1897	.1304	.0738	.0350	.0139	.0046
	5	.0319	.1028	.1746	.2023	.1789	.1272	.0746	.0365	.0148
	6	.0089	.0454	.1091	.1686	.1916	.1712	.1244	.0746	.0370
	7	.0020	.0160	.0545	.1124	.1643	.1844	.1659	.1221	.0739
	8	.0004	.0046	.0222	.0609	.1144	.1614	.1797	.1623	.1201
	9	.0001	.0011	.0074	.0271	.0654	.1158	.1597	.1771	.1602
	10		.0002	.0020	.0099	.0308	.0686	.1171	.1593	.1762
	11			.0005	.0030	.0120	.0336	.0710	.1185	.1602
	12			.0001	.0008	.0039	.0136	.0355	.0727	.1201
	13				.0002	.0010	.0045	.0146	.0366	.0739
	14					.0002	.0012	.0049	.0150	.0370
	15						.0003	.0013	.0049	.0148
	16							.0003	.0013	.0046
	17								.0002	.0011
	18									.0002
	19									
	20									

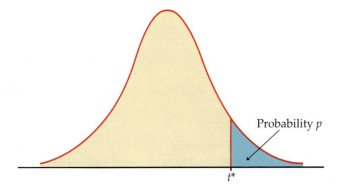

Table entry for *p* and *C* is the critical value *t** with probability *p* lying to its right and probability *C* lying between −*t** and *t**.

Probability *p*

t*

TABLE D	*t* Distribution Critical Values										
					Upper-tail probability *p*						
df	.25	.20	.15	.10	.05	.025	.02	.01	.005	.0025	.001	.0005
1	1.000	1.376	1.963	3.078	6.314	12.71	15.89	31.82	63.66	127.3	318.3	636.6
2	0.816	1.061	1.386	1.886	2.920	4.303	4.849	6.965	9.925	14.09	22.33	31.60
3	0.765	0.978	1.250	1.638	2.353	3.182	3.482	4.541	5.841	7.453	10.21	12.92
4	0.741	0.941	1.190	1.533	2.132	2.776	2.999	3.747	4.604	5.598	7.173	8.610
5	0.727	0.920	1.156	1.476	2.015	2.571	2.757	3.365	4.032	4.773	5.893	6.869
6	0.718	0.906	1.134	1.440	1.943	2.447	2.612	3.143	3.707	4.317	5.208	5.959
7	0.711	0.896	1.119	1.415	1.895	2.365	2.517	2.998	3.499	4.029	4.785	5.408
8	0.706	0.889	1.108	1.397	1.860	2.306	2.449	2.896	3.355	3.833	4.501	5.041
9	0.703	0.883	1.100	1.383	1.833	2.262	2.398	2.821	3.250	3.690	4.297	4.781
10	0.700	0.879	1.093	1.372	1.812	2.228	2.359	2.764	3.169	3.581	4.144	4.587
11	0.697	0.876	1.088	1.363	1.796	2.201	2.328	2.718	3.106	3.497	4.025	4.437
12	0.695	0.873	1.083	1.356	1.782	2.179	2.303	2.681	3.055	3.428	3.930	4.318
13	0.694	0.870	1.079	1.350	1.771	2.160	2.282	2.650	3.012	3.372	3.852	4.221
14	0.692	0.868	1.076	1.345	1.761	2.145	2.264	2.624	2.977	3.326	3.787	4.140
15	0.691	0.866	1.074	1.341	1.753	2.131	2.249	2.602	2.947	3.286	3.733	4.073
16	0.690	0.865	1.071	1.337	1.746	2.120	2.235	2.583	2.921	3.252	3.686	4.015
17	0.689	0.863	1.069	1.333	1.740	2.110	2.224	2.567	2.898	3.222	3.646	3.965
18	0.688	0.862	1.067	1.330	1.734	2.101	2.214	2.552	2.878	3.197	3.611	3.922
19	0.688	0.861	1.066	1.328	1.729	2.093	2.205	2.539	2.861	3.174	3.579	3.883
20	0.687	0.860	1.064	1.325	1.725	2.086	2.197	2.528	2.845	3.153	3.552	3.850
21	0.686	0.859	1.063	1.323	1.721	2.080	2.189	2.518	2.831	3.135	3.527	3.819
22	0.686	0.858	1.061	1.321	1.717	2.074	2.183	2.508	2.819	3.119	3.505	3.792
23	0.685	0.858	1.060	1.319	1.714	2.069	2.177	2.500	2.807	3.104	3.485	3.768
24	0.685	0.857	1.059	1.318	1.711	2.064	2.172	2.492	2.797	3.091	3.467	3.745
25	0.684	0.856	1.058	1.316	1.708	2.060	2.167	2.485	2.787	3.078	3.450	3.725
26	0.684	0.856	1.058	1.315	1.706	2.056	2.162	2.479	2.779	3.067	3.435	3.707
27	0.684	0.855	1.057	1.314	1.703	2.052	2.158	2.473	2.771	3.057	3.421	3.690
28	0.683	0.855	1.056	1.313	1.701	2.048	2.154	2.467	2.763	3.047	3.408	3.674
29	0.683	0.854	1.055	1.311	1.699	2.045	2.150	2.462	2.756	3.038	3.396	3.659
30	0.683	0.854	1.055	1.310	1.697	2.042	2.147	2.457	2.750	3.030	3.385	3.646
40	0.681	0.851	1.050	1.303	1.684	2.021	2.123	2.423	2.704	2.971	3.307	3.551
50	0.679	0.849	1.047	1.299	1.676	2.009	2.109	2.403	2.678	2.937	3.261	3.496
60	0.679	0.848	1.045	1.296	1.671	2.000	2.099	2.390	2.660	2.915	3.232	3.460
80	0.678	0.846	1.043	1.292	1.664	1.990	2.088	2.374	2.639	2.887	3.195	3.416
100	0.677	0.845	1.042	1.290	1.660	1.984	2.081	2.364	2.626	2.871	3.174	3.390
1000	0.675	0.842	1.037	1.282	1.646	1.962	2.056	2.330	2.581	2.813	3.098	3.300
z*	0.674	0.841	1.036	1.282	1.645	1.960	2.054	2.326	2.576	2.807	3.091	3.291
	50%	60%	70%	80%	90%	95%	96%	98%	99%	99.5%	99.8%	99.9%
					Confidence level *C*							

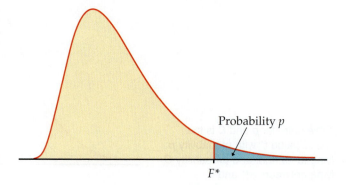

Table entry for p is the critical value F^* with probability p lying to its right.

Probability p

F^*

TABLE E   F Distribution Critical Values

		Degrees of freedom in the numerator								
	p	1	2	3	4	5	6	7	8	9
1	.100	39.86	49.50	53.59	55.83	57.24	58.20	58.91	59.44	59.86
	.050	161.45	199.50	215.71	224.58	230.16	233.99	236.77	238.88	240.54
	.025	647.79	799.50	864.16	899.58	921.85	937.11	948.22	956.66	963.28
	.010	4052.2	4999.5	5403.4	5624.6	5763.6	5859.0	5928.4	5981.1	6022.5
	.001	405284	500000	540379	562500	576405	585937	592873	598144	602284
2	.100	8.53	9.00	9.16	9.24	9.29	9.33	9.35	9.37	9.38
	.050	18.51	19.00	19.16	19.25	19.30	19.33	19.35	19.37	19.38
	.025	38.51	39.00	39.17	39.25	39.30	39.33	39.36	39.37	39.39
	.010	98.50	99.00	99.17	99.25	99.30	99.33	99.36	99.37	99.39
	.001	998.50	999.00	999.17	999.25	999.30	999.33	999.36	999.37	999.39
3	.100	5.54	5.46	5.39	5.34	5.31	5.28	5.27	5.25	5.24
	.050	10.13	9.55	9.28	9.12	9.01	8.94	8.89	8.85	8.81
	.025	17.44	16.04	15.44	15.10	14.88	14.73	14.62	14.54	14.47
	.010	34.12	30.82	29.46	28.71	28.24	27.91	27.67	27.49	27.35
	.001	167.03	148.50	141.11	137.10	134.58	132.85	131.58	130.62	129.86
4	.100	4.54	4.32	4.19	4.11	4.05	4.01	3.98	3.95	3.94
	.050	7.71	6.94	6.59	6.39	6.26	6.16	6.09	6.04	6.00
	.025	12.22	10.65	9.98	9.60	9.36	9.20	9.07	8.98	8.90
	.010	21.20	18.00	16.69	15.98	15.52	15.21	14.98	14.80	14.66
	.001	74.14	61.25	56.18	53.44	51.71	50.53	49.66	49.00	48.47
5	.100	4.06	3.78	3.62	3.52	3.45	3.40	3.37	3.34	3.32
	.050	6.61	5.79	5.41	5.19	5.05	4.95	4.88	4.82	4.77
	.025	10.01	8.43	7.76	7.39	7.15	6.98	6.85	6.76	6.68
	.010	16.26	13.27	12.06	11.39	10.97	10.67	10.46	10.29	10.16
	.001	47.18	37.12	33.20	31.09	29.75	28.83	28.16	27.65	27.24
6	.100	3.78	3.46	3.29	3.18	3.11	3.05	3.01	2.98	2.96
	.050	5.99	5.14	4.76	4.53	4.39	4.28	4.21	4.15	4.10
	.025	8.81	7.26	6.60	6.23	5.99	5.82	5.70	5.60	5.52
	.010	13.75	10.92	9.78	9.15	8.75	8.47	8.26	8.10	7.98
	.001	35.51	27.00	23.70	21.92	20.80	20.03	19.46	19.03	18.69
7	.100	3.59	3.26	3.07	2.96	2.88	2.83	2.78	2.75	2.72
	.050	5.59	4.74	4.35	4.12	3.97	3.87	3.79	3.73	3.68
	.025	8.07	6.54	5.89	5.52	5.29	5.12	4.99	4.90	4.82
	.010	12.25	9.55	8.45	7.85	7.46	7.19	6.99	6.84	6.72
	.001	29.25	21.69	18.77	17.20	16.21	15.52	15.02	14.63	14.33

Degrees of freedom in the denominator

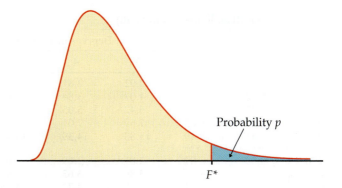

Table entry for *p* is the critical value *F** with probability *p* lying to its right.

Probability *p*

F*

TABLE E	*F* Distribution Critical Values (continued)									
			Degrees of freedom in the numerator							
10	12	15	20	25	30	40	50	60	120	1000
60.19	60.71	61.22	61.74	62.05	62.26	62.53	62.69	62.79	63.06	63.30
241.88	243.91	245.95	248.01	249.26	250.10	251.14	251.77	252.20	253.25	254.19
968.63	976.71	984.87	993.10	998.08	1001.4	1005.6	1008.1	1009.8	1014.0	1017.7
6055.8	6106.3	6157.3	6208.7	6239.8	6260.6	6286.8	6302.5	6313.0	6339.4	6362.7
605621	610668	615764	620908	624017	626099	628712	630285	631337	633972	636301
9.39	9.41	9.42	9.44	9.45	9.46	9.47	9.47	9.47	9.48	9.49
19.40	19.41	19.43	19.45	19.46	19.46	19.47	19.48	19.48	19.49	19.49
39.40	39.41	39.43	39.45	39.46	39.46	39.47	39.48	39.48	39.49	39.50
99.40	99.42	99.43	99.45	99.46	99.47	99.47	99.48	99.48	99.49	99.50
999.40	999.42	999.43	999.45	999.46	999.47	999.47	999.48	999.48	999.49	999.50
5.23	5.22	5.20	5.18	5.17	5.17	5.16	5.15	5.15	5.14	5.13
8.79	8.74	8.70	8.66	8.63	8.62	8.59	8.58	8.57	8.55	8.53
14.42	14.34	14.25	14.17	14.12	14.08	14.04	14.01	13.99	13.95	13.91
27.23	27.05	26.87	26.69	26.58	26.50	26.41	26.35	26.32	26.22	26.14
129.25	128.32	127.37	126.42	125.84	125.45	124.96	124.66	124.47	123.97	123.53
3.92	3.90	3.87	3.84	3.83	3.82	3.80	3.80	3.79	3.78	3.76
5.96	5.91	5.86	5.80	5.77	5.75	5.72	5.70	5.69	5.66	5.63
8.84	8.75	8.66	8.56	8.50	8.46	8.41	8.38	8.36	8.31	8.26
14.55	14.37	14.20	14.02	13.91	13.84	13.75	13.69	13.65	13.56	13.47
48.05	47.41	46.76	46.10	45.70	45.43	45.09	44.88	44.75	44.40	44.09
3.30	3.27	3.24	3.21	3.19	3.17	3.16	3.15	3.14	3.12	3.11
4.74	4.68	4.62	4.56	4.52	4.50	4.46	4.44	4.43	4.40	4.37
6.62	6.52	6.43	6.33	6.27	6.23	6.18	6.14	6.12	6.07	6.02
10.05	9.89	9.72	9.55	9.45	9.38	9.29	9.24	9.20	9.11	9.03
26.92	26.42	25.91	25.39	25.08	24.87	24.60	24.44	24.33	24.06	23.82
2.94	2.90	2.87	2.84	2.81	2.80	2.78	2.77	2.76	2.74	2.72
4.06	4.00	3.94	3.87	3.83	3.81	3.77	3.75	3.74	3.70	3.67
5.46	5.37	5.27	5.17	5.11	5.07	5.01	4.98	4.96	4.90	4.86
7.87	7.72	7.56	7.40	7.30	7.23	7.14	7.09	7.06	6.97	6.89
18.41	17.99	17.56	17.12	16.85	16.67	16.44	16.31	16.21	15.98	15.77
2.70	2.67	2.63	2.59	2.57	2.56	2.54	2.52	2.51	2.49	2.47
3.64	3.57	3.51	3.44	3.40	3.38	3.34	3.32	3.30	3.27	3.23
4.76	4.67	4.57	4.47	4.40	4.36	4.31	4.28	4.25	4.20	4.15
6.62	6.47	6.31	6.16	6.06	5.99	5.91	5.86	5.82	5.74	5.66
14.08	13.71	13.32	12.93	12.69	12.53	12.33	12.20	12.12	11.91	11.72

(*Continued*)

TABLE E   *F* Distribution Critical Values (continued)

| | *p* | \multicolumn{9}{c}{Degrees of freedom in the numerator} |
		1	2	3	4	5	6	7	8	9
8	.100	3.46	3.11	2.92	2.81	2.73	2.67	2.62	2.59	2.56
	.050	5.32	4.46	4.07	3.84	3.69	3.58	3.50	3.44	3.39
	.025	7.57	6.06	5.42	5.05	4.82	4.65	4.53	4.43	4.36
	.010	11.26	8.65	7.59	7.01	6.63	6.37	6.18	6.03	5.91
	.001	25.41	18.49	15.83	14.39	13.48	12.86	12.40	12.05	11.77
9	.100	3.36	3.01	2.81	2.69	2.61	2.55	2.51	2.47	2.44
	.050	5.12	4.26	3.86	3.63	3.48	3.37	3.29	3.23	3.18
	.025	7.21	5.71	5.08	4.72	4.48	4.32	4.20	4.10	4.03
	.010	10.56	8.02	6.99	6.42	6.06	5.80	5.61	5.47	5.35
	.001	22.86	16.39	13.90	12.56	11.71	11.13	10.70	10.37	10.11
10	.100	3.29	2.92	2.73	2.61	2.52	2.46	2.41	2.38	2.35
	.050	4.96	4.10	3.71	3.48	3.33	3.22	3.14	3.07	3.02
	.025	6.94	5.46	4.83	4.47	4.24	4.07	3.95	3.85	3.78
	.010	10.04	7.56	6.55	5.99	5.64	5.39	5.20	5.06	4.94
	.001	21.04	14.91	12.55	11.28	10.48	9.93	9.52	9.20	8.96
11	.100	3.23	2.86	2.66	2.54	2.45	2.39	2.34	2.30	2.27
	.050	4.84	3.98	3.59	3.36	3.20	3.09	3.01	2.95	2.90
	.025	6.72	5.26	4.63	4.28	4.04	3.88	3.76	3.66	3.59
	.010	9.65	7.21	6.22	5.67	5.32	5.07	4.89	4.74	4.63
	.001	19.69	13.81	11.56	10.35	9.58	9.05	8.66	8.35	8.12
12	.100	3.18	2.81	2.61	2.48	2.39	2.33	2.28	2.24	2.21
	.050	4.75	3.89	3.49	3.26	3.11	3.00	2.91	2.85	2.80
	.025	6.55	5.10	4.47	4.12	3.89	3.73	3.61	3.51	3.44
	.010	9.33	6.93	5.95	5.41	5.06	4.82	4.64	4.50	4.39
	.001	18.64	12.97	10.80	9.63	8.89	8.38	8.00	7.71	7.48
13	.100	3.14	2.76	2.56	2.43	2.35	2.28	2.23	2.20	2.16
	.050	4.67	3.81	3.41	3.18	3.03	2.92	2.83	2.77	2.71
	.025	6.41	4.97	4.35	4.00	3.77	3.60	3.48	3.39	3.31
	.010	9.07	6.70	5.74	5.21	4.86	4.62	4.44	4.30	4.19
	.001	17.82	12.31	10.21	9.07	8.35	7.86	7.49	7.21	6.98
14	.100	3.10	2.73	2.52	2.39	2.31	2.24	2.19	2.15	2.12
	.050	4.60	3.74	3.34	3.11	2.96	2.85	2.76	2.70	2.65
	.025	6.30	4.86	4.24	3.89	3.66	3.50	3.38	3.29	3.21
	.010	8.86	6.51	5.56	5.04	4.69	4.46	4.28	4.14	4.03
	.001	17.14	11.78	9.73	8.62	7.92	7.44	7.08	6.80	6.58
15	.100	3.07	2.70	2.49	2.36	2.27	2.21	2.16	2.12	2.09
	.050	4.54	3.68	3.29	3.06	2.90	2.79	2.71	2.64	2.59
	.025	6.20	4.77	4.15	3.80	3.58	3.41	3.29	3.20	3.12
	.010	8.68	6.36	5.42	4.89	4.56	4.32	4.14	4.00	3.89
	.001	16.59	11.34	9.34	8.25	7.57	7.09	6.74	6.47	6.26
16	.100	3.05	2.67	2.46	2.33	2.24	2.18	2.13	2.09	2.06
	.050	4.49	3.63	3.24	3.01	2.85	2.74	2.66	2.59	2.54
	.025	6.12	4.69	4.08	3.73	3.50	3.34	3.22	3.12	3.05
	.010	8.53	6.23	5.29	4.77	4.44	4.20	4.03	3.89	3.78
	.001	16.12	10.97	9.01	7.94	7.27	6.80	6.46	6.19	5.98
17	.100	3.03	2.64	2.44	2.31	2.22	2.15	2.10	2.06	2.03
	.050	4.45	3.59	3.20	2.96	2.81	2.70	2.61	2.55	2.49
	.025	6.04	4.62	4.01	3.66	3.44	3.28	3.16	3.06	2.98
	.010	8.40	6.11	5.19	4.67	4.34	4.10	3.93	3.79	3.68
	.001	15.72	10.66	8.73	7.68	7.02	6.56	6.22	5.96	5.75

Degrees of freedom in the denominator

TABLE E   *F* Distribution Critical Values (continued)

Degrees of freedom in the numerator

10	12	15	20	25	30	40	50	60	120	1000
2.54	2.50	2.46	2.42	2.40	2.38	2.36	2.35	2.34	2.32	2.30
3.35	3.28	3.22	3.15	3.11	3.08	3.04	3.02	3.01	2.97	2.93
4.30	4.20	4.10	4.00	3.94	3.89	3.84	3.81	3.78	3.73	3.68
5.81	5.67	5.52	5.36	5.26	5.20	5.12	5.07	5.03	4.95	4.87
11.54	11.19	10.84	10.48	10.26	10.11	9.92	9.80	9.73	9.53	9.36
2.42	2.38	2.34	2.30	2.27	2.25	2.23	2.22	2.21	2.18	2.16
3.14	3.07	3.01	2.94	2.89	2.86	2.83	2.80	2.79	2.75	2.71
3.96	3.87	3.77	3.67	3.60	3.56	3.51	3.47	3.45	3.39	3.34
5.26	5.11	4.96	4.81	4.71	4.65	4.57	4.52	4.48	4.40	4.32
9.89	9.57	9.24	8.90	8.69	8.55	8.37	8.26	8.19	8.00	7.84
2.32	2.28	2.24	2.20	2.17	2.16	2.13	2.12	2.11	2.08	2.06
2.98	2.91	2.85	2.77	2.73	2.70	2.66	2.64	2.62	2.58	2.54
3.72	3.62	3.52	3.42	3.35	3.31	3.26	3.22	3.20	3.14	3.09
4.85	4.71	4.56	4.41	4.31	4.25	4.17	4.12	4.08	4.00	3.92
8.75	8.45	8.13	7.80	7.60	7.47	7.30	7.19	7.12	6.94	6.78
2.25	2.21	2.17	2.12	2.10	2.08	2.05	2.04	2.03	2.00	1.98
2.85	2.79	2.72	2.65	2.60	2.57	2.53	2.51	2.49	2.45	2.41
3.53	3.43	3.33	3.23	3.16	3.12	3.06	3.03	3.00	2.94	2.89
4.54	4.40	4.25	4.10	4.01	3.94	3.86	3.81	3.78	3.69	3.61
7.92	7.63	7.32	7.01	6.81	6.68	6.52	6.42	6.35	6.18	6.02
2.19	2.15	2.10	2.06	2.03	2.01	1.99	1.97	1.96	1.93	1.91
2.75	2.69	2.62	2.54	2.50	2.47	2.43	2.40	2.38	2.34	2.30
3.37	3.28	3.18	3.07	3.01	2.96	2.91	2.87	2.85	2.79	2.73
4.30	4.16	4.01	3.86	3.76	3.70	3.62	3.57	3.54	3.45	3.37
7.29	7.00	6.71	6.40	6.22	6.09	5.93	5.83	5.76	5.59	5.44
2.14	2.10	2.05	2.01	1.98	1.96	1.93	1.92	1.90	1.88	1.85
2.67	2.60	2.53	2.46	2.41	2.38	2.34	2.31	2.30	2.25	2.21
3.25	3.15	3.05	2.95	2.88	2.84	2.78	2.74	2.72	2.66	2.60
4.10	3.96	3.82	3.66	3.57	3.51	3.43	3.38	3.34	3.25	3.18
6.80	6.52	6.23	5.93	5.75	5.63	5.47	5.37	5.30	5.14	4.99
2.10	2.05	2.01	1.96	1.93	1.91	1.89	1.87	1.86	1.83	1.80
2.60	2.53	2.46	2.39	2.34	2.31	2.27	2.24	2.22	2.18	2.14
3.15	3.05	2.95	2.84	2.78	2.73	2.67	2.64	2.61	2.55	2.50
3.94	3.80	3.66	3.51	3.41	3.35	3.27	3.22	3.18	3.09	3.02
6.40	6.13	5.85	5.56	5.38	5.25	5.10	5.00	4.94	4.77	4.62
2.06	2.02	1.97	1.92	1.89	1.87	1.85	1.83	1.82	1.79	1.76
2.54	2.48	2.40	2.33	2.28	2.25	2.20	2.18	2.16	2.11	2.07
3.06	2.96	2.86	2.76	2.69	2.64	2.59	2.55	2.52	2.46	2.40
3.80	3.67	3.52	3.37	3.28	3.21	3.13	3.08	3.05	2.96	2.88
6.08	5.81	5.54	5.25	5.07	4.95	4.80	4.70	4.64	4.47	4.33
2.03	1.99	1.94	1.89	1.86	1.84	1.81	1.79	1.78	1.75	1.72
2.49	2.42	2.35	2.28	2.23	2.19	2.15	2.12	2.11	2.06	2.02
2.99	2.89	2.79	2.68	2.61	2.57	2.51	2.47	2.45	2.38	2.32
3.69	3.55	3.41	3.26	3.16	3.10	3.02	2.97	2.93	2.84	2.76
5.81	5.55	5.27	4.99	4.82	4.70	4.54	4.45	4.39	4.23	4.08
2.00	1.96	1.91	1.86	1.83	1.81	1.78	1.76	1.75	1.72	1.69
2.45	2.38	2.31	2.23	2.18	2.15	2.10	2.08	2.06	2.01	1.97
2.92	2.82	2.72	2.62	2.55	2.50	2.44	2.41	2.38	2.32	2.26
3.59	3.46	3.31	3.16	3.07	3.00	2.92	2.87	2.83	2.75	2.66
5.58	5.32	5.05	4.78	4.60	4.48	4.33	4.24	4.18	4.02	3.87

(Continued)

TABLE E *F* Distribution Critical Values (continued)

		Degrees of freedom in the numerator								
	p	1	2	3	4	5	6	7	8	9
18	.100	3.01	2.62	2.42	2.29	2.20	2.13	2.08	2.04	2.00
	.050	4.41	3.55	3.16	2.93	2.77	2.66	2.58	2.51	2.46
	.025	5.98	4.56	3.95	3.61	3.38	3.22	3.10	3.01	2.93
	.010	8.29	6.01	5.09	4.58	4.25	4.01	3.84	3.71	3.60
	.001	15.38	10.39	8.49	7.46	6.81	6.35	6.02	5.76	5.56
19	.100	2.99	2.61	2.40	2.27	2.18	2.11	2.06	2.02	1.98
	.050	4.38	3.52	3.13	2.90	2.74	2.63	2.54	2.48	2.42
	.025	5.92	4.51	3.90	3.56	3.33	3.17	3.05	2.96	2.88
	.010	8.18	5.93	5.01	4.50	4.17	3.94	3.77	3.63	3.52
	.001	15.08	10.16	8.28	7.27	6.62	6.18	5.85	5.59	5.39
20	.100	2.97	2.59	2.38	2.25	2.16	2.09	2.04	2.00	1.96
	.050	4.35	3.49	3.10	2.87	2.71	2.60	2.51	2.45	2.39
	.025	5.87	4.46	3.86	3.51	3.29	3.13	3.01	2.91	2.84
	.010	8.10	5.85	4.94	4.43	4.10	3.87	3.70	3.56	3.46
	.001	14.82	9.95	8.10	7.10	6.46	6.02	5.69	5.44	5.24
21	.100	2.96	2.57	2.36	2.23	2.14	2.08	2.02	1.98	1.95
	.050	4.32	3.47	3.07	2.84	2.68	2.57	2.49	2.42	2.37
	.025	5.83	4.42	3.82	3.48	3.25	3.09	2.97	2.87	2.80
	.010	8.02	5.78	4.87	4.37	4.04	3.81	3.64	3.51	3.40
	.001	14.59	9.77	7.94	6.95	6.32	5.88	5.56	5.31	5.11
22	.100	2.95	2.56	2.35	2.22	2.13	2.06	2.01	1.97	1.93
	.050	4.30	3.44	3.05	2.82	2.66	2.55	2.46	2.40	2.34
	.025	5.79	4.38	3.78	3.44	3.22	3.05	2.93	2.84	2.76
	.010	7.95	5.72	4.82	4.31	3.99	3.76	3.59	3.45	3.35
	.001	14.38	9.61	7.80	6.81	6.19	5.76	5.44	5.19	4.99
23	.100	2.94	2.55	2.34	2.21	2.11	2.05	1.99	1.95	1.92
	.050	4.28	3.42	3.03	2.80	2.64	2.53	2.44	2.37	2.32
	.025	5.75	4.35	3.75	3.41	3.18	3.02	2.90	2.81	2.73
	.010	7.88	5.66	4.76	4.26	3.94	3.71	3.54	3.41	3.30
	.001	14.20	9.47	7.67	6.70	6.08	5.65	5.33	5.09	4.89
24	.100	2.93	2.54	2.33	2.19	2.10	2.04	1.98	1.94	1.91
	.050	4.26	3.40	3.01	2.78	2.62	2.51	2.42	2.36	2.30
	.025	5.72	4.32	3.72	3.38	3.15	2.99	2.87	2.78	2.70
	.010	7.82	5.61	4.72	4.22	3.90	3.67	3.50	3.36	3.26
	.001	14.03	9.34	7.55	6.59	5.98	5.55	5.23	4.99	4.80
25	.100	2.92	2.53	2.32	2.18	2.09	2.02	1.97	1.93	1.89
	.050	4.24	3.39	2.99	2.76	2.60	2.49	2.40	2.34	2.28
	.025	5.69	4.29	3.69	3.35	3.13	2.97	2.85	2.75	2.68
	.010	7.77	5.57	4.68	4.18	3.85	3.63	3.46	3.32	3.22
	.001	13.88	9.22	7.45	6.49	5.89	5.46	5.15	4.91	4.71
26	.100	2.91	2.52	2.31	2.17	2.08	2.01	1.96	1.92	1.88
	.050	4.23	3.37	2.98	2.74	2.59	2.47	2.39	2.32	2.27
	.025	5.66	4.27	3.67	3.33	3.10	2.94	2.82	2.73	2.65
	.010	7.72	5.53	4.64	4.14	3.82	3.59	3.42	3.29	3.18
	.001	13.74	9.12	7.36	6.41	5.80	5.38	5.07	4.83	4.64
27	.100	2.90	2.51	2.30	2.17	2.07	2.00	1.95	1.91	1.87
	.050	4.21	3.35	2.96	2.73	2.57	2.46	2.37	2.31	2.25
	.025	5.63	4.24	3.65	3.31	3.08	2.92	2.80	2.71	2.63
	.010	7.68	5.49	4.60	4.11	3.78	3.56	3.39	3.26	3.15
	.001	13.61	9.02	7.27	6.33	5.73	5.31	5.00	4.76	4.57

Degrees of freedom in the denominator

TABLE E   F Distribution Critical Values (continued)

Degrees of freedom in the numerator										
10	12	15	20	25	30	40	50	60	120	1000
1.98	1.93	1.89	1.84	1.80	1.78	1.75	1.74	1.72	1.69	1.66
2.41	2.34	2.27	2.19	2.14	2.11	2.06	2.04	2.02	1.97	1.92
2.87	2.77	2.67	2.56	2.49	2.44	2.38	2.35	2.32	2.26	2.20
3.51	3.37	3.23	3.08	2.98	2.92	2.84	2.78	2.75	2.66	2.58
5.39	5.13	4.87	4.59	4.42	4.30	4.15	4.06	4.00	3.84	3.69
1.96	1.91	1.86	1.81	1.78	1.76	1.73	1.71	1.70	1.67	1.64
2.38	2.31	2.23	2.16	2.11	2.07	2.03	2.00	1.98	1.93	1.88
2.82	2.72	2.62	2.51	2.44	2.39	2.33	2.30	2.27	2.20	2.14
3.43	3.30	3.15	3.00	2.91	2.84	2.76	2.71	2.67	2.58	2.50
5.22	4.97	4.70	4.43	4.26	4.14	3.99	3.90	3.84	3.68	3.53
1.94	1.89	1.84	1.79	1.76	1.74	1.71	1.69	1.68	1.64	1.61
2.35	2.28	2.20	2.12	2.07	2.04	1.99	1.97	1.95	1.90	1.85
2.77	2.68	2.57	2.46	2.40	2.35	2.29	2.25	2.22	2.16	2.09
3.37	3.23	3.09	2.94	2.84	2.78	2.69	2.64	2.61	2.52	2.43
5.08	4.82	4.56	4.29	4.12	4.00	3.86	3.77	3.70	3.54	3.40
1.92	1.87	1.83	1.78	1.74	1.72	1.69	1.67	1.66	1.62	1.59
2.32	2.25	2.18	2.10	2.05	2.01	1.96	1.94	1.92	1.87	1.82
2.73	2.64	2.53	2.42	2.36	2.31	2.25	2.21	2.18	2.11	2.05
3.31	3.17	3.03	2.88	2.79	2.72	2.64	2.58	2.55	2.46	2.37
4.95	4.70	4.44	4.17	4.00	3.88	3.74	3.64	3.58	3.42	3.28
1.90	1.86	1.81	1.76	1.73	1.70	1.67	1.65	1.64	1.60	1.57
2.30	2.23	2.15	2.07	2.02	1.98	1.94	1.91	1.89	1.84	1.79
2.70	2.60	2.50	2.39	2.32	2.27	2.21	2.17	2.14	2.08	2.01
3.26	3.12	2.98	2.83	2.73	2.67	2.58	2.53	2.50	2.40	2.32
4.83	4.58	4.33	4.06	3.89	3.78	3.63	3.54	3.48	3.32	3.17
1.89	1.84	1.80	1.74	1.71	1.69	1.66	1.64	1.62	1.59	1.55
2.27	2.20	2.13	2.05	2.00	1.96	1.91	1.88	1.86	1.81	1.76
2.67	2.57	2.47	2.36	2.29	2.24	2.18	2.14	2.11	2.04	1.98
3.21	3.07	2.93	2.78	2.69	2.62	2.54	2.48	2.45	2.35	2.27
4.73	4.48	4.23	3.96	3.79	3.68	3.53	3.44	3.38	3.22	3.08
1.88	1.83	1.78	1.73	1.70	1.67	1.64	1.62	1.61	1.57	1.54
2.25	2.18	2.11	2.03	1.97	1.94	1.89	1.86	1.84	1.79	1.74
2.64	2.54	2.44	2.33	2.26	2.21	2.15	2.11	2.08	2.01	1.94
3.17	3.03	2.89	2.74	2.64	2.58	2.49	2.44	2.40	2.31	2.22
4.64	4.39	4.14	3.87	3.71	3.59	3.45	3.36	3.29	3.14	2.99
1.87	1.82	1.77	1.72	1.68	1.66	1.63	1.61	1.59	1.56	1.52
2.24	2.16	2.09	2.01	1.96	1.92	1.87	1.84	1.82	1.77	1.72
2.61	2.51	2.41	2.30	2.23	2.18	2.12	2.08	2.05	1.98	1.91
3.13	2.99	2.85	2.70	2.60	2.54	2.45	2.40	2.36	2.27	2.18
4.56	4.31	4.06	3.79	3.63	3.52	3.37	3.28	3.22	3.06	2.91
1.86	1.81	1.76	1.71	1.67	1.65	1.61	1.59	1.58	1.54	1.51
2.22	2.15	2.07	1.99	1.94	1.90	1.85	1.82	1.80	1.75	1.70
2.59	2.49	2.39	2.28	2.21	2.16	2.09	2.05	2.03	1.95	1.89
3.09	2.96	2.81	2.66	2.57	2.50	2.42	2.36	2.33	2.23	2.14
4.48	4.24	3.99	3.72	3.56	3.44	3.30	3.21	3.15	2.99	2.84
1.85	1.80	1.75	1.70	1.66	1.64	1.60	1.58	1.57	1.53	1.50
2.20	2.13	2.06	1.97	1.92	1.88	1.84	1.81	1.79	1.73	1.68
2.57	2.47	2.36	2.25	2.18	2.13	2.07	2.03	2.00	1.93	1.86
3.06	2.93	2.78	2.63	2.54	2.47	2.38	2.33	2.29	2.20	2.11
4.41	4.17	3.92	3.66	3.49	3.38	3.23	3.14	3.08	2.92	2.78

(Continued)

TABLE E	*F* Distribution Critical Values (continued)								

		Degrees of freedom in the numerator								
	p	1	2	3	4	5	6	7	8	9
28	.100	2.89	2.50	2.29	2.16	2.06	2.00	1.94	1.90	1.87
	.050	4.20	3.34	2.95	2.71	2.56	2.45	2.36	2.29	2.24
	.025	5.61	4.22	3.63	3.29	3.06	2.90	2.78	2.69	2.61
	.010	7.64	5.45	4.57	4.07	3.75	3.53	3.36	3.23	3.12
	.001	13.50	8.93	7.19	6.25	5.66	5.24	4.93	4.69	4.50
29	.100	2.89	2.50	2.28	2.15	2.06	1.99	1.93	1.89	1.86
	.050	4.18	3.33	2.93	2.70	2.55	2.43	2.35	2.28	2.22
	.025	5.59	4.20	3.61	3.27	3.04	2.88	2.76	2.67	2.59
	.010	7.60	5.42	4.54	4.04	3.73	3.50	3.33	3.20	3.09
	.001	13.39	8.85	7.12	6.19	5.59	5.18	4.87	4.64	4.45
30	.100	2.88	2.49	2.28	2.14	2.05	1.98	1.93	1.88	1.85
	.050	4.17	3.32	2.92	2.69	2.53	2.42	2.33	2.27	2.21
	.025	5.57	4.18	3.59	3.25	3.03	2.87	2.75	2.65	2.57
	.010	7.56	5.39	4.51	4.02	3.70	3.47	3.30	3.17	3.07
	.001	13.29	8.77	7.05	6.12	5.53	5.12	4.82	4.58	4.39
40	.100	2.84	2.44	2.23	2.09	2.00	1.93	1.87	1.83	1.79
	.050	4.08	3.23	2.84	2.61	2.45	2.34	2.25	2.18	2.12
	.025	5.42	4.05	3.46	3.13	2.90	2.74	2.62	2.53	2.45
	.010	7.31	5.18	4.31	3.83	3.51	3.29	3.12	2.99	2.89
	.001	12.61	8.25	6.59	5.70	5.13	4.73	4.44	4.21	4.02
50	.100	2.81	2.41	2.20	2.06	1.97	1.90	1.84	1.80	1.76
	.050	4.03	3.18	2.79	2.56	2.40	2.29	2.20	2.13	2.07
	.025	5.34	3.97	3.39	3.05	2.83	2.67	2.55	2.46	2.38
	.010	7.17	5.06	4.20	3.72	3.41	3.19	3.02	2.89	2.78
	.001	12.22	7.96	6.34	5.46	4.90	4.51	4.22	4.00	3.82
60	.100	2.79	2.39	2.18	2.04	1.95	1.87	1.82	1.77	1.74
	.050	4.00	3.15	2.76	2.53	2.37	2.25	2.17	2.10	2.04
	.025	5.29	3.93	3.34	3.01	2.79	2.63	2.51	2.41	2.33
	.010	7.08	4.98	4.13	3.65	3.34	3.12	2.95	2.82	2.72
	.001	11.97	7.77	6.17	5.31	4.76	4.37	4.09	3.86	3.69
100	.100	2.76	2.36	2.14	2.00	1.91	1.83	1.78	1.73	1.69
	.050	3.94	3.09	2.70	2.46	2.31	2.19	2.10	2.03	1.97
	.025	5.18	3.83	3.25	2.92	2.70	2.54	2.42	2.32	2.24
	.010	6.90	4.82	3.98	3.51	3.21	2.99	2.82	2.69	2.59
	.001	11.50	7.41	5.86	5.02	4.48	4.11	3.83	3.61	3.44
200	.100	2.73	2.33	2.11	1.97	1.88	1.80	1.75	1.70	1.66
	.050	3.89	3.04	2.65	2.42	2.26	2.14	2.06	1.98	1.93
	.025	5.10	3.76	3.18	2.85	2.63	2.47	2.35	2.26	2.18
	.010	6.76	4.71	3.88	3.41	3.11	2.89	2.73	2.60	2.50
	.001	11.15	7.15	5.63	4.81	4.29	3.92	3.65	3.43	3.26
1000	.100	2.71	2.31	2.09	1.95	1.85	1.78	1.72	1.68	1.64
	.050	3.85	3.00	2.61	2.38	2.22	2.11	2.02	1.95	1.89
	.025	5.04	3.70	3.13	2.80	2.58	2.42	2.30	2.20	2.13
	.010	6.66	4.63	3.80	3.34	3.04	2.82	2.66	2.53	2.43
	.001	10.89	6.96	5.46	4.65	4.14	3.78	3.51	3.30	3.13

Degrees of freedom in the denominator

TABLE E   *F* Distribution Critical Values (continued)

Degrees of freedom in the numerator										
10	12	15	20	25	30	40	50	60	120	1000
1.84	1.79	1.74	1.69	1.65	1.63	1.59	1.57	1.56	1.52	1.48
2.19	2.12	2.04	1.96	1.91	1.87	1.82	1.79	1.77	1.71	1.66
2.55	2.45	2.34	2.23	2.16	2.11	2.05	2.01	1.98	1.91	1.84
3.03	2.90	2.75	2.60	2.51	2.44	2.35	2.30	2.26	2.17	2.08
4.35	4.11	3.86	3.60	3.43	3.32	3.18	3.09	3.02	2.86	2.72
1.83	1.78	1.73	1.68	1.64	1.62	1.58	1.56	1.55	1.51	1.47
2.18	2.10	2.03	1.94	1.89	1.85	1.81	1.77	1.75	1.70	1.65
2.53	2.43	2.32	2.21	2.14	2.09	2.03	1.99	1.96	1.89	1.82
3.00	2.87	2.73	2.57	2.48	2.41	2.33	2.27	2.23	2.14	2.05
4.29	4.05	3.80	3.54	3.38	3.27	3.12	3.03	2.97	2.81	2.66
1.82	1.77	1.72	1.67	1.63	1.61	1.57	1.55	1.54	1.50	1.46
2.16	2.09	2.01	1.93	1.88	1.84	1.79	1.76	1.74	1.68	1.63
2.51	2.41	2.31	2.20	2.12	2.07	2.01	1.97	1.94	1.87	1.80
2.98	2.84	2.70	2.55	2.45	2.39	2.30	2.25	2.21	2.11	2.02
4.24	4.00	3.75	3.49	3.33	3.22	3.07	2.98	2.92	2.76	2.61
1.76	1.71	1.66	1.61	1.57	1.54	1.51	1.48	1.47	1.42	1.38
2.08	2.00	1.92	1.84	1.78	1.74	1.69	1.66	1.64	1.58	1.52
2.39	2.29	2.18	2.07	1.99	1.94	1.88	1.83	1.80	1.72	1.65
2.80	2.66	2.52	2.37	2.27	2.20	2.11	2.06	2.02	1.92	1.82
3.87	3.64	3.40	3.14	2.98	2.87	2.73	2.64	2.57	2.41	2.25
1.73	1.68	1.63	1.57	1.53	1.50	1.46	1.44	1.42	1.38	1.33
2.03	1.95	1.87	1.78	1.73	1.69	1.63	1.60	1.58	1.51	1.45
2.32	2.22	2.11	1.99	1.92	1.87	1.80	1.75	1.72	1.64	1.56
2.70	2.56	2.42	2.27	2.17	2.10	2.01	1.95	1.91	1.80	1.70
3.67	3.44	3.20	2.95	2.79	2.68	2.53	2.44	2.38	2.21	2.05
1.71	1.66	1.60	1.54	1.50	1.48	1.44	1.41	1.40	1.35	1.30
1.99	1.92	1.84	1.75	1.69	1.65	1.59	1.56	1.53	1.47	1.40
2.27	2.17	2.06	1.94	1.87	1.82	1.74	1.70	1.67	1.58	1.49
2.63	2.50	2.35	2.20	2.10	2.03	1.94	1.88	1.84	1.73	1.62
3.54	3.32	3.08	2.83	2.67	2.55	2.41	2.32	2.25	2.08	1.92
1.66	1.61	1.56	1.49	1.45	1.42	1.38	1.35	1.34	1.28	1.22
1.93	1.85	1.77	1.68	1.62	1.57	1.52	1.48	1.45	1.38	1.30
2.18	2.08	1.97	1.85	1.77	1.71	1.64	1.59	1.56	1.46	1.36
2.50	2.37	2.22	2.07	1.97	1.89	1.80	1.74	1.69	1.57	1.45
3.30	3.07	2.84	2.59	2.43	2.32	2.17	2.08	2.01	1.83	1.64
1.63	1.58	1.52	1.46	1.41	1.38	1.34	1.31	1.29	1.23	1.16
1.88	1.80	1.72	1.62	1.56	1.52	1.46	1.41	1.39	1.30	1.21
2.11	2.01	1.90	1.78	1.70	1.64	1.56	1.51	1.47	1.37	1.25
2.41	2.27	2.13	1.97	1.87	1.79	1.69	1.63	1.58	1.45	1.30
3.12	2.90	2.67	2.42	2.26	2.15	2.00	1.90	1.83	1.64	1.43
1.61	1.55	1.49	1.43	1.38	1.35	1.30	1.27	1.25	1.18	1.08
1.84	1.76	1.68	1.58	1.52	1.47	1.41	1.36	1.33	1.24	1.11
2.06	1.96	1.85	1.72	1.64	1.58	1.50	1.45	1.41	1.29	1.13
2.34	2.20	2.06	1.90	1.79	1.72	1.61	1.54	1.50	1.35	1.16
2.99	2.77	2.54	2.30	2.14	2.02	1.87	1.77	1.69	1.49	1.22

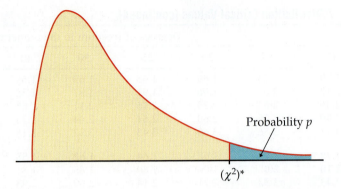

Table entry for p is the critical value $(\chi^2)^*$ with probability p lying to its right.

Probability p

$(\chi^2)^*$

TABLE F χ^2 Distribution Critical Values

df	Tail probability p											
	.25	.20	.15	.10	.05	.025	.02	.01	.005	.0025	.001	.0005
1	1.32	1.64	2.07	2.71	3.84	5.02	5.41	6.63	7.88	9.14	10.83	12.12
2	2.77	3.22	3.79	4.61	5.99	7.38	7.82	9.21	10.60	11.98	13.82	15.20
3	4.11	4.64	5.32	6.25	7.81	9.35	9.84	11.34	12.84	14.32	16.27	17.73
4	5.39	5.99	6.74	7.78	9.49	11.14	11.67	13.28	14.86	16.42	18.47	20.00
5	6.63	7.29	8.12	9.24	11.07	12.83	13.39	15.09	16.75	18.39	20.51	22.11
6	7.84	8.56	9.45	10.64	12.59	14.45	15.03	16.81	18.55	20.25	22.46	24.10
7	9.04	9.80	10.75	12.02	14.07	16.01	16.62	18.48	20.28	22.04	24.32	26.02
8	10.22	11.03	12.03	13.36	15.51	17.53	18.17	20.09	21.95	23.77	26.12	27.87
9	11.39	12.24	13.29	14.68	16.92	19.02	19.68	21.67	23.59	25.46	27.88	29.67
10	12.55	13.44	14.53	15.99	18.31	20.48	21.16	23.21	25.19	27.11	29.59	31.42
11	13.70	14.63	15.77	17.28	19.68	21.92	22.62	24.72	26.76	28.73	31.26	33.14
12	14.85	15.81	16.99	18.55	21.03	23.34	24.05	26.22	28.30	30.32	32.91	34.82
13	15.98	16.98	18.20	19.81	22.36	24.74	25.47	27.69	29.82	31.88	34.53	36.48
14	17.12	18.15	19.41	21.06	23.68	26.12	26.87	29.14	31.32	33.43	36.12	38.11
15	18.25	19.31	20.60	22.31	25.00	27.49	28.26	30.58	32.80	34.95	37.70	39.72
16	19.37	20.47	21.79	23.54	26.30	28.85	29.63	32.00	34.27	36.46	39.25	41.31
17	20.49	21.61	22.98	24.77	27.59	30.19	31.00	33.41	35.72	37.95	40.79	42.88
18	21.60	22.76	24.16	25.99	28.87	31.53	32.35	34.81	37.16	39.42	42.31	44.43
19	22.72	23.90	25.33	27.20	30.14	32.85	33.69	36.19	38.58	40.88	43.82	45.97
20	23.83	25.04	26.50	28.41	31.41	34.17	35.02	37.57	40.00	42.34	45.31	47.50
21	24.93	26.17	27.66	29.62	32.67	35.48	36.34	38.93	41.40	43.78	46.80	49.01
22	26.04	27.30	28.82	30.81	33.92	36.78	37.66	40.29	42.80	45.20	48.27	50.51
23	27.14	28.43	29.98	32.01	35.17	38.08	38.97	41.64	44.18	46.62	49.73	52.00
24	28.24	29.55	31.13	33.20	36.42	39.36	40.27	42.98	45.56	48.03	51.18	53.48
25	29.34	30.68	32.28	34.38	37.65	40.65	41.57	44.31	46.93	49.44	52.62	54.95
26	30.43	31.79	33.43	35.56	38.89	41.92	42.86	45.64	48.29	50.83	54.05	56.41
27	31.53	32.91	34.57	36.74	40.11	43.19	44.14	46.96	49.64	52.22	55.48	57.86
28	32.62	34.03	35.71	37.92	41.34	44.46	45.42	48.28	50.99	53.59	56.89	59.30
29	33.71	35.14	36.85	39.09	42.56	45.72	46.69	49.59	52.34	54.97	58.30	60.73
30	34.80	36.25	37.99	40.26	43.77	46.98	47.96	50.89	53.67	56.33	59.70	62.16
40	45.62	47.27	49.24	51.81	55.76	59.34	60.44	63.69	66.77	69.70	73.40	76.09
50	56.33	58.16	60.35	63.17	67.50	71.42	72.61	76.15	79.49	82.66	86.66	89.56
60	66.98	68.97	71.34	74.40	79.08	83.30	84.58	88.38	91.95	95.34	99.61	102.69
80	88.13	90.41	93.11	96.58	101.88	106.63	108.07	112.33	116.32	120.10	124.84	128.26
100	109.1	111.67	114.66	118.50	124.34	129.56	131.14	135.81	140.17	144.29	149.45	153.17

Answers to Odd-Numbered Exercises

CHAPTER 1

1.1 The regular price for the Smokey Grill Ribs coupon is $20, the discount price is $11.

1.3 Who: The cases are coupons; there are seven cases. What: There are six variables—ID, Type, Name, Item, RegPrice, and DiscPrice. Only RegPrice and DiscPrice have units in dollars. The data might be used to compare coupons to one another to see which are better.

1.7 (a) The cases are students. (b) Four variables: Favorite choice for online research ("Google or Google Scholar," "Library database or website," "Wikipedia or online encyclopedia," "Other"), Age (reasonable age range for first-year college students—17 to 30, etc.), Sex (M or F), and Major (could be a big list—statistics, math, engineering, English, etc.). (c) Age is quantitative, the rest are categorical. (d) The label is the number—1 to 552. (e) Who: part (a) answer; What: part (b) and (c); Why: We could look at the distribution of favorite choice across different age groups, majors, and sex.

1.9 (a) The cases are employees. (b) Employee identification number—label, last name—label, first name—label, middle initial—label, department—categorical, number of years—quantitative, salary—quantitative, education—categorical, age—quantitative.

1.11 Age: quantitative, possible values 16 to ?. Sing: categorical, yes/no. Can you play: categorical, no, a little, pretty well. Food: quantitative, possible values $0 to ?. Height: quantitative, possible values 2 feet to 9 feet.

1.17 The first-exam scores are left-skewed, the middle is around 80.

1.19 (b) Use two stems, even though one is blank. Seeing the gap is useful.

1.21 The larger classes hide a lot of detail; there are now only three bars in the histogram.

1.25 (b) The United States is a clear outlier. It has four or five times as many Facebook users as the other countries, despite having a population smaller than some of the other countries.

1.27 (b) Second class had the fewest passengers. Third class had by far the most, more than twice as many as in first class. (c) A bar graph of the percents (relative frequency) would have the same features.

1.31 (b) The distribution is somewhat right skewed. (c) There appears to be one small outlier at 2680. (d) The shape is right-skewed, the center is around 3200, the range is from 2680 to 3950.

1.33 (a) 2013 still has the highest usage in December and January. (b) The patterns are very similar, but the values for the winter months in 2014 are somewhat higher than those

in the 2013 winter months. These differences are most likely due to weather.

1.35 Opinions about least-favorite color are somewhat more varied than favorite colors. Interestingly, purple is liked and disliked by about the same percentage of people.

1.37 White is the most popular color in 2012 for North America, followed by black, silver, and gray.

1.39 (a) Four variables: GPA, IQ, and self-concept are quantitative; gender is categorical. (c) Unimodal and skewed to the left, centered near 7.8, spread from 0.5 to 10.8. (d) There is more variability among the boys; in fact, there seem to be two groups of boys—those with GPAs below 5 and those with GPAs above 5.

1.41 The distribution is unimodal and skewed to the left, with center around 59.5. Most self-concept scores are between 35 and 73, with a few below that, and one high score of 80 (but not really high enough to be an outlier).

1.43 Without Suriname: $\overline{X} = 16.29$. With Suriname: $\overline{X} = 23.96$.

1.45 The ordered list is: 2 4 5 5 5 5 6 6 7 8 10 11 12 13 16 17 19 19 24 25 32 38 49 53 ⃝208
$M = 12$. Without the outlier, the median is 11.5; with the outlier, the median is 12. The outlier does not influence the median greatly.

1.47 $M = 84$.

1.49 196.575 minutes (the value 197 in the text was rounded). The quartiles and median are in positions 20.5, 40.5, and 60.5. Based on this, $Q_1 = 54.5$, $M = 103.5$, $Q_3 = 200$.

1.53 From software: $s^2 = 157.07$, $s = 12.53$.

1.55 Without Suriname: Min = 55, $Q_1 = 75$, $M = 84$, $Q_3 = 93$, Max = 97. With Suriname: Min = 2, $Q_1 = 5.5$, $M = 11.5$, $Q_3 = 21.5$, Max = 53. The max changes drastically with the outlier removed, but otherwise, the other numbers in the five-number summary do not change drastically. This shows that generally the five-number summary is robust.

1.57 (a) $\overline{x} = 3208.44$. (b) $M = 3130.37$. (c) Because the distribution is right-skewed with a potential outlier, the median is a better measure of center.

1.59 (a) $s = 306.68$. (b) $Q_1 = 3027.64$, $Q_3 = 3286.95$. (c) Min = 2664.38 (this is the smallest value), $Q_1 = 3027.64$ (this value has 25% of the observations below it), $M = 3130.37$ (this is the middle observation, or has 50% of the observations below or above it), $Q_3 = 3286.95$ (this value has 75% of the observations below it), Max = 4213.49 (this is the largest value). (d) The five-number summary would be better for this distribution because it is right-skewed with a potential outlier.

1.61 (a) The distribution is right-skewed with a potential outlier. **(b)** The distribution is right-skewed. **(c)** Preference will vary. The only advantage of the stemplot is that it preserves the data; otherwise, the histogram is likely better. The boxplot is also fine but hides some of the details that the histogram shows.

1.63 The KPOT values are right-skewed, whereas the KSUP values are fairly symmetric. The center for KSUP is higher than the center for the KPOT. Also, the KPOT values are more spread out than the KSUP values.

1.65 (a) $\bar{x} = 122.9$. **(b)** $M = 102.5$. **(c)** The data set is right-skewed with an outlier, so the median is a better center.

1.67 (a) $IQR = 62$. **(b)** Outliers are below -26 or above 222. London is confirmed as an outlier. **(c)** The first three quarters are about equal in length, and the last is extremely long. **(d)** The main part of the distribution is relatively symmetric; there is one extreme high outlier. The minimum is about 25, the first quartile is about 70, the median is about 100, and the third quartile is about 125. There is a gap in the data from roughly 200 to about 425.

1.69 (a) $s = 8.80$. **(b)** With $n = 50$, the positions of Q_1 and Q_3 will be at 13 and 38. We find $Q_1 = 43.79$ and $Q_3 = 57.02$.

1.71 (a) Because weight is quantitative and has a decent number of observations ($n = 25$), a histogram is a good choice. Mean and standard deviation are a good starting point for numerical summaries. **(b)** Now that we see the distribution is left-skewed, the choice of using the mean and standard deviation was not a good choice. Median and quartiles would have been a better choice.

1.73 (a) With the outlier: $\bar{x} = 5.235$, $M = 4.90$. Without the outlier: $\bar{x} = 5.265$, $M = 4.905$. The values are nearly identical with and without the outlier. **(b)** With the outlier: $s = 1.406$, $Q_1 = 4.40$, $Q_3 = 5.60$. Without the outlier: $s = 1.356$, $Q_1 = 4.430$, $Q_3 = 5.600$. The values are nearly identical with and without the outlier. **(c)** Even though there is one outlier, its removal does not change the numerical summaries at all. This is partly due to the large sample and partly due to the fact that this outlier is not too far from the other observations so that removing it doesn't have a huge effect on the analysis.

1.75 Some people like celebrities and business executives have very large net worths, which will pull the mean worth making it much larger than the median (Bill Gates of Microsoft, Warren Buffett, Oprah Winfrey, etc.).

1.77 The mean is $115,909.09. Ten of the employees make less than the mean. $M = \$55,000$.

1.79 The median doesn't change, but the mean increases to $138,636.36.

1.81 The average would be 2.5 or less (an earthquake that isn't usually felt). These do little or no damage.

1.83 For $n = 2$, the median is also the average of the two values.

1.85 (a) The median of seven (sorted) points is the fourth, while the median of eight points is the average of the fourth and fifth. If these are to be the same, the added point must be equal to the fourth point of the original seven, so that the fourth and fifth points are now the same. **(b)** Regardless of the configuration of the first seven points, if the eighth point is added so as to leave the median unchanged, then in that (sorted) set of eight, the fourth and fifth points must the same. Once we add a ninth point, one of these two points will be the new middle (fifth) point, so the median will not change.

1.87 (a) Picking the same number for all four observations results in a standard deviation of 0. **(b)** Picking 10, 10, 20, and 20 results in the largest standard deviation = 5.77. **(c)** For part (a), you may pick any number as long as all observations are the same. For part (b), only one choice provides the largest standard deviation.

1.89 $\bar{x} = 5.302$ pounds and $s = 2.75$ pounds.

1.91 Full data set: $\bar{x} = 196.575$ and $M = 103.5$ minutes. The 10% and 20% trimmed means are $\bar{x} = 127.734$ and $\bar{x} = 111.917$ minutes. While still larger than the median of the original data set, they are much closer to the median than the ordinary untrimmed mean.

1.93 According to the rule, 95% of scores will fall between $\mu \pm 2$. Therefore, 95% of scores are between 212 and 364.

1.95 $z = 1.63$.

1.97 For $X = 350$, $z = 1.63$ and the proportion less than 350 is the area to the left, which is 0.9484. For the proportion greater than or equal to 350, we calculate 0.0516.

1.99 To get the top 20% of students, we need to solve for the 80th percentile. The corresponding z is 0.84. So $x = 319.92$.

1.105 (b)–(c) The following table indicates the desired ranges.

	Low	High
68%	256	320
95%	224	352
99.7%	192	384

1.107

Value	Percentile (Table A)	Percentile (Software)
150	50	50
140	38.6	38.8
100	7.6	7.7
180	80.5	80.4
230	98.9	98.9

1.109 Using the $N(153, 34)$ distribution, we find the values corresponding to the given percentiles as given here (using Table A). The actual scores are very close to the percentiles of the Normal distribution; we can conclude these scores are at least approximately Normal.

Percentile	Score	Score with $N(153, 34)$
10%	110	109
25%	130	130
50%	154	153
75%	177	176
90%	197	197

1.111 (a) Ranges are shown in the following table. In both cases, some of the lower limits are negative, which does not make sense; this happens because the women's distribution is skewed and the men's distribution has an outlier. Contrary to the conventional wisdom, the men's mean is slightly higher, although the outlier is at least partly responsible for that. **(b)** The means suggest that Mexican men and women tend to speak more than people of the same sex from the United States.

	Women	Men
68%	8489 to 20,919	7158 to 22,886
95%	2274 to 27,134	−706 to 30,750
99.7%	−3941 to 33,349	−8570 to 38,614

1.113 (a) 0.25. **(b)** 0.75. **(c)** 0.25.

1.115 (a) The mean is at point C; the median is at point B. **(b)** The mean and median are both at point A. **(c)** The mean is at point A; the median is at point B.

1.117 (a) The applet shows an area of 0.6826 between −1.000 and 1.000, while the 68−95−99.7 rule rounds this to 0.68. **(b)** Between −2.000 and 2.000, the applet reports 0.9544 (compared to the rounded 0.95 from the 68−95−99.7 rule). Between −3.000 and 3.000, the applet reports 0.9974 (compared to the rounded 0.997).

1.119 (a) 0.0808. **(b)** 0.9192. **(c)** 0.0228. **(d)** 0.8964.

1.121 (a) $z = 1.17$ or 1.18. **(b)** $z = 1.17$ or 1.18.

1.123 $z = 2$. From Table A, 2.28% qualify for membership.

1.125 For Joshua, $z = -1.02$. For Anthony, $z = -1.42$. Joshua has the higher standardized score.

1.127 $z = 1.57$. The equivalent SAT score is 1994.12.

1.129 $z = -0.46$. From Table A we get 0.3228, so about the 32nd percentile.

1.131 Scores 1125.12 and lower make up the bottom 12% of all scores.

1.133 From Table A, the quartiles have z-scores of −0.675, 0, and 0.675. Using $1498 + 316(z)$ yields scores of 1285, 1498, and 1711 (rounded to the nearest integer).

1.135 (a) $z = -0.44$. From Table A, 33% of men have low values of HDL. (Software gives 32.95%.) **(b)** $z = 1.03$. From Table A, 15.15% of men have protective levels of HDL. (Software gives 15.16%.) **(c)** 51.85% of men are in the intermediate range for HDL. (Software gives 51.88%.)

1.137 (a) The first and last deciles for a standard Normal distribution are ±1.2816. **(b)** For a $N(9.12, 0.15)$ distribution, the first and last deciles are 8.93 and 9.31 ounces.

1.139 (a) As the quartiles for a standard Normal distribution are ±0.6745, we have $IQR = 1.3490$. **(b)** $c = 1.3490$.

1.141 The deciles are shown here.

Percentile	10%	20%	30%	40%	50%
HDL level	35.2	42.0	46.9	51.1	55

Percentile	60%	70%	80%	90%
HDL level	58.9	63.1	68.0	74.8

1.143 (b) The data are roughly Normal, but there is one potential high outlier.

1.145 There is less energy used in 2014 from fossil fuels and more used from renewable sources; there was almost no change in nuclear and electric power usage.

1.147 $\bar{x} = 8.389$, $s = 1.965$. Min = 4.9, $Q_1 = 6.9$, $M = 8.2$, $Q_3 = 9.5$, Max = 15.1. The distribution is somewhat right-skewed so the five-number summary is a better description for highway fuel consumption.

1.151 (a) For car makes (a categorical variable), use either a bar graph or pie chart. For car age (a quantitative variable), use a histogram, stemplot, or boxplot. **(b)** Study time is quantitative, so use a histogram, stemplot, or box-plot. To show change over time, use a time plot (average hours studied against time). **(c)** Use a bar graph or pie chart to show radio station preferences. **(d)** Use a Normal quantile plot to see whether the measurements follow a Normal distribution.

1.155 (a) $\sigma = 7.5$.

1.157 (a) $\mu = 81.55$. $\sigma = 30.4$.

1.159 (a) Most people will "round" their answers when asked to give an estimate like this; in fact, the most striking answers are ones such as 115, 170, or 230. The students who claimed 360 minutes (six hours) and 300 minutes (five hours) may have been exaggerating. **(b)** Women seem to generally study more (or claim to), as there are none that claim less than 60 minutes per night. The center (median) for women is 170; for men the median is 120 minutes.

1.161 No, and no: There are examples of many different data sets with mean 0 and standard deviation 1. Likewise, for any given five numbers $a \leq b \leq c \leq d \leq e$ (not all the same), we can create many data sets with that five-number summary simply by taking those five numbers and adding some additional numbers in between them—for example (in increasing order): $10, \ldots, 20, \ldots, \ldots, 30, \ldots, \ldots, 40, \ldots, 50$. As long as the number in the first blank is between 10 and 20, and so on, the five-number summary will be 10, 20, 30, 40, 50.

1.163 $\bar{x} = 35.66$, $s = 41.56$, Min = 0, $Q_1 = 1$, $M = 11.5$, $Q_3 = 68$, Max = 181. On average, the band pauses for

35.66 seconds; however, the largest portion of the time, they don't pause at all. The distribution is strongly right-skewed and shows that sometimes the band pauses for as much as 181 seconds, or three minutes, before playing the final note.

1.165 Antho1 is Normally distributed. $\bar{x} = 1.630$, $s = 0.521$.

1.166 Antho2 is Normally distributed. $\bar{x} = 1.711$, $s = 0.590$.

1.167 Antho3 is right-skewed, it also has a potential high outlier. Min = 0.0546, $Q_1 = 0.4506$, $M = 0.6781$, $Q_3 = 1.1282$, Max = 6.3109.

CHAPTER 2

2.1 (a) The cases are students. (b) Number of friends and time (average amount spent on Facebook per week). (c) Both variables are quantitative because both are numeric and arithmetic operations are possible.

2.3 Cases: cups of Mocha Frappuccino. Variables: size and price (both quantitative).

2.5 (a) Tweets. (b) Click count and length of tweet are quantitative. Day of week and sex are categorical. Time of day could be quantitative (as hh:mm) or categorical (if morning, afternoon, etc.). (c) Click counts is the response. The others could all be potentially explanatory.

2.7 Some possible variables are condition of the book (with values poor, good, or excellent), number of pages, and binding type (with values hardback or paperbound), in addition to purchase price and buyback price. Cases would be individual textbooks. Here, we would likely be interested in the relationship between the buyback price and other variables.

2.9 Some possible variables are university, size, etc., in addition to the average number of tickets sold and the percentage of games won. Cases would be individual teams. Here, we would likely be interested in if there is a relationship between the average number of tickets sold and the percentage of games won.

2.11 Rating looks Normally distributed. Price also looks somewhat Normal but has a high outlier.

2.13 (c) There is no difference in the plot except for the scale on the x axis.

2.15 Answers may vary. Most will likely prefer the plot with the log transformation because it spreads out the data points and makes the plot easily to read and interpret.

2.17 (a) A negative association means that low values of one variable are associated with *high* values of the other variable. (b) A stemplot is used for only a single quantitative variable. (c) We put the response variable on the y axis and the explanatory variable on the x axis.

2.19 (b) The form is somewhat linear; the direction is positive; the strength is still weak. (c) There are a few possible low outliers for Antho3. (d) Adding a line could

be very useful because the relationship is somewhat linear. (e) Smoothing does not contribute much in describing the relationship.

2.21 (b) The form is linear; the direction is positive; the strength is very strong. (c) There is one outlier. (d) Yes, the line shows the direction and strength. (e) Smoothing does not help at all because the relationship is quite linear.

2.23 (a) For all fuel types, as highway fuel consumption increases, so do carbon dioxide emissions. (b) Vehicles with fuel type D have the largest emissions, while vehicles with fuel type E have the smallest emissions. The other two types, X and Z, have fairly similar emissions.

2.25 (b) As nondominant arm strength increases, so does dominant arm strength. There is one outlier with an extremely high nondominant arm strength. (c) Linear; positive; strong. (d) There is one outlier. (e) Yes, the relationship is linear except for the outlier.

2.27 Parents' income is explanatory, and college debt is the response. Both variables are quantitative. We would expect a negative association: for parents with lower incomes, student college debt would be high and vice versa.

2.29 (b) Smoothing does not help as the relationship is quite linear.

2.31 The relationships between calories and alcohol content are quite similar for both domestic and imported beers. Also, the outlier for the imported beers no longer is an outlier because there are several other domestic beers that have a similar alcohol content.

2.33 (b) As time increases, logcount goes down. (c) Linear; negative; strong. (d) There are no outliers. (e) The relationship is very linear.

2.35 (b) The plot is more linear than the original scatterplot. (c) The log transformed data should be preferred because it straightens out the relationship.

2.37 (b) The association is positive, linear, and strong. (c) Overall the relationship is strong, but it is stronger for women than for men. Male subjects generally have both larger lead body mass and higher metabolic rates than women.

2.39 (a) This is a linear transformation. Dollars = 0 + 0.01 × Cents. (b) $r = 0.671$. (c) They are the same. (d) Changing the units does not change the correlation.

2.41 (a) Strong and positive. (b) Strong and negative. (c) Weak and negative. (d) No linear relationship.

2.43 (a) $r = 0.298$. (b) Probably, the plot is somewhat linear. (c) No, if they were approximately equal, the correlation would be closer to 1.

2.45 For fuel type D: $r = 0.977$. For fuel type E: $r = 0.983$. For fuel type X: $r = 0.981$. For fuel type Z: $r = 0.976$. For all four fuel types: the correlations are around 0.98 and the relationship between CO_2 emissions and highway fuel consumption is linear and very strong.

2.47 **(a)** $r = 0.905$. **(b)** Yes, because the pattern is very linear. There is one outlier, but it fits the overall pattern.

2.49 There is little linear association between research and teaching—for example, knowing that a professor is a good researcher gives little information about whether she or he is a good or bad teacher.

2.51 **(a)** $r = -0.999$. **(b)** Yes, because the scatterplot is very strongly linear. **(c)** You must be careful; there can be a strong correlation between two variables even when the relationship is curved. Plot the data first!

2.53 **(a)** $r = 0.905$. **(b)** Yes, because it is quite linear; however, there is one outlier in this dataset, O'Douls, with an extremely low alcohol percent.

2.55 $r = 0.912$. Both correlations for the imported and domestic beers are quite similar, especially when the outlier O'Doul's is removed. The relationships between calories and percent alcohol for both types of beers are linear and very strong and quite similar in pattern.

2.57 **(a)** With only two points, the correlation will be 1 or -1 because they form a perfect straight line.

2.59 **(a)** $r = -0.72971$. **(b)** The correlation is not a good numerical summary for this relationship because there is a curvature in the plot.

2.61 **(a)** Almost all the data points are below the line.

2.63 Predictions for 300 and 600 are trustworthy: interpolation. Predictions from -300 and 800 are not trustworthy: extrapolation.

2.65 **(a)** $\hat{y} = 0.34475 + 0.15185 \text{Antho3}$. **(c)** The line does not fit the data well. **(d)** $\hat{y} = 0.572525$.

2.67 **(a)** $\hat{y} = 56.75012 + 20.78205 \text{FuelConsHwy}$. **(c)** A single regression line would not be a good fit for the four types of vehicles, even though the correlations were all very close. There are several different lines that need to be accounted for.

2.69 **(c)** There is a strong linear relationship. For each unit increase in nondominant, the dominant arm bone strength increases by 1.373.

2.71 Predicted bone strength is 22.854 cm^4/1000.

2.73 **(a)–(c)**

		Count = 602.8 − (74.7 × time)		
Time	Count	Predicted	Difference	Squared difference
1	578	528.1	49.9	2490.01
3	317	378.7	−61.7	3806.89
5	203	229.3	−26.3	691.69
7	118	79.9	38.1	1451.61

(d)

		Count = 500 − (100 × time)		
Time	Count	Predicted	Difference	Squared difference
1	578	400	178	31,684
3	317	200	117	13,689
5	203	0	203	41,209
7	118	−200	318	101,124

(e) The first line is a better description of the relationship.

2.75 **(b)** The relationship is linear, positive and strong. There are several outliers. **(c)** $\hat{y} = -15057 + 0.05326x$.

2.77 **(a)** $\hat{y} = 197,993$. **(b)** $\hat{y} = 202,327$. **(c)** The outliers didn't change the prediction for the median-sized state.

2.79 **(a)** $\hat{y} = 8491.907 + 0.048x$. **(b)** $r^2 = 0.942$. **(c)** 94.2% of the variation in the number of undergraduates is accounted for by the population size. **(d)** The software does not report nature of the relationship.

2.81 **(a)** There is a weak positive linear relationship but with one extreme outlier. **(b)** $\hat{y} = -7.28789 + 1.89089x$. **(c)** $r^2 = 0.7715$. **(d)** Although the x variable accounts for 77.15% of the variation in y, there is a very high outlier for x, which pulls the regression line unnaturally, accounting for the jump in R-square.

2.83 **(a)** $\hat{y} = 6.41895 + 28.34687x$. **(b)** $r^2 = 0.8193$. **(c)** The relationship between calories and percent alcohol is linear, positive, and strong; however, there does seem to be one low outlier, O'Douls, with a very low alcohol content.

2.85 **(a)** The correlations and regression lines for all four data sets are essentially the same: $r = 0.82$ and $\hat{y} = 3 + 0.5x$. For $x = 10$, $\hat{y} = 8$. **(c)** Only for data set A should regression be used.

2.87 **(a)** $y = 7$. **(b)** For each unit increase in x, y decreases by 2. **(c)** 15.

2.89 $\hat{y} = 0.896x - 1517.935$. $r = 0.982$. We can conclude that NAEP scores are steadily increasing about 0.896 point per year.

2.91 $r = 0.4$.

2.93 The residuals sum to 0.01.

2.95 The residuals are 10.58, -0.13, -7.24, -0.14.

2.97 **(a)–(b)**

Time	LogCount	Predicted	Residual
1	6.35957	6.33244	0.02713
3	5.75890	5.81121	−0.05231
5	5.31321	5.28997	0.02323
7	4.77068	4.76874	0.00195

(c) The residual plot looks random; the model using logs is much better.

2.99 (c) No. **(d)** No. **(e)** No. **(f)** California is not an outlier and does not influence the regression line using the log transformations.

2.101 (a) If the line is pulled toward the influential point, the observation will not necessarily have a large residual. **(b)** High correlation is always present if there is causation. **(c)** Extrapolation is using a regression to predict for x-values outside the range of the data (here, using 20, for example).

2.103 Internet use does not cause people to have fewer babies. Possible lurking variables are economic status of the country, levels of education, etc.

2.105 For example, a reasonable explanation is that the cause-and-effect relationship goes in the other direction: doing well makes students or workers feel good about themselves, rather than vice versa.

2.107 The explanatory and response variables were "consumption of herbal tea" and "cheerfulness/health." The most important lurking variable is social interaction; many of the nursing home residents may have been lonely before the students started visiting.

2.109 (a) It is difficult to draw the correct line by hand. **(b)** Most people tend to overestimate the slope for a scatterplot with $r = 0.6$; that is, most students will find that the least-squares line (the one without the ending dots) is less steep than the one they draw.

2.111 Each group has a positive association, but when combined, the regression slope is negative.

2.113 Sum the rows of the table. 1278 met the requirements and 751 did not meet requirements.

2.115 Divide the cell count by the table total. 861/2029 = 0.4243.

2.117 Divide the cell count by the total number of children aged 11 to 13. 417/974 = 0.4281 (which rounds to 43%).

2.119 (a) Because they want to see the effect of driver's education courses, that is the explanatory variable. The number of accidents is the response. **(b)** Driver's Ed would be the column (x) variable, and number of accidents would be the row (y) variable. **(c)** There are six cells (two columns by three rows). For example, the first row, first column entry could be the number who took driver's education and had 0 accidents.

2.121 (a) Age is the explanatory variable. Rejected is the response. With the dentistry available at that time, it's reasonable to think that as a person got older, he would have lost more teeth.

(b)

	<20	20−25	25−30	30−35	35−40	>40
Yes	0.0002	0.0019	0.0033	0.0053	0.0086	0.0114
No	0.1761	0.2333	0.1663	0.1316	0.1423	0.1196

(c)

Marginal distribution of "Rejected"

Yes	No
0.03081	0.96919

Marginal distribution of age

<20	20−25	25−30	30−35	35−40	>40
0.1763	0.2352	0.1696	0.1369	0.1509	0.1310

(d) The conditional distribution of Rejected given Age, because we have said Age is the explanatory variable. **(e)** In the table, note that all columns sum to 1. We can clearly see the proportion of rejected recruits increasing with increasing age.

	<20	20−25	25−30	30−35	35−40	>40
Yes	0.0012	0.0082	0.0196	0.0389	0.0572	0.0868
No	0.9988	0.9918	0.9804	0.9611	0.9428	0.9132

2.123 Sex is the explanatory, Lied is the response. For the males, about 55% admitted that they lied, whereas for the females, 51% admitted that they had lied. Males maybe be slightly more willing to admit that they lied than females.

2.125 (a) 50.5% get enough sleep and 49.5% do not. **(b)** 32.2% get enough sleep and 67.8% do not. **(c)** Those who exercise more than the median are more likely to get enough sleep.

2.127 3.0% of Hospital A's patients died, compared with 2.0% at B.

2.129 In general, choose a to be any number from 0 to 300, and then all the other entries can be determined.

2.131 For example, causation might be a negative association between the setting on a stove and the time required to boil a pot of water (higher setting, less time). Common response might be a positive association between SAT score and grade point average. Both of these will have a positive relationship with a person's IQ. An example of confounding might be a negative association between hours of TV watching and grade point average. Once again, people who are naturally smart could finish required work faster and have more time for TV; those who aren't as smart could become frustrated and watch TV instead of doing homework.

2.133 This is a case of confounding: The association between dietary iron and anemia is difficult to detect because malaria and helminths also affect iron levels in the body.

2.135 Responses will vary. For example, students who choose the online course might have more self-motivation or better computer skills. A diagram is shown on the right; the generic "Student characteristics" might be replaced with something more specific.

2.137 No; self-confidence and improving fitness could be common responses to some other personality trait, or high self-confidence could make a person more likely to join the exercise program.

2.139 Patients suffering from more serious illnesses are more likely to go to larger hospitals (which may have more or better facilities) for treatment. They are also likely to require more time to recuperate afterward.

2.141 People who are overweight are more likely to be on diets and so choose artificial sweeteners over sugar.

2.143 This is an observational study—students choose their "treatment" (to take or not take the refresher sessions).

2.145 (a) There is no linear relationship. (b) $\hat{y} = 110.95813 + 0.07315\text{DwellPermit}$. (c) For each new index point of dwelling permits issued, production increases by 0.07315. (d) 110.95813. This is what we would expect sales to be when the index for permits issued for new dwellings is 0. (e) $\hat{y} = 127.3444$. (f) $e = -5.34444$. (g) R-square = 1.99%.

2.147 (a) As the percent under 15 increases, the percent of the population over 65 decreases. (b) $r = -0.851$. The correlation gives a pretty good representation of the relationship; however, there is an outlier, Nunavut.

2.149 (b) The three territories have smaller percentages than any of the provinces. Additionally, two of the three territories have larger percentages of the population under 15 than any of the provinces.

2.151 Generally speaking, banks that offer RDC are more likely to have larger asset sizes, while the banks that don't offer RDC are more likely to have smaller asset sizes.

2.153 (a) The marginal totals are SBL: 1697; SME: 911; AH: 801; Ed: 319; Other: 857. By country, Canada: 176; France: 672; Germany: 218; Italy: 321; Japan: 654; UK: 475; U.S. 2069. (b) Canada: 3.84%; France: 14.7%; Germany: 4.8%; Italy: 7%; Japan: 14.3%; UK: 10.4%; U.S. 45.1%. (c) SBL: 17.5%; SME: 7%; AH: 18.7%; Ed: 19.9%; Other: 37%.

2.155 A school that accepts weaker students but graduates a higher-than-expected number of them would have a positive residual, whereas a school with a stronger incoming class but a lower-than-expected graduation rate would have a negative residual. It seems reasonable to measure school quality by how much benefit students receive from attending the school.

2.157 (a) The residuals are positive at the beginning and end, and negative in the middle. (b) The behavior of the residuals agrees with the curved relationship seen in Figure 2.34.

2.159 (a) $\hat{y} = 41.253 + 3.9331\text{Year}$; for year 25, the predicted salary is 139.58, or about $139,600. (b) Using logs: $\hat{y} = 3.8675 + 0.04832\text{Year}$. At year 25, we predict 5.0755,

or about $160,050. (c) Although both predictions involve extrapolation, the second is more reliable because it is based on a linear fit to a linear relationship. (d) Interpreting relationships without a plot is risky.

2.161 (a) $\hat{y} = 2990.42521 + 0.99216\text{Salary2014\_15}$. (b) There are two outliers in the y direction.

2.163 *Number of firefighters* and *amount of damage* are common responses to the seriousness of the fire.

2.165 There is a strong linear positive association. $\hat{y} = 2652.66 + 0.00474\text{MOE}$; $r^2 = 0.6217$. We can use MOE to get fairly good predictions of MOR.

2.167 (a) Table shown here. (b) Males: 70% admitted. Females: 56% admitted. (c) Business: 80% of males and 90% of females. For Law, 10% of males and 33% of females. (d) 75% of men apply to business school, where admission is easier. More women apply to law school, which is more selective.

Sex	Admit	Deny	Total
Male	490	210	700
Female	280	220	500
Total	770	430	1200

2.169 If we ignore "Year," Department A teaches 61.54% small classes and Department B teaches 39.62% small classes. However, in upper-level classes, A has 77.5% and B has 83.33% small classes. Additionally, 77.78% of A's classes are upper-level courses, compared to 33.96% of B's classes.

2.171 (a) People have mixed feelings on the quality of the recycled filters. (b) 55.6% of buyers think the quality is higher, while 44.3% of nonbuyers think the quality lower. It is plausible that using the filters may cause more favorable opinions.

2.173 (a) The tables are shown here.

Female *Titanic* passengers

	Class			
	1	2	3	Total
Survived	139	94	106	339
Died	5	12	110	127
Total	144	106	216	466

Male *Titanic* passengers

	Class			
	1	2	3	Total
Survived	61	25	75	161
Died	118	146	418	682
Total	179	171	493	843

(b) If we look at the conditional distribution of survival given class for females, 96.53% of first-class females survived, 88.68% survival among second-class females, and 49.07% survival among third-class females. Survival depended on class. **(c)** For males, 34.08% survival among first class, 14.62% survival among second class, and 15.21% survival among third class. Once again, survival depended on class. **(d)** Females overall had much higher survival rates than males.

CHAPTER 3

3.1 This is anecdotal evidence; the preference of a friend likely would not generalize to the entire college.

3.3 This is anecdotal evidence; the opinion of the three students does would not generalize to all students, especially if the students were picked by the newspaper because of their strong opinions in the first place.

3.5 In order to evaluate the claim, we would need data for a sample of the Millennial generation, including their brand preferences and measures of loyalties toward these brands.

3.7 Yes to both.

3.9 This is an observational study. All the soap dispensers were run until their batteries died; there were no treatments.

3.11 This is an experiment. Explanatory variable: apple form (juice or whole fruit); response variable: how full the subject felt.

3.13 The data used was likely observational data from a sample survey. Most likely, a random sample of cans of tuna was measured and compared with the amount printed on the labels.

3.15 (a) This is a sample survey if it is supposed to represent only those who did not get tickets; otherwise, it is anecdotal evidence and certainly doesn't apply to all of those who tried to get tickets to the concert. **(b)** This represents a sample survey of all those that tried to get tickets to the concert. **(c)** This also represents a sample survey of all those who tried to get tickets to the concert. It differs from part (b) because only a sample of students were selected to participate rather than allowing participants to be self-selected. **(d)** It is not an experiment, there is no treatment. **(e)** The method used in part (a) is not very useful, especially if it was determined they were upset before being interviewed. The method in part (b) also can be problematic because those who choose to respond may have strong opinions, especially negative ones. The method in part (c) is likely the best of the three in order to get an accurate representation of students' opinion on how the tickets were allocated.

3.17 (a) This is an experiment because the girls were assigned a controlled diet during the study period. The experimental units/subjects are the girls, the treatments are the two controlled diets (low and high calcium), the response is the amount of calcium retained. The factor in this instance is the same as the treatment, the diet assigned, and it has two levels—low and high calcium.

3.19 (a) This study is biased because no control group was used. Therefore, the placebo effect may be present. **(b)** To remove the bias, two treatment groups should be used, one with the aspirin and a control group that receives a placebo.

3.25 (a) Experimental units were the 30 students. They are human, so we can use "subjects." **(b)** We have only one "treatment," so the experiment is not comparative.

3.27 (a) Experimental units: people who go to the website. Treatments: description of comfort or showing discounted price. Response variable: shoe sales. **(b)** Comparative, because we have two treatments. **(c)** One option to improve: randomly assign morning and afternoon treatments. **(d)** Yes, a placebo (no special description or price) could give a "baseline" sales figure.

3.29 No, a matched pairs design would require that the two measurements are taken on the same subject—in this case, the person who visits the website—which would be impossible in this case.

3.31 (a) Shopping patterns may differ on Friday and Saturday. **(b)** Responses may vary in different states. **(c)** A control is needed for comparison.

3.33 For example, new employees should be randomly assigned to either the current program or the new one. One possible outcome would be whether the new employee is still with the company six months later.

3.35 (a) The factors are calcium dose and vitamin D dose. There are nine treatments. **(d)** Yes, there is a placebo. The group that gets 0 mg of both calcium and vitamin D serves as the placebo group.

3.39 Design **(a)** is an experiment. Because the treatment is randomly assigned, the effect of other habits would be "diluted" because they would be more-or-less equally split between the two groups. Therefore, any difference in colon health between the two groups could be attributed to the treatment.

Design **(b)** is an observational study. It is flawed because the women observed choose whether or not to take bee pollen; one might reasonably expect that people who choose to take bee pollen have other dietary or health habits that would differ from those who do not.

3.45 (a) The sample is the 772 forest owners that the survey was sent to. **(b)** The population is all forest owners from this region. **(c)** The response rate is the percentage of those who were sent the survey who returned it: 45%.

3.47 The next three labels selected are 114, 080, 094. These correspond to Malta, Iraq, and Kosovo.

3.51 (a) The population is all 5674 season ticket holders. **(b)** The sample is the 150 fans who were sent the survey. **(c)** The response rate is 65.33%. **(d)** The nonresponse rate is 34.67%.

3.55 (a) If the entire population is found in our sample, we have a *census* rather than a sample. **(b)** "Dihydrogen monoxide" is H_2O (water). Any concern about the dangers posed by water most likely means that the respondent did not know what dihydrogen monoxide was and was too embarrassed to admit it. **(c)** Honest answers to such questions are difficult to obtain, even in an anonymous survey; in a public setting like this, it would be surprising if there were any raised hands.

3.57 The population is local businesses. The sample is the 80 businesses selected. The nonresponse rate is 42.5%.

3.59 Using line 136, and labels 01−33 we select: 08, 14, 20, 09, 24, 12, 11, and 16. Complexes chosen may vary depending on labels; alphabetically, we would choose: Burberry, Crestview, Georgetown, Cambridge, Nobb Hill, Country View, Country Square, and Fairington.

3.63 Each student has a 20% chance: five out of 25 over-21 students, and three of 15 under-21 students. This is not an SRS because not every group of eight students can be chosen; the only possible samples are those with five older and three younger students.

3.65 The sample is random because the starting point is randomly selected (so every individual has an equal chance to be selected before the process begins). Once the random starting point has been selected, the rest of the sample is determined. There is no possibility of selecting students 04 and 05, which could happen in a simple random sample.

3.69 **(a)** This design would omit households without telephones or with unlisted numbers. Such households would likely be made up of poor individuals (who cannot afford a phone), those who choose not to have phones (perhaps because they use a cell phone exclusively), and those who do not wish to have their phone numbers published. **(b)** Those with unlisted numbers would be included in the sampling frame when a random-digit dialer is used.

3.71 These three proposals are clearly in increasing order of risk. Most students will likely consider that **(a)** qualifies as minimal risk, and most will agree that **(c)** goes beyond minimal risk.

3.73 It is good to plainly state the purpose of the research ("To study how people's religious beliefs and their feelings about authority are related"). Stating the research *thesis* (that orthodox religious beliefs are associated with authoritarian personalities) would cause bias.

3.81 To control for changes in the mass spectrometer over time, we should alternate between control and cancer samples.

3.83 They cannot be anonymous because the interviews are conducted in person in the subject's home. They are certainly kept confidential.

3.87 **(b)** The subjects should be told what kind of questions will be asked and how long it will take. **(d)** Revealing the sponsor could bias the poll, especially if the respondent doesn't like or agree with the sponsor. However, the sponsor should be announced once the results are made public; that way, people can know the motivation for the study and judge whether it was done appropriately, etc.

3.93 **(a)** You need information about a random selection of his games, not just the ones he chooses to talk about. **(b)** These students may have chosen to sit in the front; all students should be randomly assigned to their seats.

3.95 This is an experiment because each subject is (randomly) assigned to a treatment. The explanatory variable is the price history seen by the subject (steady prices or fluctuating prices), and the response variable is the price the subject expects to pay.

3.99 The two factors are gear (three levels) and steepness of the course (number of levels not specified). Assuming there are at least three steepness levels—which seems like the smallest reasonable choice—that means at least nine treatments. Randomization should be used to determine the order in which the treatments are applied. Note that we must allow ample recovery time between trials, and it would be best to have the rider try each treatment several times.

3.103 Use a block design: separate men and women, and randomly allocate each sex among the six treatments.

3.105 The latter method (CASI) will show a higher percentage of drug use because respondents will generally be more comfortable (and more assured of anonymity) about revealing illegal or embarrassing behavior to a computer than to a person, so they will be more likely to be honest.

CHAPTER 4

4.1 Six of the first 10 digits on line 131 correspond to "heads," so the proportion of heads is 60%. Although the average number of heads in 10 tosses is five, the actual outcome is random and can vary from sample to sample.

4.3 **(a)** We can discuss the probability (chance) the temperature would be between 30°F and 35°F, for example. **(b)** Depending on your school, student identification numbers are probably not random. For example, at the author's university, all student IDs begin with 900, which means that the first three digits are all the same and not random. **(c)** The probability of an ace in a single draw is 4/52 if the deck is well shuffled.

4.9 S = {all numbers between 0 and 168}.

4.11 0.84. Adding three probabilities and subtracting that result from 1 is slightly easier than adding the five probabilities of interest.

4.13 0.252.

4.15 0.25.

4.17 **(a)** S = {Yes, No}. **(b)** S = {0, 1, 2,..., x}, where x is a reasonable upper limit. **(c)** S = {18, 19, 20,...}. There is some leeway here with the lower and upper ends of the ages. **(d)** Answers will vary by institution.

4.19 **(a)** Not equally likely: she wins close to 60% of her matches. **(b)** Equally likely. The chance of a 3 and the chance of a 4 are both 1/6. **(c)** Probably not equally likely because it depends on the intersection. **(d)** Not equally likely: home teams win more than half their games.

4.21 (a) The probability that both of the two disjoint events occur is 0. **(b)** Probabilities must be no more than 1; $P(A$ and $B)$ will be no more than 0.7. **(c)** $P(\text{not } A) = 0.55$.

4.23 There are five possible outcomes: $S = \{$link1, link2, link3, link4, leave$\}$.

4.25 (a) 0.03. **(b)** 0.55.

4.27 (a) No. These student categories are disjoint and the probabilities sum to more than 1. **(b)** This is legitimate, but the deck would be a nonstandard one. **(c)** This is legitimate, but it represents a "loaded" die.

4.29 (a) $P(\text{some education beyond high school, but no degree}) = 0.28$. **(b)** $P(\text{at least high school}) = 0.88$.

4.31 Possible types: A+, A−, B+, B−, AB+, AB−, O+, O−. Probabilities: 0.294, 0.056, 0.084, 0.016, 0.0252, 0.0048, 0.4368, 0.0832.

4.33 (a) $P(\text{win}) = 0.006$. **(b)** $P(\text{win}) = 0.003$.

4.35 $P(\text{at least one is O-negative}) = 0.5160$.

4.37 Observe that $P(A \text{ and } B^c) = P(A) - P(A)P(B) = P(A)(1 - P(B))$.

4.39 (a) Either A or O. **(b)** $P(\text{O}) = 0.25$ and $P(\text{A}) = 0.75$.

4.41 (a) 0.578125. **(b)** 0.015625; 0.140625.

4.43 Possible values: 0, 1, 2. Probabilities: 1/4, 1/2, 1/4.

4.45 Possible values: 1, 2, 3, 4, 5, 6. Probabilities: 0.05, 0.05, 0.13, 0.26, 0.36, 0.15.

4.47 (a) 0.23. **(b)** 0.62. **(c)** 0.

4.49 (a) The possible values certainly can be negative, it is the probabilities that can't be negative. **(b)** Continuous random variables can take values from any interval, not just 0 to 1. **(c)** A Normal random variable is continuous.

4.51 Possible values: 0, 1, 2. Probabilities: 0.0361, 0.3078, 0.6561.

4.53 (a) Time is continuous. **(b)** Hits are discrete (you can count them). **(c)** Yearly income is discrete (you can count money).

4.55 (b) Each pair has probability 1/36. **(c)** For the distribution, we see that there are four pairs that add to 5, so $P(X = 5) = 4/36$. **(d)** $P(7 \text{ or } 11) = 0.22\overline{2}$. **(e)** $P(\text{not } 7) = 0.8333$.

Sum	2	3	4	5	6	7	8	9	10	11	12
Probability	1/36	2/36	3/36	4/36	5/36	6/36	5/36	4/36	3/36	2/36	1/36

4.57 (b) $P(X \geq 1) = 0.9$. **(c)** "At most, three nonword errors." $P(X \leq 3) = 0.9$, $P(X < 3) = 0.7$.

4.59 (a) The height should be 1/2. **(b)** 0.8. **(c)** 0.6. **(d)** 0.525.

4.61 This probability is essentially 1.

4.63 The possible values: \$0 and \$10. Probabilities: 0.5, 0.5. Mean: \$5.

4.65 $\mu_Y = 84$.

4.67 $\mu_X = 2.1$. $\sigma^2_X = 1.89$. $\sigma_X = 1.3748$.

4.69 As sample size gets larger, the standard deviation decreases. The mean for 1000 will be much closer to μ than the mean for 2 (or 100) observations.

4.71 $\sigma^2_X = 1.24$, $\sigma = 1.1136$.

4.73 (a) $\sigma^2_Z = 1600$, $\sigma_Z = 40$. **(b)** $\sigma^2_Z = 2304$, $\sigma_Z = 48$. **(c)** $\sigma^2_Z = 48$, $\sigma_Z = 6.928$. **(d)** $\sigma^2_Z = 48$, $\sigma_Z = 6.928$. **(e)** $\sigma^2_Z = 192$, $\sigma_Z = 13.856$.

4.75 $\mu = 2.2$.

4.77 $\sigma = 0.373$.

4.79 (a) $\sigma = \$85.48$. **(b)** Larger; the negative correlation makes the standard deviation of the sum smaller.

4.81 The exercise implies a positive correlation between calcium intake and compliance. Because of this, the variance of total calcium intake is greater than the variance we would see if there were no correlation.

4.83 (a) $\mu_X = 0.5$. $\sigma_X = 0.5$. **(b)** $\mu_Y = 2$. $\sigma_Y = 1$. **(c)** $\mu = 2$. $\sigma = 1$.

4.85 (a) Not independent. **(b)** Independent.

4.87 If one of the 10 homes were lost, it would cost more than the collected premiums. For many policies, the average claim should be close to mean.

4.89 $P(3 \text{ or } 4 \text{ or } 5) = 0.5$.

4.91 $13/48 = 0.2708$.

4.93 The addition rule for disjoint events.

4.97 (a) 0.7. **(c)** 0.3.

4.99 (a) 0.4.

4.101 (a) $A = 5$ to 10 years old; $A^C = 11$ to 13 years old; $C = $ adequate calcium intake; $C^C = $ inadequate. **(b)** $P(A) = 0.52$; $P(A^C) = 0.48$. $P(C^C|A) = 0.18$; $P(C^C|A^C) = 0.57$. **(c)** $P(\text{inadequate calcium}) = 0.3672$.

4.103 Not independent; $P(A^C|C^C) = 0.7451 \neq P(A^C) = 0.48$.

4.105 (a) 0.14. **(b)** 0.11. **(c)** 0.47. **(d)** For parts (a) and (b), use the addition rule for disjoint events. For part (c), use the general addition rule.

4.107 $L = $ "student admits to lying" and $M = $ "student is male," so $P(L) = 0.53$, $P(M) = 0.44$, and $P(M \text{ and } L) = 0.24$. Then, $P(M \text{ or } L) = P(M) + P(L) - P(M \text{ and } L) = 0.44 + 0.53 - 0.24 = 0.73$.

4.109 (a) The four entries are: 0.2684, 0.3416, 0.1599, 0.2301. **(b)** 0.5975.

4.111 The tree diagrams should be different because each branch on the tree is a conditional probability given the previous branches. So by rearranging the order of the branches, we are changing the probabilities.

4.113 $P(A|B) = 0.3142$. If A and B are independent, then $P(A|B) = P(A)$.

4.115 (a) $P(A^C) = 0.69$. (b) $P(A \text{ and } B) = 0.08$.

4.117 $P(\text{at least one offer}) = 0.9$.

4.119 $P(B|C) = 0.33$. $P(C|B) = 0.2$.

4.121 (a) Beth's brother has type aa, and he got one allele from each parent. (b) $P(aa) = 0.25$, $P(Aa) = 0.5$, and $P(AA) = 0.25$. (c) $P(AA|\text{not } aa) = 1/3$, $P(Aa|\text{not } aa) = 2/3$.

4.123 0.9333.

4.125 Close to $\mu = -2.4$.

4.127 (a) Possible values: 19 and 44. Probabilities: 0.2 and 0.8. (b) $\mu = 39$. $\sigma^2 = 100$, so $\sigma = 10$. (c) There are no rules for a quadratic function of a random variable; we must use the definitions.

4.129 (a) $P(A) = 1/36$, $P(B) = 15/36$. (b) $P(A) = 1/36$, $P(B) = 15/36$. (c) $P(A) = 10/36$, $P(B) = 6/36$. (d) $P(A) = 10/36$, $P(B) = 6/36$.

4.131 For each bet, the mean is the winning probability times the winning payout, plus the losing probability times $-\$10$. All mean payoffs equal \$0.

4.133 (a) All the probabilities are between 0 and 1 and sum to 1. (b) 0.61. (c) 0.39; 0.39.

4.135 0.00568.

4.137 $P(\text{no point}) = 1/3$. Probability of winning(losing): $1/36(1/18)$ on 4 or 10, $2/45(1/15)$ on 5 or 9, $25/396(5/66)$ on 6 or 8.

4.139 1/18.

CHAPTER 5

5.1 84% is the statistic; the population is all women in the world; the population parameter would be the true proportion of all women who experience their first street harassment before the age of 17.

5.3 It is necessary for the amounts he collected to serve as a sample from a larger population, but it is not entirely clear what this population would be. Without these assumptions, his records would be considered a population and would not represent a sampling distribution.

5.5 The centers of the sampling distributions would be the same. The spread of the sampling distribution for the SRS of 600 would be smaller than it would be for the SRS of 200.

5.7 (a) The population is the 153 English majors at your college. (b) The sample is the 30 selected to be on the committee. (c) The statistic is the number or proportion from the 30 in favor of the change. (d) The parameter would be the number or proportion of all 153 students who would favor the change.

5.9 (a) Population: all students in the United States at four-year colleges. Sample: the 17,096 students surveyed.

(b) Population: all restaurant workers. Sample: the 100 people asked. (c) Population: all 584 longleaf pine trees. Sample: the 40 trees measured.

5.11 (a) We would expect the distribution to be right-skewed, causing the mean to be larger than the median.

5.13 (a) The shape of the sampling distribution should be roughly Normal centered at 0.4. (b) The shape of the sampling distribution should be roughly Normal centered at 0.2.

5.15 (a) and (b) The mean should be "close" to 50.5. (c) The histogram theoretically should be a Normal distribution, centered at 50.5.

5.17 Population: all adults in the United States. Statistic: mean of 37 hours and 6 minutes. Likely values: answers will vary, any amount of time someone could spend during a month using mobile apps.

5.19 $\mu_{\bar{x}} = 44$. $\sigma_{\bar{x}} = 0.667$.

5.21 About 95% of the time, $\bar{x}$ should be between 81 and 83.

5.23 (a) Each sample size has $\mu_{\bar{x}} = 1$. For $n = 2$, $\sigma_{\bar{x}} = 0.7071$. For $n = 10$, $\sigma_{\bar{x}} = 0.3162$. For $n = 25$, $\sigma_{\bar{x}} = 0.2$.

5.25 (a) The standard deviation for $n = 10$ will be $\sigma_{\bar{x}} = 10/\sqrt{10}$. (b) Standard deviation *decreases* with increasing sample size. (c) $\mu_{\bar{x}}$ always equals μ. (d) The population size N does not matter as long as it is relatively large compared to n.

5.27 (a) $\mu = 120.5$. (d) The center of the histogram should theoretically be close to μ.

5.29 (a) Larger. (b) We need $\sigma_{\bar{x}} \leq 0.04$. (c) We need $n = 775$.

5.31 The mean will be 250 ml. The standard deviation will be $\sigma_{\bar{x}} = 0.1789$ ml.

5.33 (b) $P = 0.0124$. (c) P is essentially 0.

5.35 (a) $\bar{x}$ is not systematically higher than or lower than μ. (b) With large samples, $\bar{x}$ is more likely to be close to μ.

5.37 (a) $\mu_{\bar{x}} = 0.4$. $\sigma_{\bar{x}} = 0.09$. (b) 0.1335. (c) Yes, $n = 100$ is a large enough sample to be able to use the central limit theorem.

5.39 0.0051.

5.41 (a) $\bar{y}$ has a $N(\mu_Y, \sigma_Y/\sqrt{m})$ distribution and $\bar{x}$ has a $N(\mu_X, \sigma_X/\sqrt{n})$ distribution. (b) $\bar{y} - \bar{x}$ has a Normal distribution with mean $\mu_Y - \mu_X$ and standard deviation $\sqrt{\sigma_y^2 + \sigma_x^2}$.

5.43 $n = 1391$. $X = 362$. $\hat{p} = 0.26$.

5.45 (a) $n = 1500$. (c) If the choice is "Yes," $X = 1234$. (d) For "Yes," $\hat{p} = 0.8227$.

5.47 $B(20, 0.5)$.

5.49 (a) $P(X = 0) = 0.0576$ and $P(X \geq 6) = 0.0113$. (b) $P(X = 8) = 0.0576$ and $P(X \leq 2) = 0.0113$. (c) The number of "failures" in the $B(8, 0.3)$ distribution has the $B(8, 0.3)$ distribution. With eight trials, zero successes is equivalent to eight failures, and six or more successes is equivalent to two or fewer failures.

5.51 **(a)** $\mu_X = 2.4$. $\sigma_X = 1.296$. **(b)** $\mu_X = 5.6$. $\sigma_X = 1.296$. **(c)** For the means, we have $2.4 + 5.6 = 8$. The standard deviations for parts (a) and (b) are the same.

5.53 **(a)** 0.9858. **(b)** 0.7814.

5.55 **(a)** 0.1687. **(b)** 0.7199.

5.57 **(a)** Separate flips are independent. **(b)** The coin is fair. The probabilities are still $P(H) = P(T) = 0.5$. **(c)** The parameters for a binomial distribution are n and p. **(d)** This is best modeled with a Poisson distribution.

5.59 **(a)** A $B(200, p)$ distribution seems reasonable for this setting. **(b)** This setting is not binomial; there is no fixed value of n. **(c)** A $B(500, 1/12)$ distribution seems appropriate for this setting. **(d)** This is not binomial because separate cards are not independent.

5.61 **(a)** The distribution of those who say they have stolen something is $B(10, 0.2)$. The distribution of those who do not say they have stolen something is $B(10, 0.8)$. **(b)** X is the number who say they have stolen something. $P(X \geq 4) = 0.1209$.

5.63 **(a)** $\mu = 2$ will say they have stolen. $\mu = 8$ will say they have not stolen. **(b)** $\sigma = 1.265$. **(c)** If $p = 0.1$, $\sigma = 0.949$. If $p = 0.01$, $\sigma = 0.315$. As p gets smaller, the standard deviation becomes smaller.

5.65 **(a)** 13 is the smallest value of m. **(b)** Using 75%, $P(X \leq 13) = 0.9198$. **(c)** The probability will decrease. When the sample size increases, the mean will increase as well.

5.67 The probability that a digit is greater than 5 is 0.4 and 0.6 that the digit is not greater than 5. **(a)** 0.9533. **(b)** $\mu = 16$.

5.69 **(a)** $n = 4$, $p = 0.62$. **(b)** See table below. **(c)** $\mu = 2.48$.

x	0	1	2	3	4
$P(x)$	0.0209	0.1361	0.3330	0.3623	0.1478

5.71 **(a)** 0.8324. **(b)** 0.9750. **(c)** As p gets closer to 1, the probability of being within ± 0.03 of p increases because the standard deviation decreases.

5.73 **(a)** The mean is $\mu = 0.69$, and the standard deviation is $\sigma = 0.0008444$. **(b)** 68.83% to 69.17%. **(c)** It is more reasonable to assume that the population proportion has changed over time.

5.75 **(a)** $\hat{p} = 0.28$. **(b)** $\hat{p}$ is approximately $N(0.28, 0.0317)$ $P(\hat{p} \geq 0.28) = 0.5$.

5.77 **(a)** $p = 0.25$. **(b)** $P(X \geq 10) = 0.0139$. **(c)** $\mu = 5$ and $\sigma = 1.9365$. **(d)** No. The trials would not be independent because the subject may alter his or her guessing strategy based on this information.

5.79 **(a)** X has a $B(900, 1/5)$ distribution, with mean $\mu = 180$ and $\sigma = 12$ successes. **(b)** For $\hat{p}$, the mean is $\mu_{\hat{p}} = 0.2$ and $\sigma_{\hat{p}} = 0.01333$. **(c)** $P(\hat{p} > 0.24) = 0.0013$. **(d)** 208 or more successes (correct guesses) in 900 attempts.

5.81 Poisson with $\mu = 2.768$. **(a)** $P(X = 0) = 0.0628$. **(b)** $P(X \geq 3) = 0.5229$.

5.83 **(a)** $\mu = 50$. **(b)** $\sigma = 7.7071$. $P(X > 60) = 0.0793$. Software gives 0.0786.

5.85 **(a)** $\bar{x}$ is approximately Normal with mean 137 and standard deviation 0.05196. **(b)** Essentially 0.

5.87 **(a)** Approximately Normal with mean 2.07 and standard deviation 0.178. **(b)** 0.3483. **(c)** Yes, because $n = 140$ is large.

5.89 The probability that the first digit is 1, 2, or 3 is 0.602.

5.91 **(a)** $\mu_X = 3.75$. **(b)** $P(X \geq 10) = 0.000795$. **(c)** $P(X \geq 540) = 0.0213$ (0.0192 using the Normal approximation).

5.93 **(a)** $m = 15$. **(b)** $\mu = 13.52$ and $\sigma = 3.629$. **(c)** Without the continuity correction, $P(X \geq 15) = 0.3409$. With the continuity correction, we have $P(X \geq 14.5) = 0.3936$. The continuity correction is much closer.

5.95 **(a)** $\hat{p}_F$ is approximately $N(0.82, 0.01921)$ and $\hat{p}_M$ is approximately $N(0.88, 0.01625)$. **(b)** $\hat{p}_M - \hat{p}_F$ is approximately $N(0.06, 0.02516)$. **(c)** $P(\hat{p}_F > \hat{p}_M) = 0.0087$ (software gives 0.0085).

5.97 $P(Y \geq 200) = 0$.

5.99 The Poisson distribution is not appropriate because the rate is not constant and increases between the midnight and 6 A.M. period.

5.101 Y has possible values 1, 2, 3,... $P(Y = k) = (5/6)^{k-1}(1/6)$.

CHAPTER 6

6.1 $\sigma_{\bar{x}} = \$0.40$.

6.3 $\$0.80$.

6.7 The margin of error would be halved.

6.9 $n = 228$.

6.11 It is likely the 532 who responded are different from those who didn't respond so that our estimated margin of error is not a good measure of accuracy.

6.13 The margins of error are 13.067, 7.84, 4.356, and 3.92. Interval width decreases as sample size increases.

6.15 **(a)** She did not divide the standard deviation by $\sqrt{500} = 22.361$. **(b)** Confidence intervals concern the population mean, not the sample mean. **(c)** 95% is a confidence level, not a probability. **(d)** The large sample size does not affect the distribution of individual alumni ratings.

6.17 **(a)** $m = 0.2045$. (5.295, 5.704). **(b)** (5.231, 5.769).

6.19 The margin of error is 2.29 U/l and the 95% confidence interval for μ is 10.91 to 15.49 U/l.

6.21 Scenario A has a smaller margin of error. The value of σ would likely be smaller for A because we might expect less variability in textbook cost for freshman students than all students.

6.23 **(a)** $m = 22.24$. **(b)** To yield a margin of error of 15, we would need a larger sample than 2265. **(c)** $n = 4979$.

6.25 **(a)** Larger. **(b)** We would need the standard deviation to be 0.04167 hours. **(c)** $n = 886$.

6.27 **(a)** The 95% confidence interval for the mean number of hours spent listening to the radio in a week is 11.03 to 11.97 hours. **(b)** No. This is a range of values for the mean time spent, not for individual times. **(c)** The sample size is large ($n = 1200$ students surveyed).

6.29 **(a)** We can be 95% confident, but not *certain*. **(b)** We obtained the interval 53.1% to 55.1% by a method that gives a correct result 95% of the time. **(c)** The margin of error is about 0.51%. **(d)** No, confidence intervals only account for random sampling error.

6.31 $\bar{x}_{kpl} = 18.3515$ and margin of error 0.6521 kpl.

6.33 $n = 73$.

6.35 No; confidence interval methods of this chapter can only be used on an SRS.

6.37 **(a)** 0.7738 **(b)** 0.9510. **(c)** 0.99488 or about 99.5%.

6.39 H_0: $\mu = 1.4$ g/cm^2 versus H_a: $\mu \neq 1.4$ g/cm^2.

6.41 $P = 0.1164$.

6.43 **(a)** $z = 1.645$. **(b)** $z > 1.645$.

6.45 **(a)** $z = 1.75$. **(b)** P-value $= 0.0401$. **(c)** P-value $= 0.0802$.

6.47 **(a)** No. **(b)** Yes.

6.49 **(a)** Yes. **(b)** No. **(c)** Because $0.033 \leq 0.05$, we reject H_0. Because $0.033 > 0.01$, we do not reject H_0.

6.51 **(a)** P-value $= 0.038$; P-value $= 0.962$. **(b)** Suppose the null hypothesis is H_0: $\mu = \mu_0$. The smaller P-value (0.038) goes with the one-sided alternative that is consistent with the observed data.

6.53 **(a)** Hypotheses should be stated in terms of the population mean, not the sample mean. **(b)** The null hypothesis H_0 should be that there is no change. **(c)** A small P-value is needed for significance. **(d)** We compare the P-value, not the z-statistic, to α.

6.55 **(a)** H_0: $\mu = 77$ versus H_a: $\mu \neq 77$. **(b)** H_0: $\mu = 20$ seconds versus H_a: $\mu > 20$ seconds. **(c)** H_0: $\mu = 880$ ft^2 versus H_a: $\mu < 880$ ft^2.

6.57 **(a)** H_0: $\mu = \$42{,}800$ versus H_a: $\mu > \$42{,}800$, where μ is the mean household income of mall shoppers. **(b)** H_0: $\mu = 0.4$ hr versus H_a: $\mu \neq 0.4$ hr, where μ is this year's mean response time.

6.59 **(a)** P-value $= 0.9082$. **(b)** P-value $= 0.0918$. **(c)** P-value $= 0.1836$.

6.61 H_a: $\mu \neq \$48{,}127$. $Z = -2.25$. P-value $= 0.0244$.

6.63 $P = 0.09$ means there is some evidence for the wage decrease, but it is not significant at the $\alpha = 0.05$ level.

6.65 Even if the two groups (the health and safety class and the statistics class) had the same level of alcohol awareness, there might be some difference in our sample due to chance. The difference observed was large enough that it would rarely arise by chance.

6.67 Because the difference for public school students was statistically significant, we can say the mean score for them increased. The difference for private school students was not significant; that does not mean that it didn't increase, but rather that it didn't increase *enough* to be called significant.

6.69 H_0: $\mu = 0$ versus H_a: $\mu \neq 0$. $z = 4.14$. P-value $= 0.00003$ from software.

6.71 H_0: $\mu = 100$ versus H_a: $\mu \neq 100$. $z = 5.75$. P-value < 0.0001.

6.73 **(a)** $z = 1.38$. P-value $= 0.0838$. **(b)** The important assumption is that this is an SRS from the population of older students. We also assume a Normal distribution, but this is not crucial provided there are no outliers and little skewness.

6.75 **(a)** H_0: $\mu = 0$ mpg versus H_a: $\mu \neq 0$ mpg. **(b)** The mean of the 20 differences is $\bar{x} = 2.73$, so $z = 4.07$. P-value < 0.0001. We reject H_0. We conclude that $\mu \neq 0$ mpg; that is, we have strong evidence that the computer's reported fuel efficiency differs from the driver's computed values.

6.77. Smaller α means that $\bar{x}$ must be farther away from μ_0 in order to reject H_0.

6.79 With sample size $n = 50$, sample means greater than 0.3 are statistically significant.

6.81 Changing to two-sided doubles each P-value. The values are 0.7518, 0.5271, 0.3428, 0.2059, 0.1139, 0.0578, 0.0269, 0.0114, 0.0044, 0.0016.

6.83 Something that occurs "fewer than one time in 100 repetitions" must also occur "fewer than five times in 100 repetitions," so significance at the 1% level guarantees significance at the 5% level.

6.85 Any $2.576 < |z| < 2.807$.

6.87 $0.40 < P$-value < 0.50. (Software gives 0.4694.)

6.89 $0.05 < P$-value < 0.10; P-value $= 0.0602$.

6.91 In order to determine the effectiveness of alarm systems, we need to know the percent of all homes with alarm systems and the percent of burglarized homes with alarm systems.

6.93 The first test was barely significant at $\alpha = 0.05$, while the second was significant at any reasonable α.

6.95 A significance test answers only question (b).

6.97 **(a)** If SES had no effect on LSAT results, there would still be some difference in scores due to chance variation. **(b)** Knowing the effects were small tells us that the statistically insignificant test result did not occur merely because of a small sample size.

6.99 **(a)** $P = 0.2843$. **(b)** $P = 0.1020$. **(c)** $P = 0.0023$.

6.101 We expect more variation with small sample sizes than with large sample sizes, so even a large difference between $\bar{x}$ and μ_0 might not turn out to be significant.

6.107 We would need $n = 100,000$ tests.

6.109 We reject the fifth (P-value $= 0.001$) and 11th (P-value < 0.002) tests.

6.111 The one with the larger sample size will have more power.

6.113 The power for $\mu = 35$ will be higher than 0.73 because larger differences are easier to detect.

6.115 **(a)** Changing from the one-sided to the two-sided alternative decreases power. **(b)** Decreasing σ increases power. **(c)** Power increases.

6.117 0.986.

6.119 **(a)** It is better to overestimate σ somewhat because our particular random sample might have shown a smaller s than the reality. **(b)** We reject H_0 at the 5% significance level if $\bar{x} > 4.329$. **(c)** When $\mu = 4.25$, the probability of rejecting H_0 is 0.3446 (0.3464 from software). **(d)** The power of this test is not up to the 80% standard suggested in the text; he should collect a larger sample. **(e)** The needed sample size is 396.

6.121 0.7881.

6.123 **(a)** The hypotheses are "subject should go to college" and "subject should join workforce." Errors: recommending college for someone better suited for the workforce, and recommending the workforce for someone who should go to college. **(b)** We typically wish to decrease the probability of wrongly rejecting H_0.

6.125 **(a)** For example, if μ is the mean difference in scores, $H_0: \mu = 0$ versus $H_a: \mu \neq 0$. **(b)** P-value $= 0.13$, we would not reject H_0. **(c)** For example: Was this an experiment? What was the design? How big were the samples?

6.127 **(a)** For boys:

Energy (kJ)	2399.9 to 2496.1
Protein (g)	24.00 to 25.00
Calcium (mg)	315.33 to 332.87

(b) For girls:

Energy (kJ)	2130.7 to 2209.3
Protein (g)	21.66 to 22.54
Calcium (mg)	257.70 to 272.30

(c) Because the confidence interval for boys is entirely above the confidence interval for girls for each food intake, we could conclude that boys consume more of each, on average.

6.129 Most students should find that their final proportion is between 0.84 and 0.96; 85% will have a proportion between 0.87 and 0.93.

6.131 Because there is nonresponse, the accuracy is in question regardless of the small margin of error.

6.133 **(a)** 4.61 to 6.05 mg/dl. **(b)** $H_0: \mu = 4.8$ mg/dl versus $H_a: \mu > 4.8$ mg/dl, $z = 1.45$. P-value $= 0.0735$.

6.135 **(a)** The distribution is roughly symmetric. **(b)** (26.06, 37.74). **(c)** $H_0: \mu = 25$. $H_a: \mu > 25$, $z = 2.44$. P-value $= 0.0073$.

6.137 **(a)** Under H_0, $\bar{x}$ has a $N(0\%, 5.3932\%)$ distribution. **(b)** $z = 1.28$. $P = 0.1003$. **(c)** This is not significant at $\alpha = 0.05$.

6.139 Yes.

6.141 For each sample, find $\bar{x}$, then take $\bar{x} \pm 2.53$.

6.143 For each sample, find $\bar{x}$, then compute $z = \dfrac{\bar{x} - 24}{5/\sqrt{15}}$. and reject H_0 based on your chosen α.

CHAPTER 7

7.1 **(a)** $45. **(b)** 15.

7.3 ($670.105, $861.895).

7.5 **(a)** df $= 22$, $0.04 < P$-value < 0.05, which is significant at the 5% level. **(b)** df $= 8$, $0.05 < P$-value < 0.10, which is not significant at the 5% level.

7.7 From software, the 95% confidence interval is (2.0817, 26.9183).

7.9 A paired t-test is appropriate because the number of receptivity displays was recorded twice for each female when exposed to the two different videos.

7.11 $\bar{x} = -1.6$, $s = 2.9665$. df $= 4$, $t^* = 2.776$. The 95% confidence interval is $(-5.32, 2.12)$.

7.13 Although the sample size is quite large, $n = 518$, there are potential outliers, so we should not use t procedures with these data.

7.15 **(a)** df $= 14$, $t^* = 2.145$. **(b)** df $= 27$, $t^* = 2.052$. **(c)** df $= 27$, $t^* = 1.703$. **(d)** As sample size increases, the margin of error decreases. As confidence increases, the margin of error increases.

7.17 The 5% critical value for a t distribution with df $= 23$ is 1.714, reject H_0 when $t > 1.714$; the other is simply the mirror image, so reject when $t < -1.714$.

7.19 Yes, because we want $\mu > 0$, P-value $= 0.9625$.

7.21 **(a)** df $= 12$. **(b)** $2.681 < t < 3.055$. **(c)** $0.01 < P$-value < 0.02. **(d)** $t = 2.78$ is significant at the 5% level but not at the 1% level. **(e)** 0.0167.

7.23 It depends on if $\bar{x}$ is on the appropriate side of μ_0.

7.25 $\bar{X} = 28.1563$, $s = 1.4841$. df $= 15$. The 95% confidence interval is (27.37, 28.95).

7.27 **(a)** The histogram shows the data are Normally distributed, so t procedures are appropriate. **(b)** The 95% confidence interval is 36.157 ± 2.603. **(c)** (33.55, 38.76). **(d)** Inference from the confidence interval only applies to the mean, not the median. Other tools would need to be used for inference on the median.

7.29 **(a)** $H_0: \mu = 4.7\%$, $H_a: \mu \neq 4.7\%$. $\bar{x} = 4.9767$. $s = 0.03215$. $t = 14.907$. df $= 2$, $0.002 < P$-value < 0.005. **(b)** (4.8968%,

5.0566%). **(c)** For the cans and bottles to be within 0.3% of the advertised level, they need to be between 4.7% and 5.3%; it appears that Budweiser is within the standards.

7.31 (a) H_0: $\mu = 10$, H_a: $\mu < 10$. **(b)** $t = -5.2603$, df = 33, P-value < 0.0005.

7.33 (a) The distribution has two peaks, so the distribution is not Normal. The five-number summary is 2.2, 10.95, 28.5, 41.9, 69.3. **(b)** Maybe: We have a large enough sample to overcome the non-Normal distribution, but we are sampling from a small population. **(c)** $\bar{x} = 27.29$, $s = 17.7058$, df = 39; the 95% confidence interval is (21.57, 33.01). **(d)** Answers may vary.

7.35 (a) 15.833. **(b)** df = 2499; the 95% confidence interval is (1.93, 3.17). **(c)** The large number of observations eliminates any concern of skewness, but the outliers pose a risk for using t procedures.

7.37 H_0: $\mu = 45$, H_a: $\mu > 45$. $t = 5.457$. df = 49, P-value < 0.0005.

7.39 (a) H_0: $\mu = 0$, H_a: $\mu \neq 0$. $t = 5.125$, df = 15, P-value = 0.00012. **(b)** (191.6, 464.4).

7.41 (a) H_0: $\mu = 0$, H_a: $\mu \neq 0$. **(b)** $\bar{x} = 2.73$, $s = 2.8015$, $t = 4.358$, df = 19, P-value < 0.001.

7.43 (a) H_0: $\mu = 925$, H_a: $\mu > 925$. $t = 3.27$, df = 35, P-value = 0.0012. **(b)** H_0: $\mu = 935$, H_a: $\mu > 935$, $t = 0.80$, df = 35, P-value = 0.2146. **(c)** The interval is 931.3 to 945.0, which includes 935, but not 925.

7.45 (a) The differences are spread from -0.018 to 0.020. A Normal quantile plot reveals the data are approximately Normal and, therefore, t methods are appropriate. **(b)** H_0: $\mu = 0$, H_a: $\mu \neq 0$. $t = -0.347$, df = 7, P-value = 0.7388. **(c)** (-0.0117 to 0.0087). **(d)** The subjects from this sample may be representative of future subjects, but the test results and confidence interval are suspect because this is not a random sample.

7.47 (a) H_0: $\mu = 0$, H_a: $\mu > 0$. **(b)** Left-skewed, $\bar{x} = 2.5$, $s = 2.8928$. **(c)** $t = 3.8649$, df = 19, P-value = 0.0005. **(d)** (1.15, 3.85).

7.49 (2.65, 24.29). With the smaller sample sizes, the confidence interval is wider.

7.51 H_0: $\mu = 5$, H_a: $\mu > 5$. The confidence interval is (5.48, 21.46). Because 5 is not in this interval, we reject H_0. The data show evidence that the improvement is greater than five points.

7.53 SPSS and SAS give $t = 2.279$, P-value = 0.052.

7.55 (a) Hypotheses should involve μ_1 and μ_2. **(b)** The samples are not independent. **(c)** We need the P-value to be small to reject H_0. **(d)** Assuming the researcher computed the t statistic using $\bar{x}_1 - \bar{x}_2$, a positive value of t does not support H_a.

7.57 (a) We cannot reject H_0: $\mu_1 = \mu_2$ in favor of the two-sided alternative at the 5% level. **(b)** We could reject H_0 in favor of H_a: $\mu_1 < \mu_2$ if the t statistic is negative.

7.59 (a) Both distributions are Normally distributed, except the low-intensity class has a low outlier. **(b)** H_0: $\mu_H = \mu_L$, H_a:

$\mu_H \neq \mu_L$. $t = 5.30$. df = 14. P-value < 0.001. **(c)** Because the low-intensity class has an outlier, the t-test is not appropriate. **(d)** $t = 6.31$. df = 13. P-value < 0.001. Removing the outlier didn't change the results. **(e)** Because the outlier is not affecting the results, it is probably okay to report both tests.

7.61 (a) The t procedure is robust. Because $n_1 + n_2 \geq 40$, we can use the t procedures on skewed data. **(b)** H_0: $\mu_{days} = \mu_{month}$, H_a: $\mu_{days} \neq \mu_{month}$. $t = -2.42$. df = 89. $0.01 < P$-value < 0.02.

7.63 H_0: $\mu_{Brown} = \mu_{Blue}$, H_a: $\mu_{Brown} > \mu_{Blue}$. $t = 2.59$. df = 39. $0.005 < P$-value < 0.01.

7.65 52.4%. We don't know if these students were not Facebook users; because of this, the results should be viewed with caution.

7.67 (a) The data are not Normally distributed, but because neither distribution is strongly skewed or has outliers, t procedures are still appropriate. **(b)** For N group: $\bar{x} = 0.5714$, $s = 0.73$, $n = 14$. For S group: $\bar{x} = 2.1176$, $s = 1.24$, $n = 17$. **(c)** H_0: $\mu_N = \mu_S$, H_a: $\mu_N \neq \mu_S$. **(d)** $t = -4.31$, df = 13, P-value < 0.001. **(e)** (-2.32, -0.77).

7.69 (a) Taking averages on ratings is likely not appropriate. **(b)** The data are integers but the samples are large, the t procedures can be used. **(c)** McDonald's: $\bar{X} = 3.9937$, $s = 0.8930$. Taco Bell: $\bar{X} = 4.2208$, $s = 0.7331$. **(d)** H_0: $\mu_M = \mu_T$, H_a: $\mu_M \neq \mu_T$. $t = -3.48$. df = 307. P-value < 0.001.

7.71 (a) Assuming we have SRSs from each population, this seems reasonable. **(b)** H_0: $\mu_{Early} = \mu_{Late}$, H_a: $\mu_{Early} \neq \mu_{Late}$. **(c)** $SE_D = 1.0534$, $t = 1.614$, df = 199, P-value = 0.1081. **(d)** (-0.39, 3.79).

7.73 You need sample sizes and standard deviations or df and a more accurate P-value. The confidence interval could give us useful information about the magnitude of the difference.

7.75 This is a matched pairs design.

7.77 There could be things that are similar about the next eight employees who need new computers as well as the following eight, which could bias the results.

7.79 (a) The north distribution is right-skewed, while the south distribution is left-skewed. **(b)** The methods of this section seem to be appropriate because the sample sizes are relatively large, and there are no outliers. **(c)** H_0: $\mu_N = \mu_S$, H_a: $\mu_N \neq \mu_S$. **(d)** $\bar{x}_N = 23.7$, $s_N = 17.5001$, $\bar{x}_S = 34.53$, and $s_S = 14.2583$. $t = -2.629$, df = 29, $0.01 < P$-value < 0.02. **(e)** (-19.2614, -2.4053).

7.81 (a) (-1.07, 7.07). **(b)** With 95% confidence, the mean change in sales from last year to this year is between -1.07 and 7.07. Because the interval covers 0 and includes some negative values, it is possible sales have actually decreased.

7.83 (a) H_0: $\mu_B = \mu_F$, H_a: $\mu_B > \mu_F$. $t = 1.654$, df = 18, $0.05 < P$-value < 0.10. **(b)** (-0.2434, 2.0434). **(c)** We need two independent SRSs from Normal populations.

7.85 $t = 4.17$, df = 63, P-value < 0.001. Confidence interval: (14.57, 41.43). The results are similar.

7.87 s_p = 15.9617, SE_D = 4.1213. t = −2.629, df = 58, P-value = 0.0110. The 95% confidence interval is (−19.113, −2.554).

7.89 df = 55.725.

7.91 (a) df = 137.066. (b) s_p = 5.332, which is slightly closer to s_O. (c) With no assumption of equality, SE_D = 0.7626. With the pooled method, SE_D = 0.6136. (d) t = 18.74 and df = 333, for which $P <$ 0.0001, and the 95% confidence interval is (10.2827, 12.7173). The t value is larger, the confidence interval is narrower, and the P-value is smaller. (e) df = 121.503. s_p = 1.734; the standard errors are 0.2653 and 0.1995; t = 24.56, df = 333, P-value < 0.0001, and the 95% confidence interval is (4.5042, 5.2958). With the pooled procedure, t is larger, and the interval is narrower.

7.93 n = 24 guarantees the margin of error is less than 5000.

7.95 Higher; if the alternative μ is farther away from 18.5 then we will have more power.

7.97 Decrease.

7.99 0.72. 0.72.

7.101 (a) To halve the margin of error the sample size needs to be *quadrupled*. (b) The sign test is *less* powerful than the t test when the differences are close to Normal. (c) For a two-sided alternative, the power would be the *same*. (d) Increasing the sample size has no effect on the probability of a Type I error, this is determined by the choice of α.

7.103 No, the confidence interval is for the mean monthly rate, not the individual apartment rates.

7.105 (a) n = 104. (b) We would need to use the bigger sample to make sure both margin of error conditions are met.

7.107 (a) n = 18. (b) For n = 10, the power will be less than 90%. (c) For n = 10 the power is 0.80.

7.109 Answers will vary based on the choice of σ. For σ = 0.015, power = 0.09.

7.111 (a) Increase. A larger alpha gives more power. (b) 0.89.

7.113 Using a larger σ for planning the study is advisable because it provides a conservative (safe) estimate of the power.

7.115 H_0: median = 0, H_a: median > 0; P-value = 0.0898. In Exercise 7.38, we were able to reject H_0; here, we cannot.

7.117 H_0: median = 0, H_a: median > 0; P-value = 0.0002. Using a t test, we found the same conclusion.

7.119 $\bar{x}$ = 153.5, s = 14.708, $s_{\bar{x}}$ = 7.354. It would not be appropriate to construct a confidence interval because we cannot consider these four scores to be an SRS.

7.121 (b) The plot shows that t^* approaches z^* = 1.96 as the df increases. (c) The plots would be similar, but t^* would approach z^* = 1.645 as the df increases.

7.123 (a) Use two independent samples. (b) Use a matched pairs design. (c) Take a single sample of college students, and ask them to rate the appeal of the product.

7.125 (a) H_0: μ = 1.5, H_a: $\mu <$ 1.5; t = −9.974, df = 199, P-value ≈ 0. (b) (0.697, 0.963). (d) We have a large sample, so t procedures should be safe.

7.127 (a) (−3.008 to 1.302). (b) (−1.761, 0.055). (c) The centers of the intervals are the same, but the margin of error for the independent samples interval is much larger.

7.129 (a) We are looking at the average proportions across both samples. (b) H_0: μ_B = μ_W, H_a: μ_B ≠ μ_W. (c) For the first year: t = 0.982, df = 52.3, P-value = 0.3305. For the third year: t = 2.126, df = 46.9, P-value = 0.0388.

7.131 (a) $\bar{x}_1$ = −0.7, SE_1 = 2.298; $\bar{x}_2$ = 14, SE_2 = 56.125. (b) df = 13, t_1 = −0.305 (P_1 = 0.7655), and t_2 = 0.249 (P_2 = 0.8069). (c) (−5.66 to 4.26) and (−107.23, 135.23).

7.133 (a) Because the same mockingbird responded on each day. (b) 6.9774. (c) H_0: μ_1 = μ_4, H_a: μ_1 ≠ μ_4; t = 6.319, df = 23, P-value < 0.001. (d) t = −0.973, P-value = 0.3407. (e) There is a significant difference between day 1 and day 4 but not day 1 and day 5.

7.135 How much a person eats or drinks may depend on how many people he or she is with.

7.137 A two-sample t procedure was used. We assumed the data are approximately Normal. H_0: μ_C = μ_N, H_a: $\mu_C > \mu_N$; t = 0.95, df = 89. 0.15 < P-value < 0.20.

7.139 A two-sample t procedure was used. H_0: μ_C = μ_N, H_a: $\mu_C > \mu_N$; t = −0.16, df = 89, P-value > 0.25.

7.141 $\bar{X}$ = 77.76%, s = 32.6768%, (64.27% to 91.25%). This seems to support the retailer's claim.

7.143 The distributions appear similar. GPA: t = −0.91, df = 30, 0.15 < P-value < 0.20. Confidence interval: (−1.35, 0.52). IQ: t = 1.64, 0.05 < P-value < 0.10 (df = 30). Confidence interval: (−1.24, 11.48).

7.145 H_0: μ_1 = μ_2, H_a: $\mu_1 > \mu_2$; t = 3.65, P-value < 0.0005. Confidence interval: (0.7714, 2.6086).

7.147 (a) H_0: μ_B = μ_D, H_a: $\mu_B < \mu_D$; t = 2.87, P-value < 0.005. Confidence interval: (1.7, 9.7). (b) H_0: μ_B = μ_S, H_a: $\mu_B < \mu_S$; t = 1.88, P-value < 0.05. Confidence interval: (−0.24, 6.7).

7.149 No: What we have is nothing like an SRS of the population of school corporations; we have census data for your state.

CHAPTER 8

8.1 (a) n = 5013. (b) p is the (fixed, but unknown) population proportion of smartphone users who have purchased an item after using the phone to search for information. (c) X = 2657. X is the number of smartphone users from the sample who have purchased an item after using the phone to search for information. (d) $\hat{p}$ = 0.53.

8.3 (a) 0.007056. **(b)** 0.53 ± 0.0138. **(c)** (51.62%, 54.38%).

8.5 For $z = 1.34$, the two-sided P-value is the area under a standard Normal curve above 1.34 and below -1.34.

8.7 $\hat{p} = 0.75$, $z = 2.24$, P-value $= 0.0250$.

8.9 (a) $\hat{p} = 0.35$, $z = -1.34$, P-value $= 0.1802$. The results are exactly the same. **(b)** (14.1%, 55.9%).

8.11 The plot is symmetric about 0.5, where it has its maximum.

8.13 0.99.

8.15 (a) $n = 300$, $X = 109$. **(b)** $\hat{p} = 0.3633$. **(c)** $\hat{p} = 36.33\%$ is the estimate of p, the population proportion of students at your college who regularly eat breakfast.

8.17 (a) $\hat{p} = 0.461$, $\text{SE}_{\hat{p}} = 0.0157$, $m = 0.0308$. **(b)** Yes. **(c)** (0.4302, 0.4918). **(d)** We are 95% confident that between 43.0% and 49.2% of cell phone owners used their cell phone while in a store within the last 30 days to call a friend or family member for advice about a purchase they were considering.

8.19 (a) $\hat{p} = 0.84$, $\text{SE}_{\hat{p}} = 0.0232$, $m = 0.0454$. **(b)** This was not an SRS; they asked all customers in the two-week period. **(c)** (0.7946, 0.8854). **(d)** Based on the sample, we estimate with 95% confidence that between 79.5% and 88.5% of the service department's customers would recommend the service to a friend; because this was not an SRS, we must use caution in relying on this interval.

8.21 $n = 382$.

8.23 (a) Confidence cannot be bigger than 100%. **(b)** The margin of error only takes into account errors from random sampling, not errors due to bias. **(c)** The P-value gives no indication of the truthfulness of null hypothesis; rather, it gives the amount of evidence in favor of the alternative hypothesis.

8.25 $\hat{p} = 0.6548$, (0.642, 0.668).

8.27 (a) $X = 934.5$, which rounds to 935. **(b)** (0.8711, 0.9089). **(c)** (87.1%, 90.9%). **(d)** For example, parents might be conscious of violence because of recent events in the news.

8.29 (a) Values of $\hat{p}$ outside the interval 0.2482 to 0.5518. **(b)** Values outside the interval 0.29626 to 0.5074.

8.31 (a) 67,179. **(b)** (0.41682, 0.42318).

8.33 $\hat{p} = 0.43$, (0.4043, 0.4557).

8.35 (a) $m = 0.001321$. **(b)** Other sources of error are much more significant than sampling error.

8.37 (a) $\hat{p} = 0.3275$, (0.3008, 0.3541). **(b)** Speakers and listeners probably perceive sermon length differently.

8.39 (a) $\hat{p} = 0.5067$, $z = 1.34$, P-value $= 0.1802$. **(b)** (0.497, 0.516).

8.41 $n = 4269$.

8.43 The sample sizes are 71, 126, 165, 189, 196, 189, 165, 126, 71. Use $n = 196$.

8.45 $n = 153$.

8.47 Mean $= -0.1$, standard deviation $= 0.1339$.

8.49 (a) Means p_1 and p_2. Standard deviations $\sqrt{p_1(1 - p_1)/n_1}$ and $\sqrt{p_2(1 - p_2)/n_2}$. **(b)** $p_1 - p_2$. **(c)** $p_1(1 - p_1)/n_1 + p_2(1 - p_2)/n_2$.

8.51 $(-0.019, 0.225)$. We can just reverse the sign of the interval in the previous exercise.

8.53 H_0: $p_W = p_M$, H_a: $p_W < p_M$. $z = -1.65$. P-value $= 0.0495$.

8.55 686 women and 686 men.

8.57 (a) This was an experiment; although you can't assign who visits the establishment, you could randomly assign which type of server they get. There were more than 10 successes and failures in each group so assuming random assignment the guidelines are generally met. **(b)** This is an experiment; the runners were randomly assigned. However, there are less than 10 successes in each group; hence, the guidelines are not met.

8.59 (a) The guidelines are met; the number of successes and failures in each group was at least five. **(b)** The guidelines are met; the number of successes and failures in each group was at least five.

8.61 (a) RR $= 1.56$. **(b)** RR $= 1.5$.

8.63 (a) Type of college is explanatory; response is whether physical education is required. **(b)** The populations are private and public colleges and universities. **(c)** $X_1 = 101$, $n_1 = 129$, $\hat{p}_1 = 0.7829$. $X_2 = 60$, $n_2 = 225$, $\hat{p}_2 = 0.2667$. **(d)** (0.4245, 0.6079). **(e)** H_0: $p_1 = p_2$. H_a: $p_1 \neq p_2$; $\hat{p} = \dfrac{60 + 101}{225 + 129} = 0.4548$, $z = 9.39$, P-value ≈ 0. **(f)** All these counts are greater than five. We do not know if the samples were SRSs.

8.65 (0.0363, 0.1457). Among youth who stress about their health, there is between 3.6% and 14.6% more who are exergamers than not.

8.67 (a) $X_1 = 574$, $n_1 = 1063$, $\hat{p}_1 = 0.54$. $X_2 = 947$, $n_2 = 1064$, $\hat{p}_2 = 0.89$. **(b)** 0.35. **(c)** Yes. **(d)** (0.3146, 0.3854). **(e)** 35%; 31.5% to 38.5%. **(f)** A possible concern is that adults were surveyed only before Christmas.

8.69 (a) $X_1 = 776$, $n_1 = 1063$, $\hat{p}_1 = 0.73$. $X_2 = 809$, $n_2 = 1064$, $\hat{p}_2 = 0.76$. **(b)** 0.03. **(c)** Yes. **(d)** $(-0.0070, 0.0670)$. **(e)** 3%; -0.7% to 6.7%. **(f)** A possible concern is that adults were surveyed only before Christmas.

8.71 No; this procedure requires independent samples from different populations.

8.73 (a) Using software: (i) 199, (ii) 294, (iii) 356, (iv) 388, (v) 388, (vi) 356, (vii) 294, (viii) 199. **(b)** As p_1 and p_2 get closer to 0.5, the necessary sample size to achieve the same power gets larger. As p_1 and p_2 get further from 0.5, the necessary sample size to achieve the same power gets smaller.

8.75 (a) $n = 2342$, $X = 1639$. **(b)** $\hat{p} = 0.7$, SE $= 0.0095$. **(c)** (0.681, 0.718). **(d)** Yes.

8.77 We have large samples from two independent populations (different age groups). $\hat{p}_1 = 0.8161$, $\hat{p}_2 = 0.4281$, SE $= 0.0198$, $(0.3492, 0.4268)$.

8.81 (a) 1207. (b) $(0.6483, 0.6917)$. (c) About 64.8% to 69.2.

8.83 No. Many people use both; there was only one sample, not two independent samples.

8.85 (a) We have six chances to make that mistake. (b) Use $z^* = 2.65$. (c) $(0.705, 0.775)$; $(0.684, 0.756)$; $(0.643, 0.717)$; $(0.632, 0.708)$; $(0.622, 0.698)$; $(0.571, 0.649)$.

8.87 $\hat{p} = 0.375$, $SE_D = 0.01811$, $z = 6.08$, P-value < 0.0001.

8.89 $\hat{p} = 0.6575$, $(0.6337, 0.6813)$.

8.91 $H_0: p_F = p_M$. $H_a: p_F \neq p_M$; $X_M = 171$, $X_F = 150$, $\hat{p} = 0.1600$, $SE_D = 0.0164$, $z = 1.28$, P-value $= 0.2009$.

8.93 The difference becomes more significant as the sample size increases. For example, with $n = 60$, $z = 1.1$ and P-value $= 0.2713$, but with $n = 70$, $z = 1.18$ and P-value $= 0.2380$.

8.95 (a) $n = 534$. (b) $n = (z^*/m)^2/2$.

8.97. (a) $p_0 = 0.7911$. (b) $\hat{p} = 0.3897$, $z = -29.1$, P-value is tiny. (c) $\hat{p}_1 = 0.3897$, $\hat{p}_2 = 0.7930$. $z = -29.2$, P-value ≈ 0.

CHAPTER 9

9.1 (a) 61.08% of women use Instagram; 38.92% do not. (b) 43.98% of men use Instagram; 56.02% do not. (d) A greater percent of women use Instagram.

9.3 Most will probably prefer the distribution broken down by Sex, which emphasizes the difference between men and women and their Instagram usage.

9.5 Among all three fruit consumption groups, vigorous exercise is most likely. Incidence of low exercise decreases with increasing fruit consumption.

9.7 They are shown in the following table.

| Fruit | Physical Activity | | | |
	Low	Medium	Vigorous	Total
Low	51.9	212.9	304.2	569
Medium	29.3	120.1	171.6	321
High	26.8	110.0	157.2	294
Total	108	443	633	1184

9.9 df $= (r - 1)(c - 1)$. (a) df $= 6$, $0.025 < P$-value < 0.05. (b) df $= 6$, $0.025 < P$-value < 0.05. (c) df $= 2$, $0.10 < P$-value < 0.15. (d) df $= 2$, $0.01 < P$-value < 0.02.

9.11 (a) Explanatory 1: 35.7% Yes and 64.3% No. Explanatory 2: 45.2% Yes and 54.8% No. (c) Explanatory variable value 1 had proportionately fewer Yes responses.

9.13 (a) $p_i =$ proportion of "yes" responses in group i. $H_0: p_1 = p_2$, $H_a: p_1 \neq p_2$; $\hat{p}_1 = 0.357$, $\hat{p}_2 = 0.452$.

(b) $z = -1.9882$, P-value $= 0.0469$. (c) The P-values agree. (d) $z^2 = (-1.9882)^2 = 3.9529$.

9.15

| Consoles | Gamer | | |
	Adult	Teen	Total
Yes	574	489	1063
No	945	119	1064
Total	1519	608	

9.17 27% are adult who play games on consoles, 23% are teens who play games on consoles, 44.4% are adults who don't play games on consoles, 5.6% are teens who don't play games on consoles. 71.42% are adults and 28.58% are teen. 49.98% play games on consoles, 50.02% do not. For adult gamers, 37.79% play games on consoles, 62.21% do not. For teen gamers, 80.43% play games on consoles, 19.57% do not. For those who play games on consoles, 54% are adult gamers, 46% are teen gamers. For those who do not play games on consoles, 88.82% are adult gamers, 11.18% are teen gamers. Usage of game consoles by gamer type are probably the most informative. A much higher percent of teen gamers play games on consoles than adult gamers.

9.19 The expected counts are 759.14, 303.86, 759.86, 304.14.

9.21 $X^2 = 315.7770$, df $= 1$, P-value < 0.0001.

9.23 $X^2 = 315.7770 \approx z^2 = (-17.77)^2 = 315.7729$.

9.25 More and Boys: 37.23%; More and Girls: 34.13%. Never and Boys: 5.39%; Never and Girls: 7.12%. Once and Boys: 6.36%; Once and Girls: 9.77%. 48.98% are boys, 51.02% are girls. 71.36% have witnessed it more than once, 16.12% have witnessed it once, and 12.51% have never witnessed it. For boys: 76.01% have witnessed it more than once, 12.98% have witnessed it once, and 11.01% have never witnessed it. For girls: 66.90% have witnessed it more than once, 19.14% have witnessed it once, and 13.96% have never witnessed it. For those witnessing sexual harassment more than once: 52.17% are boys, 47.83% are girls. For those witnessing sexual harassment once: 39.43% are boys, 60.57% are girls. For those never witnessing sexual harassment: 43.09% are boys, 56.91% are girls. The distribution of times by sex are probably the most informative, showing that boys tend to say they have witnessed sexual harassment slightly more than girls, with boys having 10% higher in the More category than the girls.

9.27 $X^2 = 20.822$, df $= 2$, P-value < 0.0001.

9.29 This is due to rounding error.

9.31 CA: 0.5820, HI: 0.0000, IN: 0.0196, NV: 0.0660, OH: 0.2264; $0.0369 + 0.5820 + 0.0000 + 0.0196 + 0.0660 + 0.2264 = 0.9308$.

9.33 (a) The null hypothesis is that the coin is fair. The alternative is that the coin is biased. (b) $X^2 = 1.7956$. df $= 1$, $0.15 < P$-value < 0.20.

9.35 Most results will give a fairly decent randomization and should fail to reject the null hypothesis. Changing the interval likely will not change the result and should still fail to reject the null hypothesis.

9.37 $X^2 = 20.8548$, df $= 3$, P-value < 0.0005.

9.39 (a) $H_0: p_1 = p_2$, $H_a: p_1 \neq p_2$; $\hat{p}_1 = 0.8892$, $\hat{p}_2 = 0.3120$, $\hat{p} = 0.5200$, $z = 17.556$, P-value ≈ 0. (b) H_0: There is no association between being harassed online and in person, H_a: There is a relationship; $X^2 = 308.23$, df $= 1$, P-value ≈ 0. (c) $17.556^2 = 308.21$, which agrees with X^2. (d) Perhaps one girl wouldn't answer these questions.

9.41 (a) The solution to Exercise 9.39 used "harassed online" as the explanatory variable. (b) Changing to use "harassed in person" for the two-proportions z test gives $\hat{p}_1 = 0.6161$, $\hat{p}_2 = 0.0832$, $\hat{p} = 0.3603$. We again compute $z = 17.556$, P-value ≈ 0. No changes will occur in the chi-square test. (c) The test statistic will be the same regardless of which is viewed as explanatory.

9.43 100 for each face of the die.

9.45 (a) Marginal totals are Small: 57; Medium: 17; Large: 5. Allowed: 67; Not allowed: 12. (b) 10.53% of Small, 29.41% of Medium, and 20% of Large were not allowed. (c) The expected count for Large/Not allowed is too small. (d) H_0: There is no association between the size of the claim and whether or not it is allowed; H_a: There is an association between the size of the claim and whether or not it is allowed. (e) $X^2 = 3.456$, df $= 1$, $0.05 < P$-value < 0.10.

9.47 H_0: the DFW rate has not changed; $X^2 = 308.3$, df $= 2$, $P < 0.0001$.

9.49 (a) For example, among those students in trades, 320.28 enrolled right after high school, and 621.72 enrolled later. (b) For example, 39.4% of these students enrolled right after high school. Health is the most popular field with 38%. (c) $X^2 = 276.1$, df $= 5$, P-value < 0.0001.

9.51 $X^2 = 852.4330$, df $= 1$, P-value < 0.0001, $z^2 = (-29.2)^2 = 852.64 = X^2$ with rounding error.

9.53 Most results will give a fairly decent randomization and should fail to reject the null hypothesis. Changing the interval likely will not change the result and should still fail to reject the null hypothesis.

9.55 (a) Each quadrant accounts for one-fourth of the area, so we expect it to contain one-fourth of the 100 trees. (b) *Some* random variation would not surprise us; we no more expect exactly 25 trees per quadrant than we would expect to see exactly 50 heads when flipping a fair coin 100 times. (c) $\chi^2 = 10.8$, df $= 3$, and P-value $= 0.0129$.

CHAPTER 10

10.1 (a) 0.8. (b) For each serving of fruits and vegetables in a calorie-controlled diet, the expected average decrease in blood pressure goes up by 0.8. (c) 8.4. (d) $N(8.4, 3.2)$. (e) (2.0, 14.8).

10.3 (a) $t = 2.154$, df $= 19$, $0.04 < P$-value < 0.05. (b) $t = 2$, df $= 29$, $0.05 < P$-value < 0.10. (c) $t = 2$, df $= 99$, $0.04 < P$-value < 0.05.

10.5 0.7. At $x = 11.0$, the margin of error there will be larger.

10.7 (a) The parameters of the regression model are β_0, β_1, and ϵ; those given are estimates of these. (b) It should be $H_0: \beta_1 = 0$. (c) The prediction interval will be wider than the mean response interval.

10.9 Prediction intervals concern individuals instead of means. Departures from the Normal distribution assumption would be more severe here.

10.11 (a) $\hat{y} = 8.25464 + 0.11259$EDUC. The points in the residual plot look scattered and random so the assumptions are satisfied. The histogram shows the residuals are Normally distributed. (b) Yes, the log transformed data can effectively be used for inference.

10.13 Because the list was narrowed before we took our SRS, our sample really only reflects the schools that met the "academic quality" criteria and not all 500+ colleges.

10.15 (a) $19,907.16. (b) $22,388.51. (c) Larger because its in-state cost is farther from the average in-state cost.

10.17 (a) InCostAid looks somewhat linear but has several outliers. Admit also looks linear but also has a couple outliers. GradRate does looks weakly linear and may have one low outlier. OutCostAid does not have a linear relationship with AvgDebt. Baruch College is notable in each of the plots. (b) IncostAid: $s = 3665.8(P = 0.0058)$; Admit: $s = 4203.5$ $(P = 0.1882)$; GradRate: $s = 4282.8(P = 0.3356)$; OutCostAid: $s = 4369.8(P = 0.8025)$.

10.19 (a) A linear trend looks reasonable; nothing unusual. (b) $\hat{y} = -24517 + 12.83450$Year. (c) The intercept only describes what happens at when x is 0, which is far outside the range of our data. (d) The residual plot looks mostly random. (e) The residuals are Normal as shown in the Normal quantile plot. (f) Yes.

10.21 (b) The points are much closer to a straight line. (c) (0.37305, 0.43643).

10.23 Preferences vary.

Package	SS	MS	F	r^2
SPSS	Sums of Squares	Mean Square	17.096	0.149
Minitab	Adj SS	Adj MS	17.10	14.85%
Excel	SS	MS	17.09644	0.148540143
JMP	Sums of Squares	Mean Square	17.0964	0.14854

10.25 (a) The null hypothesis should test the slope β_1. (b) Sums of squares add; mean squares do not. (c) The r^2 value determines explanatory power. (d) The total df is equal to $n - 1$.

10.27 (a) Obligations are increasing linearly over time. (b) $\hat{y} = -3609.51 + 1.815$Year. (c) -0.38, 0.19, 0.76, -0.57; $s = 0.73587$. (d) $y_i = \beta_0 + \beta_1 x_i + \varepsilon_i$, $\varepsilon_i \sim N(0, \sigma)$. $b_0 = -3609.51$; $b_1 = 1.815$; $s = 0.73587$. (e) (1.10702, 2.52298).

10.29 (a) $t = 2.25$, $0.01 < P\text{-value} < 0.02$. **(b)** States with more adult binge drinking are more likely to have underage drinking; 10.24% of the variation in underage drinking can be accounted for by the prevalence of adult binge drinking. **(c)** Even though most states were used, it is assumed that sampling took place for each state; thus, we can still infer about the true unknown correlation.

10.31

Source	DF	SS	MS	F
Regression	1	5552.9	5552.9	28.22
Residual Error	23	4525	196.739	
Total	24	10077.9		

10.33 (a) 0.15619. **(b)** (0.4898, 1.1362). **(c)** It tells us what the EAFE is when there is no return in U.S. markets. **(d)** 3.377; (−10.177, 3.797).

10.35 The first plot shows nonconstant variance. The second plot also shows nonconstant variance. The third plot has no violations. The fourth plot has a nonlinear pattern.

10.37 (a) $\hat{y} = -0.01270 + 0.01796x$; $r^2 = 79.98\%$. **(b)** H_0: $\beta_1 = 0$, H_a: $\beta_1 \neq 0$; $t = 7.48$, $P\text{-value} < 0.0001$. **(c)** (0.04, 0.1142). He can't be confident he won't be arrested.

10.39 (a) H_0: $\beta_1 = 0$, H_a: $\beta_1 \neq 0$. **(b)** $t = 10.67$, $P\text{-value} < 0.0001$. **(c)** (0.83963, 1.24033). **(d)** $r^2 = 81.41\%$. **(e)** The intercept represents the tuition at $x = 0$; this is extrapolation.

10.41 (a) $\hat{y} = 4466.19453 + 1.70030x$. **(b)** The residual plot looks good; the assumptions are valid.

10.43 (a) Both distributions are also right-skewed; the five-number summaries are 0%, 0.31%, 1.43%, 17.65%, 85.01% and 0, 2.25, 6.31, 12.69, 27.88. **(b)** Only the residuals need to be Normal. **(c)** The relationship is quite scattered. **(d)** $\hat{y} = 6.24693 + 0.10634x$. **(e)** The residuals are right-skewed.

10.45 (a) 30. **(b)** The relationship is linear, positive, and strong. **(c)** House 27 is unusual and could be influential. **(d)** $\hat{y} = 9.0176 + 1.15705x$, $s = 37.34442$. **(e)** $\hat{y} = 9.43181 + 1.123x$, $s = 25.39177$. **(f)** The outlier has some influence; the first model has a much larger standard error.

10.47 (a) 8.41%. **(b)** H_0: $\rho = 0$. H_a: $\rho \neq 0$. $t = 9.12$, $P\text{-value} < 0.0001$. **(c)** Students who did not answer might have different characteristics.

10.49 (a) IBI is slightly left-skewed; $\bar{x} = 65.94$, $s = 18.28$; Forest is slightly right-skewed; $\bar{x} = 28.29$, $s = 17.71$. **(b)** A weak positive association. **(c)** $y_i = \beta_0 + \beta_1 x_i + \epsilon_i$, $\epsilon_i \sim N(0, \sigma)$. **(d)** H_0: $\beta_1 = 0$, H_a: $\beta_1 \neq 0$. **(e)** $\hat{y} = 59.91 + 0.1531\text{Forest}$, $s = 17.79$. $t = 1.92$, $P\text{-value} = 0.0608$. **(f)** The residual plot shows that there is more variation for small x. **(g)** The residuals seem reasonably close to Normal.

10.51 The first change decreases P (that is, the relationship is more significant) because it accentuates the positive association. The second change weakens the association, so P increases (the relationship is less significant).

10.53 Using area: 57.52; (23.5598, 91.4892). Using forest: 69.55; (33.2085, 105.9006). Both prediction intervals have a lot of error.

10.55 (a) Very linear. **(b)** $\hat{y} = -61.12 + 9.3187\text{Year}$; $r^2 = 98.8\%$. **(c)** (8.3562, 10.2812).

10.57 (a) 116. **(b)** $\hat{y} = 1019.8492$, for a prediction of 3.0020 m. **(c)** Prediction interval.

10.59 For $n = 15$, $t = 2.08$; for $n = 25$, $t = 2.77$. The P-values are 0.0579 and 0.0109. Finding the same correlation with more data points is stronger evidence that the observed correlation is not just due to chance.

10.61 (a) Strong, positive linear relationship with one outlier. **(b)** $\hat{y} = 1.63 + 0.0214\text{SAT}$. $t = 10.78$, $P\text{-value} < 0.0005$. **(c)** $r = 0.8167$.

10.63 (a) $a_1 = 0.02617$, $a_0 = -2.7522$. **(c)** $\bar{y} = 21.13$ and $s_y = 4.7137$.

10.65 For $n = 123$: between 0.24 and 0.28 have a P-value < 0.01; between 0.20 and 0.23 have a P-value < 0.05. For $n = 96$: between 0.22 and 0.24 have a P-value < 0.05; the others are not significant.

10.67 (a) For women: (14.72609, 33.32604). For men: (−9.46079, 42.96351). These intervals overlap quite a bit. **(b)** For women: 22.78. For men: 16.38. The women's standard error is smaller in part because it is divided by a larger n. **(c)** Choose men with a wider variety of lean body masses.

CHAPTER 11

11.1 (a) Final exam scores. **(b)** 166. **(c)** Seven. **(d)** Math course anxiety, math test anxiety, numerical task anxiety, enjoyment, self-confidence, motivation, and perceived usefulness of the feedback sessions.

11.3 (a) A small P-value indicates that at least one explanatory variable is significant. **(b)** R^2 is not obtained from squaring and adding the pairwise correlations. **(c)** The null hypothesis should be β_2.

11.5 (a) (−0.4442, 13.2442). **(b)** (0.5391, 12.2609). **(c)** (−0.816, 10.416). **(d)** (1.2288, 8.3712).

11.7 (a) $y = \beta_0 + \beta_1 x_1 + \beta_2 x_2 + \beta_3 x_3 + \beta_4 x_4 + \beta_5 x_5 + \beta_6 x_6 + \epsilon$, where $\epsilon \sim N(0, \sigma)$ and independent. **(b)** The sources of variation are model (DFM $= p = 7$), error (DFE $= n - p - 1 = 134$), and total (DFT $= n - 1 = 141$).

11.9 (a) Things seem as expected: the three anxiety variables have negative signs, and the other four variables all have positive signs. **(b)** 7 and 158. **(c)** Only Feedback usefulness tests significant ($t = 3.320$, $P\text{-value} = 0.0011$) at the 0.05 level.

11.11 (a) We need $|t| > 1.984$. So Division, November, Weekend, Night, and Promotion are all significant in the presence of all the other explanatory variables. **(b)** df $= 12$ and 110, $P\text{-value} < 0.001$. **(c)** 52%. **(d)** 15246.36. **(e)** A prediction interval is more appropriate to represent this particular.

11.13 Answers will vary. Scatterplots are visual, the correlation matrix gives the actual values.

11.15 Generally, all four plots show the same random scattering and the conditions are met for a multiple regression model.

11.17 (a) For Model 1: 200; for Model 2: 199. **(b)** $t = 3.09$, P-value $= 0.0023$. **(c)** For Gene expression: $t = 2.44$, P-value $= 0.0153$; for RB.composite: $t = 3.33$, P-value $= 0.0010$. **(d)** The relationship is still positive after adjusting for RB. When gene expression increases by 1, popularity increases by 0.204 in Model 1 and by 0.161 in Model 2 (with RB fixed).

11.19 (a) $\hat{y} = 15656 + 43.08092\text{Admit} + 75.96892\text{Grad} + 0.83610\text{InCostAid} - 0.29900\text{OutCostAid}$. **(b)** $R^2 = 43.53\%$. **(c)** $F = 3.66$, df $= 4$ and 19, P-value $= 0.0225$. **(d)** $F = 3.66$, P-value $= 0.0225$. Only InCostAid has a significant t test ($t = 2.89$, P-value $= 0.0093$), and the other three are not significant when added to the model last.

11.21 (a) (16462, 31773). **(b)** (15818, 31347). **(c)** The intervals give very similar predictions for the Ohio State University. The model using all four predictors has only a slightly narrower prediction interval.

11.23 (a) 8 and 786. **(b)** 7.84%; it is not very predictive. **(c)** Males and Hispanics consume energy drinks more frequently. Consumption also increases with risk-taking scores. **(d)** Within a group of students with identical (or similar) values of those other variables, energy-drink consumption increases with increasing jock identity and increasing risk taking.

11.25 (a) $F = 10.44$, P-value < 0.0001; $\hat{y} = 23.39556 - 0.68175x_1 + 0.10195x_2$. **(b)** 17.71%. **(c)** No violations. **(d)** $H_0: \beta_2 = 0$, $H_a: \beta_2 \neq 0$; $t = 1.83$, P-value $= 0.0696$.

11.27 (a) Budget and Opening are right-skewed. Theaters and Ratings are left-skewed. **(b)** The correlations are 0.403, 0.570, 0.625, 0.281, 0.151, and -0.022. Budget, Opening, and Theaters have the largest correlations among them (first three listed), Ratings is not highly correlated with any of the other three (last three listed).

11.29 (a) $F = 32.28$, P-value < 0.0001. USRevenue $= \beta_0 + \beta_1\text{Budget} + \beta_2\text{Opening} + \beta_3\text{Theaters} + \beta_4\text{Ratings} + \epsilon$, where $\epsilon \sim N(0, \sigma)$ and independent. **(b)** $\hat{y} = -170.40874 + 0.09252\text{Budget} + 1.91600\text{Opening} + 0.02961\text{Theaters} + 17.29923\text{Ratings}$. **(c)** The residual plot shows a slight downward trend, suggesting another model may be more appropriate. **(d)** $R^2 = 77.26\%$.

11.31 (a) ($-\$25,514,500$, $\$158,200,600$). **(b)** ($-\$28,098,300$, $\$154,239,900$). **(c)** The intervals are similar.

11.33 (a) Teaching and research are both right-skewed. Citations is left-skewed. **(b)** Teaching and research are very strongly linearly related ($r = 0.8993$). Citations does not look related to either teaching or research ($r = 0.1878$ and 0.0691, respectively).

11.35 (a) Overall $= \beta_0 + \beta_1\text{Teaching} + \beta_2\text{Research} + \beta_3\text{Citations} + \varepsilon$, $\varepsilon \sim N(0, \sigma)$ and independent. **(b)** $F = 736.38$, P-value < 0.0001, $\hat{y} = 8.16814 + 0.26432\text{Teaching} + $ 0.32800Research $+ 0.27513$Citations. **(c)** For teaching: (0.19280, 0.33583); for research: (0.26790, 0.38810); for Citations: (0.23850, 0.31176). **(d)** $R^2 = 97.74\%$; $s = 1.72272$.

11.37 (a) GINI and CORRUPT to the right, the other three to the left. CORRUPT, DEMOCRACY, and LIFE have the most skewness. **(b)** LSI seems moderately correlated with Corrupt, Democracy, and Life, ($r = 0.6974$, 0.6092, and 0.7219) but is not related to GINI much at all ($r = -0.0503$). Among the others, only CORRUPT seems to be moderately related to both DEMOCRACY and LIFE ($r = 0.7474$ and 0.6503); other relationships appear weak.

11.39 (a) Refer to your regression output. **(b)** For example, the t statistic for the GINI coefficient grows from $t = -0.42$ ($P = 0.675$) to $t = 4.25$ ($P < 0.0005$). The DEMOCRACY t is 3.53 in the third model ($P < 0.0005$) but drops to 0.71 ($P = 0.479$) in the fourth model. **(c)** A good choice is to use GINI, LIFE, and CORRUPT. All three coefficients are significant, and $R^2 = 70\%$.

11.41 (a) $F = 22.34$, P-value < 0.0001, $\hat{y} = 334.03439 + 19.50471\text{OC}$. The residual plot shows a possible outlier. **(b)** $F = 21.62$, P-value < 0.0001, $\hat{y} = 57.70419 + 6.41466\text{OC} + 53.39331\text{TRAP}$. TRAP is much more significant ($t = 3.50$, P-value $= 0.0016$) in this model than OC ($t = 1.25$, P-value $= 0.2210$).

11.43 All variables are Normal when log transformed. All pairs are positively associated: strongest between LVO+ and LVO$-$ ($r = 0.8396$) and weakest between LOC and LVO$-$ ($r = 0.5545$). Using logOC: $\hat{y} = 4.39 + 0.71\text{logOC}$, $t = 6.57$, $P < 0.0001$, $R^2 = 59.83\%$, $s = 0.36$. Using logOC and logTRAP: $\hat{y} = 4.26 + 0.43\text{logOC} + 0.42\text{logTRAP}$, $t = 2.56$, $P = 0.0162$, $t = 2.06$, $P = 0.0484$, $R^2 = 65.14\%$, $s = 0.34$. Using all three: $\hat{y} = 0.87 + 0.39\text{logOC} + 0.03\text{logTRAP} + 0.67\text{logVO}-$, $t = 3.40$, $P = 0.0021$, $t = 0.17$, $P = 0.8624$, $t = 5.71$, $P < 0.0001$, $R^2 = 84.21\%$, $s = 0.23$. The best model uses only logOC and logVO$-$: $\hat{y} = 0.83298 + 0.40589\text{logOC} + 0.68159\text{logVO}-$, $R^2 = 84.19\%$, $s = 0.23$.

11.45 Using logOC: $\hat{y} = 5.21 + 0.44\text{logOC}$, $t = 3.59$, $P = 0.0012$, $R^2 = 30.75\%$, $s = 0.41$. Using logOC and logTRAP: $\hat{y} = 5.04 + 0.06\text{logOC} + 0.59\text{logTRAP}$, $t = 0.31$, $P = 0.7618$, $t = 2.61$, $P = 0.0144$, $R^2 = 44.30\%$, $s = 0.37$. Using all three: $\hat{y} = 1.57 - 0.29\text{logOC} + 0.24\text{logTRAP} + 0.81\text{logVO}+$, $t = -2.08$, $P = 0.0468$, $t = 1.47$, $P = 0.1523$, $t = 5.71$, $P < 0.0001$, $R^2 = 74.77\%$, $s = 0.26$. The best model uses only logVO+ alone: $\hat{y} = 1.75657 + 0.7305\text{logVO}+$, $R^2 = 70.49\%$, $s = 0.27$.

11.47 (a) PCB $= \beta_0 + \beta_1\text{PCB52} + \beta_2\text{PCB118} + \beta_3\text{PCB138} + \beta_4\text{PCB180} + \epsilon$, where $\epsilon \sim N(0, \sigma)$ and independent. **(b)** $F = 1456.18$, P-value < 0.0001, $\hat{y} = 0.93692 + 11.87270\text{PCB52} + 3.76107\text{PCB118} + 3.88423\text{PCB138} + 4.18230\text{PCB180}$. All individual predictors are significant. **(c)** The residual plot shows a possible violation of constant variance. The residuals are Normal, except for two possible outliers.

11.49 (a) $F = 786.71$, P-value < 0.0001, $\hat{y} = -1.01840 + 12.64419\text{PCB52} + 0.31311\text{PCB118} + 8.25459\text{PCB138}$.

(b) $b_{118} = 0.31311$, P-value $= 0.7083$. **(c)** $b_{118} = 3.76107$, P-value < 0.0001. **(d)** When we add PCB180 to the model, it makes PCB118 useful for prediction.

11.51 TEQ $= \beta_0 + \beta_1 PCB52 + \beta_2 PCB118 + \beta_3 PCB138 + \beta_4 PCB180 + \epsilon$, where $\epsilon \sim N(0, \sigma)$ and independent. $F = 33.53$, P-value < 0.0001. Only PCB118 tests significant individually. The residual plot shows a couple potential outliers, which are also causing a slight right-skew in the Normal quantile plot.

11.53 **(a)** The correlations are all positive; the largest correlation is 0.956 (LPCB and LPCB138); the smallest 0.227 (LPCB28 and LPCB180). There is one outlier (specimen 39) in LPCB28; the latter stands out because of the "stack" of values in the LPCB126 data set that arose from the adjustment of the zero terms. **(b)** All correlations are higher with the transformed data.

11.55 A good model includes logPCB28, logPCB118, and logPCB126; $R^2 = 0.7764$. Adding more variables doesn't increase R^2 much.

11.57 **(a)** Taste: 24.53, 20.95, 16.26, 23.9. Acetic: 5.50, 5.43, 0.57, 0.66. H2S: 5.94, 5.33, 2.13, 3.69. Lactic: 1.44, 1.45, 0.30, 0.43. None of the variables show striking deviations from Normality in the quantile plots. Taste and H2S are slightly right-skewed, and Acetic has an irregular shape. There are no outliers.

11.59 $F = 12.11$, P-value $= 0.0017$, $\hat{y} = -61.49861 + 15.64777$Acetic. $R^2 = 30.20\%$. The residuals are Normally distributed, but the scatterplots show that the residuals are linearly related to both H2S and Lactic.

11.61 $F = 27.55$, P-value < 0.0001, $\hat{y} = -29.85883 + 37.71995$Lactic. $R^2 = 49.59\%$. The residuals are Normally distributed, but the scatterplots show that the residuals are linearly related to both H2S and Lactic.

11.63 $\hat{y} = -26.94 + 3.801$ Acetic $+ 5.146$ H2S with $s = 10.89$ and $R^2 = 0.582$. For Acetic: $t = 0.84$ (P-value $= 0.406$). This two-variable model is not much better than the model with H2S alone (which explained 57.1% of the variation in Taste).

11.65 $\hat{y} = -28.88 + 0.328$ Acetic $+ 3.912$ H2S $+ 19.671$ Lactic with $s = 10.13$. $R^2 = 65.2\%$. Acetic is not significant (P-value $= 0.942$); there is no gain in adding Acetic to the model with H2S and Lactic. Residuals appear to be Normally distributed and show no patterns in scatterplots with explanatory variables. It appears that the H2S/Lactic model is best.

CHAPTER 12

12.1 **(a)** ANOVA tests the null hypothesis that the *population* means are all equal. **(b)** *Experiments* are best for establishing causation. **(c)** ANOVA is used to compare *means* (and assumes that the variances are equal). **(d)** Multiple comparisons procedures are used when we wish to determine which means are significantly different, but we do not need specific relations in mind prior to looking at the data.

12.3 $x_{ij} = \mu_i + \varepsilon_{ij}$, $i = 1, 2, 3, j = 1, 2, \ldots, 20$. $\varepsilon_{ij} \sim N(0, \sigma)$; $I = 3$, $n_i = 20$. Parameters: μ_1, μ_2, μ_3, and σ.

12.5 **(a)** Yes, $80 < 2(68)$. **(b)** The estimates for μ_1, μ_2, and μ_3 are 279, 245, and 258. The estimate for σ is 75.516.

12.7 The Normal quantile plot shows the residuals are Normally distributed.

12.9 **(a)** This sentence describes *between-group* variation. **(b)** The *sums of squares* in an ANOVA table will add; that is, SST $=$ SSG $+$ SSE. **(c)** σ is a parameter, not s_p. **(d)** A small P means the means are not all the same, but the distributions may still overlap quite a bit.

12.11 **(a)** 4 and 25. In Table E, $4.18 < F < 6.49$. **(c)** $0.001 < P$-value < 0.01. **(d)** Because the P-value is small we reject H_0; however, this does not say that all pairs of group means are different, only that at least one mean is different.

12.13 Generally, the one with the largest difference between means and the smallest standard deviation will be the most significant. If this is not clear a ratio between the two can be used. So part (b) can be ruled out because it has a larger sigma than part (a) with the same means. Of the remaining two we can see that part (c) has a bigger max difference relative to its sigma (10 to 3) than part (a) does (5 to 2), so part (c) will be the most significant.

12.15 **(a)** $F = 6.8$, DFG $= 2$, DFE $= 60$. $0.001 < P$-value < 0.01. **(b)** DFG $= 7$, DFE $= 40$. MSG $= 11$, MSE $= 4.75$; $F = 2.32$, $0.025 < P$-value < 0.05.

12.17 **(a)** Response: egg cholesterol level. Populations: chickens with different diets or drugs. $I = 3$, $n_1 = n_2 = n_3 = 25$, $N = 75$. **(b)** Response: rating on seven-point scale. Populations: the three groups of students. $I = 3$, $n_1 = 31$, $n_2 = 18$, $n_3 = 45$, $N = 94$. **(c)** Response: quiz score. Populations: students in each TA group. $I = 3$, $n_1 = n_2 = n_3 = 14$, $N = 42$.

12.19 For all three situations, we have $H_0: \mu_1 = \mu_2 = \mu_3$. H_α: not all of the μ_i are equal. DFG $= I - 1 = 2$, DFE $= N - I$, and DFT $= N - 1$. The degrees of freedom for the F test are DFG and DFE. **(a)** DFG 2, DFE 72, DFT 74; $F(2, 72)$. **(b)** DFG 2, DFE 91, DFT 93; $F(2, 91)$. **(c)** DFG 2, DFE 39, DFT 41; $F(2, 39)$.

12.21 **(a)** This sounds like a fairly well-designed experiment, so the results should at least apply to this farmer's breed of chicken. **(b)** It would be good to know what proportion of the total student body falls in each of these groups—that is, is anyone overrepresented in this sample? **(c)** How well a TA teaches one topic (power calculations) might not reflect that TA's overall effectiveness.

12.23 **(a)** Both drugs cause an increase in activity level; Drug B appears to have a greater effect. **(b)** Yes; $\sqrt{17.2} < 2\sqrt{7.75}$, $s_p = 3.487$. **(c)** DFG $= 4$, DFE $= 20$. **(d)** $2.25 < F < 2.87$, $0.05 < P$-value < 0.10.

12.25 **(a)** 4 and 178. **(b)** $5 + 146 = 151$ athletes actually participated. **(c)** For example, the individuals could have been outliers in terms of their ability to withstand the water-bath pain. In either case of low or high outliers, their removal would lessen the standard deviation for their sport and move that sports mean.

12.27 $a_1 = 0.5$, $a_2 = 0.5$, $a_3 = -0.5$, $a_4 = -0.5$.

12.29 Because there are only two groups, if the ANOVA test establishes that there are differences in means, then we already know that the two means we have must be different. In this case, contrasts and multiple comparison provide no further useful information.

12.31 The power would be larger. For larger differences between alternative means, λ gets bigger, increasing our power to see these differences.

12.33 (a) Yes, $0.824 < 2(0.657)$; $s_p = 0.7683$. (b) df = 2 and 767, P-value < 0.001. (c) $\psi = \mu_2 - 0.5(\mu_1 + \mu_3)$; we test H_0: $\psi = 0$; H_a: $\psi > 0$. We find $c = 0.585$, $t = 5.99$, P-value < 0.0001.

12.35 (a) $1\mu_2 - 0.5\mu_1 - 0.5\mu_4$. (b) $1/3\mu_1 + 1/3\mu_2 + 1/3\mu_4 - 1\mu_3$.

12.37 (a) $\psi_1 = \mu_1 - 1/2(\mu_2 + \mu_4)$ and $\psi_2 = \mu_3 - \mu_2 - (\mu_5 - \mu_4)$. (b) $c_1 = -3.9$ and $c_2 = 2.35$; $SE_{c_1} = 2.1353$ and $SE_{c_2} = 3.487$. (c) The first contrast is significant ($t_1 = -1.826$, P-value $= 0.0414$), but the second is not ($t_2 = -0.674$, P-value 0.2540).

12.39 (a) $\psi = \mu_7 - 0.25(\mu_1 + \mu_2 + \mu_3 + \mu_4)$. (b) H_0: $\psi = 0$; H_a: $\psi > 0$. (c) $t = 1.894$, P-value $= 0.0302$.

12.41 The power would be larger. For larger differences between alternative means, λ gets bigger, increasing our power to see these differences.

12.43 (a) F can be made very small and the P-value close to 1. (b) F increases and P-value decreases.

12.45 (a) Cable has the highest prices; Fiber has the cheapest. (b) Yes, $40.39 < 2(26.09)$. (c) df = 2 and 44, $0.025 < P$-value < 0.05.

12.47 (a) Portals and Transactions have higher integration features than Presence. (b) An observational study. They are not imposing a treatment on the winery. (c) Yes; $2.346 < 2(2.097)$. (d) Our inference is based on sample means, which will be approximately Normal given the sample sizes. (e) $F_{(2, 190)}$; P-value < 0.001.

12.49 (a) H_0: $\mu_1 = \mu_2 = \mu_3$, H_a: not all of the μ_i are equal, $F = 5.31$, P-value $= 0.0067$. (b) The Bonferroni shows that group 2 is not significantly different from either group 1 or group 3, but group 3 is significantly different (larger) from group 1. (c) This is not appropriate. The regression assumes that group 2 (coded as 2) would have twice the effect of group 1 (coded as 1), and group 3 (coded as 3) would have three times the effect of group 1, etc. This is likely not true.

12.51 (a) See accompanying table. Pooling is appropriate, $0.621669 < 2(0.572914)$. (b) While the distributions aren't Normal, there are no outliers or extreme departures from Normality that would invalidate the results. We can likely proceed with the ANOVA.

	Score		
Level of food	N	Mean	Std dev
Comfort	22	4.8873	0.5729
Control	20	5.0825	0.6217
Organic	20	5.5835	0.5936

12.53 (a) H_0: $\mu_1 = \mu_2 = \mu_3$, H_a: not all of the μ_i are equal, $F = 8.89$, P-value $= 0.000$. There are significant differences in the number of minutes that the three groups are willing to volunteer. The Comfort group is willing to donate significantly more minutes than the Organic group. The Control group is in the middle, not significantly different from either the Comfort or Organic group. (b) The residual plot shows a possible violation of constant variance. The Normal quantile plot looks fine and shows a roughly Normal distribution.

12.55 (a) See accompanying table. (b) Yes, the largest s is less than twice the smallest s; $11.501 < 2(9.078) = 18.156$. (c) All three distributions are roughly Normal.

	Loss		
Level of group	N	Mean	Std dev
Control	35	-1.0086	11.5007
Group	34	-10.7853	11.1392
Individual	35	-3.7086	9.0784

12.57 (a) All weight loss values are divided by 2.2. (b) $F = 7.77$, df = 2 and 101, P-value $= 0.0007$. The results are identical with the transformed data.

12.59 (a) The variation in sample size is some cause for concern, but there can be no extreme outliers in a 1-to-7 scale. (b) Yes; $1.26 < 2(1.03)$. (c) $F_{(4, 405)}$, P-value < 0.0002. (d) Hispanic Americans are highest, Japanese in the middle, and the other three are the lowest.

12.61 (a) df = 2, 117. (b) P-value < 0.001. (c) Because the bargainer was the same person each time, the results would certainly not be generalizable.

12.63 (a) See accompanying table. (b) All three distributions appear to be reasonably close to Normal, and the standard deviations are suitable for pooling. (c) $F = 7.72$ (df 2 and 42), P-value $= 0.001$. (d) For Bonferroni, $t^{**} = 2.4937$ and MSD = 0.01308. The high-dose mean is significantly different from the other two. (e) High doses of kudzu isoflavones increase BMD.

	n	$\bar{x}$	s	SE
Control	15	0.2189	0.01159	0.002992
Low dose	15	0.2159	0.01151	0.002972
High dose	15	0.2351	0.01877	0.004847

12.65 (a) See accompanying table. Yes; $27.364 < 2(16.594)$. (b) $F = 7.98$ (df = 2 and 27), P-value = 0.002.

	n	$\bar{x}$	s
Control	10	601.1	27.364
Low jump	10	612.5	19.329
High jump	10	638.7	16.594

12.67 (a) $\psi_{sex} = \mu_1 - \mu_3$; $c_{sex} = -1.255$, $t = -3.6$; P-value < 0.001. (b) $\psi_{nat} = \mu_1 - \mu_2$; $c_{nat} = 0.005$, $t = 0.014$; P-value > 0.25. (c) The conclusions from the contrasts were limited to only comparing sex among Hispanics and nationalities among males. Including an Anglo female would alleviate this limited inference.

12.69 (a) See accompanying table. Pooling is not appropriate, the largest s is more than twice the smallest s; $8.6603 > 2(2.8868) = 5.7736$. (b) $F = 137.94$ (df = 5 and 12), P-value < 0.0005.

	n	Mean	s
ECM1	3	65.00%	8.66%
ECM2	3	63.33%	2.89%
ECM3	3	73.33%	2.89%
MAT1	3	23.33%	2.89%
MAT2	3	6.67%	2.89%
MAT3	3	11.67%	2.89%

12.71 (a) $\psi_1 = \mu_C - 0.25(\mu_{30\times1} + \mu_{30\times2} + \mu_{60\times1} + \mu_{60\times2})$, $\psi_2 = 0.5(\mu_{30\times1} + \mu_{30\times2}) - 0.5(\mu_{60\times1} + \mu_{60\times2})$, $\psi_3 = 0.5(\mu_{60\times1} - \mu_{60\times2}) - 0.5(\mu_{30\times1} - \mu_{30\times2})$. (b) $c_1 = 14.65$, $c_2 = 6.1$, $c_3 = -0.5$. $SE_{C1} = 4.209$, $SE_{C2} = SE_{C3} = 3.784$. (c) $t_1 = 3.481$, P-value = 0.0007; $t_2 = 1.612$, P-value = 0.1097; $t_3 = -0.132$, P-value = 0.8952. The first two are significant, the third is not.

12.73 (a) The plot shows granularity, but otherwise, independence is not violated. (b) Yes; $1.6 < 2(0.93) = 1.86$. (c) The individual Normal quantile plots for each shampoo show a lot of granularity due to using integer values but most look at least roughly Normal. (d) The Normal quantile plot for the residuals shows that the residuals are Normally distributed.

12.75 (a) $\psi_1 = 1/3(\mu_{Pyr1} + \mu_{Pyr2} + \mu_{Keto}) - \mu_{Placebo}$, $\psi_2 = 0.5(\mu_{Pyr1} + \mu_{Pyr2}) - \mu_{Keto}$, $\psi_3 = \mu_{Pyr1} - \mu_{Pyr2}$. (b) $s_p = 1.1958$. $c_1 = -12.51$, $SE_{C1} = 0.2355$; $c_2 = 1.269$, $SE_{C2} = 0.1413$; $c_3 = 0.191$, $SE_{C3} = 0.1609$. (c) H_0: $\psi_i = 0$, H_a: $\psi_i \neq 0$; $t_1 = -53.17$, P-value$_1 < 0.0005$; $t_2 = 8.98$, P-value$_2 < 0.0005$; $t_3 = 1.19$, P-value$_3 = 0.2359$. The Placebo mean is significantly higher than the average of the other three, while the Keto mean is significantly lower than the average of the two Pyr means. The difference between the Pyr means is not significant (meaning the second application of the shampoo is of little benefit).

12.77 The means increase by 5%, but everything else remains the same.

12.79 All distributions are reasonably Normal and standard deviations are close enough to justify pooling. For PRE1, $F = 1.13$, P-value = 0.329; for PRE2, $F = 0.11$, P-value = 0.895. There is no reason to believe that the mean pretest scores differ between methods.

12.81 (b) Answers will vary with choice of H_a and desired power. For example, with the alternative $\mu_1 = \mu_2$ 4.4, $\mu_3 = 5$, and $\sigma = 1.2$, three samples of size 75 will produce power 0.78.

12.83 The design can be similar, although the types of music might be different. Bear in mind that spending at a casual restaurant will likely be less than at the restaurants examined in Exercise 12.62; this might also mean that the standard deviations could be smaller. A pilot study might be necessary to get an idea of the size of the standard deviations. Decide how big a difference in mean spending you would want to detect; then do some power computations.

CHAPTER 13

13.1 $x_{ijk} = \mu_{ij} + \varepsilon_{ijk}$, $i = 1, 2$, $j = 1, 2$, $k = 1, \ldots, n_{ij}$; $\varepsilon_{ijk} \sim N(0, \sigma)$. We have $I = 2$, $J = 2$, $n_{11} = 27$, $n_{12} = 23$, $n_{21} = 19$, and $n_{22} = 31$. The parameters of the model are μ_{11}, μ_{12}, μ_{21}, μ_{22}, and σ.

13.3 (a) Two-way ANOVA is used when there are two factors (explanatory variables). (b) Each level of A should occur with all three levels of B. (Level A has two factors.) (c) The RESIDUAL part of the model represents the error. (d) DFAB $= (I - 1)(J - 1)$.

13.5 (a) A *large* value of the AB F statistic indicates that we should reject the hypothesis of no interaction. (b) The relationship is backward: *Mean* squares equal the *sum of* squares divided by degrees of freedom. (c) Under H_0, the ANOVA test statistics have an F distribution. (d) If the sample sizes are not the same, the sums of squares may not add for "some methods of analysis."

13.7 (a) There are nine cells with four observations each, so $N = 36$; DFA = 2, DFB = 2, DFAB = 4, DFE = 27. (c) $0.025 < P$-value < 0.05. (d) Interaction is not significant; the interaction plot should have roughly parallel lines.

13.9 (a) The factors are sex ($I = 2$) and age ($J = 3$). The response variable is the percent of pretend play. $N = 66$. (b) The factors are weeks after harvest ($I = 5$) and amount of water ($J = 2$). The response variable is the percent of seeds germinating. $N = 30$. (c) The factors are mixture ($I = 6$) and freezing/thawing cycles ($J = 3$). The response variable is the strength of the specimen. $N = 54$. (d) The factors are different colored tags ($I = 4$) and the type of buyer ($J = 2$). The response variable is the total dollar amount spent. $N = 138$.

13.11 (a) For those with real-time feedback, the average total cost for those not informed was only slightly larger than the average total cost for those informed, while for those without feedback, the not informed average was much larger than for those who were informed. Additionally, we can see an interaction effect. For those not informed, the lack of real-time feedback increased their spending, while for those who were informed, the lack of real-time feedback decreased their spending. **(b)** $F = 18.18$, df = 3, 190, P-value < 0.0001. **(c)** Interaction: $F = 16.18$, P-value < 0.0001; Informed: $F = 37.06$, P-value < 0.0001; Smartcart: $F = 1.16$, P-value $= 0.2828$.

13.13 (a) The same students were tested twice. **(b)** The plot shows a definite interaction; the control group's mean score decreased, while the expressive writing group's mean increased somewhat. **(c)** No, the largest s is more than twice the smallest s; $14.3 > 2(5.8) = 11.6$.

13.15 (a) The Normality assumption is for the error terms, not the measurements; however, recall from Chapter 12 that ANOVA is robust to reasonable departures from Normality, especially when sample sizes are similar. **(b)** Yes; the largest s is less than twice the smallest: $1.62 < 2(0.82) = 1.64$.

Source	DF	SS	MS	F	P-value
Age	6	31.97	5.328	4.4	0.0003
Sex	1	44.66	44.66	36.879	0.0000
Age* Sex	6	13.22	2.203	1.819	0.0962
Error	232	280.95	1.211		

13.17 (a) There appears to be an interaction effect; the lines are not parallel. **(b)** There appears to be a significant Focus main effect, so the marginal means for Focus would be useful in explaining this difference. **(c)** If each participant looked at a picture of each body type, then their responses likely would be related to each other, which violates the independence assumption.

13.19 (a) There appears to be an interaction; a thank you increases repurchase intent for consumers with a short history but not for those with long history. **(b)** Short: 6.245; Long: 7.45; No: 6.61; Yes: 7.085. Generally, the long history consumers are more likely to repurchase. Thank you is misleading, suggesting that no thanks is lower than yes thanks, but that is only true for the short history group.

13.21 (a) The plot suggests a possible interaction. **(b)** By subjecting the same individual to all four treatments, rather than four individuals, we reduce the within-groups variability.

13.23 (a) We'd expect reaction times to slow with older individuals. If bilingualism helps brain functioning, we would not expect that group to slow as much as the monolingual group. The expected interaction is seen in the plot; mean total reaction time for the older bilingual group is much less than for the older monolingual group; the lines are not parallel. **(b)** The interaction is just barely not significant ($F = 3.67$, P-value $= 0.059$). Both main effects are very significant (P-value $= 0.000$).

13.25 (a) There appears to be an interaction effect. A favorable process increases satisfaction for those with a favorable outcome but not for those with an unfavorable outcome. **(b)** No interaction effect. Favorable outcome means were only slightly higher than the unfavorable outcomes means. But both favorable process means were higher than both unfavor-

able process means. **(c)** Yes, there appears to be a three-factor interaction because the interactions in parts (a) and (b) are different.

13.27 Humor slightly increases satisfaction (from 3.60 to 3.88). A favorable process greatly increases satisfaction (from 2.74 to 4.80). A favorable outcome also increases satisfaction (from 3.21 to 4.30).

13.29 $s_p = 1.7746$. Yes, the largest s is less than twice the smallest s; $2.024 < 2(1.601) = 3.202$.

13.31 Patterns are similar for all four responses. Canada has the highest mean responses, then the United States, then France. Females are higher than males in all cases, except for credibility and purchase intention in France. Sex differences are largest for Canada, then the United States, and very little for France (except for credibility and purchase intention, where U.S. differences are the largest).

13.33 Main effect A is not significant, $0.05 < P$-value < 0.100 (df = 2, 18). Main effect B is significant, $0.025 < P$-value < 0.050 (df = 2, 18). The interaction is not significant, P-value > 0.100 (df = 4, 18).

13.35 (a) There is little evidence of an interaction. **(b)** $s_p = 0.1278$. **(c)** $\psi_1 = 1/2(\mu_{new,city} + \mu_{new,highway}) - 1/2(\mu_{old,city} + \mu_{old,highway})$, $\psi_2 = \mu_{new,city} - \mu_{new,highway}$, $\psi_3 = \mu_{old,highway} - \mu_{old,city}$. **(d)** By subjecting the same individual to all four treatments, rather than four individuals to one treatment each, we reduce the within-groups variability.

13.37 (b) There seems to be a fairly large difference between the means based on how much the rats were allowed to eat, but not very much difference based on the chromium level. There may be an interaction: the NM mean is lower than the LM mean, while the NR mean is higher than the LR mean. **(c)** L: 4.86, N: 4.871, M: 4.485, R: 5.246; LR minus LM: 0.63; NR minus NM: 0.892. Mean GITH levels are lower for M than for R; there is not much difference for L versus N. The difference between M and R is greater among rats that had Normal chromium levels in their diets (N).

13.39 (a) $s_p = 38.14$. **(b)** Yes, the largest s is less than twice the smallest s; $42.4 < 2(31.2) = 62.4$. **(c)** Sender Individual: 70.24; Sender Group: 48.65; Responder Individual: 60.46; Responder Group: 59.37. **(d)** There appears to be an interaction effect. Individuals send more to groups; groups send more to individuals. **(e)** Sender: P-value $= 0.0033$; Responder: P-value $= 0.9748$; Interaction: P-value $= 0.1522$. Only the sender effect is significant.

13.41 Yes, the iron-pot means are the highest, and the F statistic for testing the effect of the pot type is very large.

13.43 (a) $\bar{x}_{11} = 25.031$, $s_{11} = 0.0011547$; $\bar{x}_{12} = 25.028$, $s_{12} = 0$; $\bar{x}_{13} = 25.026$, $s_{13} = 0$; $\bar{x}_{21} = 25.017$, $s_{21} = 0.0011547$; $\bar{x}_{22} = 25.020$, $s_{22} = 0.002$; $\bar{x}_{23} = 25.016$, $s_{23} = 0$; $\bar{x}_{31} = 25.006$, $s_{31} = 0.0015275$; $\bar{x}_{32} = 25.013$, $s_{32} = 0.0011547$; $\bar{x}_{33} = 25.009$, $s_{33} = 0.0011547$. **(b)** Tools 3, 4, and 5 are fairly consistent across time differences, with time 2 having the largest diameters and time 1 the smallest diameters. For tool 2, however, time

1 diameters get larger than time 3, and for tool 1, time 1 diameters get the largest, bigger than both time 2 and time 3 diameters. **(c)** $F_A = 412.94$, df = 4, 30, P-value < 0.0001; $F_B = 43.60$, df = 2, 30, P-value < 0.0001; $F_{AB} = 7.65$, df = 8, 30, P-value < 0.0001. **(d)** There are differences in diameter among the five different tools; there are also differences in diameter in the different shifts, although not as large as the tool differences. There also appears to be a small interaction effect.

13.45 (a) All three F values have df 1 and 945, and the P-values are < 0.001, < 0.001, and 0.1477. Sex and handedness both have significant effects on mean lifetime, but there is no significant interaction. **(b)** Women live about six years longer than men (on the average), while right-handed people average nine more years of life than left-handed people. "There is no interaction" means that handedness affects both sexes in the same way, and vice versa.

13.47 (a) and **(b)** The first three means and standard deviations are $\bar{x}_{11} = 3.2543$, $s_{11} = 0.2287$; $\bar{x}_{12} = 2.7636$, $s_{12} = 0.0666$; $\bar{x}_{13} = 2.8429$, $s_{13} = 0.2333$. Standard deviations range from 0.0666 to 0.3437 for a ratio of 5.16—larger than we like. **(c)** $F_S = 1301.32$, df = 3, 224, P-value = 0.000; $F_W = 9.76$, df = 6, 224, P-value = 0.000; $F_{SW} = 5.97$, df = 18, 224, P-value = 0.000.

13.49 The seven F statistics are 184.05, 115.93, 208.87, 218.37, 220.01, 174.14, and 230.17, all with df = 3 and 32 and P-value < 0.0005. The only *nonsignificant* differences are between species 1 and 3 for water levels 1, 4, and 7. Therefore, for every water level, species 4 has the lowest nitrogen level and species 2 is next. For water levels 1, 4, and 7, species 1 and 3 are statistically tied for the highest level; for the other levels, species 3 is the highest, with species 1 coming in second.

13.51 (a) and **(b)** The first three means and standard deviations for Fresh are $\bar{x}_{11} = 109.095$, $s_{11} = 20.949$; $\bar{x}_{12} = 165.138$, $s_{12} = 29.084$; $\bar{x}_{13} = 168.825$, $s_{13} = 18.866$. The first three means and standard deviations for Dry are $\bar{x}_{11} = 40.565$, $s_{11} = 5.581$; $\bar{x}_{12} = 63.863$, $s_{12} = 7.508$; $\bar{x}_{13} = 71.003$, $s_{13} = 6.032$. Standard deviation ratios are quite high for both fresh and dry biomass: $108.01/6.79 = 15.9$ and $35.76/3.12 = 11.5$. **(c)** For Fresh: $F_S = 81.45$, df = 3, 84, P-value = 0.000; $F_W = 43.71$, df = 6, 84, P-value = 0.000; $F_{SW} = 1.79$, df = 18, 84, P-value = 0.004. For Dry: $F_S = 79.93$,

df = 3, 84, P-value = 0.000; $F_W = 44.79$, df = 6, 84, P-value = 0.000; $F_{SW} = 2.22$, df = 18, 84, P-value = 0.008.

13.53 Fresh: the F statistics are 15.88, 11.81, 62.08, 10.83, 22.62, 8.20, and 10.81; Dry: the F statistics are 8.14, 26.26, 22.58, 11.86, 21.38, 14.77, and 8.66; all with df = 3 and 12 and P-value < 0.003.

13.55 (a) Sex: 1 and 174; Floral: 2 and 174; Interaction: 2 and 174. **(b)** Damage to males was higher for all characteristics. For males, damage was highest under characteristic level 3, while for females, the highest damage occurred at level 2. Interaction should be significant because the distance between the means increases from floral type 1 to floral type 3. **(c)** Three of the standard deviations are at least half as large as the means. Because the response variable (leaf damage) must be non-negative, this suggests that these distributions are right-skewed.

13.57 Men in CS: $n = 39$, $\bar{x} = 7.7949$, $s = 1.5075$; men in EO: $n = 39$, $\bar{x} = 7.4872$, $s = 2.1505$; men in Other: $n = 39$, $\bar{x} = 7.4103$, $s = 1.5681$. Women in CS: $n = 39$, $\bar{x} = 8.8462$, $s = 1.1364$; women in EO: $n = 39$, $\bar{x} = 9.2564$, $s = 0.7511$; women in Other: $n = 39$, $\bar{x} = 8.6154$, $s = 1.1611$. The means suggest that females have higher HSE grades than males. For a given sex, there is not too much difference among majors. Normal quantile plots show no great deviations from Normality, apart from the granularity of the grades (most evident among women in EO). In the ANOVA, only the effect of sex is significant. Residual analysis (not shown) reveals some causes for concern; for example, the variance does not appear to be constant.

13.59 Men in CS: $n = 39$, $\bar{x} = 526.949$, $s = 100.937$; men in EO: $n = 39$, $\bar{x} = 507.846$, $s = 57.213$; men in Other: $n = 39$, $\bar{x} = 487.564$, $s = 108.779$. Women in CS: $n = 39$, $\bar{x} = 543.385$, $s = 77.654$; women in EO: $n = 39$, $\bar{x} = 538.205$, $s = 102.209$; women in Other: $n = 39$, $\bar{x} = 465.026$, $s = 82.184$. The means suggest that students who stay in the sciences have higher mean SATV scores than those who end up in the "Other" group. Female CS and EO students have higher scores than males in those majors, but males have the higher mean in the "Other" group. Normal quantile plots suggest some right-skewness in the "Women in CS" group and also some non-Normality in the tails of the "Women in EO" group. Other groups look reasonably Normal. In the ANOVA table, only the effect of major is significant.

Notes and Data Sources

CHAPTER 1

1. See **census.gov**.

2. From *State of Drunk Driving Fatalities in America 2010*, available at **responsibility.org**.

3. James P. Purdy, "Why first-year college students select online research sources as their favorite," *First Monday*, 17, No. 9 (September 3, 2012). See **firstmonday.org**.

4. Data collected in the lab of Connie Weaver, Department of Nutrition Sciences, Purdue University, and provided by Linda McCabe. For more information, see Corrie M. Whisner, et al., "Soluble maize fibre affects short-term calcium absorption in adolescent boys and girls: A randomized controlled trial using dual stable isotropic tracers," *British Journal of Nutritrion*, 112 (2014), pp. 446–456.

5. Haipeng Shen, "Nonparametric regression for problems involving lognormal distributions," PhD dissertation, University of Pennsylvania, 2003. Thanks to Haipeng Shen and Larry Brown for sharing the data.

6. From the Digest of Education Statistics at the website of the National Center for Education Statistics, **nces.ed.gov /programs/digest**.

7. See Note 4.

8. Based on Barbara Ernst et al., "Seasonal variation in the deficiency of 25–hydroxyvitamin D_3 in mildly to extremely obese subjects," *Obesity Surgery*, 19 (2009), pp. 180–183.

9. See, for example, **facebook.com/Million.Dollar .Application**.

10. From **socialbakers.com**. The website says that the data are updated daily. These data were downloaded on June 15, 2014.

11. More information about the *Titanic* can be found at the website for the Titanic Project in Belfast, Ireland, at **titanicbelfast.com**.

12. Data describing the passengers on the *Titanic* can be found at **cran.r-project.org/web/packages/titanic /titanic.pdf**.

13. See **health.gov/dietaryguidelines/2015/**.

14. Data collected in the lab of Connie Weaver, Department of Nutrition Sciences, Purdue University and provided by Linda McCabe.

15. Data from Table 1.1 in the U.S. Energy Information Administration's *July 2015 Monthly Energy Review*, available at **eia.gov/totalenergy/data/monthly/pdf/mer.pdf**.

16. From the Color Assignment website of Joe Hallock, **joehallock.com/edu/COM498/index.html**.

17. From the U.S. Environmental Protection Agency. See **www.epa.gov/sites/production/files/2015-09 /documents/2012_msw_fs.pdf**.

18. See **dupont.com/**.

19. Data provided by Darlene Gordon, Purdue University.

20. Data for 1980 to 2013 are available from the World Bank at **data.worldbank.org/indicator/IC.REG.DURS**. Data for 2013 were used for this example.

21. See, for example, **nacubo.org/Research**.

22. The data were provided by James Kaufman. The study is described in James C. Kaufman, "The cost of the muse: Poets die young," *Death Studies*, 27 (2003), pp. 813–821. The quote from Yeats appears in this article.

23. See, for example, the bibliographic entry for Gosset in the School of Mathematics and Statistics of the University of St. Andrews, Scotland, MacTutor History of Mathematics archive at **www-history.mcs.st-and.ac.uk/Biographies /Gosset.html**.

24. These and other data that were collected and used by Gosset can be found in the Guinness Archives in Dublin. See **www.guinness-storehouse.com/en/archives**.

25. These data were provided by Krista Nichols, Department of Biological Sciences, Purdue University.

26. From **beer100.com/beercalories.htm** on July 14, 2015.

27. Net worth from the *Federal Reserve Bulletin*, 100, No. 4 (2014), p. 12.

28. For more information about earthquakes, see the U.S. Geological Service website at **usgs.gov**.

29. See Noel Cressie, *Statistics for Spatial Data*, Wiley, 1993.

30. The National Assessment of Educational Progress (NAEP) is conducted by the National Center for Education Statistics (NCES). The NAEP is a large assessment of student knowledge in a variety of subjects. See **nces.ed.gov /nationsreportcard/naepdata**.

31. See the NCAA Eligibility Center Quick Reference Sheet, available at **fs.ncaa.org/Docs/eligibility_center/Quick _Reference_Sheet.pdf**.

32. Distributions for SAT scores can be found at the College Board website, **research.collegeboard.org/content /sat-data-tables**.

33. See Note 32.

34. See **stubhub.com**.

35. From Matthias R. Mehl et al., "Are women really more talkative than men?" *Science*, 317, No. 5834 (2007), p. 82. The raw data were provided by Matthias Mehl.

36. From the American Heart Association website, **www.heart.org**.

37. See **eia.gov/totalenergy/**.

38. From **nrcan.gc.ca/energy/efficiency/11938**.

39. Data from the careerbuilder.com website on July 3, 2014. See **careerbuilder.com/jobs/keyword /business-administration**.

40. See **online.wsj.com/articles/the-world-rankings -of-flopping-1403660175**.

41. Data for 2015 from **statista.com/statistics/398152 /us-twitter-user-age-groups/**.

42. The Institute of Medicine website, **www.iom.edu**, provides links to reports related to dietary reference intakes as well as other health and nutrition topics.

43. *Dietary Reference Intakes for Vitamin C, Vitamin E, Selenium and Carotenoids*, National Academy of Sciences, 2000.

44. See Note 43.

45. See **phish.net/song/divided-sky/history**.

46. Data from Tadd Colver, Department of Statistics, Purdue University.

47. Data provided by Mary Ann Lila, Director, Plants for Human Health Institute, David H. Murdock Distinguished Professor, North Carolina Research Campus, North Carolina State University.

CHAPTER 2

1. Shana M. Wilson et al., "Prediction of emotional eating during adolescents' transition to college: Does body mass index moderate the association between stress and emotional eating?" *Journal of American College Health*, 63, No. 3 (2015), pp. 163–170.

2. See Note 1.

3. See **cfs.purdue.edu/fn/campcalcium/** for information about findings from these camps.

4. See **consumerreports.org**.

5. From **consumerreports.org/cro/laundry-detergents.htm**.

6. Data for 2014 from **usgovernmentspending.com /compare_state_education_spend**.

7. These studies were conducted by Connie Weaver, Department of Nutrition Science, Purdue University, over the past 20 years. The data for this example were provided by Linda McCabe. More details concerning this particular study and references to other related studies are given in

Lu Wu et al. "Calcium requirements and metabolism in Chinese-American boys and girls," *Journal of Bone Mineral Research*, 25, No. 8 (2010), pp. 1842–1849.

8. A sophisticated treatment of improvements and additions to scatterplots is W. S. Cleveland and R. McGill, "The many faces of a scatterplot," *Journal of the American Statistical Association*, 79 (1984), pp. 807–822.

9. Data provided by Mary Ann Lila, Director, Plants for Human Health Institute, David H. Murdock Distinguished Professor, North Carolina Research Campus, North Carolina State University.

10. From **nrcan.gc.ca/energy/efficiency/11938**.

11. Stewart Warden et al., "Throwing induces substantial torsional adaption within the midshaft humerus of male baseball players," *Bone*, 45 (2009), pp. 931–941. The data were provided by Stewart Warden, Department of Physical Therapy, School of Health and Rehabilitation Sciences, Indiana University.

12. See **beer100.com/beercalories.htm**.

13. See **spectrumtechniques.com/isotope_generator.htm**.

14. These data were collected under the supervision of Zach Grigsby, Science Express Coordinator, College of Science, Purdue University.

15. See **worldbank.org**.

16. A careful study of this phenomenon is W. S. Cleveland, P. Diaconis, and R. McGill, "Variables on scatterplots look more highly correlated when the scales are increased," *Science*, 216 (1982), pp. 1138–1141.

17. Data from a plot in James A. Levine, Norman L. Eberhardt, and Michael D. Jensen, "Role of nonexercise activity thermogenesis in resistance to fat gain in humans," *Science*, 283 (1999), pp. 212–214.

18. From the Digest of Education Statistics at the website of the National Center for Education Statistics, **nces.ed.gov /programs/digest**.

19. Frank J. Anscombe, "Graphs in statistical analysis," *American Statistician*, 27 (1973), pp. 17–21.

20. From the website of the National Center for Education Statistics, **nces.ed.gov**.

21. Debora L. Arsenau, "Comparison of diet management instruction for patients with non-insulin dependent diabetes mellitus: Learning activity package vs. group instruction," Master's thesis, Purdue University, 1993.

22. See Note 19.

23. See **iom.edu**.

24. Based on a study described in Corby C. Martin et al., "Children in school cafeterias select foods containing more saturated fat and energy than the Institute of Medicine recommendations," *Journal of Nutrition*, 140 (2010), pp. 1653–1660.

25. You can find a clear and comprehensive discussion of numerical measures of association for categorical data in Chapter 2 of Alan Agresti, *Categorical Data Analysis*, 2nd ed., Wiley, 2002.

26. Edward Bumgardner, "Loss of teeth as a disqualification for military service," *Transactions of the Kansas Academy of Science*, 18 (1903), pp. 217–219.

27. Based on *The Ethics of American Youth—2012*, available from the Josephson Institute at **charactercounts.org/wp-content/uploads/2014/02/ReportCard-2012-DataTables.pdf**.

28. From M.-Y. Chen et al., "Adequate sleep among adolescents is positively associated with health status and health-related behaviors," *BMC Public Health*, 6, No. 59 (2006); available from **bmcpublichealth.biomedcentral.com/articles/10.1186/1471-2458-6-59**.

29. M. S. Linet et al., "Residential exposure to magnetic fields and acute lymphoblastic leukemia in children," *New England Journal of Medicine*, 337 (1997), pp. 1–7.

30. *The Health Consequences of Smoking: 1983*, U.S. Public Health Service, 1983.

31. OECD StatExtracts, Organisation for Economic Co-operation and Development, downloaded on June 29, 2008, from **stats.oecd.org/wbos**.

32. See **www12.statcan.gc.ca/census-recensement/index-eng.cfm**.

33. From **en.wikipedia.org/wiki/10000_metres**.

34. For an overview of remote deposit capture, see **remotedepositcapture.com/overview/rdc.overview.aspx**.

35. From the "Community Bank Competitiveness Survey," 2008, *ABA Banking Journal*. The survey is available at **nxtbook.com/nxtbooks/sb/ababj-compsurv08/index.php**.

36. The counts reported were calculated using counts of the numbers of banks in the different regions and the percents given in the ABA report.

37. *Education Indicators: An International Perspective, Institute of Education Studies*, National Center for Education Statistics; see **nces.ed.gov/surveys/international**.

38. Information about this procedure was provided by Samuel Flanigan of *U.S. News & World Report*. See **colleges.usnews.rankingsandreviews.com/best-colleges** for a description of the variables used to construct the ranks and for the most recent ranks.

39. We thank Zhiyong Cai of Texas A&M University for providing the data. The data are from work performed in connection with his PhD dissertation in the Department of Forestry and Natural Resources, Purdue University.

40. Although these data are fictitious, similar though less simple situations occur. See P. J. Bickel and J. W. O'Connell, "Is there a sex bias in graduate admissions?" *Science*, 187 (1975), pp. 398–404.

41. Condensed from D. R. Appleton, J. M. French, and M. P. J. Vanderpump, "Ignoring a covariate: An example of Simpson's paradox," *The American Statistician*, 50 (1996), pp. 340–341.

42. Lien-Ti Bei, "Consumers' purchase behavior toward recycled products: An acquisition-transaction utility theory perspective," MS thesis, Purdue University, 1993.

CHAPTER 3

1. See the news release of June 24, 2015, concerning the 2014 results for the American Time Use Survey, Table 11, at **bls.gov/news.release/pdf/atus.pdf**.

2. See **norc.uchicago.edu**.

3. Stewart Warden et al., "Throwing induces substantial torsional adaption within the midshaft humerus of male baseball players," *Bone*, 45 (2009), pp. 931–941. The data were provided by Stewart Warden, Department of Physical Therapy, School of Health and Rehabilitation Sciences, Indiana University.

4. Corby C. Martin et al., "Children in school cafeterias select foods containing more saturated fat and energy than the Institute of Medicine recommendations," *Journal of Nutrition*, 140 (2010), pp. 1653–1660.

5. Based on "Look, no hands: Automatic soap dispensers," *Consumer Reports*, February 2013, p. 11.

6. From "Did you know," *Consumer Reports*, February 2013, p. 10.

7. Bruce Barrett et al., "Echinacea for treating the common cold," *Annals of Internal Medicine*, 153 (2010), pp. 769–777.

8. For a full description of the STAR program and its follow-up studies, go to **heros-inc.org/star.htm**.

9. See Note 6.

10. Based on Gerardo Ramirez and Sian L. Beilock, "Writing about testing worries boosts exam performance in the classroom," *Science*, 331 (2011), p. 2011. Although we describe the experiment as not including a control group, the researchers who conducted this study did, in fact, use one.

11. A general discussion of failures of blinding is Dean Ferguson et al., "Turning a blind eye: The success of blinding reported in a random sample of randomised, placebo controlled trials," *British Medical Journal*, 328 (2004), p. 432.

12. Based on a study conducted by Sandra Simonis under the direction of Professor Jon Harbor from the Purdue University Department of Earth, Atmospheric, and Planetary Sciences.

13. Based on a study conducted by Tammy Younts directed by Professor Deb Bennett of the Purdue University Department of Educational Studies. For more information about Reading Recovery, see **readingrecovery.org/**.

14. Based on a study conducted by Rajendra Chaini under the direction of Professor Bill Hoover of the Purdue University Department of Forestry and Natural Resources.

15. From the Hot Rock Songs list at **billboard.com** for the week of September 5, 2015.

16. From the Hot 100 list at **billboard.com** for the week of September 5, 2015.

17. From the online version of the Bureau of Labor Statistics, *Handbook of Methods*, modified April 17, 2003, at **bls.gov**. The details of the design are more complicated than we describe.

18. For more detail on the material of this section and complete references, see P. E. Converse and M. W. Traugott, "Assessing the accuracy of polls and surveys," *Science*, 234 (1986), pp. 1094–1098.

19. From **www.census.gov/programs-surveys/cps/technical -documentation/methodology/non-response-rates.html** on January 29, 2013.

20. From **www3.norc.org/GSS+Website/FAQs/** on January 29, 2013.

21. See **pewresearch.org/about**.

22. See "Assessing the representativeness of public opinion surveys," May 15, 2012, from **people-press.org/2012/05/15**.

23. Sex: Tom W. Smith, "The *JAMA* controversy and the meaning of sex," *Public Opinion Quarterly*, 63 (1999), pp. 385–400. Welfare: From a *New York Times*/CBS News Poll reported in the *New York Times*, July 5, 1992. Scotland: "All set for independence?" *Economist*, September 12, 1998. Many other examples appear in T. W. Smith, "That which we call welfare by any other name would smell sweeter," *Public Opinion Quarterly*, 51 (1987), pp. 75–83.

24. John C. Bailar III, "The real threats to the integrity of science," *Chronicle of Higher Education*, April 21, 1995, pp. B1–B2.

25. The difficulties of interpreting guidelines for informed consent and for the work of institutional review boards in medical research are a main theme of Beverly Woodward, "Challenges to human subject protections in U.S. medical research," *Journal of the American Medical Association*, 282 (1999), pp. 1947–1952. The references in this paper point to other discussions.

26. Quotation from the *Report of the Tuskegee Syphilis Study Legacy Committee*, May 20, 1996. A detailed history is James H. Jones, *Bad Blood: The Tuskegee Syphilis Experiment*, Free Press, 1993.

27. Dr. Hennekens's words are from an interview in the Annenberg/Corporation for Public Broadcasting video series *Against All Odds: Inside Statistics*.

28. See **ftc.gov/opa/2009/04/kellogg.shtm**.

29. On February 12, 2012, the CBS show *60 Minutes* reported the latest news on this study, which was published in the *Journal of Clinical Oncology* in 2007. See **cbsnews.com/video/watch/?id=7398476n**.

30. R. D. Middlemist, E. S. Knowles, and C. F. Matter, "Personal space invasions in the lavatory: Suggestive evidence for arousal," *Journal of Personality and Social Psychology*, 33 (1976), pp. 541–546.

31. From Randi Zlotnik Shaul et al., "Legal liabilities in research: Early lessons from North America," *BMJ Medical Ethics*, 6, No. 4 (2005), pp. 1–4.

32. The report was issued in February 2009 and is available from **www.ftc.gov/sites/default/files/documents/reports /federal-trade-commission-staff-report-self-regulatory -principles-online-behavioral-advertising/p085400 behavadreport.pdf**.

CHAPTER 4

1. An informative and entertaining account of the origins of probability theory is Florence N. David, *Games, Gods and Gambling*, Charles Griffin, London, 1962.

2. See **dupont.com/**.

3. You can find a mathematical explanation of Benford's law in Ted Hill, "The first-digit phenomenon," *American Scientist*, 86 (1996), pp. 358–363; and Ted Hill, "The difficulty of faking data," *Chance*, 12, No. 3 (1999), pp. 27–31. Applications in fraud detection are discussed in the second paper by Hill and in Mark A. Nigrini, "I've got your number," *Journal of Accountancy*, May 1999, available online at **www.journalofaccountancy.com /issues/1999/may/nigrini.html**.

4. Royal Statistical Society news release, "Royal Statistical Society concerned by issues raised in Sally Clark case," October 23, 2001, at **www.rss.org.uk**. For background, see an editorial and article in *The Economist*, January 22, 2004. The editorial is entitled "The probability of injustice."

5. See **cdc.gov/mmwr/preview/mmwrhtml/mm57e618a1 .htm**.

6. See the Note 5.

7. See **bloodbook.com/world-abo.html** for the distribution of blood types for various groups of people.

8. From Statistics Canada, **www.statcan.ca**.

9. We use $\bar{x}$ both for the random variable, which takes different values in repeated sampling, and for the numerical value of the random variable in a particular sample. Similarly, s and $\hat{p}$ stand both for random variables and for specific values. This notation is mathematically imprecise but statistically convenient.

10. We will consider only the case in which X takes a finite number of possible values. The same ideas, implemented with more advanced mathematics, apply to random variables with an infinite but still countable collection of values.

11. Based on a Pew Internet report, "Teens and distracted driving," available from **pewinternet.org/Reports/2009/Teens-and-Distracted-Driving.aspx**.

12. See **pewinternet.org/Reports/2009/17-Twitter-and-Status-Updating-Fall-2009.aspx**.

13. The mean of a continuous random variable X with density function $f(x)$ can be found by integration:

$$\mu_X = \int xf(x)dx$$

This integral is a kind of weighted average, analogous to the discrete-case mean

$$\mu_X = \sum xP(X\bar{x})$$

The variance of a continuous random variable X is the average squared deviation of the values of X from their mean, found by the integral

$$\sigma_X^2 = \int (x - \mu)^2 f(x)dx$$

14. See A. Tversky and D. Kahneman, "Belief in the law of small numbers," *Psychological Bulletin*, 76 (1971), pp. 105–110, and other writings of these authors for a full account of our misperception of randomness.

15. Probabilities involving runs can be quite difficult to compute. That the probability of a run of three or more heads in 10 independent tosses of a fair coin is $(1/2) + (1/128) = 0.508$ can be found by clever counting. A general treatment using advanced methods appears in Section XIII.7 of William Feller, *An Introduction to Probability Theory and Its Applications*, Vol. 1, 3rd ed., Wiley, 1968.

16. R. Vallone and A. Tversky, "The hot hand in basketball: On the misperception of random sequences," *Cognitive Psychology*, 17 (1985), pp. 295–314. A later series of articles that debate the independence question is A. Tversky and T. Gilovich, "The cold facts about the 'hot hand' in basketball," *Chance*, 2, No. 1 (1989), pp. 16–21; P. D. Larkey, R. A. Smith, and J. B. Kadane, "It's OK to believe in the 'hot hand,' " *Chance*, 2, No. 4 (1989), pp. 22–30; and A. Tversky and T. Gilovich, "The 'hot hand': Statistical reality or cognitive illusion?" *Chance*, 2, No. 4 (1989), pp. 31–34.

17. Based on a study discussed in S. Atkinson, G. McCabe, C. Weaver, S. Abrams, and K. O'Brien, "Are current calcium recommendations for adolescents higher than needed to achieve optimal peak bone mass? The controversy," *Journal of Nutrition*, 138, No. 6 (2008), pp. 1182–1186.

18. Based on a study described in Corby C. Martin et al., "Children in school cafeterias select foods containing more saturated fat and energy than the Institute of Medicine recommendations," *Journal of Nutrition*, 140 (2010), pp. 1653–1660.

19. Based on *The Ethics of American Youth--2012*, available from the Josephson Institute, **charactercounts.org/wp-content/uploads/2014/02/ReportCard-2012-DataTables.pdf**.

20. See **nces.ed.gov/programs/digest**. Data are from the 2012 *Digest of Education Statistics*.

CHAPTER 5

1. See the 2015 press release from the *Student Monitor*, at **www.studentmonitor.com**.

2. 2015 study conducted by Dr. Beth Livingston and graduate assistants Maria Grillo and Rebecca Paluch, Cornell University ILR School in partnership with Hollaback!

3. K. M. Orzech et al., "The state of sleep among college students at a large public university," *Journal of American College Health*, 59 (2011), pp. 612–619.

4. Findings can be found at **www.nielsen.com/us/en/insights/news/2014/smartphones-so-many-apps--so-much-time.html**.

5. Haipeng Shen, "Nonparametric regression for problems involving lognormal distributions," PhD dissertation, University of Pennsylvania, 2003. Thanks to Haipeng Shen and Larry Brown for sharing the data.

6. Findings from a 2015 DMR article titled "By the numbers: 60 amazing Snapchat statistics."

7. Statistical methods for dealing with time-to-failure data, including the Weibull model, are presented in Wayne Nelson, *Applied Life Data Analysis*, Wiley, 1982.

8. Statistics are from Pew Research Center's article titled "6 new facts about Facebook," posted February 3, 2014, on **www.pewresearch.org/**.

9. From the grade distribution database of the Indiana University Office of the Registrar, **gradedistribution.registrar.indiana.edu**.

10. Diane M. Dellavalle and Jere D. Haas, "Iron status is associated with endurance performance and training in female rowers," *Medicine and Science in Sports and Exercise*, 44, No. 8 (2012), pp. 1552–1559.

11. Results of this and other questions from this survey can be found at **www.mumsnet.com/surveys/pressure-on-children-and-parents**.

12. Results are from S. Rinehart et al., "Sexual harassment and sexual violence experiences among middle school youth," presented at the 2014 American Educational Research Association annual meeting.

13. U.S. Department of Education, National Center for Education Statistics, "The Condition of Education 2015" (NCES 2015-144), High School Coursetaking, 2015.

14. S. A. Rahimtoola, "Outcomes 15 years after valve replacement with a mechanical vs. a prosthetic valve: Final report of the Veterans Administration randomized trial," American College of Cardiology, **content.onlinejacc.org/article.aspx?articleid=1126703**.

15. Based on the article "E-retailers beat stores in customer satisfaction study," posted on February 19, 2015, by Internet Retailer.

16. The results of this 2012 survey can be found at **www.theaa.com/newsroom/news-2012/streetwatch-october-2012-fewer-potholes.html**.

17. The results of this 2012 survey can be found at **josephsoninstitute.org**.

18. Results from the *Global News* article "New regulations about illegal downloading go into effect," posted January 2, 2015, and found at **globalnews.ca/news/1752246/new-regulations-about-illegal-downloading-go-into-effect/**.

19. "The Wireless Report 2014" can be found at **www.ditchthelabel.org/the-wireless-report-2014/**.

20. A summary over time can be found at **www.gallup.com/poll/1588/children-violence.aspx**.

21. A summary of Larry Wright's study can be found at **www.nytimes.com/2009/03/04/sports/basketball/04freethrow.html**.

22. Barbara Means et al., "Evaluation of evidence-based practices in online learning: A meta-analysis and review of online learning studies," U.S. Department of Education, Office of Planning, Evaluation, and Policy Development, 2010.

23. Dafna Kanny et al., "Vital signs: Binge drinking among women and high school girls—United States, 2011," *Morbidity and Mortality Weekly Report*, January 8, 2013.

24. Information was obtained from "Price comparisons of wireline, wireless and internet services in Canada and with foreign jurisdictions," Canadian Radio-Television and Telecommunications Commission, April 6, 2012.

25. This information can be found at **www.census.gov/topics/population/genealogy/data/2000_surnames.html**.

CHAPTER 6

1. Noel Cressie, *Statistics for Spatial Data*, Wiley, 1993. The significance test result that we report is one of several that could be used to address this question. See pp. 607–609 of the Cressie book for more details.

2. The 2014–2015 statistics for California were obtained from the California Department of Education website, **dq.cde.ca.gov**.

3. Based on information reported in "How America pays for college 2015," found online at **news.salliemae.com/files/doc_library/file/HowAmericaPaysforCollege2015FNL.pdf**.

4. See Note 3. This total amount includes grants, scholarships, loans, and assistance from friends and family.

5. Average starting salary taken from the January 2015 salary survey by the National Association of Colleges and Employers.

6. See **www.thekaraokechannel.com/**.

7. These annual surveys can be found at **www.apa.org/news/press/releases/stress/index.aspx**.

8. C. M. Weaver et al., "Quantification of biochemical markers of bone turnover by kinetic measures of bone formation and resorption in young healthy females," *Journal of Bone and Mineral Research*, 12 (1997), pp. 1714–1720.

9. Average starting salary taken from the spring 2015 salary survey by the National Association of Colleges and Employers.

10. Euna Hand and Lisa M. Powell, "Consumption patterns of sugar-sweetened beverages in the United States," *Journal of the Academy of Nutrition and Dietetics*, 113, No. 1 (2013), pp. 43–53.

11. See the 2015 press release from the *Student Monitor*, at **www.studentmonitor.com**.

12. Alyssa Brown, "Americans' life outlook best in seven years," Gallup News Service, January 16, 2015. Found at **www.gallup.com/**.

13. The vehicle is a 2002 Toyota Prius previously owned by the third author.

14. Regional cost-of-living rates are often computed using the Department of Labor, Bureau of Labor Statistics, metropolitan-area consumer price indexes. These can be found at **www.bls.gov/cpi**.

15. See Note 10.

16. M. Garaulet et al., "Timing of food intake predicts weight loss effectiveness," *International Journal of Obesity*, 1 (2013), pp. 1–8.

17. Giacomo DeGiorgi et al., "Be as careful of the company you keep as of the books you read: Peer effects in education and on the labor market," National Bureau of Economic Research, working paper 14948 (2009).

18. Seung-Ok Kim, "Burials, pigs, and political prestige in neolithic China," *Current Anthropology*, 35 (1994), pp. 119–141.

19. These data were collected in connection with the Purdue Police Alcohol Student Awareness Program run by Police Officer D. A. Larson.

20. National Assessment of Educational Progress, *The Nation's Report Card*, Mathematics & Reading Assessments 2015.

21. Matthew A. Lapierre et al., "Background television in the homes of U.S. children," *Pediatrics*, 130, No. 5 (2012), pp. 839–846.

22. Sogol Javaheri et al., "Sleep quality and elevated blood pressure in adolescents," *Circulation*, 118 (2008), pp. 1034–1040.

23. Victor Lun et al., "Evaluation of nutritional intake in Canadian high-performance athletes," *Clinical Journal of Sports Medicine*, 19, No. 5 (2009), pp. 405–411.

24. R. A. Fisher, "The arrangement of field experiments," *Journal of the Ministry of Agriculture of Great Britain*, 33 (1926), p. 504, quoted in Leonard J. Savage, "On rereading

R. A. Fisher," *Annals of Statistics*, 4 (1976), p. 471. Fisher's work is described in a biography by his daughter: Joan Fisher Box, *R. A. Fisher: The Life of a Scientist*, Wiley, 1978.

25. The editorial was written by Phil Anderson. See *British Medical Journal*, 328 (2004), pp. 476–477. A letter to the editor on this topic by Doug Altman and J. Martin Bland appeared shortly after. See "Confidence intervals illuminate absence of evidence," *British Medical Journal*, 328 (2004), pp. 1016–1017.

26. A. Kamali et al., "Syndromic management of sexually-transmitted infections and behavior change interventions on transmission of HIV-1 in rural Uganda: A community randomised trial," *Lancet*, 361 (2003), pp. 645–652.

27. T. D. Sterling, "Publication decisions and their possible effects on inferences drawn from tests of significance—or vice versa," *Journal of the American Statistical Association*, 54 (1959), pp. 30–34. Related comments appear in J. K. Skipper, A. L. Guenther, and G. Nass, "The sacredness of 0.05: A note concerning the uses of statistical levels of significance in social science," *American Sociologist*, 1 (1967), pp. 16–18.

28. For a good overview of these issues, see Bruce A. Craig, Michael A. Black, and Rebecca W. Doerge, "Gene expression data: The technology and statistical analysis," *Journal of Agricultural, Biological, and Environmental Statistics*, 8 (2003), pp. 1–28.

29. Erick H. Turner et al., "Selective publication of antidepressant trials and its influence on apparent efficacy," *New England Journal of Medicine*, 358 (2008), pp. 252–260.

30. Robert J. Schiller, "The volatility of stock market prices," *Science*, 235 (1987), pp. 33–36.

31. Corby K. Martin et al., "Children in school cafeterias select foods containing more saturated fat and energy than the Institute of Medicine recommendations," *Journal of Nutrition*, 140 (2010), pp. 1653–1660.

32. Data from Joan M. Susic, "Dietary phosphorus intakes, urinary and peritoneal phosphate excretion and clearance in continuous ambulatory peritoneal dialysis patients," MS thesis, Purdue University, 1985.

33. Mugdha Gore and Joseph Thomas, "Store image as a predictor of store patronage for nonprescription medication purchases: A multiattribute model approach," *Journal of Pharmaceutical Marketing & Management*, 10 (1996), pp. 45–68.

CHAPTER 7

1. Average hours per week obtained from "The Total Audience Report, 4th Quarter 2014," Nielsen Company (2015).

2. C. Don Wiggins, "The legal perils of 'underdiversification'—a case study," *Personal Financial Planning*, 1, No. 6 (1999), pp. 16–18.

3. Data provided by Bill Berezowitz and James Malloy of GE Healthcare.

4. Brent Stoffer and George W. Uetz, "The effects of social experience with varying male availability on female mate preferences in a wolf spider," *Behavioral Ecology Sociobiology*, 69 (2015), pp. 927–937.

5. Go to **www.futurity.org/fried-food-taste-without-all-the-fat/** for more information.

6. These recommendations are based on extensive computer work. See, for example, Harry O. Posten, "The robustness of the one-sample *t*-test over the Pearson system," *Journal of Statistical Computation and Simulation*, 9 (1979), pp. 133–149; and E. S. Pearson and N. W. Please, "Relation between the shape of population distribution and the robustness of four simple test statistics," *Biometrika*, 62 (1975), pp. 223–241.

7. The standard reference here is Bradley Efron and Robert J. Tibshirani, *An Introduction to the Bootstrap*, Chapman Hall, 1993. A less technical overview is in Bradley Efron and Robert J. Tibshirani, "Statistical data analysis in the computer age," *Science* 253 (1991), pp. 390–395.

8. From "Insolvency Statistics in Canada 2013—Annual report" available at **www.ic.gc.ca/eic/site/bsf-osb.nsf/eng/br03221.html**.

9. This announcement can be found at **epa.gov/fueleconomy/labelchange.htm**.

10. Based on the scatterplot found at **newsroom.uber.com/nyc/what-does-a-typical-new-york-uberx-partner-earn-in-a-week/**.

11. Statistics are from the article "6 new facts about Facebook," posted February 3, 2014, on **www.pewresearch.org/**.

12. A description of the lawsuit can be found at **www.cnn.com/2013/02/26/business/california-anheuser-busch-lawsuit/index.html**.

13. See Note 1.

14. Christine L. Porath and Amir Erez, "Overlooked but not untouched: How rudeness reduces onlookers' performance on routine and creative tasks," *Organizational Behavior and Human Decision Processes*, 109 (2009), pp. 29–44.

15. The vehicle is a 2002 Toyota Prius previously owned by the third author.

16. Information regarding Instagram can be found at **locowise.com/tools.php**.

17. Sujata Sethi et al., "Study of level of stress in the parents of children with attention-deficit/hyperactivity disorder," *Journal of Indian Association for Child and Adolescent Mental Health*, 8, No. 2 (2012), pp. 25–37.

18. James A. Levine, Norman L. Eberhardt, and Michael D. Jensen, "Role of nonexercise activity thermogenesis in resistance to fat gain in humans," *Science*, 283 (1999), pp. 212–214. Data for this study are available from the *Science* website, **www.sciencemag.org**.

19. These data were collected in connection with a bone health study at Purdue University and were provided by Linda McCabe.

20. Based on Praveetha Patalay et al., "Equivalence of paper and computer formats of a child self-report mental health measure," *European Journal of Psychological Assessment*, advance online publication, doi:10.1027/1015-5759/a000206.

21. Data provided by Joseph A. Wipf, Department of Foreign Languages and Literatures, Purdue University.

22. Summary information can be found at the National Center for Health Statistics website, **www.cdc.gov/nchs/nhanes.htm**.

23. Detailed information about the conservative *t* procedures can be found in Paul Leaverton and John J. Birch, "Small sample power curves for the two sample location problem," *Technometrics*, 11 (1969), pp. 299–307; in Henry Scheffé, "Practical solutions of the Behrens-Fisher problem," *Journal of the American Statistical Association*, 65 (1970), pp. 1501–1508; and in D. J. Best and J. C. W. Rayner, "Welch's approximate solution for the Behrens-Fisher problem," *Technometrics*, 29 (1987), pp. 205–210.

24. This example is adapted from Maribeth C. Schmitt, "The effects of an elaborated directed reading activity on the metacomprehension skills of third graders," PhD dissertation, Purdue University, 1987.

25. See the extensive simulation studies in Harry O. Posten, "The robustness of the two-sample *t* test over the Pearson system," *Journal of Statistical Computation and Simulation*, 6 (1978), pp. 295–311.

26. M. Garaulet et al., "Timing of food intake predicts weight loss effectiveness," *International Journal of Obesity*, advance online publication, January 29, 2013, doi:10.1038/ijo.2012.229.

27. This study is reported in Roseann M. Lyle et al., "Blood pressure and metabolic effects of calcium supplementation in normotensive white and black men," *Journal of the American Medical Association*, 257 (1987), pp. 1772–1776. The individual measurements in Table 7.5 were provided by Dr. Lyle.

28. J.D. Vescovi and T. Goodale, "Physical demands of womens Rugby Sevens matches: Female athletes in motion (FAiM) study," *International Journal of Sports Medicine*, advance online publication, doi:10.1055/s-0035-1548940.

29. Elizabeth F Beach and Valerie Nie, "Noise levels in fitness classes are still too high: Evidence from 1997–1998 and 2009–2011," *Archives of Environmental & Occupational Health* 69, No. 4 (2014), pp. 223–230.

30. Y. Charles Zhang and Norbert Schwarz, "How and why 1 year differs from 365 days: A conversational logic analysis of inferences from the granularity of quantitative expressions," *Journal of Consumer Research* 39 (August 2012), pp. S212–S223.

31. Karel Kleisner et al., "Trustworthy-looking face meets brown eyes," *PLoS ONE* 8, No. 1 (2013), e53285, doi:10.1371/journal.pone.0053285.

32. Reynol Junco, "Too much face and not enough books: The relationship between multiple indices of Facebook use and academic performance," *Computers in Human Behavior*, 28, No. 1 (2012), pp. 187–198.

33. C. E. Cryfer et al., "Misery is not miserly: Sad and self-focused individuals spend more," *Psychological Science*, 19 (2008), pp. 525–530.

34. Grant D. Brinkworth et al., "Long-term effects of a very low-carbohydrate diet and a low-fat diet on mood and cognitive function," *Archives of Internal Medicine*, 169 (2009), pp. 1873–1880.

35. These reports can be found at **www.qsrmagazine.com/reports**.

36. Samara Joy Nielsen and Barry M. Popkin, "Patterns and trends in food portion sizes, 1977–1998," *Journal of the American Medical Association*, 289 (2003), pp. 450–453.

37. Gordana Mrdjenovic and David A. Levitsky, "Nutritional and energetic consequences of sweetened drink consumption in 6- to 13-year-old children," *Journal of Pediatrics*, 142 (2003), pp. 604–610.

38. David Han-Kuen Chu, "A test of corporate advertising using the elaboration likelihood model," MS thesis, Purdue University, 1993.

39. M. F. Picciano and R. H. Deering, "The influence of feeding regimens on iron status during infancy," *American Journal of Clinical Nutrition*, 33 (1980), pp. 746–753.

40. Average starting salary taken from the spring 2015 salary survey by the National Association of Colleges and Employers.

41. The data were obtained on August 24, 2006, from an iPod owned by George McCabe, Jr.

42. The method is described in Xiao-Hua Zhou and Sujuan Gao, "Confidence intervals for the log-normal mean," *Statistics in Medicine*, 16 (1997), pp. 783–790.

43. See the 2015 press release from the *Student Monitor*, at **www.studentmonitor.com**.

44. Data from Wayne Nelson, *Applied Life Data Analysis*, Wiley, 1982, p. 471.

45. This city's restaurant inspection data can be found at **www.jsonline.com/watchdog/dataondemand/**.

46. Braz Camargo et al., "Interracial friendships in college," *Journal of Labor Economics*, 28 (2010), pp. 861–892.

47. Based on Loren Cordain et al., "Influence of moderate daily wine consumption on body weight regulation and metabolism in healthy free-living males," *Journal of the American College of Nutrition*, 16 (1997), pp. 134–139.

48. B. Wansink et al., "Fine as North Dakota wine: Sensory expectations and the intake of companion foods," *Physiology & Behavior*, 90 (2007), pp. 712–716.

49. Douglas J. Levey et al., "Urban mockingbirds quickly learn to identify individual humans," *Proceedings of the National Academy of Sciences*, 106 (2009), pp. 8959–8962.

50. Morgan K. Ward and Darren W. Dahl, "Should the devil sell Prada? Retail rejection increases aspiring consumers' desire for the brand," *Journal of Consumer Research*, 41, No. 3 (2014), pp. 590–609.

51. Anne Z. Hoch et al., "Prevalence of the female athlete triad in high school athletes and sedentary students," *Clinical Journal of Sports Medicine*, 19 (2009), pp. 421–428.

52. This exercise is based on events that are real. The data and details have been altered to protect the privacy of the individuals involved.

53. Based loosely on D. R. Black et al., "Minimal interventions for weight control: A cost-effective alternative," *Addictive Behaviors*, 9 (1984), pp. 279–285.

54. These data were provided by Professor Sebastian Heath, School of Veterinary Medicine, Purdue University.

CHAPTER 8

1. The actual distribution of X based on an SRS from a finite population is the *hypergeometric distribution*. Details regarding this distribution can be found in Sheldon M. Ross, *A First Course in Probability*, 8th ed., Prentice Hall, 2010.

2. From **pewinternet.org/2014/08/06/future-of-jobs**.

3. Results of the survey are available at **slideshare.net/duckofdoom/google-research-about-mobile-internet-in-2011**.

4. Details of exact binomial procedures can be found in Myles Hollander and Douglas Wolfe, *Nonparametric Statistical Methods*, 2nd ed., Wiley, 1999.

5. See A. Agresti and B. A. Coull, "Approximate is better than 'exact' for interval estimation of binomial proportions," *American Statistician*, 52 (1998), pp. 119–126. A detailed theoretical study is Lawrence D. Brown, Tony Cai, and Anirban DasGupta, "Confidence intervals for a binomial proportion and asymptotic expansions," *Annals of Statistics*, 30 (2002), pp. 160–201.

6. See, for example, **pilatesmethodalliance.org**.

7. See **pewinternet.org/Reports/2013/in-store-mobile-commerce.aspx**.

8. Heather Tait, *Aboriginal Peoples Survey, 2006: Inuit Health and Social Conditions*, Social and Aboriginal Statistics Division, Statistics Canada, 2008. Available from **statcan.gc.ca/pub**.

9. See **southerncross.co.nz/about-the-group/media-releases/2013.aspx**.

10. See **commonsensemedia.org/sites/default/files/full_cap-csm_report_results_1-7-13.pdf**.

11. See "National Survey of Student Engagement, the College Student Report," available online at **nsse.iub.edu/index.cfm**.

12. This survey and others that study issues related to college students can be found at **nelliemae.com**.

13. See Note 11.

14. Information about the survey can be found online at **saint-denis.library.arizona.edu/natcong**.

15. From **pewinternet.org/2015/04/09/teens-social-media-technology-2015**.

16. See Alan Agresti and Brian Caffo, "Simple and effective confidence intervals for proportions and differences of proportions result from adding two successes and two failures," *American Statistician*, 45 (2000), pp. 280–288. The plus four interval is a bit conservative (true coverage probability is higher than the confidence level) when p_1 and p_2 are equal and close to 0 or 1, but the traditional interval is much less accurate and has the fatal flaw that the true coverage probability is *less* than the confidence level.

17. J. M. Tanner, "Physical growth and development," in J. O. Forfar and G. C. Arneil, *Textbook of Paediatrics*, 3rd ed., Churchill Livingston, 1984, pp. 1–292.

18. Based on T. A. Brighton et al., "Low-dose aspirin for preventing recurrent venous thromboembolism," *New England Journal of Medicine*, 367, No. 21 (2012), pp. 1979–1987. The analysis in the published manuscript used a slightly more complicated summary, called the hazard ratio, to compare the treatments.

19. Nicolas Gueguen and Celine Jacob, "Clothing color and tipping: Gentlemen patrons give more tips to waitresses with red clothes," *Journal of Hospitality & Tourism Research*, 38, No. 2 (2014), pp. 275–280.

20. Edward Bumfardner, "Loss of teeth as a disqualification for military service," *Transactions of the Kansas Academy of Science*, 18 (1903), pp. 217–219.

21. B. J. Bradley et al., "Historical perspective and current status of the physical education requirement at American 4-year colleges and universities," *Research Quarterly for Exercise and Sport*, 83, No. 4 (2012), pp. 503–512.

22. Erin K. O'Loughlin et al., "Prevalence and correlates of exergaming in youth," *Pediatrics*, 130 (2012), pp. 806–814.

23. From a Pew Internet Project Data Memo by Amanda Lenhart et al., dated December 2008. Available at **pewinternet.org**.

24. The report, dated May 18, 2012, is available from **pewinternet.org/Reports/2012/Future-of-Gamification/Overview.aspx**.

25. From the Pew Research Center's Project for Excellence in Journalism, *The State of the News Media 2012*, available from **stateofthemedia.org/?src=prc-headline**.

26. See **iom.edu**.

27. Based on a study described in Corby C. Martin et al., "Children in school cafeterias select foods containing more saturated fat and energy than the Institute of Medicine recommendations," *Journal of Nutrition*, 140 (2010), pp. 1653–1660.

28. From **pewinternet.org/~/media//Files/Reports/2013 /PIP_SocialMediaUsers.pdf**.

29. From **forbes.com/sites/ericsavitz/2013/01/11/totally -pwned-2012-u-s-video-game-retail-sales-tumble-22**.

30. From the Entertainment Software Association website at **theesa.com**.

31. See Note 12.

32. See S. W. Lagakos, B. J. Wessen, and M. Zelen, "An analysis of contaminated well water and health effects in Woburn, Massachusetts," *Journal of the American Statistical Association*, 81 (1986), pp. 583–596, and the following discussion. This case is the basis for the movie *A Civil Action*.

33. This case is discussed in D. H. Kaye and M. Aickin (eds.), *Statistical Methods in Discrimination Litigation*, Marcel Dekker, 1986; and D. C. Baldus and J. W. L. Cole, *Statistical Proof of Discrimination*, McGraw-Hill, 1980.

CHAPTER 9

1. From a Pew Research Institute article by Monica Anderson, "Young adults more likely to say vaccinating kids should be a parental choice," February 2, 2015.

2. When the expected cell counts are small, we prefer a test based on the exact distribution rather than the chi-square approximation, particularly for 2×2 tables. Many statistical software systems offer an "exact" test as well as the chi-square test for 2×2 tables.

3. D.-C. Seo et al., "Relations between physical activity and behavioral and perceptual correlates among midwestern college students," *Journal of Americal College Health*, 56, No. 2 (2007), pp. 187–197.

4. From P. Strazzullo et al., "Salt intake, stroke, and cardiovascular disease: A meta analysis of prospective studies," *British Medical Journal*, 339 (2009), pp. 1–9. The meta-analysis combined data from 14 study cohorts taken from 10 different studies.

5. N. R. Cook et al., "Long term effects of dietary sodium reduction on cardiovascular disease outcomes: Observational follow-up of the trials of the hypertension prevention (TOHP)," *British Medical Journal*, 334 (2007), pp. 1–8.

6. Catherine Hill and Holly Kearl, *Crossing the Line: Sexual Harassment at School*, American Association of University Women, Washington, DC, 2011.

7. For an overview of remote deposit capture, see **remotedepositcapture.com/overview/rdc.overview.aspx**.

8. From the Community Bank Competitiveness Survey, 2008, *ABA Banking Journal*. The survey is available at **nxtbook.com/nxtbooks/sb/ababj-compsurv08/index.php**.

9. The sampling procedure was designed by George McCabe. It was carried out by Amy Conklin, an undergraduate honors student in the Department of Foods and Nutrition at Purdue University.

10. The analysis could also be performed by using a two-way table to compare the states of the selected and not-selected students. Because the selected students are a relatively small percent of the total sample, the results will be approximately the same.

11. See the M&M Mars website at **us.mms.com/us/about /products** for this and other information.

12. See **nhcaa.org**.

13. These data are a composite based on several actual audits of this type.

14. Data provided by Professor Marcy Towns of the Purdue University Department of Chemistry.

15. Based on *The Ethics of American Youth—2008*, available from the Josephson Institute at **charactercounts.org**.

16. From the Survey of Canadian Career College Students Phase II: In-School Student Survey, 2008. This report is available from **files.eric.ed.gov/fulltext/ED514952.pdf**.

CHAPTER 10

1. Data based on Michael L. Mestek et al., "The relationship between pedometer-determined and self-reported physical activity and body composition variables in college-aged men and women," *Journal of American College Health*, 57 (2008), pp. 39–44.

2. M. Van Praag et al., "The higher returns to formal education for entrepreneurs versus employees," *Small Business Economics* 40 (2013), pp. 375–396.

3. Information regarding bone health can be found in "Osteoporosis: Peak bone mass in women," last reviewed in June 2015 and available at **www.niams.nih.gov/Health _Info/Bone/Osteoporosis/bone_mass.asp**.

4. The data were provided by LindaMcCabe and were collected as part of a large study of women's bone health and another study of calcium kinetics, both directed by Professor Connie Weaver of the Department of Foods and Nutrition, Purdue University.

5. This annual report can be found at **www.kiplinger.com**.

6. Data available at **www.ncdc.noaa.gov**.

7. Data sampled from **www.jcmit.com/memoryprice.htm**.

8. C.U. Eze et al., "Relationship between sonographic umbilical cord size and gestational age among pregnant women in Enugu, Nigeria," *African Health Sciences*, 14, No. 2 (2014), pp. 334–338, doi:10.4314/ahs.v14i2.7.

9. The method is described in Chapter 2 of M. Kutner et al., *Applied Linear Statistical Models*, 5th ed., Irwin, 2004.

10. National Science Foundation, Division of Science Resources Statistics, *Federal Science and Engineering Support to Universities, Colleges, and Nonprofits: Fiscal Year 2013*. Detailed Statistical Tables NSF 15-327, Arlington, VA, 2015. Available at **www.nsf.gov/statistics/2015/nsf15327/**.

11. L. Cooke et al., "Relationship between parental report of food neophobia and everyday food consumption in 2–6-year-old children," *Appetite*, 41 (2003), pp. 205–206.

12. Toben F. Nelson et al., "The state sets the rate: The relationship among state-specific college binge drinking, state binge drinking rates, and selected state alcohol control policies," *American Journal of Public Health*, 95, No. 3 (2005), pp. 441–446.

13. These data can be found in the report titled "Grade inflation at American colleges and universities," at **www.gradeinflation.com**.

14. Rates can be found in various "Annual Return of Key Indices" reports available at **www.lazardnet.com**.

15. Andrew Lepp et al., "The relationship between cell phone use, academic performance, anxiety, and satisfaction with life in college students," *Computers and Human Behavior*, 31 (2014), pp. 343–350.

16. These are part of the data from the EESEE story "Blood Alcohol Content," found on the text website, **www.macmillanhighered.com/launchpad/ips8e**.

17. Tuition and fees for 2008 and tuition for 2014 were obtained from **www.findthebest.com**. Tuition rates for 2000 from the "2000–2001 Tuition and Required Fees Report," University of Missouri.

18. M. Mondello and J. Maxcy, "The impact of salary dispersion and performance bonuses in NFL organizations'" *Management Decision*, 47 (2009), pp. 110–123. These data were collected from **www.cbssports.com/nfl/playerrankings/regularseason/** and **content.usatoday.com/sportsdata/football/nfl/salaries/team**.

19. Selling price and assessment value available at **php.jconline.com/propertysales/propertysales.php**.

20. Matthew P. Martens et al., "The co-occurrence of alcohol use and gambling activities in first-year college students," *Journal of American College Health*, 57 (2009), pp. 597–602.

21. Based on Dan Dauwalter's master's thesis in the Department of Forestry and Natural Resources at Purdue University. More information is available in Daniel C. Dauwalter et al., "An index of biotic integrity for fish assemblages in Ozark Highland streams of Arkansas," *Southeastern Naturalist*, 2 (2003), pp. 447–468. These data were provided by Emmanuel Frimpong.

22. James T. Fleming, "The measurement pf children's perception of difficulty in reading materials," *Research in the Teaching of English*, 1 (1967), pp. 136–156.

23. G. Geri and B. Palla, "Considerazioni sulle più recenti osservazioni ottiche alla Torre Pendente di Pisa," *Estratto dal Bollettino della Società Italiana di Topografia e Fotogrammetria*, 2 (1988), pp. 121–135. Professor Julia Mortera of the University of Rome provided valuable assistance with the translation.

24. Z. Xuan et al., "Tax policy, adult binge drinking, and youth alcohol consumption in the United States," *Alcoholism: Clinical and Experimental Research*, 37, no. 10 (2013), pp. 1713–1719.

25. Alexandra Burt, "A mechanistic explanation of popularity: Genes, rule breaking, and evocative gene-environment correlations," *Journal of Personality and Social Psychology*, 96 (2009), pp. 783–794.

CHAPTER 11

1. This data set is similar to those used at Purdue University to assess academic success.

2. M.I. Núñez-Peña et al., "Feedback on students performance: A possible way of reducing the negative effect of math anxiety in higher education," *International Journal of Educational Research*, 70 (2015), pp. 80–87.

3. Katharine Kelley et al., "Estimating consumer spending on tickets, merchandise, and food and beverage: A case study of a NHL team," *Journal of Sport Management*, 28 (2014), pp. 253–265.

4. Based on Leigh J. Maynard and Malvern Mupandawana, "Tipping behavior in Canadian restaurants," *International Journal of Hospitality Management*, 28 (2009), pp. 597–603.

5. Kathleen E. Miller, "Wired: Energy drinks, jock identity, masculine norms, and risk taking," *Journal of American College Health*, 56 (2008), pp. 481–489.

6. From a table entitled "Largest Indianapolis-area architectural firms," *Indianapolis Business Journal*, June 15, 2014.

7. The data were obtained from the Internet Movie Database (IMDb), **www.imdb.com**, on August 14, 2014.

8. The 2015 table of 200 top universities can be found at **www.timeshighereducation.co.uk**.

9. The results were published in C. M. Weaver et al., "Quantification of biochemical markers of bone turnover by kinetic measures of bone formation and resorption in young healthy females," *Journal of Bone and Mineral Research*, 12 (1997), pp. 1714–1720. The data were provided by Linda McCabe.

10. This data set was provided by Joanne Lasrado of the Purdue University Department of Foods and Nutrition.

11. These data are based on experiments performed by G. T. Lloyd and E. H. Ramshaw of the CSIRO Division of Food Research, Victoria, Australia. Some results of the statistical analyses of these data are given in G. P. McCabe, L. McCabe,

and A. Miller, "Analysis of taste and chemical composition of cheddar cheese, 1982–83 experiments," CSIRO Division of Mathematics and Statistics Consulting Report VT85/6; and in I. Barlow et al., "Correlations and changes in flavour and chemical parameters of cheddar cheeses during maturation," *Australian Journal of Dairy Technology*, 44 (1989), pp. 7–18.

CHAPTER 12

1. Statistics from "The Infinite Dial 2014" survey by *Edison Research* and *Triton Digital* posted in March 2014.

2. Based on Stephanie T. Tong et al., "Too much of a good thing? The relationship between number of friends and interpersonal impressions on Facebook," *Journal of Computer-Mediated Communication*, 13 (2008), pp. 531–549.

3. This rule is intended to provide a general guideline for deciding when serious errors may result by applying ANOVA procedures. When the sample sizes in each group are very small, this rule may be a little too conservative. For unequal sample sizes, particular difficulties can arise when a relatively small sample size is associated with a population having a relatively large standard deviation.

4. Penny M. Simpson et al., "The eyes have it, or do they? The effects of model eye color and eye gaze on consumer ad response," *Journal of Applied Business and Economics*, 8 (2008), pp. 60–71.

5. Discussion on this an other tests can be found in M.H. Kutner et al., *Applied Linear Models*, 5th ed., McGraw-Hill/Irwin, 2005.

6. Bryan Raudenbush et al., "Pain threshold and tolerance differences among intercollegiate athletes: Implication of past sports injuries and willingness to compete among sports teams," *North American Journal of Psychology*, 14 (2012), pp. 85–94.

7. Based on "Don't bring me down: A study of the perceived emotional impact of positive, negative, and neutral content on Facebook," Thesis (2015), Isis Lopez, University of Texas at Brownsville.

8. Several different definitions for the noncentrality parameter of the noncentral F distribution are in use. When $I = 2$, the λ defined here is equal to the square of the noncentrality parameter δ that we used for the two-sample t test in Chapter 7. Many authors prefer $\phi = \sqrt{\lambda/I}$. We have chosen to use λ because it is the form needed for the SAS function PROBF.

9. Woo Gon Kim et al., "Influence of institutional DINESERV on customer satisfaction, return intention, and word-of-mouth," *International Journal of Hospitality Management*, 28 (2009), pp. 10–17.

10. Eileen Wood et al., "Examining the impact of off-task multi-tasking with technology on real-time classroom learning," *Computers & Education*, 58 (2012), pp. 365–374.

11. Sangwon Lee and Seonmi Lee, "Multiple play strategy in global telecommunication markets: An empirical analysis," *International Journal of Mobile Marketing*, 3 (2008), pp. 44–53.

12. F. Madhumita, "A study of changes to the websites of British Columbia wineries between 2004 and 2012," MS Dissertation (2013), University of British Columbia.

13. Kendall J. Eskine, "Wholesome foods and wholesome morals? Organic foods reduce prosocial behavior and harshen moral judgments," *Social Psychological and Personality Science*, 2012, doi:10.1177/1948550612447114.

14. Adam I. Perlman et al., "Massage therapy for osteoarthritis of the knee: A randomized dose-finding trial," *PLoS ONE*, 7, No. 2 (2012), e30248, doi:10.1371/journal.pone.0030248.

15. Jeffrey T. Kullgren et al., "Individual- versus group-based financial incentives for weight loss," *Annals of Internal Medicine*, 158, No. 7 (2013), pp. 505–514.

16. Corinne M. Kodama and Angela Ebreo, "Do labels matter? Attitudinal and behavioral correlates of ethnic and racial identity choices among Asian American undergraduates," *College Student Affairs Journal*, 27, No. 2 (2009), pp. 155–175.

17. Christie N. Scollon et al., "Emotions across cultures and methods," *Journal of Cross-cultural Psychology*, 35 (2004), pp. 304–326.

18. Jesus Tanguma et al., "Shopping and bargaining in Mexico: The role of women," *Journal of Applied Business and Economics*, 9 (2009), pp. 34–40.

19. Adrian C. North et al., "The effect of musical style on restaurant consumers' spending," *Environment and Behavior*, 35 (2003), pp. 712–718.

20. The experiment was performed in Connie Weaver's lab in the Purdue University Department of Foods and Nutrition. The data were provided by Berdine Martin and Yong Jiang.

21. The data were provided by James Kaufman. The study is described in James C. Kaufman, "The cost of the muse: Poets die young," *Death Studies*, 27 (2003), pp. 813–821. The quote from Yeats appears in this article.

22. Data provided by Jo Welch of the Purdue University Department of Foods and Nutrition.

23. Steve Badylak et al., "Marrow-derived cells populate scaffolds composed of xenogeneic extracellular matrix," *Experimental Hematology*, 29 (2001), pp. 1310–1318.

CHAPTER 13

1. Based on a student project of Stefannie Garcia, Stephanie Morgan, Jeremy Sanders, Taylor Hooper, and Natalie Rowe titled "The effect of scarcity on consumer purchase intentions," University of New Orleans, 2014.

2. We present the two-way ANOVA model and analysis for the general case in which the sample sizes may be unequal. If the sample sizes vary a great deal, serious complications can arise. There is no longer a single standard ANOVA analysis. Most computer packages offer several options for the computation of the ANOVA table when cell counts are unequal. When the counts are approximately equal, all methods give essentially the same results.

3. Sara N. Bleich and Julia A. Wolfson, "Trends in SSBs and snack consumption among children by age, body weight, and race/ethnicity," *Pediatric Obesity*, 23 (2015), pp. 1039–1046.

4. Rick Bell and Patricia L. Pliner, "Time to eat: The relationship between the number of people eating and meal duration in three lunch settings," *Appetite*, 41 (2003), pp. 215–218.

5. Karolyn Drake and Jamel Ben El Hine, "Synchronizing with music: Intercultural differences," *Annals of the New York Academy of Sciences*, 99 (2003), pp. 429–437.

6. Example 13.10 is based on a study described in P. D. Wood et al., "Plasma lipoprotein distributions in male and female runners," in P. Milvey (ed.), *The Marathon: Physiological, Medical, Epidemiological, and Psychological Studies*, New York Academy of Sciences, 1977.

7. Koert van Ittersum et al., "Smart shopping carts: How real-time feedback influences spending," *Journal of Marketing*, 77 (2013), pp. 21–36.

8. Alex L. Jones et al., "Miscalibrations in judgements of attractiveness with cosmetics," *The Quarterly Journal of Experimental Psychology*, (2014), doi:10.1080/17470218.2014.908932.

9. Gerardo Ramirez and Sian L. Beilock, "Writing about testing worries boosts exam performance in the classroom," *Science*, 331 (2011), pp. 211–213.

10. Felix Javier Jimenez-Jimenez et al., "Influence of age and gender in motor performance in healthy adults," *Journal of the Neurological Sciences*, 302 (2011), pp. 72–80.

11. Tomas Brodin et al., "Ecological effects of pharmaceuticals in aquatic systems impacts through behavioural alterations," *Philosophical Transactions of the Royal Society B*, (2014), doi:10.1098/rstb.2013.0580.

12. Sarah J. Gervais et al., "My eyes are up here: The nature of the objectifying gaze toward women," *Sex Roles*, 69 (2013), pp. 557–570.

13. Vincent P. Magnini and Kiran Karande, "The influences of transaction history and thank you statements in service recovery," *International Journal of Hospitality Management*, 28 (2009), pp. 540–546.

14. Brian Wansink et al., "The office candy dish: Proximity's influence on estimated and actual consumption," *International Journal of Obesity*, 30 (2006), pp. 871–875.

15. Data based on Brian T. Gold et al., "Lifelong bilingualism maintains neural efficiency for cognitive control in aging," *Journal of Neuroscience*, 33, No. 2 (2013), pp. 387–396.

16. Annette N. Senitko et al., "Influence of endurance exercise training status and gender on postexercise hypotension," *Journal of Applied Physiology*, 92 (2002), pp. 2368–2374.

17. Willemijn M. van Dolen, Ko de Ruyter, and Sandra Streukens, "The effect of humor in electronic service encounters," *Journal of Economic Psychology*, 29 (2008), pp. 160–179.

18. Jane Kolodinsky et al., "Sex and cultural differences in the acceptance of functional foods: A comparison of American, Canadian, and French college students," *Journal of American College Health*, 57 (2008), pp. 143–149.

19. Gad Saad and John G. Vongas, "The effect of conspicuous consumption on men's testosterone levels," *Organizational Behavior and Human Decision Processes*, 110 (2009), pp. 80–92.

20. Klaus Boehnke et al., "On the interrelation of peer climate and school performance in mathematics: A German-Canadian-Israeli comparison of 14-year-old school students," in B. N. Setiadi, A. Supratiknya, W. J. Lonner, and Y. H. Poortinga (eds.), *Ongoing Themes in Psychology and Culture* (Online Ed.), International Association for Cross-Cultural Psychology.

21. Data provided by Julie Hendricks and V. J. K. Liu of the Department of Foods and Nutrition, Purdue University.

22. Lijia Lin et al., "Animated agents and learning: Does the type of verbal feedback they provide matter?" *Computers and Education*, 2013, doi:10.1016/j.compedu.2013.04.017.

23. Tamar Kugler et al., "Trust between individuals and groups: Groups are less trusting than individuals but just as trustworthy," *Journal of Economic Psychology*, 28 (2007), pp. 646–657.

24. Based on A. A. Adish et al., "Effect of consumption of food cooked in iron pots on iron status and growth of young children: A randomised trial," *Lancet*, 353 (1999), pp. 712–716.

25. Based on a problem from Renée A. Jones and Regina P. Becker, Department of Statistics, Purdue University.

26. For a summary of this study and other research in this area, see Stanley Coren and Diane F. Halpern, "Left-handedness: A marker for decreased survival fitness," *Psychological Bulletin*, 109 (1991), pp. 90–106.

27. Data provided by Neil Zimmerman of the Purdue University School of Health Sciences.

28. See I. C. Feller et al., "Sex-biased herbivory in Jack-in-the-pulpit (*Arisaema triphyllum*) by a specialist thrips (*Heterothrips arisaemae*)," in *Proceedings of the 7th International Thysanoptera Conference*, Reggio Callabrio, Italy, pp. 163–172.

Index

Acceptance sampling, 396
ACT college entrance examination, 73, 380, 604–605
Adequate Calcium Today (ACT) study, 545–546
Adjusted R^2, 631
Aggregation, 145
Alternative hypothesis. *See* Hypothesis, alternative
American Community Survey (ACS), 201
Analysis of variance (ANOVA)
  one-way, 643–685
  regression, 583–587, 596, 613–614, 616
  two-way, 697–713
Analysis of variance table
  one-way, 658–663
  regression, 586, 596, 613–614, 616
  two-way, 708–709
Anonymity, 206
Applet
  Central Limit Theorem, 300–302, 311
  Confidence Interval, 347–348, 360, 403
  Correlation and Regression, 104, 106, 135
  Law of Large Numbers, 220, 251
  Mean and Median, 31, 50
  Normal Approximation to Binomial, 322
  Normal Curve, 62, 72
  One-variable statistical calculator, 16
  One-Way ANOVA, 661, 687
  Probability, 217, 220, 291–292, 335
  Simple Random Sample, 187, 192, 202, 292
  Statistical Power, 400, 401
  Statistical Significance, 383
  t Statistic, 423
  Two-variable statistical calculator, 123
Association, 80–81, 84, 532–533
  and causation, 131, 133, 149–150
  negative, 89, 96
  positive, 89, 96
Attention deficit hyperactivity disorder (ADHD), 429–430
Available data, 165, 170

Bar graph, 10, 23, 530
Bayes's rule, 273–274
Behavioral and social science experiments, 209–211
Behavioral Risk Factor Surveillance System (BRFSS), 604
Benchmarking, 89
Benford's law, 226, 249, 338

Bias *see also* Unbiased estimator
  in a sample, 190, 198–201
  in an experiment, 174–175, 185
  of a statistic, 287–289, 290
Binomial coefficient, 327, 333
Binomial distribution. *See* Distribution, binomial
Binomial setting, 312, 332, 14-1
Block, 183–184, 185
Bonferroni procedure, 391, 678–680
Bootstrap, 424–425, 15-2. *See also* Chapter 16
Boston Marathon, 27, 17-40
Boxplot, 34, 46
  modified, 37, 46
  side-by-side, 37, 46, 646, 649
Brown-Forsythe test. *See* Modified Levene's test
Buffon, Count, 218

Canadian Internet Use Survey (CIUS), 14-24
Capability, 17-34, 17-36
Capture-recapture sampling, 199
Case, 2, 7, 609
Categorical data. *See* Variable, categorical
Causation, 131, 133, 148–152
Cause-and-effect diagram, 17-4
Cell, 137, 145, 699
Census, 167, 170, 636
Census Bureau, 8, 338, 380
Center of a distribution, 28–31, 46, 54
Centers for Disease Control and Prevention, 214, 336, 604
Central limit theorem, 298–301, 313, 325, 328, 335
Chi-square distribution. *See* Distribution, chi-square
Chi-square statistic, 534, 543
  and the z statistic, 540–541
  goodness of fit test, 547
Classes in a histogram, 15
Clinical trials, 207
Coefficient of determination, 662. *See also* Correlation, squared multiple
Coin tossing, 217, 221, 238–239, 291, 312–313, 331, 335, 339
College Alcohol Study (CAS), 604
Column variable. *See* Variable, row, and Variable, column
Common response, 149–150, 152
Complement of an event. *See* Event, complement
Conditional distribution. *See* Distribution, conditional

Conditional probability. *See* Probability, conditional
Confidence interval, 346–348, 356
  behavior, 352–353
  bootstrap, 16-13–16-16, 16-31–16-35
  cautions, 355–356
  for multiple comparisons, 680
  for odds ratio, 14-10, 14-19
  for slope in a logistic regression, 14-9, 14-18
  relation to two-sided tests, 375–377
  simultaneous, 680
  t for a contrast, 674
  t for difference of means, 437–439, 454
    pooled, 449
  t for matched pairs, 421
  t for mean response in regression, 570–571, 578
  t for one mean, 410–412, 425–426
  t for regression parameters, 568, 578, 612–613, 616
  z for one mean, 348–352
  z for one proportion
    large sample, 486, 500
    plus four, 489
  z for difference of proportions
    large sample, 506–507, 519
    plus four, 509–510
Confidence level, 347, 356
Confidentiality, 204, 206–207, 211
Confounding, 149–150, 152, 169, 170, 388, 419
Consumer Report of Eating Share Trends (CREST), 632, 14-23
Consumer Reports National Research Center, 319
Consumers Union, 85, 16-35
Continuity correction, 325–326, 333, 15-7
Contrast, 650, 670–677, 685
Control chart, 17-7, 17-17
  individuals chart, 17-40
  p chart, 17-51–17-56
  R chart, 17-23, 17-35
  s chart, 17-12–17-17
  $\bar{x}$ chart, 17-8–17-12, 17-14, 17-18
Control group, 174, 310
Correlation, 100–104
  and regression, 115, 118
  based on averaged data, 131, 133
  between random variables, 257, 261
  bootstrap confidence interval, 16-35–16-37
  cautions about, 123–133
  nonsense, 131
  inference for, 593–596
  population, 594

Correlation (*continued*)
  properties, 102–103, 104
  squared, 116, 118, 585, 596
  squared multiple, 615. *See also*
      Coefficient of determination
  test for, 594, 596
Count, 9. *See also* Frequency
  distribution of, 310–314, 321–322,
      328–333
Critical value, 378, 379
  of chi-square distribution, 535, Table F
  of *F* distribution, 585–586, Table E
  of standard Normal distribution,
      349, 410, Table A
  of *t* distribution, 409–410, Table D
Current Population Survey, 289
Cumulative proportion, 61, 70
  standard Normal, 63–64, Table A

Data, 2
  Anecdotal, 164, 170
  Available, 156, 170
Data mining, 132–133
Decision analysis, 396–400
Degree of Reading Power, 437, 16-42
Degrees of freedom, 40
  approximation for, 436, 447, 453
  of chi-square distribution, 535
  of chi-square test, 535
  of *F* distribution, 585
  of one-way ANOVA, 659
  of *t* distribution, 409, 425
  of two-way ANOVA, 704, 708–709
  of regression ANOVA, 584–586, 613–614
  of regression *t*, 568, 570, 572, 594,
      612–613
  of regression s^2, 562, 612
  of two-sample *t*, 436, 447, 449
Deming, W. Edwards, 17-40
Density curve, 51–54, 69, 240, 243
Density estimation, 68–69
Design, 171–185. *See also* Experiment
  block, 183–184, 185
  repeated-measures, 707
  sampling, 188–200
Direction of a relationship, 88, 96
Disjoint events. *See* Event, disjoint
Distribution, 23, 46
  bimodal, 69
  binomial, 312–318, 332, 14-2, Table C
    formula, 326–328, 333
    Normal approximation, 321–324, 332
    use in the sign test, 472–473
  bootstrap, 16-24–16-29
  of categorical variable, 9
  chi-square, 535, Table F
  conditional, 140, 145, 528, 537
  describing, 20, 23
  examining, 18
  exponential, 300
  geometric, 340

F, 585–586, Table E
  joint, 138, 145
  jointly Normal, 594
  marginal, 139, 145
  noncentral *F*, 682
  noncentral *t*, 467, 474
  Normal, 56–57, 69
    for probabilities, 242–243
    standard, 60, 63, 70, Table A
  Poisson, 328–332, 333, 551
  population, 291, 294
  probability. *See* Probability distribution
  of quantitative variable, 11–16
  sampling. *See* Sampling distribution
  skewed, 18, 23
  symmetric, 18, 23
  t, 409–410, Table D
  trimodal, 69
  tails, 18
  uniform, 71, 240, 243, 554
  unimodal, 18
  Weibull, 305–307
Distribution-free procedure, 470. *See also*
      Chapter 15
Double-blind experiment, 181–182, 185
Dual X-ray absorptiometry scanner, 364,
      431–432, 475, 17-38–17-39

Equivalence testing, 420–422
Estimation, 250–251
Ethics, 163, 203–211
Excel, 3, 178–179, 191–192, 417, 445,
      487, 508, 563, 609, 629, 664,
      713, 17-22
Expected value, 248. *See also* Mean of a
      random variable
Expected cell count, 533, 543, 547, 550
Experiment, 167–168, 170
  block design, 183–184, 185
  cautions about, 181–182
  comparative, 173–174, 185
  completely randomized, 180
  matched pairs, 182–183, 185
  principles, 177
  units, 171, 185
Explanatory variable. *See* Variable,
      explanatory
Exploratory data analysis, 9, 16, 23, 163
Extrapolation, 110, 118
Event, 223, 232
  complement of, 224, 232
  disjoint, 224, 232
  empty, 266
  independent, 229, 232
  intersection, 271, 274
  union, 264, 274

F distribution. *See* Distribution, *F*
F test
  one-way ANOVA, 661, 667
  regression ANOVA, 586, 614

for collection of regression coefficients,
      631–632, 635
  for standard deviations, 665–666
  two-way ANOVA, 709
Facebook, 23, 308–309, 428, 456, 522,
      648–650, 661–663, 670–673, 686,
      687–688, 694, 696, 15-31, 16-3–16-5
Factor, experimental, 172, 185, 644,
      698–702
Federal Aviation Administration (FAA), 309
Fisher, Sir R. A., 385, 400, 585
Fitting a line, 108–109
Five-number summary, 34, 46
Flowchart, 17-4–17-5
Form of a relationship, 88, 96
Frequency, 15, 23
Frequency table, 15

Gallup-Healthways Well-Being Index,
      359–360
Gallup Poll, 335–336
Genetic counseling, 277
Genomics, 388
General Social Survey (GSS), 167,
      197, 211
Goodness of fit, 545–550
Google, 9, 485
Gosset, William, 48, 409, 16-10

Histogram, 14, 23
Hypothesis
  alternative, 363–364, 370, 379
    one-sided, 364, 379
    two-sided, 364, 379
  null, 363, 379
Hypothesis testing, 399-400. *See also*
      Significance test

Independence, 218–219
  in two-way tables, 532–533, 543–544
  of events, 228–229, 232
  of random variables, 257–258, 261, 274
Indicator variable. *See* Variable, indicator
Inference, statistical. *See* Statistical
      inference
Influential observation, 127–129, 133,
      566, 624
Informed consent, 204, 205–206, 211
Institutional review board (IRB),
      204–205, 211
Instrument, 5
Interaction, 701, 703–707
Intercept of a line, 108
  of least-squares line, 112, 118, 557
Internet Movie Database (IMDb), 637
Intervention, 169, 170
Intersection of events, *See* Event,
      intersection
Interquartile range (IQR), 36, 46
iPod, 422, 470–471

Jitter, 87
JMP, 416, 441, 446, 469, 493, 499, 509, 513, 517, 528, 532, 545, 549, 552, 564, 580, 622, 623, 666, 683, 689, 14-4, 14-14, 15-8, 15-12, 15-15, 15-20, 15-30, 15-31

Karaoke Channel, 358
Kerrich, John, 218
Key characteristics of a data set, 4, 7
Key characteristics of data for relationships, 83
Kruskal-Wallis test, 15-26–15-31

Label, 2, 7
Law of large numbers, 250–252, 253, 261
Law School Admission Test (LSAT), 390, 476
Leaf, in a stemplot, 11, 23
Leaning Tower of Pisa, 604
Least significant difference, 678
Least squares, 111, 611–612
Least squares regression line, 112, 118, 555, 577
Level of a factor, 172, 185, 698–701
Line, equation of, 108
    least-squares, 112, 118, 611–612
Linear relationship, 88, 96
Linear transformation. See Transformation, linear
Logarithm transformation. See Transformation, logarithm
Logistic regression, 632–633. See also Chapter 14
Logit, 14-5
Lurking variable. See Variable, lurking

Main effect, 701, 703–707, 714
Major League Baseball (MLB), 15-3
Mann-Whitney test, 15-5, 15-8
Margin of error, 287, 289, 291, 352
    for a difference in two means, 437, 449, 454
    for a difference in two proportions, 508, 519
    for a single mean, 349, 353, 356–357, 411, 426
    for a single proportion, 486, 500
Marginal means, 705, 714
Matched pairs design, 182–183, 185
    inference for, 419–420, 426, 472–473, 15-17
McNemar's test, 554
Mean, 28, 46
    of binomial distribution, 318, 332
    of density curve, 55, 69
    of difference of sample means, 434
    of difference of sample proportions, 506
    of Normal distribution, 56
    of random variable, 246–248, 261
    rules for, 253–254, 261

of sample mean, 296–297, 307
of sample proportion, 320, 332, 500, 584
trimmed, 51
versus median, 31
Mean square
    in one-way ANOVA, 660, 667
    in two-way ANOVA, 709
    in multiple linear regression, 613
    in simple linear regression, 584–586
Median, 30, 46
    inference for, 472–473, 15-9, 15-23, 16-28–16-29
    of density curve, 55, 69
Mendel, Gregor, 230
Meta-analysis, 542
Minitab, 315, 395, 417, 422, 441, 463, 466, 493, 499, 509, 514, 517, 529, 548, 563, 595, 627, 630, 665, 684, 690, 713, 14-11, 14-14, 14-16, 14-17, 14-20, 14-21, 15-8, 15-20, 15-24
Minnesota Multiphasic Personality Inventory (MMPI), 5
Mode, 18, 23
Model selection, 629
Modified Levene's test, 665–666
Mosaic plot, 143, 531, 534
Motorola, 17-2
Multiple comparisons, 650, 677–681, 15-51

National AIDS Behavioral Surveys, 335
National Assessment of Educational Progress (NAEP), 70–71, 381
National Association of Colleges and Employers (NACE), 354, 358, 464
National Center for Education Statistics, 119–120, 166
National Collegiate Athletic Association (NCAA), 16-11
National Endowment for the Humanities, 432
National Enquirer, 458
National Football League, 601
National Health and Nutrition Examination Survey (NHANES), 372, 434, 704
National Hockey League (NHL), 617–618
National Longitudial Survey of Youth (NLSY), 574
National Oceanic and Atmospheric Administration (NOAA), 581
National Public Radio (NPR), 360
National Science Foundation (NSF), 597
Neyman, Jerzy, 399–400
Nielsen Company, 294, 411, 428
Noncentrality parameter
    for t, 468, 474
    for F, 682
Nonparametric procedure, 470, 472–473. See also Chapter 15

Nonresponse, 196, 200
Normal distribution. See Distribution, Normal
Normal distribution calculations, 61–66
Normal probability plot. See Normal quantile plot
Normal quantile plot, 66–67, 70
Normal scores, 66
Null hypothesis. See Hypothesis, null

Observational study, 168
Odds, 633, 14-2, 14-18
Odds ratio, 633, 14-7, 14-10, 14-18
Outcomes, 171, 185
Out-of-control rules, 17-23–17-26
Outliers, 19, 23, 15-1
    $1.5 \times IQR$ criterion, 35–36
    regression, 127–129, 133, 574–575

Parameter, 282, 290
Pareto chart, 17-18, 17-53–17-54, 17-57
Pearson, Egon, 399
Pearson, Karl, 218
Percent, 9
Percentile, 32
Permutation tests, 15-2, 16-41–16-50
Pew Research Center survey, 198, 308, 428, 484, 485, 501, 521, 522, 527, 14-19, 15-15, 16-55
Pie chart, 11, 23
Placebo effect, 174
Plug-in principle, 16-9, 16-10
Pooled estimator
    of population proportion, 512, 519
    of ANOVA variance, 654, 660, 703
    of variance in two samples, 448
Population, 189, 200
Population distribution. See Distribution, population
Power, 392, 400
    and Type II error, 399
    increasing, 395
    of one-way ANOVA, 681–685
    of t test
        one-sample, 465–467
        two-sample, 467–469
    of z test, 391–395
    of z test for a single proportion, 498–499
    of z test for comparing two proportions, 516–517
Prediction, 107, 110, 118
Prediction interval, 572–573, 578, 613
Probability, 216–217, 219
    conditional, 267–268, 269
    equally likely outcomes, 227
    finite sample space, 225–226
Probability distribution, 236, 241
    mean of, 246–248, 261
    standard deviation of, 255–256, 261
    variance of, 255–256, 261

Probability histogram, 237, 243
Probability model, 221, 232
Probability rules, 223–224, 232
    addition, 224, 232, 264, 266, 275
    complement, 224, 232, 264, 275
    general, 264–275
    multiplication, 228–229, 232, 264–265, 268, 275
Probability sample. See Sample, probability
Process capability indices, 17-40–17-47
Proportion, 9
    distribution of, 319–321, 322–323
    inference for a single proportion, 483–501
    inference for comparing two proportions, 505–517
    population, 283
    sample, 283, 319, 484, 500
P-value, 366, 379

Quartiles, 32–33, 46
    of a density curve, 55, 69

R, 315, 329, 330, 331, 332, 16-9, 16-11, 16-14, 16-18, 16-34, 16-38, 16-45
Randomization
    consequences of, 177
    experimental, 175–176, 185
    how to, 177–180
Random digits, 180–181, 192–193, 200, 284, Table B
Random number generator, 375
Random phenomenon, 217, 219
Random variable, 235–236, 243
    continuous, 239–242, 243
    discrete, 236, 243
    mean of, 248, 261
    standard deviation of, 256, 261
    variance of, 256, 261
Randomized comparative experiment, 177, 185
Randomized response survey, 279–280
Ranks, 15-4, 15-14
Rate, 6
Regression, 107–117
    and correlation, 115, 118
    cautions about, 123–133
    deviations, 88, 560, 577, 610
    interpretation, 113
    least-squares, 111, 611–612
    logistic, 632–633, Chapter 14
    model conditions, 567
    model selection, 627–631
    multiple, 608–615
    multiple logistic, 632–633, 14-16–14-18
    nonlinear, 576–577
    simple linear, 556–576
Regression equation, population, 608, 615

Regression line, 107, 117
    population, 557, 577
Relative risk, 518, 519
Reliability, 313
Resample, 424. See also Chapter 16
Residual, 123–124, 133, 561, 577, 612, 616, 653
    plots, 125, 133, 566, 577–578, 599, 690
Resistant measure, 30, 46
Response bias, 198, 200
Response rate, 189
Response variable. See Variable, response
Robust, 30, 423–424, 426, 442, 15-1
Roundoff error, 125, 138, 139
Row variable. See Variable, row, and Variable, column
Rugby sevens, 455

Sallie Mae, 350
Sample, 189, 200
    cautions about, 196–199, 200
    design of, 189, 200
    multistage, 195–196, 200
    probability, 194, 200
    simple random (SRS), 191–193, 200
    stratified, 193–194, 200
    systematic, 202
Sample size, choosing
    confidence interval for a difference in means, 462–463
    confidence interval for a difference in proportions, 514–515, 519
    confidence interval for a mean, 353, 461–463
    confidence interval for a proportion, 494–495, 500
    one-way ANOVA, 681–684
    power for a proportion, 498–499
    power for a difference in proportions, 516–517
    t test, one-sample, 465–467
    t test, two-sample, 467–469
Sample space, 221, 232
    finite, 225
Sample survey, 167–168, 170, 188–200
Sampling distribution, 281, 284–287, 290
    of difference of means, 434
    of regression estimators, 567
    of sample count, 314, 322, 332
    of sample mean, 298, 307
    of sample proportion, 285, 319–321, 322, 332
Sampling variability, 287–288
SAS, 445, 587, 619, 626, 628, 631, 664, 710, 711
SAT college entrance examination, 73, 344, 604–605, 618–619, 14-26
Scatterplot, 86, 96
    adding categorical variables to, 93
    smoothing, 94, 96
Shape of a distribution, 11, 23

Shewhart, Walter, 17-7, 17-32
Sign test, 472–473, 491–492, 549–550
Significance level, 367, 383–385
Significance, statistical, 367–369, 379
    and Type I error, 398
Significance test, 361–370
    chi-square for two-way table, 534–535, 543
        relation to z test, 540–541
    chi-square test for goodness of fit, 547, 550
    chi-square test for logistic regression slope, 14-10, 14-19
    F test in one-way ANOVA, 660–662
    F test in regression, 585–586, 596, 614
    F test for a collection of regression coefficients, 631–632, 635
    F test for standard deviations, 665–666
    F tests in two-way ANOVA, 709
    Kruskal-Wallis test, 15-26–15-31
    Mann-Whitney test, 15-5
    relationship to confidence intervals, 375–377
    sign test for matched pairs, 472–473, 491–492
    t test for a contrast, 674
    t test for correlation, 594, 596
    t test for one mean, 413, 425
    t test for matched pairs, 419–420
    t test for two means, 440, 454
        pooled, 449
    t test for regression coefficients, 568, 578
    t tests for multiple comparisons, 678
    use and abuse, 384–389
    Wilcoxon rank sum test, 15-3–15-14
    Wilcoxon signed rank test, 15-17–15-24
    z test for one mean, 372, 379
    z test for one proportion, 491, 500
    z test for logistic regression slope, 14-10, 14-19
    z test for two proportions, 511–512, 519
Simple random sample. See Sample, simple random
Simpson's paradox, 143, 145, 160
Simulation, 284
Simultaneous confidence intervals, 680
68–95–99.7 rule, 57–58, 70
Skewed distribution. See Distribution, skewed
Slope of a line, 108
    of least-squares line, 112, 118, 561
Small numbers, law of, 252
Spread of a distribution, 32, 38, 46, 54
Spreadsheet, 3. See also Excel
SPSS, 417, 446, 530, 549, 562, 621, 657, 658, 671, 676, 680, 14-14, 14-17, 15-9, 15-20
Standard & Poor's 500-Stock Index, 416, 598

Standard deviation, 38, 46. *See also* Variance
  of binomial distribution, 318, 332
  of density curve, 55, 69
  of deviations in ANOVA, 652, 702
  of deviations in regression, 560, 577, 611, 615
  of difference between sample means, 434
    pooled, 448
  of difference between sample proportions, 515, 519
  of Normal distribution, 57
  of Poisson distribution, 329, 333
  of random variable, 256, 261
  of regression intercept and slope, 590–591
  of sample mean, 297, 307
  of sample proportion, 485
  properties, 40
  rules for, 257–258, 261
Standard error, 408
  bootstrap, 16-5, 16-8
  for regression prediction, 592, 596
  of a contrast, 674
  of a difference in sample means, 436–437
    pooled, 448
  of a difference in sample proportions, 508, 519
  of a sample mean, 408, 425
  of a sample proportion, 486
  of mean regression response, 592, 596
  of regression intercept and slope, 591, 596
Standard Normal distribution. *See* Distribution, standard Normal
Standardized observation, 59, 70
Statistic, 282, 290
Statistical inference, 282, 290, 341–343
  for non-Normal populations, 470–473. *See also* Chapter 15
  for small samples, 444–447
Statistical process control, Chapter 17
Statistical significance. *See* Significance, statistical
Stem-and-leaf plot. *See* Stemplot
Stemplot, 11, 23
  back-to-back, 12
  splitting stems, 13
  trimming, 13
Strata, 195, 200. *See also* Sample, stratified
Strength of a relationship, 88, 96. *See also* Correlation
StubHub! 69, 16-11–16-12, 16-22
Student Monitor, 283, 291
Subjects, experimental, 171, 185
Subpopulation, 557, 608

Sums of squares
  in one-way ANOVA, 659–660
  in two-way ANOVA, 708–709
  in multiple linear regression, 613–614
  in simple linear regression, 583–584
Survey of Study Habits and Attitudes (SSHA), 382
Systematically larger, 15-9
Symmetric distribution. *See* Distribution, symmetric

t distribution. *See* Distribution, *t*
t inference procedures
  for contrasts, 674
  for correlation, 594, 586
  for matched pairs, 419–420
  for multiple comparisons, 678
  for one mean, 411, 413
  for regression coefficients, 568, 578, 612–613, 616
  for regression mean response, 570, 578
  for regression prediction, 572, 578
  for two means, 437, 440
  for two means, pooled, 449
  robustness of, 423–424, 442–443
Tails of a distribution. *See* Distribution, tails
Test of significance. *See* Significance test
Test statistic, 364–365
Testing hypotheses. *See* Significance test
The Times Higher Education Supplement, 637–638
Three-way table, 145
Ties, 15-10, 15-22
Time plot, 21, 23
Titanic, 24, 52, 146, 161, 16-12, 16-22
Transformation
  linear, 44–45, 46, 254
  logarithm, 91, 96, 470–471, 574–575
  rank, 15-4
  square root, 671–672
Treatment, experimental, 171, 174, 185
Tree diagram, 271–273, 275
Tuskegee study, 208
Twitter, 75, 244, 522
Two-sample problems, 433
Two-way table, 136, 145
  data analysis for, 136–145
  inference for, 525–543
  models for, 543
  relationships in, 81, 528
Type I and II errors, 396–397

Uber, 428
Unbiased estimator, 287
Undercoverage, 196, 200
Unimodal distribution. *See* Distribution, unimodal

Union of events, 264–265
Unit of measurement, 3, 43
Unit, experimental, 171, 185
U.S. Agency for International Development, 15-26
U.S. Department of Education, 336

Value of a variable, 2, 7
Variability, 32, 287–288
Variable, 2, 7
  categorical, 3, 7, 10, 11
  column, 137, 145
  dependent, 83
  explanatory, 82, 84
  independent, 83
  indicator 14-3
  lurking, 129–130, 133, 172
  quantitative, 3, 7, 11
  response, 82, 84
  row, 137, 145
Variance, 38, 46
  of a difference between two sample means, 434
    pooled, 448
  of a difference between two sample proportions, 507, 519
  of a random variable, 255–256, 261
  a pooled estimator, 448, 454
  rules for, 257–258, 261
  of a sample mean, 297
Variation
  among groups, 658, 667
  between groups, 647, 658, 667
  common cause, 17-7
  special cause, 17-7
  within group, 647, 658, 667
Venn diagram, 224
Voluntary response, 190–191

Wald statistic, 14-10, 14-19
Wall Street Journal, 458
Whiskers, 35
Wilcoxon rank sum test, 15-3–15-16
Wilcoxon signed rank test, 15-17–15-24
Wording questions, 198, 200
World Bank, 28
World Database of Happiness, 638

z-score, 59, 70
z statistic
  for one proportion, 491, 500
  for two proportions, 512, 519
  one-sample for mean, 371
  two-sample for means, 435
    pooled, 449

Table entry for p and C is the critical value t^* with probability p lying to its right and probability C lying between $-t^*$ and t^*.

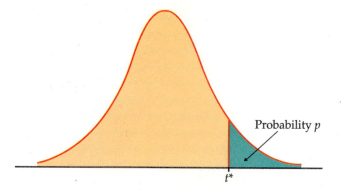

Probability p

t^*

TABLE D   t Distribution Critical Values

					Upper-tail probability p							
df	.25	.20	.15	.10	.05	.025	.02	.01	.005	.0025	.001	.0005
1	1.000	1.376	1.963	3.078	6.314	12.71	15.89	31.82	63.66	127.3	318.3	636.6
2	0.816	1.061	1.386	1.886	2.920	4.303	4.849	6.965	9.925	14.09	22.33	31.60
3	0.765	0.978	1.250	1.638	2.353	3.182	3.482	4.541	5.841	7.453	10.21	12.92
4	0.741	0.941	1.190	1.533	2.132	2.776	2.999	3.747	4.604	5.598	7.173	8.610
5	0.727	0.920	1.156	1.476	2.015	2.571	2.757	3.365	4.032	4.773	5.893	6.869
6	0.718	0.906	1.134	1.440	1.943	2.447	2.612	3.143	3.707	4.317	5.208	5.959
7	0.711	0.896	1.119	1.415	1.895	2.365	2.517	2.998	3.499	4.029	4.785	5.408
8	0.706	0.889	1.108	1.397	1.860	2.306	2.449	2.896	3.355	3.833	4.501	5.041
9	0.703	0.883	1.100	1.383	1.833	2.262	2.398	2.821	3.250	3.690	4.297	4.781
10	0.700	0.879	1.093	1.372	1.812	2.228	2.359	2.764	3.169	3.581	4.144	4.587
11	0.697	0.876	1.088	1.363	1.796	2.201	2.328	2.718	3.106	3.497	4.025	4.437
12	0.695	0.873	1.083	1.356	1.782	2.179	2.303	2.681	3.055	3.428	3.930	4.318
13	0.694	0.870	1.079	1.350	1.771	2.160	2.282	2.650	3.012	3.372	3.852	4.221
14	0.692	0.868	1.076	1.345	1.761	2.145	2.264	2.624	2.977	3.326	3.787	4.140
15	0.691	0.866	1.074	1.341	1.753	2.131	2.249	2.602	2.947	3.286	3.733	4.073
16	0.690	0.865	1.071	1.337	1.746	2.120	2.235	2.583	2.921	3.252	3.686	4.015
17	0.689	0.863	1.069	1.333	1.740	2.110	2.224	2.567	2.898	3.222	3.646	3.965
18	0.688	0.862	1.067	1.330	1.734	2.101	2.214	2.552	2.878	3.197	3.611	3.922
19	0.688	0.861	1.066	1.328	1.729	2.093	2.205	2.539	2.861	3.174	3.579	3.883
20	0.687	0.860	1.064	1.325	1.725	2.086	2.197	2.528	2.845	3.153	3.552	3.850
21	0.686	0.859	1.063	1.323	1.721	2.080	2.189	2.518	2.831	3.135	3.527	3.819
22	0.686	0.858	1.061	1.321	1.717	2.074	2.183	2.508	2.819	3.119	3.505	3.792
23	0.685	0.858	1.060	1.319	1.714	2.069	2.177	2.500	2.807	3.104	3.485	3.768
24	0.685	0.857	1.059	1.318	1.711	2.064	2.172	2.492	2.797	3.091	3.467	3.745
25	0.684	0.856	1.058	1.316	1.708	2.060	2.167	2.485	2.787	3.078	3.450	3.725
26	0.684	0.856	1.058	1.315	1.706	2.056	2.162	2.479	2.779	3.067	3.435	3.707
27	0.684	0.855	1.057	1.314	1.703	2.052	2.158	2.473	2.771	3.057	3.421	3.690
28	0.683	0.855	1.056	1.313	1.701	2.048	2.154	2.467	2.763	3.047	3.408	3.674
29	0.683	0.854	1.055	1.311	1.699	2.045	2.150	2.462	2.756	3.038	3.396	3.659
30	0.683	0.854	1.055	1.310	1.697	2.042	2.147	2.457	2.750	3.030	3.385	3.646
40	0.681	0.851	1.050	1.303	1.684	2.021	2.123	2.423	2.704	2.971	3.307	3.551
50	0.679	0.849	1.047	1.299	1.676	2.009	2.109	2.403	2.678	2.937	3.261	3.496
60	0.679	0.848	1.045	1.296	1.671	2.000	2.099	2.390	2.660	2.915	3.232	3.460
80	0.678	0.846	1.043	1.292	1.664	1.990	2.088	2.374	2.639	2.887	3.195	3.416
100	0.677	0.845	1.042	1.290	1.660	1.984	2.081	2.364	2.626	2.871	3.174	3.390
1000	0.675	0.842	1.037	1.282	1.646	1.962	2.056	2.330	2.581	2.813	3.098	3.300
z^*	0.674	0.841	1.036	1.282	1.645	1.960	2.054	2.326	2.576	2.807	3.091	3.291
	50%	60%	70%	80%	90%	95%	96%	98%	99%	99.5%	99.8%	99.9%

Confidence level C